AQUACULTURE

PRINCIPLES
AND
PRACTICES

Cage farm of the Svanoy Foundation at Svanoy, Norway (photograph: Ola Sveen).

An aerial view of the fish farm complex of the Fish Culture Research Institute at Szarvas, Hungary (courtesy of Ferrence Müller).

AQUACULTURE
PRINCIPLES
AND
PRACTICES

T.V.R. PILLAY

Former Programme Director
Aquaculture Development and Coordination Programme
Food and Agriculture Organization of the
United Nations, Rome, Italy

Fishing News Books

First published 1990

British Library Cataloguing in Publication Data
Pillay, T.V.R.
    Aquaculture.
    1. Aquaculture.
    I. Title
    639.8
ISBN 0−85238−168−9

Fishing News Books
A division of Blackwell Scientific
    Publications Ltd
Editorial Offices:
Osney Mead, Oxford OX2 0EL
    (Orders: Tel. 0865 240201)
25 John Street, London WC1N 2BL
23 Ainslie Place, Edinburgh EH3 6AJ
3 Cambridge Center, Suite 208,
    Cambridge, MA 02142, USA
54 University Street, Carlton,
    Victoria 3053, Australia

Set by Setrite Typesetters Ltd
Printed and bound in Great Britain by
the University Press, Cambridge

Dedication

This book is dedicated to the memory of the late Dr Sunder Lal Hora, who introduced me to aquaculture and with whom I wrote my first book on the subject, and to the Svanoy Foundation of Norway that encouraged me to write this book during my association with them.

Contents

Part II Aquaculture practices

Preface

Fishery administrations and private sector industries now accord a place of importance to aquaculture, albeit after several decades of hesitation or downright scepticism. Though for farmers in many Asian countries aquaculture has been a way of life for centuries, its status in the context of global food production, aquatic resource management and socioeconomic development of rural areas remained until recently a matter of debate. The scenario has changed radically with changes in world fisheries and the spectacular success of certain types of aquaculture enterprises. Development and donor agencies now consider it a priority area, and several scientific and technical institutions are presently involved in research on a number of aspects of aquaculture. Aquaculture workshops, symposia, conferences and expositions have become very frequent. All these have contributed to the recognition of some of the basic needs and problems of this new and emerging industry. The two world conferences on aquaculture, viz, the Food and Agriculture Organization (FAO) Technical Conference on Aquaculture held in Kyoto (Japan) in 1976[1] and the subsequent World Conference on Aquaculture in Venice (Italy) in 1981,[2] highlighted the importance of (i) technology development and improvement through research, (ii) training of personnel and (iii) information dissemination, in strategies for rapid and orderly development of the sector.

Many present-day aquaculture practices are based on biological studies with only limited involvement of other concerned disciplines. This major handicap is now being increasingly understood, and aquaculture has come to be recognized as a multidisciplinary science, although expertise in the associated disciplines continues to be scarce. Farm management, which is an interdisciplinary science in itself,

has yet to be developed for application in aquaculture. There are very significant communication gaps, and access to existing experience and information are extremely difficult. It is believed that large-scale application of the present technologies – despite all their deficiencies – will result in much greater production, if only sufficient numbers of adequately trained and experienced personnel are available.

The two world conferences, and many later international consultations, have repeatedly emphasised the crucial importance of training in development strategies. While different categories of personnel are needed in an aquaculture development programme, the key personnel are technicians at the middle management level, with a broad background of the more important disciplines involved, and practical experience of relevant technologies. They may not be experts in all the disciplines, but their background would help in providing necessary leadership in planning and executing aquaculture development programmes. While planning and organizing training of aquaculture personnel of this type on a regional basis under the United Nations Development Programme/ Food and Agriculture Organization (UNDP/ FAO) projects, I came to recognize the need for a comprehensive book covering the various aspects involved, for educational, as well as technology development and information programmes. The problems faced in ensuring appropriate instruction in different institutions, and collecting adequate information on culture technologies, further highlighted the usefulness of such a basic reference. This book is an attempt to fill this need, and to present a global picture of aquaculture as practised. As the main objective of the book is to present subsistence and commercial aquaculture, only culture systems of major species groups that presently contribute to production or are likely to do so in the near future, are described. Due to space limitations many of the descriptions have had to be abbreviated, and data, especially economic data that are not very recent, omitted. My exposure to aquaculture in developing countries for about three decades has prompted me to lay greater emphasis on practical low-level technologies, that have resulted in cost-effective production.

Though all major aspects of the subject are covered, this book is essentially intended for those with a biological background who wish to obtain a basic understanding of the principles and practices of aquaculture, as an interdisciplinary science. In a rapidly expanding field like aquaculture there are limitations to the up-to-dateness of global information that can be achieved in a book like this. However, any deficiencies in this regard may not affect its intended use. As the emphasis is on practical application, exhaustive reviews of research are not attempted, although major findings of

importance are included. Inland and marine aquaculture are not treated separately as in other publications, as I believe that the technological differences, if any, are not fundamental, nor dictated so much by salinity conditions. The distinction made between freshwater fish culture and mariculture has become largely obsolete and aquaculture has now to be considered as an integrated science. It is hoped that this book will be of some assistance in this direction too.

Rome   T.V.R. Pillay
August, 1988

1. FAO (1976) Report of the FAO Technical Conference on Aquaculture, Kyoto, Japan, 28 May–2 June, 1976. *FAO Fish Rep.*, **188**.
2. Bilio M., Rosenthal H. and Sinderman C.J. (Eds) (1986) *Realism in Aquaculture: Achievements, Constraints, Perspectives*. Summary Statement, World Conference on Aquaculture, Venice, Italy, 21–25 September, 1981, 577–86.

Acknowledgements

A number of people have assisted in various ways in the preparation of this book. I am indebted to all of them, and particularly to the following colleagues who reviewed specific chapters or sections of the book and suggested changes or additions: Imre Csavas, Regional Aquaculturist, FAO Regional Office for Asia and the Far East, Bangkok (Chapters 4 and 6); Josef Kovari, Former FAO/ADCP Aquaculture Engineer, Budapest (Chapters 4 and 6); Dr John Halver, University of Washington, Seattle (Chapter 7); Dr Nikola Fijan, University of Zagreb (Chapter 9); Dr Eugene Shang, University of Hawaii, Honolulu (Chapter 13); Dr V.R.P. Sinha, Director, Central Institute of Fisheries Education, Bombay (Chapter 15); M.M.J. Vincke, Senior Aquaculturist, FAO, Rome (Chapter 15); Michael B. New, EEC/ASEAN Aquaculture Development and Co-ordination Project, Bangkok (Chapter 15).

Appropriate credits for illustrations and tabulated data are given in the text. Special mention should, however, be made of the help of Nikola Fijan, Michael New. Josef Kovari and Ola Sveen (Svanoy Foundation, Norway) for providing several photographs. Some of the photographs contributed by the late Marcel Huet on an earlier occasion have also been included.

The main sources of documented information used in preparing the book have been listed at the end of each chapter, though citation in the text had to be kept to the minimum.

I would like to record my deep appreciation of the assistance given by Amanda Vivarelli, who typed the manuscript, and by my wife Dr Sarojini Pillay who helped edit the manuscript and prepare the index.

Part I

Principles of Aquaculture

1
Basis of Aquaculture

1.1 Scope and definition

The word 'aquaculture', though used rather widely for over a decade to denote all forms of culture of aquatic animals and plants in fresh, brackish and marine environments, is still used by many in a more restrictive sense. For some, it means aquatic culture other than fish farming or fish husbandry, whereas others understand it as aquatic farming other than mariculture. It is also sometimes used as a synonym for mariculture. However, the term aquaculture is sufficiently expressive and all-inclusive. It only needs a clarification that it does not include the culture of essentially terrestrial plants (as, for example, in hydroponics) or of basically terrestrial animals. However, when it needs to be used to denote (i) the type of culture techniques or systems (e.g. pond culture, raceway culture, cage culture, pen culture, raft culture), (ii) the type of organism cultured (e.g. fish culture or fish husbandry, oyster, mussel, shrimp or seaweed culture), (iii) the environment in which the culture is done (e.g. fresh water, brackish water, salt water or marine aquaculture or mariculture) or (iv) a specific character of the environment used for culture (e.g. cold-water or warm-water aquaculture; upland, low land, inland, coastal, estuarine), the use of restrictive terms would probably be more appropriate.

While aquaculture is generally considered a part of fisheries science, there is now a tendency to denote the distinction between the two by using the term 'fisheries and aquaculture', be-cause of some of the basic differences in development and management.

1.2 Cultural and socio-economic basis

Man depended on hunting and gathering for subsistence until the neolithic period. Fishing developed as part of this basic subsistence activity, but has witnessed considerable technological advances in modern times in methods of capture and utilization of aquatic products. Fish production from the sea increased at a rapid rate with the expansion of fishing fleets, development of efficient methods of fishing and improvements in processing and transportation of catches. Although new fishery resources were discovered, intensive fishing efforts began to show their effects on the resource base, and the increase in production, particularly of the more valuable products, has steadily declined. Overfishing and depletion of stocks have become a living reality and the need to enhance or create new stocks by human intervention has begun to be recognized.

Over the years, human societies have adopted forms of cultivation, pastoralism and ranching that were expected to stabilize production and bring it under greater human control. For various reasons, this type of evolution in the basic forms of food production has been too slow to occur in respect of living aquatic resources. Agriculture and animal husbandry probably developed from a need to adopt more productive means to feed increasing populations. In the case of fishery resources, the need to increase production was sought to be met by discovering new resources and by adopting more efficient methods of hunting and utilization. Further, unlike in agriculture, common access rights prevailed for most of the resources. Conditions have, however, changed rather drastically in recent years. The methods so far widely adopted to obtain increased production are often proving to be counter-productive. Restrictions in access rights, brought about by the new laws of the sea, have affected the fishing industries of many nations. Increasing demands in foreign and domestic markets for some of the favoured species like shrimps, salmons, eels, sea basses and sea breams and their decline or lack of potential for expansion of natural production, have created a situation where adoption of methods of farming and ranching have become logical and inevitable. Since most forms of aquaculture can be undertaken within national jurisdiction, there are fewer chances of international conflicts relating to rights and ownership in culture fisheries, except possibly in ranching operations.

There are also other concurrent factors that have promoted enhanced attention to aquatic farming. One is the recognized need in many countries to achieve greater self-reliance in food production and greater balance of international trade. Saving or earning of foreign exchange has also become an inevitable need for economic development. Further, as will be discussed in Chapter 3, aquaculture has shown its potential to increase rural employment and improve the nutrition and income of rural populations, particularly in developing countries. The labour-intensive nature of certain types of farming and the opportunities for waste recycling and integration with crop and animal farming, have made development agencies consider aquatic farming as particularly appropriate to developing areas.

Aquatic farming is also of special significance in fish marketing strategies. Production can be organized according to market demand, in respect of quantity, preferred size, colour, preservation and processing, etc. In many markets there is a special demand for fresh or chilled fish and it may not be easy for the fishing industry adequately to satisfy such a demand. Harvesting from farms can be regulated to meet this demand and make available the product during off-seasons in order to maintain regular supplies. The species can be grown to the size most preferred by consumers, when size restrictions have to be observed in capture fisheries.

1.3 Biological and technological basis

The rationale of aquaculture is not limited merely to socio-economic and marketing advantages. There are also scientific principles that weigh very much in favour of aquatic farming of fish and shellfish. It is a relatively efficient means of producing animal protein and can compare very favourably with poultry, pork and beef in the economies of production, when appropriate species and techniques are adopted. Poikilothermic (cold-blooded) animals, especially fish, have relatively low energy requirements, except for metabolism and maintenance of body functions, as they use little energy for maintenance of body temperature and normal locomotion. Since their body weight is nearly the same as of the water they inhabit, loss of energy in supporting themselves and swimming is minimal. Little energy is used by cold-blooded animals for thermoregulation. These advantages result in higher growth rates and greater production per unit area, taking full benefit of the three-dimensional nature of water bodies. Filter-feeding sessile shellfish, such as oysters and mussels, spend very little energy in obtaining their food. Fish are highest on the comparative list in terms of gross body weight gain and high in terms of protein gain per unit of feed intake (Hastings and Dickie, 1972). When fed balanced diets under favourable environmental conditions, the feed conversion ratio (wet weight gain per unit of dry feed intake) has been found to be in the range 1:1 to 1:1.25. The protein efficiency ratio (weight gain per unit of protein intake) is either equal to or higher than that for poultry and higher than for swine, sheep and steers (Hastings and Dickie, 1972). Fish are able to utilize high levels of protein in the diet, whereas in poultry almost one-half of the amino acids are deaminated and lost for protein synthesis. A weanling pig may lose as much as two-thirds of the amino acids through deamination.

The absolute economics of a culture system depend very much on the species, production technology and market conditions. Basically, low trophic feeders can generally be raised at lower costs than those which are high in the food chain and which thus require a higher proportion of proteins, particularly animal proteins. However, the latter species usually fetch higher prices in the market place and compensate for the higher production costs. As will be discussed in Chapter 3, aquaculture offers the option to produce low- or high-cost products, and it is up to the farmer to decide which. However, it has to be remembered that many types of proteins that are not consumed by man can be upgraded through aquaculture to produce highly acceptable and well-relished products. Very often, waste products of capture fisheries and animal and crop farming form the main basis of aquaculture feeds. Also, much of present-day aquaculture is based on the natural fertility of soil and water, supplemented by organic or inorganic fertilizers and the plentiful energy of the sun.

In certain situations, the application of aquaculture technologies is an inevitable necessity and not a matter of choice. The case in point is of species or populations that have been decimated by overfishing or environmental perturbations. Culture techniques have to be used to prevent the extinction of species that are ecologically or economically important to the environment. The diminishing salmon stocks in river systems of countries in the northern hemisphere and their slow rehabilitation through environmental improvements and repopulation with hatchery-produced smolts are probably a good example of the role of fish propagation. Similarly, recreational fisheries and aquaria are largely dependent on the application of culture techniques.

Irrigation and hydropower development projects, as well as land reclamation, have seriously affected fishery resources in many areas. At the same time, some of these projects have resulted in the creation of vast reservoirs that require the development of new fishery resources to compensate for the losses incurred. The potential for the application of culture techniques in developing fishery resources has been clearly demonstrated in many countries such as the USSR (Volgogradskoya and Tzimljanskoye reservoirs), China (Taihu Lake), India (Damodar Valley Corporation and Mettur reservoirs) and the USA (TVA reservoirs).

1.4 Role in fishery management

The foregoing discussions have indicated the rationale for the increasing emphasis given to aquaculture in fishery development and management programmes. While the current emphasis would appear to be in enhancing the production of high-valued species for export, its benefits in overall fishery management are also being slowly recognized. Export-oriented farming has clearly been responsible for attracting investment from the private sector and for starting several supporting industries like feed and equipment manufacture. Because of its possible role in improving foreign trade, governments in many countries are now offering incentives, including financial support, for the aquaculture sector. Industry and scientific institutions are devoting attention to research and development for the handling, preservation and presentation of aquaculture products. Even though the enthusiasm is restricted to a small number of export products, the benefits of progress can certainly trickle down to the production of other species, as sooner or later the need for diversification will be recognized by most enterprises. Even now the newly established supporting industries can be of benefit to other types of aquaculture as well.

A major element in fishery management in many countries is to prevent any increase in, and possibly even reduce, fishing pressure in the intensively fished foreshore areas. Aquaculture would probably be the only means of maintaining the overall supplies, if fishing restrictions affect the landings. Sizeable increases in production can also be expected through aquaculture under favourable conditions.

Reduction in fishing pressure in developing countries often involves the displacement of large numbers of small-scale fishermen, who are unable to obtain a reasonable income, even when unrestricted fishing is allowed. Many of these fishermen and their families are reluctant to leave their traditional homes and change to professions unrelated to fisheries. Efforts are therefore made in some areas to assist these surplus fishermen to become aquafarmers. According to some social scientists, the fisherman, who is essentially a hunter, looks down with some contempt on those who adopt land- or coast-based production methods, devoid of the excitements of open-water hunting and the prestige that is believed to go with it. However, in many areas of the world there are large numbers of part-time fishermen farmers. Further, the origins of some of the present-day aquaculture systems, such as cage culture, are to be found in the fishermen's practice of holding live fish for marketing. That, in course of time, led to fattening before sale and then to techniques of rearing from fingerlings or fry stages. Numerous oyster farmers and some of the present-day cage-farmers of yellow tail, groupers and sea basses are former fishermen.

Conflicts can arise between capture and culture fishery sectors, but with appropriate planning these two activities can be harmonized to provide an integrated development policy and programme. For many years, extraction and reforestation have formed the basic elements in the management of forest resources which in many ways is the terrestrial analogue of fisheries, and there is no strong reason why such a development cannot be achieved in fishery management. Ways of harmonizing the two sectors will be discussed further in Chapter 4.

1.5 References

Dill W.A. and Pillay T.V.R. (1968) Scientific basis for the conservation of non-oceanic living aquatic resources. *FAO Fish. Tech. Paper*, **82**.

Hastings W.H. and Dickie L.M. (1972) Feed formulation and evaluation. In *Fish in Nutrition*. (Ed. by J.E. Halver, pp. 327–74. Academic Press, New York.

Jhingran V.G. (1982) *Fish and Fisheries of India*, 2nd edn. Hindustan Publishing Corporation (India), Delhi.

Pillay T.V.R. (1973) The role of aquaculture in fishery development and management. *J. Fish. Res. Board Can.*, **30**(12), 2202–17.

Pillay T.V.R. (1977) *Planning of Aquaculture Development — An Introductory Guide*. Fishing News Books, Oxford.

Pillay T.V.R. (1983) Return to the sea — not as a hunter but as a farmer. *Impact of Science on Society*, **3/4**, 445–52.

Pillay T.V.R. and Wijkstrom U.N. (1980) Aquaculture and small-scale fisheries development. Symposium on the development and management of small-scale fisheries. *Proc. IPFC 19th Session, Kyoto 1980*, 978–87.

Pollnac R. (1978) *Socio-cultural Aspects of Implementing Aquaculture Systems in Marine Fishing Communities*. Anthropology working paper, no. 29, University of Rhode Island, Providence.

Tapiador D.D. *et al.* (1977) Freshwater fisheries and aquaculture in China. *FAO Fish. Tech. Paper*, **168**.

2
History of Aquaculture and its Present State

Despite the clear rationale for the adoption of farming as a natural evolution from hunting and gathering, the technological advances needed to achieve such a complete transformation of fishing to farming are enormous. Even though the contribution of culture to total fishery production is likely to increase steadily, and in certain cases exceed production from hunting, it is unlikely to reach the levels of human control comparable with crop and animal farming on the land, in the foreseeable future. However, what can be expected is an integrated or harmonized development of the two sectors, as in the enhancement and management of forest resources.

Large-scale aquatic farming is a relatively recent development, but small-scale aquatic farming existed in inland areas in some countries from ancient times, most likely from the time of evolution to pastoralism and land cultivation.

2.1 Origins and growth of aquaculture

Most publications on aquaculture refer to the long history of fish culture in Asia, ancient Egypt and in central Europe. The *Classic of Fish Culture* believed to have been written around 500 BC by Fan Lei, a Chinese politician-turned-fish-culturist, is considered proof that commercial fish culture existed in China in his time, as he cited his fish ponds as the source of his wealth! (Ling, 1977). Later writings of Chow Mit of the Sung Dynasty (*Kwei Sin Chak Shik* in 1243 AD) and of Heu (*A Complete Book of Agriculture* in 1639 AD) describe in some detail the collection of carp fry from rivers and in the latter publication, methods of rearing them in ponds. Even though stews or storage ponds for eels and other fish existed in Roman times and later in monastic houses in the Middle Ages, and a 2500 BC bas-relief of fish in Egypt is believed to be of tilapia raised in a pond, the earliest form of fish culture appears to be of the common carp (*Cyprinus Carpio*), a native of China. It was introduced into several countries of Asia and the Far East by Chinese immigrants, and to Europe during the Middle Ages for culture in monastic ponds. From there it spread to many countries. From the 6th century AD the common carp lost its pre-eminence in China. This is said to have been due to the identity with the name of the Tang Dynasty Emperor 'Lee', which is also the name of the common carp in Chinese (Ling, 1977). Since the name of Emperor Lee was considered sacred, it was inconceivable that lee could be cultured and caught for eating. So they looked for other species of carp and that is how the culture of the so-called Chinese carps (grass, silver, bighead and mud carp) came into being. Irrespective of whether it is fact or fiction, this and probably also the practical problems of separating larvae of different species of carp caught from rivers gave rise to the celebrated system of polyculture. Until very recent times, carp culture in ponds remained the mainstay of aquaculture in China, but for the introduction of tilapia (*Tilapia mossambica*) from Vietnam and the development of simple methods of oyster cultivation in certain foreshore areas of the coast. However, a number of other fish have been added to the species combinations, with the expectation of increasing productivity in polyculture ponds.

Fig. 2.1 Fan Lei Park in Wuxi, China, in memory of Fan Lei who is believed to have written there the *Classic of Fish Culture*.

While the Chinese immigrants were the focal points for most of the developments of fish farming in Southeast Asia, indigenous systems of Indian carp culture seem to have existed in eastern parts of the Indian subcontinent in the 11th century AD. Fish culture was practised in Indo-China for many centuries and the early systems of pen and cage culture of catfish appear to have indeed originated in Cambodia, present-day Kampuchea. Probably starting as a means of holding fish alive before marketing, flow-through culture from fry to market size with artificial feeding developed in the course of time. Variations of this system came to be practised in Indonesia for carps and in Thailand for the catfish *Pangasius*.

The earliest brackish-water farming in Southeast Asia appears to have originated in Indonesia in the island of Java during the 15th century AD. It is believed that the culture of the milkfish (*Chanos chanos*) and other brackish-water species in embanked coastal areas (tambaks) originated under the influence of the Hindu rule, and by the 18th century there were over 80 000 acres (32 389 ha) of ponds. The early tambaks are reported to have

been constructed by convicts who were sent to the coastal areas to work on salt marshes and to guard the coastal fires.

As mentioned earlier, the history of aquaculture in Europe starts from the Middle Ages with the introduction of common carp culture in monastic ponds. Common carp attained a social and religious significance as the chosen food to be eaten on special occasions, as for example Christmas, in certain areas. However, there was also a certain amount of prejudice against it in some Western countries, particularly because of the lack of acceptance of its culinary properties, and it was considered a pest because its feeding habits gave rise to soil erosion and muddying of water, particularly water used for game fishing. Despite this, carp culture continued and flourished in almost all East European countries and from there it was introduced into the present-day Israel. In recent times, the polyculture of Chinese carp has also been adopted in many of these countries.

The propagation of trout, which has a fairly long history, originated in France and the monk Don Pinchot, who lived in the 14th century, is credited with the discovery of the method of artificial impregnation of trout eggs (Davis, 1956). Being a sport fish and of more widely accepted culinary properties, trout culture spread to almost all continents in the course of time. Even though early efforts were focused on repopulating natural water bodies for improving sport fishing, pond culture and other forms of intensive culture gradually developed to produce fish for the market. Commercial trout culture in fresh water on a fairly large scale developed in countries like France, Denmark, Japan and recently in Italy and Norway. During this period, the culture of the Atlantic salmon also became a commercial success and with the development of cage farming of salmon and trout in Norwegian fjords, salmonid culture achieved a remarkable boost in production and public attention.

The British introduced trout in their colonies in Asia and Africa, mainly to develop sport fisheries. The early development of fish culture in North America was centred on the propagation of salmon and trout, and to a lesser extent on the black bass. Starting in the 18th century, trout hatcheries were established in government stations mainly for release of fry into open waters, but in the course of time the private sector started commercial production of consumption fish. Slowly the practice of trout propagation for release in open waters or, more recently for farming, spread to the temperate and semi-temperate areas of Central and South America.

When tracing the history of fish culture, one has to take into account the rather ancient practice of breeding and rearing ornamental fish, such as goldfish by the Japanese and the Chinese. The spread of tilapia, a native of the African continent, to several countries in all parts of the world is a remarkable phenomenon. Even though there was resistance to its introduction in many countries and it was considered as a pest by some, its culture spread far and wide, especially in developing tropical countries. Tilapia culture was considered by many as an easy means of producing cheap proteins for the masses. Research and experimentation have in recent years found solutions to some of the problems of culturing tilapia, and commercial-level farming has developed in certain areas.

The oldest form of coastal aquaculture is probably oyster farming, and the Romans, Greeks and Japanese are believed to be the earliest oyster farmers. Oyster culture in intertidal stretches is said to have been practised in Japan around 2000 years ago. Aristotle mentions the cultivation of oysters in Greece and Pliny gives details of Roman oyster farming from 100 BC. The culture of other molluscs, like mussels and clams, which follow methods similar to oyster farming appear to have developed much later.

From a historical point of view, the only other culture system that needs mentioning is the large-scale farming of seaweeds, which is of relatively recent origin. The earliest text book of seaweed culture appears to have been published in Japan in 1952. After the Second World War, culture of edible seaweeds expanded and intensified considerably and spread to other countries like Korea, Taiwan and mainland China.

2.2 Present state of aquaculture

Although traditionally fish farming was part of rural life in certain areas, the present day aquaculture has a much greater significance in socio-economic development and natural resource management. Despite some temporary increases in a few areas, the total world capture fishery production appears to have plateaued around 70 to 75 million tons, with the production of preferred species remaining stationary or in some cases diminishing. The fraction of the total catch utilized for human consumption has increased from 58 to 70 per cent, due to increased demand and increased utilization through new or improved processing techniques and marketing of value-added products. The most optimistic estimates of total catch of conventional species from the wild are around 100 million tons, and any significant increase due to harvesting of new unconventional species for food is considered unrealistic due to problems of consumer acceptance, harvesting technology and costs.

If the hoped for 100 million tons catch is obtained, about 70 million tons can be expected to become available for human food at the current rate of utilization. Even if this can be increased, the maximum total catch used for human consumption cannot be expected to surpass 80 million tons. On the other hand, it is estimated that about 100 to 140 million tons of edible fishery products will be required to meet the demand of the projected world population by the year 2000. There is thus a deficit of approximately 20 to 60 million tons to be made up, and the only major means presently known for this is an accelerated development of aquaculture.

As indicated in Chapter 1, aquaculture has been historically a small-scale activity. Some spectacular successes have been achieved in large-scale commercial farming, but in the minds of many it is still a development for the future. It is true that aquaculture contributes probably not more than 15 per cent of what the capture fishery contributes on a global basis, despite the fact that in certain areas and sectors the volume of production and economic significance are much greater. Culture technologies are far from perfect and research efforts to develop and improve technologies are in the very early stages. The promotional efforts of dedicated institutions and individuals have created the recognition of its potentials and provided the climate suitable for testing theories and practices on a much larger scale than had been possible before. Being a new and emerging industry, there are bound to be more mistakes made than in established ones, for conceptual, technological or managerial reasons. But, as pointed out by Rosenthal and Murray (1986), 'under good management many of the aquaculture systems have surpassed all expectations of a decade ago'. If one prefers to take a negative approach, there are also systems, considered as real breakthroughs, which have failed to take off. It has also been demonstrated that aquaculture programmes have a relatively longer gestation period, in comparison with fishing or other forms of food production. Even when tested technologies are adopted, the construction of physical facilities (particularly pond farms), solution of site specific problems, the building up of the productivity of the system and, above all, attainment of skills by workers take considerable time. Lack of allowance for such time-lags has often resulted in the premature termination of many enterprises.

In a discussion of the worldwide state of aquaculture, individual successes and failures can serve only as indicators. The type of statistics that will be needed for an appraisal of the situation are unfortunately not available. In the absence of suitable mechanisms for the collection of aquaculture statistics in most countries, The Food and Agriculture Organization of the United Nations (FAO) has been making estimates of world production at frequent intervals, based largely on data provided by various governments. The estimates of total production were 5.0 million tons in 1973, 6.1 million tons in 1975, 8.7 million tons in 1980, over 10.5 million tons in 1983, about 10.6 million tons in 1985 and over 13.2 million tons in 1987. It is difficult to determine the accuracy of the estimates, as different types of computations have been used in certain countries, such as total production based on average yield per unit area, conversion of processed products to wet weight of harvests, isolation of harvests of cultivated species from total landings in large water bodies containing resident species, etc. It is also likely that the productions from many small-scale farming operations have been overlooked, as government institutions may have had no records of them and their harvests may not reach major markets. The possibility of some of the increases in estimates being due to better coverage also cannot be ruled out. Nevertheless, the available figures clearly show the main trends in production and demonstrate convincingly that aquaculture is a growth industry, indicating that some of the forecasts of future production may not be unattainable. It is believed that a production of over 26 million tons by the turn of the century can be achieved if the observed rate of increase is maintained and the necessary technical, financial and policy supports become available.

The regional distribution and composition of world aquaculture production for 1983, 1985 and 1987 are given in Tables 2.1 and 2.2 respectively. Asia is the largest aquaculture producer, followed by Europe. The rates of increase in production in North America and in Africa are remarkably higher, although their overall contributions still remain rather small. Available data for South America do not seem to permit any justifiable conclusions on development trends in the region. Analysis of the composition of production figures shows that major increases are due to expansion of finfish

Table 2.1 Regional Distribution of Aquaculture Production (in million tons).

Region	1983	1985	1987
Africa	43 865	61 100	62 502
Asia	8 642 377	8 928 800	11 131 302
Europe	1 277 853	1 136 600	1 340 181
Near East			23 834
North America	316 203	392 600	449 993
South America	220 478	68 200	200 104
Total	10 500 776	10 587 300	13 207 916

Table 2.2 Composition of Aquaculture Production (in million tons).

Product	1983	1985	1987
Finfish	4 671 244	4 717 500	6 793 441
Molluscs*	3 301 948	2 798 600	2 672 394
Crustaceans[†]	133 800	265 700	574 906
Seaweeds	2 393 784	2 777 200	3 139 473
Others		28 300	27 702
Total	10 500 776	10 587 300	13 207 916

* (Oysters, mussels, etc.)
[†] (Shrimps, prawns, crayfish, etc.)

and crustacean culture, for which there is more widespread consumer demand.

Extensive, semi-intensive and intensive systems of production are adopted according to local conditions. Extensive systems are characterized by low inputs, maximum use of natural processes for the production of food, low density of stock and low harvest per unit of area under culture. In countries having large-scale operations, a gradual evolution towards semi-intensive and intensive systems can be observed. This involves higher stock densities, hatchery production of seed where feasible, greater human control of environmental conditions, at least supplementary feeding, and higher yields per unit area. Social and economic changes seem to require the adoption of semi-intensive systems in many areas to make aquaculture a viable industry.

2.3 References

Davies, H.S. (1956) *Culture and Diseases of Game Fishes*. University of California Press, Berkeley.

Hickling C.F. (1962) *Fish Culture*. Faber and Faber, London.

Hora S.L. and Pillay T.V.R. (1962) Handbook on fish culture in the Indo-Pacific region. *FAO Fish. Biol. Tech. Paper*, **14**.

Ling S.W. (1977) *Aquaculture in Southeast Asia — A Historical Review*. University of Washington Press, Seattle.

Nash C.E. (1987) *Future Economic Outlook for Aquaculture and Related Assistance Needs*. ADCP/REP/87/25, FAO of the UN, Rome.

Pillay T.V.R. (1973) The role of aquaculture in fishery development and management. *J. Fish. Res. Board Can.*, **30**(12), 2202–17.

Pillay T.V.R. (1976) The state of aquaculture 1976. In *Advances in Aquaculture*, (Ed. by T.V.R. Pillay and W.A. Dill), pp. 1–10. Fishing News Books, Oxford.

Pillay T.V.R. (1985) Some recent trends in aquaculture development. In *Status and Prospects on Aquaculture Worldwide*, pp. 61–4. Aquanor '85, Trondheim.

Rosenthal H. and Murray K.R. (1986) System design and water quality criteria. In *Realism in Aquaculture: Achievements, Constraints, Perspectives*, (Ed. by M. Bilio, H. Rosenthal and C.J. Sindermann), pp. 473–93. European Aquaculture Society, Bredene.

Schuster W.H. (1952) Fish culture in brackishwater ponds of Java. *Indo-Pac. Fish. Com. Spec. Publ.*, **1**.

United Nations Development Programme–FAO (1988) *Aquaculture and Coordination Programme Aquaculture Minutes* **1**.

3
National Planning of Aquaculture Development

Most of existing aquaculture has developed through isolated and uncoordinated efforts. But in view of the new roles that it is expected to have in national economic development and the rapid expansion envisaged, it is most desirable for a country to have a national plan, defining objectives, policies and strategies that are most suited for achieving the selected goals and targets. The need for clearly defined policies and plans for aquaculture in both developing and industrially advanced countries has been widely recognized in recent years, irrespective of whether the country adopts a centrally planned or market economy. Based on such macro-plans, specific development projects or plans can be formulated by the private or the public sector, after detailed feasibility studies, including site surveys and studies of technical and economic viability in the proposed areas.

Some of the basic considerations in overall planning are discussed in this chapter, with a view to pin-pointing its role in the fisheries sector and in the national economy. Many of these will come in for further discussion in some of the succeeding chapters dealing with actions involved in establishing and operating aquaculture farms and related activities. An attempt is made to outline the various steps involved in designing a national plan and methods of revising and updating it. Being a new and emerging industry in many countries, a national plan is a prerequisite for the orderly development of aquaculture and this accounts for the inclusion of it in this book, before descriptions of technological aspects.

The objectives of aquaculture development depend on the socio-economic conditions of the country and on environmental suitability. National priorities may differ very significantly between countries, but in the majority of situations aquaculture can have an important role, as for example in:

(1) increasing food production, especially of animal proteins, and achieving self-sufficiency in aquatic product supplies,

(2) producing food near consuming centres in rural areas, thus contributing to improvement of human nutrition,

(3) supplementing or replacing capture fishery production of over-exploited fish and shellfish stocks,

(4) generating new sources of employment in rural areas, including part-time employment of farmers and small-scale fishermen, and arresting the migration of people from rural to urban areas,

(5) overall development of rural areas through integrated projects, including aquaculture,

(6) earning foreign exchange through export or saving foreign exchange through import substitution,

(7) using waste lands productively and using organic wastes for food production and environmental management,

(8) creating and maintaining leisure-time activities, including sport fishing and home and public aquaria,

(9) promoting agro-industrial development, which could include processing and marketing of fishery products, feeds and equipment for aquaculture, and seaweed culture for the production of marine colloids, pearl oyster culture, etc.

3.1 National priorities and aquaculture development

Although aquaculture planning has necessarily to be based on the priorities and existence of conditions where aquaculture can make a significant contribution, in the majority of countries increased food production and attainment of self-sufficiency will form an integral part of economic development policy. As mentioned earlier, integrated rural development, which could include rural aquaculture, will also have high priority in many countries. In either case and, for that matter, in any form of aquaculture planning, the first step should be an examination of the state and contribution of capture fisheries and projected future production against expected demand. A global estimate of supply and demand by the end of the century is attempted in Chapter 1. Similar estimates on a national basis may be made to identify in as precise a manner as possible the role of aquaculture in national fishery production. For various reasons it will be preferable to harmonize aquaculture production within the framework of overall fishery production and prevent unnecessary competition in the market place. This does not mean that aquaculture should not take advantage of its inherent strengths of product quality and regularity of supplies. Production can be planned within the limits of demand, so that both sectors will have fair markets. It will also be desirable to select species for which there is no capture fisheries or the landings of which are insufficient to meet consumer demand.

One other means of harmonizing aquaculture development with capture fisheries is by the provision of opportunities for part-time or full-time employment to excess fishermen. Coastal fisheries in many countries are presently over-exploited and the small-scale fishermen are often unable to make a living because of dwindling catches and the high cost of fishing. By rehabilitating excess fishermen in aquaculture projects, fishing pressure in coastal waters can be reduced and catch per unit of effort increased, to make fishing more profitable.

Based on the estimates of current and future capture fisheries production and the projections of demand for domestic consumption and export, the aquaculture production needed to fill the gaps in demand and supply can be determined. This of course, will only show in general terms what is likely to be absorbed in the market, as assessed largely on the basis of existing conditions. In aquaculture, market-oriented production is the general practice, unlike the production-oriented marketing in capture fisheries. So there is a need to obtain basic information on consumer preferences and demand, both within the country and in export markets. There may also be circumstances in which less acceptable products may have to be produced to meet national priorities, such as a cheaply produced fish for feeding needy sections of the population, supported by strong promotional activities. This option should also be taken into account, where appropriate, in setting targets of production.

3.2 National resources

The feasibility of achieving the required production will naturally depend on a number of factors including natural, technical and human resources, legal and environmental conditions and funding and financing arrangements. Land, water and climatic conditions are probably the most important natural resources to be assessed. Though detailed site surveys cannot generally be done in the overall planning stage, identification and mapping of potential areas have to be attempted. Among land areas, priority will naturally be given to those that are presently not productively used, as for example in agriculture. However, in some areas where economic agriculture is not possible due to soil or other conditions, even agricultural land may be suitable sites for pond farms. In the assessment of sites for aquaculture, careful consideration has to be given to possible environmental impacts. There is often a tendency to describe all lands unused by man directly as potential sites for aquaculture, and estimate production potential of the country based on total areas of such lands, as for example coastal swamps.

Many of the coastal swamps are under mangroves, the ecological importance of which has received special attention in recent years. Dense mangroves along the margins of estuaries, lagoons and bays prevent soil erosion and aid silt accretion (Fig. 3.1). The marginal zones that are inundated regularly by tides contribute substantially to the productivity of the waters and form the nursery grounds of many marine and some inland species of fish and shellfish. Conversion of such areas into fish ponds is likely to affect the capture fishery resources and adversely affect the environment. In fact, such areas should not be considered for pond farms, for reasons other than ecological. In the first place, it would be too expensive and difficult to clear the tall mangroves that grow along the margins and the dense growths behind it. In addition, pond farms built along these marginal zones will be more exposed to natural disasters like high tidal storms, typhoons and cyclones. On the other hand, the back mangroves, characterized by less dense and bushy growths and only occasional tidal inundation, contribute little to aquatic productivity and are more easily cleared and converted into pond farms (Fig. 3.2), so only these areas should be considered when assessing available land for aquaculture. Similarly, when considering coastal waters, other uses of such areas should

be taken into account such as for navigation or recreation, besides possible problems of environmental management, including water quality control and waste disposal.

For land-based aquaculture, the availability of adequate quantities of water of appropriate quality is an obvious requirement. Natural bodies of water, man-made irrigation or multipurpose systems and ground water are major resources to be assessed, besides rainfall which is not always a reliable direct source of supply. The quantity of water required will depend on the type of culture system to be adopted as, for example, stagnant ponds, flow-through systems, cages or pens, etc. Though technologies for recirculation of water exist, large-scale use of it other than for hatcheries seems too expensive at present. As well as the source of water, the means of supplying it to the aquaculture facilities is also important, because of the costs and availability of energy for pumping.

Assessment of suitable areas for development will lead to the determination of the systems of culture or species of animals or

Fig. 3.1 Dense mangroves along the margins of creeks in the Niger delta, Nigeria.

Fig. 3.2 Back-swamps of red mangroves in the Niger delta, Nigeria.

plants that can be cultured. Although many cultured species are gradually being acclimatized to grow in different temperatures and salinity ranges, there are still tolerance limits and ranges within which they will grow fast or reproduce. So, agro-climatic conditions are a determinant in species selection. As mentioned earlier, consumer preference and acceptance will also have to be considered.

3.3 Technology and human resources

The existence of tested technologies or the ability to develop or adapt existing technologies to suit local conditions, is an important aspect to be taken into account in a planning exercise. As problems of aquaculture are very often site-specific, even well-established technologies have to be adapted or modified for local application and tested to determine their economic viability. This would require at least minimum research capability and facilities. Large-scale developments often necessitate a regular health inspection and disease diagnostic programme and this could form a part of the research

establishment or of an extension service attached to it.

The development of appropriate human resources should form part of any national aquaculture development plan. The primary importance of hands-on experience in successful farming has been shown all over the world. While traditionally the skills have been passed from father to son, in modern large-scale rapid development programmes it may not be feasible to follow this practice. Organized, institutionalized training may be needed for farm managers and technicians. Since aquaculture is interdisciplinary by nature, specialized training programmes will be required for major categories of personnel. Through analysis of farm performance data, basic guidelines for farm management have to be developed and farm managers must be trained in the use of relevant methods, including decision-making, as described in Chapter 14.

Universities and specialized research institutes will be the major sources of research personnel. Depending on the organization of aquaculture in the country (small-scale, large or industrial-scale) a suitable extension machinery with the appropriate number of adequately trained and experienced extension personnel will have to be built up. Extension officers have to be well-trained technicians with necessary field experience and proficiency in extension methodologies. A combination of institutionalized and on-the-job training may, therefore, be required for extension personnel.

3.4   Legal and environmental factors

For an orderly development of aquaculture, as for any industry, a suitable legal framework is essential. As a first step, it is necessary to examine existing laws and see how far they are applied to, or applicable to aquaculture. In many cases, it may be found that there is no accepted legal definition of aquaculture or any of the associated terms like fish culture, mariculture, etc. in the country. Fishery laws do not normally apply to aquaculture, nor can it be brought under the existing regulations relating to agriculture and animal husbandry. Even though it is realized that specialized regulations are necessary to meet the specific needs of aquaculture, there is a growing body of opinion in favour of bringing it under the same legally protected and financially assisted footing as agriculture. In fact this has already been done in some countries, and seems to have helped aquaculturists gain access to greater incentives and scientific and extension support. Aquaculture has, in many cases, to be closely associated with agriculture and animal husbandry in integrated rural development programmes. However, there are many who feel that aquaculture should continue to be a part of fisheries and legally come under it because of its closeness or identity in the secondary and tertiary phases of the industry (harvesting handling, processing and marketing). The need and potential for harmonizing farming with fishing on a national or regional basis are also arguments in favour of this. In ranching or enhancement of natural stocks with hatchery-raised young, the need for harmonized fishing restrictions becomes specially important.

One of the first problems that an aquaculture entrepreneur faces is in obtaining the right to establish and operate a farm in a suitable area. As mentioned earlier, generally wastelands unsuitable for agriculture are utilized for land-based aquaculture, and foreshore or protected coastal areas for other types of open-water farming. Often the ownership of these is vested in the state or local communities. Whether they can be acquired on outright purchase or long-term lease will depend on existing laws. The duration of the lease, maximum extent of land or water area that can be leased or sold to individuals or private enterprises and the environmental regulations that limit the use of the sites are major considerations. Just as existing regulations may be restrictive, the lack of specific regulations governing the use of such areas for aquaculture or the interpretation and application of regulations promulgated for other purposes (without any reference whatsoever to aquaculture) may become a major impediment.

In cases where regulations for sale or lease of sites for aquaculture exist, the procedures may be long and complicated, often requiring the clearance and permission of many departments and authorities, as for example land revenue for property rights; irrigation department for water use; agriculture department for water use in relation to agriculture in neighbouring lands; department of environment for environmental impacts; fisheries department for regulations on fishery production, culture of selected indigenous or exotic species and marketing. In order to facilitate aquaculture development, these procedures have to be streamlined and simplified, so that the necessary decisions will be made in a reasonable period of time, based on a proper understanding of the case. Successive project reports, government inspections (or public hearings where applicable) and departmental clearances before issue of permission can not only be frustrating to entrepreneurs, but also very expensive and time-consuming. In order to harmonize the various interests involved and to avoid conflicts among users, it may often be necessary to resort to zoning of areas. The development of a unified approach and the establishment of a suitable administrative authority at the regional or federal level have been suggested as appropriate for effective consideration and accommodation of all relevant interests and for early decisions on aquaculture proposals.

For various reasons it may be reasonable and necessary to limit areas to be used for aquaculture or the number and magnitude of operations to be permitted. But this should be based on relevant data on socio-economic and environmental impacts, sanitation and fish health, markets, etc., and measures may have to be taken to prevent the misuse of permits and speculative deals.

A legislative safeguard is required for the successful operation of aquaculture in order to prevent the introduction and spread of infectious diseases. This would entail the regulation of movement of uninspected ova and fry and institution of quarantine measures where feasible. As far as possible, eggs and fry should be obtained only from hatcheries and farms that are regularly monitored by fish health specialists and certified by competent authorities as free from disease. Indiscriminate introduction of non-indigenous species has to be controlled, but the rules should not be arbitrary and must be based on scientific evidence of adverse effects.

In order to facilitate investments in the sector, an aquaculture plan may have to include possible incentives. Besides offering the usual tax holiday for the run-in-period offered to new industries, the types of subsidies and loans given to agriculture could be extended to aquaculture as an incentive in the early stages.

As in other types of industry, there is currently much interest and some movement towards foreign investment and joint venture operations, particularly for the farming of species that have export markets. Review of statutes and regulations relating to foreign investments, admissible percentage of equities, partnership arrangements and general administrative procedures should be carried out with a view to attracting foreign participation and reducing red tape in the establishment of new ventures.

Aquaculture may, in some cases, require special market regulations. The farmer, as well as the consumer, will benefit from the availability of products throughout the year. For this purpose, it may be necessary to permit the sale of farmed products when there is a closed season in the capture fishery. Similarly, it may be an advantage for the farmer to grow the fish or shellfish to a size below the minimum size allowed to be fished in capture fishery, to meet specific market demands. Unless legal provision is made to exempt culture fishery products from such restrictions, aquaculture industry will lose some of its intrinsic advantages.

If aquaculture is expected to become a sizeable industry in the country, it will be advisable to consider the need for a regular compulsory health inspection programme for all major communicable diseases through certified inspectors with adequate diagnostic facilities. Legal provision may also be necessary to build up a suitable insurance programme for aquaculture. Most forms of aquaculture are presently classed as high-risk activities because of a number of hazards which are beyond the control of the operators, such as adverse weather conditions, changes in water supply, natural calamities like typhoons, cyclones and floods and epidemics of mortality due to communicable diseases. Despite major problems in arranging insurance cover, an aquaculture development plan should aim towards establishing a suitable insurance facility in the private or public sector.

3.5 Organization of aquaculture

As indicated earlier, aquaculture can be organized at different levels in order to meet specific developmental objectives. Major support services, such as research, training and extension are generally organized by the State, even though all of them can as well be organized in the private sector. However, when the national policy is to establish aquaculture in the form of small-scale operations as an integral part of rural development, the State has necessarily to take the responsibility for these, or at least take a leading role, and enlist the cooperation of non-governmental agencies including cooperatives. Similarly, greater involvement of government may be needed in making available credit on reasonable terms to small producers and in promoting the production and distribution of inputs such as feeds and fertilizers, as well as farm equipment. In the absence of effective cooperative organizations, government assistance may be required for making appropriate arrangements to market products within the country and to export to foreign markets.

On the other hand, large-scale enterprises can ideally be organized on a vertically integrated basis. Hatchery and seed production, grow-out, feed manufacture, processing and marketing of products can all be integrated into one unit, if the necessary natural resources like land, water and energy are available in adequate quantities. Restrictions relating to land ownership or use of foreshore or inland waters would naturally affect the size of farms and therefore the possibilities of such integrated enterprises, conforming to minimum economic size requirements. There is, of course, the possibility of the owners of a large number of small areas combining their resources to establish and operate a large enterprise on a cooperative basis. It is also possible to organize production on a large number of small holdings, with a central managing or coordinating agency providing the necessary finance, technical assistance, essential inputs and marketing services. Despite the large number of small units of production, the enterprise could then function with economic efficiency and effective overall management.

3.5.1 Aquaculture for rural development

Irrespective of the economic or other benefits of large-scale aquaculture operations, greater emphasis is laid on small-scale farming in developing countries. This is largely because of the opportunities it offers for part- and full-time employment, which help in sustaining peasants and fishermen in rural areas, reducing the drift of populations to urban centres.

If aquaculture has to be developed for socio-economic benefits, it has to be planned for that purpose. In view of the importance given to this subject, it will be useful to consider it in some detail.

The small size of a farm alone cannot be expected to yield the desired social benefits. The most important factor is that in this type of aquaculture the focus of development has to be the farmer or the community, and not the aquaculture product *per se*. So, development has to be designed on the basis of the social, economic and behavioural patterns of the community involved.

In the vast majority of people-oriented aquaculture projects, the immediate target groups are likely to be the poorest of the poor, including landless labourers and marginalized peasants. It is generally accepted that if these groups are really to benefit, the project should be based on an adequate understanding of the needs, desires, behaviour and capacities of the people and their indigenous institutions. The basic needs of the community are considered to be food, clothing, primary housing, household equipment, sanitation, water supply, cheap mass transport, elementary education and extension services, basic medicine, simple health services, etc. It should be obvious that an aquaculture development programme by itself cannot provide these needs directly, or in most cases even indirectly. It has to be integrated with, or complemented by, other development activities. Hence the need for integrated community development programmes aimed at the required production and consumption in terms of essential goods and services, where aquaculture could play a major role.

While the ideal people-oriented aquaculture programme is an integrated one, it should be recognized that sectoral programmes can also yield major socio-economic benefits, such as improvement in the availability of protein food, enhanced income contributing to improved purchasing power of the population and better standards of living. Employment opportunities generated through aquaculture development, including production, processing, transport and marketing, can be expected to control, to some extent, the drift of rural people to urban areas. Large-scale development of aquaculture can also eventually lead to better communications in rural areas, as they are needed also for proper management of aquaculture production and distribution.

3.5.2 Aquaculture for social benefit

Whether a project is intended to meet fully or partly the socio-economic needs of a community, it is necessary to design it carefully to provide the expected outputs. On the assumption that the potential for aquaculture development in the area is established, priority has to be given to the study of the community. It should aim at identifying the basic needs to be fulfilled and those that can be met through an aquaculture programme. For example, if an increase in family income is needed to afford the specified basic necessities, the project has to be designed to yield at least that minimum income required. A knowledge of the level of human, economic and social infra-structure development, and the cultural and political context in which the programme has to be implemented, is necessary for appropriate project design.

The technology or the farming system to be adopted will have to be carefully selected, not only on the basis of the agro-climatic and hydrological conditions of the area, but also on the skills and educational background of the target population and their socio-cultural system. As it is often not practicable for those concerned with the project design to live long enough among the target community to learn all that needs to be learnt, adequate flexibility should be built in. It should be possible to make necessary project changes later, based on field tests and the results of the early-phase activities.

Participation of the local community

The need for the participation of the local community in planning and implementation in rural development projects is a widely accepted ideal. Where agriculture and allied industries are organized through cooperatives or communes, there are established mechanisms for broad participation in decision making and benefit sharing. Even though the ideal is seldom reached, people get opportunities to express their views and influence decisions. But in the majority of developing countries most people belong to what is called atomistic societies, characterized by extreme individualism, great reliance upon household members, and general distrust of others outside the household. Tribal, communal or plantation communities have more cohesive social organization, but these form only a minority in the global context. Government-sponsored cooperatives in unregimented societies have a rather poor performance record in most countries. Success rates tend to be higher when the members have reached a certain socio-economic status. So the option is either to concentrate on individuals or family units or to form or seek the intervention of non-governmental voluntary agencies.

Many of the basic needs of a community could be factored into individual or family needs and an activity that meets these needs, and leads to improvement in their standard of living in the aggregate, may constitute social benefits to the community as a whole. The improvements in the economic well-being of the individuals and families can be expected to result in greater political clout and assertiveness to demand from the State, social services that they themselves cannot develop. Such an approach, however, may sometimes lead to the accumulation of benefits of development in a few hands, and thus alter the social structure. The assumption that the success of receptive and progressive individuals will motivate the rest of the community to adopt the productive activities may not always prove true.

The alternative of close involvement of a non-governmental voluntary agency has the potential to reach the community and its individuals more easily and motivate them to adopt development activities. However, the effectiveness of such agencies will depend largely on their organization, objectives and the motivation and dedication of their workers. In cases where no such agencies exist, it will be necessary to promote their formation. The need for such agencies is greater, and their involvement more valuable, when the project activities are intended to benefit rural women. In traditional rural societies, government agents, especially men, may find it very difficult to reach and establish rapport with the women's groups. Suitably trained women volunteers would have a better chance of achieving this.

Ideally, people's participation should be spontaneous and by the free will of the community. However, in practice it has often to be achieved through effective education, persuasion and demonstration of benefits. Participation is needed not only at the initial decision-making stage, but also during implementation, including decisions on benefit sharing. Educating the target group for proper understanding and appreciation of the development programme and national policies and procedures relating to financing and credit are essential to enable meaningful participation.

Employment opportunities

Generally, small-scale aquaculture projects provide more employment opportunities per unit of capital invested than larger farms. In addition, they have the advantage of being more widely distributed geographically and locally owned, enabling improved income distribution among the population. The preference has therefore to be for small-scale farms in people-oriented aquaculture. However, the size and production should be adequate for the targeted income to be earned by individuals or families and should conform to the minimum economic size of the particular type of farming.

When aquaculture is developed as an additional or part-time activity, the system selected has to be compatible with, or complementary to, the normal vocation of the target group. Crop and animal farmers may find it comparatively more easy to integrate fish culture with their on-going farming activities and obtain increased production at minimum cost. When a small-scale fisherman wants to undertake aquaculture on a part-time basis to supplement his income, he will normally have to base it near his dwelling, as for example on the sea coast, estuary, river or lake. Although land-based aquaculture may be feasible in some areas, cage farming and raft culture of molluscs may be easier to adopt, not only because of the technology but also the ease with which attitudes to such activities can be influenced. Since these systems of farming have a closer association with capture fishery environments and practices, the fisherman who is basically a hunter may find it easier to adapt to such farming approaches.

3.5.3 Industrial-scale aquaculture

When a target production has to be achieved and maintained for meeting substantial local or foreign market demand, in the majority of cases the answer may be large-scale production. Small-scale aquaculture can often make a greater impact on local consumption, but maintaining regular supplies to distant markets can prove more difficult and expensive for small holders. As well as the economies of scale in production, economic arrangements for storage, transport and processing are inherent strengths of large operations. It also becomes possible to be more self-reliant in the supplies of inputs. The volume of products may allow the owners to have their own processing and marketing arrangements. It also becomes possible to introduce mechanization in many of the operations if necessary and so save labour, increasing cost-effectiveness. The magnitude of operations often justifies and enables the maintenance of in-house expertise and in some cases even problem-oriented research and on-the-job training of field personnel.

Experience gained so far has shown the crucial importance of effective management in the successful implementation of aquaculture. Large-scale enterprises make it possible to recruit and maintain experienced managers and thus improve the chances of success. Small-scale aquaculture has often to depend on special credit arrangements organized by government institutions. Larger enterprises can in most cases depend on existing industrial or agro-industrial financing, including raising of capital through public financing and sales of shares in the open market.

Another advantage of larger aquaculture enterprises is the potential to maintain more efficient security arrangements for the farm stock. Poaching and wanton killing of stocks are problems faced in varying degrees in almost all countries, but when the farm is small and production consequently small, the cost of keeping an adequate security staff or installing safety equipment may prove prohibitive.

While there are a number of advantages of the type mentioned above in large-scale operations, there are also some disadvantages. An important one is the need to start with tested technologies suited to such operations. Many of the traditional practices, which can easily be adopted in small farms, may not be suitable for large farms, as for example reliance on wild caught fry for rearing, maintenance of algal pastures for feeding in grow-out facilities, etc.

The greater capital outlay needed and the requirements of raising such capital on the open market bring in a number of limitations. Besides the normal investment criteria such as expected rate of return, payback period, degree of risk, etc., the investor will have to compare financial benefits from aquaculture with those from alternative ventures. Generally speaking, one can say that the larger the farm, the longer the start-up period. Even when the construction and operation are phased, it may take some time before the shareholders can expect reasonable returns on their investments. It may well be that production and therefore incomes are low in the early stages. The possibility of compensating low profits from primary production by higher profits from marketing, and in the case of export-oriented aquaculture, the incentives offered by the State, may to some extent offset these disadvantages.

One other problem with regard to large-scale aquaculture that needs to be mentioned is the disposal of higher quantities of waste water or pollutants. It is preferable to recycle the wastes, but when this is not possible, suitable waste treatment facilities will have to be provided. Spread of diseases through the discharge of farm effluents must also be avoided through suitable effluent treatment.

3.5.4 Investment requirements

A national development plan has necessarily to assess the magnitude of investment needed to achieve the targeted production and determine possible sources. As aquaculture procedures are diverse and depend on the systems and species cultured, uniform models for estimating investment costs and carrying out project analysis cannot be expected. In any case, for planning purposes only rough estimates of costs and returns of sample operations will be needed. The most useful data for such estimates would be commercial or pilot-scale aquaculture carried out within the country. In the absence of such data, nearest approximations, based on experience in more or less similar circumstances in comparable countries, have to be used. Methods of calculating financial and economic feasibility of projects, including internal rates of return are described in Chapter 13. An appropriate method has to be used to determine the potential viability of the selected aquaculture systems in the country. To estimate the impact of aquaculture enterprises on the national economy, the methods of determining direct and secondary benefits and costs described in Chapter 13 may prove useful. It will also be useful to determine the contribution that the aquaculture sector can make to the gross national product (GNP) and foreign exchange earnings or savings of the country, in specified periods of time. These estimates could form the basis for proposals for financing various actions required under the plan.

3.5.5 Plan reviews and revisions

The plan period may, in many cases, depend on the period of the overall 'national economic development plan' if one exists. Where none exists, a suitable period for an aquaculture plan has to be determined, based on the activities proposed and their phasing. Many such plans have selected a period of 5 to 10 years, with details of development activities and budgetary provisions on an annual basis. A suitable mechanism for coordination and review of plan activities in both public and private sectors would be very valuable. Any long-term development plan would need periodic reviews to determine the relevance of each activity, appropriateness and effectiveness of its implementation, budgetary requirements, benefits of the projects including the removal of remaining constraints and modification of implementational strategies. The need for such reviews in the case of projects formulated on the basis of limited data and experience has been emphasized earlier in connection with socially oriented small-scale aquaculture. Being a new and fast developing science, considerable changes in technologies can be expected to occur in short periods of time. This is also true for market conditions, particularly in respect of export products, and this may necessitate changes in technology or the nature of the products. So, provision for flexibility in plan implementation and the ability to revise project objectives should the need arise are essential features of a realistic plan.

3.6 References

FitzGerald W.J. Jr (1982) *Aquaculture Development Plan for the Territory of Guam*. Department of Commerce, Government of Guam.

Freshwater Institute, Winnipeg (1973) Summary report of the Government/Industry Policy Development Seminar on Aquaculture, 1973.

Gerhardsen G.M. (1979) Aquaculture and integrated rural development, with special reference to economic factors. In *Advances in Aquaculture*, (ed. by T.V.R. Pillay and W.A. Dill), pp. 10–22. Fishing News Books, Oxford.

Hawaii State Center for Science Policy and Technology Development (1978) *Aquaculture Development for Hawaii: Assessment and Recommendations*. Honolulu, Department of Planning and Economic Development, State of Hawaii.

Mayo R.D. (1974) *A Format for Planning a Commercial Model Aquaculture Facility*. Kramer, Chin and Mayo Inc., Technical Report no. 30, Seattle.

National Research Council (1978) *Aquaculture in the United States – Constraints and Opportunities*. National Academy of Science, Washington.

NOAA (1975) *NOAA Aquaculture Plan*. National Marine Fisheries Service and Office of Sea Grant, Washington.

Pillay T.V.R. (1977) *Planning of Aquaculture Development – An Introductory Guide*. Fishing News Books, Oxford.

Pritchard G.I. (1976) Structured aquaculture development with a Canadian perspective. *J. Fish. Res. Board Can.*, **33**(4–1), 855–70.

Smith L.J. and Peterson S. (1982) *Aquaculture Development in Less Developed Countries – Social, Economic and Political Problems*. Westview Press, Boulder.

UNDP/FAO (1975) *Aquaculture Planning in Africa*. Report of the First Regional Workshop on Aquaculture Planning in Africa, ADCP/REP/75/1.

UNDP/FAO (1976a) *Aquaculture Planning in Asia*. Report of the Regional Workshop on Aquaculture Planning in Asia, ADCP/REP/76/2.

UNDP/FAO (1976b) *Planificacion Sobre Acuicultura en Amèrica Latina*, ADCP/REP/26/3.

Webber H.H. (1971) The design of an aquaculture enterprise. *Proc. 24th Session Gulf Carib. Fish Inst.*, 117–25.

Webber H.H. and Riordan P.F. (1979) Problems of large-scale vertically-integrated aquaculture. In *Advances in Aquaculture*, (ed. by T.V.R. Pillay and W.A. Dill), pp. 27–34. Fishing News Books, Oxford.

4
Selection of Sites for Aquaculture

The history of aquaculture projects all over the world has led to the conclusion that the right selection of sites is probably the most important factor that determines the feasibility of viable operations. Even though, after many years of painful efforts and of new technology, some farms on poor sites have been turned into productive units, there are many that have been abandoned after considerable investment of money and effort. So there is no gainsaying the basic importance of selecting suitable sites for successful aquaculture. At the same time it has to be recognized that compromises have often to be made, as ideal sites may not always be available, and conflicts of land and water use will have to be resolved. In many situations good, irrigated agricultural land may be the best site for pond farms for fish culture, but national priorities in cereal food production may make it unavailable for aquaculture, ir-respective of economic or other advantages. On the other hand many countries, particularly in Asia, are now giving higher priority to aqua-culture and farmers are utilizing rice fields increasingly for fish and shrimp culture.

Although site selection will generally be based on the species to be cultured and the technology to be employed, under certain cir-cumstances the order may have to be reversed. If it is decided to bring under culture certain sites, the selection may be oriented to deter-mining the species that can best be cultured there and the most suitable technologies to be used for that purpose, if indeed the site is primarily suited for aquaculture. Limitations in any of the three factors, namely site character-istics, species and appropriate technology, obviously restrict choice of the others. How-

ever, as mentioned earlier, in the large majority of cases the species to be cultured would have been determined in advance, based on market requirements and consumer preferences.

4.1 General considerations

Although many of the factors to be investigated in the selection of suitable sites will depend on the culture system to be adopted, there are some which affect all systems, such as agro-climatic conditions, access to markets, suitable communications, protection from natural dis-asters, availability of skilled and unskilled labour, public utilities, security, etc. (see Chapter 3). It may be possible to find solutions when these factors are unfavourable and pre-sent problems, but it would involve increased investment and operating costs and would affect profitability. In the case of small-scale aqua-culture, it is necessary to determine that the selected site has easy access to materials that cannot be produced on the farm and that the necessary extension services are available.

All available meteorological and hydro-logical information about the area (generally available from meteorological and irrigation authorities), such as range and mean monthly air temperature, rainfall, evaporation, sun-shine, speed and direction of winds, floods, water table, etc., have to be examined to assess their suitability.

In land-based aquaculture, the most com-monly used installations are pond farms and hatcheries. Since most such farms have earthen ponds, soil characteristics, the quality and quantity of available water and the ease of filling and drainage, especially by gravity, are basic considerations. For fresh-water pond farms, the land available consists mainly of swamps, unproductive agricultural land, valleys, stream and river beds exposed due to changes of water flow, etc. (figs 4.1–4.3). Land elevation and flood levels have to be ascertained. The maximum flood level in the last ten years or the highest astronomical tide (in the case of brackish-water sites) should not be higher than the normal height of the dikes that will be constructed for the farm. It will be advantageous to select land with slopes not steeper than 2 per cent. The area should be sufficiently extensive to allow future expansion and preferably of regular shape to facilitate farm design and construction.

The nature of the vegetation indicates the soil type and elevation of the water table. Obviously dense vegetation, particularly tall trees, make clearing more difficult and expensive. Land under grass or low shrubs is much better suited in this respect. However, in areas exposed to strong winds and cyclonic or similar weather conditions, sufficiently tall vegetative

Fig. 4.1 A swampy area reclaimed into a fish farm in Indonesia.

Fig. 4.2 A fish farm under construction in saline soil area, in Egypt.

Fig. 4.3 A fish farm in a valley in Northern Cameroon.

cover around the farm can serve as effective wind breakers. High ground-water level may create problems in farm operation, as drainage will become difficult and expensive. The use of mechanical equipment for pond construction will also become inconvenient.

Among the other important general factors to be considered are the existing and future sources of pollution and the nature of pollutants. In this connection, information on development plans for the neighbourhood areas will be necessary. It will be useful to ascertain the past use of the site, if any. Croplands that have been treated for long periods with pesticides may have residues that are harmful to fish and shellfish. If the site is located adjacent to croplands that are sprayed from air or land, there is the risk of contamination occurring directly or through run-off water. Similarly, the possible effects of discharges from the pond farms into the waterways and irrigation systems in the neighbouring area should be considered. This can greatly influence the attitudes of the neighbourhood communities to the proposed farming and hence their future cooperation.

When a hatchery is planned in connection with a pond-rearing facility, the selection of its site depends on the location of the nursery and rearing ponds. The more important consideration is the unrestricted availability of good quality water, such as from springs, tube wells, reservoirs, etc. If earthen nursery ponds are to be constructed alongside the hatcheries, it is necessary to ensure the quality of the soil for pond construction and pond management. In many modern hatcheries, fry rearing is mostly done in tanks and troughs, with as much control over ambient conditions as possible. So the main consideration is the availability of essential utilities like electricity. The situation is very similar for the selection of sites for raceway farms. When the raceways are made of cement concrete the main consideration is the availability of adequate quantities of good quality water and essential utilities.

The choice of sites for integrated aquaculture – such as fish culture combined with crop and livestock farming – is governed by factors other than their mere suitability for aquaculture. Land available for integrated aquaculture is generally agricultural land, even if it is somewhat less productive. A satisfactory irrigation system is likely to have been developed for agriculture, in which case water and soil management can be expected to be easier.

Since integrated farming is based on the recycling and utilization of farm wastes, problems of pollution can be expected to be minimal.

4.2 Land-based farms

Sites generally available for coastal pond farms are tidal and intertidal mud flats in protected areas near river estuaries, bays, creeks, lagoons and salt marshes including mangrove swamps. The traditional and, in many cases, the most economical method of water management for a coastal farm is through tidal flow and so one of the essential pieces of information is the tidal amplitude and its fluctuations at the site. The tidal range along the shore line may be more easily obtained from tide tables or other sources, but in estuaries and other water bodies away from the coast the figures will be different: the mean tidal level generally becomes higher, the duration of the ebb tide becomes longer and the flood tide shorter. The diurnal tidal range, that is the differences in height between the mean higher high and the mean lower low waters, becomes less. In order to determine the relation between tidal levels and ground elevation at the proposed coastal farm site, tide measurements will have to be made on the site with a tide gauge or tide staff over a period of time. The relationship of tides between the nearest port and the tide gauge placed at the site has to be determined first for this purpose. The tide curves and other necessary tidal data at the site can be calculated from the highest astronomical tide (HAT), mean high water springs (MHWS), mean high water neaps (MHWN), mean low water neaps (MLWN) and mean low water springs (MLWS).

The construction of ponds in areas reached only by the high spring tides would require excavation, leading to high construction costs and problems in disposal of the excess soil. If the dikes are made higher than necessary to deposit excess earth, the productive water area in the farm will be reduced. Excavation may also affect efficient drainage using tidal energy. Further, the removal of fertile top soil, which is important to induce the growth and maintenance of benthic food organisms in coastal ponds, will result in the loss of much time in reconditioning the pond bottom to stimulate such growths. However, in certain mangrove areas, particularly those under the red mangrove *Rhizophora*, the top layer may contain peat or a very dense mass consisting of rootlets of mangroves, which in any case will have to be excavated to make the pond bottom productive.

The selection of suitable sites, based on tidal fluctuations and elevation, is shown in fig. 4.4. A tidal fluctuation of around 3 m is considered ideal for coastal ponds. However, it has to be remembered that if the tidal energy can be replaced by other forms of energy for water management, the limitations indicated would not apply. As mentioned earlier, the main consideration then would be the cost involved and the economics of operation. Gedrey *et al.* (1984) estimate that the construction and operation of a farm with a pumped water supply system can be more economical than that of a tidal water farm.

4.2.1 Soil characteristics

The quality of soil is important in pond farms, not only because of its influence on productivity and quality of the overlying water, but also because of its suitability for dike construction. The ability of the pond to retain the required water level is also greatly affected by the characteristics of the soil. It is therefore essential to carry out appropriate soil investigations when selecting sites for pond farms. Such investigations may vary from simple visual and tactile inspection to detailed subsurface exploration and laboratory tests. Because of the importance of soil qualities, detailed investigation are advisable, particularly when large-scale farms are proposed. Sandy clay to clayey loam soils are considered suitable for pond construction. To determine the nature of the soil, it is necessary to examine the soil profile, and either test pits will have to be dug or soil samples collected by a soil auger at regular distances on the site. To obtain samples, rectangular pits (1.0–2.0 m deep, 0.8 m wide and 1.5 m long) are recommended. If available, a standard core sampler or soil auger of known capacity (e.g. 100 cm^3) can be used for collecting samples of soil from each soil horizon.

Texture and porosity are the two most important physical properties to be examined. Soil texture depends on the relative proportion of particles of sand, silt and clay. The size

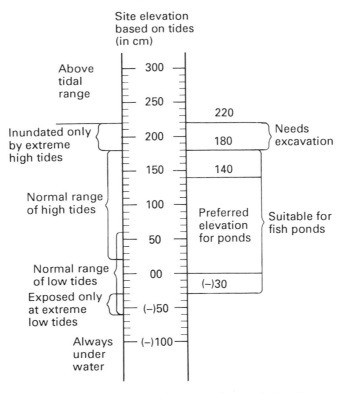

Fig. 4.4 An example of site selection based on tidal range and ground elevation.

limits and some general characteristics of the soil constituents are given in Table 4.1. By touch and feel one can roughly determine the texture. A sample of the soil should be kneaded in the hand (to make it somewhat drier, if it is wet and sticky; if the sample is dry add some water to make it moist but not sticky). If the kneaded sample can be rolled into a bar (about 6 mm thick) and bent to form a ring around the thumb, without any cracks, the soil must be clayey. If it cannot be made into a bar and remains separate with visible grains when dry, the sample is sandy. If the sample does not fall into either of these categories it can be classified as silty or loamy. Sand grains can be felt distinctly, even when not readily visible in loamy soils. Silty soils feel like flour or dough between the fingers. There are, of course, intermediate categories depending on the proportions of the constituents.

Because of their cohesive properties, the fine-textured soils (clay, silty clay, clay loam, silty clay loam and sandy clay) are more suitable for pond farms. They have a greater surface area and can therefore absorb more nutrients and retain and release them for organic production in ponds; they are also less subject to erosion and other damage. The soil structure or the arrangement of soil particles is of special importance in determining the compactness, and therefore the porosity, of the soil. Light-textured soils, particularly in close proximity to open drains can cause high seepage and percolation. Pond farms built on such soils may, however, improve in the course of time due to the blocking of interstitial pores by organic sediments produced in the pond, or introduced with the water supply or derived from manuring. Puddling is an efficient means of sealing ponds. In this process, fine particles clog the most permeable parts and in due course the bottom of the pond may be completely sealed. Compaction of soil by mechanical means during pond construction can also assist in reducing seepage. Suitable linings like polyethylene sheets have been used on pond

Table 4.1 Diameter and characteristics of soil constituents (size fractions).

Soil constituent	Diameter of particles	General characteristics
Sand	2.0−0.05 mm	Individual particles feel gritty when the soil is rubbed between the fingers. Not plastic or sticky when moist.
Silt	0.05−0.002 mm	Feels smooth and powdery when rubbed between the fingers. Not plastic or sticky when moist.
Clay	<0.002 mm	Feels smooth, sticky and plastic when moist. Forms very hard clods when dry. Particles may remain suspended in water for a very long period of time.

bottoms and water supply channels to prevent seepage with some success. But it is difficult to prevent damage to the lining and it often turns out to be too expensive for practical use. It may also greatly reduce the contribution of the pond bottom to natural productivity in the pond, even if the initial and continuing costs of the lining are acceptable.

Generally, the soil on sites selected for coastal pond farms is alluvial. It is usually porous with varying masses of fine roots of mangroves and other swamp vegetation. The preferred soils are clay, clayey loam, silty clay loam, silt loam and sandy clay loam. Sandy clay loam is the best for diking.

4.2.2   Acid sulphate soils

As mentioned earlier one of the major problems in site selection for coastal pond farms in the tropics is the prevalence of acid sulphate soils or cat-clays. Even though such soils are also found in fresh-water swamps, the problem is more pronounced in brackish-water areas. The highly acidic conditions inhibit the production of fish and fish food organisms. Elements, particularly iron and aluminium, are released into the water in toxic quantities which render phosphorus unavailable, causing severe phosphorus deficiency for algal growth. Sudden fish kills during rains after long dry periods are a common phenomenon due to leaching of extremely acidic water from surrounding dikes into ponds built on such soils.

Acid sulphate soil results from the formation of pyrite which is fixed and accumulated by the reduction of sulphate from salt water. The process involves bacterial reduction of sulphate to sulphide, partial oxidation of sulphide to elemental sulphur followed by interaction between ferrous or ferric iron with sulphide and elemental sulphur. A sufficient supply of sulphate and iron; high concentrations of metabolizable organic matter; and sulphate-reducing bacteria (*Desulfovibrio desulfuricans* and *Desulfo maculatum*) in an anaerobic environment alternated with limited aeration, are the factors that give rise to sulphate soils.

In mangrove swamp areas, the most favourable conditions for pyrite formation exist in the zones between the mean high water and mean low water levels which have limited periodic aeration due to tidal fluctuation. There is less pyrite in the better drained parts of the marshes which are aerobic most of the time.

The reclamation of mangrove swamps for pond farms with drainage results in the exposure and oxidation of pyrite and causes acidic conditions. Ferrous iron (Fe_2) is released during atmospheric oxidation of pyrite under moist conditions at an optimum moisture content of 30−40 per cent. At low pH, oxidizing bacteria convert ferrous iron to ferric iron (Fe_3). It can remain in solution in appreciable amounts only at pH values in the range 3−3.5 and is a more effective oxidant for pyrite and elemental sulphur than free oxygen. At higher pH, almost all ferric iron is hydrolysed and precipitated as ferric hydroxide. Basic ferric sulphate is also formed during pyrite oxidation. Elemental sulphur is oxidized to sulphuric acid by bacteria.

The most harmful effect of pyrite oxidation lies in the excessive amount of sulphuric acid produced, which if not neutralized by exchangeable bases creates strongly acid conditions. In selecting sites for pond farms, one has to take into account not only the existence of acid sulphate soils but also the potential for acid conditions to develop as a result of drainage after construction. The levels of pyrite and acid-neutralizing components such as calcium carbonate from mineral deposits and metal cations have to be considered. The use of combined criteria, as for example sedimentary relationships and sulphur sources, land form, vegetation and soil characteristics, has been suggested as a basic approach for recognition and prediction of potential and actual acid sulphate soils. Although it is desirable to have both field and laboratory investigations, it is considered possible to use with confidence certain simple criteria. Potential and existing acid sulphate soils are generally found in mangrove swamps and marshy back swamps, on the seaward side of river deltas and on marine and estuarine plains (Figure 4.5). Tidal brackish-water vegetation with dense rooting systems are usually related to accumulation of pyrite. Association with the red mangrove (*Rhizophora*), *Nipa* and *Melaleuca* stands is a fair indication of potential acid soils. Soils that

are likely to become acidic have a high organic matter content, such as the fibrous roots of mangroves, and a grey subsoil with dark grey to black specks or mottles with partially decomposed matter.

The detection of actual sulphate soils is easy. They can be recognized by the pale yellow mottles of the top soil, overlying pyritic subsoil. The older acid sulphate soil shows the red-brown ferric hydroxide. Their pH is generally below 4. A comparatively easy method of estimating the extent of acid and non-acid soil layers is by implanting stakes coated with red-lead paint in the soil profile. Hydrogen sulphide generated in the layer with active sulphate reduction turns the red-lead marking black within about a week, leaving on the stake a record of the upper limit of the present sulphide accumulation.

As will be described later in Chapter 6 on construction and maintenance of pond farms, it is possible to minimize the harmful effects of acid soils, but it is time-consuming and expensive. However, in many tropical areas, the available sites for pond farms may almost all have such poor soils and there may be little choice. In such cases sites that can more readily be reclaimed should be selected. Basically, reclaiming consists of removing the source of acidity by oxidizing the pyrite from the pond bottom and flushing it out of the 10–15 cm

Fig. 4.5 Back-swamps with secondary growth of mangroves – potential sites for aquafarms (photograph: H.R. Rabanal).

deep surface soil and preventing further diffusion of acids, aluminium salts and ferrous salts from the subsoil. Acid and toxic elements are also leached and removed. If this is feasible, the farm can be made suitable for aquaculture within a period ranging from 3 to 5 years, depending on the extent of the problem.

4.3 Open-water farms

Open-water aquaculture includes mollusc culture in shallow salt- and fresh-water areas, seaweed farming in coastal seas and pen and cage culture in sea and fresh-water bodies. As is obvious, in selecting sites for such systems of culture the main considerations are the hydrographic and climatic conditions. In spite of some limited success in extending certain types of aquaculture to deeper and more exposed coasts, the most suitable and preferred areas continue to be sheltered bays, estuaries, lagoons, straits, lakes and reservoirs, protected from strong winds and rough seas. While moderate currents and water flows are necessary to maintain water quality and removal of waste products from farm sites, frequent storms and turbulent seas will make it difficult to practise most types of aquaculture. Winds will directly affect culture installations above water, whereas waves affect both the submerged structures and the animals under culture. In most cases low current velocities are preferred.

In systems like the ones for bottom culture of molluscs, the nature of the sea or river bed is important. Suitable stable substrates are needed for the attachment of the animals. Most modern open-water culture is of the off-bottom type, where the water conditions and quality are more important.

Since mollusc culture is based largely on natural food organisms that the molluscs filter from the environment, it is essential to select sites with high primary production. Though some experimental work has been done on artificial feeding of certain molluscs, in commercial farming production is dependent on the growth of plankton or algae. In order to make natural food available to the animals the current velocity should not exceed 5 cm/s.

Even though controlled reproduction and hatchery production of seed are possible in mollusc farming, in many places aquaculturists depend on wild spat for culture. In such cases, it is advisable to select sites where there is an abundance of spat. A breeding population of the species nearby is, of course, necessary, but it does not necessarily follow that the spat will settle in the immediate neighbourhood. The larvae may be carried away by currents, so sufficient shelters and suitable current speeds are necessary to keep the larvae in the area. Field observations, supplemented by experimental spat setting, may be a necessary basis for a decision on site suitability.

In the farming of seaweed such as laver fertilizers are used to increase growth, but naturally fertile areas are still selected as in open-water situations fertilization can only be a complement to natural productivity. Movement of water prevents the increase of pH which can be caused by the consumption of carbon dioxide in seaweed-growing areas. Therefore it is necessary to select sites with an adequate current. A current of about 10–30 cm/s is considered suitable, depending on the content of nutrients in the water. Waters deficient in nutrients should have a current of 30 cm/s and those rich in nutrients about 10 cm/s. Since periodic exposure of leafy thalli is important for growth in some seaweeds, it is necessary to select a place with a tidal range of 1–1.5 m or more.

4.4 Water quantity and quality

The availability of water of appropriate quality is important for all systems of aquaculture, but the quantity is particularly important for land-based systems. It is therefore necessary to investigate, as thoroughly as possible, the extent and seasonality of water sources as well as liability to pollution. Since predictions have to be made of long-term water conditions, it is desirable to have data for a reasonably long period of time. In areas with controlled irrigation, reliability of supplies can generally be expected. This together with the availability of cheap electricity has made water management fairly easy for fish farmers in Southern China, in spite of dense stocks of fish and heavy loading of manures in pond farms. On the other hand, when rain-fed or ground-water ponds are used, as in Eastern India, water levels in the ponds become dangerously low due to seepage and evaporation in summer months, when the ponds have generally the maximum biomass of fish. Access to other reliable sources of water, such as rivers, streams, lakes and reservoirs or even tube wells which can yield enough water are essential for the enterprise to succeed. Loss of water due to seepage and evaporation varies considerably. For example, the average loss in Europe is reported to be about 0.4−0.8 cm per day, whereas in tropical regions it may be as much as 2.5 cm per day. When ground water is the major source of water supply, the effect of pumping on the water table and possible land subsidence have to be considered.

The need to investigate the elevation and ranges of tides for coastal aquaculture has already been referred to. This is most important when tidal movements have to be depended on for filling and draining the ponds. The constant flushing of newly constructed ponds to leach out toxic elements from the soil has also been mentioned. It is believed that if pumping were to be used for water management, the costs of construction of dikes and sluice gates would be minimized and the ponds could be constructed and operated without disturbing the acid soils, allowing a non-acidic layer of sediment to deposit on the bottom. In the long run, this may be more economical, despite the increased energy costs. However, it will be necessary to make rough calculations of the comparative costs before finally selecting the site and deciding on the system of management to be adopted.

The temperature of the water will be an important criterion as to whether the species selected can be cultured on the site. Although in hatcheries and in systems with a recirculating water supply the temperature can be controlled, it is extremely difficult, if not impossible, to do so at affordable cost in large pond farms. Industrial waste heat can to a certain extent be used to raise temperature in aquaculture areas, but very often practical problems of quality of heated water or irregularity in availability limits their use, except in well controlled environments or where the animals can stand considerable variations in temperature.

Salinity and variations thereof are also important environmental factors which have to be taken into account. Some species have wide salinity tolerance limits and it has been noted that some fresh-water fish grow faster in slightly saline water and some brackish-water fish faster in fresh water. However, they still have their limits of tolerance. Even if they survive, their growth and reproduction may be affected. For example, the common carp (*Cyprinus carpio*) can grow well in salinities up to 5 ppt, but at 11.5 ppt the salinity becomes lethal. Similarly, the tiger shrimp (*Penaeus monodon*) can tolerate 0.2 to 40 ppt salinity, but grows well only between 10 and 25 ppt.

As will be discussed in Chapter 6, salinity and water temperature are important considerations in deciding on the sites for hatcheries. Not only do these require higher water quality but the levels of salinity and water temperature required for spawning and larval rearing may differ from those needed for grow-out to market size. This may make it sometimes necessary to select separate sites for hatcheries and grow-out farms for certain species.

High turbidity of water caused by suspended solids can affect productivity and fish life. It will decrease light penetration into the water and thus reduce primary production. This would naturally also affect secondary production. In certain cases, oxygen deficiency has also been reported as a result of a sudden increase in turbidity. The suspended solids may clog the filter-feeding apparatus and digestive organs of planktonic organisms. The gills of fish may be injured by turbid water. Although the effect will depend on the species and the nature of the suspended matter, pronounced effects are seen when the water contains about 4 per cent by volume of solids. The use of turbid water in hatcheries should be avoided, as it can greatly affect the hatching and rearing of larvae.

If it becomes necessary to select sites with highly turbid water, which the candidate species cannot tolerate, suitable methods of reducing turbidity have to be adopted. The use of settling tanks, different types of filters and repeated application of gypsum (200 kg per 1000 m^3 initially, followed if necessary by an additional application of 50 g per 1000 m^3) have been recommended. All these will involve higher capital or operational cost, but in cases where there are no alternatives the possibility of absorbing the costs will have to be examined in feasibility studies. Improvements in drainage from catchment areas, often the cause of high turbidity, may also be considered.

Among other water quality criteria of importance in site selection are acidity and alkalinity. The most suitable pH of water for aquaculture farms is considered to lie in the range 6.7–8.6 and values above or below this inhibit growth and production, although the extent of their effect will depend on the species concerned and environmental conditions such as the concentration of carbon dioxide or the presence of heavy metals like iron.

The prevalence of low pH in brackish-water areas and the problems of improving soil and water quality in farms built in such areas have been described earlier. Water of low pH is also common in fresh-water areas with soils low in calcium and rich in humic acids. Acid water with a pH range of 5.0–5.5 can be harmful to the eggs and fry of most fish and the adults of many. Acidity reduces the rate of decomposition of organic matter and inhibits nitrogen fixation, thereby affecting the overall productivity.

The most common method of correcting low pH is by liming to neutralize the acidity. The dose will depend on the pH value and the chemical composition of the water, especially the concentration of calcium bicarbonate $[Ca(HCO_3)_2]$. It also will depend on the type of lime applied. The relative quantities of quick lime (calcium oxide, CaO), slaked lime or agricultural lime (calcium hydroxide, $Ca(OH)_2$) and limestone (calcium carbonate, $CaCO_3$) required will be in proportions of 1:1.5:2 respectively. The actual dosage has to be determined by titrating the water to neutrality and calculating the equivalent amount of lime to be added. The additional costs involved will have to be taken into account before selecting sites with acid water.

High pH, indicating excessive alkalinity, can also be harmful. However, it should be noted that in productive water pH may reach higher values of 9 to 10 due to the uptake of carbon dioxide during photosynthesis in the daily pH cycle. This is why it will be better to take pH measurements before daybreak to determine their suitability for aquaculture. A pH level of 11 may be lethal to fish.

Toxic substances in water supplies can affect aquaculture, particularly in hatcheries.

Liebmann (1960) summarizes the threshold levels of toxicity and maximum permissible concentration of toxic substances in indoor fish hatcheries, as shown in Table 4.2.

4.5 Sources of pollution and user conflicts

As indicated earlier, it is essential to investigate any existing or potential sources of pollution and the nature of pollutants that are likely to affect the water supply to the proposed farm. Thorough local enquiries will be needed, as the situation at the time of the site selection studies may not represent conditions at other times of the year. Therefore data for previous years should also be examined as far as possible. Certain types of organic and harmless wastes

Table 4.2 Threshold of toxicity and maximum permissible concentration of toxic substances in the water supply of indoor fish hatcheries.

Substance	Threshold concentration (mg/l)	Maximum permissible concentration (mg/l)
Ammonia	0.2–2.0	0.05
DDT	0.02–0.1	absent
Calcium bisulphate	30–60	
Calcium chloride	7000–12 000	
Potassium chloride	700–5 200	
Potassium sulphate	800–1 000	
Magnesium chloride	5000–15 000	20
Magnesium nitrate	10 000	15
Magnesium sulphate	30 000	50
Manganese (nitrate, chloride, sulphate)	75–200	5
Copper (compounds)	0.08–0.80	0.005
Sodium bicarbonate	5000	
Sodium carbonate	200–500	
Sodium chloride	7000–15 000	
Cadmium	3–20	0.003
Ozone	0.02	
Mercury	0.1–0.9	
Rotenone	0.01–0.012	absent
Sulphides	0.4–4.0	0.1
Hydrogen sulphide	1.0	0.1
Iron (compounds)	0.9–2.0	0.01
Phenol	6–17	0.0005
Formaldehyde	15–30	
Tannin	15	5
Paraquinone	0.1–10	
Chlorine	0.05–0.4	absent
Carbolineum	7	
Zinc (compounds)	0.1–2.0	0.005

can be used to increase the productivity of aquaculture farms. The use of waste heat in temperate and cold climates has already been referred to. Sewage effluents and properly treated animal wastes can be used successfully to fertilize aquaculture farms in order to increase the growth of food organisms. However, it will be necessary to incorporate such uses at the design stage of waste disposal (in order to render the wastes readily usable for aquaculture purposes) as well as the aquafarm (to provide for controlled use of the waste material in the appropriate form and doses to enable its safe use). The likelihood of discharges from facilities used for intensive aquaculture polluting public water bodies and spreading communicable diseases from farmed stocks to wild stocks should also be considered. Though these can be prevented in well designed and managed farms, there is still the possibility of such arguments being used by neighbouring communities who are not very appreciative of the use of the selected site.

In open-water aquaculture, particularly cage and pen culture of fish and stick and raft culture

of molluscs in lagoons, estuaries, bays, fjords, etc., there is a likelihood of the organic load from metabolic wastes of cultured organisms and unused feeds accumulating, sometimes giving rise to a high biological oxygen demand and accumulation of toxic gases. The pattern of water flow may also be altered. It will therefore be necessary to consider ways of preventing this and avoiding conflicts with other uses of the area, such as navigation, recreation and fishing.

Some of the major considerations in reclaiming mangrove swamps for aquaculture have been discussed earlier. From what is known of mangrove ecology and the effect of reclamation, it would appear that if properly planned, clearing of mangroves retaining a belt of at least 50 m along the coast ensures that their ecological functioning is unimpaired. It has been suggested that clearing of mangroves should be done without changing the general morphology of the area, leaving for every hectare of mangrove cleared at least three hectares untouched, for conservation purposes.

Conflicts may arise with agriculture, as for

Fig. 4.6 Aerial view of a sheltered fjord used for cage culture in Norway.

example rice farming, in areas where for economic reasons rice fields may be converted into fish ponds. However, if national priorities require that they be used for rice cultivation, the possibility of integrated rice field aquaculture could be considered. In areas where crop/livestock/fish integrated farming is possible, conflicts with agriculture communities can be minimized by adopting such practices that will add to the income of the farmers.

With the expansion of aquaculture, many governments have brought in systems of licensing to regulate the enterprise. Where no such unified regulations exist, very often the prospective farmer has to obtain permits and clearances for his project from a number of agencies (see Chapter 3). Naturally these legal and administrative matters will also be major considerations in the selection of sites for aquaculture.

4.6 References

Boyd C.E. (1979) *Water Quality in Warmwater Fish Ponds.* Auburn University, Alabama.

Breemen N. Van (1976) Genesis and solution chemistry of acid sulfate soils in Thailand. *Agric. Res. Rep. (Versl. Land-bowkd, Onders)*, 848, PUDOC, Wageningen.

Coulter J.K. (1973) The management of acid sulfate and pseudo-acid sulfate soils for agricultural and other uses. *Proceedings International Symposium on Acid Sulfate Soils, Wageningen*, **1**, 255−74.

FAO/UNDP (1982) *Report of Consultations/ Seminar on Coastal Fishpond Engineering*, 1982, Surabaya, Indonesia FAO/UNDP South China Sea Fisheries Development and Coordination Programme, SCS/GEN/82/42.

FAO/UNDP (1984) *Inland Aquaculture Engineering.* ADCP/REP/84/21, FAO of the UN, Rome.

Gedrey R.H., Shang Y.C. and Cook H.L. (1984) Comparative study of tidal and pumped water supply for brackishwater aquaculture ponds in Malaysia. In *Malaysia − Coastal Aquaculture Development.* FAO (Field Document), Rome.

Hepher B. and Pruginin Y. (1981) *Commercial Fish Farming.* John Wiley and Sons, New York.

Huet M. (1986) *Textbook of Fish Culture*, 2nd edn. Fishing News Books, Oxford.

Imai T. (Ed.) (1978) *Aquaculture in Shallow Seas: Progress in Shallow Sea Culture.* (Translated from Japanese.) A.A. Balkema, Rotterdam.

Liebmann H. (1960) *Handbuch der Frischwasser- und Abwasser-Biologie.* R. Oldenburg, München.

New M.B. (1975) The selection of sites for aquaculture. *Proc. World Maricul. Soc.*, **6**, 379−88.

Poernomo A. and Singh V.P. (1982) Problems, field identification and practical solutions of acid sulfate soils for brackishwater ponds. In *Report of Consultations/Seminar on Coastal Fishpond Engineering*, pp. 49−61. FAO/UNDP South China Sea Fisheries Development and Coordination Programme SCS/GEN/82/42.

Rickard D.T. (1973) Sedimentary iron sulfide formation. *Proceedings International Symposium on Acid Sulfate Soils, Wageningen*, **1**, 28−65.

Singh V.P. (1980) The management of fishponds with acid sulfate soils. *Asian Aquaculture*, **3**(4), 4−6.

5
Selection of Species for Culture

Jhingran and Gopalakrishnan (1974) include about 465 species, belonging to 28 families of plants and 107 families of animals, in a catalogue of cultivated aquatic organisms. It would probably be possible to culture almost all aquatic organisms, but the main consideration is whether it is worth the effort and how far they can contribute to the main objectives of aquaculture (see beginning of Chapter 3). The availability of a large number of aquaculture species adapted to different environmental conditions is an advantage, as it will often be possible to choose from locally occurring species and avoid the introduction of exotic ones for culture. However, this also means that aquaculture misses the advantages that crop and animal production have had in agriculture: of concentrated research on a few species that has led to the development of advanced technologies of production, and of selected high-yielding strains and hybrids. The history of agricultural research indicates the time and effort that are needed to develop such technologies. The science of aquaculture (as distinct from traditional practices), which is relatively new, will probably require a longer period of time to reach that level of advancement if efforts have to be shared among so many species. It has to be remembered that long traditional experience and scientific research, so far, have actually succeeded in domesticating, in the sense of animal husbandry practices, only a small number of species like the trout, common carp and salmon. However, one can clearly see a tendency towards limiting the number of species in large-scale commercial aquaculture, unlike in aquaculture research where an increasing number of species are still being investigated.

Despite the value of limiting species for culture for speedy technological advancement, it has to be recognized that there is a real need to have species suited for different environmental conditions and economic circumstances. Species have to be selected according to the objectives of culture, for example increasing protein supplies to the poor, export to earn foreign exchange or waste recycling in a polyculture system.

5.1 Biological characteristics of aquaculture species

A major characteristic that determines the suitability of a species for aquaculture is the rate of growth and production under culture conditions. Although certain slow-growing species may be candidates for culture because of their market value, it is often difficult to make their culture economical. Through the use of heated water growth rates of many species can be improved, but commercial grow-out using such methods has not yet proved very successful. In principle, a faster growth rate, as obtained in many tropical species, allows them to grow to marketable size in a shorter time, making it possible to have more frequent harvests. The size and age at first maturity is also an important consideration, as it will be preferable to have them reach marketable size before they attain first maturity, so that most of the feed and energy are used for somatic growth. Early maturity would ensure easier availability of breeders for hatchery operations, but early maturity before the species reaches marketable size will also be a great handicap, as in the case of tilapia species.

It is certainly preferable to culture a species

that can be bred easily under captive conditions. This would permit hatchery production of seed in adequate quantities. If it is a species that matures more than once a year, it should be possible to have several crops of seed and possibly adults, if other conditions are suitable. High fecundity can be an advantage, as also frequency of spawning; however small-sized eggs and small larvae make hatching operations more difficult. A shorter incubation period and larval cycle often contribute to lower mortality of larvae and greater survival in hatcheries.

Larvae that would accept artificial feeds would be easier to rear in hatcheries. The raising of live foods is comparatively more difficult and often expensive.

In cases where controlled breeding techniques have not been perfected, the aquaculturist may have to depend on seed available from the wild. But as has been experienced in many situations, it proves to be an unreliable source in large-scale farming, as their abundance in nature depends on a number of unpredictable factors. Further, large-scale collection of wild spawn and fry has given rise to conflicts with commercial fishermen, who ascribe the decline in catches of the concerned species to the removal of early stages, despite the lack of any scientific evidence. So, even from a public relations point of view, it is better to select species that can be propagated in hatcheries and to start hatchery production as early as possible.

In modern aquaculture, feeding is one of the major elements of cost of production and may amount to 50 per cent or more. Nutritional requirements of aquaculture species are discussed in Chapter 7. In most traditional aquaculture practices, herbivorous or omnivorous species have been preferred as they feed on natural food organisms in water, the growth of which can be enhanced through fertilization and water management. In such cases, the cost of feeding will be relatively low and because of this, species low in the food chain are preferable for the production of low-priced products. However, even with such species, supplementary feeding with artificial feedstuffs has to be adopted in intensive culture systems. The feed efficiency in relation to growth and productivity then becomes an important criterion. Some of the low trophic level feeders can also be highly

selective in their feeding habits, as in the case of filter-feeders that require plankton of a particular size and shape. The need to grow the species to market size within a limited season or period often makes it necessary to resort to artificial feeding.

Carnivorous species generally need a high protein diet and are therefore considered to be more expensive to produce, even though the costs will depend largely on local availability and price of the required feedstuffs. To compensate for feeding costs, most carnivorous species command higher market prices. Such species generally have greater export markets and therefore attract substantial investments.

Species that are hardy and can tolerate unfavourable conditions will have the advantage of better survival in relatively poor environmental conditions that may occur occasionally in culture situations. The temperature and oxygen concentration can fluctuate in ponds and other enclosures and deterioration of the water quality may occur unavoidably. In such situations, hardier species will obviously fare better. Besides the possible effects of poor water quality on the candidate species, it is also necessary to consider the influence of the species on the environment. Soil erosion that may be caused by the feeding habits of carp has been referred to in Chapter 2.1. Species that easily escape into natural bodies of water and upset their ecology would need special protective measures, leading to higher costs and environmental concern.

In intensive and semi-intensive culture, dense populations are confined in a limited space. In such cases, behaviour patterns of species in confinement are of special significance. Increases in transmission of disease, cannibalism in the early stages and accumulation of waste products are related to overcrowding. Species that have better resistance to such unfavourable conditions are better candidates for culture.

5.2   Economic and market considerations

Economic considerations are as important or even more important to an aquaculturist than biological factors in the selection of species to be cultured. Many of the relevant factors have already been referred to in Chapter 3 when discussing national priorities and investment requirements. The availability of proven technologies of culture, backed by economic viability, should guide an investor or an aquaculturist in the selection of a species or a culture system. Despite the scarcity of this type of information and the variability of economic returns of enterprises, it is of such crucial importance that even incomplete information from actual commercial or pilot operations would be useful in validating available experimental results.

Consumer acceptance and availability of markets for the species are very intimately interlinked with the economics of raising them. There are several instances where culture techniques were in existence for many years but never resulted in any large-scale production until new or improved markets developed, whether for domestic consumption or for export. Markets can, of course, be developed in places where none existed for a species, but this would require very considerable time and effort. Public and/or private organizations will have to undertake very intensive promotional activities to achieve this in a reasonable period of time.

The above considerations appear to be the main reasons for the widespread interest in introducing exotic species. The concerned species are generally those for which established culture technologies exist and the economics of production and marketability have been demonstrated.

5.3 Introduction of exotic species

The advantages of limiting the number of aquaculture species and the scarcity of really domesticated species for culture have been referred to at the beginning of this chapter. The economic and market considerations that create interest in the introduction of exotic species, have also been mentioned in the previous section. Considering the natural geographic ranges of distribution of proven species, there is a strong argument for introduction and transplantation of exotic species where necessary. However, the problem very often is how to decide whether it is necessary and, if so, what procedures and precautions should be taken to prevent possible undesirable consequences.

History reveals that several indiscriminate introductions and transplantations have been made in the past for establishing sport and commercial fisheries, for ornamental purposes and for biological control. Some of them have had detrimental effects on the local fauna and have contributed to the spread of communicable diseases. There is no gainsaying the need for preventing such consequences by following appropriate procedures and effective national regulations. However, expanding aquaculture may find it very difficult to avoid the introduction or transplantation of species, or selected strains of local species, for experimentation or commercial production. Munro (1986) lists some of the aquaculture species that have already colonized outside their historical distributional range: tilapia species, cyprinids (common carp, Chinese carps), rainbow trout, walking catfish, Japanese and European oysters and fresh-water crayfish (*Pacifastacus* sp.). The majority of them have been introduced for valid reasons, but it is most doubtful whether any of these or other successful introductions have been preceded by detailed screening procedures.

Turner (1949) suggested criteria to be considered in introducing new species. The species should

(a) fill a need, because of the absence of a similar desirable species in the locality of transplantation;
(b) not compete with valuable native species to the extent of contributing to their decline;
(c) not cross with native species and produce undesirable hybrids;
(d) not be accompanied by pests, parasites or diseases which might attack native species; and
(e) live and reproduce in equilibrium with its new environment.

The basic logic of these criteria is still valid and organizations such as the American Fisheries Society (Anonymous, 1973) and the International Council for the Exploration of the Sea (ICES, 1972 and 1979) have tried to strengthen the arguments for critical evaluation and propose methods of obtaining basic data for predicting the consequences of introductions. A close scrutiny of the rationale for introductions and the advantages and disadvantages of the candidate species has to be followed by a preliminary assessment of impacts before a decision is made to introduce the species for testing. If it is decided to proceed, thorough experimental studies should be carried out and results critically evaluated to make the final decision for general introduction or transplantation. ICES (1979) recommends the following procedures for the investigations:

(a) A brood stock should be established in an approved quarantine environment. The first progeny of the introduced species, not the original import, can be transplanted to the natural environment if no diseases or parasites become evident. The quarantine period will be used to provide an opportunity for the observation of disease and parasites. In fish, brood stock should be developed from stocks imported preferably as eggs or possibly juveniles, to allow sufficient time for observation in quarantine.
(b) All effluents from quarantine units are to be disinfected in an approved manner.
(c) A continued study should be made of the introduced species in its new environment.

While it is relatively less difficult to determine whether the imported stock brings in parasites or diseases, prediction of ecological effects based on controlled experiments has many limitations. It is therefore important that monitoring of the effects of introductions should be carried out on a long-term basis, in order to adopt necessary measures as early as possible.

5.4 Common aquaculture species

As mentioned at the beginning of this chapter, there are several species of finfish, shellfish and plants that are used in experimental or commercial aquaculture. But the bulk of present-day production is based on a smaller number of species. The more important of them, along with species for which appreciable progress has been made in developing culture technologies, are listed below. The list is not claimed to be exhaustive and is based on gross evaluations as it was found not feasible to use precise criteria. It does not include ornamental and bait fish, purely game fish or amphibians.

Family	Species	Common name
Finfish		
Acipenseridae	*Huso huso*	Beluga − sturgeon
	Acipenser ruthenus	Sterlet sturgeon
	Acipenser guldenstadti	Russian sturgeon
	Acipenser nudiventris	Thorn sturgeon
	Acipenser stellatus	Starred sturgeon, sevrjuga
	Acipenser baeri	Siberian sturgeon
	Acipenser transmontanus	White sturgeon
Chanidae	*Chanos chanos*	Milkfish
Channidae	*Channa marulius*	Murrel, snakehead
(= Ophicephalidae)	*Channa punctatus*	Murrel, snakehead
	Channa striatus	Murrel, snakehead
	Channa maculatus	Murrel, snakehead
	Channa micropeltes	Murrel, snakehead
Heterotidae	*Heterotis niloticus*	Heterotis
Salmonidae	*Salmo gairdneri*	Rainbow trout
	Salmo trutta	Brown trout
	Salvelinus fontinalis	Brook trout
	Salmo salar	Atlantic salmon

Family	Species	Common name
	Oncorhynchus gorbuscha	Pink salmon
	Oncorhynchus nerka	Sockeye salmon
	Oncorhynchus kisutch	Coho salmon
	Oncorhynchus keta	Chum salmon, dog salmon
	Oncorhynchus tschawytscha	Chinook salmon
Plecoglossidae	*Plecoglossus altivelis*	Ayu
Coregonidae	*Coregonus albula*	Lake whitefish
	Coregonus lavaretus	Common whitefish
Anguillidae	*Anguilla anguilla*	European eel
	Anguilla japonicus	Japanese eel
Characidae	*Colossoma brachypomus*	Pirapitinga
	Colossoma macropomum	Tambaqui
	Colossoma mitrei	Pacu
Cyprinidae	*Aristichthys nobilis*	Bighead
	Catla catla	Catla
	Cirrhina mrigala	Mrigal
	Ctenopharyngodon idella	Grass carp
	Cyprinus carpio	Common carp
	Hypophthalmichtys molitrix	Silver carp
	Labeo rohita	Rohu
	Labeo calbasu	Calbasu
	Mylopharyngodon piceus	Black carp
	Osteochilus hasseltii	Nilem
	Puntius gonionotus	Tawes
Siluridae	*Silurus glanis*	Wels
Ictaluridae	*Ictalurus punctatus*	Channel catfish
Claridae	*Clarias batrachus*	Catfish, Asian
	Clarias lazera, gariepinus	African catfish
	Clarias macrocephalus	Catfish, Asian
Pangasidae	*Pangasius larnaudi*	Catfish
	Pangasius pangasius	Catfish
	Pangasius sutchi	Catfish
Scophthalmidae	*Scophthalmus maximus*	Turbot
Helostomidae	*Helostoma temmincki*	Kissing gourami
Osphronemidae	*Osphronemus goramy*	Gourami
	Trichogaster pectoralis	Siamese gourami, sepat siam
Mugilidae	*Mugil brasiliensis*	Grey mullet
	Mugil curema	Grey mullet
	Mugil capito	Grey mullet
	Mugil auratus	Grey mullet
	Mugil saliens	Grey mullet
	Mugil chelo	Grey mullet
	Mugil grandisquamis	Grey mullet
	Mugil falcipinnis	Grey mullet
	Mugil cephalus	Grey mullet
	Mugil parsia (dussumieri)	Grey mullet
	Mugil tade	Grey mullet
	Mugil macrolepis	Grey mullet
	Rhinomugil corsula	Freshwater mullet

Family	Species	Common name
Carangidae	*Seriola quinqueradiata*	Yellowtail
	Trachinotus carolinus	Pompano, Florida pompano
	Trachinotus falcatus	Atlantic permit
	Trachinotus goodei	Permit
Esocidae	*Esox lucias*	Pike
	Lucioperca lucioperca	Pike perch
Siganidae	*Siganus canaliculatus (= oramin)*	Rabbit fish
(= Teuthidae)	*Siganus rivulatus*	Rabbit fish
	Siganus lurida	Rabbit fish
	Siganus vermiculatus	Rabbit fish
Centropomidae	*Lates calcarifer*	Sea-bass, Asian sea-bass
Serranidae	*Dicentrarchus labrax*	Sea-bass, Mediterranean sea-bass
	Epinephelus tauvina	Estuarine grouper, greasy grouper
	Epinephelus akaara	Red grouper
	Morone saxatilis	Striped bass
Sparidae	*Pagrus major*	Red porgy, Red sea-bream
	Sparus aurata	Gilthead sea-bream
Cichlidae	*Tilapia andersonii*	Tilapia
	Tilapia aurea	Tilapia
	Tilapia hornorum	Tilapia
	Tilapia melanotheron	Tilapia
	Tilapia mossambica	Tilapia
	Tilapia nilotica	Tilapia
	Tilapia spilurus	Tilapia
	Tilapia rendalli	Tilapia
	Tilapia zillii	Tilapia
Tetraodontidae	*Fugu rubripes*	Pufferfish
	Fugu vermicularis	Pufferfish
Crustaceans		
Penaeidae	*Penaeus aztacus*	Brown shrimp
	Penaeus duorarum	Pink shrimp
	Penaeus indicus	Indian shrimp, white shrimp
	Penaeus japonicus	Kuruma shrimp
	Penaeus monodon	Tiger shrimp
	Penaeus orientalis (= chinensis)	Oriental shrimp
	Penaeus merguiensis	Banana shrimp
	Penaeus penicillatus	Red-tailed shrimp
	Penaeus kerathurus	Mediterranean shrimp, triple-grooved shrimp
	Penaeus schmitti	Southern white shrimp
	Penaeus semisulcatus	Green tiger shrimp, bear shrimp
	Penaeus notialis	Shrimp

Family	Species	Common name
	Penaeus setiferus	White shrimp, common shrimp
	Penaeus stylirostris	Blue shrimp
	Penaeus vannamei	White shrimp
	Metapenaeus monoceros	Shrimp
	Metapenaeus brevicornis	Shrimp
	Metapenaeus ensis	Shrimp
Palaemonidae	*Macrobrachium rosenbergii*	Giant freshwater prawn
Nephropsidae	*Homarus americanus*	American lobster
	Homarus gammarus	European lobster
Astacidae	*Astacus astacus*	European noble crayfish
	Procambarus clarkii	Red swamp crayfish
	Pacifastacus leniusculus	Signal crayfish, American crayfish
	Pacifastacus acutus	White river crayfish
	Oreonectes immunis	Paper-shell crayfish
	Cherax tenuimanus	Freshwater crayfish
	Cherax destructor	Freshwater crayfish
Scyllidae	*Scylla serrata*	Swimming crab
Portunidae	*Portunis trituberculatus*	Blue crab
	Neptunus pelagicus	Blue crab
	Paralithodes camchatica	King crab

Molluscs

Family	Species	Common name
Arcidae	*Anadara granosa*	Blood cockle, blood clam
Mytilidae	*Mytilus edulis*	Mussel
	Mytilus galloprovincialis	Mussel
	Mytilus crassitesta	Black mussel
Aviculidae	*Perna perna*	Blue mussel
	Perna viridis	Green mussel
	Perna indica	Brown mussel
	Perna canaliculus	Green mussel
Ostreidae	*Crassostrea angulata*	Portuguese oyster
	Crassostrea rhizophora	Mangrove oyster
	Crassostrea eradilie	Slipper oyster
	Crassostrea tulipa	Mangrove oyster
	Crassostrea brasiliana	Mangrove oyster
	Crassostrea belcherii	Mangrove oyster
	Crassostrea virginica	American oyster
	Crassostrea plicatula	Chinese oyster
	Crassostrea rivularis	Chinese oyster
	Crassostrea gigas	Japanese oyster, Pacific oyster
	Crassostrea commercialis	Sydney rock oyster
	Crassostrea glomerata	Auckland rock oyster
	Ostrea edulis	Flat oyster, European oyster
	Ostrea chilensis	Chilean oyster

Family	Species	Common name
Pectenidae	*Patinopecten yessoensis*	Deepsea scallop, giant ezo scallop
	Argopecten irradians	Bay scallop
	Pectinopecten maximus	European king scallop
	Chlamys tigerina	European tiger scallop
	Chlamys farreri	Chinese scallop
	Chlamys nobilis	Chinese scallop
Mercenaridae	*Mercenaria mercenaria*	Hard clam, quahog
Veneridae	*Meretrix meretrix*	Big clam
	Meretrix lusoria	Clam
	Tapes (= Ruditapes) phillippinarum	Small-necked clam
	Venerupis japonica	Japanese little-neck, Manila clam
Haliotidae	*Haliotis discus hannai*	Abalone
	Haliotis rufescens	Red abalone

Aquatic plants/seaweeds

Family	Species	Common name
Chlorophyceae	*Enteromorpha compressa*	Green algae
	Caulerpa racemosa	Green algae
	Monostroma sp.	Green algae
Laminariaceae	*Laminaria japonica*	Kombu, brown algae
Lessoniaceae	*Undaria pinnatifida*	Kelp, brown algae, wakame
	Undaria undarioides	Wakame
	Undaria peterseniana	Wakame
Bangiaceae	*Porphyra angusta*	Nori, red algae
	Porphyra haitanensis	Nori, red algae
	Porphyra kuniedai	Nori, red algae
	Porphyra tenera	Nori, red algae
	Porphyra pseudolinearis	Nori, red algae
	Porphyra yezoensis	Nori, red algae
Gelidiaceae	*Gelidium amansii*	—
Solieriaceae	*Eucheuma cottonii*	Red algae
	Eucheuma edule	Red algae
	Eucheuma muricatum (= spinosum)	Red algae
Gracilariaceae	*Gracilaria gigas*	Red algae
	Gracilaria confervoides	Red algae

5.5 References

Anonymous (1973) Position of American Fisheries Society on introductions of exotic aquatic species. *Trans. Am. Fish Soc.*, **102**, 274–6.

Chazari E. (1984) *Piscicultura en Agua Dulce*, (Ed. by M.A. Porrua). Secretaria de Pesca, Mexico.

Elton C.S. (1972) *The Ecology of Invasions by Animals and Plants*. Chapman Hall, London.

FAO (1977) Control of the spread of major communicable fish diseases. *FAO Fish. Rep.*, **192**. FAO of the UN, Rome.

ICES (1972) *Report of the Working Group on Introductions of Non-indigenous Marine Organisms*. ICES Cooperative Research Report, **32**.

ICES (1979) *Report of the Working Group on Introductions of Non-indigenous Marine Organisms*. ICES/CM, 1979/E22.

Jhingran V.G. and Gopalakrishnan V. (1974) Catalogue of cultivated aquatic organisms. *FAO Fish. Tech. Paper*, **130**.

Mann R. (Ed.) (1979) *Exotic Species in Aquaculture*. MIT Press, Cambridge.

Munro A.L.S. (1986) Transfers and introductions: do the dangers justify greater public control? In *Realism in Aquaculture: Achievements, Constraints, Perspectives*, (Ed. by M. Bilio, H. Rosenthal and C.J. Sinderman). European Aquaculture, Society, Bredene.

Rosenthal H. (1980) Implications of transplantations to aquaculture and ecosystems. *Marine Fish. Rev.*, 1–14.

Thorpe J. (Ed.) (1980) *Salmon Ranching*. Academic Press, London.

Turner H.J. (1949) *Report on Investigations of Methods of Improving the Shellfish Resources of Massachusetts*. Department of Conservation, Division of Marine Fisheries, Commonwealth of Massachusetts.

Webber H.H. and Riordan P.F. (1976) Criteria for candidate species for aquaculture. *Aquaculture*, **7**, 107–203.

Welcomme R.L. (1981) Register of international transfers of inland fish species. *FAO Fish. Tech. Paper*, **213**.

6
Design and Construction of Aquafarms

Some of the basic information required for designing an aquaculture farm would have been collected at the time of determining the feasibility of the project. However, further investigations will usually be needed for designing the most appropriate lay-out, construction methods and operation. The design of the farm and its construction are as important as the selection of the site in ensuring the success of the project, both technically and economically. As indicated earlier, the ideal sites may not always be available. Deficiencies of the site will in most cases have to be made up by suitable designs for construction and operation. Though engineering designs may be available to meet the requirements of aquaculture in almost any adverse conditions, the economics and practicality of using them for commercial aquaculture render them of little help. In fact, the designs normally used in water or irrigation engineering works cannot be used without very considerable modifications for aquaculture constructions, because of the costs involved. This applies especially to pond farms which account for a very good proportion of present-day aquaculture.

As pond farm design is so site-specific, one cannot conceive of a design that can be of universal use. However, some of the major design features can be defined on the basis of the site physiography, the source and nature of the water supply, type of stock enclosures to be used, organisms to be cultured and the techniques of management, including feeding or food production and harvesting methods. The detailed investigations mentioned earlier should be directed towards obtaining the basic information required to determine the appropriate design features.

6.1 Inland and coastal pond farms

6.1.1 Data for pond farm design

Since the majority of aquaculture installations at present are land-based pond farms, we may first consider the procedures for designing those. Despite the similarity of basic principles involved, it will be convenient to consider inland fresh-water pond farms and coastal brackish-water or salt-water pond farms separately, mainly because of the differences in operational details.

As already indicated, the investigations prior to farm design will depend on the extent of information collected during the preliminary feasibility studies. The meteorological data relating to mean monthly temperature, rainfall, evaporation, humidity, sunshine and wind speed and direction should already be available. A contour map (scale 1:25 000 to 1:50 000) of the area will be most useful in determining the catchment area of the site and its relative location. A soil or geological map, if available, would be useful in studying the subsoil at the site.

Detailed investigations may be necessary with regard to water sources, soil characteristics and topography of the site. Topographic maps, if available, are likely to be of a small scale, which would not allow all the relevant features to be reflected. Therefore a new or updated map will have to be prepared showing the nature of the ground relief and its characteristics, such as differences in elevation, location and measurements of boundaries or fences, physical facilities if any (such as buildings, roads, canals, bridges), etc. It will assist in determining the direction of water movement, location of water control structures and quantity of earthwork needed. There are a number of methods used for surveying the land, such as

(a) gridding,
(b) plane tabling,
(c) cross-section method with transverse survey,
(d) radiating lines method with transverse survey and
(e) tachymetry.

Among these, tachymetry is relatively rapid in field surveys and more versatile in that it can be used for surveying all types of areas. Methods like gridding and plane tabling are more suited for relatively flat land, and the others cited above are especially useful for hilly terrains (Kovary, 1984). For field surveying a temporary bench mark with a convenient datum should be established. The location of this bench mark should be marked on the contour map and all the elevations of embankments, canals, ponds, structures, buildings, etc. set out from it. The contour map, which should show any structures observed or measured on the land, should preferably be scaled at 1:1000 to 1:5000, with contour lines of $10-25$ cm vertical spacing, so complete pond drainage can be designed and earthwork volume estimated with the required accuracy. If the proposed construction is an extension of an existing farm, the cross- and longitudinal-sections of the adjacent ponds, drains and channels should be obtained.

Soil quality

The soil characteristics of importance in site selection have been described in Chapter 4. Based on the results of feasibility investigations, the extent of further soil samplings required will have to be determined.

One or two sample stations to each 2 to 5 ha of site should be adequate if the soil conditions are uniform. If not, more sampling stations will be needed. The minimum depth of a bore hole for soil sampling is suggested to be 2 m below the deepest intended excavation of the project area. For the building of special structures, such as large water towers, greater depths of boring, commensurate with the size of the

structures, will be needed. The soil tests should be to estimate

(a) seepage loss,
(b) under-seepage conditions and the hazard of piping failure,
(c) stability of dikes constructed with the soil,
(d) the degree of compaction needed,
(e) the permissible flow velocity in the earthen supply channels and
(f) the foundation requirements of the structures.

Soil on potential borrow sites within economical hauling distance should be studied to determine the nature of the soil available for building embankments. The embankments for the farm have to be built with cohesive soils that have adequate plasticity (generally designated by the plasticity index − a measure of the interaction between water and the cohesive plastic components present in the soil), as for example soil with a plasticity index above 15 per cent. Such soils should be checked for their susceptibility to long-term changes in permeability caused by atmospheric factors, such as the development of stable density or aggregation of particles. The losses that are likely to occur due to under-seepage and infiltration have to be determined using standard methods. To estimate long-term losses through seepage it is necessary to take into account the sediment content of the water supply, which along with decaying debris, pond wastes, algal growths, etc., would cause natural sealing or colmatation in the course of time.

While embankment stability can be determined by standard methods of soil mechanics, the assessment of the possible long-term performance of structures is more difficult. Due to their relatively small size and the practice of repeated draining and filling, there is the greater possibility of entire embankments of farms becoming desiccated, causing cracks to develop and entry of water into the embankment at times of rain or pond filling. The soil will then swell, but the extent of swelling at any particular point will depend not only on the swelling potential of the soil, but also the magnitude of the confining pressure by the surrounding, especially overlying, soil masses. Repeated drying and wetting, and therefore shrinking and swelling, will produce a stable density distribution, with higher densities in the interior of the cross-section. Szilvassy (1984) describes the adverse effects of drying and rewetting fish ponds. The cracks formed by drying facilitate the entrance of water into the body of the embankment. The crack faces are saturated and the moisture penetrates into the interior by capillary action. The saturated parts become almost impervious to air and the air in the pores comes under the combined pressure of the capillary action and the hydrostatic pressure of underwater parts. This pressure on the confined air leads to spalling and subsequent sudden liquefaction of unprotected slopes. If water flows through the cracks, the liquefying soil will be scoured at a faster rate, resulting in the development of gully or tunnel erosion, which is often the cause of failure of small embankments.

Besides careful exploration of the surface layer of the area where the ponds and water supply canals are planned, the soils along the canal traces should be investigated also for their hydraulic properties to estimate slope inclinations and the allowable (non-scouring) velocity of flow in the canal. The sequence of soil strata down to the first impervious layer should be determined as accurately as possible. If the soil is impervious at least to 0.6 m thickness below the designed deepest bottom level in the ponds or the drainage channels, no further exploration may be needed. In view of the difficulties in obtaining fully undisturbed soil samples for laboratory tests, field permeability studies are recommended in the vicinity of each exploratory borehole by the infiltration method.

The buildings and other structures on fish farm sites are generally small, and so the loads acting on the foundation are not likely to be large. In cases where these are to be built on newly filled sites, special care should be taken to avoid damage due to future soil subsidence. The standard sounding methods used by building engineers should be applied.

While there is no gainsaying the importance of careful soil studies in planning aquaculture farms, it has also to be remembered that laboratory tests for design values of soil strength are costly. Even when done, the engineer has to use his judgement to decide whether to use it for the type of constructions involved in a pond farm with low dikes and dams. Because of this, in countries like Hungary, aquaculture engineers use special practical guidelines based on local experience for the construction of dikes, levees and dams lower than 3 m in height and retaining less than 3 million m^3 water (Szilvassy, 1984).

Special features of soils on coastal sites, especially mangroves have already been discussed in Chapter 4. The presence of large quantities of organic matter in the soil, particularly mangrove roots, is a special problem to be reckoned with in pond design and construction in coastal areas. There is a growing body of opinion in favour of leaving the pond beds undisturbed without any excavation and, depending on the flocculation and settling of sediments brought in with tidal water, to build up a thick top layer on the bed to reduce acid soil problems. In that case, soil to build the embankments has to be obtained from outside the pond limits. If a mechanical means of construction is planned, the necessary cohesive soil should be available within reach of drag-line excavators or similar equipment, working from the embankment base. If manual means of construction is the choice, it may be possible to cut soil into blocks and transport them on rafts or flat-bottom boats at high tides to the pond site. Besides the comparative costs, the construction time has also to be taken into account in making decisions.

Water supply

The chemical properties of the water source for the farm and the sources of pollution, if any, would already have been studied during site selection. Very often, further information would be needed on the quantity of water required, at the design stage. For a fish pond with an average depth of 1.5 m the amount of water required to fill it initially is 15 000 m^3/ha. Loss through seepage and evaporation varies considerably between areas. In an arid climate, the average loss during the growing season could amount to 1–2 cm/day or more. With proper management, the total minimum quantity required for filling and topping under such a situation is estimated to be between 35 000 and 60 000 m^3/ha per year. The size of the farm should naturally depend on the quantity of water available during the period of operations. When the source of supplies is a stream, data will be needed on the stages and flow rates to be anticipated at the diversion point in the periods of pond filling and for compensation of water losses. It has been recommended that flow rates should be designed for 80 per cent probability.

In areas exposed to floods, data on design floods and discharges will be required. Water control agencies can generally provide values for probability of occurrence of the design flood, but in cases where such values are not available, it has been suggested that 1 per cent probability of occurrence (that is once in a hundred years) should be adopted as the design flood for the spillway of a dam. In the case of smaller dams with a design volume less than 1 million m^3 and of ponds with a water area less than 20 ha built farther away from human settlements, where the dike failure would not cause other losses, a flood of 3 per cent probability may be adopted as the design flood. The runoff of the water catchment area of the site should also be calculated to determine the capacity of the farm reservoir or ponds. Data on the peak values of monthly evaporation and rainfall are necessary to estimate water demand.

Estimates of the annual volume of sediment entering the ponds would be necessary to determine desilting requirements; or in cases where it is planned to build up a top layer of silt, to estimate the time it will take for it to be accomplished. Again, where the water turbidity is undesirably high and separate sedimentation tanks are required to reduce it, this information is essential. One of the problems in ponds filled from natural bodies of water is the entry of extraneous fish and other organisms in the egg or larval stages with the water, even when the inlets are protected by small-meshed screens. Filtration of such water to remove pests and predators is extremely difficult and expensive. In special circumstances, when considered essential, sand or other filters may be designed according to the size and quantity of sediments.

The use of waste water, including sewage effluents, to irrigate and increase productivity of ponds is an age-old practice and fish culture is used now in many places as an efficient means of recycling organic wastes. Reference has already been made to the use of heated water effluents from power stations in temperate and cold climates. The main problem with the use of wastes is the possible development of anoxia in ponds, due to excessive organic loading and contamination with toxicants and heavy metals. The risk of transmitting bacterial and viral pathogens through the use of domestic wastes has received some attention. It has been shown that under conditions existing in fish ponds, actual reduction of pathogens occurs. Due to high photosynthetic rates, such ponds have high dissolved oxygen contents and high pH values, which increase the rate of disinfection of coliforms. Investigations have not yet found evidence of the transmission of any human bacterial diseases through fish. Even though fish do not suffer from enterobacterial infections, the possibility still remains that fish can harbour bacteria in their alimentary tract, tissues and mucus and hence serve as passive vectors of pathogens. Experimental studies made on artificially infected fish have shown that by holding them in clean water for an adequate period of time they can be cleansed of pathogenic vibrios. Depuration is often practised in waste-water aquaculture. So, if the use of waste water is planned, necessary facilities will have to be included in the farm design. Similarly, possible measures should be adopted to avoid incorporation of toxic substances. This can best be done at source. Detergents are often difficult to exclude from domestic and municipal wastes, but at least their concentration should be kept under permissible limits. The lethal limits of detergent for common carp is reported to be 10 ppm ABS (alkylbenzene sulphonate), but even sublethal concentrations can affect their growth (Hepher and Pruginin, 1981). The short duration of the grow-out period in aquaculture reduces the risk of accumulation of heavy metals from waste water, unless the concentration is very high. Experience so far seems to show that even when there is some accumulation, it is generally within accepted standards for safe use.

Public attitudes to eating products grown in waste water, particularly sewage effluents, can be a problem and solutions have to be found on the basis of socio-cultural ambience and should include public education and product promotion. In modern aquaculture, only pre-treated wastes are used. In some cases, the use of wastes is avoided in the final grow-out stages and when there is possible exposure to waste water at that stage the product is depurated for an adequate period before marketing. These are some of the measures that could help in meeting consumer concerns.

Salinity and tidal flows in coastal farms

For designing coastal pond farms the most important data needed are the seasonal variations in salinity of the water available and access to fresh water to reduce salinity when required. When the ponds have to be filled using tidal energy, detailed studies are needed to determine the stage/duration/frequency relationship necessary for engineering designs. Continuous data for as long a period as possible from the site will be necessary to verify calculated values from available tide tables and observations during feasibility studies. For designing proper water management in tide-fed ponds, it is necessary to determine the ground elevation, which actually approximates the tidal levels of mean lower high water or of mean high water at neap tide. If possible, the measurements should be made when the lowest critical tides of the year occur (which can be found from the tide tables). Alternatively, the measurements should be taken during the lowest and highest tides of the month. The days with the lowest tides should be selected, and the O datum or mean lower low water (MLLW) noted. A fifteen-day observation during the dry season for the mean high water and another fifteen-day observation at the height of the rainy season for mean low water, are considered sufficient to ascertain whether the pond-system will be drainable during rainy season and whether the desired depth can be maintained. Measurements may best be done in front of the area where the main gate of the farm is likely to be constructed. On the tide gauge, which can be a measuring stake driven into the ground, the point at which the water level was lowest should be marked. The O datum level, correlated with the lowest water level, should also be marked. This will serve as the base line for determination of all elevations in the farm system. A bench mark can be established by running a level to a permanent marker near the site to be developed. The elevation of points within the area can be measured as reckoned from the datum plane. It is considered uneconomical to excavate more than 50 cm for pond construction. If this is needed it will be better to resort to pumping, rather than depend on tides for water supply and drainage.

Location of hatcheries and availability of other inputs

The essential data required for hatchery design would become available through some of the investigations mentioned earlier in this chapter. Decisions as to whether a hatchery, together with nursery facilities, should be established in the same farm complex or in a different locality have to be made on the basis of the site conditions, water quality requirements, ease of operation, security, etc. As mentioned in Chapter 4, in certain types of coastal aquaculture, as for example shrimp culture, the need for unpolluted high salinity water for hatchery operation may make it necessary to site hatchery installations nearer to the sea, rather than in the brackish-water areas where the grow-out ponds may be located. Similarly for the giant fresh-water prawn (*Macrobrachium rosenbergii*) which requires saline water for spawning and larval development, the hatchery may have to be situated away from the fresh-water pond farms used for grow-out. However, in some circumstances it may be more economical to transport the necessary salt water to the inland farm site rather than maintain two separate units. In the case of salmonid culture, especially of the trout, the low temperatures required for spawning, hatching and larval development may make it necessary to establish the hatchery at high elevations with cold water, and grow-out farms at lower elevations with higher temperatures for faster growth. Smolt production for salmon in fresh-water installations may have to be done in different locations and the smolt transported and acclimatized for salt-water culture or for sea ranching.

The other input production facility that may be considered for inclusion in the farm design is for feed. For this, as well as for processing of farm products, the main requirements to be investigated are suitable land for the necessary constructions, clean water supplies and electricity.

The availability of skilled and unskilled labour in the area is an important factor in deciding on construction which would require adequate maintenance and careful operations. In many developing countries, priority is given to aquaculture development because of its potential to generate employment, and so there is a definite preference for the use of manual labour in construction and day-to-day operations. At the same time, it will be necessary to achieve cost-effectiveness and profitability. So, it will be necessary to obtain comparative information on costs of construction and maintenance, using mechanical equipment against manual labour. Besides the actual costs, the time it takes to construct the farm and bring it under production by these two methods and its economic consequences should also be considered.

6.1.2 Design and construction of pond farms

Considering the fact that the construction of farm facilities forms the major capital investment in pond farms and the operational efficiency of the facilities will largely determine the success of the project, it is fully justified and necessary to devote adequate attention to their design and construction. As mentioned earlier, pond farm designs are site-specific and so it is difficult to detail all possible variations. The aquaculturist will have to work closely with the design engineer to arrive at an economically acceptable design that will meet the operational requirements of the species and the culture technologies. Planned construction is feasible in most projects, except when existing undrainable ponds, tanks or mining pits have to be used, as is the practice in some of the South Asian and South East Asian countries. Even then, there may often be the possibility of designing a proper farm, incorporating the existing water bodies for easier management.

Size and shape

The size of a farm has to be determined on the basis of a number of factors, including quantity of water and extent of land available, technology to be followed (e.g. extensive, semi-intensive or intensive farming), production and income required to make the enterprise economically viable, and access to markets, manpower and equipment.

Even though the design of the farm will depend on several factors, there are some basic principles which are generally followed. Whether a hatchery is incorporated in the farm or not, there is usually a series of nursery ponds for growing larvae to fry stage, another series of rearing ponds to rear fry to the fingerling or yearling stage and a final series of production or stock ponds. Many farms, particularly in tropical areas, may not have the transitional rearing ponds, and fry may directly be introduced into the stock or production ponds for grow-out to marketable size. In farms incorporating hatchery operations, there is a need for brood-stock ponds to rear selected brood-stock, and in some cases also spawning ponds. Depending on the harvesting system to be adopted, there may also be a need for market ponds for holding the harvests before marketing. Instead of earthen ponds, tanks or raceways may be used for fry rearing in certain types of culture, such as of salmonids and shrimps. Tanks for culture of fry can very well be incorporated in the design of a pond farm, particularly in conjunction with a hatchery. Details of tank and raceway design will be discussed in a later section. In temperate and cold climates, wintering ponds or indoor wintering facilities may be needed.

It is possible to use some of the ponds mentioned above for more than one purpose, depending on the seasonality of operations. Spawning ponds can often be used for fry nursing after suitable preparation, and sometimes also as market ponds. Properly designed rearing ponds can, after the fry season, be used as production ponds. Thus, economy can be effected in pond area and most of the ponds can be brought under operation for a major part of the year, if the farming technology permits it. Because of these possibilities, it will not be very meaningful to suggest a particular ratio between different types of ponds. The estimated number of fry required and the number of crops of fry that can be raised to meet the requirements of the farm will decide the total area to be assigned for nursery purposes. The production target of the farm, based on markets and technology, will decide the area to be set apart for production ponds. For small-scale fish culture in tropical areas based on quick-growing species, it has been suggested that each farm should have a multiple of twelve production ponds, so that each month an equal number of ponds can be drained or harvested, ensuring a regular supply of fish for sale each month of the year (Maar *et al.*, 1966).

The size of ponds would vary according to the intensity of culture operations, but ranges of 0.05–2.00 ha for nursery ponds and 0.25–10.00 ha for production or stock ponds have been suggested. Spawning ponds could be 0.01 ha. Smaller ponds would result in a larger area covered by embankments and water supply channels. In intensive culture systems, there is an obvious preference for smaller ponds, ranging in size from 1 to 5 ha as against 3 to 10 ha in extensive systems, as small ponds allow greater control. Larger ponds take longer to fill or drain, under given water conditions. This may mean, in certain situations, sizeable loss of production time. Similarly, moderate-sized ponds facilitate safe harvesting, as too much crowding in harvesting sumps and handling can result in fish loss. It has been suggested that the harvesting of a pond should not take more than a day. This again points to the need for less extensive production ponds. However, it costs more per unit area to construct smaller ponds, because of the cost of the additional embankments and water supply structures needed.

There appears to be a greater preference for rectangular-shaped ponds in fresh-water farms. This is mainly to facilitate harvesting with seines of manageable length or through draining to a sump using the regular slope of the pond bottom. The lengths of drainage and feeder canals required will also be less. From the point of view of cost of construction, square-shaped ponds are considered preferable, as the ratio of water area to the length of embankment will be higher, but if the slope of the site selected is high it may be necessary to construct rectangular ponds, to enable easy drainage. In cases where fish culture is combined with animal production or cultivation of vegetables, fruit trees, etc., as in southern China, the cost of construction of the main embankments would not be a major consideration, as the farmers need wide land areas near the ponds for animal or plant production. Many of the new farms there have square ponds, but others are rectangular in shape. Some East European countries, particularly Hungary, have tried different shapes of ponds, such as radial ponds, all of which drain into a central sump.

The layout of coastal pond farms is largely dependent on the farming procedures. Some of the typical layout designs will be described later, but there appears to be no special shape preference in newly designed ponds, though most of them are rectangular. The shape of the traditional farms largely follow the land contours and many of them have irregular shapes. In modern designs this is generally avoided and embankments are straight where possible.

Layout of farms

The conventional classification of fish pond design into barrage ponds, contour ponds and paddy ponds can still be used to describe the major types of pond layout. The barrage ponds are constructed in flat or gently sloping valleys, or abandoned river beds, by putting a low dam at a suitable site (fig. 6.1). The dam has to be built at the narrowest point to reduce construction costs. The sides of the ponds are formed by the slopes of the valley and a series of ponds can be built on the site. The source of water is a stream or river nearby. A spillway has to be built to avoid flooding of the ponds. A feeder canal from the stream will be necessary to regulate the water supply. Suitable drainage has to be provided to prevent flooding and consequent loss of stock and damage to the pond structures.

Contour ponds (fig. 6.2) are also generally located near a stream, canal, river or reservoir and in a valley, the bottom having a slightly sloping contour. The farm is situated on one side of the valley only and floods pass through the other side. A weir diverts the water for intake through a gate to a supply canal, from which each pond can be filled and drained separately. The dikes should be built to carry the design flood safely. Such a layout is possible

Fig. 6.1 A barrage type of pond farm.

only in sufficiently wide valleys or river beds.

Paddy ponds (fig. 6.3) are constructed on relatively flat areas surrounded by a dike. Such sites make it possible to use much better layout designs, including separate supply and drainage channels, seepage and pond drains, harvesting sumps, etc. Most of the sites selected for carp, tilapia and catfish culture in fresh water and brackish water, or salt-water finfish and shrimps in coastal areas, would be suitable for this type of pond.

Fig. 6.2 A contour type of pond farm.

Dike design and construction

The most important constructions in a pond farm are the dike system and the water control

Fig. 6.3 A paddy type of pond farm (photograph: Y.A. Tang).

structures. The pond bottoms may or may not be excavated, depending on the topography and soil conditions, but the supply and drainage channels and harvesting pits have to be excavated.

As indicated earlier, the constructional details of the dike will depend upon the nature of the soil to be used, water depth required in the ponds, wave action and possible erosion, etc. Fig. 6.4 illustrates cross-sections of some typical dikes. Since cost-efficiency is the major consideration, it is necessary to determine the steepest slope inclination of the dikes that will ensure stability of the structure on a long-term basis. Where the soil conditions warrant it, the economics of lining the dike slopes with bricks, rip-rap, wood, etc. should be determined, taking into account long-term maintenance costs and the security provided. The freeboard has to be determined according to wave action and design flood levels. Table 6.1 lists the recommended side slopes, top width and freeboard of dikes (Kovari, 1984b). It is recommended that small ponds be located, if topography permits, with their long axis parallel to the prevailing winds, in order to provide maximum aeration. Large ponds may have the long axis at right angles to the prevailing winds, as the winds blowing over a long stretch of water may create higher waves and greater erosion of the dike. A minimum of 3 m top width will be required for embankments to be used by vehicles, but when heavy vehicles are to be used, the main embankments may have to be as wide as 6 m. Adjustments may be needed if the design includes a water supply channel on the crest of the dike. Secondary and tertiary dikes can be narrower and lower, according to the water depth in the enclosed ponds.

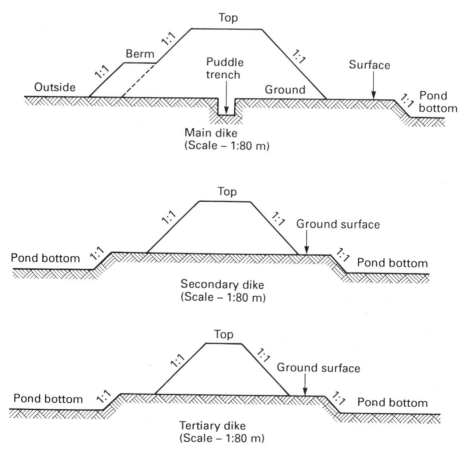

Fig. 6.4 Cross-sections of some typical types of dikes.

Table 6.1 Recommended side slopes and top width of pond dikes.

Type of soil	Inside slope	Outside slope	Water depth in the pond (m)	Top width of dike (m)	Freeboard (m)
Sandy loam	1:2–1:3	1:1.5–1:2	0.50	0.50	0.40
Sandy clay	1:1.5	1:1.5	0.50–0.80	0.50–1.00	0.40–0.50
Firm clay	1:1	1:1	0.80–1.20	1.50	0.50
With brick lining inside	1:1–1:1.5	1:1.5–1:2	1.20–2.00	2.00–2.50	0.50
With concrete lining inside	0.75–1:1	1:1.5–1:2	2.00–3.00	2.50–4.00	0.50–0.60

The depth of water to be maintained in a pond depends very much on the climatic conditions and culture practices. The recommended depth of trout ponds is 1 m at the intake, sloping to 1.5 or 2 m at the outflow. A depth of about 1 m is preferred in tropical and subtropical carp culture ponds. Beside minimizing wide fluctuations of water temperature, it assists in reducing the growth of rooted aquatic weeds which are a major problem in fertilized ponds, particularly in the tropics. However, shallower ponds will be preferable during the growing period in temperate climates, to make use of the higher water temperature for enhanced production. Because of these differences, wide variations occur in the depths and size of different types of ponds in culture systems. A range of average water depth of 0.4–1.5 m for nursery ponds and 0.8–3.0 m for production or stocking ponds have been recorded. In fresh-water fish culture, spawning ponds may have an average depth of 0.4–1 m and holding or market ponds 1.2–2.0 m. The water area of nursery ponds varys between 0.05 and 2 ha and of production or stocking ponds between 0.25 and 10.0 ha. Spawning ponds are smaller, ranging from 0.01 to 0.5 ha and holding or market ponds from 0.10 to 1.0 ha.

It is a common practice to provide an impervious core of soils of high cohesiveness, with shells of less cohesive soils on both sides or only on one side (fig. 6.5). The most important causes of deterioration of dike slopes are erosion due to wind and wave action and burrowing by aquatic animals and the feeding habits of fish like carp that rout around pond dikes. A proper grass cover is necessary to protect the exposed parts of the dike. Quick-growing and spreading varieties of grass are preferred. In East European ponds, a 4 m wide reed belt is recommended for larger stocking ponds (greater than 10 ha) to protect the sections of the dike exposed to wave action. In farms elsewhere, rip-rap placed in thicknesses of 0.25 to 0.5 m is preferred. In countries like China, blocks of concrete or bricks are used to line the slope. Lime stabilization has also been suggested as a means to improve soil compaction in dikes to resist erosive action.

The general principles of design and construction are very similar for fresh-water and coastal pond farms. However, the nature of the terrain and dependence on tidal water, as well as the practice of utilizing benthic algal pastures as food for the cultivated stock, necessitate some changes in the details of design and construction. The type of soil found in tidal areas selected for coastal farms has already been described (see Section 4.22, Acid Sulphate Soils). The construction of large embankments and heavy concrete structures on soft ground creates special problems. The fibrous or peaty top soil in mangrove areas (with roots sometimes constituting more than 50 per cent of the soil) is an additional problem. The need for, and methods of, easy drainage to leach out the acid contents of cat-clay soils have already been discussed. A system of open canals along the natural waterways will be required for this purpose. The canals should have a depth approximating the tidal level of the mean lower low water, in order to allow drainage of the maximum amount of percolated tidal flow over permeable stratum. Two to three years or more may be required to bring the pH value of acid soils to an acceptable level.

The depth of water in coastal fish ponds also varies as indicated earlier. Because of the need for easy drainage and constant changes of water in present culture practices, it is necessary to maintain lower depths of 30–60 cm. It is held

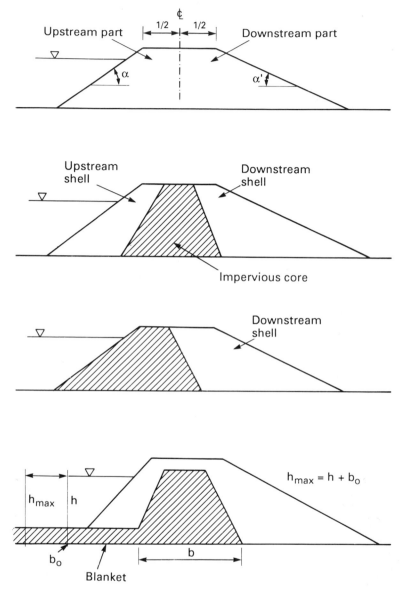

Fig. 6.5 Cross-sections of dikes with impervious cohesive soils (from Z. Szilvassy, 1984).

that production, even in coastal ponds, can be increased through increasing the depth of water, but present culture practices make it impractical to do this. When pumping proves to be a feasible means of water management in such farms, a greater depth of water should become possible.

A typical design of the perimeter dike or main embankment of a coastal pond farm located in an estuarine area is shown in fig. 6.6. The dikes are aligned along the river banks on the seaward side. If the farm is located in a mangrove area it is advisable to maintain a belt (50–100 m wide) of mangroves to protect it from waves and currents. The soft foundation soil in such swamps makes it necessary to allow

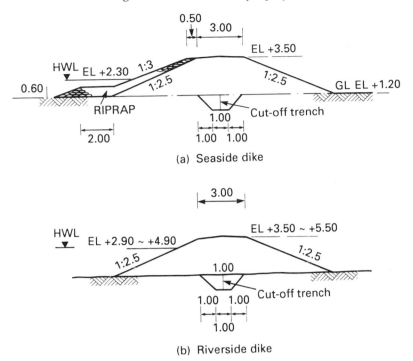

Fig. 6.6 Cross-section of a typical perimeter dike: (a) seaside dike (b) riverside dike (from Tang, 1979).

for a period of natural consolidation, before it becomes stable. High volume changes and surface cracks may occur in the process of drying. The height of such embankments is generally limited to 2.0 m, with a freeboard, after shrinkage and settlement, of 0.6–1.0 m above the design flood water level. Usually an allowance of 15–20 per cent is made for shrinkage due to settling.

A puddle trench of 0.5–1 m depth and 0.5–1 m width is considered essential to prevent seepage under the dike. For shrimp ponds in Southeast Asia, the following slopes are recommended (ASEAN, 1978):

2:1 when dike is above 4.26 m and exposed to wave action,
1:1 when dike height is less than 4.26 m and tidal range is above 1 m,
1:2 when the tidal range is 1 m or less and the dike height is less than 1 m.

The perimeter dike of the farm should be 0.5 m above the highest tide or flood level recorded in the locality. The berm built on the inside of the dike should be slightly above the water line, in order to minimize the effects of wave action on the dikes. Holes made by burrowing animals damage the dikes in coastal ponds; incorporation of bamboo screens or plastic film in the puddle trench helps to minimize this.

To facilitate the drainage and harvesting of fish, the pond bottom should have a minimum slope of 0.1–0.2 per cent towards the outlet. In inland fresh-water ponds, a harvest sump is constructed near the outlet in the deepest part of the pond as a long trench or in some other convenient shape, about 50 cm deeper than the surrounding area and with sloping sides to facilitate netting.

Harvesting sumps can also be constructed outside the pond, and a combined sump can be made for a number of ponds. The recommended bottom area for the harvesting (cropping) sump is around 40 m²/ha, and the depth 0.6–1 m. A width of 10–25 m would be convenient for the use of nets. The external harvesting sumps are connected to outlet sluices of the ponds and fresh water has to be introduced into the external sumps at the time of harvesting. The bottom should be at least 30 cm deeper than the deepest point of the pond and an additional differential elevation of 20 cm is necessary between the two ends of the harvesting pit. In order to avoid rapid silting, the sump may be constructed 5–10 m away from the main dike of the pond. Low levees made of sandstones, gravel, bricks or concrete may be built around the sump to prevent silting.

In coastal farms using tidal flow for water management, it is common to have a central canal from which tide water is taken in through a pipe and fed into a set of two or three ponds through a common catching pond (Fig. 6.7). It is connected to the rearing ponds through sluice gates. The catching pond and the central canal serve the same purpose as the harvesting sumps in fresh-water ponds. For harvesting from nursery ponds, the catching ponds are particularly useful. The central canal becomes more important for harvesting from rearing or stocking ponds. The habit of many brackish-water fish to swim against the current is used to capture them. In fact, the elaborate system of 'lavorieri' or traps in Mediterranean lagoon farming is based on this behaviour. However, in new coastal farms, especially those meant for shrimp culture, separate feeder and drainage canals and harvesting sumps are provided. Harvesting sumps are usually located at the pond outlet.

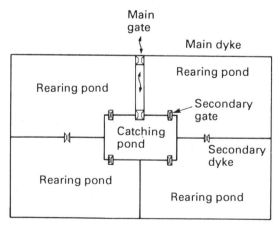

Fig. 6.7 Layout of a rearing system, with a catching pond.

Water supply and drainage

One of the most important factors that govern the success of an aquaculture operation is proper water management. Although traditionally fish farming has been done in some areas in 'undrainable' ponds, the ability to intensify operations will be greatly restricted in such waters. So the adequacy of available water should be a major criterion in selecting the site. The actual quantity required for the ponds will depend on the soil and climatic conditions, but as a rule of thumb, one may calculate it at the rate of 19000–23000 m^3 for a 1 ha pond with an average depth of 1.5 m, which would include an extra 25–50 per cent more to compensate for evaporation and seepage. To have a more accurate assessment of the yearly water requirement the following formula can be used (Kovari, 1984b):

$$Q_r = V_f + V_{rf} + L_e + L_s + L_c - V_{ra} \ (m^3)$$

or

$$Q_r = \frac{V_f + V_{rf} + L_e + L_s + L_c - V_{ra}}{86400 \times T} \ (\ell/s)$$

where

Q_r = annual water requirement (m^3 or ℓ/s)
V_f = $A \times h$ = the pond volume to be filled (m^3)
A = average water surface area of pond (m^2)
h = average water depth of pond (m)
V_{rf} = $N_o \times V_f$ = the pond volume to be refilled (m^3)
N_o = number of refillings a year
L_e = $A \times E$ = water loss from evaporation (m^3)
E = mean annual evaporation (m)
L_s = $A \times T \times S$ = seepage loss in the pond (m^3)
S = seepage coefficient (m/day)
L_c = $A_c \times 1.2 \times E$ = transmission loss in earthen canal (m^3)
A_c = water surface area of feeder canal (m^2)
V_{ra} = $A_{eff} \times R_a$ = water inflow from rainfall to pond (m^3)
A_{eff} = total area of pond including dikes exposed to rain (m^3)
R_a = mean annual rainfall (m)

T = operational time in days.

The water supply and drainage system have to be designed to convey the required quantities. Different designs have been adopted, obviously based on different criteria and requirements. In many designs, the same canals are used for feeding and drainage of water, so as to economize on space and construction costs. In others, it is considered essential to have separate feeder and drainage canals, as well as inlets and outlets, for operational safety and efficiency. It is generally considered necessary to locate the inlets and outlets on opposite sides of a pond, but in some farm designs the inlet is located near the outlet and the harvesting sump, so as to facilitate the supply of water to the sump when the pond is drained for harvesting.

The quantity of water conveyed through a canal depends on the area of the cross-section of the water passing through (referred to sometimes as the 'wet cross-section') and the speed of the current. This can be calculated by the equation $Q = F \times V$, where Q is the water quantity transported in m^3, F is the wet cross-section in m^2 and V is the speed of water current in m/s (Woynarovich, 1975). If the bottom of the canal is 1 m wide and the slope is 1:1.5, the wet cross-section under different water depths will be as shown in Table 6.2.

If the bottom of the canal has a slope of 0.1–0.2 m in 1000 m, the speed of water will be about 0.3–0.5 m/s. Unless such current is maintained, rapid siltation may take place in the canals. On the other hand, faster flow may result in erosion. In areas where the soil quality is poor, lining with suitable reinforced plastic films has been successfully employed to reduce erosion and seepage from the feeder canals. It is, however, more common to construct brick or cement concrete lined canals, where earthen canals are not feasible. In such cases, a higher velocity of water flow can be maintained and so the width of the canal can be smaller, for example 0.4–0.5 m, with a bottom slope of 0.5–1 m per 1000 m. When a feeder canal is built on the crest of the dikes, it is necessary to construct it with bricks or cement concrete.

The principles of designing water supply and drainage systems in coastal ponds are essentially the same as for inland fresh-water ponds.

Table 6.2 Water flow through canals at different water depths.

Water depth (m)	Wet cross-section m^2 (approx)	Water transported	
		ℓ/s	m^3/day
0.1	0.11	33	2 800
0.2	0.26	78	5 700
0.3	0.43	129	11 100
0.4	0.64	192	16 500
0.5	0.87	260	24 400

* Calculations based on a water current speed of 0.3 m/s.

The only major differences are caused by the tidal fluctuations, when tidal energy has to be used for filling and draining the ponds, and by the prevalance of acid sulphate soils. The tidal range data at the site will have to be used in estimating the duration and quantity of water that the farm can extract at the site at high water for feeding the ponds to the level required. Similarly, an estimate of the quantity of water that can be drained during low tides has also to be made. The size of the feeder canals and the size and number of water intakes will depend on the tidal supply. The duration of low tides and their amplitude will determine the quantity of water that can be drained from the pond. The acid sulphate soils that occur in coastal swamp areas make it necessary to drain the seepage water from the ponds, without allowing it to contaminate the water in the feeder channels. The level of water inside the pond should be maintained at a higher level than water outside the pond, to ensure that the acidic water does not stagnate there. It will also be advisable to construct a berm near the water's edge to catch acidic run-off during rains, so preventing it from washing into the ponds.

A coastal fish farm generally has a main canal and subsidiary canals for water supply and drainage. The main canal distributes from the main water gates to the subsidiary canals and from there to individual ponds. The flow is in the reverse direction for drainage. There are many types of water control structures in use in fresh-water and coastal fish farms. The inlets may be anything from a simple pipe to a concrete sluice. A turn-down pipe, open sluice or

monk, is used as outlet structures. Probably
the most versatile water control structure is the
monk (fig. 6.8) which can be used for inlets and
also for outlets. One major advantage is that
by adjusting the stoplog and fish screens, the
operator can release the top or bottom layer of
water from the pond. The monk consists of a
vertical tower with three pairs of grooves for
housing screens and stoplogs, and a horizontal
conduit passing through the dike, both of which
may be made of concrete, brick or a combination of the two. In recent years, monks made
of fibreglass, plastic and even non-corrosive
metal have been used. The selection of material
has to be made on the basis of long-term costs,
including maintenance and economic life of the
structure. The height of the tower part of the
monk depends on the highest allowable water
level and the size of the pipe used under the
dike. The opening in front of the tower need
not be more than 40 cm wide for ponds measuring up to 5 ha. Stoplogs or flash boards are
inserted into two of the grooves in the monk
tower. The space between the boards can be
filled tightly with wet clay, barnyard manure or
compost, to prevent leakage of water. Another
means of preventing lateral seepage is by attaching a rubber liner (the inner tube of an
automobile tyre can be used) to the board to
provide a water-tight seal. The third groove is
for a suitable screen to prevent debris from
entering the pond and fish from escaping. It is
advisable to construct wings as support for high
monks. Since they are heavy structures, particularly those built with bricks or concrete,
strong bases will be needed. A base 30−50 cm
deep and 30 cm wide on each side will have to
be constructed with boulders and cement mor-
tar. In the case of soft soil, the base may have
to be 60−90 cm deep, and 50−60 cm wider
than the actual size of the structure.

Another commonly used water control structure is the open sluice, (figs 6.9 and 6.10). It
is especially useful where the discharges are
higher than those which normal monks are
capable of carrying or in catching ponds serving
two adjacent ponds to facilitate the passage of
fish or fry. The closing mechanism in open
sluices can be stoplogs or vertical lift gates.
Open sluice gates are commonly used on
coastal fish farms in Asia. In order to avoid
poor performance due to defective workman-

Fig. 6.8 Monks made of concrete (above) and wood
(below). Outlet pipe can be seen behind the wooden
monk (from Huet, 1986).

Fig. 6.9 Concrete sluice gate for a tidal pond farm in the Philippines. Note the bamboo screen.

Fig. 6.10 Wooden sluice gate in a coastal pond farm in the Philippines. The attached net is for harvesting by draining.

ship and to reduce construction costs, many farmers prefer a wooden sluice although its economical life will be less than that of concrete structures.

In catfish farms in the southern USA, the most popular water regulatory system is the turn-down pipe, located at the lowest point of the base of the dike. It serves as an overflow and drainpipe. The water levels can be adjusted by pivoting the pipe, in order to increase or decrease the quantity of water flowing out of the pond (fig. 6.11). Besides providing a screen over the end of the pipe inside the pond to prevent loss of fish and obstruction of water flow by aquatic animals, it will be desirable to provide a special anti-seep collar around the drainpipe inside the dike to prevent water from seeping along the pipe and causing leaks. Some turn-down pipes are constructed with a double-sleeve device that permits water to be drained from the bottom of the pond rather than the surface (Lee, 1973). This will rectify the main disadvantage of turn-down pipes, of not being able to drain the bottom water (low in oxygen and containing a higher percentage of metabolites). The size of pipe to be used should be selected on the basis of the size of the pond, the speed at which drainage has to be done and the rate at which the pond is to be filled. The higher the diameter of the pipe, the greater the

water flow capacity. Doubling the diameter of the pipe will result in an increase of over four times in the water flow capacity. Generally, an 11 cm pipe will be adequate for small ponds of 1−2 ha, but pipes of 16−32 cm are recommended for 6−8 ha ponds (Lee, 1973).

A water control structure of special importance for farms susceptible to flooding is a spillway. This serves to bypass the floods reaching the farm, without damage to the ponds, so preventing the stock of fish from escaping. Spillways are also useful in farms built on level ground, when there is a large watershed area and there is a likelihood of surplus water caused by rainfall or by filling. A wide variety of spillway designs are available. Unlined spillways (fig. 6.12) with fish screens between piers are relatively simple to design and construct.

When intensive aquaculture is practised, some form of aeration system becomes essential to enhance oxygen transfer and the dissolving of organic carbon in the water. Gravity aeration is often achieved through weirs and splash boards in ponds and raceways. Simple surface aerators like open impeller or centrifugal pumps and paddle wheels are commonly used to break up or agitate the water and increase the surface area available for oxygen transfer. Different types of aerators used in carp ponds are described in Chapter 15.

Fig. 6.11 Pond with a turn-down pipe drain (from Stickney, 1979 − by permission of John Wiley & Sons Inc).

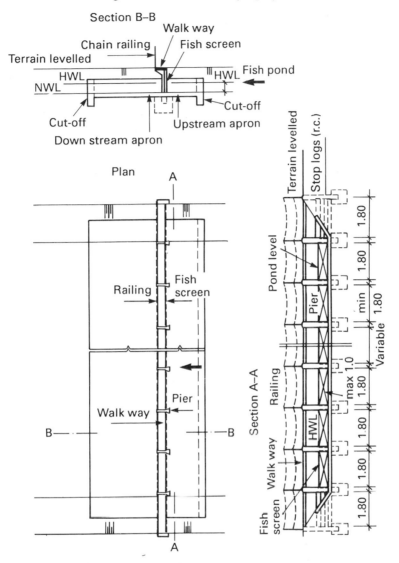

Fig. 6.12 An unlined spillway (from Elekes, 1984).

Methods of construction

In planning construction generally, there is an option to use mechanical equipment or manual labour for much of the work involved. From an economic point of view, mechanical methods of construction have many advantages. The construction period can be greatly reduced, the need for recruitment and supervision of a large labour force can be minimized and in a majority of cases more efficient structures can be achieved. Table 6.3 shows the comparative figures for earthwork in a tidal fish farm in the Philippines.

However, under certain socio-economic situations it may be necessary to select labour-intensive methods in order to generate employment in rural areas. Also, small homestead-type fish farms can probably be constructed equally or more efficiently with manual labour. On certain swamp-land areas, particularly in peaty soils in tidal lands, the use of manual labour may prove efficient. For example, it may be possible to adopt the technique of cutting earth

Table 6.3 Comparison of mechanical and manual methods for earthwork in tide-fed fish farm construction, based on a 1974 case study (from Tang, 1979).

Item	Mechanical method	Manual method
Cost of contruction (US\$/m^3)		
Perimeter dike	1.15	1.38
Main canal	0.62	0.77
Partition dike	0.62	0.46
Levelling (US\$/ha)	200.00	600.00
Labour requirements (man-days/million m^3)		
Skilled labour	80 000	
Unskilled labour		440 000
Construction period		
Skilled labour (500 man-days/million m^3)	160 days	
Unskilled labour (1000 man-days/million m^3)		440 days

Fig. 6.13 Bulldozer used for pond construction (courtesy of J. Kovari).

into blocks and loading them on to rafts or flat-bottomed boats for transport at high tide to the embankment site. The embankment can be built at low tide, placing the blocks the same way as bricks for building walls and compacting them mechanically or manually to make the embankment watertight. Nevertheless, overall experience so far would indicate the need to use mechanical equipment, where feasible, for construction of larger farms.

The bulldozer (fig. 6.13) is probably the most versatile earthmoving equipment for inland fresh-water farm construction as it can be used for clearing, grubbing, stripping, excavating, diking and levelling. However, the earth will have to be compacted to prevent erosion. The economical length of haul for a bulldozer is generally between 20 and 50 m. Another piece of equipment especially preferred for embankment construction is a scraper (fig. 6.14), which can be used for stripping, excavating, diking as well as compacting. The economical length of haul of a scraper is generally between 100 and 1000 m. As a scraper does not move very easily on heavy clay, a tractor must be used to push along the cutting haul. Hydraulic power shovels and hoes can be useful, particularly in excavating trenches, drainage and feeder canals.

For coastal fish farm construction, a small dragline excavator (figs 6.15 and 6.16) with a bucket capacity of 0.3–0.5 m^3, or a hydraulic excavator, has been found to be convenient for operation and handling. A crawler tractor, with a lower ground contact pressure, is very suitable for trimming the soil for profile formation and also for pond bottom levelling. The main constraint is in the haulage of earth in areas where mass movement of earth is required. Multiple handling has to be resorted to, as no other means of truck transportation is possible in swamps.

A wide variety of compactors, such as sheepfoot, steel wheel and rubber tyred rollers and platform and vibratory compactors, are available; but, in swampy soil conditions, it may be difficult to use them. In such cases it has been recommended that the dikes be constructed in layers and that the dragline travels on them to effect proper compaction.

Construction materials

A point that needs to be emphasized in pond farm construction is the choice of construction materials. As cost and availability of materials differ so much between areas one cannot suggest a uniform standard of materials. The guiding principle, however, should be cost-effectiveness, where durability and maintenance costs are important. Table 6.4 gives the values of durability and maintenance costs of the commonly used construction materials in pond farms.

Fig. 6.14 Tractor-driven scrapers being used in pond construction (courtesy of J. Kovari).

Schedule and sequence of construction

It is necesssary to plan the construction work very carefully to avoid waste of effort, funds and efficiency of the structures. Based on project financing, availability of labour and equipment and climatic conditions, the schedule of construction and farm operation should be determined in advance. Catfish farms in the USA are often constructed during summer and autumn, allowing the soil to settle during rainy seasons in late autumn and winter. Obviously it is preferable to complete the construction and start farming in the shortest possible time, as the investment would then start giving returns early enough. Where a longer construction period is unavoidable, as when manual methods are employed, the possibility of constructing the farm in sections, in order to start production while construction work of the rest of the farm continues, should be considered.

In order to plan the construction work properly, a detailed contour map will have to be prepared. This usually can be done only after the site has been cleared. Clearing presents greater problems and takes more time in marshy areas, particularly mangroves. In such cases, the clearing of the area for the perimeter dike may be done first, based on the available topographic map, and the rest of the area cleared and mapped as the work progresses. It is generally easier to fell trees and remove

Table 6.4 Durability and maintenance costs of materials commonly used in the construction of pond farms.

Material	Durability (years)	Maintenance cost (% of material cost)
Reinforced concrete (1:2:4)	20–30	100
Stone rubble in 1:5 cement mortar	10–15	150
Brick masonry in 1:5 cement mortar	5–10	250
Wood	5–8	300–400

Fig. 6.15 Dragline excavator, operated from pontoons in tidal lagoons in Italy, for dike construction (photograph: Carlo Mozzi).

Fig. 6.16 Dragline being used for dike construction (courtesy of J. Kovari).

dense brush after the perimeter dike is constructed, as the ground can then be dried to support heavy equipment for cutting, such as chain saws, pluckers for uprooting bushes and small trees and winches for pulling the trees and brush.

The general features of the farm, including the boundary of the site, layout, number and size of ponds, main and subsidiary dikes, water supply and drainage, location of water control structures, etc., have to be shown on the detailed map to aid construction work. The area reserved for the hatchery (if one is planned at the pond site), auxiliary buildings for storage of dry feed and chemicals, fish handling, preservation and storage, workshops and space for storage of nets and equipment and the approach road and other utilities should also be shown on the map. If feed production is to be done on the farm site, adequate space for housing the equipment, feed ingredients and storage of processed feeds has to be provided.

The sequence of construction work of the farm has to be decided in advance and followed, in order to achieve good quality construction. If flooding of the area during construction is likely, the drainage channels should be excavated before starting construction of the dikes. All the outlets should be constructed before commencing on the dikes. For the actual construction of the dikes it is necessary to estimate the quantity of earth required (taking into account also the packing coefficient of the soil, usually 20−50 per cent) and decide whether the pond area should be excavated for obtaining the earth and for levelling the pond bottom. The quantity of earth required per hectare for the construction of dikes for a 4 ha pond is estimated to be 2500−4000 m^3 (Pruginin and Ben-Ari, 1959). As far as possible, all organic matter, including roots, should be removed from the soil used for dike construction, as rotting organic matter will weaken the dike. Similarly, the humus should be removed from the base of the dike to bind the dike to the base properly and avoid seepage. The need to preserve the top soil in ponds built on cat-clay soils and allow a layer of silt to settle on it to reduce hazards of acidity in pond water has already been discussed (see Section 6.1.1). To attain the necessary height of the dike, it may be necessary to compensate for the subsidence by recapping it two or three times. It may also be necessary to make a large berm between the toe of the dike and the drainage canal, to offset the weight of the dike.

6.2   Tank and raceway farms

As will be evident from the preceeding section, pond farms, although comparatively less expensive to construct and operate, are affected by too many external factors over which the aquaculturist has very little control. Because of this, it is not generally possible to employ a highly intensive technology in pond farm culture. Tank and raceway farms attempt to bring greater human control in operations and facilitate highly intensive farming.

6.2.1   Tank farms

Tanks can be made of concrete, fibreglass, marine plywood, metal or other hard substances (figs 6.17−6.19). Durable materials that are free from toxic paints or chemicals only are used. Fibreglass is a popular material for tank construction as it is light, strong and inert to fresh and salt water. It can be moulded into most desired shapes and is strongest in tension loading, which is usually the stress experienced in circular tank walls. Fibreglass tanks are generally circular in shape. Sectional metal tanks can readily be obtained in the market in many places and can easily be erected or dismantled. Circular tanks are very commonly used for nursery and grow-out purposes. Besides being easy to assemble and install, the water supply and drainage in such tanks can be organized in such a way as to create a vortex that will sweep most of the detritus and other waste material out of the system. Ready-made plastic-coated metal tank sections can be bought to make tanks of the required size. They are bolted together and sealed with waterproof cement or similar material. The base is screeded in waterproof cement and slopes to a central drain, from which a pipe of suitable size carries discharges to the main drainage pipe; the latter collects discharges from all the tanks and conveys them to the final discharge point. The water level in the tank is controlled by a vertical pipe which is moveable and fitted in the main drain pipe, its height above the base of the tank being thus adjustable. The outflow is usually screened by a vertical, cylindrical plastic or metal mesh of the required size that projects above the water surface. A screened overflow pipe of adequate size is fitted into the upper wall of the tank. In order to protect the stock from predatory birds and other animals, the tanks are covered with suitable netting or metal screens.

Many variations on the arrangement of the water supply system and protective devices are possible in circular tank farms, including regular aeration and recirculation of water where

Fig. 6.17 An out-door tank farm. Tanks are made of cement concrete.

Fig. 6.18 An out-door tank farm. Tanks are made of fibreglass.

Fig. 6.19 An indoor tank farm. Tanks are made of plastic-covered metal.

it is necessary. Circular cement concrete tanks in Chinese fish breeding farms have water inlet nozzles arranged on the walls in such a way as to cause a regular circulation in the tanks (fig. 6.20). Circular tanks used for catfish culture in the USA range up to 6 m in diameter and 0.8 m in depth, with a fall of approximately 5 cm from the circumference to the centre drain, making it easier to clean. In fact they are to some extent self-cleaning.

Rectangular tanks are also used and they are approximately 8 m long, 1 m wide and 75 cm deep. The bottom may slope towards one end or towards the middle, to facilitate cleaning and draining (fig. 6.21). In tanks that drain at one end, water enters at the opposite end and flows the length of the tank, whereas in tanks that drain in the middle, water enters at each end and flows towards the middle. One advantage of a rectangular tank is that it is comparatively easier to harvest fish from it than from a circular tank. Rectangular tanks can be arranged in stacks four or five high, in which case they are made slightly smaller, with a length of 4.5−6 m, a width of 1.5−1.8 m and a depth of 40−45 cm. Such systems can be arranged indoors and it should be possible to install the necessary equipment for controlling water temperature and thus use the tanks for year-round production in colder climates.

Rectangular tanks are easy to construct, but

circulation of water is often characterized by what may be called 'dead' areas, where metabolic products can build up and also cause oxygen depletion. In such tanks, solid waste products can build up at the bottom, unless water velocities are maintained high enough to remove them. It is, of course possible to incorporate suitable designs for better circulation, but it becomes more complex and expensive to maintain.

As mentioned earlier, tanks can be built of different materials and in different shapes and sizes. Though not very common, there are large cement concrete tanks of area 200−300 m² used for rearing salmon, trout, shrimp, etc. In Japan and Taiwan, cement concrete tanks measuring up to 0.2 ha are used for eel culture. To enable a high density culture, suitable aeration equipment is provided. A more recent type of facility for aquaculture, in many ways similar to tank farms, is the silo, which has been tried largely in the USA (fig. 6.22). Essentially it is a deep tank, with water pumped down the centre through a pipe. Water flows upward in the culture tank, outside the centre pipe, and discharges into a trough constructed around the outside of the tank at the top. The flow rates are high, but higher densities of fish can be grown in a silo − as much as 136 kg/m³ or 27.5 kg/m³ per second of water flow (Buss *et al.*, 1970).

Fig. 6.20 Circular tanks in a Chinese fish breeding station. Inlet nozzles (arrow marks on the walls show their positions) ensure proper water circulation.

Fig. 6.21 A tank farm with cement-concrete rectangular tanks arranged on the side of the water supply system. Note the rails on the side of the tanks for transporting feed and the silos on the right for feed storage (courtesy of J. Kovari).

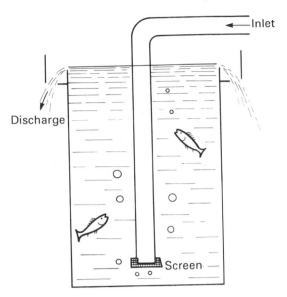

Fig. 6.22 Design of a silo tank (from Wheaton, 1977 — by permission of John Wiley & Sons Inc.).

Tank farms built in areas with limited water supply have often to resort to purification of water through biological or other filters and recycle them for repeated use. The costs involved have been rather prohibitive for commercial grow-out, but it is often possible to use the system in hatchery and nursery tanks as described in Section 6.5.2, Reconditioning and recirculation of water.

6.2.2 Raceway farms

Raceways are designed to provide a flow-through system to enable rearing of much denser populations of animals. An abundant flow of good quality, well-oxygenated water is essential to provide respiratory requirements and to flush out the metabolic wastes, particularly ammonia. The specific flow rate required to meet the oxygen consumption of organisms and the flushing out of metabolites can be determined on the basis of the temperature and oxygen concentration of the inflow water and the oxygen consumption and ammonia/nitrogen excretion of the organisms in the raceway.

Raceways are obviously smaller in size and occupy much less space than ponds. Although earthen raceways are sometimes used, the large majority of them are made of reinforced concrete, or cement blocks (figs 6.23 and 6.24). Earthen raceways can be lined with plastic material to reduce loss of water through seepage. Just as for pond farms, site selection for a raceway farm has to be done with special care. Naturally, the most important consideration is the water supply. The main sources of water are springs, streams, deep wells or reservoirs. For trout, which is probably the most common species cultivated in raceways, there is generally a preference for spring water of uniform temperature. For raceway farming of catfish a supply of 79 ℓ/s (1250 gal/min) is required for every 0.4 ha (acre) of raceway (Lee, 1973). In channels 3 m (10 ft) wide at the bottom and 1.2 m (4 ft) deep with a 1:1 side slope, a flow of 2.5 ℓ/s (530 gal/m) is often recommended. If such a flow is maintained, the water in a raceway segment of 30 m (100 ft) will be completely exchanged in about 1 hour.

In designing a raceway it is preferable to make use of the contour of the land. A slope of 1–2 per cent is preferred so that water flowing in at one end can be removed at the other. Each segment of a raceway can be about 30 m long, 2.5–3 m wide at the bottom and 1–1.2 m deep. A raceway farm consists of 15 to 20 segments or more. Many of them are constructed with side slopes of 1:1 or 1:0.5.

It is generally advisable to have a water supply reserve for emergencies. A storage reservoir near the beginning of a raceway system from where water can flow into the raceway by gravity would be most useful, in case there is pump failure. Raceways should be built straight and avoid curves, to ensure uniform flow. As many raceways as necessary can be built alongside each other. Many have a dozen or more rows. Very often the segments are built at different levels. The general practice of discharging water through a series of raceways carries with it the risk of unhygienic conditions developing in the lower level segments. However, when there are not too many segments and the water flow is sufficiently fast, the risk is not very significant, except when there are infective diseases in the upper raceways. In order to meet this contingency, it is necessary to have water control structures to cut off the affected segment and discharge the water through a separate drainage channel. Then

Fig. 6.23 A raceway farm under construction (courtesy of J. Kovari).

Fig. 6.24 A cement concrete raceway farm used for raising trout in Italy.

there should also be a separate feeder channel for each segment, very much as in a pond or tank.

It is important to have water control structures or weirs to regulate the flow and depth of water. They could also serve to aerate the water as it flows through. Commonly used materials for such structures are reinforced concrete, reinforced concrete blocks, wood, sheet metal and culverts with flashboards. They should permit discharge of water from the bottom of raceways and include screens to prevent loss of stock. Removal of water from the bottom helps in flushing out metabolic wastes and water low in oxygen. Other means of achieving the flow of bottom water is through a syphon arrangement or by constructing a vertically adjustable wide baffle on the upstream side of the weir, extending down into the water, so that water flows under it rather than over it. It is essential to adjust the rate at which water is pumped or flows into a raceway in order to prevent overflow or emptying. For cleaning raceway bottoms in emergencies, a suitable suction device can be used.

6.3   Cage farms

Holding or rearing fish in cages is a traditional practice in some Asian countries and appears to have originated almost two centuries ago in Kampuchea, from where it spread to Indonesia and Thailand and in recent times in a more advanced form to several other countries. Coche (1979) summarizes the historical evolution of the concept. It was a general practice in The Great Lakes area of Kampuchea to hold commercially valuable fish in bamboo cages to be sold alive. The cages were trailed in water behind a fishing boat for transport to the markets. Since this often took a long time and some of the catches were of smaller size, the fishermen began feeding them with trash fish and kitchen refuse. The fish grew well in the cages and as a result their market value increased considerably. This naturally led to longer-term rearing of catfish in Thailand, carp in sewage-fed canals in Java (Indonesia) and later on yellowtail in Japan and groupers and sea bass in Hong Kong and Singapore. Through recognition of the value of cage farms in aquaculture and the opportunities they offer for productive use of open waters, cage culture has attracted considerable research and development efforts in most parts of the world. In the last two decades it has become a major source of aquaculture production, particularly of high valued species like salmon, trout, sea bass and groupers. Several types and designs of cages and cage farms have been developed and are available commercially.

6.3.1   Types of cages and layout of cage farms

It is obviously difficult to describe the various designs of cages presently available. Detailed descriptions of different types of cages are given in Beveridge (1987). Although there are submersible and rigid-walled cages in use, the majority consist of a floating unit, a framework and a flexible mesh-net suspended under it. There are different methods of floatation and mooring, placement and attachments of individual cages in a farm, means of approach and handling of cages. The floating unit can consist of empty barrels, styrofoam polyethylene pipes, or ready-made pontoons of plastic and metal. The buoy units are often built into a framework, the material of which can be impregnated wood, bamboo spars, galvanized scaffolding or welded aluminium bars. Nylon is commonly used for the net, but weldmesh or even woven split bamboo are also used. Cage flotillas provide safer working conditions and enable storage of feed on site, as well as installation of automatic feeders. The diversity of materials used shows that the design of the cages and cage farms should be based on conditions prevailing at the selected site. Reasonably sheltered areas, with sufficient water movement to effect adequate mixing and aeration, are selected as sites for cage farms. The occurrence of typhoons, hurricanes and cyclones in the area and the vulnerability of the site to these are also major considerations in the design of cage farms. Polluted sites are generally avoided. In cold climates, areas that receive safe heated water effluents are preferred, as higher water temperatures generally improve growth and productivity.

Unused feed and fish faeces fall from the bottom of floating net cages on to the floor of the water body. Accumulated wastes decompose and cause oxygen depletion or generation of methane or other toxic gases under anaerobic conditions. Cages also increase deposition of silt on the bottom of the site. It is therefore necessary to have enough movement of clean water below the floating cages, and if the movement is not enough to clear them, provision has to be made for regular mechanical removal with suction or slush pumps and disposal of the waste at safe distances. Though the determination of precise carrying capacity of waterbodies is difficult, at least emperical estimates should be used to avoid overcrowding of cages.

Extensive testing of materials for construction of cages, supporting framework, floats, sinkers, walkways, etc. has been carried out. Despite differences in technical efficiency, different types of materials continue to be used

Fig. 6.25 A modern cage farm in Svanoy Island, Norway (photograph: Ola Sveen).

Fig. 6.26 A cage farm in Norway. Note the hexagonal cages with wooden framework moored alongside a walkway from a jetty.

depending on availability and cost. The most sophisticated designs appear to be used for sea cage farms, especially in Norway (fig. 6.25) and Scotland. Cage size can be anything up to 1000 m³, but is normally between 100 and 500 m³. A simple unit holds a net of four vertical sides and rectangular cross-section, but the more popular ones are circular in cross-section. When timber is used as framework it is not easy to have a perfectly circular shape and an approximation is achieved with six- or eight-sided structures. A commonly used cage in Norway has an eight-sided floating framework of timber, which is impregnated to reduce rotting. They are linked together by flexible joints to reduce the rigidity of the structure (fig. 6.26). Planks 12 cm × 5 cm or larger form the sides of the section and are spaced 30 cm apart by wooden slats nailed across the top and bottom. The slats on the top should be positioned close together, as the framework has also to serve as the walkway around the net. Expanded polystyrene or other flotation material is inserted between the two layers of slats and held in place by nails driven through the timber. The joint between two sections of the frame is formed by bolting strips of heavy rubberized machine belts. For further safety, all the sections are securely fastened together by suitable nylon rope or reinforced plastic piping. In every alternate corner of the frame a loop is provided to attach the anchoring devices. Inside the collar, four 120 cm long laths are nailed to each of the eight units. A nylon net is stretched between the laths to prevent leaping fish from escaping.

Another common type of cage system used in Norway employs rectangular cages suspended from a rectangular float (as in fig. 6.25). The float consists of four PVC pontoons in iron frames. The two frames, made of 25 mm galvanized iron tubes, are connected to a wooden frame made of 5 cm × 10 cm impregnated planks with galvanized bolts and nuts. The wood frame is made in two sizes (4 m × 1 m and 5 m × 1 m), depending on the length of the two types of elements. On top, the wooden frames are covered by 2.5 cm × 12.7 cm impregnated planks. The elements of the two different lengths are joined together to form a rectangular float with a network of 4 m × 4 m square openings. The bag net, equipped with

Fig. 6.27 A farm in Norway, which uses a different design of floating six-sided cages. Flotation is provided by six inflatable rubber buoys kept in place by six fibreglass poles radiating from a steel plate above the cage, looking like an inverted umbrella frame.

headline and leadline, is fixed to the rafts with brass hooks screwed into the wooden frame. Fence poles are erected from pedestals. The raft is usually moored by its four corners and can easily be towed.

When welded tubular metal or PVC or fibreglass tubing is used for the framework, there is greater flexibility in shapes and sizes of cages (fig. 6.27). Besides cost and safety of the structures, a major consideration in designing cages is the ease with which they can be handled. Obviously, large cages, though cheaper to buy and install, are not very convenient in this respect and would need a larger labour force or special mechanical equipment to handle. Cages with an underwater net volume of more than 1000 m^3 are not recommended; the preferred size is between 200 and 500 m^3. It is common practice to have double netting: the outer one serving as a predator net to protect the inner one and the fish stock in it.

There are many ways of arranging cages in a cage farm. Where possible, it is preferable to moor cages to a jetty with direct access to a quay, in order to facilitate work and reduce labour costs (fig. 6.28). However, environmental and site conditions may require them to be located farther away from the coast, in which case a work boat will be needed for access (fig. 6.29). In either case, the cages should be installed on the sides of a central walkway to facilitate day-to-day work on the farm. In many modern cage farms, feed dispensers are installed above each cage; in others, manual feeding is done. Mooring blocks have to be sufficiently heavy and are usually made of concrete with heavy galvanized bolts. Cages should be attached to the mooring by chain or, for lighter structures, by nylon rope.

While the arrangement of cages in a battery is the most common practice, in cases where infection of diseases is feared they may be moored separately and workers use boats to attend to feeding and the care of the cages and fish stock.

Most of the presently available cages are designed for use in protected areas like bays, fjords and lakes. In order to utilize more open waters and high seas, special cages with a flexible rubber framework have been recently developed. Some of these designs have twin rubber booms for increased stability in rough

Fig. 6.28 A series of cages moored to a jetty along the shore (courtesy of Ola Sveen).

seas. It is claimed that the low weight of the cage reduces strain on the moorings and contributes to all-weather capability and high durability. Other types of very large cages made of high strength galvanized steel, fitted with all accessories including even independent power supply, have been tried and are claimed to be capable of withstanding very severe storms. If such cages prove technically and economically successful, an enormous expansion of unpolluted potential sites for cage farming can be expected.

6.3.2   Submersible cages and cage maintenance

One of the advantages of a cage farm is that it can generally be towed away to a different location for harvesting or if unfavourable weather or other environmental conditions occur. In areas subject to typhoons or cyclones, submersible types of cages can be useful. Such cages are used in Japan for yellowtail rearing.

Fig. 6.29 A cage farm moored in Johore Straits, Singapore. Note the floating house of the caretaker and work boats.

They can withstand wind and waves much better than floating cages and can often also be used in open sea areas. The headropes of the cages, constructed on more or less the same design as floating cages, are attached to taut mooring ropes suspended from plastic surface buoys. Under normal conditions, the mooring ropes will be only about 2–5 m from the surface, but in rough weather the ropes are dropped to 10 m or more. By attaching weights of about 10 kg at each corner of the cage bottom, the shape of the cage is maintained. Additional weights may be added if the currents are too strong (Fujiya, 1979). The net is raised to the surface at feeding time, and the feeds conveyed through the feeding passage attached to the top of the net. A more sophisticated version of this type of cage uses variable buoyancy synthetic rubber floats that can be filled or emptied with compressed air or sea water from the surface.

A different type of submersible cage has been designed for use in hurricane-affected seas in the Caribbean. It is a spindle-shaped collapsible net cage held in position by circular PVC rings of different diameters (the largest rings in the middle and progressively smaller rings towards the ends) (figs 6.30 and 6.31). It looks very much like an enlarged and modified version of a fyke net. There are funnel-shaped pockets through which fish in the cage can be fed. Under normal weather conditions, the cage floats with the top above the surface, but when there is a hurricane warning it can be sunk to the bottom by increasing the weights and removing the floats. When the hurricane has passed, the cage can be raised again by removing the extra weights and replacing the floats. The spindle shape helps in rotating the cage and exposing the submerged parts to the sun for drying and removal of fouling organisms on the net.

Two of the major problems for cage farms are fouling of cage materials, particularly nets and mesh, and susceptibility to easy poaching. Fouling makes the net heavier and prevents easy exchange of water. Antifouling coating, which does not harm the fish, is a solution, but the most practical way at present is regular change of nets. Clean nets are installed at regular intervals (the frequency of change will depend on the rate of fouling at the site) and the old nets cleaned and dried for further use if they are sufficiently strong. Generally, the economic life of nets in cages is two to three years, depending on local conditions. Constant watch has to be kept of deterioration in the framework and other structures of the farm,

Fig. 6.30 Framework of a collapsible net cage. Note the PVC rings and the central beam.

Fig. 6.31 Submersible cages under operation in Martinique.

and repairs or replacements have to be in proper time to avoid unnecessary risks. Besides fouling, cage farms have to be protected from floating debris, rotten tree branches and drift wood. Unless they are exceptionally heavy, no damage may be done to the strong frameworks.

However, they can tear the nets sometimes and any such damage should immediately be repaired. Even though floating booms are sometimes erected with old netting slung beneath them to protect the cage, they can hinder routine operations.

There are different types of alarms that can be used to warn against poachers, but there appears to be no better means of looking after a cage farm than by the owner or a watchman on the spot, possibly with the help of trained dogs. In cage farms in Singapore, there is a floating house for the operator attached to the cage complex and a good number of guard dogs, for round-the-clock watch and constant care of the cages.

6.4 Pens and enclosures

Pens and enclosures can in some ways be considered as transitional structures between ponds and cages, in so far as environmental and stock control are concerned. While enclosures or pens continue to be used as in the culture of yellowtail in Japan, milkfish in the Philippines and salmon in Norway, attempts to introduce these systems have not met with much success in many other countries. This can probably be ascribed to the difficulties in the use of intensive techniques and in some cases the rather high costs of embankments and water management, such as through pumping. Experience seems to indicate that the success of enclosures for culture, to a large extent, depends on the hydrological conditions of the site. The design of the structures and operational procedures have to be based on adequate knowledge of water quality, floods, waves and currents, prevalence of predatory animals, etc.

Probably the simplest and relatively most efficient type of enclosure used for aquaculture is the one formed by damming a bay, cove, fjord or arm of the sea, estuary or river (fig. 6.32). Sites are selected where the barriers can be constructed across narrow sections, or channels, in order to reduce costs and increase the ease of operation. Most of the perimeter of the enclosure is formed by the natural shoreline. When the blind end of a water area is enclosed, there may be only one or one series of barriers, but in enclosures that permit direct flow-through there may be two or two series of

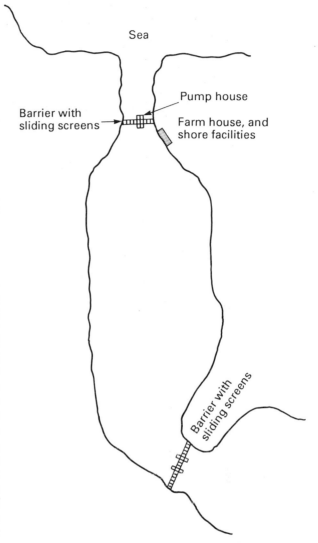

Fig. 6.32 Diagram of an enclosed fjord.

barriers: one upstream and another downstream. The dams are constructed with stones, earth or concrete, depending on the availability of materials and exposure of the site to storms and other natural disasters. They hold screens consisting of vertical aluminium or galvanized metal bars, with about 1 cm gaps between them to allow free flow of water. The screens, which are removable, can be located in the concrete by moulded guideways. Two guideways are provided, one behind the other for each screen, to enable a duplicate to be placed in position before removal for cleaning. The screens prevent the escape of the fish stock. For proper management these enclosures have to be relatively small (2−7 ha), although there are much larger enclosures in Japan, measuring up to 120 ha or more, which should strictly be called ranches or reserves, rather than farms. The depth of water in different portions of the enclosure differs and the circulation of water in the deepest parts may be insufficient to prevent the accumulation of waste and organic matter, giving rise to oxygen depletion in parts of the enclosure. Suction pumps have been used successfully for removing such accumulations and additional flow-through has been created by propellers mounted on floats. Obviously this increases the cost of operation.

Another type of enclosure or pen is the one formed by net barriers to partition off areas of an open water body, such as the intertidal areas of the sea or foreshore areas of lakes and reservoirs. Different designs of enclosures have been constructed. But, generally, the enclosure is formed on one side by the shore and the other three by a wall of nylon netting hung from poles driven into the sea. In many such enclosures, concrete or stone walls (about 3 m wide) are built on each side where it joins the shore, to provide adequate support for the net. Around the rest of the perimeter, heavy posts of impregnated timber or concrete piles are driven into the bed (at least about 3 m), extending for about 2 m out of the water at all times (during floods and high tide). Net barriers may be hung from steel cables strung between the poles or the concrete or steel piles. To prevent the lateral movements of these piles some are anchored fore and aft, to large anchor blocks using strong steel cables. The nets are generally made of knotless nylon netting material. In some cases, two walls of netting are used, the outer one to protect the enclosure from floating debris and to prevent the escape of fish if the inner wall gets damaged. However, this has been found to be a hindrance to free flow-through of water and now in most enclosures only one net wall is used. Though not so common as nylon nets, galvanized wire mesh or chain links are also being used as barriers. At the bottom of the poles or pilings under water, the net barrier is fixed by a rope along the sea bed for about 1 m, until it terminates in a lead line. Normally the net embeds itself in the sand or silt at the bottom, forming a good seal. As a further precaution to prevent escape of fish, heavy rubble may be piled at the bottom of the net or in some cases a net bottom may be provided.

A unique system of pen culture has developed in the shallow eutrophic Laguna de Bay in the Philippines and in lakes in China. Using bamboo scaffolding, enclosures of different sizes have been made in the lakes (fig. 6.33). Because of the shallow nature of the lakes, enclosures can be fairly easily constructed. The average depth of Laguna de Bay at low water is only 3 m and at high water 5 m. The netting is taken above the surface to prevent the escape of fish by jumping (fig. 6.34). An improved floating net enclosure has been developed, where the net enclosure is held in place by concrete block sinkers (about 500 kg in weight), with a series of small weights on the foot rope which is secured to a chain link between the sinkers. The net is kept afloat by floats attached to the headropes. There is a horizontal net at the top of the enclosure to prevent fish from jumping. To spread the load exerted by water movement and wave action, a lattice work of nylon rope is provided. The enclosures are generally assembled on land, then taken to the site on barges where they are installed by attaching them to the sinkers. Such floating enclosures are now used in lakes for tilapia and milkfish. It is reported that a similar system of pen culture of the giant fresh-water prawn has developed in the Songkhla Lake in Thailand. Several circular pens consisting of a bamboo framework and perlon nets have been built in the fresh-water portion of the lake by small-scale farmers.

6.5 Design and construction of hatcheries

Methods of seed production in aquaculture differ considerably with the species under culture and the state of technology, as well as with the level of operation (as, for example, extensive or intensive and the number or frequency of crops). Where the techniques for artificial propagation are still to be developed or perfected, or where it is feasible and economical to collect eggs, larvae or fry from natural

Fig. 6.33 A fish pen made of bamboo in Tahoe Lake in China.

Fig. 6.34 A fish pen in Laguna de Bay in the Philippines. Note the bamboo scaffolding (courtesy of Y.A. Tang).

breeding areas, sophisticated hatchery systems are not often used (see Chapter 14). However, even in such cases it is generally accepted that, eventually, hatchery production of seed will be necessary to stabilize and ensure regular supplies and introduce breeding techniques to raise improved seed for better growth and production.

As is only to be expected, there are different types of hatchery facilities in use, depending on the species, locality and investment capabilities of the aquaculturists. However, the basic requirements are about the same: there has to be the necessary facilities for holding or rearing an adequate brood stock, spawning or stripping and fertilization of ova, incubation of fertilized ova and rearing of larvae to the required stage for transfer to nurseries or other culture facilities.

6.5.1 Source and supply of water

The selection of a suitable site for a hatchery is very important for its successful operation. Although, for various reasons, it is preferable to have it located near the grow-out farm, often a different site may have to be selected because of the water quality and quantity requirements. There are also cases where the hatchery forms an independent enterprise or is meant to produce seed for a number of grow-out farms. In principle, surface or ground water can be used in hatcheries, if it satisfies the necessary water quality criteria. Surface water from streams, rivers, lakes and the open sea may be relatively less expensive to utilize. However, very often there will be the need to filter the water and where there is a high content of silt it may be necessary to have a settling tank. Generally, sand or gravel filters with backflushing will make the water suitable for the hatchery. In salmon and trout hatcheries, where stricter water quality conditions are maintained, spring or borewell water is preferred, to eliminate the risk of contamination. As far as possible the source of water, or at least the entire course of the water supply system from the intake, should be under the control of the hatchery manager. Well water often has an excess of gas, which can cause gas bubble disease, but through adequate aeration (fig. 6.35) before use in the hatchery, this problem can be overcome. A large reservoir of properly aerated water from a spring can be a suitable source for a controlled water supply to a hatchery.

Water temperature is of special importance in a hatchery system, as the maturation of the brood stock, spawning, development of fertilized ova and growth of larvae are all directly

Fig. 6.35 Deepwell water being aerated through a tower for hatchery use in Poland (photograph: J. Walugo).

affected by it. Spring water has often the advantage of constant temperature conditions. The temperature to be maintained in different units of the hatchery installation will depend on the requirements of the propagated species. It may be necessary or desirable to have provisions for regulating the temperature, as for example by mixing cold and warm water from separate supply lines or by the provision of an in-line heat exchanger, including a thermostatically controlled boiler which can be bypassed if heat control is not needed. While 20−30°C is generally the temperature requirement of warm water fishes, trout and salmon hatcheries maintain a temperature between 7 and 15°C. In tropical shrimp hatcheries, a temperature ranging between 20 and 29°C is suitable, but for the majority of species a temperature not lower than 25°C is considered optimal. In the giant fresh-water prawn hatchery a higher temperature of 30−31°C is recommended for better growth and survival. In oyster hatcheries a temperature of about 29°C is maintained.

Dissolved oxygen and pH are other important properties of the water for hatcheries. The lowest safe level of dissolved oxygen for trout hatcheries is about 5 ppm, but a higher concentration of 7 ppm is preferred. According to Wickins (1981), salmonids and warm water crustacea should not be exposed to levels of dissolved oxygen below 5 mg/ℓ for more than a few hours. Equivalent levels for eels and carp range from 3 to 4 mg/ℓ. In other warm water species of fish and shrimp, slightly lower oxygen contents may be adequate. Oxygenation of water in a hatchery is relatively simple and is generally achieved by the manipulation of water flow from the source or the use of appropriate aerating devices. In salt-water hatcheries, maintenance of the required salinity can be important, although many species are quite tolerant of fluctuations within limits. In order to be able to regulate salinity when required, salt-water hatcheries generally maintain supplies of fresh water as well as sea water.

6.5.2 Reconditioning and recirculation of water

Where the availability of good quality water is limited, hatcheries have to resort to reconditioning and recirculation. In certain circumstances, it may also be considered necessary to reduce risks of infection by pathogens and parasites through continued use of water from external sources. When the water has to pass through a series of tanks, it has often been the practice in hatcheries to pump the water through an aerator, after it has passed through a certain number of tanks, before further distribution. Naturally, there are intrinsic dangers in such simple systems of recirculation. Though oxygen can be replenished through aeration and most of the carbon dioxide dissipated, the removal of metabolic products like ammonia

will involve more complex systems, which besides reaeration and mechanical filtration may involve biological treatment. Recent designs of semi-closed systems employ one or more by-pass treatment units, such as for denitrification, oxygenation, ozonization, etc. In principle, such recirculation should make it economically feasible to grow warm water species in temperate climates by reducing the cost of heating. In practice, there are many constraints to its application in commercial aquaculture, but it can be used in hatchery situations, when essential.

Ammonia can be removed by nitrifying bacteria. Ammonia is first converted primarily by *Nitrosomonas* to nitrous acid, then by *Nitrobacter* to nitric acid. The acid combines with an available base to form nitrites and then nitrates. Nitrates are harmless in the recirculating system. Even in prolonged exposures in culture systems, no toxic effects have been reported below 100 mg/ℓ nitrate nitrogen (Wickins, 1981). Nitrite toxicity is influenced by water chemistry, but it has been suggested that concentrations in hard fresh water should not exceed 0.1 mg/ℓ nitrite nitrogen, and in sea water 1.0 mg/ℓ nitrite nitrogen.

There are several systems and designs employed in waste-water treatment for hatchery use as well as for intensive aquaculture. Many of them have been described in recent literature (for example Tiews, 1981). The most commonly used and relatively more economic treatment would appear to be biofiltration, which may incorporate downflow filters (e.g. trickling filters), upflow filters or horizontal flow filters. Several types of filter media are in use, such as sand, gravel, oyster shells, plastics, anthracite, activated carbon, diatomaceous earth and their combinations. Besides serving as strainers, they provide surface area for biological growth. Through biological growth and oxidation, ammonia is converted into nitrite and nitrate. The nitrate may be further combined with ions in water to form salts or reduced to nitrogen gas through a denitrification process. According to Liao and Mayo (1974), with a retention time of about 30 minutes and a hydraulic loading rate of 1.0−3.7 ℓ/s per m^2, about 48 per cent of the initial ammonia load can be removed.

A rotating biofilter process is employed as a secondary treatment. The basic unit consists of a half-cylinder tank, through the ends of which a horizontal shaft is mounted. As waste water flows through the cylinder, the discs, which are half submerged, are rotated. A layer of micro-organisms grows on the discs and acts as the aerobic biochemical agent to remove the dissolved wastes from the water. By arranging the discs in a series of stages, the rate of oxidation of the organic materials is increased by improved residence time. A surface area loading of $0.06-0.1$ m^3/day per m^2 will be needed to achieve a better than 95 per cent removal efficiency for biological oxygen demand (BOD) decrease and nitrate nitrogen.

Ion exchange is an efficient and reliable means of ammonia removal, but it is much more expensive than biological filtration. Clinoptilolite, one of the zeolites used in water treatment, is an effective natural ion exchange material for the removal of ammonia from hatchery water. It has a high selectivity for ammonia. The minimum depth of an ion exchange bed is $0.61-0.76$ m with a flow rate in the range $1.4-3.4$ ℓ/sec per m^2 (Liao, 1981). The regeneration is achieved by passing $5-10$ per cent brine solution through the exchange bed, with a flow of $0.68-1.36$ ℓ/sec per m^2.

Although in principle it is possible to use aquatic plants, including algae, to remove metabolites and nutrients from water through assimilation or biological conversion there are practical difficulties, particularly for use in connection with a hatchery system. Another means of reconditioning water is through the use of an activated sludge process, in which biological oxidation occurs in the fluidized suspension of the sludge. Mention may also be made of the use of ozone, which is a very strong oxidizing agent and oxidizes the organics in the water and at the same time serves as a disinfectant. However, its use in hatcheries has not progressed beyond the experimental stage.

In all these systems there is the need for a regular supply of make-up water, at a rate of at least $5-7$ per cent of the total volume, to replace the losses due to filter back-flushing, draining of the sludge and evaporation.

6.5.3 Hatchery equipment

Besides the installations for an adequate supply of water of the required quality, the other general equipment required in a hatchery includes tanks for brood stock; implements for collecting and handling breeders, for stripping and fertilization; spawning tanks where necessary, jars, troughs, tanks or other containers, net cages or 'hapas' (mesh cloth tanks) for incubating and hatching fertilized eggs; food dispensers; larval rearing tanks and aeration systems.

The brood tanks are very similar to the ones described in Section 6.2. There are a variety of implements used for collection, handling and transport of breeders in the hatchery. The main consideration is to avoid damage during collection and transport. In the case of large finfish, special hammocks have proved very efficient (fig. 6.36). Scoop nets are commonly used for catching brood stock for spawning. The simple tools used for stripping and fertilization are referred to in Chapter 15.

In shrimp hatcheries, circular maturation tanks of about 12 m^3 capacity made of fibreglass or cement concrete are used. The substrate consists of a layer of gravel (about 10 cm thick) separated from a $5-10$ cm thick second layer of coral sand by a permeable synthetic cloth to prevent gravel from mixing with sand (AQUACOP, 1983). Concentric perforated plastic pipes fitted in a PVC tube (10 cm diameter) are embedded in the gravel. Water

Fig. 6.36 Hammock used for transporting large brood fish.

that flows into these pipes passes through the sand, preventing the accretion of wastes and sediments in the sand. Water is discharged through two concentric tubes, allowing the bottom water to be drained first. Water exchange is achieved two or three times a day.

Spawning tanks may be built in the open, of concrete, with adequate arrangements for water circulation, as in the case of Chinese carp hatcheries (fig. 6.37). In shrimp hatcheries, indoor tanks of concrete or fibreglass are commonly used for spawning (fig. 6.38). Cylindroconical, 150 ℓ capacity figreglass tanks have been found to be very convenient (AQUACOP, 1983). The overflow passes through a concentrator provided with a 100 μm mesh that retains the eggs. A perforated plastic plate fitted over the conical bottom prevents the spawners from eating the eggs accumulated.

Different types and designs of incubators are used for hatching fertilized eggs, ranging from improvised earthen and polyethylene jars to sophisticated batteries of jars and troughs. To a certain extent, the degree of sophistication depends on the species, the size of the eggs and the magnitude of operations. However, the general principle involved is the provision of a regulated flow of good quality water of the required temperature, for the development and hatching of fertilized eggs and prevention of infections that will affect the hatching rate.

Troughs made of wood, concrete, aluminium, plastic or fibreglass are commonly used in many types of hatcheries (figs 6.39 and 6.40). The size varies, but an average size may be about 3 m × 0.5 m × 0.25 m. They are generally screened at the intake to prevent the entry of detritus and at the outflow to prevent the escape of larvae. By the use of vertically adjustable screens, the depth of water in the trough can be regulated. As an alternative, the water depth can be regulated by adjustable elbow pipes.

In the case of trough-type incubators, such as the ones used in trout and salmon hatcheries, there are egg baskets fitted in the tray. The perforations are of the shape and size to retain the eggs, but allow the hatchlings to fall through to the water below in the trough (fig. 6.41). In order to allow aeration of the eggs, water is forced upward through the perforations.

Special glass jar incubators (known as Zuger jars, Weiss jars or Zug-Weiss jars and Mac-Donald jars) and plexiglass or other plastic funnels (figs 6.42 and 6.43) as well as less expensive sieve-cloth funnels and even earthen jars are used for incubating non-adhesive eggs (figs 6.44 and 6.45). Even in cases like the common carp, these devices can be used after removing the sticky layer.

In oyster hatcheries, mature male and female oysters are spawned individually in battery jars containing filtered water of the required temperature (about 27°C). The spawned eggs and milt are filtered out and the ova fertilized. Larval tanks are provided for the development and hatching of the ova and for rearing through various larval stages.

Different types of larval rearing troughs, tanks and pools are in use and some types are readily available from manufacturers. The basic requirements are proper circulation and drainage of water to keep them well supplied with clean oxygenated water and prevent accumulation of waste products. For rearing carp larvae, circular cement pools built on the ground are commonly used. For hatching and larval rearing in Indian carp culture, meshed cloth tanks (hapas) fixed on the pond bottom by means of stakes (fig. 6.46) are widely used. It is common to have double tanks (fig. 6.47), the inner smaller tank with fine mesh holds the fertilized eggs for hatching. The hatched larvae fall into the outer larger cage, leaving behind the egg shells and debris. The inner cage can easily be removed after hatching. Rectangular cement cisterns, with an adequate water supply and drainage, are also used in many places.

6.5.4 Layout and accessories

Although some of the simple and improvised hatchery systems such as the carp hatcheries in China and India are built in the open, modern hatcheries are installed indoors. In some cases the larval rearing may be carried out in tanks, pools or ponds outdoors, but where water temperature has to be controlled they are generally provided with at least a protective roofing.

There are many ways of arranging the installations in a hatchery. Some have maturation, spawning and fertilization, hatching and larval rearing in different sections. Others have most of them in one area, particularly when there are limitations of space. An important factor

Fig. 6.37 Cement concrete spawning tanks in a Chinese hatchery.

Fig. 6.38 Indoor shrimp spawning.

Fig. 6.39 An example of simple hatchery troughs made of wood.

Fig. 6.40 Troughs made of fibreglass and aluminium used in a modern carp hatchery in Hungary.

Fig. 6.41 Trough-type incubators with egg baskets, used in a salmon hatchery (courtesy of Ola Sveen).

that would determine the arrangement of the equipment is the convenience of water distribution and drainage. The main supply channel or pipe could run along one wall of the hatchery building or along its centre. The latter is more common, when the tanks and incubators can be installed on either side of the supply system, with drains along the walls. The size of the water supply system should be adequate to carry one and a half times the quantity of water required for operation of the hatchery. When siting supply lines, it will be advisable to take into account possible future expansion of the hatchery. An overhead reservoir or a head tank from which the water flows by gravity will be very useful in maintaining a constant water pressure in the system and consequently a uniform flow in the hatching units. It also serves as a stand-by in times of power failure.

In order to save space and reduce the use of water, it is possible to stack the troughs or trays used for incubation and larval rearing, one above the other. Battery incubators are available from manufacturers. They consist of vertically stacked troughs, each having an egg tray and a cover. Each trough can be pulled out separately for inspection. The water flow will pass downwards through each vertical trough stack, trickling through each one from top to bottom.

As indicated earlier, control of water temperature is an important factor in all stages of hatchery operation. If the natural supplies of water need to be heated or cooled for maintaining the required temperature, it will be necessary to install equipment for this purpose. Electrical heating and cooling would probably be the best, but economic considerations and availability may make it necessary to look to other sources such as heated water from industries and power stations. Most hatcheries would require some type of aeration system. Air blowers and air stones are commonly used to provide the aeration that will be needed to meet the extra oxygen requirements in brood stock and nursery tanks.

Suitable means of dispensing feed in maturing and larval tanks have to be provided. Different types of automatic feeders are available (see Chapter 7), but in many hatcheries manual feeding is still the common practice. In crustacean and mollusc hatcheries, there is also the need to have facilities, for algal culture to feed the larvae, besides an artificially lit room for maintaining algal cultures. Large concrete or fibreglass tanks can be used to grow the algal food required. The algal tanks can be outdoors, but if kept indoors there has to be adequate lighting. The tanks should also be provided with proper aeration through air stones or other devices. Shrimp hatcheries may also require similar tanks for hatching Artemia cysts, for feeding the larvae.

Besides suitable storage space for feeds, it is desirable to have, in large hatcheries, some laboratory space for routine tests and examinations.

Fig. 6.42 Glass jar incubators.

Fig. 6.43 Plexiglass funnel incubators.

Fig. 6.44 Sieve-cloth incubators used for hatching carp eggs in Nepal (photograph: E. Woynarovich).

Fig. 6.45 Earthen jars used for carp egg hatching in Nepal (photograph: E. Woynarovich).

Fig. 6.46 Hapas (cloth tanks) used for rearing carp larvae in India.

Fig. 6.47 A double hapa for hatching carp eggs.

6.6 References

Alcantara L.O. (1982) *Variations of Fish Pond Layouts for Different Types of Brackishwater Fish Pond Management*, pp. 83–6. SCS/GEN/82/42, FAO/UNDP South China Sea Fisheries Development and Coordination Programme, Manila.

AQUACOP (1983) Constitution of broodstock, maturation, spawning, and hatching systems for Penaeid Shrimps in the Centre Oceanologique du Pacifique. In *Handbook of Mariculture, Vol. 1, Crustacean Aquaculture*, (Ed. by J.P. McVey) pp. 105–27. CRC Press, Boca Raton.

ASEAN (1978) *Manual on Pond Culture of Penaeid Shrimp*. ASEAN National Coordinating Agency of the Philippines, Manila.

Bandie M.J. *et al.* (1982) *Present Status of Brackishwater Fish Ponds in East Java, Indonesia with*

Emphasis on Engineering Related Problems, pp. 104–7. SCS/GEN/82/42, FAO/UNDP South China Sea Fish. Dev. Coord. Progr. Manila.

Beveridge M.C.M (1987) *Cage Culture*. Fishing News Books, Oxford.

Breemen N. van (1973) Soil forming process in acid sulfate soils. *Proc. Int. Symp. Acid Sulfate Soils, Wageningen*, **1**, 66–130.

Buss K., Graft D.R. and Miller E.R. (1970) Trout culture in vertical units. *Progr. Fish Cult.*, **32**(2), 86–94.

Chen T.P. (1976) *Aquaculture Practices in Taiwan*. Fishing News Books, Oxford.

Coche A.G. (1979) A review of cage fish culture and its application in Africa. In *Advances in Aquaculture*, (Ed. by T.V.R. Pillay and W.A. Dill) pp. 428–41. Fishing News Books, Oxford.

Cruz C.R. de la (1982) *Equipment and Facilities for Coastal Fishpond Construction, Maintenance and Repair*, pp. 156–68. SCS/GEN/82/42, FAO/ UNDP South China Sea Fisheries Development and Coordination Programme, Manila.

Davis H.S. (1956) *Culture and Diseases of Game Fishes*. University of California Press, Berkeley.

Edwards, D.J. (1978) *Salmon and Trout Farming in Norway*. Fishing News Books, Oxford.

Elekes K. (1984) Principles of designing inland fish farms. In *Inland Aquaculture Engineering*, pp. 105–23. ADCP/REP/84/21 FAO of the UN, Rome.

Fujiya M. (1979) Coastal culture of yellowtail (*Seriola quinqueradiata*) and sea bream (*Pagrus major*) in Japan. In *Advances in Aquaculture*, pp. 453–8. (Ed. by T.V.R. Pillay and W.A. Dill). Fishing News Books, Oxford.

Hepher B. and Pruginin Y. (1981) Commercial Fish Farming. John Wiley & Sons, New York.

ISI (1970) *Classification and Identification of Soils for General Engineering Purposes*. Indian Standards Institution, New Delhi.

Kovari J. (1984a) The organization and supervision of fish farm construction. In *Inland Aquaculture Engineering*, pp. 275–94. ADCP/REP/84/21, FAO of the UN, Rome.

Kovari J. (1984b) Preparation of plans and cost estimates and tender documents. In *Inland Aquaculture Engineering*, pp. 125–203. ADCP/REP/ 84/21, FAO of the UN, Rome.

Lee J.S. (1973) *Commercial Catfish Farming*. The Interstate Printers and Publishers, Danville.

Leitritz E. and Lewis R.C. (1976) Trout and salmon culture (hatchery methods). *Fish. Bull.*, **164**.

Liao P.B. (1981) Treatment units used in recirculation systems for intensive aquaculture. In *Aquaculture in Heated Effluents and Recirculation Systems*, Vol. I (Ed. by K. Tiews), pp. 183–97. Schriften der Bundesforschungsanstalt für Fischerei, Berlin.

Liao P.B. and Mayo R.D. (1974) Intensified fish culture combining water reconditioning with pollution abatement. *Aquaculture*, **4**, 61–85.

Maar A., Mortimer M.A.E. and Lingen J. Van der (1966) Fish culture in Central East Africa. *FAO Fisheries Series*, **20**, FAO, Rome.

Matsui I. (1979) *Theory and Practice of Eel Culture* (Translated from Japanese). Amerind Publishing Co, New Delhi.

Merritt F.S. (1968) *Standard Handbook for Civil Engineers*, 2nd edn. McGraw-Hill, New York.

Meske Ch. (1979) Fish culture in a recirculating system with water treatment by activated sludge. In *Advances in Aquaculture*, (Ed. by T.V.R. Pillay and W.A. Dill), pp. 527–31. Fishing News Books, Oxford.

Milne P.H. (1972) *Fish and Shellfish Farming in Coastal Waters*. Fishing News Books, Oxford.

Poernomo A. and Singh V.P. (1982) Problems, field identification and practical solutions of acid sulfate soils for brackish-water fish ponds. In *Report of Consultation/Seminar on Coastal Fish Pond Engineering*, pp. 49–61. 1982 Surabaya (Indonesia) SCS/GEN/82/42FAO/UNDP South China Sea Fisheries Development and Coordination Programme, Manila.

Pruginin Y. and Ben-Ari A. (1959) Instructions for the construction and repair of fish ponds. *Bamidgeh*, **11**(1), 25–8.

Rosenthal H. (1981) Recirculation systems in Western Europe. In *Aquaculture in Heated Effluents and Recirculation Systems*, Vol. II (Ed. by K. Tiews), pp. 305–15. Schriften der Bundesforschungsanstalt für Fischerei, Berlin.

Silas E.G., *et al.* (1985) Hatchery production of Penaeid prawn seed: *Penaeus indicus*. *CMFRI Spec. Publ. Cochin*, **23**.

Stevenson J.P. (1980) *Trout Farming Manual*. Fishing News Books, Oxford.

Szilvassy Z. (1984) Soils engineering for design of ponds, canals, and dams in aquaculture. In *Inland Aquaculture Engineering*, pp. 79–101. ADCP/ REP/84/21, FAO of the UN, Rome.

Tang Y.A. (1979a) Planning, design and construction of a coastal milkfish farm. In *Advances in Aquaculture*, (Ed. by T.V.R. Pillay and W.A. Dill), pp. 104–17. Fishing News Books, Oxford.

Tang Y.A. (1976b) Physical problems in fish farm construction. In *Advances in Aquaculture*, pp. 99–104. (Ed. by T.V.R. Pillay and W.A. Dill). Fishing News Books, Oxford.

Terzaghi K. and Peck R.B. (1967) *Soil Mechanics in Engineering Practice*. John Wiley & Sons, New York.

Tiews K. (Ed.) (1981) *Aquaculture in Heated Effluents and Recirculation Systems*, Vols I and II.

Schriften der Bundesforschungsanstalt für Fischerei, Berlin.

Wheaton F.W. (1977) *Aquacultural Engineering.* John Wiley & Sons, New York.

Wickins J.F. (1981) Water quality requirements for intensive aquaculture: a review. In *Aquaculture in Heated Effluents and Recirculating Systems*, Vol. I (Ed. by K. Tiews), pp. 17–37. Schriften der Bundesforschungsanstalt für Fischerei, Berlin.

Woynarovich E. (1975) *Elementary Guide to Fish Culture in Nepal.* FAO of the UN, Rome.

Woynarovich E. and Horvath L. (1984) The artificial propagation of warm-water finfishes — a manual for extension. *FAO Fish. Tech. Paper*, **201**.

7
Nutrition and Feeds

7.1 Feeding habits and food utilization

When discussing the criteria for selection of fish and shellfish for culture in Chapter 5, the importance of feeding habits and feed efficiency in terms of growth and production has been pointed out. Most forms of traditional animal aquaculture rely largely on the production of foods through natural processes, or by fertilization and water management in enclosed areas. To a certain extent, this practice is still followed in extensive and semi-intensive pond farming, but supplementary feeding is resorted to for ensuring adequate availability of food to dense stocks and for enhanced growth and production. In the case of mollusc culture, live foods continue to be the source of nutrition, even though experimental studies are under way to develop inert feeds. In nature, one can observe distinctly different feeding habits among fish and shellfish species, such as those that feed on zoo- and phytoplankters, filamentous algae, macrophytes, benthos, detritus, molluscs and other smaller animal species, etc. Many of them feed on more than one type of food or even a number of them. Fish generally use one or more sensory systems for acquiring feed, such as visual detection, sound and water turbulence and chemical stimuli released by food. Of these, the visual stimuli are best understood and include the properties of size, movement, shape and colour contrast.

Digestion involves the conversion of the three major nutrients (proteins, carbohydrates and lipids) which occur as macromolecules in nature into sizes that pass through the walls of the alimentary canal and are absorbed into the bloodstream. Proteins are converted into amino acids or polypeptide chains of a few amino acids, carbohydrates into simple sugars and lipids into glycerols and fatty acids. This is made possible through the activity of enzymes. Digestibility ranges from 100 per cent for glucose to as little as 5 per cent for raw starch or 5–15 per cent for plant material containing cellulose. Digestibility of most natural proteins and lipids ranges over 80–90 per cent. Indigestible materials are eventually voided as faeces.

The enzyme amylase catalyses the digestion of starch and together with dextrinases produces maltose. Maltase hydrolyses maltose to give the final product of starch digestion, glucose. Most fish have amylase; in plant-eating fish such as tilapia it may be present in all parts of the digestive tract, whereas in carnivorous fish it may be found only in the pancreas, pyloric caeca and intestines.

Cellulase and cellobiase are the enzymes involved in digesting cellulose. Cellulase hydrolyses cellulose to disaccharide cellobiose, which is then acted upon by cellobiase producing the final breakdown product, glucose. Very few fish have cellulase activity, but the microflora in their intestines may serve as a source of cellulases.

Protein digestion in fish begins in the stomach and is catalysed by pepsin and acid pH ranging from 1 to 4. Pepsin is synthesized in the gastric gland in the inactive form called pepsinogen. Hydrochloric acid, produced by another enzyme-controlled reaction between sodium chloride and carbonic acid, activates the pepsinogen. Pepsin attacks most proteins where the linkages are formed by aromatic and acidic amino acids, such as phenylalanine, tyrosine, tryptophan and asparatic and glutamic acids.

Trypsin and chymotrypsin are involved in the alkaline digestion of proteins. These enzymes are generally synthesized and stored in the pancreatic cells as inactive forms, viz. trypsinogen and chymotrypsinogen. These are then transported mainly to the intestines and pyloric caeca, or in some cases to the liver. In the intestines, trypsinogen is converted to the active form trypsin by the enzyme enterokinase. Trypsin in turn activates chymotrypsinogen into chymotrypsin. Trypsin is specific for peptide linkages which come from basic amino acids: arginine and histidine. Chymotrypsin attacks linkages with aromatic amino acids: phenylalanine, tyrosine and tryptophan.

Carboxypeptidases (A and B) hydrolyse the C-terminal peptide of proteins. This is found in the pancreas, pyloric caeca and intestines of fish. Carboxypeptidase A is not active towards proteins with aromatic C-terminal amino acids: phenylalanine, tyrosine and tryptophan, while carboxy-peptidase B acts preferentially on these with lysine and arginine. Amino peptidase hydrolyses the amino terminal peptide of polypeptidia, releasing one amino acid at a time from the N-terminal end. Most fish have also lipase enzymes that hydrolyse ester linkages in triglyceride and produce glycerol and fatty acids.

The effectiveness of digestive enzymes is influenced by temperature and pH. In general, the reaction rate increases with temperature until the enzymes begin to denature around 50−60°C. However the range of pH within which they function is very limited, often as little as 2 pH units. In the case of channel catfish, which is probably representative of many teleosts, the pH in the stomach ranges between 2 and 4, becoming alkaline (pH 7−9) below the pylorus, decreasing to 8.6 in the upper intestine, and finally nearing neutrality in the hind gut (Page *et al.*, 1976).

Absorption of amino acids, peptides and simple carbohydrates in fish have not been studied much, but presumably they diffuse through or are transported across the gut epithelium into the bloodstream. Digested food, particularly protein, is not fully available to the fish even after it has been absorbed. Amino acids may be used as absorbed for building new

tissue. But if digested food has to be oxidized for energy, deamination (removal of the amino group) which requires an input of energy (a process known as specific dynamic action) would have to occur first. Fish which have not grown due to low temperature or due to low levels of feeding would deaminate most or all of their amino acids. Those reared at high temperatures or having very high metabolic rates due to high activity levels would also do likewise. On the other hand, fish having rapid growth and high protein intake would deaminate a relatively small portion of the digested protein.

The energy for deamination need not come from amino acids, but may be preferentially taken from carbohydrate or lipid, if available. This 'protein-sparing' action accounts for the addition of limited amounts of inexpensive carbohydrate in the diet of fish, which helps in reducing feed costs. The calorie-to-protein ratio (kcal:g) can be applied in diets containing adequate energy and protein. Optimal ratios for catfish diets are reported to be between 6.5 and 8.3 kcal of digestible energy per g protein.

7.2 Energy metabolism

From among the two types of energy, viz. heat energy utilized for maintaining body temperature and the free energy available for biological activity and growth, the latter is more important for poikilothermal animals like fish. Free energy is needed for maintenance, growth and reproduction. Seaweeds and other plants can obtain it directly from the sun and water and synthesize the complex molecules that constitute its structural parts. Animal species have to depend on the oxidation of the complex molecules contained in the food that they eat for energy requirements. The complex molecules are broken down during digestive processes to simpler molecules and are absorbed into the body, where oxidation occurs and energy is released. The biological process of energy utilization is known as metabolism and the rate at which it is utilized is referred to as the metabolic rate.

Energy metabolism in cold-blooded animals, such as fish, is different from mammals and birds in that they do not expend energy to maintain a body temperature different from its environment, as is the case for warm-blooded animals, and the excretion of waste nitrogen requires less energy than in homeothermic land animals.

The metabolic rate in fish, which is probably the most studied aquaculture animal group, is influenced by temperature, age or size, activity and seasonal and diurnal fluctuations of body function. It is also affected by oxygen or carbon dioxide concentration, pH and salinity of the water. The energy requirements necessary for all metabolic functions can be calculated for each species. As, for example, carp utilize 25 cal/dec^2/h at 25°C. Approximately 70 per cent of this is used for maintenance and growth and the remaining 30 per cent is lost to the environment. As the body temperature of the fish is maintained at or near the environmental water temperature, the heat that is produced is lost to the environment. The biological partition of energy is shown in fig. 7.1.

Energy is also lost in faeces, urine and gill excretions, besides the small amounts of heat lost from external body surface.

7.3 Energy requirements and sources

As indicated at the beginning of Section 7.2, the energy requirements of fish and other cultured organisms are generally supplied by carbohydrates, proteins and fats. Most of the available information on nutrient requirements of aquaculture species is based on researches on a small number of these, viz. trouts, salmon, channel catfish, common carp, grass carp, eel, plaice, gilthead bream, red sea-bream and yellowtail. Work on penaeid shrimps and the giant fresh-water prawn has shown considerable similarities with the fish species studied, although there are some differences. Very little work has been done on the nutritional needs of molluscs, as culture has been based on fil-

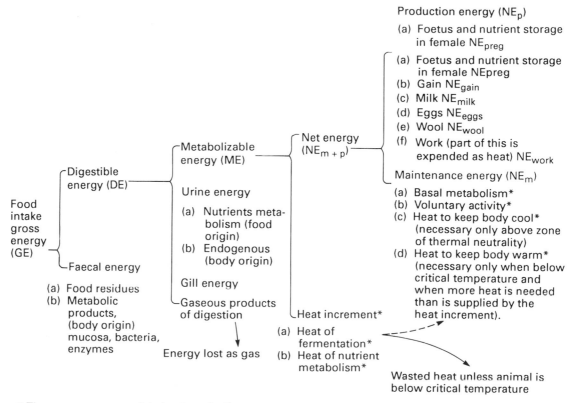

* These processes result in heat production

Fig. 7.1 Partition of feed energy (from: Smith, 1980).

ter feeding of naturally occurring phytoplankton and similar organisms. However, efforts are presently underway to develop encapsulated or fine-particulate feeds, which should lead to a better understanding of their feed requirements.

Energy requirements have in most cases been derived primarily from experimentation, in which fish were fed rations varying in calorific value. The ration yielding the best growth was assumed to be the most satisfactory calorific value for the species concerned.

7.3.1 Carbohydrates

Carbohydrates are the most abundant and relatively least expensive source of energy in animal aquaculture. These may consist of easily digested sugars to complex cellulose which is difficult to digest. Based on results of research on carnivorous species, doubts have been expressed on the value of carbohydrates in fish feeds, but practical experience in fish culture shows that digestible carbohydrate can be an energy source if kept in proper balance with other nutrients. The ability to assimilate starches depends on enzymatic activity (production of amylase). In herbivores, amylase occurs through the entire digestive tract. Up to levels of 25 per cent in the diet, it can be as effective an energy source as fat for several species of fish, such as channel catfish, rainbow trout and plaice (Cowey and Sargent, 1972). Stickney and Shumway (1974) have shown the presence of cellulase activity associated with cellulolytic microflora in several species of brackish-water fish and fresh-water catfish. Metabolizable energy values of carbohydrates may range up to 3.8 kcal/g for easily digestible sugars, whereas for undigestible cellulose it may be near zero. Values for raw starch range from 1.2−2.0 kcal/g. When processed in high moist temperatures, for making pelleted feed, starch gelatinizes and its digestibility therefore improves. When digested, the products of hydrolysis are assimilated into the bloodstream, where their known function is to provide energy. Therefore they have a protein sparing action (see the end of Section 7.1) and any excess is partially stored in the liver as glycogen and partially converted into visceral and muscular fat.

Successful fish feeds contain a certain amount of carbohydrates, as for example 20 per cent for cold-water fish feeds and 30 per cent for warm-water fish feeds. Besides providing energy, they have the physical function of texturizing manufactured feeds and acting as a binder in the formulation of pellets. Cereal grain products are generally used as 'fillers' to complete feed formulae. Formulae for expanded pellets often contain up to 50 per cent of whole cereal grains, to achieve the floating properties.

7.3.2 Proteins

Dietary protein is the main source of nitrogen and essential amino acids in animals. It is also the most expensive source of energy in artificial diets. In nature, carnivorous fish consume food which are about 50 per cent protein. They have a very efficient system for excretion of waste nitrogen from protein, which is catabolised for energy. Therefore high-protein diets are not harmful but, being expensive, it is necessary to keep the proportion of protein down to optimum levels necessary for good growth and feed conversion. Protein has a metabolizable energy value of about 4.5 kcal/g in fish, which is comparatively higher than that of mammals and birds.

Gross protein requirements of a number of cold-water and warm-water finfish have been determined (Table 7.1). The requirements are the highest in the initial feeding of fry and decrease as fish size increases. For maximum growth, young fish require between 40 and 60 per cent of their diet as proteins, which is much higher than the requirements of terrestrial animals. However, most of the wet weight gain in lean fish is in the form of muscle tissue, unlike in terrestrial animals where there is considerable deposition of both fat and protein. Salmonids continue to need, from young to adult stages, higher levels of 40−60 per cent protein in their diets. But other species like the milkfish (*Chanos chanos*) appear to make a rapid transition to a diet of algae, containing 10−20 per cent protein in its natural environment. The protein component of this material is digested and the amino acids absorbed, while most of the undigestible cellulose is excreted.

Protein requirements are influenced by water

Table 7.1 Estimated dietary protein requirement of certain fish (from Castell, *et al.*, 1986).

Species	Crude protein level in diet for optimal growth (g/kg)	Reference
Rainbow trout (*Salmo gairdneri*)	400−600	Satia, 1974; Zeitoun *et al.*, 1976
Common carp (*Cyprinus carpio*)	380	Ogino and Saito, 1970
Chinook salmon (*Oncorhynchus tshawytscha*)	400	De Long *et al.*, 1958
Eel (*Anguilla japonica*)	445	Nose and Arai, 1972
Plaice (*Pleuronectes platessa*)	500	Cowey *et al.*, 1974
Gilthead bream (*Chrysophrys aurata*)	400	Sabaut and Luquet, 1973
Grass Carp (*Ctenopharyngodon idella*)	410−430	Dabrowski, 1977
Red sea bream (*Chrysophrys major*)	550	Yone, 1976
Yellowtail (*Seriola quinqueradiata*)	550	Takeda, 1975

temperature, body size, stocking density, oxygen levels and the presence of toxins. As water temperature declines, the body temperature of fish also declines and consequently the metabolic rate is reduced. The most favourable temperature for a given species is the one at which the difference between maintenance requirement and voluntary food intake is greatest and at which optimum efficiency of growth occurs (Smith, 1980). Chinook salmon (*Oncorhynchus tshawytscha*) need food containing 40 per cent protein in water temperatures of about 8°C for optimum growth, whereas in temperatures of about 14°C the same fish will need food containing 55 per cent protein (De Long *et al.*, 1958). At lower temperatures, foods containing more than 40 per cent protein produce stress due to an excess of ammonia released from gills. Channel catfish (*Ictalurus punctatus*) show optimum growth at 20°C on a 35 per cent protein diet, whereas at 25°C they need a 40 per cent protein diet to achieve optimum growth (Dupree and Sneed, 1966). However, work by Slinger *et al.* (1977) and Cho and Slinger (1978), does not confirm the results relating to temperature effects. The greater absolute need for protein at higher temperatures might be satisfied through

increased consumption of the lower protein diets.

Readily digested high-protein materials have higher metabolizable energy (ME) values for fish than other mono-gastric animals. Similarly, protein has more net energy for fish than it has for mammals or birds. Smith *et al.* (1978) showed that less than 5 per cent of the ME is lost as heat increment in fish. Fish are among the most efficient of all animals in converting feed energy into high quality protein.

Available information does not seem to support the general view that omnivorous and herbivorous aquaculture species require less protein in their diets. As, for example, the juveniles of herbivorous grass carp (*Ctenopharyngodon idella*) require levels of protein similar to salmon and trout. Phytoplankton and zooplankton contain high percentages of protein (40−60 per cent) and there is reason to believe that the protein requirements of plankton-feeding species are also similarly high. The real difference between species of different feeding habits would appear to be in the ability to digest carbohydrates. Most of the carnivores, like trout and yellowtail, have a limited ability to digest complex carbohydrates.

Juveniles and adults of most cultured crustaceans have protein requirements in the range of 30–50 per cent of their dry diet weight. Like fish, they also require much higher levels of protein than terrestrial animals. However, there are differences in their nitrogen metabolism. One major difference is based on the habit of moulting. The crustacean exoskeleton consists of a mineral-organic matrix. Chitin, one of the primary compounds, is composed of glucosamine units (an amine group and glucose). Growth occurs when the old exoskeleton is partly resorbed, then shed and a new one is grown in its place. Prior to moulting, they produce high levels of ammonia, indicating the resorption of the old exoskeleton. Even when the exuvia is eaten, substantial losses of nitrogen occur. Nitrogen balance can be used to evaluate the amino acid and nitrogen requirements. Moulting (ecdysis) affects the animal's nitrogen balance.

Crustaceans appear to have a limited ability to store protein, unlike terrestrial vertebrates (Maynard and Loosi, 1969). Recent studies show that both carbohydrates and lipids can be used to spare dietary protein in crustaceans. There are wide variations in the protein requirements of shrimp species.

Little work has been done on defining the nutritional requirements of molluscs, probably because of the successful use of algae as food and the lack of successful microencapsulated diets. However, as pointed out earlier, the protein content of phyloplankton species is generally high (above 40 per cent), although it varies with environmental conditions.

Studies of Langton *et al.* (1977) on the relationship between protein consumed and nitrogen retained by the clam *Tapes japonica* showed that they are directly correlated. However, the efficiency dropped from 48 to 36 per cent when the nitrogen intake doubled from approximately 50 to 100 mg. Its nitrogen retention efficiency thus appears to be similar to that of other aquatic animals.

Amino acids

Ingested proteins are first split into smaller fragments by pepsin or by trypsin or chymotrypsin from the pancreas in fish. These peptides are then further reduced by the action of carboxypeptidase and amino peptidase, which hydrolyses off one amino acid at a time, beginning at each end of the polypeptide chain. The free amino acids released into the digestive system are then absorbed through the walls of the gastro-intestinal tract into the bloodstream, where they are then resynthesized into new tissue proteins, or catabolysed for energy or fragmented for further tissue metabolism.

Amino acids are described as the building blocks of proteins, and about twenty-three of them have been isolated from natural proteins. Ten of these amino acids are considered indispensable for fish, crustaceans and molluscs; these are arginine, histidine, isoleucine, leucine, lysine, methionine, phenylalanine, threonine, tryptophan and valine. Alanine, aspartic acid, cystine, glutamic acid, glycine, proline, serine and tyrosine are considered the non-essential amino acids, at least in the case of trout and channel catfish. In the case of molluscs, such as mussels, proline may be essential. While the type of amino acids required are similar to those of other animals, the quantity required is much higher.

Cowey (1979) summarized the available information on amino acid requirements of certain species of cultured fish, as shown in Table 7.2. The comparative amino acid requirements (as percentage of protein) of four selected fish were given by Ketola (1980) (Table 7.3). The data suggest that differences exist between species in their requirements of certain amino acids. This creates difficulties in practical diet formulation for species for which amino acid requirements are not clearly known. The solution suggested is to provide for the maximum observed requirements, despite possible additional costs.

Several investigations have shown the potential for supplementing amino acid deficient proteins with limiting amino acids in diets for salmonids. Casein supplemented with six amino acids gave feed conversion ratios similar to isolated fish proteins as a dietary source in Atlantic salmon. It was demonstrated that soybean meal supplemented with five or more amino acids, including methionine and lysine, was a superior protein source to soybean meal alone, for rainbow trout. However, this does not appear to be the case with young carp and channel catfish. Generally speaking, diets containing fish meal, meat and bone meal,

Table 7.2 Some quantitative essential amino acid requirements of certain species of fish in g/kg dry diet, (reproduced with permission of Heenemann Verlagsgesellschaft).

	Chinook salmon[1]	Japanese eel[2]	Carp[2]	Channel catfish[3]	Gilthead bream[4]	Rainbow trout[5]
Arginine	24	17	16		<10.4	12
Histidine	7	8	8			
Isoleucine	9	15	9			
Leucine	16	20	13			
Lysine	20	20	22	12.3	20	
Methionine	16*	12*	12*		16[†]	
Phenylalanine	21[‡]	22[‡]	25[‡]			
Threonine	9	15	15			
Tryptophan	2	4	3		2.4	
Valine	13	15	14			

* In the absence of cystine
[†] Methionine + cystine
[‡] In the absence of tryosine
[1] Mertz (1972)
[2] S. Arai and T. Nose (private communications)
[3] Wilson *et al.*, (1977)
[4] Luquet and Sabaut (1974)
[5] Kaushik (1977)

Table 7.3 Comparative amino acid requirements, as percentage of protein (from Ketola, 1980).

	Arg	His	Iso	Leu	Lys	Met + Cys	Phe + Try	Thr	Trp	Val
Japanese eel	4.2	2.0	3.8	4.7	5.1	4.8	5.8	3.8	1.1	3.8
Chinook salmon	6.0	1.8	2.3	4.0	5.0	3.8	5.3	2.3	0.5	3.3
Channel catfish	4.3	1.5	2.6	3.5	6.8	2.9	4.5	2.2	0.5	3.0
Common carp	4.3	2.1	2.5	3.6	5.7	3.1	6.5	3.9	0.8	3.6

yeast and soybean can be improved by supplementing with cystine (10 g/kg) and tryptophan (5 g/kg) together. It is reported that fish meal can be entirely replaced, without reduction in food conversion rate, in diets for rainbow trout, by a mixture of poultry by-product meal and feathermeal together with 17 g lysine HLC/kg, 4.8 g DL-methionine/kg and 1.44 g DL-tryptophan/kg.

The absolute amino acid requirements of crustaceans have yet to be defined. Table 7.4 presents the amino acid composition of commonly used proteins in crustacean purified diets. It is, however, to be remembered that processing and digestibility affect their actual availability to the animals.

As mentioned earlier, radiotracer work indicates that molluscs have the same requirements for essential amino acids as fish and crustaceans. The algae they consume are rich in essential amino acids and, in satisfactory culture conditions, their requirements are generally fulfilled adequately.

7.3.8 Lipids and essential fatty acids

Lipids are a group of fat-soluble compounds occurring in the tissues of plants and animals and broadly consist of fats, phospholipids, sphingomyelins, waxes and sterols. Fats are the fatty acid esters of glycerol and are the principal form of energy storage. They contain more

Table 7.4 Amino acids commonly used in crustacean diets, in percentage of total weight of amino acids,* (from Castell *et al.* 1986).

Amino acid	Brine shrimp		Wheat gluten		Soy protein		Casein		Egg albumin		Shrimp meal[†]	
	d	c	a	d	d	b	c	a	c	a	d	a
Asp	10.1	9.2	3.5	4.8	11.6	7.0	7.1	6.8	9.3	10.7	11.3	11.5
Thr[‡]	4.9	4.6	2.5	3.3	4.1	3.9	4.9	3.8	4.0	4.3	4.2	4.6
Ser	5.2	4.8	4.6	4.9	5.3	5.6	6.3	4.7	8.2	6.2	4.3	4.5
Glu	14.6	14.2	35.5	30.4	19.7	21.5	22.4	21.3	16.5	13.6	16.6	16.2
Pro	4.7	5.2	13.1	8.9	5.2	10.5	10.6	10.2	3.8	3.6	3.7	5.4
Gly	4.9	5.3	3.5	6.2	4.3	1.8	2.0	1.8	3.6	3.6	4.9	8.9
Ala	5.2	6.9	2.7	3.6	4.4	2.9	3.2	2.9	7.6	6.1	5.8	6.2
Cys[§]	1.3	2.2	1.2	1.4	1.2	0	0.3	0	2.8	1.5	1.1	0
Val[‡]	5.3	5.4	4.4	4.2	4.2	6.1	7.2	6.7	8.8	7.4	6.1	6.2
Met[‡§]	2.3	2.7	1.4	1.4	1.3	1.0	2.8	3.0	5.3	3.9	2.8	0.4
Iso[‡]	5.1	5.3	3.9	3.0	4.7	5.0	6.1	5.4	7.0	5.4	4.5	4.6
Leu[‡]	8.6	8.0	7.2	6.2	7.6	9.2	9.2	9.2	9.9	8.5	7.8	7.4
Tyr	4.6	4.5	3.6	4.2	3.5	5.4	6.3	5.3	4.1	3.7	4.1	2.4
Phe[‡]	5.3	4.7	5.2	3.7	5.1	4.7	5.0	4.5	7.2	5.7	4.7	4.6
Lys[‡]	7.4	7.6	1.7	3.3	6.3	7.7	8.2	7.8	6.5	7.1	8.2	5.9
His[‡]	2.2	1.8	2.3	2.0	2.4	2.9	3.1	3.0	2.9	2.6	2.3	2.5
Arg[‡]	6.8	6.5	3.8	5.3	7.4	3.6	4.1	3.6	6.0	6.1	7.9	8.3
Trp[‡§]	0	1.0	0	0	0	1.1	1.7	0	1.2	0	0	0

* Many of the sources can vary by more than 20% total protein.
a Bodega Marine Lab 79F diet ingredients; analysis by University of California at Davis Medical Center.
b ICN.
c Gallagher and Brown, 1975.
d Deshimaru and Shigueno 1972.
[†] Purified
[‡] Essential for the lobster, *Homarus* (Gallagher and Brown, 1975).
[§] Partially or totally destroyed by acid hydrolysis.

energy per unit weight than any other biological product — it is estimated that they provide 8.5 kcal metabolizable energy (ME) per gramme. Natural diets may contain as much as 50 per cent fat. Phospholipids are the esters of fatty acids and phosphatic acid. These are the main constituent lipids of cellular membranes, determining the hydrophobic or hydrophylic properties of the membrane surfaces. Sphingo-myelins are present in the brain and nerve tissue compounds. Waxes are fatty acid esters of long-chain alcohols and can be metabolized for energy. Sterols are polycyclic, long-chain alcohols and are components of several hormone systems, especially those related to sexual maturation and reproductive functions. The protein-sparing function of lipids has already been referred to.

Fatty acids are described as saturated when they contain no double bonds, and unsaturated when they contain one (mono-unsaturated) or more (polyunsaturated) double bonds. They are composed of carbon, hydrogen and oxygen and are generally acylic, unbranched molecules containing an even number of carbon atoms. The polyunsaturated fatty acids (PUFA) are divided into three families, named after the shortest chain length fatty acid representing each, namely oleic, linoleic and linolenic acids. The omega (ω) system of nomenclature is used to identify the families. (In some recent literature ω is replaced by *n*.) Those belonging to oleic family are referred to as ω9, linoleic as ω6 and linolenic as ω3 fatty acids. An abbreviated form is used to refer to the structure of the fatty acid, for example linoleic acid is expressed as 18:2ω6, where 18 is the number of carbon atoms in the fatty acid molecule, 2 is the number of double bonds and ω6 is the location of the first double bond relative to the methyl (CH_3) end of the molecule.

Fish in general contain more ω3 than ω6 PUFA, but fresh-water fish appear to have higher levels of ω6 fatty acids than marine species. The requirement for ω3 fatty acids in diets may, therefore, be greater in salt-water species. Besides salinity, there are other environmental factors that affect the fatty acid composition, particularly PUFA of fish: temperature is an important factor and the effects of temperature on fatty acid composition have been clearly demonstrated. There is a general trend towards a higher content of long-chain PUFA at lower temperatures. The ω6/ω3 ratio decreases with decrease in temperature. If this trend in fatty acid composition can be taken as clues to the essential fatty acid (EFA) requirements, the ω3 requirements of fish grown at lower temperatures would be greater, and conversely those grown in warmer waters may do better with a mixture of ω6 and ω3 fatty acids (Halver, 1980). The melting point of fat, related to the degree of unsaturation, has an important bearing on digestibility. Liquid fats are more readily digested and used by fish, whereas high-melting-point fats are not effectively utilized. Fats that solidify at relatively low environmental temperatures may be poor lipid sources for aquaculture diets, especially during cold weather, as the temperature of the animal will be about the same as the environment. Other significant factors that affect EFA are depth, season, diet and reproduction and possibly also genetic variation.

The known EFA requirements of a number of species of fish are presented in Table 7.5. Most of the estimates of EFA requirements for crustaceans are suggestive rather than conclusive. There are very few reports on the effects of purified fatty acids in semi-purified test diets for crustaceans. Kanazawa *et al.* (1977) found that both 18:2ω6 and 18:3ω3 improved the growth of *Penaeus japonicus* compared to diets containing 18:1ω9 as the sole lipid. It has also been shown that, even though the chain elongation/desaturation ability of the prawn (*P. japonicus*) was less than that of rainbow trout, they were able to convert 18:3ω3 into 20:5ω3 and 22:6ω3 at a faster rate than marine fish like red sea-bream (Kanazawa

et al., 1979b). Fresh-water crustaceans will probably have a requirement for one or both of ω3 and ω6 fatty acids, depending on culture temperature. Herbivores will probably be more capable of utilizing the 18 carbon ω3 and ω6 fatty acids than carnivores (Castell and Tiews, 1980).

In addition to EFA, crustaceans require other dietary lipids. The sterols are very important in many essential hormonal functions, besides being membrane lipids. The level of sterol required varies from 0.5−2 per cent of the dry weight diet or 5−30 per cent of the dietary lipid. Phospholipid phosphatidyl choline has been observed to have growth-promoting properties. The addition of 1 per cent lecithin from the short-necked clam (*Tapes* sp.) to a semi-purified test diet resulted in optimum growth of *P. japonicus* in studies made by Kanazawa *et al.* (1979a). Similarly, optimum growth and survival of juvenile lobsters (*Homarus americanus*) were obtained when 7−8 per cent lecithin from soybean lipids was added to the casein/albumin-based test diet (Conklin *et al.*, 1980, D'Abramo *et al.*, 1981). Though the precise role of dietary lecithin is not yet fully known, it is believed that it may have an important role in the transport of lipids in crustacean haemolymph.

As mentioned earlier, there is very little information on the nutritional requirements of molluscs and this applies also to the requirements of lipids and EFA. Marine molluscs tend to have relatively high levels of 20:5ω3 and 22:6ω3. The content of 18:2ω6, 20:4ω6 and 18:3ω3 tend to be higher in fresh-water than in marine molluscs and higher in warm-water than in cold-water species (Ackman *et al.*, 1974; Hoskin, 1978). Based on these similarities, it is predicted that the range of EFA requirements of molluscs will be similar to those of finfish and crustaceans. But unlike them, the ability to synthesize sterols *de novo* from acetate and mevalonate appears to vary from species to species in molluscs.

The lipid contents of most commercial aquaculture diets are less than 10 per cent, mainly due to processing problems. Though the optimum levels have not yet been determined, higher levels do not appear to result in higher growth. Excessive dietary lipid levels can cause nutritional diseases like fatty liver. From the

Table 7.5 Essential fatty acid requirements (as a percentage of the diet) reported for various species of finfish (after Castell *et al.*, 1986).

Species	EFA Requirements*				References
	18:2ω6	18:3ω3	20:4ω6	20:5ω3 +22:6ω3	
Cold, fresh-water fish					
Rainbow trout	–	1.0			Castell *et al.*, 1972a,b,c
(*Salmo gairdneri*)	–	1.0		or 1.0	Yu and Sinnhuber, 1972
Chum salmon	–	1.0		or 1.0	Takeuchi *et al.*, 1979
(*Oncorhynchus keta*)					
Coho salmon	–	1.0			Yu and Sinnhuber, 1979
(*Oncorhynchus kisutch*)					
Warm, fresh-water fish					
Common carp	1.0	+1.0		or 0.5–1.0	Takeuchi and Watanabe, 1977
(*Cyprinus carpio*)					
Eel	0.5	+0.5			Takeuchi *et al.*, 1980
(*Anguilla japonica*)					
Tilapia	1.0	–	or 1.0	–	Kanazawa *et al.*, 1980
(*Tilapia zillii*)					
Marine-fish					
Turbot	–	–	–	0.8–1.0	Gatesoupe *et al.*, 1977a,b,
(*Scophthalmus maximus*)	–	3.7	–	or 0.6	Leger *et al.*, 1979
Red sea bream	–	–	–	2.0	Fujii and Yone, 1976
(*Chrysophrys major*)					

* In most cases, the ω3 or ω6 fatty acids of 20 or 22 carbon chain length were more effective than those with 18 carbons.

point of view of product quality also this may not be desirable as high lipid levels may cause greater deposit of visceral fat.

7.3.4 Vitamins

Vitamins are a chemically diverse group of organic substances that are either not synthesized by organisms or are synthesized at rates insufficient to meet the organisms' needs. They constitute only a minute fraction of the diet and are more catalytic in their function, but are critical for the maintenance of normal metabolic and physiological functions. They can be classified into two groups: water-soluble and fat-soluble vitamins. Water-soluble vitamins include eight members of the vitamin B complex: thiamin, riboflavin, pyridoxine, pantothenic acid, niacin, biotin, folic acid and vitamin B_{12}. They include the essential nu-

tritional factors choline, inositol, ascorbic acid and vitamins with less-defined activities for fish: *p*-aminobenzoic acid, lipoic acid and citrin. The fat-soluble group comprises vitamins A, D, E and K.

The information available on vitamin nutrition of aquatic animals is limited. The leaching of water-soluble vitamins from test diets, before the animals feed on them, has been a major problem in determining their requirements. This is more so in the case of crustaceans because of their slow feeding habits, and in the case of molluscs because they feed on algae and other natural food. Much of the available information is based on the work done on salmonids, but some data are also available for a few other species, which are presented in Table 7.6. If natural food organisms are available to cultured animals, as in extensive pond culture, prepared feed may not need any vitamin supplements. On the contrary, in inten-

Table 7.6 Vitamin requirements (in mg/kg dry diet) (from Halver, 1980).

Vitamin	Rainbow trout	Brook trout	Brown trout	Atlantic salmon	Chinook salmon	Coho salmon	Common carp	Channel catfish	Eel	Sea bream	Turbot	Yellowtail
Thiamin	10–12	10–12	10–12	10–15	10–15	10–15	2–3	1–3	2–5	*	2–4	*
Riboflavin	20–30	20–30	20–30	5–10	20–25	20–25	7–10	*	*	*	*	—
Pyridoxine	10–15	10–15	10–15	10–15	15–20	15–20	5–10	*	*	2–5	*	*
Pantothenate	40–50	40–50	40–50	*	40–50	40–50	30–40	25–30	*	*	*	*
Niacin	120–150	120–150	120–150	*	150–200	150–200	30–50	*	—	*	—	—
Folacin	6–10	6–10	6–10	5–10	6–10	6–10	—	*	*	—	*	—
Cyanocobalamin	*	*	*	*	0.015–0.02	0.015–0.02	—	*	—	*	*	—
Myo-inositol	200–300	*	*	*	300–400	300–400	200–300	*	—	300–500	—	—
Choline	*	*	*	*	600–800	600–800	500–600	*	—	—	*	—
Biotin	1–1.5	1–1.5	1.5–2	—	1–1.5	1–1.5	1–15	*	*	*	*	*
Ascorbic	100–150	*	*	*	100–150	50–80	30–50	30–50	*	*	—	*
Vitamin A	2000–2500 IU	*	*	—	*	*	1000–2000 IU	*	—	—	—	*
Vitamin E†	*	*	*	40–50	*	*	80–100	*	—	—	—	*
Vitamin K	*	*	*	—	*	*	*	*	—	—	—	—

Fish fed at reference temperature with diets at about protein requirement.
* Denotes a requirement, the level of which has not been established.
† Requirement directly affected by amount and type of unsaturated fat fed

sive farming, where natural food items do not contribute much to dietary intake, the addition of adequate quantities of vitamins will be essential. Insufficient information often makes it difficult to decide with precision the quantities which should be added. Hypervitaminosis is rare in fish, although it is possible at very high levels (for example excess vitamin A causes enlargement of liver and spleen, abnormal growth and bone formation and epithelial keritonization), particularly with the fat-soluble vitamins. Generally, vitamin levels in prepared feeds are sufficiently high that even with processing, storage and leaching losses, the remaining levels meet requirements.

The deficiency symptoms for most of the water-soluble vitamins have been described (Halver, 1979) (Table 7.7). The symptomatic results of these deficiencies were observed in fish that had been grown on test diets devoid of an individual vitamin. The functions of these vitamins are also beginning to be understood. For example, thiamin deficiency produces characteristic symptoms, including nervous system damage. As thiamin is involved in carbohydrate metabolism, thiamin needs were expected to be correlated with dietary levels of carbohydrates, and this has been shown to be so in common carp. Riboflavin also has a variety of coenzyme functions in carbohydrate metabolism, and its absence from the diet produces symptoms generally associated with the eye. Dietary deficiencies of folacin and vitamin B_{12} in fish are expressed as anaemias. Folic acid primarily functions in purine synthesis, and deficient purine synthesis strikes most directly at nucleoprotein production in blood cell formation, causing a macrocytic, normochromic anaemia. Severe choline deficiency impairs lipid metabolism and leads to fatty livers. Choline is usually needed in much larger amounts than water-soluble vitamins, because of its role in phospholipid synthesis needed for cell membrane structure. Myo-inositol is also required in larger amounts, as it is also probably needed as a constituent of phospholipids. Vitamin C deficiencies in fish often cause spinal deformities and slow wound repair, because of slow collagen formation.

As has been pointed out earlier, there is very little definite information available at present on vitamin requirements of crustaceans and molluscs. Even where requirements have been demonstrated, the deficiency symptoms have been defined only as reduced growth. No evidence has been found to show the physiological role of any of the fat-soluble vitamins by crustaceans. A dietary source of thiamin and pyridoxine, as well as inositol, has been shown to promote growth in *Penaeus japonicus*. Similarly, choline also appears to be necessary for improved growth and survival. The American lobster (*Homarus americanus*) has an apparent requirement for choline. Lack of choline-containing phospholipids results in a characteristic deficiency syndrome, in which juvenile mortalities are associated with unsuccessful molts (D'Abramo *et al.*, 1981a, b).

For the Japanese shrimp (*P. japonicus*) the required dietary levels of vitamins have been tentatively established as thiamin 120 mg/kg, pyridoxine 120 mg/kg, choline 600 mg/kg, inositol 2000 mg/kg and ascorbic acid 10 000 mg/kg (Castell *et al.*, 1986). A vitamin C requirement has also been indicated in the species.

7.3.5 Minerals

Minerals are required by all animals either in their elemental form or incorporated into specific compounds, for various biological functions such as the formation of skeletal tissue, respiration, digestion and osmoregulation. Of the twenty-six naturally occurring essential elements described for animals, only nine have been shown to be required by finfish. Very little is known of the mineral requirements of crustaceans and molluscs. One of the main difficulties in determining the quantitative mineral requirements of aquatic animals is their ability to absorb inorganic elements from their external environment in addition to their diets.

Since water generally contains an abundance of minerals, supplementation of diets may not be necessary, except in the case of those that are required in relatively high concentrations, especially in fresh-water fish. The available information on the mineral requirements of fish is summarized in Table 7.8 and information on deficiency symptoms in Table 7.9.

Calcium and phosphorus are closely related in metabolism, especially in bone formation and the maintenance of acid-base equilibrium.

Table 7.7 Vitamin deficiency symptoms in fish (after Halver, 1979).

Vitamin	Deficiency symptoms
Thiamin	Anaemia, anorexia, ataxia (terminal), convulsions (when moribund), corneal opacities, degeneration of vestibular nerve nucleus, fatty liver, hemorrhage of midbrain or medulla, loss of equilibrium, melanosis in older fish, muscle atrophy, paralysis of dorsal and pectoral fins, rolling, whirling motion, vascular degeneration, weakness.
Riboflavin	Anorexia, cloudy lens, darkened skin, dim vision, discoloured iris, hemorrhage in eyes, nares or operculum, incoordination, lens cataract, mortalities, photophobia, xerophthalmia.
Pyridoxine	Anorexia, ascites, ataxia, convulsions, flexing of opercles, hyperirritability, indifference to light, microcytic hypochromic anaemia, rapid jerky breathing, spasms, weight loss, nervous disorders, increased mortalities, rapid onset of rigor mortis.
Folic acid	Anaemia, anorexia, ascites, dark coloration, erythropenia, exophthalmia, fragility of caudal fin, lethargy, macrocytic anaemia, pale gills, poor growth.
Pantothenic acid	Anorexia, clubbed gills, flared opercula, gill exudate, general 'mumpy' appearance, lethargy, necrosis of jaw, barbels and fins, prostration, poor weight gain.
Inositol	Anaemia, bloated stomach, poor growth, anorexia, skin lesions.
Biotin	Anaemia, anorexia, blue slime disease, colonic lesions, contracted caudal fins, dark coloration, erythrocyte fragmentation, mortalities, muscle atrophy, poor growth, spastic convulsions.
Choline	Anaemia, poor food conversion, poor growth, vascular stasis and hemorrhage in kidney and intestine.
Nicotinic acid (niacin)	Anaemia, anorexia, colonic lesions, edema of stomach and colon, incoordination, jerky movements, muscle spasms, lethargy, photophobia, swollen gills, tetany, skin hemorrhage, high mortality.
p-Aminobenzoic acid	No significant change in growth, appetite or survival.
Cobalamin (B_{12})	Anorexia, erratic haemoglobin and erythrocyte counts, fragmentation.
Ascorbic acid	Anorexia, impaired collagen production, impaired wound healing, lordosis with dislocated vertebrae and focal hemorrhage, poor growth, scoliosis with hemorrhage in severe cases, twisted deformed hyaline cartilage in gill filaments and sclera of eyes.
Vitamin A	Ascites, edema, exophthalmos, hemorrhagic kidneys.
Vitamin D	Reduced conversion.
Vitamin E (tocopherol)	Anaemia, ascites, ceroid in liver, spleen, kidney, clubbed gills, epicarditis, exophthalmia, microcytic anaemia, mortalities, pericardial edema, poor growth, red blood cell fragility.
Vitamin K	Anaemia, coagulation time prolonged.

Table 7.8 Mineral requirements of certain finfish, in percentage or amount per kg feed (after Castell *et al.*, 1986).

Species	Ca (%)	P* (%)	Mg (%)	Fe (mg)	Cu (mg)	Zn (mg)	Mn (mg)	I (μg)	Se (mg)
Rainbow trout (*Salmo gairdneri*)	0.02	0.7−0.8	0.05−0.07	−	3	15−30	12−13	−	0.15−0.38
Atlantic salmon (*Salmo salar*)	0.03	0.6	−	R	R	R	R	R	0.1
Chinook salmon (*Oncorhynchus tshawytscha*)	−	−	−	−	−	−	−	0.6−1.1	−
Chum salmon (*Oncorhynchus keta*)	−	0.5−0.6	−	−	−	−	−	−	−
Catfish (*Ictalurus punctatus*)	0.03	0.6−0.7	0.04−0.05	−	3	15−30	12−13	−	−
Tilapia (*Tilapia nilotica*)	−	0.9	−	−	−	−	−	−	−
Eel (*Anguilla japonica*)	0.27	0.30	0.04	170	−	−	−	−	−
Red sea bream (*Pagrus major*)	0.34	0.56−0.6	NR	150	−	−	−	−	−

* Inorganic.
  R: Required
  NR: not required.

Table 7.9 Mineral deficiency symptoms in certain finfish (after Castell *et al.*, 1986).

Mineral	Deficiency symptoms
Calcium	Poor growth and feed efficiency,[1,8] high mortality.
Phosphorus	Skeletal abnormalities,[5,7,8] poor growth and feed efficiency and bone mineralization.[1,3,5,7,8]
Magnesium	Renal calcinosis,[1] loss of appetite,[1,8] poor growth,[1,8] high mortality, sluggishness, skeletal abnormalities.
Iron	Hypochromic microcytic anaemia.[2,7,9]
Copper	Poor growth[7]
Manganese	Poor growth,[1] short and compact body,[1] abnormal tail growth.[7]
Iodine	Thyroid hyperplasia.[6]
Zinc	Cataract,[1] caudal fin and skin erosion,[1,7] growth depression.[1]
Selenium	Muscular dystrophy,[3] exudative diathesis.[3]

[1] *Salmo gairdneri*, [2] *Salmo fontinalis*, [3] *Salmo salar*, [4] *Oncorhynchus keta*, [5] *Ictalurus punctatus*, [6] *Oncorhynchus tshawytscha*, [7] *Cyprinus carpio*, [8] *Anguilla japonica*, [9] *Chrysophrys major*.

While fish can obtain calcium from food and also from the environment through gills and fins in fresh water, phosphorus has to come mainly from food, as both fresh and salt waters are generally deficient in phosphates. Almost the entire store of calcium (99 per cent) and most of the phosphorus (80 per cent) in the body of fish are in the bones, teeth and scales. The remaining small portions are widely distributed throughout the organs and tissues. Calcium is present in body fluids in non-diffusible form bound to protein and in a diffusible fraction largely as phosphate and bicarbonate compounds. It is this diffusible fraction that is of significance in calcium phosphorus nutrition (Chow and Schell, 1980).

Ionized calcium in the extracellular fluids and in the circulatory system participates in muscle activity and osmoregulation. Phosphorus in combination with proteins, lipids, sugars, nucleic acids and other compounds are vital exchange currencies in life processes, and are distributed throughout the organs and tissues.

Among the feed ingredients in common use, fish meal is rich in both calcium and phosphorus, but feedstuffs of plant origin usually lack calcium, and phosphorus though abundant, occurs predominantly in the form of phytin or phytic acid, which are generally not readily absorbed.

Magnesium is closely associated with calcium and phosphorus in distribution and metabolic activities. While the bulk of the magnesium is stored in the skeleton, the rest (40 per cent) is distributed throughout the organs and muscle tissues and extracellular fluids. This fraction plays a vital role as enzyme co-factors and as an important structural component of cell membranes.

Among the trace elements of importance in fish nutrition, mention has to be made of cobalt. Studies made in the USSR seem to indicate that addition of cobalt chloride and/or cobalt nitrate to the feed, or the addition of cobalt chloride to the water of fish ponds, enhances growth and haemoglobin formation in common carp. Cobalt probably satisfies the requirement of bacteria that synthesize vitamin B_{12}, which the fish can utilize.

As mentioned earlier, there is very limited information available on mineral requirements and deficiency symptoms in crustaceans and molluscs. It was reported that the best growth of *P. japonicus* was obtained with diets containing 1.04 per cent phosphorus and 1.24 per cent calcium (Kitabayashi *et al.*, 1971). Other research (Deshimaru and Yone, 1978) showed that 2 per cent phosphorus, 1 per cent potassium and 0.2 per cent trace element supplementation in a purified diet are essential for maximum growth of this species and that calcium, magnesium and iron supplementation are not essential. The best growth and survival of the American lobster (*Homarus americanus*) was achieved with a diet containing a Ca/P ratio of 0.51 (0.56 per cent calcium and 1.1 per cent phosphorus) (Gallagher *et al.*, 1978).

7.4　Live foods

7.4.1　Nature and source of live foods

Aquaculture animals have to obtain all their nutritional requirements except for part of the mineral requirements, through the foods they consume. In nature, most of them subsist on live foods consisting of plants and animals obtained from the environment, but some do ingest and possibly utilize detritus along with associated organisms. As mentioned at the beginning of Section 7.1, these foods are generally rich in essential nutrients. Table 7.10 gives an example of the nutrient composition of algal pastures grown in milkfish ponds in brackish-water ponds, where culture is largely based on natural foods. There is a difference of opinion on the food value of bacteria, although fair quantities can be found in the alimentary tract of aquaculture species, particularly in detritus and periphyton feeders. Experience in aquaculture seems to show that most adult finfish and crustaceans can be weaned to accept inert foods, even though there are advantages in providing some live food. But larval stages of many of these species have to depend entirely on live food. The initial source of food for many larval organisms is phytoplankton. This is probably associated with the size of the larvae at hatching. After a certain period of time, the larvae of most species, except molluscs, can be fed exclusively on zooplankton or other animal species, or a combination of plant and animal matter.

It is possible to obtain both types of food from nature. But it will be more convenient to culture algae under controlled conditions for hatchery use. In nursery and grow-out ponds, they are generally produced as a result of the biological cycle initiated by mineral nutrients in solution. Using the sun's heat and light they transform the inorganic matter and carbonic acid in solution into organic matter, in the form of vegetable tissues consisting of plankton and periphyton. Of particular interest in pond culture of bottom-feeding fish is the production of benthic algal complexes which also include animal species and bacteria associated with detritus.

Light penetration is an important factor in photosynthesis and, therefore, in the growth of

Table 7.10 Composition of four major groups of algae and their relative nutritive value as milkfish food (after Tang and Hwang, 1967).

Group of algae	Number of samples	Total composition						Digestive coefficient[a,b]				Digestible protein[c] (%)	Total digestible nutrients[d] (%)
		Total dry matter (%)	Crude protein (%)	Crude fat (%)	Nitrogen-free extract (%)	Fibre (%)	Mineral matter (%)	Crude protein (%)	Crude fat (%)	Nitrogen-free extract (%)	Fibre (%)		
Chaetomorpha													
Fresh form	15	8.54:	2.82	0.91	1.50	1.22	2.09	3	72	87	21	0.09	3.12
Detrital form	15	10.72:	3.46	0.38	3.21	0.98	2.69	66	89	85	37	2.28	6.13
Phytoflagellates[e]													
Fresh form	5	11.89:	3.91	1.32	5.61	0.42	0.72	81	91	78	23	3.17	10.41
Diatoms[f]													
Fresh form	15	12.87:	2.89	0.94	2.25	0.27	6.52	87	96	84	19	2.51	6.48
Filamentous blue-green algae[g]													
Fresh form	15	9.86:	2.32	0.21	1.52	0.70	5.11	69	86	81	38	1.60	3.49

[a] Digestive coefficient:

$$\frac{\text{the amount of a class of organic nutrient in the feed} - \text{the amount of that class of organic nutrient in faeces}}{\text{the amount of that class of organic nutrient in the food}} \times 100.$$

[b] The water temperature during digestion experiments ranged from 29 to 33°C and the salinity from 24 to 27 ppt.

[c] Digestible protein: the percentage of protein in the food × digestion coefficient of protein.

[d] Total digestible nutrients: the sum of digestible protein, fibre, nitrogen-free extract and fat × 2.25.

[e] Centrifuged from the pond water where *Chlamydomonas* and *Chilomonas* flagellates bloomed predominantly.

[f] Furnished as the diatom sludge.

[g] Collected from the pond bottom where the dominant genera, *Oscillatoria* and *Lyngbya*, grew.

aquatic algae and macrophytes. Since aquaculture is generally done in shallow waters, the light intensity at the bottom usually exceeds 1 per cent incident, which is the accepted compensation depth for aquatic plants. Even on ponds with highly turbid water, some photosynthetic activity takes place and this is further enhanced by water circulation.

Among the major nutrients required by plants are phosphorus and nitrogen, primarily PO_4 and NO_3. Nitrogen is removed from water as nitrates (NO_3) by plants. Nitrogenous wastes are excreted by animals and nitrogenous compounds are released during the bacteriological decomposition of plant and animal matter. They are eventually transformed into ammonia, which undergoes nitrification to nitrate through a nitrite (NO_2) as a result of the action of aerobic bacteria. Phosphorus is an important major nutrient because it plays a key role in photosynthesis and intermediary metabolism and forms a constituent of nucleic acid and proteins. Available carbon is also of major importance as its deficiency is reflected in decreased production. The ratio of carbon: nitrogen:phosphorus required by most species of phytoplankton is about 106:16:1.

Algal culture

The culture of algae for feeding larvae and postlarvae through fertilization and water management has been in practice in traditional finfish culture. In recent years there has been greater interest in intensive forms of algal culture, largely because of the need for live foods in rearing larvae in crustacean culture, which has expanded rapidly. Algal culture facilities now form an integral part of many shrimp and prawn hatcheries. Where controlled breeding of molluscs such as oysters is practised, algal culture to feed the larvae forms probably the most important activity.

Several species of microalgae are cultured for experimental purposes in laboratories or for commercial use in special tanks or batteries of large flasks or carboys. A partial list of commonly cultured algal species is given by Fox (1983) as follows, under the headings of their classes.

Bacillariophyceae
Skeletonema costatum, Thalassiosira pseudomonas, Thalassiosira fluviatilis, Phaeodactylum tricornutum, Chaetoceros calcitrans, Chaetoceros curvisetus, Chaetoceros simplex, Ditylum brightwelli, Scenedesmus

Haptophyceae
Isochrysis galbana, Isochrysis sp., *Dicrateria inornata, Cricosphaera carterae, Coccolithus huxley*

Chrysophyceae
Monochrysis sp.

Prasinophyceae
Pyraminimonas grossii, Tetraselmis suecica, Tetraselmis chuii, Micromonas pusilla

Chlorophyceae
Dunaliella tertiolecta, Chlorella autotrophica Chlorococcum sp., *Nannochloris atomus, Chlamydomonas coccoides, Brachiomonas submarina*

Chryptophyceae
Chroomonas salina

Cyanophyceae
Spirulina

According to Imai (1978) the following are the main algal forms that form the food of shellfish larvae.

Food organisms	Species of shellfish
Carteria sp.	*Mercenaria mercenaria*
Chaetoceros simplex	*Haliotis discus*
Chlamydomonas sp.	*Mercenaria mercenaria*
Chlamydomanas sp.	*Mytilus edulis*
Chlorella sp.	*Mercenaria mercenaria*
Chlorella sp.	*Crassostrea virginica*
Chlorococcum sp.	*Mercenaria mercenaria*
	Crassostrea virginica
Chromulina pleiades	*Crassostrea virginica*
Cryptomonas sp.	*Crassostrea virginica*
Cyclotella sp.	*Mercenaria mercenaria*
	Crassostrea virginica
Dicrateria inornata	*Crassostrea virginica*
Dicrateria sp.	*Mercenaria mercenaria*
Dunaliella euchlora	*Mercenaria mercenaria*
	Crassostrea virginica
Dunaliella sp.	*Mercenaria mercenaria*
	Crassostrea virginica
Hemiselmis refescens	*Crassostrea virginica*
Isochrysis galbana	*Mercenaria mercenaria*
	Crassostrea virginica
	Ostrea edulis
Monas sp.	*Crassostrea gigas*
	Pinctada martensii
	Haliotis gigantea
	Mactra sachalinensis
	Ostrea edulis
	Ostrea lurida
	Pteria penguin
Monochrysis lutheri	*Mercenaria mercenaria*
	Crassostrea virginica
	Ostrea edulis
	Ostrea lurida
	Mactra sachalinensis
	Patinopecten yessoensis
Olisthodiscus sp.	*Mercenaria mercenaria*
Platymonas sp.	
(= *Tetraselmis maculata*)	*Haliotis discus hannai*
Platymonas sp.	*Mercenaria mercenaria*
	Crassostrea virginica
Phaeodactylum tricornutum	*Mercenaria mercenaria*
	Crassostrea virginica
Pyramimonas grossi	*Crassostrea virginica*
Rhodomonas sp.	*Mercenaria mercenaria*
Skeletonema costatum	*Mercenaria mercenaria*
Stichococcus sp.	*Mercenaria mercenaria*

Although considerable work has been done on mass cultivation of algae, the continuous and consistent production of large quantities of the desired species is still uncertain. Experience shows that it is not easy to manipulate the nutrient levels, nutrient proportions, detention time and mixing required to effect species control in induced blooms of algae. The rapid growth of algae affects the aquatic environment to such an extent that the continued production of those algae is arrested and other species able to grow under the altered environment take over. Despite these limitations, aquaculturists have to use the available techniques to produce the microalgae that appear essential for the culture of larvae or adults of certain species.

The techniques presently employed differ somewhat according to species. But in all cases, an enriched medium is used and the optimum temperature, lighting and aeration are maintained in order to obtain dense growths. There are several formulations and variations of media used and no attempt is made here to describe them all. Relevant references listed at the end of this chapter may be consulted for a comprehensive review of these. The media used for culture of certain algal species will be described in Part II, with reference to specific technologies.

Except in laboratory-scale culture, completely pure cultures are not expected. However, efforts are made to maintain as pure a culture of the desired species as necessary and feasible. Where required, it is possible to obtain pure cultures from collections of commercial or non-profit organizations. An alternative would be to isolate them from local collections from natural sources. Isolation is rather complex and difficult but can be done successfully under laboratory conditions. Fox (1983) describes some of the successful techniques employed. All of them involve capillary pippette isolation and the maintenance of stringent aseptic conditions.

For the species mentioned above, the most common culture medium used in mass production is filtered surface seawater enriched with essential growth nutrients. An alternative is a synthetic seawater medium, consisting of distilled water, growth nutrients and artificial sea salts. A variety of inorganic and organic nutrients are used in different types of media. Fox (1983) cites Guillard's F/2 medium (Guillard, 1975) (Table 7.11) as one that has received extensive use and is suitable for the growth of most algae. The macronutrients in this medium include nitrate, phosphate and silica. Inorganic micronutrients include ferric chloride, the chelate EDTA (to keep essential trace elements in solution) and a number of trace elements. Organic micronutrients include the vitamins thiamin (B_1), cyanocobalamin (B_{12}) and biotin. They are not considered essential in all mass cultures of algae. The inclusion of silica is of special importance in the culture of diatoms.

As in the case of enriched seawater media, there are also several formulations of artificial seawater media in use. Fox (1983) cites Gates and Wilson's NH medium, the composition of which is given in Table 7.12, as a proven medium for the culture of marine algae. Several brands of synthetic sea salts are available in

Table 7.11 Enriched seawater media (from Guillard, 1975).

Additive	Concentration (μm/l)				
	F/2	H/2	F/2 beta	ES	SWM
Inorganic macronutrients					
$NaNO_3$	880	—	880	660	500−2000
$NH_4 Cl$	—	500	—	—	—
$NaH_2 PO_4$	36.3	36.3	36.3	—	50−100
Na_2 glycerophosphate	—	—	—	25.0	—
$Na_2 SiO_3.9H_2 O$	54−107	54−107	54−107	—	200
Inorganic micronutrients					
FeEDTA	—	—	—	7200	2.0
$FeCl_3.6H_2 O$	11.7	11.7	11.7	1.8	—
Na_2 EDTA	11.7	11.7	11.7	26.9	48.0
$CuSO_4.5H_2 O$	0.04	0.04	0.04	—	0.3
$ZnSO_4.5H_2 O$	0.08	0.08	0.08	0.80	35.0
$CoCl_2.6H_2 O$	0.05	0.05	0.05	0.17	0.30
$MnCl_2.4H_2 O$	0.90	0.90	0.90	7.30	10.0
$Na_2 MoO_4.2H_2 O$	0.03	0.03	0.03	—	5.0
Boron	—	—	—	185	400
Organic micronutrients					
Thiamin HCl (B_1)	100 μg	100 μg	100 μg	20 μg	—
Nicotinic acid	—	—	—	—	0.1 mg/ℓ
Ca pantothenate	—	—	—	—	0.1 mg/ℓ
p-Aminobenzoic acid	—	—	—	—	10 μg/ℓ
Biotin	0.5 μg	0.05 μg	0.5 μg	0.8 μg	1.0 μg/ℓ
i-Inositol	—	—	—	—	5.0 mg/ℓ
Folic acid	—	—	—	—	2.0 μg/ℓ
Cyanocobalamin	0.5 μg	0.5 μg	0.5 μg	1.6 μg	1.0 μg/ℓ
Thymine	—	—	—	—	3.0 μg/ℓ
Tris	—	—	—	0.66 μg	0−5000
Glycylglycine	—	—	—	—	5000
Soil extract	—	—	—	—	50 ml/ℓ
Liver extract	—	—	—	—	10 mg/ℓ

Table 7.12 Gates and Wilson's NH artificial seawater medium.

Additive	Concentration
NaCl	24.0 g/ℓ
KCl	0.6 g/ℓ
MgCl$_2$.6H$_2$O	4.5 g/ℓ
MgSO$_4$.7H$_2$O	6.0 g/ℓ
CaCl$_2$	0.7 g/ℓ
K$_2$HPO$_4$	10.0 mg/ℓ
KNO$_3$	10.0 mg/ℓ
Vitamin B$_{12}$	1.0 µg/ℓ
Thiamin HCl	10.0 mg/ℓ
Biotin	0.5 µg/ℓ
Sulphides*	1.0 ml/ℓ
Vitamin mix 8[†]	0.1 ml/ℓ
Metals T[‡]	5.0 ml/ℓ
Adenine sulphate	1.0 mg/ℓ
Tris	0.1 g/ℓ
NaEDTA	10.0 mg/ℓ

* Sulphides: 0.2 g NH$_4$Cl, 0.1 g KH$_2$PO$_4$, 0.04 g MgCl$_2$.6H$_2$O, 0.2 g NaHCO$_3$, 0.15 g Na$_2$SiO$_3$.9H$_2$. Make up to 1 ℓ with distilled water.

[†] Vitamin mix: 20 mg thiamin-HCl, 50 µg biotin, 5 µg vitamin B$_{12}$, 0.25 mg folic acid, 1.0 mg PABA, 10 mg nicotine acid, 80 mg thymine, 50 mg choline, 100 mg inositol, 0.8 mg patrescine, 0.5 mg riboflavin, 4.0 mg pyridoxine, 2.0 mg pyridoxine, 26 mg orotic acid. Make up 100 ml with distilled water.

[‡] Metals T: 1% solutions; 2.5 ml Fe Tartrate (5 mg Fe), 3.0 ml H$_3$BO$_3$ (5.1 mg B), 0.1 ml H$_2$ScO$_3$ (1 mg Se), 0.12 ml NH$_4$VO$_3$ (0.5 mg V), 0.11 ml K$_2$CrO$_4$ (0.2 mg Cr), 0.37 ml MnCl$_2$ (1.0 mg Mn), 0.83 ml TiO$_2$ (5.0 mg Ti), 5.0 ml Na$_2$SiO$_3$ (5.0 mg Si), 0.4 ml ZrOCl$_2$ (2.0 mg Zr), 0.15 ml BaCl$_2$ (1.0 mg Ba). Make up to 100 ml with distilled water.

non-sterile powdered form and can be used as an additive; they should be sterilized and homogenized prior to use.

As only small amounts of nutrients are needed at any particular time for algal culture, concentrated nutrient stock solutions are maintained so that the problems of frequently weighing small amounts can be avoided. Not only is time thus saved in preparing media, but the possibility of contamination due to frequent handling of reagents is reduced.

An algal culture system generally consists of three or four main stages, starting with and maintaining a stock culture, from which cultures are made at regular intervals in small flasks (of about 50 ml volume), followed by culture in larger carboys (of about 12 ℓ volume) or tanks of 300 ℓ or more capacity.

Stock cultures can be maintained in small screw-top test tubes that can be autoclaved for sterilization, using low-level enrichment media for maintenance, rather than heavy growth. While constant illumination may be needed for flagellate stocks, a 12-hour photoperiod is considered enough for diatoms. The incident light level required for stock cultures is 750–1000 lux (measured horizontally), which can be provided by two 30–40 W cool white fluorescent bulbs placed in front of the stock culture tubes. A temperature of about 24°C is maintained. After about a month the stock cultures should be transferred under aseptic conditions to create new culture lines.

In the second culture phase, aliquots (2.0 ml) of the stock culture are used to inoculate autoclaved small (about 125 ml) Erlenmeyer flasks. A light intensity of about 1500 lux is necessary. Though aeration may not be necessary, the flasks should be shaken to reduce shading.

A 4-day-old flask culture is used as inoculum for the next phase of culture. Figure 7.2 illustrates the type and possible arrangements of carboys for algal culture. They are generally 12–20 ℓ in capacity with an autoclaveable stopper fitted with an air supply line and a screw-top inoculation tube. The latter enables more or less aseptic inoculation of cultures and the aeration pipe reduces settling of cells on the bottom, assuring homogeneity of nutrients, exposure to greater light intensity and provision of small amounts of carbon dioxide for growth. A battery of carboys can be arranged on a shelf as shown in fig. 7.2, with large 40 W, cool white fluorescent bulbs placed behind the carboys. Guillard's F medium, diluted to half strength, is reported to give the high growth rates required for this phase.

After four days of growth in the carboys, the final phase of culture can be started. For this, large fibreglass tanks are used. While for laboratory use 300 ℓ tanks may be suitable, for commercial farms larger tanks of about 1 ton capacity will be necessary. Circular or rectangular tanks, painted white, are often preferred. Illumination is provided by a series

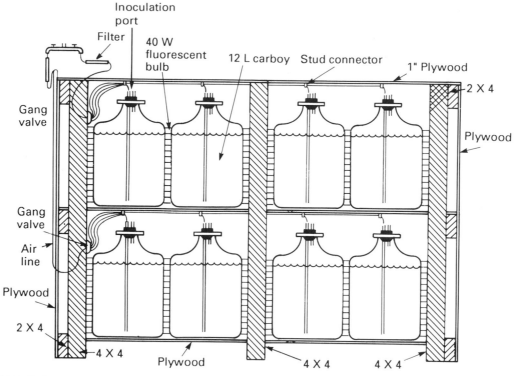

Fig. 7.2 Carboy culture shelf (after: Fox, 1983).

of 40 W fluorescent bulbs suspended directly above the tank. Suspended plastic bags have also been successfully used, but would require increased illumination. Constant illumination and aeration with air stones or other devices for adequate circulation are necessary.

Several methods are available for harvesting algae. High-density cultures can be concentrated by chemical flocculation or by centrifuging. Many flocculants cause the cells to settle at the bottom. Others, in combination with an electrical charge, can keep them floating. Harvesting is done by siphoning off the supernatant or by skimming cells off the surface as applicable. Centrifugation of algal cultures can be performed with a standard dairy cream separator. The culture has to be transferred mechanically or by pumping into the bowl of the separator. The rate of flow of the culture from the bowl to the centrifuge head is adjusted according to the species of algae and the centrifugation rate of the separator. The algae deposit on the wall of the centrifuge head as a thick paste during centrifugation, and this has to be removed and resuspended in water.

Where possible, live algae are directly pumped into larval tanks. When necessary, concentrated cultures are frozen for storage; thawed algae are diluted and supplied to larval tanks through small peristaltic pumps. Living algae are considered best for feeding larvae, although frozen algae may be more convenient to handle.

Many aquaculture farms use much less elaborate methods for producing live foods, predominently algae, in outdoor or indoor tanks, using suitable fertilizers as sources of nutrients. Cultures may not always be of one species, but if the biomass produced meets the requirements of larval and/or adult feeding, it is likely to be much more economical. Large outdoor tanks of 8–40 tons capacity, with supply lines for fresh and sea water and compressed air, have been used successfully for growing selected species such as *Chaetoceros*. A commercial hydroponic fertilizer is added at the time of inoculation at the rate of about 20 g per cubic metre. The fertilizer is placed in bags hung in the tank. In about three to four days, an average density of 1.2×10^6 cells per ml can be expected to develop.

The 'green water' method of algal production is practised in some *Macrobrachium* culture centres (fig. 7.3). Green water is a mixed phytoplankton culture, with a predominance of *Chlorella* spp. Large indoor or outdoor tanks are used for its culture. The tank is fertilized once a week with a solution of a mixture of 4 parts of urea to 1 part of NPK (15:15:15) agriculture fertilizer at the rate of 185 g per 10 m^3 water. In order to control the growth of filamentous algae, a small number of tilapia (at the rate of one per 400 ℓ water) is held in the tank to graze on them. Tilapia also help to fertilize the water. Copper sulphate is applied at the rate of 0.6 ppm once a week to control the growth of rotifers. Green water develops at a salinity below 12 ppt, and up to three-day-old

green water can be used at the same salinity level for rearing larvae.

Live animal food

Larval and adult stages of many aquaculture species grow well on live animal foods, especially zooplanktonic organisms. The two major sources for such organisms are collections from natural waters or culture under controlled conditions. The production of plankton in ponds and other impoundments by organic and inorganic fertilization will be discussed in a later section.

Besides the small-scale collection of plankton from natural waters using standard or modified plankton nets, large-scale plankton fishing is done in certain areas. Propeller pumps and net cones have been used successfully for fishing in the sea in Norway. Two net cones, one inside the other, are used for filtering. The outside cone has a bigger mesh (about 250 μm).

Fig. 7.3 Indoor tanks used for 'green water' method of algal production in a prawn hatchery in Martinique.

Plankton goes through the inner cone, but is retained by the outer one. A hose from the tip of the outer cone leads the plankton to the larvae-rearing units which, as far as possible, are located near by. Another method is based on the phototactic reaction of many plankton species. They are attracted by underwater lights directly into net cages containing larvae to be fed, or pumped away by an air lift. Fishing for plankton from natural waters may be more economical, but obviously there are many uncertainties in the quality and quantity of plankton that can be obtained. So in any large-scale aquaculture, particularly in hatchery production of larvae, culturing plankton will be more dependable.

Brine shrimp

Among live animal foods used in aquaculture, the brine shrimp *Artemia salina* has received considerable attention in recent years, largely due to its expanding use in crustacean hatcheries. In most cases, *Artemia* is used as freshly hatched or frozen nauplii, starting from dry cysts. Dry cysts are commercially available and sources of supply have increased. Most of the *Artemia* cysts are collected from accumulations along the shores of salt lakes and as a by-product of salt production. Until recently, the commercial exploitation of *Artemia* has been restricted to only a few places. Because of increasing demand and high market prices, *Artemia* has now been transplanted into new environments where they have started yielding appreciable quantities of cysts.

Nutritionally, there may be better live animal food for larvae than *Artemia*, but the ease and speed with which the cysts can be hatched make their use very convenient in hatcheries. Studies have shown that nauplii from cysts from different geographical regions do not give equally good results in larval rearing. Differences between strains, modes of harvesting, handling and hatching may all affect their nutritive values. Cysts that accumulate on the shores of salt lakes may be subjected to repeated hydration and dehydration due to rainfall and other atmospheric conditions before they are collected, which may take up to several weeks. The size of nauplii from such cysts and their energy contents may therefore be

below normal. It has also been observed that transplanted *Artemia* have reduced cyst production. Solar salt works and ponds in wet tropics have been used for full-cycle *Artemia* culture, but the economics of combining it with commercial salt production has not yet been established. It is thought that the evaporation ponds of salt works can be used simultaneously for *Artemia* production.

Because of the variations in hatching rate, length and weight of nauplii and nutrient contents, the cysts have to be selected carefully. The size of nauplii that the larvae can ingest is also an important factor in selecting the strain. It is reported that more than fifty strains have been registered. Liao *et al.* (1983) have compiled data on the performance of cysts from different sources (Table 7.13), which could provide some guidance in the selection of sources of cysts. Hatching efficiency and hatching percentage are probably the most important characteristics, but a shorter hatching rate ensures better cysts and greater reliability of supplies of nauplii on a timely basis. Also, heavier individual weights and a greater hatching output provide higher nutritional values.

Hatching of *Artemia* cysts and production of nauplii for feeding larvae in aquaculture farms is relatively easy. Most types of hatchery tanks can be used for the purpose. Some aquaculturists prefer cylindroconical containers, with water circulation to keep the cysts in suspension; others recommend funnel-shaped containers including heat-sealed plastic bags that are aerated from the bottom. As far as possible, natural sea water with a temperature of 25–30°C and pH of 8–9 should be used. High dissolved oxygen contents up to saturation should be maintained in the tanks. For optimal results, a continuous illumination of about 1000 lux, should be maintained. This can be accomplished in a 75 ℓ tank by suspending two 60 W fluorescent light bulbs at a distance of about 20 cm. A density of less than 10 g cysts per ℓ water is recommended for hatching. The nauplii should be harvested soon after hatching, for feeding larvae of crustaceans or fish, in the instar I stage when they have the highest calorific content. After the second moulting, that generally takes place within twenty-four hours after hatching at 25°C, the weight and calorific value of the individual nauplii decreases by over 20

Table 7.13 Data relating to performance of artemia cysts, (after Liao *et al.*, 1983).

Source	Hatching efficiency (nauplii/g)	Hatching percentage (%)	Hatching rate T_o (h)	T^{90} (h)	Individual dry weight (µg)	Individual energy content (10^{-3} joule)	Hatching naupliar biomass (mg/g cysts)	Output naupliar energy (joule/g cysts)
San Francisco Bay,	267 200	71.4	15.0	20.5	1.63	366	435.5	9 780
California	259 200	–	16.4	23.2	–	–	–	–
	249 600	–	25.8	37.6	–	–	–	–
San Pablo Bay, California	259 200	84.3	13.9	20.1	1.92	429	497.7	11 120
Macau, Brazil	304 000	82.0	15.7	23.7	1.74	392	529.0	11 917
	182 400	–	16.0	29.1	–	–	–	–
	297 600	–	16.4	21.9	–	–	–	–
Philippines	214 000	78.0	14.7	22.0	1.68	382	359.5	8 175
Great Salt Lakes, Utah	106 000	43.9	14.1	21.7	2.42	541	256.6	5 735
Shark Bay, Australia	217 600	87.5	20.3	28.1	2.47	576	537.5	12 534
Chaplin Lake, Canada	65 600	19.5	14.3	33.0	2.04	448	133.8	2 937
Buenos Aires, Argentina	193 600	62.8	16.1	22.6	1.72	379	333.0	7 337
Lavalduc, France	182 400	75.8	19.5	30.5	3.08	670	561.8	12 221
Tien-Tsin, China	129 600	73.5	16.0	37.2	3.09	681	400.5	8 826
Margerita Di Savoia, Italy	137 600	77.2	18.7	25.3	3.33	725	458.2	9 976
Artemia Reference Centre, Belgium	211 000	45.7	18.0	32.2	1.78	403	375.6	8 503

Cysts are from different sources, hatched under standard conditions (35 ppt, 25°C).

per cent. Harvesting of nauplii can be done by siphoning or by a 125 micron screen net, after an interruption of aeration in the tank. Empty shells float at the surface, while the nauplii concentrate in the lower part of the container. The concentration of nauplii can be increased by utilizing their phototactic properties. The upper part of the container may be covered with a black plastic sheet and the lighting directed to reach the lower part of the tank. To improve the flotation of cyst shells, the salinity of the tank water may be increased up to 35 ppt, shortly before harvesting, by addition of saturated brine or crude salt (which does not harm the nauplii).

The harvested nauplii should be washed thoroughly on a 125 µm screen prior to feeding to larvae, in order to prevent contamination of larval tanks with glycerol, hatching metabolites and bacteria.

To improve the hatching rate, decapsulation of the cysts is recommended by Sorgeloos *et al.* (1977). The hard shells of the cysts can be removed by short exposure to a hypochlorite solution. Decapsulation also has the advantage of easy separation of nauplii, disinfection of cysts and a lower threshold for light stimulation at the onset of hatching. The main disadvantage is that extra water circulation with devices such as air/water lifts will be needed in hatching containers, because the cysts have a tendency to settle out of suspension due to the increased buoyancy caused by the loss of chorion.

Decapsulation consists of a series of consecutive treatments which include hydration of the cysts, exposure to hypochlorite solution and washing and deactivation of chlorine residues. Hydration is done in fresh or sea water not exceeding 35 ppt salinity at a temperature of about 25°C. Full hydration is reached in 1−2 hours, after which the cysts are transferred to a hypochlorite solution. Either liquid bleach NaOCl or bleaching powder $Ca(OCl)_2$ can be used. The weight of active product per volume of the solution required for both of these is the same, namely 0.5 g per g cysts and 14 ml solution per g cysts. The pH of the solution is stabilized by the addition of CaO or Na_2CO_3 and then aerated for about 10 minutes. It is then stored overnight for precipitation and cooling. The supernatant can be siphoned off the next day for decapsulation.

The same type of containers with an aeration system, as used for the hatching of cysts, can be used for decapsulation also. The hydrated cysts are transferred to the tank containing the solution and kept in continuous suspension. As the chorions dissolve after exposure to the solution, foam accumulates and a gradual change in the colour of the cysts occurs. When large quantities of cysts (500 g or more) are treated, the temperature of the bath should be monitored, and if it goes above the optimum it should be brought down by the addition of ice. The lethal temperature for the cysts is about 40°C. Complete decapsulation can be achieved in 10–15 minutes, when the cysts should be filtered off and washed well on a 120 μm screen with fresh or sea water, until the smell of chlorine is completely removed.

In order to deactivate toxic residues that may remain adsorbed to the decapsulated cysts, they should then be dipped a couple of times in a bath of chloric or acetic acid. After deactivation in the baths for less than half a minute, the cysts should be washed again with tap or sea water, when they will be ready for incubation. Decapsulated cysts can also be used directly for feeding larvae, but then the volume of the food will be at least 50 per cent less than of hatched nauplii.

Hatched instar stages of nauplii of most strains of *Artemia* can, if necessary, be stored under refrigeration (0–4°C) in aerated containers for up to 48 hours, with minimal energy losses.

Artemia can be cultivated from nauplius to adult stage under controlled conditions using extensive methods in earthen ponds or intensively, in air/water lift-operated raceway-type tanks (Sorgelos *et al.*, 1983). Some farmers in tropical countries like the Philippines and Brazil have raised adult *Artemia* in earthen ponds to feed larvae of fish or shrimps. Lime and crude salt are added to sea water to provide an acceptable medium for their culture. Weekly additions of organic and inorganic fertilizers enhance phyloplankton growth to feed the *Artemia*. Production of more than 10 g brine shrimp per square metre per day has been reported.

In the intensive system, raceway-type rectangular or elongated oval tanks with central partitioning and air/water lifts can be used. The height/width ratio of the tank should be close to 1, and the water depth is not allowed to exceed 1 m to enable optimal water circulation. The central partitioning along the length of the tank leaves enough space at both ends to ensure easy water circulation. Heaters and thermostats can be directly immersed in the culture tank in order to maintain the optimal temperature of 25–30°C. Water evaporation can be reduced by using an insulated cover and the darkness thus created is also conducive to faster growth of *Artemia*.

Natural sea water enriched with bicarbonate (2 g $NaHCO_3$/ℓ) or artificial sea water is used as the culture medium. The artificial sea water is prepared according to the following recipe:

Evaporated sea salt	31.08 g/ℓ	(tap water)
$MgSO_4$	7.74 g/ℓ	(dissolved in hot water)
$MgCl_2$	6.09 g/ℓ	
$CaCl_2$	1.53 g/ℓ	
KCl	0.97 g/ℓ	(dissolved in hot water)
$NaHCO_3$	2.00 g/ℓ	(dissolved in hot water)

The cultures are inoculated with freshly hatched nauplii at a rate of about 10 000/ℓ. Several types of small size feed items that do not dissolve in the culture medium can be used, such as micronized rice bran, spray-dried *Spirulina*, dried algae, yeast, etc. Since *Artemia* filter feeds continuously, it is necessary to maintain constant feed densities throughout the day for best results. An automatic food dispensing device would be most useful. An alternative is to hold aerated suspensions of food in special containers, from which the food is introduced at preset time intervals by an electric clock-activated air pump, which triggers food distribution for preset time periods.

Solid wastes from the tank have to be removed from about the fourth day of culture, at least every other day. Dissolved oxygen and the pH are monitored regularly. If oxygen levels drop below 2 mg/ℓ, aeration rates are increased; if the pH drops below 7.5, more $NaHCO_3$ is added.

After two weeks of culturing, the larvae will reach an average length of 8 mm, yielding a wet weight of around 5 kg *Artemia* per cubic metre of culture media. Harvesting of preadults is facilitated by turning off the aeration. When the dissolved oxygen level in the medium drops to critical levels after about 30 minutes, the

Artemia concentrate at the surface, from where they can easily be scooped out.

Much higher quantities of *Artemia* can be produced in flow-through systems with continuous renewal of culture water and removal of all waste products. This is possible in areas where sufficiently warm sea water or brine can be had inexpensively.

Live *Artemia* are also a better food for postlarvae of crustaceans and finfish than most artificial feeds. Adult *Artemia* are omnivorous and can feed on protozoa, micro-algae, yeast or bacteria, as well as a variety of artificial feeds. Unused feed materials, when disintegrated, become fertilizers and are recirculated for the production of their natural food items in cultures. Thus the cost of culturing can be greatly reduced.

Preadult *Artemia* are known to be more nutritive than freshly hatched nauplii. During growth, their protein contents increase from about 42 to 60 per cent and the fat contents decrease from about 20 to less than 10 per cent of its dry weight. Nauplii are deficient in histidine, methionine, phenylalanine and threonine, but the adults are rich in all amino acids (Tobias *et al.*, 1980). Being an extremely hardy species that can tolerate salinities ranging from 5 to 150 ppt, water temperatures from 6 to 35°C and dissolved oxygen even less than 1 ppm, it is becoming a favoured live food in some countries.

Rotifers

Some of the species belonging to the class Rotifera are considered to have greater nutritive value than *Artemia* for larvae of marine species of fish and crustacea. In fact, in countries like Japan, the success of mass production of marine fish larvae is largely dependent on the availability of rotifers. *Brachionus plicatilis* is one of the commonly used species. Its average size is 250−260 μm, although there are strains which differ markedly from this in size, probably due to genetic reasons. It reproduces asexually under favourable conditions by laying one or two large eggs (80 to 100 μm × 110 to 130 μm) which hatch into amictic females. The factors that induce such parthenogenetic reproduction appear to be high population densities, supplies of the right type of food, water temperature, stable salinity, light penetration, pH values and lack of contamination by other species, besides genetic characteristics. In the sexual reproduction phase, mictic females bearing one to six small eggs which hatch into males are formed from the large eggs of an amictic female. Then the mictic female is fertilized by a male to produce one or two dormant eggs. A dormant egg has to undergo a period of dormancy before it can be hatched into an amictic female. From the amictic female asexual reproduction continues.

In mass culture it is necessary to keep the species in the asexual reproduction phase, by excluding factors that induce sexual reproduction. Sexual reproduction is the source for resting eggs, which can be stored at low temperatures for a longer period of time.

There are different strains of *B. plicatilis* characterized by specific ranges of size. The growth rate depends on, among other factors, the strain characteristics, as for example the large size strain (300−350 μm) grows more slowly (about 0.40 μm per day) whereas the smaller sized strain (150−200 μm) grows faster (0.69 μm per day). Different types of feeds are used for growing *Brachionus*, including micro-algae, bacteria, yeast and small organic particles. Marine *Chlorella* and bread yeast are considered to be the best foods.

Marine *Chlorella* grow at salinities between 0 and 35 ppt, with an optimum range of 10−20 ppt. The optimum temperature for growth is 25−35°C. Under high temperatures, there is a likelihood of contamination by other organisms such as Protozoa. *Chlorella* stock cultures are maintained in small flasks (250 ml) containing a medium such as Walne's medium (Table 7.14) at 20°C, 2000 lux light intensity and 12 h dark/ 12 h light photoperiod. Subcultures are made every fortnight to maintain algal vigour.

To raise mass cultures of *Chlorella*, 1 ℓ carboys containing the same medium as the stock culture can be used. About 100 ml stock culture is inoculated and kept with vigorous aeration at 23−30°C and 6000−9000 lux light intensity. The same photoperiod as for the stock culture is maintained. In four to five days, when maximum growth has been achieved, this 1 ℓ culture is inoculated into a large glass container with 10 ℓ boiled sea water enriched with agri-

Table 7.14 Formula of Walne medium for algae culture.

Additive	Concentration
Stock A	
$FeCl_3.6H_2O$	1.30 g/ℓ
$MnCl_2.4H_2O$	0.36 g/ℓ
H_3BO_3	33.60 g/ℓ
EDTA (Na salt)	45.00 g/ℓ
$NaH_2 PO_4.2H_2O$	20.00 g/ℓ
$NaNO_3$	100.00 g/ℓ
Trace metal solution	1.0 ml/ℓ

Make up to 1 ℓ with distilled water. Add 1–2 ml Stock A per ℓ sea water, depending on the species cultured.

Stock B		
Vitamin B_{12} (cyanocobalamin)	10	mg/100 ml
Vitamin B_1 (thiamin)	200	mg/100 ml

Make up to 100 ml with distilled water. This solution should be acidified to pH 4.5 before autoclaving. Add 0.1 ml Stock B per ℓ sea water.

Stock C	
$Na_2SiO_3.5H_2O$	4.0 g/100 ml

Make up to 100 ml with distilled water. Add 2 ml Stock C per ℓ sea water for diatom culture only.

Trace metal solution	
$ZnCl_2$	2.1 g/100 ml
$CoCl_2.6H_2O$	2.0 g/100 ml
$(NH_4)_6Mo_7O_{24}.4H_2O$	0.9 g/100 ml
$CuSO_4.5H_2O$	2.0 g/100 ml

Make up to 100 ml with distilled water. Acidify with sufficient concentrated HCl to obtain a clear solution.

cultural fertilizer (150 m/ℓ ammonium sulphate, 7.5 mg/ℓ urea and 25 mg/ℓ calcium superphosphate). Media containing 300 mg/ℓ ammonium sulphate and 50 mg/ℓ calcium superphosphate have also been used successfully. The culture is exposed to light and aeration with an air stone. After four to five days, the culture is sealed up in a 500 ℓ plastic outdoor tank containing 200 ℓ filtered sea water enriched with the same fertilizer. After four to five days, a further 250 ℓ filtered sea water are added to the tank and fertilized with the same

fertilizer. The scaling up can be continued with further inoculations.

Though the basic methodology for the culture of *B. plicatilis* is generally the same as for most live aquatic animals, there are variations in the details of procedures. In one system, tanks of 0.5–3 ton capacities are used. One tank is inoculated with *Chlorella* and after the *Chlorella* densities reach 1×10^7 cells per litre, *Brachionus* is inoculated at a density of 10–20 individuals per ml. When the *Chlorella* is used up and the water becomes clear, bread yeast is added twice daily at a rate of 1 g per 10^6 rotifers. After about five to seven days, when the density of *Brachionus* would have exceeded about 100 per ml, most of the culture can be harvested. The remaining culture is removed to another tank and a new batch of *Brachionus* raised. This procedure is repeated using a series of tanks. Although this method is somewhat labour-intensive, it is simple and reliable.

In another system, a number of large 200 ton *Chlorella* tanks, and several smaller 40 ton rotifer culture tanks are used. *Chlorella* is initially grown in the rotifer tanks and when the algal density reaches $1–2 \times 10^7$ cells per ml, the rotifer is inoculated at a density of 10–12 individuals per ml. The same procedure is followed as for the previous system, and when the density reaches above 100 individuals per ml, part of the culture (generally between one-fifth and one-third) is harvested. The same quantity of *Chlorella* as before is then added. The remaining rotifers in the tank multiply and large quantities can be harvested daily. The culture can be maintained at peak productivity only up to a maximum of about 30 days, because of changes in water quality. After this period, the cultures have to be completely harvested and the tanks cleaned to start new cultures.

A third system of rotifer culture is based on the use of bread yeast as the sole source of food. A large concrete tank containing 2 tons of sea water is supplied with a 200 g yeast suspension and inoculated with rotifers, at a rate of 10–20 individuals per ml. Thereafter, twice a day a fresh-water suspension of bread yeast is provided at the rate of 1 g yeast per 10^6 rotifers. After a week to ten days, the rotifers can be completely or partially harvested, when the population density would have exceeded about 100 per ml.

Rotifers are harvested using a 75 μm plankton net and rinsed with sea water before they are used for larval feeding. Survival of the rotifers in larval tanks depends on the salinity: if the salinity is the same as in the culture tank, they can survive for about a day, but a variation of plus or minus 15 ppt could reduce survival to 50 per cent.

It is generally believed that, in due course, *Brachionus* along with an artificial diet would be able to relieve the present dependency of many hatcheries on *Artemia*. *Brachionus* can be frozen and stored for feeding crustacean larvae, even though fish larvae appear to accept only live ones. Besides *B. plicatilis*, there are other species of rotifers, some of them occurring in fresh waters, like *Brachionus rubens*, which are suited for mass culture and can be kept amictic under culture conditions and fed with *Scenedesmus* grown in fertilized outdoor ponds. The population growth is dependent on food concentrations (ranging from 250 to 450 mg/ℓ). A culture with 500 rotifers per ml has been obtained when 25 per cent of the culture volume was replaced by fresh food suspension every 12 hours.

Copepods and cladocerans

Even though *Brachionus* is a very valuable live food, some of the early larval stages of fish may be unable to ingest it because of its larger size, and would therefore prefer smaller copepod larvae instead. In the natural environment, the vast majority of fish larvae show preference for copepods in the larval and adult stages. A growing population of a copepod species may often be able to produce all the sizes of food needed by a fish larva from the first feeding to metamorphosis. Running water systems will have to be used to avoid deterioration of the medium in copepod cultures. Mesh-bottomed floating trays have been used as culture units for *Tisbe* species. When suspended in larval tanks nauplii just fall from the culture basket. A production of 132 000 nauplii per day from a 200 mℓ basket, which works to approximately 10 nauplii per ml in a 200 ℓ rearing tank has been recorded.

Cladocerans have the advantages of high reproduction rates, wide environmental tolerance and the ability to thrive on phytoplankton and organic wastes for mass rearing. The most common cladocerans cultured as food are *Daphnia* spp., especially fresh water forms. Small tanks, pools or even earthen ponds can be used for raising *Daphnia*. Phytoplankton has to be raised in the medium to feed *Daphnia* by the application of appropriate fertilizers, such as H_3PO_4, followed by regular refertilization with nitrogen. Densities of up to 500 *Daphnia* per ml have been raised by phosphate fertilization. Aeration will be a substantial advantage. As with rotifers, daphnids also consume more food per unit of time than algae. The growth rate and population density can be greatly increased by providing such food. Using rice bran as feed, 160–250 ℓ cultures of *Daphnia magna* have produced population densities of 1000–3000 per ℓ. There are several records of very high production rates of *Daphnia* in well fertilized and managed media.

Besides the microfauna mentioned earlier in this chapter, there are also other species cultured as live animal food, though only on a smaller scale. These include Protozoa and Oligochaete worms such as Tubifex, and Chironomid larvae. Flat trays filled with soil and manure and covered with water can attract Chironomid flies to deposit eggs in them. A production of about 0.3 kg/m^2 per week has been reported.

7.4.2 Pond fertilization for production of live foods

The nutrient value of live foods for aquaculture species and the need for essential nutrients in the media for the growth of food organisms have been referred to earlier in this chapter. The traditional practice of fertilizing fish ponds is based on this knowledge and the relative economics of fertilizing compared with the use of processed or unprocessed artificial feeds. Even today, the majority of pond culture of herbivorous and omnivorous species is based on food production through fertilization, sometimes combined with supplementary feeding with easily available feedstuffs. In many tropical developing countries, where the priority is to produce less expensive species to feed low-income populations, use of artificial feeds may not be feasible as it is likely to raise the cost of production, making the product beyond the reach of the majority of the population. Further, in most of these countries, manufacture of aquaculture feeds is not well developed and import of large quantities of feed is not practical. Furthermore, the design and operation of pond farms in tropical Asia are also best suited for the growth of natural food organisms. Viewed from the point of view of input/output ratios, such types of farming have to be considered as intensive, even when no supplementary feedstuffs are used.

The production and maintenance of a crop of live food organisms in an aquaculture pond or similar enclosure is rather complex, even though fish farmers in Asia have been practising it for centuries and have made it into an art. There are many factors that affect the growth of live food which can be controlled only to a limited extent in large ponds and enclosures. The interaction between the underlying soil and the pond water is one of the factors that distinguish algae and zooplankton culture methods described earlier in the chapter, from pond raising of live food. Fertilizers introduced in a pond ecosystem are intended to support and modify the food chain or food web, where each link depends for its food supply on the lower trophic levels. The first links are generally bacteria and algae. Phytoplankton is the autotrophic link which produces organic matter, whereas the other organisms are heterotrophs that consume organic matter. Besides the fertilizer that is added by the farmer, the pond usually also receives a certain amount of additional fertilization through water introduced from outside sources, for replenishment or maintenance of water circulation in the ponds and through rainfall. A third important factor is that the aquaculture species is raised in the same media where its food organisms are grown, and so continuous grazing by a progressively increasing biomass of the culture species takes place. Besides these, the seasonal variations of temperature, photoperiods, pH and density-related interactions influence the growth of live food in ponds. Consequently, the application of fertilizers has to be adjusted according to needs determined on the basis of regular monitoring. Often rough and ready methods are followed by farmers, which sometimes result in the development of unfavourable conditions.

The fertilizers used in aquaculture are inorganic or organic in nature, or a combination of both. Probably because of the large bulk to be handled when using organic manures and the variability of its chemical composition, industrially advanced countries have not adopted its use very widely, except in a few instances. On the other hand, there is greater reliance on organics in developing tropical countries.

Inorganic fertilizers

Though considerable experience in the application of inorganic fertilizers has been accumulated by studies made in Eastern and Western Europe and North America, no standardized fertilizer has been evolved similar to the formulae for culture media in controlled live food production discussed in Section 7.4.1. This is only to be expected, in view of the differences in climatic, soil and hydrological conditions that affect fertilizer requirements.

Inorganic fertilizers are in simple inorganic compound form containing at least one of the following primary nutrients: nitrogen, phosphorus and potassium (NPK). They may also include nutrients like calcium, magnesium and sulphur and trace elements like copper, zinc, boron, manganese, iron and molybdenum. Generally, aquaculturists use commercially available agricultural fertilizers. The composition of some of the commonly used inorganic fertilizers is given in Table 7.15 (Boyd, 1979).

While NPK fertilizers seem to be favoured in North America, West European aquaculturists lay greater stress on phosphate fertilizers and East Europeans on both nitrogen and phosphorus fertilizers. The main problem with phosphate fertilizers is that phosphorus compounds are not easily soluble in water and are absorbed by the bottom soil or mud and often converted to insoluble compounds. These are released only when microorganisms change them into assimilable forms, so even though present in the pond, phosphorus is not always available for algal growth. Daily application would be a solution, but the quantities required being small, there are practical difficulties in distributing them evenly in ponds. Mixing the daily requirement of phosphates with organic ma-

Table 7.15 Composition of some of the commonly used inorganic fertilizers (after Boyd, 1979).

Material	N (%)	P_2O_5 (%)	K_2O (%)
Ammonium nitrate	33–35	–	–
Ammonium sulphate	20–21	–	–
Calcium metaphosphate	–	62–64	–
Calcium nitrate	15.5	–	–
Ammonium phosphate	11–16	20–48	–
Muriate of potash	–	–	50–62
Potassium nitrate	13	–	44
Potassium sulphate	–	–	50
Sodium nitrate	16	–	–
Superphosphate (ordinary)	–	18–20	–
Superphosphate (double or triple)	–	32–54	–

nure such as pig manure for application has been found to be beneficial, not only for better distribution, but also because inorganic phosphates are gradually converted into organic phosphate compounds in manures. For this purpose, phosphate fertilizer should be kept mixed with manure for a few days before application. A dose of 0.2–0.5 mg per litre of dissolved P_2O_5 is considered adequate to maintain a high rate of production of plankton algae. Fish ponds in the USSR are generally fertilized with 15–20 kg P_2O_5 per ha. Phosphate fertilizers can have the effect of increasing the nitrogen content of phytoplankton through the fixation of water-soluble nitrogen by nitrogen-fixing bacteria and blue-green algae (Martyshev, 1983).

Nitrogen fertilizers are easily soluble in water and can therefore become readily available for organic production in ponds. A commonly used nitrogen fertilizer is ammonium sulphate, containing 20–21 per cent nitrogen. Urea, which is an organic compound that has to decompose into an inorganic form for absorption by algae, is also used in many aquaculture farms. Liquid ammonia, which is a solution of NH_3 in water containing about 20 per cent N by volume, is a cheaper source of nitrogen fertilization in some areas. However, special precautions have to be taken in handling, because of its strong odour and possible danger to human health due to inhalation or bodily contact. The usefulness of nitrogen fertilizers has been demonstrated in a number of places. Contrary results reported

from Europe have been ascribed to the climatic differences and production pattern in West European fish culture. A recommended dose of application is 30–40 kg ammonium sulphate bi-weekly or 15–20 kg weekly, together with phosphorus fertilizer. Application of nitrogen fertilizers alone is reported to result in the suppression of nitrogen-fixing bacteria under certain conditions. But if there is a vigorous development of phytoplankton, the compounds are reduced at a rapid rate and will not have any suppressing effect on nitrogen-fixing bacteria such as *Azobacter*.

As in the case of nitrogen, there are differences of opinion on the value of potassium fertilizers in ponds. Certain types of soils contain large quantities of potassium, but others such as peaty soils contain little. A large content of potassium in the soil does not ensure its availability in the pond, since most of it is in the form of stable alumosilicate minerals. Exchangeable potassium and water-soluble potassium are present only in small quantities. The rate of potassium fertilizer used varies considerably from 30 to 100 kg/ha, depending on the soil and water conditions.

Calcium fertilizers have a definite role in pond fertilization, but how much of it is a direct contribution to fertility and how much is ameliorative is difficult to assess. Calcium is an essential element of aquatic flora and fauna. It causes precipitation of colloidal humus, reducing its absorption capacity and thereby releasing previously absorbed nutrients into the water. Calcium is added to ponds rich in organic matter and to ponds with acidic soil and water.

Calcium carbonate ($CaCO_3$) and unslaked lime (CaO) are commonly used in ponds. The need for liming to correct acidity in brackish-water swamp soils has been described in Chapter 4. According to Martyshev (1983) a dose of 30–5000 kg lime per ha is applied as fertilizer in the USSR, depending on the conditions in the pond. He quotes norms in other European countries as 1000–2000 kg/ha in Germany, 200 kg/ha in France, 600–700 kg/ha in The Netherlands and 500 kg/ha in Yugoslavia. For a good growth of plankton, the total hardness and total alkalinity of the pond water should not be less than 10 mg/ℓ. Waters with values above 20 mg/ℓ are reported to produce consistently adequate quantities of phytoplankton after inorganic fertilization. Liming is therefore required when the total alkalinity or hardness is below 20 mg/ℓ. Dark-coloured water containing large quantities of humic substances will also require liming to clear the water and improve light penetration for photosynthesis.

For correction of water quality, much higher doses of lime than mentioned above will be required. Higher acidity requires greater quantities of lime to neutralize, depending not only on pH, but also on the chemical composition of the water, especially the concentration of calcium bicarbonate [$Ca(HCO_3)_2$] and its relation with carbon dioxide and carbonates. In Asian brackish-water ponds, with soils of pH value around 5, treatment with 3 tons of agricultural lime (calcium hydrozide $Ca(OH)_2$) per ha are recommended. The lime has to be worked into the pond soil after dewatering. Though agricultural lime will raise soil pH, its effect may not be adequate to maintain the pH of brackish-water ponds because of its low solubility in salt water. Natural carbonates that contain a minimum of about 4 per cent magnesium (such as dolomite, mollusc shells or coral) are more soluble at the pH of sea water and will aid in maintaining optimum alkalinity and pH levels of the pond water. It is therefore desirable for salt-water ponds to have a supply of such types of lime to maintain water quality.

It has been demonstrated that a combination of inorganic fertilizers gives the best results in pond fertilization. Different combinations of the major fertilizers are in common use in many areas. Combinations of the primary fertilizers should be based on the specific requirements of the pond. A dose of 50–60 kg/ha of NPK (16:20:4) fertilizer combination has given satisfactory results in many situations, particularly in fresh waters. When the pond soil has high potassium contents, the potassium can be omitted from the fertilizer.

Studies made in the USSR seem to show that the addition of certain trace elements promotes plankton production. The application of a fertilizer consisting of 10 kg/ha cobalt, 1200 kg/ha ammonium nitrate and 200 kg/ha superphosphate results in a marked increase in the biomass of plankton and benthos (Martyshev, 1983). It has also been shown that cobalt is absorbed and retained by pond silt for a long period of time and utilized in the biological cycle. Among the other trace elements studied, boron and molybdenum have proved to be beneficial. The addition of these at the rate of $0.07 \, g/m^3$ and $0.01 \, g/m^3$ respectively is reported to result in an increase of zooplankton biomass of about 1.5 to 2.0 times. However, the accumulation of these trace elements in the species of fish or shellfish cultured has not been studied, and so their use in commercial aquaculture is not widely accepted.

The common practice is to apply about 50 per cent of the total fertilizer requirements initially in preparation for the release of the stock, so that a standing crop of the desired food organisms will develop. Where liming is required to improve the soil or water conditions, this has to be done two to three weeks before the application of nitrogen and phosphorus fertilizers. As chemical fertilizers are rapidly utilized and large doses inhibit bacterial growth, it is preferable to apply them after the ponds are filled with water. Fertilizers applied on the pond bottom may also be adsorbed by the bottom soil and be used by rooted plants rather than plankton. In temperate climates, the initial application is usually in spring, when temperature conditions are optimal for rapid growth of microorganisms and when major rearing activities begin. In tropical climates, where growth occurs throughout the year, the main consideration is the commencement of a new crop of the aquaculture species. After the required bloom of micro-organisms has developed in the ponds and the stock of young ones introduced for rearing, further fertilization is intended only to maintain the density of the required food organisms.

The method of application of the fertilizer is important. Distribution should be as even as possible, to ensure full utilization and prevent loss by precipitation or release into the atmosphere that may happen when it is applied in a limited area. One method of application recommended is dissolving the fertilizer in a suitable container and distributing the solution over the pond surface from a boat. Another method is to apply the fertilizer in a dry powdered form, with a suitable blower. Hepher and Pruginin (1981) described a method of using currents caused by winds in ponds to distribute fertilizers. The fertilizer is deposited on the windward side of the pond at a spot 2−3 m away from the bank, preferably when the wind is blowing. The dissolving fertilizer is carried away by the current generated by the wind and distributed throughout the pond.

In Southeast Asian brackish-water ponds, fertilization is performed to develop the algal complexes known as 'lab lab' and 'lumut'. The 'lab lab' complex predominantly consists of benthic blue-green algae (Myxophyceae) and diatoms (Bacillariophyceae), whereas the 'lumut' complex is composed primarily of filamentous green algae and associated forms of life. Though organic manures are preferred for growing such algal complexes, inorganic fertilizers can also be used either independently or mixed with organics. The dosage recommended is 50−100 kg of 18:46:0 (NPK) or 100−150 kg of 16:20:0 (NPK) fertilizer, depending on soil conditions. These are applied on the dry pond bottom and 3−5 cm of water let into the pond soon after treatment. After one week, the same amount of fertilizer is applied again and the water level is raised to 10−15 cm. Fertilization is repeated after two weeks and the water level raised to 20−25 cm. The level of water is topped up to make up for the loss by evaporation. The best growth of 'lab lab' occurs in water salinities around 25 ppt or higher. The maximum water level is about 40 cm.

'Lumut' grows best in low to medium salinity ranges, above 25 ppt and at water depths of 40−60 cm. Soft mud bottoms with a pH of 6.8−7.5 are considered most favourable for its growth. In cases where the pH is below 6.5, liming should be done in such a way as to incorporate it in the soil. The pond bottom is dried for about three days, after which sufficient water is let in to wet the soil. The pond bottom is then seeded with the desired species of green algae (*Oscillatoria, Lyngbia, Phormidium, Spirulina, Microcoleus*, etc.) by sticking young filaments into the mud. Then the pond is filled to a depth of 20 cm. Three to seven days after planting the pond is fertilized with 16:20:0 (NPK) fertilizer at the rate of 18−20 g/m^3 water. The fertilizer can be broadcast over the pond or dissolved into the water from a submerged platform (about 10 cm below the surface). After a week, the water level is raised to 40 cm. Starting with the second week, a weekly application of the fertilizer at the rate of 9−10 g/m^3 water is recommended for the duration of the culture.

The above fertilizer treatment and water management not only result in the production of green algae, but a series of associated organisms including bacteria, protozoans, diatoms, nematodes, small crustaceans, etc., which contribute to the live food resources of the pond.

Organic fertilizers

The use of organic fertilizers or manure in aquaculture is an ancient practice and, despite its drawbacks, continues to be used by aquaculturists as an efficient and economical means of increased production in aquaculture ponds. In fact, the limited research that has been carried out on the subject has served to highlight its advantages in recent years. Some of the relative advantages of organic fertilizers in aquaculture are similar to those demonstrated in terrestrial farming. They improve the soil structure and fertility, particularly in newly constructed ponds deficient in deposits of silt and other organic substances. They promote the growth of zooplanktonic organisms, which form the nutritious and preferred food of many aquaculture species. Manures often facilitate the utilization of chemical fertilizers, when appropriate fertilizer practices are adopted. Proper manuring generally has a longer lasting effect on pond production. In rural areas of most Third World countries, organic fertilizers are more easily available and relatively less expensive to use. When pond culture is integrated with crop and animal production, as will be described in Chapter 11, efficient recycling of farm wastes and overall production economies become possible.

One of the major problems in the use of organic fertilizers is the extreme variability of composition. In the case of animal manures, the quality depends not only on the animal species, but also on the nature of its food, handling, storage, climatic conditions, etc. Another problem is the effect of over-fertilizing with organics. Rapid decomposition of the manure, and the increased bacterial population associated with it, can result in oxygen depletion and fish mortality. Over-manuring is also believed to promote the incidence of diseases.

The most commonly used organic fertilizers are animal manure, green manure, composts and domestic sewage, besides organo-chemical manures. The composition of fresh manure from a number of animal species is quoted by Martyshev (1983) from the *Agronomist Handbook on Manures* (in Russian), apparently based on conditions in the USSR (Table 7.16). The nutrient contents of different animal manures used in China are shown in Table 7.17 (FAO, 1977). The most common manure in use in China is undoubtedly from pigs and the composition of pig manure is given in Table 7.18 (FAO, 1977). In South Asia and Israel, cow dung is commonly used for pond fertilization. Its composition usually ranges from 78–79 per cent water, 0.5–0.7 per cent N, 0.1–2 per cent P and 0.5 per cent K with an organic matter of 17 per cent and C:N:P ratio of 17:1:0.2 (see also Table 7.16).

Animal manures applied in a pond in fine particulate or colloidal state stimulate heterotrophic growth of bacteria by providing the necessary surface area for their attachment and facilitating their mineralization. The mineral fraction is also directly available for photosynthesis, but apparently because of limited light penetration in manured ponds, phytoplankton production is reduced. However, the production of zooplanktonic organisms is generally more rapid and they feed on the nannoplankton and bacteria produced in the pond. Heterotrophic production is at a maximum level at the soil/water interface in fish ponds and this is conducive to abundant benthic growth. In brackish-water ponds, growth of benthos is of special significance as many of the species grown in such ponds are primarily benthic feeders.

Various methods of application of animal manures are adopted. The traditional way of scattering or dumping them in heaps in ponds is not very commonly practised now. Manure may be applied on the dried pond bottom, particularly in new ponds that have little silt

Table 7.16 *Composition of fresh manure* from various animal species.

Components	Mixed dung	Horse dung	Cattle dung	Sheep dung	Pig dung
Water	75.0	71.3	77.3	64.6	72.4
Organic matter	21.0	25.4	20.3	31.8	25.0
Total nitrogen (N)	0.50	0.58	0.45	0.83	0.45
Proteinic nitrogen	0.31	0.35	0.28	—	—
Ammoniacal nitrogen	0.15	0.19	0.14	—	0.20
Phosphorus (P_2O_5)	0.25	0.28	0.23	0.23	0.19
Potassium (K_2O)	0.60	0.63	0.50	0.67	0.60
Calcium (CaO)	0.35	0.21	0.40	0.33	0.18
Magnesium (MgO)	0.15	0.14	0.11	0.18	0.09
Sulphuric acid (SO_3^{2-})	0.10	0.07	0.06	0.15	0.08
Chlorine (Cl^-)	—	0.04	0.10	0.17	0.17
Silicic acid	—	1.77	0.85	1.47	1.08
Iron and aluminum sesquioxides (R_2O_3)	—	0.11	0.05	0.24	0.07

Table 7.17 Nutrient content of animal manures (from FAO, 1977).

Source	N (%)	P_2O_5 (%)	P (%)	K_2O (%)	K (%)
Buffalo	0.30	0.25	0.11	0.10	0.08
Sheep	0.70	0.60	0.26	0.30	0.25
Poultry	1.63	1.54	0.68	0.85	0.71
Rabbit	1.72	2.96	1.30	–	–

Table 7.18 Nutrient content of pig manure (from FAO, 1977).

Source	Organic matter (%)	N (%)	P_2O_5 (%)	P (%)	K_2O (%)	K (%)
Fresh manure	15.00	0.60	0.40	0.18	0.44	0.37
Urine	2.00	0.30	0.12	0.05	1.00	0.83
Air-dried manure	34.32	2.12	0.98	0.43	2.45	2.03
Litter manure	34.00	0.48	0.24	0.11	0.63	0.52

deposition. The widely recommended method is frequent (if possible daily) application of manure in an easily dispersible form, such as liquid cowshed manure or powdered poultry waste. In large ponds it will be possible to disperse manure from mesh baskets towed around from a powered boat. It is also possible to mix the manure with water and disperse it from a small work boat. Diluted manure can be applied uniformly on the pond surface with a pump fitted to the boat. Different doses are suggested, but it is difficult to determine the most suitable one because of the many unknown variables, including the precise composition of the manure, the organic content of the pond and the climatic conditions. Since no standardized treatments are available, dosages have to be worked out in each area, based on properly monitored trials. Woynarovich (1975) suggested the application of about 1 ton of organic manure per hectare per year in freshwater ponds that are deficient in organic matter at the bottom. When distributed at the bottom before inundation, it can be expected to contribute to the development of a rich bottom fauna. In Eastern Europe the best application rate is reported to be 5 tons/ha in stagnant or mildly flowing water. In China, animal manure is often composted with plant materials before application. Where it is used directly, most farmers these days keep it in fermentation tanks for a few days before introduction into fish ponds. The rate of fertilization is about 3.7 tons/ha per year, but in addition they may also apply 60 kg crushed plant material and other agricultural or processing wastes (FAO/UNDP, 1979). In southern parts of China, a high rate of 5.6–10 tons/ha per year, in three applications, was reported by Tapiador *et al.*, (1977).

Composts made from cow dung and green plants are generally used in the USSR for fertilizing nursery ponds. Special waterproof pits are dug, in which layers of green grass are placed, alternated with layers of dung. The compost pile is covered with a layer of unslaked lime (about 70 kg), flooded with liquid dung manure and covered with earth. The usual ratio of green plants, dung and liquid manure is 4:2:1. For every 100 kg plants, 100 litres of water are added so that the decomposition process is accelerated. Such compost is applied at the rate of 7.5 tons/ha. It is thoroughly sieved in an 8 mm mesh sieve to remove all undecomposed materials and then mixed with water and applied from a boat at the rate of 3000–3500 ℓ/ha on the first day, followed by one-quarter to one-fifth of the quantity on the second day. The remaining liquid is diluted and applied twice a day in the morning and evening.

In China, concrete pits are sometimes used for composting (fig. 7.4). The pits are usually circular, measuring 2.5 m in bottom diameter, 1.5 m deep and 3.0 m in top diameter (Delmendo, 1980). Each pit is filled with a layer of a mixture of river silt and rice straw (7.5 tons and 0.15 tons respectively), pig or cow manure (1 ton) and aquatic plants or green manure crops (0.75 tons) in 15 cm layers. The top is covered with mud and a water column 3–4 cm deep is kept at the hollowed surface to create anaerobic conditions and thus minimize losses. The compost is turned over in six to ten weeks, after which it is ready for use. In the first turning over, 20 kg superphosphate are added and thoroughly mixed with the organic material, adding water to ensure moist conditions. The chemical composition of the compost as a percentage of wet weight is 0.30 N, 0.30 P, 0.25 K and the organic matter 7.8 to 10.3. The carbon nitrogen ratio is 15–20:1

Fig. 7.4 Cement concrete pits used for composting silt, rice straw and animal manure in China (From FAO Soils Bulletin, 40).

(FAO, 1977). Compost is applied at the rate of 5–10 tons/ha in three applications (the first being the largest) in six to eight months of the fish rearing period.

In certain parts of South Asia, composting is done in a corner of the pond in bamboo enclosures. Plant matter such as leaves, grass cuttings and aquatic vegetation is composted in layers about 30 cm high with 7.5 cm layers of manure in between and dusted liberally with superphosphate and lime (Hora & Pillay, 1962). For a rapid decomposition of vegetable matter, 25 kg nitrate of soda to about 1000 kg compost have to be applied. To maintain the humidity of the compost heap, water is sprinkled on to it. The compost has to be turned at intervals of about five weeks. When composted in the pond corner, farmers often expect the manure to diffuse into the pond, but this is a long process. It has

to be removed from the pit and spread in the pond. Further, it has the disadvantage that anaerobic decomposition may produce methane which may accumulate in the pond. Generally, a rate of 5 tons of compost per ha of pond area is applied in small lots. But, as indicated earlier, the dosage varies very considerably. A very important pre-condition for the effective use of compost, or for that matter any fertilizer, is the absence of dense growth of macrovegetation. The fertilizer can easily be utilized by such vegetation, stimulating their rapid growth and consequent choking of the pond. As a result the growth of microorganisms will be minimized.

As will be discussed in Chapter 29, in integrated farming of ducks and fish the droppings of ducks fall directly into the ponds from duck houses built on the ponds, or are washed into the ponds from enclosures on the banks of the ponds. Fresh duck manure contains about 57 per cent water and 26 per cent organic matter. On average, 100 kg contain about 10 kg carbon, 1.4 kg P_2O_5, 1 kg N, 0.6 kg potash (K_2O), 1.8 kg calcium and 2.8 kg other materials (Woynarovich, 1980). Chicken manure is also an efficient fertilizer and contains a high percentage of organic matter (26 per cent with a low water content (56 per cent). It is high in nitrogen (1.6 per cent) P_2O_5 (1.5 per cent) K_2O (0.9 per cent) and Ca (2.4%). Fresh as well as dry powdered chicken manure has been used very successfully for live food production in fish ponds.

Human sewage is also used for fertilizing aquaculture ponds in some countries. It has been a traditional practice in China and has been adopted in parts of India, Malaysia, other Asian countries and Europe, experimentally or on a small scale. Even in China and Hong Kong, the system of using raw sewage is gradually disappearing. Night soil is now stored in properly designed closed fermentation chambers for a four-week period to destroy pathogenic organisms, before application in ponds. Another method is to subject sewage to anaerobic digestion in a biogas plant, to kill parasitic micro-organisms. In some biogas plants a mixture of night soil and animal manure is used together with 10−15 per cent grass and crop residues. The mixture may sometimes have 10−30 per cent night soil, but the percentage can also be much less, as low as 10 per cent. The carbon/nitrogen ratio varies between 1:15 and 1:25 and the ratio of solids to liquid between 1:15 and 1:20 (FAO, 1978).

In certain parts of East and North India, the use of treated sewage for fertilizing fish farms is expanding, particularly in areas near urban centres where the disposal of sewage is an ever increasing problem. Treatment generally consists of sedimentation, dilution and storage. Sedimentation may be done in two stages: a primary stage to settle most of the heavier solids and a secondary stage to increase mixing and homogenization as well as improve natural purification processes. It has been estimated that about 33 per cent of the BOD can be reduced by the sedimentation process. Before application, the sewage is diluted with water to maintain a proper dissolved oxygen balance.

During storage, microbic digestion occurs, reducing pathogenic organisms if present in the sewage.

Green manuring is another method of pond fertilization which is adopted in certain areas, particularly where the pond soils are of poor quality. Leguminous plants which fix atmospheric nitrogen are probably the best green manure. Some farmers grow such plants as an alternative crop on dried pond bottoms. After they are fully grown, the plants are cut and ploughed into the soil before the ponds are filled with water. This improves the fertility of the soil and water very considerably. Green manuring is also sometimes done to increase the nitrogen content of the water, using different types of plant matter.

Although the value of live foods in meeting the nutritional requirements of aquaculture organisms is well recognized, the practical problems of maintaining a steady supply of adequate quantities of the required foods at appropriate times are often too difficult to solve. It is not easy to ensure the quality and quantity of natural food to provide the essential components like proteins, carbohydrates, lipids, vitamins, etc., in the required proportions. The quantities of food needed to feed the increasing biomass of the cultured organisms vary so much with climatic and hydrobiological conditions, that it becomes almost impossible to regulate food production to synchronize with requirements through known fertilization or management practices. Because of these, an aquaculturist has to resort to artificial feeds for intensified production. Depending on the nature of the culture operations and local conditions, he has to select 'supplementary feeding' or 'complete feeding' with artificial feeds.

7.5 Artificial feeds

7.5.1 Supplementary feedstuffs

Different types of feedstuffs have been used in traditional aquaculture, ranging from kitchen wastes and foliage in homestead-type fish farming to fishery and agro-industrial by-products like oil cakes, wheat and rice bran, mill wastes, brewery waste, bean residues, silkworm pupae, poultry wastes, slaughterhouse wastes (blood and entrails), trash fish, fish offal, etc. Table

7.19 gives the composition of some of the commonly used supplementary feeds. It is, however, quite obvious that any precise determination of the supplementary feeding rates is difficult, because of the variability of natural food production in the farm and indeed also the variation in nutrient levels in the supplementary feedstuffs.

The feedstuffs are often provided as mixtures in order to improve the quantity and food conversion ratios. Hickling (1962) gives the food conversion (the amount of feed required to produce a unit of weight gain) of a number of such supplementary diets as follows:

fresh sardine, mackerel, scad, dried silkworm pupae	5.5
liver, sardine, silkworm pupae	4.5
silkworm pupae, silkworm faeces, grass, soybean cake, pig manure, night soil	4.1
raw silkworm pupae, pressed barley, *Lemna* and *Gammarus*	2.55
67 per cent groundnut cake, 33 per cent manioc leaves	3.5
50 per cent manioc leaves, 50 per cent ground rice	11.0
manioc leaves and fresh manioc root	26.8
fish flour, rice flour	2.5–3
meat flour, potato	3.5–4
fresh silkworm pupae, wheat flour	10.4
fish flour, soybean cake, yeast	1.7–2.8
fish flour, cotton seed meal, yeast	1.56–3.4

Martyshev (1983) gives several combinations of feedstuffs used in the USSR for feeding carp. In the preparation of feed mixtures, animal products are generally mixed with products of plant origin. In addition to higher aquatic and terrestrial plants, fresh or dried algae (such as *Chlorella*) are added to improve the protein and trace element content, as well as to serve as a binder for the mixture.

Besides the feedstuffs referred to above, processed compound feeds are also used in aquaculture to supplement natural food produced in ponds or other enclosed waters. In such cases, the formulations are generally aimed at providing additional protein and fat. Feeding costs often amount to 40–60 per cent of the overall cost of production in most forms of intensive aquaculture when complete feeds are used. The supplementation of natural food with less expensive supplementary feeds (as feedstuffs or prepared feeds), is meant to reduce this major cost. However, based on studies of common carp feeding, Hepher and Pruginin (1981) pointed out the difficulties in deciding on the composition of supplementary feeds in ponds with different natural productivities and standing crops of fish. It is obvious that, in practice, it will be almost impossible to have separate supplementary diets for every stage of the constantly changing standing crop condition. Approximate composition and quantity of feeds have to be used, based on experience or field trials.

7.5.2 Types of processed feeds

The main purpose of using processed feeds is to ensure that the animals under culture receive a balanced diet that meets their nutritional requirements. The use of processed feeds also reduces uncertainties in the quality and availability of food, characteristic problems with unprocessed feedstuffs. As compound feeds are formulated according to nutrient specifications, the product quality can be kept uniform, even when ingredient substitution has to be made at times of shortages. As mentioned earlier, processed feeds may be formulated to serve as complete feeds to meet all the nutritional needs of the animal or as supplementary feeds to augment the major nutritional elements. Supplementary diets are usually less expensive than the complete diets.

From the point of view of composition, three types of processed feeds can be recognized: purified, semi-purified and practical. Purified diets are made with synthetic amino acids, fatty acids, carbohydrates of precisely known composition and chemically pure vitamins and minerals. Naturally, such diets are comparatively expensive and are used only for research purposes. Semi-purified diets contain natural ingredients in as pure a form as is available, as for example casein, corn oil, fish oil, etc. These are commonly used as test diets in nutritional studies to determine the efficiency of different levels of dietary components in terms of food conversion or growth. Lovell (1980) quotes examples of such test diets (Table 7.20). The

Table 7.19 Average composition (in percentage by weight) of some of the supplementary feedstuffs used in aquaculture.

	Dry matter	Crude protein	Crude fat	Carbohydrate (nitrogen-free extract)	Crude fibre	Ash	True protein
Fresh plant material							
Ipomoea reptans	7.5	2.1	0.2	2.9	0.9	1.4	–
Sweet potato leaves and stem (*Ipomoea batata*) (Congo)	13.0	1.6	0.4	6.8	2.3	1.6	
(*Ipomoea batata*) vine (China)	12.4	2.08	0.67	5.96	2.43	1.26	
Tapioca leaves (Congo) (*Manihot utilissima*)	27.3	8.8	0.9	6.2	9.8	1.7	
Guinea grass	23.0	2.9	0.2	10.3	6.6	3.0	
Cynodon datylon (land grass)	22.4	4.89	0.78	10.40	4.17	2.0	
Maize (Europe)	87.0	9.9	4.4	69.2	2.2	1.3	9.4
Oats (Europe)	87.0	10.4	4.8	58.4	10.3	3.1	9.5
Barley (Europe)	85.0	9.0	1.5	4.5	2.6	8.5	6.8
Oil cakes							
Soybean cake (China)	89.9	40.9	3.51	35.69	4.34	5.46	
Groundnut cake (China)	88.55	39.51	3.56	33.36	3.55	8.57	
Coconut cake	90.0	21.2	7.3	44.2	11.4	5.9	19.7
Palm kernel cake	89.0	13.1	10.0	54.9	7.7	3.3	
Mustard seed cake (China)	89.8	24.64	1.06	41.66	7.10	15.34	
Cotton seed cake (decorticated)	90.0	41.1	3.0	26.4	7.8	6.7	39.6
Cotton seed cake (China)	91.3	36.58	4.99	33.41	8.31	8.01	
Bran							
Rice bran (China)	89.0	13.68	17.9	37.02	6.84	13.56	
Fine rice bran (Malaysia)	89.2	11.4	6.8	45.4	14.1	11.5	
Coarse rice bran (Malaysia)	90.5	6.2	2.7	37.8	33.1	10.7	
Wheat bran (China)	87.2	11.33	2.64	58.25	8.87	5.51	
Wheat bran (Europe)	85.1	15.0	3.2	54.1	7.5	5.3	
Cotton seed bran	92.6	3.38	0.91	46.14	37.01	5.23	
Animal products							
Trash fish	28.0	14.2	1.5	–	–	10.7	
Blood meal	86.0	81.0	0.8	1.5	–	2.7	71.9
Cattle liver	25.0	21.2	0.6	–	–	1.0	
Small clams (flesh)	15.93	13.20	0.77	–	–	1.20	
Small shrimps (dry)	82.80	55.45	5.52	4.37	–	17.65	
Silkworm pupae, fresh	35.4	19.1	12.8	2.3	–	1.2	
Silkworm pupae, dried	90.0	55.9	24.5	6.6	–	1.9	
Silkworm pupae, dried and defatted	91.1	75.4	1.8	8.4	–	5.6	
Miscellaneous							
Soybean curd residue	10.75	2.38	0.41	5.39	2.19	0.38	
Brewers grain (dried)	89.7	18.3	6.4	45.9	15.2	3.9	17.4

Table 7.20 Test diets for fish (from Lovell, 1983).

Ingredient	Channel catfish diet	Test diet H440 for salmonids
Vitamin-free casein	29	38
Gelatin	6	12
Dextrin	30	28
Cellulose flour	20.25	8
Fish oil	3	3
Soybean oil	3	–
Corn oil	–	6
Carboxymethylcellulose	3	–
Mineral mixture*	4	4
Vitamin mixture[†]	1.5	1
Calcium propionate	0.25	–

* A salt mixture supplying all essential minerals for laboratory animals.
[†] Should provide the recommended vitamin allowances presented by NRC (1973 and 1977) for warm-water fishes and for cold-water fishes respectively.

third category of diets, practical diets, comprise formulations of available ingredients and would consist to a large extent of natural products such as fish meals, oil seed meals, cereal grains, etc. They are aimed at fulfilling the nutritional needs of the animal under culture, at the minimum cost.

Another classification of artificial compound feeds into moist and dry feeds, is based on the consistency or water content. Moist or wet feeds are preferred as practical or test diets, because they can be prepared without exposure to heat as in hard pelleting and drying processes that result in nutrient losses. Grinding and mixing machines only may be needed. Many species prefer moist diets and show better feed conversion ratios and growth on such diets. These diets are, however, susceptible to spoilage if not used immediately after preparation. Freezing will be required to prevent spoilage in storage by the action of microorganisms and deterioration of oxygen-sensitive nutrients, such as ascorbic acid. Dry feeds, on the other hand, have the advantage of easy storage and transport. They contain only about 8–11 per cent water and are relatively more water-stable. By the use of appropriate binders and processing techniques, the water stability can be adjusted to suit the feeding habits of the species.

Similarly, the density of the pellets can be adjusted to make them float or sink in water, by using extrusion techniques of processing.

The size and shapes of pelletted dry feeds are easily adjusted by the use of suitable dies. Different sizes of pellets, with different nutrient composition, are needed to feed successive growth stages of a species. Sizes vary from finely ground granules and encapsulated forms for first feeding, through crumbles and small pellets up to large pellets of about 1 cm diameter and 1.5 cm length for adults. The composition is related to the nutrient requirements of each life stage. Starter and fry feeds have a relatively higher protein content and in many cases contain animal proteins. For example, reasonably good trout feeds have a protein content of about 50 per cent, of which 75 per cent is recommended to be of animal origin. Grower feed has a lower protein content of 45 per cent (with 70 per cent animal protein) and finishing feed has 40–45 per cent (with 60 per cent animal protein). Brood fish are fed on special high-protein diets.

Other types of practical feeds that need to be considered are medicated feeds incorporating antibiotics and curative, prophylactic or growth-promoting drugs and pigment-fortified feeds containing pigment compounds such as carotenoids. When the cultured species does not have access to sufficient natural pigment sources in aquatic flora and fauna, it is often necessary to include a suitable source in their diets to obtain the natural coloration to make it acceptable to the consumer. Meals from various crustaceans such as shrimps and lobsters are commonly used as pigment supplements. Astaxanthin or canthanxanthin is often added in salmonid diets to impart an attractive red colour to trout and salmon grown in cages.

7.5.3 Ingredients

Most of the feed ingredients used in animal feed manufacture are believed to be potential ingredients for aquaculture feeds. The International Network of Feed Information has descriptions of over 18000 feed ingredients (Harris, 1980). Göhl (1980) has compiled information on the nutritive value of tropical feedstuffs used for animal feeding which are of importance to developing countries. Only a few of these feedstuffs have so far been evaluated as ingredients for aquaculture feeds and so only a much smaller number are presently used in formulations. These materials include feed grains, oil cakes and meals, animal by-products including fish meal and a number of agricultural waste products. As has been indicated earlier, the quality of anyone of these feed materials varies between and within countries and regions. Therefore it becomes necessary to determine the composition of ingredients from each source for formulation of diets. The available information on the composition of a number of common ingredients has been published by FAO/UNDP (1983). Some of these are utilized as supplementary feeds, without any processing.

Based on experience in feed formulation of salmonids, fish meal has become a major constituent of most aquaculture diets, especially of carnivorous species. Because of the relatively high cost of the product and shortage of supplies, substitutes have been explored. Other animal proteins, such as feather meal and blood meal, have unique patterns in essential amino acid composition. So, they have to be used in combination with other protein sources to balance the amino acid composition. It is reported that fish meal in a standard diet (70 per cent) for rainbow trout could be entirely replaced by a mixture of poultry by-product meal and feather meal on an isonitrogenous basis, if the lacking amino acids were supplemented in isolated form. Tiews *et al.* (1976) found that one-quarter of the fish meal protein could be substituted with methionine or with a mixture of meat and bone meal and blood meal, with the remaining dietary protein coming from poultry by-product and feather meal.

Among the plant protein sources, soybean meal is probably one of the most commonly used ingredients in fish feed. Enrichment with limiting amino acids greatly improves its nutritive value. Up to about one-quarter of the fish meal in trout feeds can be replaced by soybean meal, if it is enriched with methionine.

Another protein source studied as a replacement for fish meal is single cell protein (SCP) such as alkane yeast, methanol yeast, ethanol yeast, etc. These yeasts can be included in rainbow trout and eel feeds, as replacements for fish meal on an isonitrogenous basis, to a level of about 30 per cent, without any significant adverse effect on growth. A combination of these yeasts with fish meal was found to be superior to fish meal alone in carp diets. The importance of micro-algae as a protein source has been referred to in Section 7.4. Although methods of large-scale production are not perfected, these algae can be grown in moderate quantities. When dried, they are non-toxic and can provide all the protein needed for aquaculture species. Lysine supplement may be needed for *Chlorella* and methionine for *Scenedesmus* and *Spirulina*. Some experiments have been carried out on the dietary use of bacterial proteins (for example, *Methanomonas* spp.), which have high protein contents, with encouraging results.

Leaf protein concentrate (LPC) is another source of protein for aquaculture feeds. Juices extracted from the fleshy parts of plants are heated to coagulate the proteins and then dried. Even though the protein content may be lower than some other sources, such as soybean meal, the quality of the protein is generally higher. In certain plants (as for example those belonging to *Leucaena* and *Mimosa*), LPC may contain toxic substances, but it is presumed that LPC from alfalfa and other legumes can be used safely. The LPC from rye grass at levels up to 48 per cent of total dietary protein has been successfully tested on carp and trout (Ogino *et al.*, 1978).

By-catches from fishing industries are a major raw material for fish meal manufacture and have been used directly as feed for cultured fish. Fresh trash fish has a high feed value and is superior to fish meal in protein content and quality. When mixed and properly processed with less expensive ingredients like rice bran, and supplemented with vitamins, trash fish forms an excellent feed in aquaculture. However, special care has to be taken in the dispensation of the feed to avoid deterioration of the water quality. As trash fish is likely to deteriorate and lose nutrients if not used when fresh, it will be necessary to freeze it for later use. Trash fish should be pasteurized before use to prevent the recycling of fish diseases.

The cost of preserving and processing trash fish can be reduced by the use of fish silage, which is now becoming common in some countries such as Norway. Fish silage is made by adding 3−4 per cent of an acid to the fresh fish, preferably minced. Organic or inorganic acids can be used. Formic acid is commonly used, but sulphuric or propionic acid can also be used. The pH of the mixture is brought down below 4 and this inhibits bacterial decay. The enzymes in the minced fish continue to act and reduce the mixture to a slurry. An antioxidant is added to prevent the fats becoming rancid and the liquid can then be stored in tanks for up to six months. When properly prepared, fish silage contains practically all the nutrients of the raw fish. For feeding fish, the silage is mixed in equal proportions with a commercial feed meal containing vitamins and a binding agent, and is passed through a simple perforated extruder to produce moist pellets. If the content of low melting point fats in the mixture is low, fish oil can be added in required quantities before extrusion.

Though considerable importance is given to protein content in selecting ingredients for aquaculture feeds (not only because of its role in fish nutrition, but also because of its high cost), the need for lipids has also to be taken into full consideration. As described in Section 7.3.3, fish and other aquaculture animals have a relatively high requirement for lipids to serve as a source for polyunsaturated fatty acids (PUFA) and as an energy source. Besides improving the overall digestibility of diets, they also help in smoothing the pelleting process and reduce dustiness in milling. Animal and vegetable fats, as well as soap stocks, are readily available to feed manufacturers. Fish oils and most raw vegetable fats have a high degree of unsaturation. The PUFA content of fats from a number of animal and plant sources is given in Table 7.22. According to the lipid content of the other ingredients in the diet, the need for the addition of fats to achieve essential PUFA levels can be determined.

Where supplementation of vitamins is necessary commercially available vitamin premixes may be used, or the necessary premixes may be prepared using suitable diluents or ingredients with high vitamin contents. Algal meal and brewers yeast are excellent sources of vitamins. Commercially available vitamin premixes are usually meant for terrestrial animals. Table 7.23 gives the composition of vitamin premixes made specifically for fish feeds (Hastings, 1979). Table 7.24 lists sources of minerals that could be used as supplements to diets (Harris, 1980) Mineral premixes are also commercially available.

7.5.4 Feed formulation and feed formulae

The primary objective of feed formulation is to provide the species under culture with an acceptable diet that meets its nutritional requirements at different stages of its life, so as to yield optimum production at minimum cost. As has already been pointed out, our present knowledge of the nutritional needs of aquatic animals is restricted to only a few species, and feed formulation for other species has to be approximations on this basis. Although it is most likely that all aquaculture species can be weaned to consume processed compound feeds, there is still a lot to be known about the physical and chemical properties of diets preferred by

Table 7.22 Polyunsaturated fatty acid (PUFA) content of fats from a number of plant and animal sources.

Fat	PUFA (%)
Animal or fish source	
Lard	11.8
Beef tallow	4.2
Atlantic cod	42.8
Atlantic herring	14.6
Rainbow trout	31.0
Common carp	22.5
Shrimp	41.6
Vegetable source	
Rice	50.0
Maize	58.2
Wheat	60.5
Groundnut	31.0
Sesame	40.5
Cottonseed	50.7
Soybean	57.6
Sunflower	63.8
Safflower	73.8
Palm	9.3
Olive	9.0
Coconut	2.0

Table 7.23 Fish feed vitamin premixes for use in (a) confinement (tanks, cages, etc.) and (b) pond culture, ingredients per kg (from Hastings, 1978).

Ingredient	(a)	(b)
B_1 (Thiamin)	0.8 g	2.4 g
B_2 (Riboflavin)	1.5 g	1.5 g
B_3 (Pantothenic acid)	2.5 g	2.5 g
B_5 (Nicotinic acid)	12.0 g	12.0 g
B_6 (Piridoxine)	1.0 g	0.5 g
B_{12} (Cobalamine)	1.0 g	1.0 g
Choline chloride	150.0 g	50.0 g
Folic acid	0.4 g	0.4 g
Inositol	40.0 g	–
Biotin	0.06 g	0.06 g
Para-amino-benzoic acid	0.15 g	–
Ascorbic acid	20.0 g	20.0 g
Vitamin A-acetate	4 000 000 IU	2 000 000 IU
Vitamin D_3	200 000 IU	100 000 IU
Vitamin E	7.0 g	7.0 g
K_3 (Menadion-bisulphate)	2.5 g	2.5 g
Buthyl-hydroxytoluene	5.0 g	5.0 g
Iron (carbonate)	2.5 g	2.5 g
Manganese	3.5 g	3.5 g
Zinc	1.4 g	1.4 g
Copper sulphate	0.1 g	0.1 g
Cobalt	0.4 g	0.4 g
Iodine	0.1 g	0.1 g

several species. Flavour, colour, odour, texture and water stability are important characteristics related to acceptance and consumption. Better knowledge of preferences and feeding behaviour would greatly improve the choice of appropriate ingredients and feed preparation processes, as well as the methods of feed dispensing.

For formulating the basic composition of a feed, the main information needed is the levels of crude protein, energy, specific amino acids, crude fibre and ash required. The energy level may be in terms of metabolizable energy (ME) or digestible energy (DE). Most complete practical diets have to be supplemented with a vitamin premix, at levels in excess of the dietary requirement. All levels, with the exception of energy, are determined on the basis of chemical tests on samples of a feedstuff. Even though they correlate well with biological methods of feed evaluation, such as growth studies, tissue levels, etc., they are subject to errors due to variability in composition. For example, the proximate composition of fish meal made from spawning fish is different from that made from immature fish. Usually, the lipid levels increase before spawning and decrease afterwards, changing the percentage compositions of protein, ash and carbohydrates. Many plant feedstuffs also show compositional variations according to season, locality and environment. However, formulations have often to be based on average values.

Another problem to be considered is seasonality in the availability of ingredients. It may not always be possible to store large quantities of ingredients, and so it becomes necessary to vary ingredient composition according to availability. In many developing countries where feedstuffs are scarce, one cannot expect sufficient priority to be given to their use in aquaculture. Even when a particular feedstuff is available for purchase, increased prices may make it necessary to use substitutes in order to obtain a least-cost ration. It therefore follows that there are no fixed formulae for feeds and

Table 7.24 Sources of supplementary minerals and their potency (from Harris, 1980).

Nutrient	Common source	Composition or potency	Remarks
Calcium*	Feeding bone meal	26% Ca	Also contains 18% protein and 11% phosphorus.
	Feeding bone meal	29% Ca	Also contains 12% protein and 14% phosphorus.
	Bone char	27% Ca	Also contains 13% phosphorus but no protein.
	Tricalcium phosphate	13% Ca	10% P.
	Dicalcium phosphate	24% Ca	20% P.
	Monocalcium phosphate	16% Ca	12% P.
	Ground limestone	24–36% Ca	Balance likely to be carbonate and magnesium.
	Calcium carbonate	40% Ca	
	Oyster and other marine shells	38% Ca	Shells contain on average 96% $CaCO_3$.
Phosphorus	Bone meals and Ca phosphate	(see above)	
	Rock phosphate	14% P	Rock phosphate is 75–80% tricalcium phosphate. Not advised unless guaranteed to contain less than 1% fluorine.
	Defluorinated rock	18% P	Should not contain more than 1 part fluorine to 100 parts phosphorus.
Iodine	Potassium iodide	76% I	Potassium and sodium salts may be used interchangeably.
	Sodium iodide	84% I	
	Potassium iodate	59% I	
	Iodized salt		Stabilized iodine should be used. Amounts of iodine differ but 0.02% and 0.05% are commonly sold.
Iron	Ferric oxide	35% Fe	
	Ferrous sulphate	20% Fe	Originary copperas, commercial grade.
	Reduced iron	80–100% Fe	May be 20% ferric oxide.
Cobalt	Cobalt sulphate	34% CO	May be administered as a drench, as cobaltized salt, or as an ingredient in the ration.

* As sources of calcium these products are useful in direct proportion to the calcium they contain.

that they have to vary according to availability of ingredients, composition and costs. Least-cost feeds have to be formulated, but in doing so careful consideration has to be given to the quality of the nutrient content in substitute ingredients. Changes of ingredients or their proportions may also affect the physical characteristics and palatability of the diets. So feed costs should be estimated not on the basis of the price of ingredients alone, but also on the proportion of feed costs in the overall production cost per unit weight of fish.

The type of feed required and the methods of processing will influence formulation. Extruded or floating-type feeds must contain an appreciable quantity of starch for satisfactory gelatinization and expansion. Most such feeds contain 20–25 per cent cereal grain such as corn, wheat or sorghum. Again, depending on the storage conditions and duration of storage, a certain amount of loss of vitamins has to be expected. In order to cover such loss, it will be necessary to add higher levels of vitamins than are nutritionally required.

The type of culture system also has a determining role in feed formulation. For example, if the feed is for use in semi-intensive pond culture, certain vitamin and mineral supplements can be omitted from the formula, as these nutrients are likely to be available to the animal from the natural food organisms growing in the ponds. On the other hand, in intensive culture in cages, raceways and tanks, where natural foods are limited, the diet should contain all the required nutrients in adequate quantities and proportions. Similarly, when the feed is meant to be used as supplementary feed, to augment the major nutritional elements provided by natural food, the formulations will have to be based on the quantitative assessment of additional requirements of these elements. The problems of formulating supplementary feeds were mentioned in Section 7.5.1.

Since protein is the most expensive portion of an animal diet, it is usually computed first in diet formulation. The first step consists of balancing the crude protein and energy levels. Then the levels of indispensable amino acids should be assessed to ensure that the animal's dietary levels in this respect are met. Except in the case of unconventional protein supplements, if the feedstuff has the required dietary levels of arginine, lysine, methionine and tryptophan it is most likely that the other six indispensable amino acids are above required levels (Hardy, 1980). In cases where the formulation is low in amino acids, necessary alterations have to be made by the addition of ingredients with high levels of the required amino acids. After this is done, a final check will be necessary to ensure that the balance of protein and energy levels is not altered.

The most commonly used methods for balancing crude protein levels are the square method and algebraic equations. For example, to balance a supplementary feed to contain 25 per cent protein, using only two ingredients – fish meal (50 per cent protein) and rice bran (8 per cent protein) – a square is constructed as shown below. The desired protein level of the feed (25 per cent) is inserted in its centre. The two feedstuffs, along with their protein content, are placed on each corner at the left-hand side of the square and the levels of protein of each feedstuff are subtracted from the desired protein level of the feed. The differences are placed

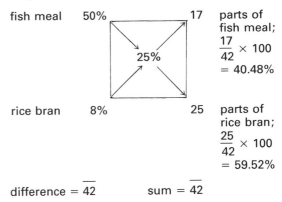

fish meal 50%   17   parts of fish meal; $\dfrac{17}{42} \times 100 = 40.48\%$

25%

rice bran 8%   25   parts of rice bran; $\dfrac{25}{42} \times 100 = 59.52\%$

difference = $\overline{42}$ sum = $\overline{42}$

on the corners of the square diagonally opposite the feedstuff, ignoring plus or minus signs (see diagram). The difference between the percentages of protein in the rice bran and the protein required in the feed under formulation show the proportion of fish meal needed. The difference between the protein percentage of fish meal and of the feed being formulated show the proportion of rice bran required. These proportions can be expressed on a percentage basis, as 40.48 per cent fish meal and 59.52 per cent rice bran, or as a ratio of 17 parts:25 parts.

Algebraic equations can be used to arrive at the same percentages as follows:

assume

$$x = \text{fish meal in kg per 100 kg feed}$$
$$y = \text{rice bran in kg per 100 kg feed}$$
$$x + y = 100 \text{ kg feed} \qquad (7.1)$$
$$0.50\,x + 0.08\,y = 25 \text{ kg protein per } 100 \text{ kg feed} \qquad (7.2)$$

Multiply equation (7.1) by 0.08
$$0.08\,x + 0.08\,y = 8 \qquad (7.3)$$
subtract equation (7.3) from (7.2)
$$0.42\,x = 17$$
$$x = \frac{17}{0.42} = 40.48 \text{ kg or \%}$$

from equation (7.1) one can derive
$$y = 100 - x$$
$$y = 100 - 40.48 = 59.52 \text{ kg or \%}$$

In actual practice, more than two ingredients are generally used in feed formulations. The use of both the square method and algebraic equations is illustrated in the following example to balance a diet containing 30 per cent protein using fish meal (60 per cent protein), soybean meal (51 per cent protein), rice bran (8 per cent protein) and corn meal (10 per cent protein) in the proportions 2 parts soybean meal: 1 part fish meal and 1 part rice bran to 1 part corn meal.

Using the square method as in the above example, the desired protein level (30 per cent) is placed in the middle of the square. The ingredients are separated into two groups and the protein level of each group calculated according to the proportion specified:

protein source: fish meal = 1 × 60 = 60%
Soybean meal = 2 × 51 = 102%
$$\text{average} = \frac{162}{3} = 54\%$$

energy source: rice bran = 1 × 8 = 8%
corn meal = 1 × 10 = 10%
$$\text{average} = \frac{18}{2} = 9\%$$

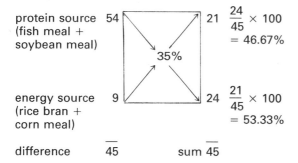

protein source (fish meal + soybean meal) 54   21 $\frac{24}{45} \times 100 = 46.67\%$

35%

energy source (rice bran + corn meal) 9   24 $\frac{21}{45} \times 100 = 53.33\%$

difference   45   sum 45

The protein and energy sources, along with their calculated protein contents are placed on each corner at the left-hand side of the square. After diagonal subtraction, the calculation proceeds as shown below. The final figure for the protein source is divided into $\frac{2}{3}$ soybean meal and $\frac{1}{3}$ fish meal:

protein source = 46.67 per cent
$$\text{fish meal} = 46.67 \times \frac{1}{3} = 15.56\%$$
$$\text{soybean meal} = 46.67 \times \frac{2}{3} = 31.11\%$$

and that for the energy source into $\frac{1}{2}$ rice bran and $\frac{1}{2}$ corn meal:

energy source = 53.33%
$$\text{rice bran} = \frac{53.33}{2} = 26.665\%$$
$$\text{corn meal} = \frac{53.33}{2} = 26.665\%$$

If the algebraic equation method is used, the ingredients are separated into two groups and the protein level of each group calculated according to the proportion required:

assume x = protein source in kg per 100 kg feed
y = energy source in kg per 100 kg feed
$$x + y = 100 \text{ kg feed} \qquad (7.4)$$
$$0.54\,x + y = \text{protein per 100 kg feed} \qquad (7.5)$$

multiply equation (7.4) by 0.09 to give equation (7.6)
subtract equation (7.6) from equation (7.5)
$$0.45\,x = 21$$
$$x = \frac{21}{0.45} = 46.67\%$$
$$\text{fish meal} = 46.67 \times \frac{1}{3} = 15.56\%$$
$$\text{soybean meal} = 46.67 \times \frac{2}{3} = 31.11\%$$

from equation (7.4) one can derive
$$y = 100 - x$$
$$y = 100 - 46.67 = 53.33\%$$
$$\text{rice bran} = \frac{53.33}{2} = 26.665\%$$
$$\text{corn meal} = \frac{53.33}{2} = 26.665\%$$

Very often, more complicated formulations than the ones described above will become necessary. As, for example, the diet may have to be formulated to contain specified amounts of certain feedstuffs and concentrations of various nutrients, as in the following example (Lim, 1982).

An all-plant diet has to be made using soybean meal and peanut meal in the ratio 1:1 as protein source, with the following nutrient specifications:

crude protein	32%
crude fat	12%
sulphur-containing amino acids	1.2%
available phosphorus	0.45%

The other ingredients to be included in the diet are:

cottonseed meal	12%
distiller's dried solubles	7.5%
starch	12%
fish oil	3%
carboxymethyl cellulose	2%
vitamin mix	0.5%

The composition of the available ingredients is listed in Table 7.25. The formulation proceeds as follows. A work sheet, as shown below, is made up and the required feed ingredients and amounts filled in first. The nutrients that will be furnished by these ingredients are calculated as a percentage (or kg per 100 kg feed) according to available data on their nutrient composition. By adding up each item, the total amounts of each nutrient supplied by the

feedstuffs are obtained. By subtracting these amounts from the level of nutrients required in the formulated feed, the additional amounts of nutrients needed and the quantities of other ingredients to provide those nutrients are determined. As soybean meal and peanut meal do not contain any available phosphorus, dicalcium phosphate should be added. As dicalcium phosphate contains 18 per cent available phosphorus, the required quantity per 100 kg is $0.45/0.18 = 2.5$ kg. This leaves an amount of $100-39.5 = 60.5$ kg per 100 kg (or per cent) of other ingredients. If this is to be supplied by soybean meal and peanut meal in equal proportions, the nutrients will be provided by 2.44 kg protein, 1.51 kg fat and 0.77 kg sulphur containing amino acids per 100 kg. Since these do not meet the full requirements for fat and sulphur-containing amino acids, animal fat and methionine have to be added. The same results can be obtained by using algebraic equations.

Linear programming in feed formulation is a mathematical procedure to obtain the optimum solution to specified objectives and involves the use of computers. It is of special importance in least-cost ration formulation, as there are

Ingredient	Amount kg	% or kg/100 kg of feed			
		Protein	Fat	Available phosphorus	Sulphur-containing amino acids
Starch	12.0	—	—	—	—
Distiller dried solubles	7.5	2.00	1.10	—	0.09
Cotton seed meal	12.0	4.32	0.38	—	0.17
Fish oil	3.0	—	3.00	—	—
Carboxy methyl cellulose	2.0	—	—	—	—
Vitamin mix	0.5	—	—	—	—
Total	37.0	6.32	4.48	0	0.26
Specifications for feed	100.00	32.00	12.00	0.45	1.20
Additional nutrients needed	63.0	25.68	7.52	0.45	0.94
Dicalcium phosphate	2.5	—	—	0.45	—
Soybean meal	27.05	12.71	0.54	—	0.39
Peanut meal	27.05	12.71	0.81	—	0.30
Fat	6.15	—	6.15	—	—
Methionine	0.25	0.25	—	—	0.25
Total nutrients supplied	100.00	31.99	11.98	0.45	1.20

Table 7.25 Composition, in percentage or kg per 100 kg, of ingredients available to formulate an all-plant diet (Lim, 1982).

Ingredient	Protein	Fat	Available phosphorous	Sulphur-containing amino acids
Soybean meal	47.0	2.0	–	1.45
Peanut meal	47.0	3.0	–	1.10
Cottonseed meal	36.0	3.2	–	1.45
Distiller dried solubles	26.0	14.6	–	1.20
Starch	–	–	–	–
Wheat short	15.0	7.1	–	0.38
Catfish waste meal	49.0	25.0	2.8	2.30
Fish meal	58.3	10.3	2.8	2.74
Fish oil	–	100	–	–
Animal fat	–	100	–	–
Dicalcium phosphate	–	–	18	–
Methionine	100.0	–	–	100

many ingredients that can provide the necessary protein levels in a ration and there may be a need to change ingredients according to availability, price and quality. While there may be several possible solutions to achieve a given set of specifications, when the cost factor is taken into account there will be only one formulation that costs the least. The elaborate calculations involve the simultaneous solution of a number of linear equations and can best be done with a computer. It is widely practised in livestock and poultry feed manufacture and has been used in the manufacture of fish feeds.

For formulation of least-cost rations, the computer should have the ration specifications together with the nutrient requirements of the animal. The critical nutrients such as methionine and lysine are specified as minimum, range, ratio or exact amounts. Nutrients, such as trace elements and vitamins, which are supplied at constant levels need not be indicated. Relevant specifications of the feedstuffs to be used should be put in, along with the composition and price of various available feedstuffs to be included in the formula. Since the accuracy of the formula will depend on the nutrient composition of the feedstuffs to be used, the values should be as accurate as possible. Prices used should be those at the time and point where the feed is made, in order to obtain a realistic and effective least-cost ration. By expressing these various data inputs as constraints or restrictions, a series of linear equations are formed. Successful solution of the various simultaneous linear equations leads to an optimal solution. When erroneous ingredient data are used or when unreasonable restrictions are placed on nutrient requirements or ingredient usage, the computer may report that a solution is not feasible. This normally happens only when there are errors in data input. It is therefore essential to check the information thoroughly and input only precise data.

7.5.5 Feed preparation

Although the number of feed mills specialized in aquaculture feeds (figs 7.5 and 7.6) and the number of animal feed mills which also produce aquaculture feeds are steadily increasing, there are large areas of the world that are not yet served by commercial feed milling for aquaculture. Either the existing demand does not justify large-scale manufacture or the farmers have not yet recognized the economics of using artificial feeds. Because of these, aquaculturists who wish to use such feeds will have to resort to on-farm production of feeds, as import of manufactured feeds is often difficult or impossible.

It is possible to prepare feeds on a small scale on the farm using inexpensive equipment and locally available ingredients (figs 7.7 and 7.8). For example, Pascual (1982) describes the procedure for small-scale preparation of a feed for the tiger shrimp (*Penaeus monodon*) juveniles. The suggested equipment includes weighing scales, sieves, a mixer of 5 or 10 kg

Fig. 7.5 A fish-feed mill in Szarvas, Hungary.

capacity, a meat grinder, corn meal or coffee grinder, a steamer or a big cauldron and bamboo basket for steaming, a saucepan for gelatinizing starch and a drier. The ingredients in the relevant formula are finely ground and passed through a sieve of nylon mesh (420 μ/cm^2). Weighed ingredients are mixed thoroughly and oil is added and mixed again for a few minutes. The starch, gelatinized with water, is added to the mixture and mixed well to form a dough which is passed through a grinder using a $1-3$ mm die. The extruded feed is cut into small sizes (0.5 cm) and then steamed to make water-stable pellets. The steamed product is dried overnight in an oven at 60°C. Such pellets can be made once a week to meet the needs of the farm.

The feed requirements in large-scale aquaculture may make this type of feed production not feasible and large-scale milling may become necessary. However, the basic processes involved are essentially the same and consist of particle size reduction, premixing of micronutrients, mixing of all components of the diet, pelleting and cooling and then sacking. These processes are illustrated in fig. 7.9. Coarse ingredients are ground in a hammer mill or other type of grinder such as an attrition roller or cutter mill. Besides reducing particle size and facilitating handling of ingredients, grinding improves feed digestibility, acceptability, mixing properties and pelletability.

The ground ingredients may be sieved to separate materials into the required sizes. For example, feeds sifted through 177 µm meshes can be used for making feeds for fry of certain species like the catfish. An excess of dust in the ground feed may be controlled by adding a spray of oil or a semi-moist ingredient, such as condensed fish solubles or fermentation solubles, on feeds entering the grinder.

The next process, namely mixing, is performed to achieve uniformity of composition in the whole feed material according to the required formula. This may involve both scattering of particles and blending. Mixing can be either a batch or continuous process. The best-suited technique for formula feeds appears to be continuous mixing of proportions by weight or volume. There are many types of mixers

Fig. 7.6 Inside view of a fish feed mill in Pirassununga, Brazil.

used in feed mills such as vertical mixers, continuous and non-continuous ribbon mixers and liquid mixers. Accurate mixing requires the addition of ingredients in a tested sequence from batch to batch. Usually, large-volume ingredients are added first and then the smaller amounts. Total mixing time is based on the composition of the formula.

The general process of pelleting involves passing a feed mixture through a conditioning chamber where 4−6 per cent water (usually as steam) may be added. The water provides lubrication for compression and extrusion and

in the presence of heat causes some gelatinization of raw starch present in ingredients of plant origin, resulting in adhesion. Within a few seconds of entering the pellet mill, the feed goes from an air-dry condition to 15−16 per cent moisture at 80−90°C. The feed temperature may increase further to nearly 92°C, during subsequent compression and extrusion through the die, due to friction. Pellets discharged from a pelletizer to a screen belt of a horizontal tunnel drier or a vertical screened hopper are air cooled in about ten minutes and dried to below 13 per cent moisture.

According to Hastings and Higgs (1980), finished pellets contain practically all the nutrients in the ingredients as compounded. The loss of thermolabile vitamins can be compensated for by the extra supplementation in the vitamin premix used.

The cooled pellets may be ground on corrugated rolls and sifted into various sizes of granules and crumbles (fig. 7.10). The crumbles are suitable for feeding small fish and are more easily consumed by sight-feeders. They are, of course, superior to meal rations and less expensive to manufacture than small size pellets.

Two qualities of the pellets are of special importance and have been referred to earlier: hardness and water stability. Moderately hard pellets are easily consumed by many species, but for some there is the danger of overfeeding with hard pellets, causing swelling and rupture of the stomach. The feed may not be digested properly and may cause fermentation and gas formation in the stomach; the fish may then float upside down.

Hardness of a pellet does not necessarily correlate with water stability. In order to prepare more water-stable pellets and thereby improve feed conversion, a number of measures can be adopted. Before pelleting, the mixed feed may be ground through a 2 mm screen to an effective size of about 125 μm. Organic flour such as rice dust, wheat endosperm or other binders may be added, replacing about 5 per cent of some non-essential ingredient, if the formula is deficient in binding material. The addition of sufficient dry steam to condition the soft feed to a temperature of 85−90°C would cause gelatinization of raw starch. The pellet mill should be operated at its optimum

Fig. 7.7 Small-scale production of compound feeds on a farm in China.

rated amperage for maximum compression and extrusion.

As mentioned earlier, feeds may be pelleted by the extrusion process to form floating pellets, which are suited for certain types of culture. One advantage of this type of pellet is that the fish can be observed while feeding and the amount of feed regulated according to the amount eaten. The processing of floating pellets consists of (i) conditioning the feed which is in a meal form to contain 25–30 per cent water, (ii) conveying it by auger into a pressure cylinder, (iii) injecting steam to increase gelatinization of raw starch and (iv) then extruding to atmospheric pressure, almost exploding the material through holes in the die plate at the end of the cylinder. The extruding ribbon is cut by a rotating knife outside the die plate and the pellets are then dried at about 120°C to a moisture content suitable for storage. Following oven drying, a standard pellet cooler is used to lower product temperature after internal moisture becomes less than 13 per cent.

Fig. 7.8 Solar drying of pelleted feeds in a farm in Central African Republic.

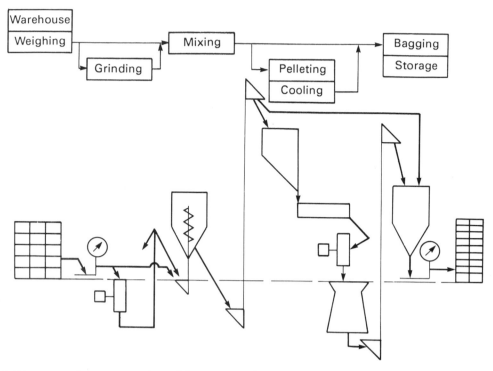

Fig. 7.9 Diagrammatic representation of feed milling (from: Raven and Walker, 1980).

Fig. 7.10 Pelleted feeds of different sizes (from: Stickney, 1979).

The previous high temperatures may partially destroy heat-labile vitamins and decrease the availability of some amino acids. Instead of over-fortifying the formula, as for hard pelleting, the necessary additional additives may be sprayed on to expanded pellets after extrusion.

The expansion process is more expensive and there is some evidence to show that fish fed on floating pellets contain relatively more liver and body fat (Hastings and Higgs, 1980), probably because of increased digestibility of the carbohydrates in the ration.

Therapy and chemoprophylaxis of a great majority of bacterial diseases and many endoparasitic invasions can be achieved efficiently through the use of medicated feeds. For this purpose antibiotics, sulphonamides, nitrofuran and antiparasitic compounds, as well as some of the disinfectants, can be added to the feed. In order to avoid misuse of such diets and hazards to human and animal health, it is necessary to follow proper standards and ensure that only permitted compounds are used.

Astaxanthin and canthaxanthin are the pigments commonly used to give attractive coloration to salmon and trout. They are generally added as water-dispersible gelatine beadlets in concentrations of 100 mg carotenoid per kg. A high dietary lipid content of the diet is known to improve the utilization of carotenoids. Carotenoids are labile compounds and are prone to degradation by heat, acids, alkalis and oxidation. They can be stabilized somewhat by the use of antioxidants.

Except for off-flavours that may develop due to environmental conditions, there is as yet no consensus regarding the organoleptic qualities of cultured aquatic animals as against wild ones. It has to be remembered that the commonly used feed formulations have been developed to provide optimum growth and feed conversion and not texture or flavour.

The treatments involved in feed processing generally help in increasing the nutritional value of feeds. Heat treatment, for example, improves the nutritional value of soybean meal by destroying the trypsin inhibitor present and by increasing the utilization of the essential elements. Digestibility is improved by partial cooking. Similarly, heat treatment increases the nutritional value of cereal grains by gelatinizing starches and improving digestibility. Grinding also increases the nutritional value by reducing the particle size and thereby facilitating digestion. Pelleting improves palatability and steam conditioning improves digestibility. The heat produced during compaction of the pellet may also destroy thermolabile toxic factors that occur in some plant products. Table 7.26 gives details of the effect of processing on toxins and inhibitors in a number of feedstuffs.

Salmonella in feedstuffs like meat meal is killed by pelleting, but aflatoxins, produced by *Aspergillus flavus*, are not inactivated by normal pelleting procedures. It is necessary to prevent mould growth. Materials susceptible to contamination by aflatoxins, such as corn, peanut meal, cottonseed meal, copra and fish meal should be monitored routinely for the presence of aflatoxins.

Storage of both raw materials and processed feeds needs special care as both may undergo deteriorative changes during storage. Loss and deterioration of raw materials often occur as a result of insect infestation, which is greatly facilitated by high ambient temperature, relative humidity and moisture content of the feed

Table 7.26 Toxins and inhibitor destruction by processing (from Walker, 1980).

Feedstuff	Inhibitor	Deactivation process
Cottonseed meal	Gossypol: cyclopropane fatty acids; phytate	Add iron salts; rupture pigment gland
Soybean meal	Trypsin inhibitor	Heat, autoclaving
Linseed meal	Crystalline water-soluble substance	Water treatment
Raw fish	Thiaminase	Heat
Alfalfa meal	Saponins; pectin methyl esterase	Limit amount fed
Rye	5-*N*-alkyl resorcinols	Limit amount fed
Sweet clover	Dicoumarol	
Wheat germ	Unidentified	Heat
Rapeseed	Isothiocyanate; thyroactive materials	

ingredient. Most feed materials undergo some chemical changes that alter their flavour and nutritive value. Besides eating the feedstuffs, insects also accelerate these changes by secretion of enzymes such as lipase. Fats in feedstuffs often break down during storage. Recontamination of feedstuffs by adventitious microorganisms is another major hazard to be avoided. Fungi grow at moisture contents of 15–20 per cent and cause spoilage of feedstuffs in storage. They produce mycotoxin, raise the temperature and moisture content and cause mustiness. The highly toxic and carcinogenic aflotoxins produced by *A. flavus*, are perhaps the most important mycotoxins contaminating feedstuffs.

Rancidity is another important type of deterioration in storage and is caused mainly by oxidation of lipids, besides some hydrolysis and ketone formation. Lipid oxidation can be inhibited by the addition of antioxidants such as ethoxyquin, butylated hydroxytoluene

(BHT) and butylated hydroxyanisole (BHA). The permitted levels are 150 ppm ethoxyquin or 200 ppm BHT or BHA.

In a feed mill, special care has to be taken to store both feed materials and processed feeds in as cool and dry conditions as possible, raised off the ground on pallets. Warehouses or silos should be constructed in such a way that the interior can be kept cool and dry with adequate ventilation. Large silos are used for storage of pellets and some of these are equipped with conveyors or worm screws to move feed conveniently (fig. 7.11). Large silos are equipped with dust collectors to collect the fines which accumulate from repeated storage of pellets. These fines can be repelleted or used for feeding mixed with fresh pellets.

Most commonly, pellets are packed in polyethylene-lined sacks made of multilayered paper or other material. Feed can be stored right on the edge of the farm in hoppers, so that outlet chutes can be swung over the water

Fig. 7.11 Feeds stored in silos near a fish farm in Hungary. Note the rail cart used to carry feeds around the farm (courtesy of J. Kovari).

to discharge pellets directly into a boat for distribution. It is advisable for the farmer to buy at a time, feeds to last only a few months, as long-term storage inevitably results in some deterioration. Properly stored, the feed will keep well for 1−3 months in summer and 2−4 months in winter. It is necessary to protect the feed from moisture and dampness and feeding hoppers should be placed at sufficient distances from water to avoid splashed water entering them.

7.5.6 Larval feeds

Dependence on live foods for feeding larvae of a number of aquaculture species has been referred to in Section 7.4. One important constraint to the formulation and preparation of appropriate artificial diets for larvae is the lack of suitable techniques for determining their nutritional requirements. Measurements of food intake, weight increment, digestibility, etc., are extremely difficult at the larval stage. So, the only solution at present appears to be the extrapolation of data for juveniles and young adults, which is obviously less reliable. Because of this, aquaculturists very often resort to feeding with a combination of live food and compound feeds, which generally gives better results.

Meals, crumbles and small pellets are used as starter feeds for certain species like salmonids and channel catfish, but many others do not accept such feeds. Pollution of water in larval tanks due to accumulation of disintegrated feed is a common problem when intake is slow. Experience shows that prepared feeds can be used for larval rearing of a number of finfish only if the level of acceptance is good and the particle intake per unit time per litre of water is high enough to prevent rapid disintegration of the feed. Flake diets, extruded diets and micro-encapsulated diets have been tried to overcome some of the problems, particularly water stability.

Flake diets can be formulated with commercially available feedstuffs. Readily metabolizable vegetable binders are used to achieve the water stability needed to optimize the residence time of the flake in water. A double drum dryer unit, consisting of steam-heated or electrically heated rollers, is used to dry homogeneous wet suspensions of the feed into thin platelets. They can be reduced to smaller particle sizes without reducing the basic stability. According to Meyers (1979) a major advantage is the comparative ease with which definable formulations can be made using natural or synthetic products. Flavours and colours can easily be introduced as required. The main disadvantage is the relatively high cost per unit weight of product. It is necessary to avoid overheating in the drying process and thereby loss of nutrients.

Extruded larval diets can have good water stability, if readily digestible vegetable binders are used. Extrusion can be run at temperatures between 60 and 100°C. Small feed particles of various shapes can be extruded using material of different densities. It is claimed that specific shapes or particle configurations provide some irregular or erratic motion that simulates prey movement and thus attracts the larvae to feed on them. Low-temperature extrusion and elimination of post-extrusion drying protect heat-labile components and lower costs of production.

Another type of larval feed under investigation is a micro-encapsulated diet. The microcapsule consists of a liquid or particulate dietary component, enclosed within a suitable shell or wall. The selection of the wall matrix depends on the material to be capsulated. It can be made of a biodegradable polymer, such as modified gelatin or zein, so that the nutrients within the capsule can be released by enzymic processes of the animal or by microflora present in its gut. It is claimed that a whole range of components can be encapsulated to constitute a complete diet, facilitating a wide range of sizes of nutritionally diverse capsules to feed different growth stages of aquaculture species.

One major advantage of the encapsulated diet is that the specified nutritional requirements of the larva, if known, can be met with a high degree of precision, since there will be minimum nutrient loss due to leaching. Its use will facilitate the maintenance of better water quality, which is of special importance in intensive culture conditions. Unlike natural or live foods which may not be nutritionally complete, an encapsulated diet could have consistent nutrient composition, be free from contaminants and have a good shelf life.

Chow (1980) describes simple methods of preparing an encapsulated egg diet. According to this author, the whole egg of chicken contains all the necessary nutrients (48.8 per cent protein, 43.2 per cent fat, 0.2 per cent calcium and 0.9 per cent phosphorus, with a gross energy of 5830 kcal/kg and metabolizable energy of 4810 kcal/kg) required during the first ten days of life of most species of fish. Egg yolk alone, which is often used in feeding larvae, is nutritious, but as a diet for very young fish its high energy/protein ratio may result in an inadequate intake of protein necessary for maximum growth. The suggested method is to encapsulate the whole egg (white and yolk).

Cracked whole egg is beaten vigorously with a fork or homogenized with a mechanical blender. Boiling water (approximately 150 ml for each egg) is poured rapidly into the homogenate with constant stirring. A fine opalescent suspension is obtained, which may be made up to the desired volume with cold water. The suspension can be introduced directly into larval tanks. The opalescent protein coat of the microcapsule reflects enough light to attract larvae to it. Whole-egg diet can also be made in larger particles by controlled stirring to feed fry or the homogenate made into a custard for feeding adult fish such as eels. When egg diets are made for fry or adults for longer term feeding, vitamin supplementation may be needed as they lack in water-soluble vitamins, especially ascorbic acid. If the protein content has to be lowered, finely ground carbohydrate ingredients like wheat flour or cassava flour can be added.

A larval feed that is often used in shrimp hatchery work in India is a crustacean tissue suspension. Small size shrimps of low commercial value are processed into wet tissue suspensions. Different feed particle sizes and dosages are used for successive larval stages. Larval survival rates vary according to culture conditions (Hameed Ali, *et al.*, 1982). The survival rates can possibly be improved by using dried crustaceans ground into free-flowing powder of appropriate particle size (Tacon, 1986).

7.5.7 Feeding techniques

While some species start feeding on prepared feeds from the larval stage, soon after yolk absorption, others can be started on it from only the juvenile or fry stage. In order to ensure cost-effective use of feeds, feeding levels have to be decided on in advance. Feed efficiency is usually expressed in one of two ways:

$$\text{feed conversion ratio} = \frac{\text{feed intake}}{\text{weight gain}}$$

or

$$\text{conversion efficiency} = \frac{\text{weight gain}}{\text{feed intake}} \times 100$$

These are often reported as ratios of dry weight of feed to wet weight of the animal, and this explains the anomalous values sometimes quoted in the literature, such as a ratio of 1:1 or even less.

Feeding levels depend to a large extent on the system of culture, nature of the feed, temperature conditions and feeding behaviour of the species. Generally speaking, feed requirements, in volume or weight, increase with the size of the animal, but the requirement per unit of weight decreases with increase in its size. For example, the feed requirements of a 1 kg fish, will be less than of ten fish of 100 g each (weighing in total one kg).

Ideally, feeding levels should be based on the dietary energy requirements or the metabolizable energy value of the feeds. In the absence of such information, the daily feed requirements are often calculated as percentages of the live weight of the animal cultured. This is valid when using complete feeds, but in the cases where the prepared feed is used as a supplement to natural food, calculations of artificial feeding levels become much less precise. The aquaculturist has to fall back on trial-and-error methods and practical experience.

Feeding rates

Rapidly growing fry of about 0.25 g generally require as much as 10 per cent of their weight daily, but as they grow and reach about 4 g in size, the ration can be reduced to about 5 per cent of the body weight per day. Many farmers overfeed at the fry stage for better growth and survival, but then special precautions have to be taken to avoid fouling of the water. The frequency of feeding decreases from about eight times a day for fry weighing less than 1.5 g to four times a day for advanced fry.

Culturists often maintain feeding charts to guide in daily feeding in grow-out farms. It is generally held that feeding in excess of twice a day has no beneficial effect on growth rate or food conversion efficiency; in most cases once daily would be adequate. The total amount of feed to be provided is generally estimated by multiplying the weight representing the desired growth by the conversion rate of the feed used. However, the calculation of the daily feeding rates has to be based on a percentage of the estimated biomass to be fed, making appropriate adjustments for daily increments in the biomass. The animals can also be fed *ad libitum* till they become satiated. When this is not practised, adjustments of feeding rate can be made on the basis of biomass estimated from measurements of small samples at regular intervals. If the feed conversion ratio for a species is known and can be assumed to be constant, the daily changes needed in the feeding rate can be calculated on an assumed specific growth rate. Studies on the effects of feeding frequency on the growth of fish seem to indicate that there is an optimum frequency, above which additional feedings produce no advantage. This frequency seems to be related also to the size of the stomach, since species with smaller stomachs require more frequent feeding for maximum growth.

The ration has to be adjusted according to the water temperature, which affects the metabolic rate of the animal. According to Stickney (1979), channel catfish should be fed no more than 1 per cent (percentage of body weight to feed) daily, when the temperature is above 32°C. Between 21 and 32°C, they are generally fed 3 per cent of the body weight per day and this amount is reduced to 2 per cent at 16–21°C and 1 per cent at 7–16°C. Many culturists do not feed during winter, and below 7°C the fish may not consume much feed, except on warmer days. In tropical countries, where the fluctuations in water temperature are not so great, the adjustments in the amount of feed could also be less.

Feed dispensing

The most common method of feed dispensing is hand-feeding (fig. 7.12). Though it is labour-intensive, it has the advantage that the farmer can observe his stock regularly, especially if feeding takes place at the surface. In pond farms, feeding is usually done in a specific place, which can be marked out by poles to enable regular checking for left-over feed (fig. 7.13). If there is an accumulation of left-over feed, provision of fresh feed is stopped until the old feed is used up. Some farmers prefer to dispense feed at the harvesting sump, as the feeding fish will stir up the deposited silt and keep the sump clean. The animals generally learn to congregate near the feeding spot at the usual feeding time. In cage culture of salmon and trout, wet feed is almost always fed by hand.

Several mechanical aids for hand feeding are available, such as hand-operated blowers and 'disc throwers'. A boat can be used for hand feeding in large ponds and other enclosures. Blower tanks facilitate the dispensation of prepared feeds, as well as feedstuffs such as cereals. The impeller in the tank blows the feed through a slanted pipe to a distance of about 10 m into the pond or enclosure. When mounted on a truck or towed by a tractor, the blower can be propelled by the motor of the vehicle. The actual quantity of feed applied can be measured with the help of an auger, which transfers the feed to the blower. With each turn of the auger, a certain amount of feed is transferred and by counting the rotations of the auger on a meter, the amount can be calculated. Wet-feed dispensers are sometimes used, particularly in sea cage farming of salmon and trout. Wet feed, prepared with a blender, is put into a hopper and the rotating arm at the bottom expresses the feed through a slit. An alternative arrangement consists of an electrically-driven screw at the bottom of the hopper, which expels the food through a nozzle in a continuous 'worm'. Spokes projecting from a revolving wheel cut the 'worm' into wet pellets of required size as it emerges from the nozzle. The pellets drop into a long plastic drain pipe, which extends out into the sea enclosure.

Certain types of feeds, such as dough for eels, are presented on submerged platforms

Fig. 7.12 Hand-feeding salmon in a cage farm in Norway (photograph: Ola Sveen).

Fig. 7.13 Bamboo frame used to hold chopped grass for feeding grass carp in a pond farm in China.

from which the animal can feed with minimum feed loss. In some countries, stationary feeders consisting of a feed hopper fitted over a submerged shallow circular plate are used. Feed falls, by the pressure of the feed and gravity, into the circular plate from where the animal feeds. As the feed from the container is used up, more feed from the hopper falls to fill it.

Different types of automatic feeders are used for prepared feeds in larval tanks and grow-out areas (fig. 7.14). Electrically-activated timing devices make it possible to dispense set amounts of feed at given intervals. This enables a certain frequency and duration of feeding, which can be adjusted according to the varying demands related to life stages, seasons, water temperature, oxygen content, etc. There are different patterns of dispersal. The feed may drop directly into the water from the storage bin or be dispersed widely by falling on to a rotating disc, which can fling the pellets out over the water. In another model, electrical impulses

are received by a vibrator situated just inside the opening at the bottom of a conical feed hopper. Underneath the hopper is a small plastic disc which is an integral part of the vibrator. Pellets fall by gravity onto the disc, which vibrates each time impulses are received. They are then thrown off the disc into the water and the vibrating action also ensures that the pellets in the hopper are kept freely running. Other models distribute feeds from a shore-based storage bin by means of a screw conveyer or auger rotating within a tube which extends over the water. The pellets fall into the water through perforations along the bottom of the tube as they are conveyed out by the screw. The animals feed from a wide area, along the conveyor and at the end of the conveyor run. Other large feed distribution systems use pneumatic tubes through which pellets are blown.

Mechanisms using compressed air to distribute pellets over the water have been used successfully in raceway culture. Mobile mech-

Fig. 7.14 Automatic feeders in a hatchery in Svanoy Island, Norway (photograph: Ola Sveen).

anical feeding systems such as blowers are used to dispense feed over large areas by means of a nozzle at the end of a boom reaching out over the water as the feeder moves along the water's edge (fig. 7.15). Dispensing feed in a stream of water ejected through a nozzle has also been found to be effective. Nevertheless, many farmers prefer to distribute only 70–80 per cent of the feed by an automatic device and the rest is fed by hand to observe the feeding behaviour of the stock and ensure that the feed is made available to all the animals as required.

An increasing number of aquafarms in industrially advanced countries are using computer-controlled automatic feeders, where the quantity of feed and frequency of feeding are regulated through appropriate programs that take into account several variables like stock density, growth rate, conversion rates, temperature conditions, mortality, harvest time, etc.

Other types of mechanical feeders in use are demand feeders, which are activated by the animals themselves by means of a rod or plate hanging from the feeder into the water (fig. 7.16). Movement of the rod or plate (referred to also as bait) releases a small amount of pellets from a hopper. They can be less expensive than automatic feeders, even though the number of such feeders needed in a farm may be relatively higher. They can be easily fabricated from inexpensive materials on the farm itself. An electric supply, so such feeders can be installed in rural farms that have no electricity. Fish very soon learn to activate the feeder and generally do so only when they want to feed, although in high-density culture conditions inadvertent activation can happen. Generally, demand feeders avoid wastage of feed and the creation of anoxia due to bacterial decomposition of unused feed.

Demand feeders permit *ad libitum* feeding, but there is a difference of opinion as to whether this helps in improving feed conversion ratios. There appears to be no conclusive proof one way or the other, but under experimental

Fig. 7.15 A blower tank for dispensing feeds.

Fig. 7.16 An electrically operated demand feeder.

Fig. 7.17 A simple hand-made demand feeder used in a tilapia farm in Central African Republic.

conditions it has been observed that uniform distribution of the feed is impeded by competition, the more aggressive animals being able to activate the feeder more frequently and obtain greater amounts of feed. This leads to uneven growth and size composition of the stock. It is expected that the weaker ones would get their chance to feed after the stronger ones have fed to satiation, but in practice this does not always appear to happen.

Different models of demand feeders are manufactured commercially but, as indicated earlier, simple ones can be made on a farm (fig. 7.17). A demand feeder consists essentially of a feed container with a conical bottom with an opening of about 5 cm diameter in the centre, through which hangs a rod or metallic chain of 12–14 mm thickness. A ball or inverted cone is mounted on the rod or chain in such a way as to more or less block the opening of the container. A small plate can be welded at the end of the rod or chain to facilitate activation by the fish. When the rod or the chain is moved aside by the fish, the feed falls into the water through the gap produced. If the capacity of the feeder is small, frequent refilling of pellets will be needed. This can be done manually or from feed distribution trucks, at regular intervals.

7.6 References

Ackman R.G., Epstein S. and Kelleher M. (1974) A comparison of lipids and fatty acids of the ocean quahaug, *Arctica islandica*, from Nova Scotia and New Brunswick. *J. Fish Res. Board Can.*, **31**, 1803–11.

Austreng E. (1976) Fat and protein in diets for salmonid fishes. I. Fat content in dry diets for salmon parr (*Salmo salar* L.). *Meld. Norq. Labdvr Hogsk*, **55**(5). (In Norwegian with English Summary.)

Berka R. (1973) A review of feeding equipment in fish culture. *EIFAC Occasional Paper*, **9**, FAO of the UN, Rome.

Boyd C.E. (1979) *Water Quality in Warmwater Fish Ponds*. Auburn University Experiment Station, Alabama.

Castell J.D. *et al.* (1972a) Essential fatty acids in the diet of rainbow trout (*Salmo gairdneri*): growth, feed conversion and some gross deficiency symptoms. *J. Nutr.*, **102**, pp. 77–86.

Castell J.D. *et al.* (1972b) Essential fatty acids in the diet of rainbow trout (*Salmo gairdneri*): physiological symptoms of EFA deficiency. *J. Nutr.*, **102**, 87–92.

Castell J.D., Lee D.J. and Sinnhuber R.O. (1972c) Essential fatty acids in the diet of rainbow trout (*Salmo gairdneri*): lipid metabolism and fatty acid composition. *J. Nutr.*, **102**, 93–100.

Castell J.D. *et al.* (1986) Aquaculture nutrition. In *Realism in Aquaculture: Achievements, Constraints, Perspectives*, (Ed. by M. Bilio, H. Rosenthal and C.J. Sindermann), pp. 251–308. European Aquaculture Society, Bredone. Belgium.

Castell J.D. and K. Tiews (Eds.), (1980) Report of the EIFAC, IUNS and ICES Working Group on standardization of methodology in fish nutrition research. *EIFAC Tech. Paper*, **36**.

Cho C.Y. and Slinger S.J. (1978) *Effect of Ambient Temperature on the Protein Requirements of Rainbow Trout and on Fatty Acid Composition of Gill Phospholipids*, pp. 24–35. 1977 Annual Report, Fish Nutrition Laboratory, University of Guelph.

Chow K.W. (1980) Microencapsulated egg diets for fish larvae. In *Fish Feed Technology*, pp. 355–61. ADCP/REP/80/11, FAO of the UN, Rome.

Chow K.W. and Schell W.R. (1980) The minerals. In *Fish Feed Technology*, pp. 105–8. ADCP/REP/80/11, FAO of the UN, Rome.

Conklin D.E. *et al.* (1980) A successful purified diet for the culture of juvenile lobsters: The effects of lecithin. *Aquaculture*, **21**, 243–9.

Cowey C.B. (1979) Protein and amino acid requirements of finfish. In *Finfish Nutrition and Fishfeed Technology*, Vol. I (Ed. by J.E. Halver and K. Tiews), pp. 3–16. Schriften de Bundesforschungsanstalt für Fischerei, Berlin.

Cowey C.B. and Sargent J.R. (1972) Fish nutrition. *Adv. Mar. Biol.*, **10**, 383–492.

Cowey C.B. *et al.* (1974) Studies on the nutrition of marine flatfish. The effect of dietary protein content on certain cell components and enzymes in the liver of *Pleuronectes platessa*. *Mar. Biol.*, **28**, 207–13.

D'Abramo L.R. *et al.* (1981a) Successful artificial diets for the culture of juvenile lobsters. *Proc. World Maricult. Soc.*, **12**(1), 325–32.

D'Abramo L.R. *et al.* (1981b) Essentiality of dietary phosphatidylcholine for the survival of juvenile lobsters. *J. Nutr.*, **111**, 425–31.

Dabrowski K. (1977) Protein requirements of grass carp fry (*Ctenopharyngodon idella*). *Aquaculture*, **12**, 63–73.

Delmendo M.N (1980) A review of integrated livestock–fowl–fish farming systems. *ICLARM Conf. Proc.*, **4**, 59–71.

De Long D.C., Halver J.E. and Mertz E.T. (1958) Nutrition of salmonid fishes VI. Protein requirements of chinook salmon at two water temperatures. *J. Nutr.*, **65**, 589–99.

Deshimaru O. and Shigueno K. (1972) Introduction to the artificial diet for prawn, *Penaeus japonicus*. *Aquaculture*, **1**, 115–33.

Deshimaru O. and Yone Y. (1978) Studies on a purified diet for prawn XII. Optimum level of dietary protein for prawns. *Bull. Jap. Soc. sci. Fish.*, **44**, 1395–7.

Dupree H.K. and Sneed K.E. (1966) Response of channel catfish fingerlings to different levels of major nutrients in purified diets. *US Dept. Interior-Fish Wildlife Serv. Tech. Paper*, **9**.

Edwards D.J. (1978) *Salmon and Trout Farming in Norway*. Fishing News Books, Oxford.

FAO (1977) China: recycling of organic wastes in

aquaculture. *FAO Soils Bull.*, **40**.

FAO (1978) China: azolla propagation and small-scale biogas technology. *FAO Soils Bull.*, **41**.

FAO/UNDP (1979) *Aquaculture Development in China.* ADCP/REP/79/10, FAO of the UN, Rome.

FAO/UNDP (1980) *Fish Feed Technology.* ADCP/REP/80/11, FAO of the UN, Rome.

FAO/UNDP (1983) *Fish Feeds and Feeding in Developing Countries.* ADCP/REP/83/18, FAO of the UN, Rome.

Fox J.M. (1983) Intensive algal culture techniques. In *CRC Handbook of Mariculture Vol. I. Crustacean Aquaculture*, (Ed. by J.P. McVey), pp. 15–41. CRC Press, Boca Raton.

Fuji M. and Yone Y. (1976) Studies on nutrition of red sea bream XIII. Effect of dietary linolenic acid and omega 3 polyunsaturated fatty acids on growth and feed efficiency. *Bull. Jap. Soc. sci. Fish.*, **42**, 583–8.

Gallagher M.L. and Brown D.W. (1975) Composition of San Francisco Bay brine shrimp (Artemia Salina) *J. Agric. Food Chem.*, **23**, 630–32.

Gallagher M.L. *et al.* (1978) Effect of varying calcium phosphorus ratios in diets fed to juvenile lobsters (*Homarus americanus*). *Comp. Biochem. Physiol.*, **60**(A), 467–71.

Gatesoupe F.J. *et al.* (1977a) Alimentation lipidique du turbot (*Scophthalmus maximus* L.) I. Influence de la longueur de chaine des acides gras de la serie e. *Annl. Hydrobiol.*, **80**, 247–54.

Gatesoupe F.J. *et al.* (1977b) Alimentation lipidique du turbot (*Scophthalmus maximus* L.) II. Influence de la supplimentation en esters mighyliqucs de l'acide linoleinque et de la complimentation in acedes gras de la seric 9 sur la croissance. *Annl. Hydrobiol.*, **8**, 89–97.

Gohl B. (1980) Tropical feeds. *FAO Anim. Prod. Health Ser.*, **12**.

Guillard R.R.L. (1975) Culture of phytoplankton for feeding marine invertebrates. In *Culture of Marine Invertebrate Animals*, p. 29. Plenum Press, New York.

Halver J.E. (1979) Vitamin requirements of finfish. In *Finfish Nutrition and Fish Feed Technology*, Vol. I (Ed. by J.E. Halver and K. Tiews), pp. 45–58. Schriften der Bundesforschungsanstalt für Fischerei, Berlin.

Halver J.E. (1980) Lipids and fatty acids. In *Fish Feed Technology*, pp. 41–53. ADCP/REP/80/11, FAO of the UN, Rome.

Halver J.E. and Tiews K. (Eds.) (1979) *Finfish Nutrition and Fishfeed Technology*, Vols I and II. Schriften der Bundersforschungsanstalt für Fischerei, Berlin.

Hameed Ali K., Dwivedi S.N. and Alikunhi K.H. (1982) A new hatchery system for commercial rearing of penaeid prawn larvae. *Bull. Central Inst. Fish. Education, Bombay*, 2–3.

Hardy R. (1980) Fish feed formulation. In *Fish Feed Technology*, pp. 233–9. ADCP/REP/80/11, FAO of the UN, Rome.

Harris L.E. (1980) Feedstuffs. In *Fish Feed Technology*, pp. 111–70. ADCP/REP/80/11, FAO of the UN, Rome.

Hastings W.H. (1979) Fish nutrition and fish feed manufacture. In *Advances in Aquaculture* (Ed. by T.V.R. Pillay and W.A. Dill) pp. 568–74. Fishing News Books, Oxford.

Hastings W.H. and Dickie L.M. (1972) Feed formulation and evaluation. In *Fish Nutrition*, (Ed. by J.E. Halver), pp. 327–74. Academic Press, New York.

Hastings W.H. and Higgs D. (1980) Feed milling process. In *Fish Feed Technology*, pp. 293–320. ADCP/REP/80/11, FAO of the UN, Rome.

Hepher B. and Pruginin Y. (1981) *Commercial Fish Farming.* John Wiley & Sons, New York.

Hickling C.F. (1962) *Fish Culture.* Faber and Faber, London.

Hora S.L. and Pillay T.V.R. (1962) Handbook on fish culture in the Indo-Pacific region. *FAO Fish. Biol. Tech. Paper*, **14**.

Hoskin G.P. (1978) Alterations in lipid metabolism of molluscs due to dietary changes. *Comp. Path.*, **4**, 25–57.

Imai T. (Ed.) (1978) *Aquaculture in Shallow Seas: Progress in Shallow Sea Culture* (Translated from Japanese) A.A. Balkema, Rotterdam.

Kanazawa A. (1984) *Feed Formulation for Penaeid Shrimp, Seabass, Grouper and Rabbitfish Culture in Malaysia.* FAO Field Document-2 (FI:DP/MAL/77/008:61–78.

Kanazawa A. *et al.* (1977) Essential fatty acids in the diet of prawn 1. Effects of linoleic and linolenic acid in growth. *Bull. Jap. Soc. sci. Fish.*, **43**, 1111–14.

Kanazawa A., Teshima S. and Endo M. (1979a) Requirements of prawn, *Penaeus japonicus* for essential fatty acids. *Mem. Fac. Fish Kagoshima Univ.*, **28**, 27–33.

Kanazawa A., Teshima S. and Ono K. (1979b) Relationship between essential fatty acid requirements of aquatic animals and the capacity for bioconversion of linolenic acid to highly unsaturated fatty acids. *Comp. Biochem. Physiol.*, **63B**, 295–8.

Kanazawa A. *et al.* (1980) Requirements of *Tilapia zillii* for essential fatty acids. *Bull. Jap. Soc. sci. Fish.*, **46**, 1353–6.

Kaushik S.J. (1977) *Influence de la Salinité sur le Metabolisme azote le Besoin en Arginine chez la Truite Arcen-ciel.* These, L'Université De Bretagne Occidentale.

Ketola H.G. (1980) Amino acid requirements of fishes: a review. *Proceedings 1980, Cornell Nutrition Conference*, pp. 43–7. Cornell University, Ithaca, New York.

Kitabayashi K. *et al.* (1971) Studies on formula feed for kuruma prawn. 1: On the relationship among glucosamine, phosphorus and calcium. *Bull. Tokai. reg. Fish. Res. Lab.*, **65**, 91–107.

Langton R.W., Winter J.E. and Roels O.A. (1977) Effect of ration size on growth and growth efficiency of the bivalve mollusc, *Tapes japonica. Aquaculture*, **12**, pp. 283–92.

Lannan J.E., Smitherman R.O. and Tchobanoglous G. (Eds) (1983) *Principles and Practices of Pond Aquaculture − A State of the Art Review.* CRSP in Pond Dynamics/Aquaculture, Oregon State University, Newport.

Lee J.S. (1973) *Commercial Catfish Farming.* The Interstate Printers and Publishers, Danville.

Leger C. *et al.* (1979) Effect of dietary fatty acids differing by chain lengths and series on growth and lipid composition of turbot, *Scophthalmus maximus. L. Comp. Biochem. Physiol.*, **643B**, 345–50.

Liao, I-C, H-M Su and J.H. Lin (1983) Larval foods for prawns. *Handbook of Mariculture.* Crustacean Aquaculture. **1**, 43–69. Ed. J.P. McVey. CRC Press Inc., Boca Raton.

Lim C. (1982) *Fish Feed Formulation.* UNDP/FAO Network of Aquaculture Centres, AQU-TRAIN/NACA/037.

Lovell R.T. (1980) Practical fish diets. In *Fish Feed Technology*. pp. 333–50 ADCP/REP/80/11, FAO of the UN, Rome.

Luquet P. and Sabaut J.J. (1974) Nutrition azotee et croissance chez la durade et la truite. *Actes de Colloques, Colloques Sur l'Aquaculture, Brest*, **1**, 243–53.

Martyshev F.G. (1983) (B.R. Sharma, transl.) *Pond Fisheries*, Russian translation series. A.A. Balkema, Rotterdam.

Maynard L.A. and Loosi J.K. (1969) *Animal Nutrition.* McGraw-Hill, New York.

Mertz E.T. (1972) The protein and amino acid needs. In *Fish Nutrition*, (Ed. by J.E. Halver), pp. 106–43. Academic Press, New York.

Meyers S.P. (1979) Formulation of water-stable diets for larval fishes. In *Finfish Nutrition and Fishfeed Technology*, Vol. II (Ed. by J.E. Halver and K. Tiews), pp. 13–20. Schriften de Bundersforschungsanstalt für Fischerei, Berlin.

Nose T. (1979) Diet compositions and feeding techniques in fish culture with complete diets. In *Finfish Nutrition and Fish Feed Technology*, Vol. I (Ed. by J.E. Halver and K. Tiews), pp. 283–96. Schriften de Bundersforschungsanstalt für Fischerei, Berlin.

Nose T. and Arai S. (1972) Optimal level of protein in purified diet for eel, *Anquilla japonica. Bull. Freshwater Fish. Res. Lab. Tokyo*, **22**(2), 145–55.

NRC (1973) Nutrient requirements of trout, salmon and catfish. National Research Council, National Academy of Sciences, Washington, **11**.

NRC (1977) Nutrient Requirements of Warmwater Fishes. National Research Council, National Academy of Sciences, Washington, **12**.

Ogino C. and Saito K. (1970) Protein nutrition in fish. I. The utilization of dietary protein by young carp. *Bull. Jap. Soc. sci. Fish.*, **36**, 250–54.

Ogino C., Cowey C.B. and Chiou J.Y. (1978) Leaf protein concentrate as a protein source in diets for carp and rainbow trout. *Bull. Jap. Soc. sci. Fish.*, **44**, 49–52.

Page J.W. *et al.* (1976) Hydrogen concentration in the gastrointestinal tract of channel catfish. *J. Fish Biol.*, **8**, pp. 225–8.

Pascual F.P. (1982a) *Economical and Practical Feed Formulation.* UNDP/FAO Network of Aquaculture Centres, AQU-TRAIN/NACA/060.

Pascual F.P. (1982b) Nutrition and feeding of sugpo (*Penaeus monodon*). UNDP/FAO Network of Aquaculture Centres, AQU-TRAIN/NACA/060.

Price Jr. W.N., Shaw K.S. and Danberg K.S. (1976) *Proceedings of the First International Conference on Aquaculture Nutrition.* University of Delaware, Newark.

Raven P. and Walker G. (1980) Material flow in feed manufacturing. In *Fish Feed Technology*, pp. 289–92. ADCP/REP/80/11, FAO of the UN, Rome.

Sabaut J.J. and Luquet P. (1973) Nutritional requirements of the gilthead bream, *Chrysophrys aurata.* Quantitative protein requirements. *Mar. Biol.*, **18**, 50–54.

Satia B.P. (1974) Quantitative protein requirements of rainbow trout. *Prog. Fish. Cult.*, **36**, 80–85.

Slinger S.J., Cho C.Y. and Holub B.J. (1977) Effect of water temperature on protein and fat requirements of rainbow trout (*Salmo gairdneri*). *Proceedings University of Guelph Nutrition Conference for Feed Manufacturers*, 1–5.

Smith R.R. (1980) Nutritional bioenergetics in fish. In *Fish Feed Technology*, pp. 21–7. ADCP/REP/80/11, FAO of the UN, Rome.

Smith R.R., Rumsey G.L. and Scott M.L. (1978) Heat increment associated with dietary protein, fat, carbohydrate and complete diets in salmonids. Comparative energetic efficiency. *J. Nutr.*, **108**, 1025–32.

Solberg S.O. (1979) Formulation and technology of moist feed − moist pellets. In *Finfish Nutrition and Fishfeed Technology*, (Ed. by J.E. Halver and K. Tiews), pp. 41–5. Schriften der Bundersforschungsanstalt für Fischerei, Berlin.

Sorgeloos P. *et al.* (1977) Decapsulation of Artemia cysts: A simple technique for the improvement of the use of brine shrimp in aquaculture. *Aquaculture*, **12**(4), 311−5.

Sorgeloos P. *et al.* (1983) The use of brine shrimp *Artemia* in crustacean hatcheries and nurseries. In *CRC Handbook of Mariculture, Vol. I. Crustacean Aquaculture*, (Ed. by J.P. McVey), pp. 79−96. CRC Press, Boca Raton, Florida.

Steffens W. (1981) Protein utilization by rainbow trout and carp: A brief review. *Aquaculture*, **23**, 337−45.

Stein J.R. (1973) *Handbook of Phycological Methods, Culture Methods and Growth Measurements*. Cambridge University Press, London.

Stickney R.R. (1979) *Principles of Warmwater Aquaculture*. John Wiley & Sons, New York.

Stickney R.R. and Shumway S.E. (1974) Occurrence of cellulase activity in the stomach of fishes. *J. Fish. Biol.*, **6**, 779−90.

Tacon A.G.J. (1986) *Larval Shrimp Feeding*. ADCP/MR/86/23, FAO of the UN, Rome.

Takeda M. (1975) The effect of dietary calorie to protein ratio on the growth, feed conversion and body composition of young yellow tail *Bull. Jap. Soc. sci. Fish.*, **41**, 443−7.

Takeuchi T. and Watanabe T. (1977) Studies on nutritive value of dietary lipids in fish. VIII. Requirements of carp for essential fatty acids. *Bull. Jap. Soc. sci. Fish.*, **43**, 541−51.

Takeuchi T., Watanabe T. and Nose T. (1979) Studies on nutritive value of dietary lipids in fish XVIII. Requirements for essential fatty acids of chum salmon (*Oncorhynchus keta*) in freshwater environment. *Bull. Jap. Soc. sci. Fish.*, **45**, 1319−23.

Takeuchi T. *et al.* (1980) Studies on nutritive value of dietary lipids in fish XXI. Requirements of eel *Anguilla japonica* for essential fatty acids. *Bull. Jap. Soc. sci. Fish.*, **46**, 345−53.

Tang Y.A. and Hwang T.L. (1967) Evaluation of the relative suitability of various groups of algae as food of milkfish in brackishwater ponds. *FAO Fish. Rep.*, **44**(3), 365−72.

Tapiador D.D. *et al.* (1977) Freshwater fisheries and aquaculture in China. *FAO Fish. Tech. Paper*, **168**.

Tiews K., Gropp J. and Koops H. (1976) On the development of optimal rainbow trout pellet feeds. *Arch. Fisch. Wiss.*, **27**, 1−29.

Tobias W.J. *et al.* (1980) International study on Artemia XIII. A comparison of production data of 17 geographical strains of Artemia in the St. Croix artificial upwelling mariculture system. In *The Brine Shrimp Artemia*, Vol. 3 (Ed. by G. Persoone *et al.*). Universa Press, Wetteren.

Vanhaecke P. and Soregeloos P. (1980) International study on Artemia IV. The biometrics of Artemia strains from different geographical origin. In *The Brine Shrimp Artemia*, Vol. 3 (Ed. by G. Persoone, *et al.*). Universa Press, Wetteren.

Vanhaecke P. and Soregeloos P. (1983) International study on Artemia XIX. Hatchery data for 10 commercial sources of brine shrimp cysts and re-evaluation of the hatching efficiency concept. *Aquaculture*, **30**, 43−52.

Walker G. (1980) Effects of processing on the nutritional value of feeds. In *Fish Feed Technology*, pp. 321−4. ADCP/REP/80/11, FAO of the UN, Rome.

Wilson R.P., Harding D.E. and Garling D.L. (1977) Effect of dietary pH on amino acid utilization and the lysine requirements of fingerling Channel Catfish. *J. Nutr.*, **107**, 166−70.

Woynarovich E. (1975) *Elementary Guide to Fish Culture in Nepal*. FAO, Rome.

Woynarovich E. (1980) Raising ducks on fish ponds. *ICLARM Conf. Proc.*, **4**, 129−34.

Yamada R. (1983) Pond production systems − feeds and feeding practices in warmwater fish ponds. In *Principles and Practices of Pond Aquaculture*, (Ed. by J.E. Lannan *et al.*), pp. 117−44. Oregon State University, Newport.

Yone Y. (1976) Nutritional studies of red sea bream. In *Proceedings of the First International Conference on Aquaculture Nutrition*, (Ed. by K.S. Price, W.N. Shaw and D.S. Danberg), pp. 39−64. University of Delaware, Newark.

Yu T.C. and Sinnhuber R.O. (1972) Effect of dietary linolenic acid and decosahexaenoic acid on growth and fatty acid composition of rainbow trout (*Salmo gairdneri*). *Lipids*, **7**, 450−54.

Yu T.C. and Sinnhuber R.O. (1979) Effect of dietary omega 3 and omega 6 fatty acids on growth and feed conversion efficiency of coho salmon (*Oncorhynchus kisutch*). *Aquaculture*, **16**, 31−8.

Zeitoun I.H. *et al.* (1976) Quantifying nutrient requirements of fish. *J. Fish. Res. Board Can.*, **33**, 167−72.

8
Reproduction and Genetic Selection

As discussed in Chapter 5, one of the major criteria in selecting a species for culture is the existence of either suitable techniques for controlled breeding or easy availability of spawn, larvae or juveniles from natural breeding grounds. Even when culture can be initiated using 'wild seed', it is essential to achieve controlled reproduction as early as possible, to ensure timely availability of young ones in adequate numbers for large-scale rearing. It is also a basic need in the domestication of the animal and for taking advantage of the benefits of genetic selection and hybridization that have contributed so much to terrestrial agriculture and animal husbandry.

Controlled breeding will obviously be possible only if there is adequate knowledge of the factors governing reproduction of the animal and its breeding behaviour. Lack of such knowledge has hampered the progress of aquaculture of several important species. The extensive culture of Chinese carps, Indian carps, mullets, milkfish, sea-bass, sea-bream, penaeid shrimps, oysters and mussels has been based until recently on 'seed' obtained from natural breeding. Despite advances made in techniques of controlled or semi-controlled breeding, the techniques have not been sufficiently perfected or adapted for large-scale production of seed, with the result that the aquaculturist has still to depend partially or entirely on natural seed resources. There are also species like the eels for which no propagation technique has so far been developed, even though some progress has been made in maturing and spawning under laboratory conditions.

Among the aquaculture species, finfish as a group has received greater research attention

in controlled reproduction. The reproductive cycles of almost all fish are regulated by environmental stimuli. Appropriate sensory receptors convey the environmental stimuli to the brain in the form of neural inputs. This neural information, on reaching the hypothalamus, causes the release of hypothalamic peptides known as releasing hormones, which in turn stimulate the pituitary gland to release the gonadotropic hormone(s), which act on the gonads. The gonads in turn produce the sex steroid hormones which are responsible for the formation of gametes, as well as for the regulation of secondary sexual characteristics, nuptial coloration and breeding behaviour. This pattern of reproductive mechanism provides the basis for methods of induced reproduction, namely the provision of appropriate environmental stimuli and the administration of hormones for maturation and release of gametes.

8.1 Reproductive cycles

The large majority of aquaculture species are seasonal breeders, although some breed intermittently or continuously. Seasonal breeding is generally related to climatic seasons. For example, most fresh-water fish of temperate zones spawn in spring and early summer, but the salmonids spawn in autumn. Rainy season and flood waters are associated with the spawning of fresh-water fishes of tropical and subtropical regions of Latin America and Africa. Obviously the fishes integrate their own reproductive functions with environmental cycles. The breeding season appears to coincide with environmental conditions that are most conducive to the survival of the offspring. These favourable factors, that act as cues for a suitable breeding season, affect the central nervous system and through it the pituitary and the gonads. Photoperiod, temperature and rainfall are important factors involved in regulation of the reproductive cycles.

Mechanisms of reproductive timing vary very considerably among species. For example, in salmonids that spawn in the autumn, gradually increasing photoperiods followed by short photoperiods or decreasing photoperiods have a major role in regulating the cycle. Temperature has an important role in the reproductive cycle of cyprinid species. Gonadal recrudescence takes place in Indian carps during the period of the year when both photoperiod and temperature are increasing. Changes in the volume and velocity of water, flooding of shallow areas and dilution or replacement of water are also considered to be important factors. Warm temperatures and long photoperiods appear to affect also the reproductive cycle of Chinese carps. A review of available information would appear to show that in the majority of cases gonadal recrudescence is regulated chiefly by seasonal variations in photoperiod and temperature, while spawning may be controlled by temperature and/or rainfall.

The age of sexual maturity varies widely between species. For example, tilapia species become mature within a few months, whereas others may take a few years. The same fish may mature earlier in a warm climate and much later in colder climates; examples of this are the common carp and the Chinese carps. The common carp, which takes three to four years to mature in Europe, takes only a year to attain maturity in tropical regions. Chinese carps that take five to seven years to mature in Europe become mature in one to three years in tropical and subtropical conditions.

Some species have only one spawning season, during which they may spawn several times. Others may have two or more spawning seasons. Some species of finfish exhibit well developed parental care, which may consist of incubating fertilized eggs in the buccal cavity of the parent, or guarding the eggs and larvae during development. Many of the species that exhibit parental care lay eggs in nests made of plant or other available material or in hollows dug out on the bottom.

Some of the species like the Chinese and Indian carps that are essentially riverine spawners would not spawn in the confined waters of fish ponds or other enclosures. Their gonads develop only up to a certain stage and then remain dormant until resorption sets in. They have however, been observed to spawn in special types of ponds (called bundhs in India) that have a flow of fresh rainwater, inundating shallow marginal areas where the conditions are favourable for the fish to breed. The simulation of conditions in natural spawning grounds may serve to induce certain fish to breed in confined areas. The provision of nest-building material for nest-breeding species and the provision of artificial substrates for the attachment of eggs required for certain species are also believed to induce spawning.

8.2 Control of reproduction

In aquaculture, the main purpose of controlled reproduction is to achieve sexual maturation and spawning at the time of the year which is normal to that species. As mentioned earlier, some species will not breed in the confined waters of an aquaculture facility. In other cases, maturation and spawning are unpredictable, because of the culture conditions or environmental factors. Controlled reproduction can also be of considerable importance in advancing or retarding the spawning period as required. This can help in making available young ones at appropriate times or of appropriate sizes. A higher level of reproduction control would involve development of the capability to mature and spawn a species at any time of the year, in order to enable continuous production and marketing throughout the year.

The two major types of control that are possible, consist of (i) manipulation of the reproductive cycle and (ii) induction of gonadal gamete release (ovulation and spermiation). The reproductive cycle is manipulated so as to have gametes available when needed. This may be initiated in the juvenile stage, or advanced or retarded in the adult stage. Altered gonadal gamete release can be achieved by hormonal supplementation, manipulation of environmental factors or the use of special selected strains.

In oviparous animals, embryos are dependent on the egg yolk for their nutritional requirements. Vitellogenesis, or the process of yolk deposition in oocytes, is a seasonal or cyclic phenomenon. All stages of it, starting with the mobilization of lipid from storage sites, the synthesis in the liver of a female-specific glycolipophosphoprotein, vitellogenin, and its eventual deposition in oocytes are known to be gonadotropin-dependent.

The interaction between the brain, pituitary gland, testis and ovary largely mediates the influence of environmental factors on the reproductive development of finfish. The thyroid and interrenal may also have a less important role. The substance formed by the nucleus lateralis tuberis in the hypothalamus, which is responsible for such influence is the gonadotropin-releasing factor or releasing hormone. In the case of mammalian luteinizing hormone (LH) and follicle stimulating hormone (FSH), the releasing activities for these two hormones have been shown to be present in the same peptide, which consists of a chain of ten amino acids (Schally and Kastin, 1972). The molecule is referred to as LH-RH. The presence of LH-RH has been demonstrated in certain species of fish (Crim *et al.*, 1978) and it has also been demonstrated that mammalian LH-RH or its analogues in large doses bring about the release of gonadotropin.

Even though attempts have been made with salmonids, the induction of a completely new reproductive cycle has not yet been successful. Chronic administration of gonadotropic hormones can, however, initiate a normal reproductive cycle and assure its progress. By pellet implantation of hormones, it has been possible to advance normal spawning by one year in pink salmon. The release of gametes can be advanced by a single dose of hormone. Similarly, it has been demonstrated that hormone injections can induce late ovulations, as in brown trout, when maturity is blocked by adverse environmental conditions.

As mentioned earlier, the two major environmental factors that affect maturation and spawning are the photoperiodic regime and temperature. Although any definitive conclusions regarding the independent influence of photoperiodism have not been possible, there is enough evidence of the combined effect of these in several species. When, by manipulation of these factors, early maturation is achieved, egg-laying can more easily be synchronized by hormonal injection. This helps in predicting ovulation more precisely and in avoiding ageing of ova, which may occur at high summer temperatures. There is considerable experimental evidence of the independent role of temperature in maturation and spawning. It is believed that spawning is timed to ensure that gametes are released into water whose temperature is within the appropriate stenothermal conditions for embryonic development. While the precise mechanisms by which temperature regulates reproductive development are not known, it is presumed that it acts as a triggering mechanism at the hypothalamic level or alternatively exerts a generalized stimulatory effect on metabolic rate. The influence of rainfall on the spawning of certain species, as referred to earlier, is also ascribed to the combined effect of temperature and photoperiod, plus the dilution of inhibitory elements in the water.

Another means of reproductive control, oriented to spreading egg production over the year, is through the use of selected strains for early or late spawning. Strains have been developed that spawn for much longer periods than normal for the species. There is also the possibility of using in a farm several strains, reproducing at different times of the year, in order to ensure the availability of young throughout the year.

8.3   Induced reproduction

As explained in the previous section, the hypothalamus regulates the reproductive functions of the pituitary gland. The correct combination of environmental factors required for maturation, ovulation and spawning, brings about an accelerated release of gonadotropin from the pituitary into the bloodstream. Ng and Idler (1978 a, b) and Idler and Ng (1979) have isolated two gonadotropic hormones: one with a low carbohydrate content that induces vitellogenesis and the other which is rich in carbohydrates, inducing maturation and ovulation. The surge of gonadotropins that occurs brings about maturational changes culminating in the act of spawning. Environmental conditions required for the initiation of oocyte maturation, ovulation and spawning are much more complex than those for gametogenesis.

Very often under culture conditions, the required environmental conditions may not be available, or may not persist for a sufficient length of time for spontaneous maturation to occur. This has led to the development of induced reproduction or hypophysation techniques (Houssay, 1931; Von Ihering, 1935 and 1937). By the injection of pituitary homogenates (fig. 8.1), the natural gonadotropin surge is simulated, by-passing to some extent the environmental variables of temperature, rainfall, photoperiod, etc. Besides the advantage of regulating the time of spawning, it enables the adoption of other methods of artificial propagation, including hand-stripping (fig. 8.2), fertilization, incubation, hatching and larval rearing. While hypophysation techniques have been demonstrated to be effective in a large variety of fish species, its major contribution in respect of aquaculture technologies, since its first field application in Brazil in 1935, has been in the inducement of spawning in fishes that do not ordinarily breed under conditions of confinement or do so only under specific environmental conditions. It has now become a common practice in many countries and utilized widely in the reproduction of finfish, despite the fact that the relevant mechanisms are not fully understood and little standardization of the techniques has been achieved.

Vitellogenesis in decapod crustacea, particularly Penaeid shrimps and lobsters, has been shown to be mediated by hormones. Male shrimps mature fully under captive conditions and spermatophores can be seen through the carapace. Female shrimps often do not mature fully, even though maturing eggs can be found in the ovaries. The maturation process seems to be inhibited by a gonad-inhibiting hormone (GIH) secreted by the medulla terminalis ganglionic x-organ (MTGX) and stored in the sinus gland. The y-organ, which secretes the moult hormone crustecdysone, also has an influence on maturation. The ablation (surgical removal) of eye stalks, which have the glands containing the inhibitory hormone, has been shown to accelerate vitellogenesis in many crustaceans. Besides environmental factors like temperature, photoperiod, salinity and pH, the state of nutrition of brood animals seems to be an important factor in the maturation and spawning of shrimps.

8.3.1 Hypophysation

A more detailed description of the techniques of induced spawning, including environmental control employed for the breeding of important aquaculture species, can be found in Part II. Only some of the common features of induced spawning, with special reference to finfish, will be discussed here.

The mammalian gonadotropic hormones, LH and human chorionic gonadotropin (HCG), are effective in inducing maturation and ovulation in fishes. Although a number of species have been induced to breed by the administration of HCG or a combination of HCG and mammalian pituitary extract, there are certain refractory breeders, like the Indian and Chinese carps, where fish pituitary homogenates or extracts are needed to induce spawning. There are reports of successful breeding of even these species, by using HCG under certain circumstances. The Chinese carps, which have been bred two or three times by administration of fish pituitary extract, will respond positively to injections of HCG. Bhowmick (1979) has reported on the use of crude HCG for induced spawning of one species of Indian carp, *Labeo rohita*. It has, however, been reported that repeated injections of HCG could induce a 'drug resistance effect' related to the production of antibodies against foreign proteins. Nevertheless, it would appear that homogenates and extracts of whole pituitary glands and partially purified fish gonadotropins are more potent in inducing maturation and ovulation in fishes than mammalian gonadotropins, and can be used extensively in commercial fish culture.

While the administration of the appropriate hormone is basic to the success of induced breeding, the condition of the brood fish and the environmental conditions are also equally important. The large number of failures in induced breeding can often be traced to poor condition of the brood fish, including their health and nutrition and stage of gonadal development, as well as to environmental conditions in spawning tanks or enclosures.

Fig. 8.1 Injection of pituitary homogenates to induce spawning.

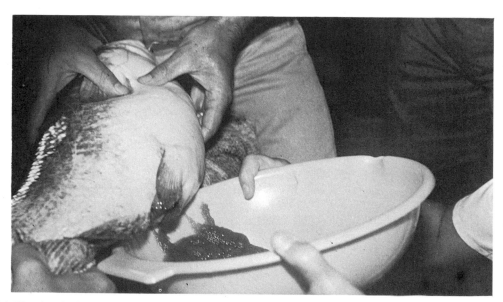

Fig. 8.2 Hand-stripping of a mature female fish.

Chinese farmers believe that it is more difficult to breed wild Chinese carps, as well as carp that have attained maturity for the first time. They prefer to rear spent fish in special holding ponds, fed on a special protein diet, for future breeding.

The identification of sex is another important requirement for successful induced breeding. Many species do not have distinctive and permanent sex characteristics. When there are no secondary sex characteristics, detailed morphometric characteristics will have to be used to separate sexes, particularly in the pre-puberty stages. After sexual differentiation, it may be possible to distinguish the sexes through examination of the gonads. This will involve the use of endoscopy or biopsy, which is difficult to use on a large scale. Siphoning of eggs and their examination under a microscope, to determine the stage of maturity of females, have been described by Chen *et al.* (1969) but the adoption of this method in large-scale breeding work is not always very practical. Other possible methods, such as the use of serum markers or detection of circulating vitellogenin, are also difficult to use in the field. Aquaculturists have therefore to depend largely on practical experience and field obser-vations to distinguish the sexes and determine the stage of maturity of breeders. Brood female fish ready for spawning are identified by the well-rounded and soft abdomen and swollen genital opening. The male releases a few drops of thick milt when its abdomen is slightly pressed.

As indicated earlier, several species of fish respond to injections of HCG and other mam-malian hormones, and these hormones are commercially available to aquaculturists. Many species, which are more difficult to spawn under confinement, need injections of fish pituitary for maturation and spawning. There are dif-ferences of opinion regarding the species-specificity of the pituitary, but aquaculturists generally prefer to use the glands of the same or closely related species. It is recommended that pituitaries from phylogenetically close donors should be used, when there is a choice. However, common carp is considered a uni-versal donor and its pituitary is being used very widely for both experimental and commercial breeding purposes for several species. Salmon pituitary is also used for breeding a number of species. Though commercially available on a limited scale, a large majority of aquaculturists have to depend on local arrangements for the collection and preservation of the glands. Glands of the recipient species or of other proven donor species are used. Fractionation and purification of teleost gonadotropins are still in experimental stages. Though potent gonadotropic preparations have been made from fish pituitaries by means of chemical/ ethanol fractionation, they have not been used widely in spawning refractory fish.

Glands extracted from catches of the selec-ted mature donor species are preserved in alcohol or acetone or frozen for storage. Freshly collected glands are first desiccated in absolute ethyl alcohol (changing the preservative sev-eral times) and then stored in fresh alcohol at room temperature or under refrigeration. The glands remain active for a period of about two years. Instead of alcohol, the glands can be desiccated in acetone, changing it several times as for alcohol. The desiccated glands are dried in vacuum and stored in that condition or sealed in vials and stored in frozen condition. Acetone-dried glands retain their activity for 6–10 years. The glands can also be preserved by quick freezing, but the most common method of preservation is acetone drying.

Though a number of methods of preparing pituitary homogenates and extracts have been tried, the most commonly accepted method is extraction with distilled water or saline sol-lution. The glands are macerated in a small volume of water or saline solution and brought up to the desired volume. Distilled water, common salt solution (0.3–1 per cent) and physiological saline can be used, as they all seem to give equivalent results. The hom-ogenate can be used as such for injection, or filtered or centrifuged to obtain filtrate or supernate which can be injected. Extraction with trichloracetic acid (TCA) at low concen-trations of 1.25–2.5 per cent for short time-periods of 3–6 hours, is reported to provide more complete extraction and better results. But this practice has not received wide accept-ance, probably because of the specific require-ments of concentration and extraction time. It is reported that higher concentrations and/or longer extraction, can result in denaturation of glycoproteins.

As pituitary extracts are subject to rapid enzymic deterioration, they have to be pre-pared fresh every time fish are to be bred. This is obviously inconvenient. Methods of pre-serving extracts have been tried with some success. One method involves the extraction of pituitary glands in a small volume of distilled water, and refrigerating it for 24–48 hours, after which glycerine is added to make a 2:1

ratio with water. The suspension is again re-frigerated for 24–48 hours, centrifuged and the supernate stored under refrigeration in air-tight vials. Another method consists of grinding acetone-dried pituitaries, sieving them through 40–60 mesh/mm^2 sieves and storing in sealed vials at 5°C. Both these techniques are aimed at achieving homogeneous preparations of uniform potency.

Despite its wide use, the dosage frequency and latency period of pituitary administration remains more or less at a trial-and-error stage, and sometimes leads to poor results. This is mainly on account of the variations in the gonadotropic content of the pituitary material used and the stage of sexual maturity of the brood fish, besides the environmental con-ditions and the stress to which the breeders are subjected. The mode of injection (intra-peritoneal or intramuscular) also appear to af-fect results. Development of an acceptable method of assessing gonadotropic content should greatly assist in determining practical dosages. Though several biological units have been proposed, none seems to have gained wide acceptance.

8.3.2 Gametes and fertilization

Injection of pituitary homogenate or extract is usually given into the dorsal muscles above the lateral line and below the anterior part of the dorsal fin, or the dorsal part of the caudal peduncle. Injections into the body cavity are considered less efficient. The required quantity of the gland is generally administered in two to four doses (one or more preparatory injections followed by one or more final doses). As indi-cated earlier, suitable environmental con-ditions, besides pituitary injection, will be needed for ovulation to take place. Tem-perature, high dissolved oxygen levels and lack of stress are some of the important require-ments. The process of ovulation takes some time, depending on the species and environ-mental conditions. Maturation of the ovum starts when its nucleus starts to migrate from the centre toward the micropyle and undergoes hydration by absorbing fluids. Ovulation starts with the disappearance of the nuclear mem-brane and ends with the first meiotic division. At the same time, the follicle which attaches

the eggs to the wall of the ovary splits and releases the eggs into the cavity of the ovary, from where it can freely flow through the geni-tal opening.

Many of the fish that are treated with gon-adotropic hormones start to spawn in the pres-ence of active males after normal ovulation. The eggs are fertilized by the male breeders and the fertilized eggs can be collected easily for hatching. Where such induced spawning does not occur, it becomes necessary to strip the sex products from the females and males and artificially fertilize them. Ripe ova remain-ing unspawned for long periods after ovulation become over-ripe and do not develop normally. It is also not uncommon for the phenomenon of 'plugging' to occur in gravid females sub-jected to overdoses of hormone. In such cases, natural spawning cannot be accomplished, since a mass of ovarian eggs forms a plug at the urinogenital opening, preventing the free flow of eggs. Stripping will be necessary to obtain eggs from such fish. Stripping and artificial fertilization are necessary also for fish with sticky eggs like the common carp. Such eggs will have to be treated to dissolve the sticky layer, so that they can be incubated in suitable incubators. A quick washing with a weak tannin solution after the eggs have swollen will be effective in removing the stickiness of the eggs. Common salt and carbamide (urea) solution can also be used for removing the sticky layer.

The ovulated egg which has undergone the first meiotic division will have the second mei-otic division when the sperm penetrates it, ending in the extrusion of the second polar body. Further embryonic development leading to the formation of the first somatic cell com-pletes the process of fertilization. The time available for the ripe egg to become fertilized is rather limited in most fresh-water fish, as the eggs swell rapidly in water and this results in the closure of the micropyle. The time available for common and Chinese carps is about 45–60 seconds. In saline solution the eggs seem to remain fertilizable for longer periods, up to several minutes.

The sperm, which is immotile in the testis, becomes motile on contact with the medium in which fertilization takes place. The duration of the activity of spermatozoa varies with the species, but is generally not longer than a couple of minutes. In the males of most species, dense semen having highly motile spermatozoa can be obtained without hormone injection. Administration of pituitary extracts brings about thinning of the seminal plasma and would facilitate spermiation. Relatively large numbers of spermatozoa are needed to fertilize an egg. For example, the requirement of a trout egg is reported to be 10 000–300 000 spermatozoa and of a carp egg 13 000–30 000. This is due to the fact that the spermatozoon can penetrate at only one place, i.e. the micropyle, and the distance that can be covered by a trout spermatozoon during its life span (2 mm) is often less than the circumference of the ovum which is about 15–20 mm. The probability of it reaching the micropyle is therefore low, if the motility is less. The number of spermatozoa compensates for the low motility. It is necessary to take special care in regulating the quantity of water added to the sexual products during fertilization. If too much water is added, many of the sperms will not be able to reach the micropyle. On the other hand, if sufficient water is not added, the micropyle of an egg may get blocked by other eggs, due to crowding, preventing the sperm from entering it.

8.4 Preservation of gametes

In many species, the maturation of gonads in the two sexes is not synchronous. Males often show testicular recrudescence earlier during the season. Because of this, ripe males occur during the beginning of the season, when the females are not yet mature and ready for spawning. The reverse situation occurs during the end of the breeding season. Under such circumstances, it will be most advantageous to have a suitable means of preserving the gametes for artificial fertilization, when needed. Methods of gamete preservation would also help in the initiation of genetic selection programmes, by providing easy access to a reserve of genetic material of known and desired qualities.

Cryopreservation with liquid nitrogen, used widely in the preservation of cattle and livestock sperm, has been tried for the preservation of a number of species of fish. Blaxter (1955) reported successful fertilization of fresh eggs with cryopreserved (−79°C) sperm of *Clupea*

harengus. Sections of ripe testis were stored in 80 per cent sea water containing 12.5 per cent glycerol as a protector, and the mixture frozen quickly or slowly at 1°C/min to −30°C, then quickly to −79°C (using dry ice). Besides the sperm of rainbow trout, spermatozoa of the common carp, Chinese and Indian carps and grey mullet are among the cultivated species which have been subjected to cryopreservation, which consists of cooling and storing at sub-zero temperatures of liquid nitrogen (−196°C), using dimethyl sulphoxide, glycerine, ethyl glycol or other cryoprotectants and diluents (Harvey and Hoar, 1979). Attempts at cryopreservation of ova have not been as successful as for sperm. Zell (1978) reported the first successful cryopreservation of unfertilized ova and zygotes of salmonid fish. Ova frozen in liquid nitrogen at −20°C for 5 minutes proved to be fertile, and zygotes frozen at −50°C survived the exposure. All subsequent attempts have failed. While it is difficult to predict possible advances in cryopreservation of fish gametes, it would appear that the results so far indicate only the feasibility of short-term preservation of semen or the prolongation of embryonation.

8.5 Use of sex steroids for sex reversal

In certain situations and species, it will be advantageous to restrict fertility. A well-known example is the cichlid tilapia, which attains maturity at an early age and breeds repeatedly at short intervals, overpopulating ponds and other rearing facilities. This results in stunted populations, as energy is expended for reproduction rather than growth. Among the techniques that can be employed for restricting fertility is the application of hormones to produce monosex populations. Androgenic and oestrogenic steroids are used for masculinization of genotypic females and feminization of genotypic males (Jalabert *et al.*, 1974; Guerrero, 1975, 1979; Shelton *et al.*, 1978). Genotypic female fry of the species of *Sarotherodon* (= *Tilapia*), when fed on methyltestosterone and ethynyltestosterone have become males. Similarly, monosex female tilapia have been produced by treatment with oestrone, ethynyloestradiol and stilboesterol. While the feasibility of sex reversal by steroid administration has been demonstrated, the percentage of fish that underwent sex change in any treated group varied greatly. Since the presence of even a small percentage of the opposite sex in a population is sufficient to initiate uncontrolled breeding, the value of the results achieved so far becomes less significant. Similar experiments to produce monosex fish have been conducted with salmonids and other species. Sex inversion of the protogynous species of *Epinephelus* (*E. tauvina*) has been accelerated to produce male brood stock from three-year-old females, by oral administration of methyltestosterone. Production of all-female eggs is now a common practice in a number of rainbow trout hatcheries (see Chapter 16.1.2). The initial functional males required for fertilizing ova from normal female brood stock are obtained by sex reversal, by treating with 17 methyltestosterone through immersion or incorporation in starter feed in the fry stage.

8.6 Genetic selection and hybridization

The use of genetically selected strains and hybrids has contributed very substantially to modern agriculture and animal husbandry. But aquaculture has so far benefited very little from efficient breeding and selection programmes. Among the many reasons for this are the delays in the development of suitable techniques for controlled reproduction of many farmed species and the paucity of genetic expertise among aquaculturists. Genetic improvements usually require long-term experimentation with a large number of individuals and generations, and so considerable time may elapse before useful results become available. Moreover research on farming technologies has not reached that level in most cases, when the only way to improve production is by genetic improvement of the stocks. Except in a few cases, the present technologies are too inefficient to benefit from the use of selected stocks.

In traditional aquaculture, certain strains have evolved as a result of environmental or farming conditions without much conscious effort by the aquaculturist, as in the case of the common carp, or as a result of the rule-of-thumb selection. These more or less accidental strains can seldom be used with confidence for commercial farming.

There is no doubt that effective selective breeding programmes are expensive and require more facilities than are presently available in most aquaculture farms or even institutions that can function as central stations for breeding and distribution of aquaculture species. Though the economic benefits of selection programmes have been worked out for domestic animals, comparable evaluations are few in aquaculture.

As pointed out earlier, the number of domesticated species used in aquaculture for food is limited (unlike in the culture of ornamental species), but the number is steadily growing. Opportunities to increase the production properties and adaptation to a new environment of species through selection can therefore be expected to expand in the future. Kirpichnikov (1966) gives some of the main aims of fish selection as follows:

(1) To increase the growth rate by better utilization of food (physiological decrease of food expenditure per unit of growth increment);
(2) to increase the growth rate by fuller utilization of natural food in ponds and higher consumption of feed mixtures;
(3) to increase resistance to oxygen deficiency, to high or low temperature, to higher salinity or to other deviations from the normal environmental conditions;
(4) to improve resistance to infectious diseases and to infestation of parasites (to develop new breeds resistant to particular diseases);
(5) to improve the nutritive properties of fish (to increase the calorie content, to decrease the proportional weight of non-edible parts, to decrease the bone content, to increase or decrease the fat content, etc.).

Other aims may include speeding up of sexual maturation, the ability to reproduce at relatively low temperatures and the slowing down of maturation to prevent early switching over of metabolism to develop sex products, affecting growth and resulting in prolific reproduction.

The relative advantages of a fish in genetic breeding schemes are brought out by Skjervold (1976), using salmonids as an example. Among the most important of these are:

- Very high fertility, leading to high 'litter' of considerable importance in selection work.
- External fertilization, which makes it possible to have several combinations of matings and the production of many 'litters' of large numbers of half-siblings.
- The high fertility of females, which enables:

  (a) some types of family selection, even when the heritability of the selected trait is low, as large family groups will in practice result in rather accurate estimation of the breeding value;
  (b) progeny testing among females, which can be carried out by using mixed sample of semen, or semen from sires of known breeding value;
  (c) the improved possibility of estimating non-additive genetic components, because of the combination of high female fertility and external fertilization;
  (d) artificial manipulation of chromosome content, which is facilitated by external fertilization, and
  (e) easier hybrid production due to high female fertility and the remarkable ability for crossing as observed in nature.

The main disadvantages of fish for genetic breeding are the rather long generation interval of many species, particularly in cold climates, and the hierarchies that may develop in fish populations and which may contribute to size variations.

Genetic gains through selection, as in the case of salmonids, are dependent on selection differentials and genetic variance of the relevant traits, which are inversely related to the length of the generation interval. Gjedrem (1983) presents average estimates of heritability and a coefficient of variation ($CV = (\sigma_p/x) \times 100$) (see Section 8.6.1) of several species, as shown in Table 8.1.

The high phenotypic variance of the body weight of adults, together with the medium heritability as seen in Table 8.1, show that there is a large genetic variance for this trait for these species. Though mortality shows low heritability, resistance to specific diseases shows medium to high heritability. Similarly, the meat quality traits show some genetic variation, though it is low in the dressing percentage. Age at sexual maturation shows medium heritability in rainbow trout but a high one in the Atlantic salmon. The conclusion is that the possibility of

Table 8.1 Average values of coefficients of variation (CV) and heritabilities (h_s^2) based on the sire component for economically important traits (superscripts give number of estimates involved) (from Gjedrem, 1983).

Economically important trait	Rainbow trout CV	h^2	Atlantic salmon CV	h^2	Common carp CV	h^2	Channel catfish CV	h^2	Tilapia CV	h^2	Oyster CV	h^2	Prawns CV	h^2
Body weight, juveniles	33[7]	0.12[4]	78[1]	0.08[1]		0.15[1]	46[1]	0.42[4]	26[5]	0.04[1]		0.36[3]		0.12[3]
Body weight, adults	22[7]	0.17[2]	27[2]	0.36[3]	22[1]	0.36[2]	27[3]	0.49[4]						
Body length, juveniles	14[3]	0.24[3]	23[1]	0.14[2]				0.12[1]	8[1]	0.06[1]		0.47[2]		
Body length, adults	9[2]	0.17[2]	8[2]	0.41[4]			8[2]	0.61[3]						
Mortality/ resistance		0.14[1]		0.11[4]	28[1]									
Carcass traits														
Meatiness	20[1]	0.14[1]	19[1]	0.16[1]										
Meat colour	23[1]	0.06[1]	16[1]	0.01[1]										
Fat (%)	10[1]	0.47[1]						0.14[1]	8[1]	0.23[3]				
Dressing (%)	6[1]	0.01[1]	4[1]	0.03[1]					2[2]	0.00[2]				
Age at maturation		0.18[1]		0.41[2]										

genetic gains from selection programmes in aquatic animals is high and compares favourably with terrestrial animals and plants.

8.6.1 Methods of genetic selection

As pointed out at the beginning of the last section, the development of some of the earlier strains of common carp and trout has not always been as a result of planned selection. Carp farming in different regions has led to the establishment of strains which appear to have adapted to the general climatic conditions under which they are grown, and which grow faster than the wild strains. In rainbow trout, the usual practice has been to pick the best-looking fish from a stock to be the parents of the next generation. These common-sense approaches cannot be depended upon in a cultivation programme to achieve genetic improvement.

Most economically important characteristics of cultivated organisms are measurable and their variation within a population usually takes the form of a 'normal' distribution (Purdom, 1972). Such a distribution of measurements occurs because the magnitude of a characteristic is determined by a large, often variable, number of factors, some of which are environmental and some genetic. The separation of environment and heredity has been one of the main aims of studies on population genetics. The reliable models of the inheritance of the measurable characteristics can be used for predicting and controlling the gains from genetic selection within cultivated species.

The variation of a character between individuals can be expressed as 'variance', the mean square deviation of individual values from the mean. This is called the phenotypic variance (V_P) of a sample or population and is the sum of two components, the environmental variance (V_E) and the genotypic variance (V_G). Hence $V_p = V_E + V_G$. The proportion of phenotypic variance that is genetic (V_G/V_p) is approximately equal to the value of 'heritability' which measures the proportion of additive genetic inheritance in the phenotypic variance. It is a measure of the degree to which multiple genes control resemblance between offspring and their parents in the face of a particular set of modifying environmental factors. Heritability

can be used to predict selection gains in the formula $R = h^2S$, where R is the response, measured as the change in the mean from one generation to the next, and S is the 'selection pressure', or the difference between the mean of the selected parents and the mean of the population from which they were chosen.

As indicated, the ratio of V_G to V_p is only an approximate measure of h^2. More reliable values can be obtained through parent/offspring correlations or by the use of the above formula in a selection. Though laborious and time-consuming, it is essential to have an indication of the magnitude of h^2 before starting an extensive selection programme.

The primary reason for desiring estimates of heritability is to enable prediction of results expected from a given level of selection. The effectiveness of selection depends upon:

(1) heritability of the attribute (h^2),
(2) degree of variation in the attribute (σ_p), and
(3) intensity of selection applied in (1).

According to Falconer (1981), the anticipated response to selection (R) can be stated algebraically as

$$R = i\sigma_p h^2$$
where
R = mean of offspring from selected parents minus mean of all adults before selection

$$i = \frac{\text{mean of group selected minus}}{\sigma_p}$$
minus mean of all adults before selection

σ_p = phenotypic standard deviation of the attribute
h^2 = heritability estimate for a particular attribute

Mass selection

Mass selection, or individual selection, is based on characteristics of the individuals under selection as opposed to selection based on the performance of their relatives. It is one of the simplest and most common methods employed in breeding programmes, where the characteristic to be improved is easy to measure. It can be used efficiently in selection for growth rate and to some extent for age at maturity.

As stated above, response in mass selection (R) is determined by the general equation

$$R = i\sigma h^2 = Sh^2$$

where

S = selective differential (the difference in a certain trait between the individuals selected and the population as a whole),

h^2 = heritability of the differences (the share of additive genetic variation in the general variation of the character), and

i = intensity of selection.

The high fecundity of cultivated fish causes high selective potential and intensity of selection compared to domestic animals and poultry. In performing mass selection in fish breeding, the selection severity coefficient or the rejection rigidity factor (V) is calculated by the equation:

$$V = \frac{100\,n}{N}\%$$

where

n = the number of individuals selected and
N = the total number of fish grown.

In fig. 8.3 the intensity of rejection is plotted against its severity on a semi-logarithmic scale. The curve obtained shows that there is a sharp increase in the intensity of selection with decrease in the severity coefficient within the range 100−10 per cent. A further decrease in V (down to 1 per cent) results in a considerably lower increase in i; with further decrease in V (0.1−0.01) there is hardly any increase in i. For fish with high fecundity, selection gives best results when the severity of selection is 1−0.1 per cent.

Response to selection is directly proportional to the heritability of the character (h^2). In many cases, a rather accurate estimate of the value of heritability of the character can be obtained from the equation:

$$h^2 = \frac{R}{S}$$

To obtain the estimate, selection has to be conducted in several successive generations.

For increasing response in mass selection, the values of i, σ and h^2 have to be increased. The value of i can be raised in fecund fish by

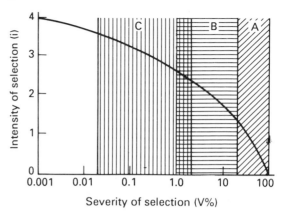

Fig. 8.3 Relation between intensity of selection and its severity in cattle (A), poultry (B) and fish (C) (adapted from Kirpichnikov, 1971).

increasing the number of individuals and, through this, increasing the severity of selection. Variability should relate only to genetic and not environmental variation, as the latter is conditioned by the non-additive genetic variation. To increase h^2 values, non-related individuals have to be crossed. Outbreeding increases the degree of heterozygosity, i.e. increase in genetic variation, but inbreeding results in higher homozygosity. A sufficient number of fish should be available every year for crossing, to enable selection of fish from different crossings for breeding purposes. Another method of increasing genetic variation is by speeding up the process of mutation through irradiation and chemical mutagens.

Non-hereditary variation can be depressed by following special rearing methods such as growing spawners under favourable conditions for maturation, simultaneous crossings, incubation of all eggs under identical physico-chemical conditions, growing larvae and young ones under conditions that do not promote food competition, avoiding the mixing of stocks grown in different ponds or enclosures, and by selection at the age when the animal is more susceptible to improvement by selection.

It is necessary to point out here that a long period of one-way selection for certain characteristics may bring about changes in other morphogenetically or genetically correlated characters. There are many observed examples of correlated responses in selection of non-selected characters, such as physiological and biochemical factors (Steffens, 1964), growth rate (Moav and Wohlfarth, 1967), fecundity, etc.

Genotypic selection

Individual or mass selection can only be used on traits that can be recorded on live animals and is not very efficient for traits with low heritability. In such cases, other types of selection have to be resorted to. The two types of genotype selection that have applications in aquaculture are family selection and progeny testing.

Family selection and sib-selection

Family selection is of special interest in selection for characteristics of low heritability, such as survival, meat quality and age at maturation. Use of full and half sib families in a selection programme has the advantage that the generation interval will not be increased, compared to individual selection. However, a disadvantage is that usually each family has to be reared in separate tanks, as it is generally difficult to mark newly hatched larvae or fry. This will introduce environmental and tank effects on characteristics, such as body weight, between families. Because of this, Falconer (1981) recommends a combination of individual and family selection.

In family selection, several families are grown under identical conditions to determine the ones to be maintained for breeding. To obtain separate progeny (family), either one male/female pair or a small group of spawners can be used. The response equation is essentially the same as in mass selection:

$$R_f = i_f \sigma_f h^2{}_f$$

The intensity of selection appears to be lower than in mass selection, as it is not possible to grow such a large number of families. Similarly, a reduction can be observed in the standard deviation, as this denotes the variation in the family and not individual variation. However, the heritability is much higher.

If the individuals have to be sacrificed for examination, the brothers and sisters of the individuals from the best families can be maintained for breeding. This is known as sib-selection. Kirpichnikov (1971), in his description of the methods of family selection, underlines the importance of carrying out crossings, egg incubation, larval rearing and grow-out of families separately, under as identical timings and conditions as possible. The main disadvantage of family selection is the practical difficulty in simultaneously growing many families under identical conditions. Marking of individuals will reduce some of the problems, as communal growing will then become possible. Fin clipping and cold or hot branding have been used in many large-scale selection programmes. Molluscs can be marked more easily on their shells, whereas in crustaceans moulting habits make marking difficult.

Progeny testing

Progeny testing enables the assessment of the breeding qualities of separate spawners or pairs of spawners and the selection of the best for further selection work. However, progeny testing will increase the generation interval very markedly. For example, in carp breeding it requires one or two years, which would mean a slowing down of selection work by 20 to 30 per cent.

Three methods of progeny testing are applicable in aquaculture. The first method is testing of pairs, without testing males and females separately (fig. 8.4a). The second is to test spawners belonging to one sex, as for example females only as shown in fig. 8.4b, and the third is the testing of both females and males (complete diallele crossing) (fig. 8.4c).

The equation to measure selection response is the same as in family selection. Intensity of selection is limited by the number of families. Variability of family means (σ_f) is also the same in both cases. Heritability of family means ($h^2{}_f$) in progeny testing is high, as in family testing. This may occur only if the breeding conditions of all progenies are practically identical or if breeding proceeds with a three- or four-fold reiteration.

By comparing the response values in the two equations ($R = Sh^2$ and $R_f = S_f h^2{}_f$), it should be possible to determine which method

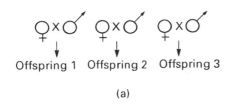

Offspring 1   Offspring 2   Offspring 3

(a)

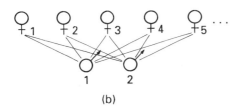

(b)

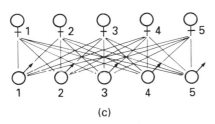

(c)

Fig. 8.4 Progeny testing in fish breeding a = Comparison of pairs without testing sexes separately, b = Testing spawners belonging to one sex only, c = Testing of both sexes (diallele crossing), (from Kirpichnikov, 1971).

is more efficient in a given programme. If Sh^2 is greater than S_fh^2, mass selection is more effective than selection for relatives. The selection for relatives is only advantageous when the increase in heritability is not accompanied by a proportionally greater drop in the selection differential.

Combined selection

Even though mass selection has been found to be more efficient than selection for relatives in fish, the advantages of the latter in selection for certain characteristics like fat content has been demonstrated. For characters like weight, it appears possible to follow a combined selection programme, which may consist of:

(1) performing mass selection among fingerlings or two-year-old fish with a great severity coefficient,

(2) progeny testing of males through to maturity, since males often mature earlier than females,

(3) family selection with a simultaneous breeding of five to ten families and

(4) repeated performance of mass selection in the best families.

Such a scheme allows summing up of the efficiency of mass selection and selection for relatives in a relatively short time.

Cross-breeding

Cross-breeding is another well known means of genetic improvement which has application in aquaculture. Heterosis or hybrid vigour enables an offspring to surpass its parents for one or more traits. On the other hand, inbreeding depression caused by mating of closely related parents has a deleterious effect on the progeny.

The inbreeding measure is the coefficient of inbreeding, incorporating the degree of the animal's homozygosity. It shows what part of the genes in a group of individuals are in the homozygous state. Close inbreeding, especially sib mating (brothers and sisters and parents with offspring), causes homozygosity to increase rapidly, to as much as 0.9−0.95 (expressed as fractions of one) or even more. In most animals, inbreeding results in inbreeding depression, characterized by a drop in viability, growth rate and often fertility. Such depression has been observed by many workers in common carp, brook trout and other fish species. Outbreeding, on the other hand, is accompanied by heterosis in growth rate and viability, especially when fish from different highly inbred groups are crossed. The main types of crossings undertaken are:

(1) Commercial crossing, directed towards breeding of the first-generation hybrids for commercial purposes. Only the first generation, that has the heterosis of productive qualities or incorporates the advantageous characteristics of both the parental forms, is used. They are maintained for further reproduction.

(2) Synthetic or distant crossing, in which distant parents, including those of intergeneric origin, are crossed to develop a new breed,

in the course of long selection. It may attempt to combine the qualities of parents of several breeds, species or even genera. Such crossings to produce new breeds should ensure the preservation and perfection of the productive qualities of the breed, the preservation of genetic variability and the prevention of inbred depression.

Distant outbreeding is indispensable in the selection of aquaculture species. The aims of such crossings are as follows (Kirpichnikov, 1971):

(1) an overall increase in genetic variability, resulting in an increase in selection response,
(2) achievement of a combination of characteristics of two or three breeds or two (rarely three) species,
(3) improvement of the productive quality of the local breed by making use of the few valuable traits of another breed,
(4) increase in the viability of the breed by introducing genes responsible for resistance to environmental factors and diseases.

Kirpichnikov (1971) describes different cross-breeding patterns to achieve these aims.

Reproductive crossing is suggested when valuable properties from both parents are to be combined in the hybrids. It can be done with complete fertility of the hybrid and requires only meticulous selection in the subsequent generations.

Introductory crossing will be advantageous when one or only a few characteristics from a breed have to be incorporated in the hybrid. Each generation of the hybrid has to be crossed with the local breed and so there is the risk of losing the useful characteristics of the improved breed in back crossing, particularly in the case of polygenic inheritance of properties selected. This type of crossing is of considerable use in selection for increased resistance to certain diseases, which is often dependent on the presence of one or a few genes. These genes can be preserved by means of proper selection in each generation.

Absorptive crossing differs from introductory crossing only in that its purpose is a nearly complete substitution of the local breed of genotype by the genotype of the improved breed. Only some features of the local breed, such as viability, are preserved.

Alternate crossing is the most complicated system which is most useful when a combination of many characteristics from two breeds with polygenic inheritance is required. It allows the preservation of high genetic variability through a number of generations. Selection efficiency is kept at a high level owing to this variability and does not reach a plateau. The main problem of obtaining new hybrid breeds by means of crossing (interspecific or intergeneric) is their complete or partial sterility, which takes a lot of time-consuming work to overcome.

A number of breeding systems have been proposed to utilize completely the advantages associated with heterogeneous crossings. Parallel breeding of two or more groups within a breed is possible when working with slowly maturing species, without intermingling, allowing a moderate inbreeding among each and carrying out selection in each generation. In breeding in groups for family selection, a large number of crossings from different groups are carried out for each generation. The parents producing the best offspring are used for subsequent commercial crossings. This system suffers from the drawback that the genetic variability gradually decreases during family selection. Moav and Wohlfarth (1967) recommend that a reserve group of a sufficient number of individuals should be kept for each group when selecting two groups marked by certain genes. In case of a drop in genetic variability, additional gene pools can be introduced into the exhausted group.

Another possible system is alternate inbreeding and outbreeding. After two or three generations of close inbreeding, the evaluation of hybrids from different inbred lines is performed. The best combinations are used for commercial rearing and among the offspring new inbred lines are established. Linear selection involving inbreeding for superior ancestors and top cross, where crossing is done between the best inbred individuals (say males) and individuals from the outbred population (say females) to preserve the genetic variability, are other methods that have applications in aquaculture.

One of the most complicated techniques of breeding is reciprocal recurrent selection (RRS), where the combining capacity of the parents from each of the two groups is evaluated by means of a cross with parents from the other group. The individuals thus selected are reproduced without recrossing and their offspring again tested for combining potential.

The basic feature of all the systems described is the utilization of heterosis in crossing individuals from different groups, lines and breeds. Along with this, moderate to very close inbreeding is employed. The most appropriate system would naturally depend on the species and the traits that are of importance in commercial culture. Gjedrem (1985), however, proposed a cross-breeding scheme for fish farming as summarized below:

(1) Test all possible crosses between different strains or species for the economic traits in question and select crosses that are likely to give useful results. It may be better to use strains with very different origins and which, in combination, have favourable traits.
(2) Develop inbred lines and test the crosses under natural conditions to find the most valuable cross for farming.
(3) Start an RRS programme to ensure continuous genetic improvement, utilizing both general and specific combining abilities simultaneously.

Chromosomal manipulation

As mentioned earlier, the sex of fish that are not differentiated into males or females at hatching can be controlled by the use of sex steroids at the time the gonads are differentiating. While direct masculanisation is frequently easier, feminisation has also been possible in some species. Methods employed for commercial production of all-female rainbow trout by using sex-reversed functional males and genetic females is summarised in Section 15.2. An alternate method of producing monosex stock is to induce sterility, and this can be done by the administration of high doses of sex steroids or by chromosomal manipulation.

Chromosomal manipulation for inducing polyploidy, gynogenesis and androgenesis has been studied with a view to controlling sex, as well as for rapid inbreeding. Manipulation becomes feasible during the nuclear cycles of cell division, and since fertilisation in fish is external, artificial means can be employed to either gamete before fertilisation, or to the fertilised egg at any period during the formation of the zygote.

The chromosome number can be increased by subjecting the egg to a pressure or temperature shock shortly after fertilization. The normal expulsion of one set of maternal chromosomes is prevented, and so after fusion of the chromosomes from the sperm, the developing embryo contains three sets of chromosomes instead of the normal two sets. The extra set of chromosomes in the triploid individuals interferes with gonad development. Such induced triploidy is also useful for producing individuals with increased heterozygosity.

Gynogenesis, the parthenogenetic development of eggs after activation with genetically inert spermatozoa, is a very effective means of achieving relatively rapid inbreeding. It can be used to generate all-female stocks and for gene-transfer.

By exposing milt to a very high dose of atomic radiation (for example by using the radioactive isotope cobalt-60), the chromosomes of the sperm cells are destroyed. The milt is kept on ice at 0°C during radiation and can be stored thus for several days without loss of vitality. When the irradiated milt is mixed with eggs, the sperm cells penetrate the eggs but play no further part in the development of the egg. The embryo develops from the egg material only, without any male chromosomes. Since the egg is haploid (with only a single set of chromosomes), the developing embryos are also haploid. Though most of them die at or soon after hatching, there will usually be some gynogenetic diploids (with a double set of chromosomes), as a result of spontaneous diploidization of the female chromosome complex. To increase the frequency of diploidization of the female chromosome complex, temperature shock can be used. Treatment of eggs with low and high temperatures at the time of meiotic divisions results in disturbances in the process, such as disintegration of the spindle, due to which none of the chromosome sets can form the polar body, or the return of the polar body into the plasma of the ovum. The output of genetic diploids varies greatly, being high under favourable conditions according to the strength and duration of the temperature shock and the stage at which it is administered.

Interspecific and intergeneric hybridization

Several interspecific and intergeneric hybridizations have been carried out for different purposes. In many cases the offspring have proved to be viable and fertile, combining some of the desired qualities. Several species of trout and salmon, cyprinids, cat fish and sturgeons have been used in such crossings, for special traits. Transplantation of nuclei of one species (common carp) into the cytoplasm of another species (crucian carp) has been tried to bring together superior characteristics of the two species. Of late, interest in the production of monosex hybrids through interspecific crossing has increased, for improving the culture of the species of tilapia that reproduce rapidly and overpopulate ponds. As will be described in Chapter 19, selected species were crossed for the purpose of obtaining all-male progeny (Pruginin, 1968; Pruginin *et al.*, 1975). The progeny consisted of a high percentage (98–100 per cent) of male offspring. However, commercial production of all-male hybrids has been difficult to maintain over a long period of time due to contamination of pure brood stock lines (Lovshin, 1982). There has also been considerable interest in breeding mutant forms of tilapia, such as the red tilapia which has better consumer acceptance.

8.7   References

Adiyodi K.G. and Adiyodi R.G. (1970) Endocrine control of reproduction in decapod crustacea. *Biol. Rev.*, **45**, 121–65.

Bhowmick R.M. (1979) Observations on the use of human chorionic gonadotropin prepared in the laboratory in induced spawning in the major carp *Labeo rohita*. Paper presented at the symposium on Inland Aquaculture, Barrackpore, India. *Tech. Paper 16.*

Billard R. (1986) The control of fish reproduction in aquaculture. In *Realism in Aquaculture; Achievements, Constraints, Perspectives*, (Ed. by M. Bilio H. Rosenthal and C.J. Sinderman), pp. 309–49.

Blaxter J.M.S. (1955) Herring rearing 1. The storage of herring gametes. *Mar. Res.*, **3**, 1–12.

Brown M.E. (Ed.) (1957) *The Physiology of Fishes*, Vol. I. Academic Press, New York.

Caillouet C.W. Jr (1972) Ovarian maturation induced by eyestalk ablation in pink shrimp, *Penaeus duorarum. Proc. Third Annual Workshop World Maricult. Soc.*, **3**, 205–25.

Chen F.Y., Chow M. and Sim B.K. (1969) Induced spawning of the three major Chinese carps in Malacca, Malaysia. *Malays. Agr. Jour.*, **47**(2),
211–38.

Crim J.W., Dickhoff W.W. and Gorbman A. (1978) Comparative endocrinology of piscine hypothalamic hypophysiotropic peptides; distribution and activity. *Amer. Zool.*, **18**, 411–24.

Donaldson E.M. (1973) Reproductive endocrinology of fishes. *Am. Zool.*, **13**, 909–27.

Donaldson E.M., Hunter G.A. and Dye H.M. (1981) Induced ovulation in coho salmon (*Oncorhynchus kisutch*), II. Preliminary study of the use of LH-RH and two high potency LH-RH analogues. *Aquaculture*, **26**, 129–41.

Falconer D.S. (1981) *Introduction to Quantitative Genetics*, 2nd edn. Longman, London.

FAO (1971) *Seminar/Study Tour in the USSR on Genetic Selection and Hybridization of Cultivated Fishes.* Lectures, REP/FAO/UNDP (TA), 2926.

Fostier A. and Jalabert B. (1982) Physiological basis of practical means to induce ovulation in fish. In *Reproductive Physiology of Fish*, (Compiled by C.J.J. Richter and H.J.T. Goos), pp. 164–73. PUDOC, Wageningen.

Fujino K. (1986) Impact of genetic factors on aquaculture and stock management. In *Realism in Aquaculture: Achievements, Constraints, Perspectives*, (Ed. by M. Bilio, H. Rosenthal and C.J. Sindermann), pp. 421–48. European Aquaculture Society, Bredene.

Gjedrem T. (1983) Genetic variation in quantitative traits and selective breeding in fish and shellfish. *Aquaculture*, **33**, 51–72.

Gjedrem T. (1985) Improvement of productivity through breeding schemes. *Geo-Journal*, **10**(3), 233–41.

Guerrero R.D. III, (1975) Use of oral androgens for the production of all-male *Tilapia aurea* (Steindachner). *Trans. Am. Fish. Soc.*, **104**, 342–8.

Guerrero R.D. III (1979) Culture of male *Tilapia mossambica* produced through artificial sex reversal. In *Advances in Aquaculture*, (Ed. by T.V.R. Pillay and W.A. Dill), pp. 166–8. Fishing News Books, Oxford.

Harvey B.J. and Hoar W.S. (1979) *The Theory and Practice of Induced Breeding in Fish.* IDRC-TS21 e, IDRC, Ottawa.

Hines R.S. *et al.* (1974) Genetic differences in susceptibility to two diseases among strains of the common carp. *Aquaculture*, **3**, 187–97.

Hoar W.S. (1969) Reproduction. In *Fish Physiology*, Vol. 3 (Ed. by W.S. Hoar and D.J. Randall), pp. 1–72. Academic Press, New York.

Houssay B.A. (1931) Action sexuelle de l'hypophyse sur les poissons et les reptiles. *C.R. Soc. Biol. Paris*, **106**, 377–8.

Ibrahim K.H. (1969) *Techniques of Collection, Processing and Storage of Fish Pituitary Gland.* FAO/ UNDP Regional Seminar on Induced Breeding of Cultivated Fishes, Bombay.

Idler D.R. and Ng T.B. (1979) Studies on two types

of gonadotropins from both salmon and carp pituitaries. *Gen. Comp. Endocrinol.*, **38**, 421–40.

Jalabert B. *et al.* (1974) Determinisme du sexe chez *Tilapia macrochir* et *Tilapia nilotica*: action de la methyltestosterone dans l'alimentation des alevins sur la differentiation sexuelle; proportion des sexes dans la descendance des males 'inverses'. *Ann. Biol. Anim. Biochim. Biophys.*, **14**, 729–39.

Kirpichnikov V.S. (1966) Goals and methods of carp selection. *Izv. gosud. nauchno-issled. Inst. ozer. rech. ryb. khoz.*, **61**, 40–61.

Kirpichnikov V.S. (Ed.) (1966) Selective breeding of carp and intensification of fish breeding in ponds. *Bull. Stat. Sci. Res. Inst. Lake Fisheries, Leningrad.* (Translated from Russian by Israel Program for Scientific Translations, Jerusalem, 1970.)

Kirpichnikov V.S. (1968) Efficiency of mass selection and selection for relatives in fish culture. *FAO Fish. Rep.* **44**(4), 179–94.

Kirpichnikov V.S. (1971) *Methods of fish selection.* In FAO seminar/study tour in the USSR on genetic selection and hybridization of cultivated fishes. Lectures. Rep. FAO/UNDP(TA), **2926**, pp. 202–16.

Lovshin L.L. (1982) Tilapia hybridization. In *The Biology and Culture of Tilapias*, (Ed. by R.S.V. Pullin and R.H. Lowe-McConnell). *ICLARM Conf. Proc.*, **7**, 279–307.

Moav, R. (1979) Genetic improvement in aquaculture industry. In *Advances in Aquaculture*, (Ed. by T.V.R. Pillay and W.A. Dill), pp. 610–22. Fishing News Books, Oxford.

Moav R. and Wohlfarth G.W. (1967) Breeding schemes for the genetic improvement of edible fish. *Prog. Rep. Israel–U.S. Fish Wildl. Serv.*, 1964–65.

Ng T.B. and Idler D.R. (1978a) 'Big' and 'little' forms of plaice vitellogenic and maturational hormones. *Gen. Comp. Endocrinol.*, **34**, 408–20.

Ng T.B. and Idler D.R. (1978b) A vitellogenic hormone with a large and small form from salmon pituitaries. *Gen. Comp. Endocrinol.*, **35**, 189–95.

Pickford G.E. and Atz J.W. (1957) *The Physiology of the Pituitary Gland of Fishes.* New York Zoological Society, New York.

Pruginin Y. (1965) *Report to the Government of Uganda on the Experimental Fish Culture Project in Uganda 1962–64.* REP/FAO/UNDP (TA) 1960.

Pruginin Y. (1968) Culture of carp and *Tilapia* hybrids in Uganda. *FAO Fish Rep.*, **44**(4), pp. 223–229.

Pruginin Y. *et al.* (1975) All male broods of *Tilapia nilotica* × *T. aurea* hybrids. *Aquaculture*, **6**, 11–21.

Purdom C.E. (1972) Genetics and fish farming. *Lab. Leaflet (N.S.), Min. Agric. Fish and Food*, **25**.

Richter C.J.J. and Goos H.J.Th. (Compilers) (1982) *Reproduction Physiology of Fish.* PUDOC, Wageningen.

Schally A.V. and Kastin A.J. (1972) Gonadotropin releasing hormone – one polypeptide regulates secretion of luteinizing hormone and follicle stimulating hormone. *Science*, **173**, 1036–8.

Shehadeh Z.H. (1976) Induced breeding techniques; a review of progress and problems. *EIFAC Tech. Paper*, **25**, 72–89.

Shelton W.L., Hopkins K.D. and Jensen G.L. (1978) Use of hormones to produce monosex tilapia for aquaculture. In *Symposium on Culture of Exotic Fishes*, (Ed. by R.O. Smitherman, W.L. Shelton and J.H. Grover), pp. 10–33. Fish Culture Section, American Fisheries Society, Auburn.

Simon R.C. (1970) Genetics and marine aquaculture. In *Marine Aquiculture*, (Ed. by W.J. McNeil) pp. 65–74. Oregon State University Press, Corvallis.

Skjervold H. (1976) Genetic improvement of salmonids for fish culture. In *Aspects of Fish Culture and Fish Breeding*, (Ed. by E.A. Huisman), pp. 7–27. Miscellaneous paper, 13 Landbouwhogeschool, Wageningen.

Stanley J.G. and Thomas A.E. (1978) Absence of sex reversal in unisex grass carp fed methyltestosterone. In *Symposium on Culture of Exotic Fishes*, (Ed. by R.O. Smitherman, W.L. Shelton and J.H. Grover), pp. 194–9. Fish Culture Section, American Fisheries Society, Auburn.

Steffens W. (1964) Vergleichende Anatomisch-physiologische Untersuchungen und Wild und Teich Karpfen (*Cyprinus carpio* L.). *Z. Fisch.*, **12**(8/10), 725–800.

Sundarararj B.I. (1981) *Reproductive Physiology of Teleost Fishes.* ADCP/REP/81/16, FAO of the UN, Rome.

Tcherfas N.B. (1971) Natural and artificial gynogenesis of fish. In *Seminar/Study Tour in the USSR on Genetic Selection and Hybridization of Cultivated Fishes*, pp. 274–91. Lectures, REP. FAO/UNDP (TA), (2926).

Thorgaard G.H. (1986) Plaidy manipulation and performance. *Aquaculture*, **57**, 57–64.

Von Ihering R. (1935) Die Wirkung von Hypophyseninjektion auf den Laichakt von Fischen. *Zool. Anz.*, **111**, 273–9.

Von Ihering R. (1937) A method for inducing fish to spawn. *Prog. Fish-Cult.*, **34**, 15–6.

Wilkins N.R. and Gosling E.M. (Ed.) (1983) *Genetics in Aquaculture.* Proceedings of an international symposium. Elsevier Science Publishers, Amsterdam.

Woynarovich E. and Horvath L. (1980) The artificial propagation of warm-water finfishes – a manual for extension./*FAO Fish.Tech. Paper*, **201**.

Zell S.R. (1978) Cryopreservation of gametes and embryos of salmonid fishes. *Ann. Biol. Anim. Biochim. Biophys.*, **18**, 1089–99.

9
Health and Diseases

9.1 Health and diseases in aquaculture

Diseases of aquaculture species caused by parasites and infectious pathogens have attracted the attention of veterinarians and fish biologists from the early days of aquaculture investigations. A number of prophylactic and curative measures have also been suggested, although many of the chemicals have not been cleared for use in some countries. As will be discussed later in this chapter, there are a number of diseases for which there are, as yet, no known remedies. With increasing investments in aquaculture and closer examination of factors that contribute to the risks faced by an aquaculturist, the concept of integrated health protection measures has developed in recent years. Similarly, experience of fish farming in the tropics has brought into focus the public health aspects of fish farm development and the possible role of aquatic farming in the spread of communicable human diseases. The extensive introduction and transplantation of aquaculture species occurring at present have clearly shown the need for regional and international cooperation in controlling the spread of communicable diseases and implementation of mutually acceptable measures for the purpose. Thus fish health and disease control are now viewed from different angles, that include environmental protection and pollution control, human health and epidemiology, site selection and culture technologies, monitoring and sanitation of aquaculture facilities, diagnosis and treatment of diseases of cultured species, avoidance of nutritional diseases, prevention of epidemics of mortality in culture facilities, formulation and implementation of regulatory measures to control national and international spread of communicable diseases, development of disease-resistant strains through genetic selection and hybridization and individual and mass immunization of cultured species.

Undoubtedly the research, development and regulatory measures needed for an integrated health management programme in aquaculture involve considerable expertise, organization and expense. Both the State and the aquaculture sector will have to share the responsibility for the successful implementation of such a programme. Being probably the most important risk factor in an aquaculture enterprise, such a programme will have direct relevance to the development of a risk insurance system to protect the farmer from unavoidable losses. While the need for these measures are readily recognized, the low magnitude of the industry at present and uncertainties about the extent to which it can develop and contribute to the overall national economies have prevented its realization in most countries. There is now however, a greater recognition of its importance among aquaculturists. For example, in Asia disease problems were considered to be of only secondary importance when extensive farming was the most common practice. With the adoption of semi-intensive and intensive systems, the occurrence of several forms of diseases and consequent mortalities have significantly increased. Similarly, improved expertise and facilities in disease diagnosis have led to the identification of several heretofore unknown pathogens and disease conditions. Consequently, greater efforts are now being made to diagnose and control disease conditions in the region.

9.1.1 Factors affecting fish health

Fish health or the health of aquaculture organisms has to be conceived as a state of physical well-being. The importance of proper nutrition for rapid growth and the prevention of nutritional deficiencies have been discussed in Chapter 7. Adequate nutrition is also vital for the overall health and vigor needed to cope with a variety of disease agents. Nutritional deficiency symptoms associated with vitamin imbalances are well-documented (see Table 7.7). However, imbalances in vitamin content of fish diets are not the only causes of nutritional diseases. Thyroid tumours, liver degeneration, visceral granuloma, anaemia and pigmentation impairment can be caused by other forms of nutritional imbalances. High levels of starch may give rise to symptoms of diabetes in trout and enlarged liver in channel catfish. Freedom from disease is an essential element of physical well-being, but physical and environmental stress have also significant roles in the maintenance of healthy conditions. Many of the potential pathogens of aquaculture species are normally found in the aquatic environment, but in spite of their presence disease may not occur. Obviously, disease is essentially the result of interaction between the species, the disease agent and the environment. So the three major factors of significance are the susceptibility of the species to the pathogens present, the virulence of the pathogenic agent and the environmental conditions that may trigger epizootics. Despite the individual importance of each one of these factors in the maintenance of good health or avoidance of disease, it should be emphasized that it is the balance between these factors that determines the state of health. Even in the presence of all three factors, the interaction may be such that no disease occurs. But a disturbance in any of the factors, leading to disruption of the relationship, can give rise to disease.

Susceptibility of the host

The susceptibility or the resistance of the culture species to the action of the disease agent is governed by its physical barriers, its exposure experience and its age. Among the physical barriers are the skin, scales, exoskeleton or shells and mucous membranes which limit the entry of toxic, infectious and parasitic agents. The physiological defences that keep the body from being over-run include the white blood cells that engulf pathogens, avoidance mechanisms, detoxification of chemicals from water or

diet by the liver, storage of certain metals by the bones and local tissue reactions. The overall nutritional well-being is the source of the host's physiological ability to defend itself. The immune system and its specific activity against biological agents such as viruses, bacteria and parasites forms an important means of disease resistance. Populations with previous exposure to specific disease agents will generally not be as readily susceptible as those on a first encounter. For this reason and also because of the fragility of their defence system, young ones are more susceptible to diseases than older ones, except that the spawners may experience additional stress because of their reproductive functions. The species specificity of certain disease agents is also a factor of importance in understanding health hazards.

Once the pathogen has established itself within or on the host under favourable conditions, the infection may take one of three routes:

(1) the pathogen proliferates, eventually causing mortality of the host;
(2) the defences of the host surmount the infection and eliminate the pathogen from its system; or
(3) a carrier state develops, whereby a balance between the host and the pathogen may persist generally, with no evident disease symptoms.

From an aquaculture point of view, the greatest concern is the rapid multiplication of the pathogen within the host and the danger of transfer to other individuals of the host population, which may result in an uncontrollable epizootic. During the incubation period (which is the interval between the penetration or establishment of the pathogen in the host and the appearance of the first symptoms of the disease) the host will often be shedding the pathogen. If the host recovers after this initial stage, or after any of the later stages of the infection, without entirely eliminating the pathogen, a carrier condition exists. A carrier can disseminate the pathogen into the surrounding environment or can harbour it in a latent state without shedding. So, even after the clinical stage of infection, some individuals recovering from the disease continue to disseminate pathogens in a manner similar to those that are chronically ill.

As the transfer of infection can occur without the manifestation of disease symptoms, the infections may often be difficult to identify and can be passed unnoticed from individual to individual or even generation to generation. Until the population experiences particularly stressful conditions, which exacerbate disease symptoms, no infection may be suspected. The problem of carrier states in aquaculture species remains one of the most crucial ones for the aquaculturist.

When a disease outbreak is encountered, the pattern of losses, the size of hosts affected and the duration of the epizootic provide valuable information. Sudden, explosive mortalities often implicate acute environmental problems, such as oxygen deficiency, the presence of lethal concentrations of toxicants or lethal levels of temperature. The appearance of a few sick individuals, unusual behaviour or loss of appetite can indicate the beginnings of infectious disease. A disease is generally due to the inability of the host to adjust adequately to environmental stress and consequent dominance of the pathogen, and so the aquaculturist should act quickly when losses occur in typical patterns. A balance between the host and the pathogen should be restored by resolving environmental problems and by effective therapeutic treatment. Timely action is the essence of success in controlling epidemics of mortality in aquaculture, but it needs considerable skill to correct adverse environmental conditions in time to prevent major losses.

The type of aquacultural practice adopted has a decisive role on the susceptibility of the culture species. As indicated earlier, a high density of stocks and the use of restricted spaces like cages, tanks and raceways lead to closer contact between individuals as well as environmental stress. Higher stock densities also mean the use of larger quantities of concentrated feeds and/or fertilizers. This leads to denser growth of plankton and benthos which may include intermediate hosts of disease agents. The environmental and disease risks related to overloading of ponds and enclosures with organic manures are very considerable. The use of heated water effluents from industries and sewage effluents also has built-in risks; so an aquaculturist has to be prepared for quick and effective action, when an adverse situation develops.

While some of the aquaculture practices are conducive to diseases, there are others that are effective in controlling them. For example, the practice of regular drying of fish ponds and application of lime on the pond bottom helps to kill parasites and many other infectious disease agents.

Pathogen

Biological agents are probably the most common cause of disease initiation and are the primary focus of attention in infectious diseases. As mentioned earlier, potential pathogens are always present in the aquatic environment. They may include viruses, bacteria, fungi, protozoans, parasitic crustaceans, helminths and other worms. The virulence or pathogenicity of the agent is the relevant factor in the determination of health hazards. It depends upon the physical or biochemical attributes of the agent. Bacteria with flagella or with capsules are generally better equipped to invade the host and resist adverse conditions. Some bacteria are able to elaborate toxins, which cause haemorrhage or affect the nervous system of the host. Enzymes such as chitinase enable bacteria to erode chitinous membranes. Parasites, on the other hand, attach themselves to the host through special organs of attachment, such as suckers.

Penetration into the host is the first step for a microbial agent to multiply and invade the vital organs of its host. This normally happens through ingestion, rupture of the skin, transgression of gill lamellae or penetration of the egg membrane. The specific point of entry may have a decisive role on the virulence of the microbe. Wounds in the skin are common entrance points for some of the bacterial and viral infections, which in turn invite fungal secondary invaders such as *Saprolegnia* sp. Other routes of entry are usually (i) the gills, where the pathogens can either enter the body through the delicate and thin epithelium, or establish themselves on them as in the case of protozoan infection with *Schizamoeba salmonis* or *Ichtyobodo necator* (*Costia necatrix*) and (ii) the digestive tract, where protozoans like *Ceratomyxa shasta* may become numerous enough to weaken the fish. Some bacteria may penetrate the intestinal lining under certain conditions. Eventually the pathogen may return to the aquatic medium when shed by the host.

The host/pathogen relationship generally undergoes several stages of development. The incubation period is when the pathogen multiplies but the host does not yet show clinical signs of disease. The incubation period may range from a day or two for virulent pathogens to prolonged periods of several months. After this asymptomatic period, specific and non-specific signs of disease become evident. Whether the host dies or survives will depend on its ability to resist the infection. During an epidemic, some of the infected animals may not exhibit clinical signs at all and become carriers, capable of transmitting the disease agent or initiating a future epizootic. Animals that recover from a disease may be completely free of the disease agent or continue to be asymptomatic carriers. In many instances, a disease condition may involve more than one pathogen or the infection by one primary agent may create conditions suitable for a secondary agent to gain access. Bacterial infections often follow the establishment of a parasite or of a virus. It is not uncommon to find a wide array of disease and parasitic problems occurring simultaneously.

Environment

The environment plays a crucial role in disrupting the balance between the host and the pathogen. In many situations, the culture animals live a healthy normal life in the presence of pathogens; but when environmental stresses occur and the balance tips in favour of the disease, the pathogen gets the upper hand and disease conditions ensue.

As the primary environmental parameters required would have been adequately considered in selection of the site and species, the relevant stress factor would normally be environmental disturbances that extend the adaptive responses of the animal beyond the normal range or affect the normal functioning to such an extent that chances of survival are significantly reduced. Morphological, biochemical and physiological disturbances occur in different stages and are characterized by a variety of metabolic conditions, such as anoxia, fright, forced exertion, anaesthesia, temperature changes and injury. Though the effect of stress is the alteration of host biochemistry in order to increase the probability of survival of the host, some of the resulting metabolic changes contribute also to increased susceptibility to infection.

Of the physical factors, temperature is one that has an effect on a number of other variables in the environment. Temperatures above or below the tolerance limits of the host animal create stress. Increased metabolic rate caused by high temperature results in higher oxygen demand. However, dissolved gases, including oxygen, generally decrease in solubility with increasing temperature. Also the solubility of toxic compounds increases with increasing temperature, creating unfavourable conditions.

As well as the environmental effect on the host, the effect of temperature on the pathogen is also an important factor to be considered. As for example, a rise in temperature generally accelerates to a certain limit all the biological processes of the causative agent, above which its viability is lowered, sometimes causing its death. Similarly, lowering of temperature decreases the biological processes to a certain minimum, below which the organism may not survive. Pathogenic organisms of the same genus in the same host may react differently to a change in temperature.

The minimum water quality conditions necessary to maintain fish health are:

dissolved oxygen	5 ppm
pH range	6.7–8.6 (extremes 6.0–9.0)
free total CO_2	3 ppm or less
ammonia	0.02 ppm or less
alkalinity	At least 20 ppm (as $CaCO_3$)

Obviously there are differences in the tolerance limits of different species, but these values provide a general guideline. Levels of tolerance of other elements are chlorine: 0.003 ppm, hydrogen sulphide: 0.001 ppm, nitrite (NO_2): 100 ppb in soft water, 200 ppb in hard water and total suspended and settleable solids: 80 ppm or less.

Pesticide pollution is one of the common causes of environmental stress in aquaculture situations. The maximum pesticide concentrations that may be tolerated by fish, without noticeable effects, and recognized by the Environmental Protection Agency of the USA are listed in Table 9.1.

Even though it is not a hazard to the aquaculture species itself, the development of off-flavour is a phenomenon that seriously affects the economics of culture. The earthy or musty taste of fish grown in affected ponds would make them unmarketable. The cause of off-flavour is reported to be a compound called geosmin produced by actinomycetes and a number of blue-green algae of the genus *Oscillatoria* (such as *O. princeps*, *O. agardhi*, *O. tenuis*, *O. prolifica*, *O. limosa*, and *O. muscorum*). All these organisms grow on mud that is high in organic matter. The organic matter decomposes, causing the reduction of the mud. These organisms grow well on the interface between the reduced mud and the oxidized water layer above it. The off flavour generally disappears when the fish are held in clean water (preferably running water) for one to two weeks.

Another source of off-flavour in fish is industrial wastes. The odour and taste of these wastes are usually concentrated in the fat deposits of the fish's body. The most important chemicals that impart off-flavours are phenols, tars and mineral oils. Chlorinated phenols, such as *o*-chlorophenol and *p*-chlorophenol, impart a distinct flavour to carp even in low concentrations of 0.015 and 0.06 mg/ℓ, respectively. Eels and oysters are even more sensitive and develop off-flavour when the water contains as little as 0.001 mg/ℓ *o*-chlorophenol. A concentration of 5−14 mg/ℓ mineral oil, or less if in suspension, also imparts a distinct flavour.

9.1.2 Integrated health management

In Section 9.1, the relevance of a concept of integrated health management in aquaculture was referred to. The discussion of the factors affecting fish health underline the need for such an approach to reduce levels of risk and accelerate the development of the emerging aquaculture industry. Such a management programme that involves (i) the implementation of appropriately planned guidelines for prevention, control and eradication of diseases, (ii) correction of disease-causing and disease-spreading conditions in farms and (iii) the adoption and implementation of policies and regulations by the State requires a high degree of cooperative effort. In its basic philosophy and approach, such a programme may not be very different from the integrated pest control programmes in agriculture, which are in operation in many countries. The basic difference would appear to be the infancy of the science, lack of proper organization of the aquaculture industry and some of the complications caused by the watery medium in which aquaculture crops are raised. In integrated pest management, all classes of pests and their interrelationships are considered together and the protection of the crop is seen as an important element in the overall management of an agro-ecosystem. It extends the concept of integrated control as a combination of all available management

Table 9.1 Maximum permissible pesticide concentrations which may be tolerated by fish.

Pesticide	Concentration
Organochlorine pesticides	
Aldrin	0.01
DDT	0.003
Dieldrin	0.005
Chlordane	0.004
Endrin	0.003
Lindane	0.02
Toxaphene	0.01
Organophosphate insecticides	
Diazinon	0.002
Dursban	0.001
Malathion	0.008
Parathion	0.001
TEPP	0.3
Carbamate insecticides	
Caebaryl	0.02
Zectran	0.1
Herbicides, fungicides and defoliants	
Aminotriazole	300.0
Diquat	0.5
Diuron	1.5
2,4−D	4.0
Silvex	2.0
Simazine	10.0
Botanicals	
Pyrethrum	0.01
Rotenone	10.0

techniques, implying the coordination of relevant control measures into a unified programme which seeks optimal control. In practice, this means using appropriate techniques to keep disease agents below a certain level that causes an unacceptable economic loss. Selection of species and strains, immunization, environmental manipulation, nutrition and chemotherapy are techniques that are used for this purpose. However, there comes a point when the economic threshold of disease control has been reached and further efforts to control or eliminate the pathogens will not compensate the cost involved in terms of the benefits derived. Exceptions would be diseases that have to be eradicated under any circumstances, at any cost, due to dangers involved not only to the individual farms concerned, but also to others in the industry and community.

In planning fish health management measures, a distinction has to be made between the control of obligate pathogens, which affect only the particular species under culture, and infectious diseases which are caused by facultative (opportunistic) pathogens usually present in the environment. The control of obligate pathogens is based on identifying the sources of infection, breaking the connection between such sources and susceptible aquaculture stocks and on reducing the susceptibility of the exposed stocks. A broader strategy that integrates all the available management techniques is required for the control of diseases caused by facultative pathogens. The economic threshold for diseases caused by facultative pathogens is often difficult to determine. It can be done only on the basis of experience in assessing the economic impact of a disease and an understanding of the merits of additional control measures that can be adopted.

Prevention can be considered as the cornerstone of a health protection programme, but is often more complex than the control of an existing disease. This would involve a reliable identification of the disease and its carriers, adequate knowledge of the transmission mechanisms, the development of effective methods of preventing the access of pathogens and their carriers into culture facilities and the provision of environmental conditions conducive to the maintenance of a healthy condition among culture species.

Health inspection and disease monitoring

Timely information on the health status of aquaculture stock and the environmental conditions is essential for an effective disease control programme. Such information will enable a quick response to disease outbreaks and reduce the mortality rate in infected stock by the provision of prompt therapy. If the disease problem is attributable to poor management, immediate corrective measures can be undertaken. When necessary, appropriate steps can be taken to prevent the introduction of disease agents from outside sources through transfer of eggs or adults or exposure to other contaminated sources.

Ideally, every large aquaculture enterprise should have at least one trained person and the basic facilities to undertake regular health and environmental monitoring of the farm. Prompt on-site diagnosis of disease permits the immediate application of chemotherapy or remedial measures to control or eradicate the disease. Where such on-site diagnosis is not feasible due to economic reasons or lack of trained personnel, a cooperative or State-organized system of health inspection would be a possible alternative. A competent person should inspect the facilities at regular intervals and gather the relevant information. The minimum number of inspections required is suggested as twice a year. The inspection should be conducted using procedures that will detect the presence of parasitic, viral and bacterial pathogens. Samples of all lots of stock in the farm should be inspected. A sample is recommended to contain sixty individuals of each lot, except in the case of brood stock where the number may have to be reduced. The inspector should have access to laboratory facilities to study the samples in as detailed a manner as is necessary. For the identification of viral agents or viral diseases, the facilities of a laboratory that maintains tissue culture and specific antisera will be needed. Collection and preservation of specimens for detailed examination will have to be done as required by the diagnostic laboratories. In situations where a viral epizootic is suspected, only a relatively few specimens (between fifteen and twenty) exhibiting the typical symptoms will be required for examination.

The value of organized health inspection and disease monitoring is manifold, despite the costs involved. By detecting predisposing conditions for an outbreak of disease in advance, preventive actions can be taken and unnecessary losses avoided. Infections diagnosed in the early stages can more easily be treated. It may also be possible to isolate infected individuals from the stock and prevent spread of the disease to others in the same farm or those in neighbouring farms.

As will be discussed later, the control of communicable diseases of aquaculture species, nationally as well as internationally, will have to be based on legislative regulations that require appropriate reporting procedures and certification at source of stock and eggs, for freedom from infection, by competent personnel. A regular health inspection of the farm may also enhance its acceptability for participation in a risk-insurance programme.

Disease treatment

The eradication of a disease, where it occurs, requires a programme that will remove infected stocks, prevent re-infection, reduce stress and maintain optimal conditions. Chemotherapy or the use of therapeutants generally gives only a temporary advantage over the pathogen. If conditions are not improved, disease can recur when, due to stress or other reasons, the animal again becomes susceptible to infection. The best treatment for disease in aquaculture is therefore the adoption of sound husbandry practices and the avoidance of pathogens.

Chemical treatment to prevent disease may have the effect of reducing or eliminating pathogens or of controlling populations of heterotrophic microorganisms which may act as facultative pathogens of animals under stress. However, such treatment may also have negative effects on biological filters in controlled recirculating systems, particularly on nitrifying bacteria, and may adversely affect algae in the culture systems, besides leaving undesirable or harmful residues in cultured animals. (Sindermann, 1986). Further, some countries have restrictions on the use of chemicals to treat animals raised for food. For example, the United States Food and Drug Administration has approved the use of only salt, acetic acid and sulphamerazine on food fish and has restricted the use of oxytetracycline in trout, salmon and catfish. Some commonly used chemicals, like malachite green, are likely to be carcinogenic to farm workers who handle it. Most of the available information on chemotherapy is based on experience in fresh-water fish farming. Table 9.2 gives data on drugs used in pond fish culture in the USSR.

The chemical characteristics of the water supply in the farm will affect the toxicity and efficacy of chemotherapeutic treatment and this has to be assessed before any large-scale treatment. It has been recommended that before using a chemical, it should be tested on a small number of sick animals. As the tolerance of sick animals to chemical treatment will often be less than of healthy ones, it may become necessary to adjust treatment levels if they are weak or in poor condition. Before starting treatment, it is necessary to ensure that the facilities are as clean as possible, because organic matter will absorb some of the treatment chemical and reduce its effectiveness. Reduction of stock density, suspension of feeding a day or two before treatment, monitoring of dissolved oxygen levels before and during treatment, supplemental aeration and treatment during the coolest part of the day are some of the other precautions suggested for effective treatment.

Continuous feeding of antibiotics at low levels in the diet as a prophylactic measure against outbreaks of bacterial disease is a practice that is discouraged. At such low levels, the antibiotic may serve only to kill those bacteria most sensitive to the drug, and lead to the development of drug-resistant strains. So antibiotic treatment should be done only when actually needed, and then only at the prescribed treatment levels and for the required period.

Commercial medicated feeds, if available, are easy to use and in certain circumstances less expensive. Medicated feed can also be made on the farm. Suspending the drug in a suitable oil such as cod liver oil and coating the daily ration of feed pellets with the mixture of oil and drug in a mixer would be an easy way of making medicated feeds.

The local application of drugs and injections have limited use in disease control in aquaculture — bath and dip treatments are much

Table 9.2 Main drugs used for disease control in pond culture in the USSR (from Bauer *et al.*, 1973).

Disease	Measures to be taken	Drugs	Dose	Time of exposure
Infectious dropsy of carp	Injections	Levomycetin	For spawners: 10–20 mg per kg fish	Spring and autumn
	In the feed	Levomycetin	100–300 mg per kg feed	Two or three times in early summer
	Baths	Methylene blue	50–900 mg/ℓ	2–11 hours
	In the feed	Methylene blue	1 g per kg feed	7 days, three times, with 3–4 days intervals
	Disinfection of ponds	Chloride of lime	0.3–0.5 tons/ha	Spring and autumn
		Quicklime	2.5 tons/ha	Spring and autumn
Inflammation of the swim bladder	In the feed	Methylene blue	For spawners: 3 g per kg feed	Spring and autumn for 15 days (3 days of curative feeding followed by 2 days of normal feeding)
			For underyearlings: 20–30 mg per fish	Four courses of 10 days each, with intervals of 5–8 days
			For two-year-olds: 35–40 mg per fish	2–4 courses
Branchiomycosis	On the water	Copper sulphate	2–3 kg/ha	Once a month beginning in May
	Disinfection of ponds	Quicklime	150–200 kg/ha	Prophylactically twice a month; daily in case of an outbreak.
Costiasis	Disinfection of ponds	Quicklime	2.5 tons/ha	Spring and autumn
	Disinfection of ponds	Chloride of lime	0.3–0.5 tons/ha	
		Quicklime	2.5 tons/ha	
	Baths	Sodium chloride	5%	5 minutes, three times, with 5-day intervals
Chilodonelliasis	Baths	Copper sulphate	8:100000	Several minutes
		Formalin 40%	1:4000	8–10 minutes
		Sodium chloride	5%	5 minutes
	Treatment of fish in the pond	Copper sulphate	8:1000000	30 minutes
		Sodium chloride	0.1–0.2%	2–3 days, 1–3 times
		Malachite green	0.5–1.0 g/m³	4–5 hours
		Basic brilliant green	0.1–0.2 g/m³	24–48 hours
	Disinfection of ponds	Quicklime	2.5 tons/ha	Spring and autumn
Trichodiniasis	Baths	Sodium chloride	5%	5 minutes
		Ammonia	0.1–0.2%	1 minute
	Treatment of fish in the pond	Mixture of copper and iron sulphate	7:1000000	

(cont.)

Table 9.2 (*cont.*)

Disease	Measures to be taken	Drugs	Dose	Time of exposure
Ichthyophthiriasis	Baths	Malachite green	$0.5-1.0 \text{ g/m}^3$	4–5 hours, two or three times
		40% formalin	1:2000	3–7 minutes, once
		Mixture of sodium and magnesium chlorides	3.5:1.5	3–10 days
	Treatment of fish in the pond	Trypaflavin	0.001%	10 hours
	Baths	Mercury nitrate	$2 \text{ mg}/\ell$	2–3 hours
		Chloride of lime and copper sulphate	0.001%	15–30 minutes, once
			0.0008%	15–30 minutes, once
	Treatment of fish in the pond	Mercury nitrate	$0.1-0.3 \text{ ml}/\ell$	1 day, once
		Basic brilliant green	$0.1-0.2 \text{ g/m}^3$	1–2 days, once or twice at a temperature to 12°C
		Basic violet K	$0.1-0.2 \text{ g/m}^3$	
Apiosomiasis	Treatment of fish in the pond	Brilliant green	$0.05-1.0 \text{ g/m}^3$	1–2 days
Whirling disease of trout	In the feed	Osarsol	0.01 g per kg of fish	3 days
			0.02 g per kg of fish	3 days, with an interval of 7 days; repeat three or four times
	Treatment of the pond bottom	Calcium cyanamide	1 kg/m^3	Once
Coccidiosis	Disinfection of the pond bottom	Chloride of lime and quicklime	0.5 and 2.5 tons/ha	Spring and autumn
	In the feed	Furazolidone	0.2–0.3 mg per under-yearling	Three times, with 1-day intervals
Dactylogyrosis and gyrodactylosis	Baths	Sodium chloride	5%	5 minutes
		Ammonia	0.1–0.2%	0.5–1 minute
		Potassium permanganate	1:100 000	15–30 minutes
	Treatment of fish in the pond	Ammoniate cupric sulphate $Cu(NH_3)_4SO_4$	$0.1-0.3 \text{ mg}/\ell$	Four times, with 48-hour intervals
	Baths	Chloride of lime	10 g/m^3	1 day
		40% formalin	1:5000	25 minutes
	Disinfection of ponds	Chloride of lime	0.5 tons/ha	
		Quicklime	2.5 tons/ha	

Disease	Method	Agent	Dose	Application
Trematodoses (sanguinicoliasis, diplostomatosis, posthodiplostomatosis)	Destruction of intermediate hosts – molluscs	Chloride of lime	0.5 tons/ha	On the wet pond bottom after draining the pond
		Quicklime	2.5 tons/ha	As above
		Chlorophos	0.1–1% solution	As above
Bothriocephalosis of carp	Dehelminthization with the food	Kamala underyearlings	0.1 g per fish	Two or three times every other day
		spawners	0.5–1.0 g per fish	As above
	Individual dehelminthization	Phenothiazine underyearlings	0.8 g per fish	As above
		Kamala spawners	0.5–1.0 g	Once
		Filixic acid	60–200 mg	Twice every other day
Argulosis	Baths	Potassium permanganate	0.001%	30 minutes
		Chlorophos	0.1 g/ℓ	1 hour
		Aqueous suspension of hexachlorane	1:1 000 000	Several hours
Lernaeosis	Baths	Chlorophos	0.1 g/ℓ	1 hour
		Potassium permanganate at: 15–20°C	1:50 000	1.5–2 hours
		21–30°C	1:100 000	
	Treatment of fish in the pond	Chlorophos	0.5 mg/ℓ	Up to 20°C, once every 2 weeks / Above 20°C, once a week
Piscicolosis	Baths	Sodium chloride	2.5%	1 hour
		Cupric chloride	0.005%	15 minutes

more applicable and common. Dips of short duration, varying from a few seconds to 5 minutes, depending on the chemical and the concentration, are recommended for certain infections, especially of brood stocks. Though very effective, such treatment can be stressful. After treatment, the animal should be rinsed in clean water and released in water free from parasites and pathogens. Where fresh water is available and adequate oxygen levels can be maintained, baths of up to an hour's duration can be given. High concentrations of chemicals can be used, but care has to be taken to avoid overdoses and over-long contact times.

In pond farms, lower concentrations of the chemical are usually used and allowed to dissipate in the ponds. Generally, the water level in the pond is lowered to reduce the amount of chemical to be used. There are, of course, possible risks to pond biota and oxygen levels. For example, the degradation of formalin uses up 1 ppm oxygen for each 5 ppm of formalin, and it acts as an algicide. In running water ponds and raceways, a system of flush treatment is possible. The required amount of the chemical is added at the inlet and allowed to flush through the pond or raceway.

Sanitation

The maintenance of sanitary conditions in an aquaculture facility is of the utmost importance in preventing the outbreak of disease. This is obviously very much tied in with sound culture practices and sometimes it becomes difficult to separate the two. Monitoring of the water supply is an effective and essential means of controlling diseases. Actual disinfection of supplies is often quite expensive and is generally possible only in hatcheries. Three acceptable methods of disinfection are recommended: ozonation, ultra-violet irradiation and chlorination. When a facility is affected by infectious diseases and the necessary treatment has been applied or the stock destroyed as the disease is incurable, disinfecting the facility and maintaining sanitary conditions on a continued basis are especially important.

The main goal of a sanitation programme is to prevent the spread of pathogens of cultured species. Egg disinfection strives to prevent the transmission of pathogen from the parent stock to the progeny and transfer from the hatchery to the rearing areas. Sanitary measures can help in confining pathogens of an infected stock to one part of the farm and prevent them from spreading infection to other parts.

The roles of health inspection and infection monitoring to prevent the spread of disease have already been referred to. When introduction or transplantation of aquaculture species is essential and proper certification of eggs, fry or adults to be transplanted, based on regular health inspection is not available, it will be necessary to consider quarantine measures. This involves retention of newly imported stocks in quarantine facilities for prescribed periods to ascertain whether they are carriers of disease agents. Quarantining is most relevant when the main purpose is to prevent the introduction of communicable diseases that have never been recorded in a country or region. It has been used more widely in research situations, rather than in commercial farming, because of the need for sophisticated holding facilities with capabilities for total disinfection of effluents. Quarantine facilities must be separated by physical barriers or located away from the farm. All effluents from the quarantine must be fully disinfected. Chlorination of effluents at 200 mg/ℓ total chlorine for two hours is recommended. Lower concentrations may be effective in certain facilities, particularly if a longer exposure is used. These levels may apply only to small water volumes, as use on large volumes of water could represent serious environmental hazards. It is obvious that effective quarantine measures are relatively very expensive and regulatory agencies often have difficulty in justifying such expenses against the economic benefits to be derived from the introduction or transplantations that can be regulated through these measures.

Immunization

A relatively new technique of disease prevention in fish is through immunization with vaccines. Licensed vaccines are now available against an increasing number of diseases. Though they do not give absolute protection from infection, they do help to combat infections, especially when the specific diseases cause repeated problems.

Protection from a disease is sought to be achieved through the development of specific resistance of the cultured organism to the causal agent. The mechanism of antibody production is a crucial element in such an acquired resistance. An antibody is a specific immunoglobulin (modified protein) which is produced in response to and reacts specifically with an antigen. An antigen is any foreign substance which is capable of stimulating the formation of antibodies and reacting with the produced antibodies, under suitable conditions. Vaccines or bacterins contain antigens that are generally attenuated or killed disease agents. When administered to a host, they stimulate the production of specific antibodies or non-specific resistance to that particular disease agent. The chances of survival of the host, when infected by the pathogen, are greatly enhanced by the immunization achieved through the production of antibodies. The immunological responses of fish are generally similar to those of terrestrial vertebrates, including phagocytosis and the elaboration of specific immunoglobin antibodies. Prophylactic immunization of several species of fresh-water fishes has been attempted, using this characteristic. Vaccination has been shown to be cost-effective in salmonid species that have a sensitive immune response system that can be stimulated by immunization. Attempts have also been made to immunize salt-water-held salmon against vibrio and other bacterial infections. It has, however, to be remembered that there are many other diseases of salmonids and other species for which vaccines have yet to be developed.

Vaccination can be done by a variety of methods, of which the most successful ones for fish immunization appear to be immersion/spray-shower vaccination and injection. Immersion is specially suited for small fish (1–4 g), whereas spray-shower vaccination is more convenient for fish larger than 4 g. The immune response is affected by temperature conditions, higher temperatures contributing to a rise in specific antibody production. Under optimal conditions, it takes 2–4 weeks for protective immunity to develop after antigenic stimulation. A vaccinated fish can retain immunity for well in excess of 300 days, if temperatures are favourable. Immersion and spray-shower vaccinations have the advantages of rapid administration and cost-effectiveness. Where it is specially needed, vaccination can also be done by injection. Intra-peritoneal injection is preferred because of the rapid development of protection and ease of administration.

Genetic resistance to disease

Based largely on experience in genetic breeding for disease resistance in agriculture, there is considerable optimism concerning the possibilities of developing strains of fish and other aquaculture animals that can resist certain infections. Fish are known to adapt to disease in nature and these traits of resistance can be measured experimentally. However, Warren (1983) pointed out some of the problems involved in the development of disease-resistant strains of fish, based on experience with salmonids. Brook trout strains, selected for their resistance to furunculosis, were reported to have acquired greater susceptibility to bacterial gill diseases during selection. Some strains of US West Coast steelhead trout, resistant to bacterial kidney disease (BKD), were also the strains most susceptible to *Vibrio anguillarum* infections. It is suggested that the loss of genetic diversity in a selection process makes it difficult to develop strains of fish that are resistant to several diseases at the same time. In a programme of selection for disease resistance, survival after a challenge by the particular disease will have a high trait value. But the surviving disease-resistant animal may develop a carrier state. This is harmful, especially if the disease carrier is to be introduced into new areas where the disease would not otherwise occur. However, it is believed that by maintaining a high level of genetic diversity in a stock and by developing hybrid vigour, the ability to withstand the stress of infectious diseases can be enhanced.

Farm disinfection

As pointed out above, when a disease occurs the aquaculturist often has no choice but to destroy the stock and disinfect the rearing facilities before starting operations with new uninfected stock. The disposal of the infected stock and disinfection are of special importance if further rearing in the facility is to succeed.

Experience seems to indicate that the control of certain diseases can be achieved only through disinfection and eradication of contaminated stocks. The two situations where disinfection becomes impractical are (i) when the probability of reinfection from nearby open waters or farms is unavoidably high and (ii) when the economic loss due to the disease is less than the cost of disinfection. When economic loss is assessed, not only the loss sustained by the particular facility but also potential losses to other facilities in the vicinity should be considered, if not by the individual farmer at least by the regulatory agencies.

Obviously it is easier to destroy the stock and disinfect small, well-controlled facilities like hatcheries, tank farms and raceways. Earthen pond farms are considerably more difficult to disinfect (fig. 9.1). The priority consideration for a commercial operator is the maximum utilization of the stocks in the farm in such a way as not to aggravate the disease problem or contribute to its spread to non-enzootic areas. Though diseased animals should not be sold to another farmer for rearing, marketable animals can in most cases be sold for human consumption, if it is confirmed that the product (after processing or cooking) does not cause any health hazards to the consumers.

If direct utilization is not an acceptable option, the stocks should be destroyed through burial or incineration.

For hatchery and raceway disinfection, chlorine is favoured by several agencies. A concentration of 200 ppm of available chlorine is recommended. If the chlorinated water will enter water bodies containing fish upon leaving the farm, it will be necessary to inactivate the chlorine by neutralization with commercial sodium thiosulphate (at the rate of 1.5 g for every litre of 200 ppm chlorine solution). For disinfecting hatcheries, an exposure of about one hour is recommended at the concentration of 200 ppm of available chlorine. For fish ponds, it may be necessary to super-chlorinate, to establish a chlorine residual of 5–10 ppm, and have an exposure period of 12–24 hours. It will take one to two days for the residual to drop to 0, and only after that should any new stock be introduced into the ponds.

As chlorine dissipates rapidly and is inactivated by organic matter, it may be advisable to maintain in hatcheries a concentration of 100 ppm or more for several hours, after the initial treatment at 200 ppm for one hour. In large hatcheries and raceways, treatment may have to be done in sections, but it should be done in such a way that no fresh contaminated water

Fig. 9.1 Disinfecting a fish pond by liming, after draining and ploughing the dried bottom (courtesy of Nikola Fijan).

flows through parts of the system after they have been disinfected and that dilution of the chlorine solution does not occur.

When dealing with whirling disease, drained earthen ponds should be disinfected by applying slaked lime at the rate of about 2 tons per hectare of wet pond bottom. Several treatments may be required to disinfect earthen ponds thoroughly, because most chemicals are less effective in mud.

For bacteria, viruses and non-specific protozoa in ponds, Finlay (1978) recommends a solution of 1 per cent sodium hydroxide and 0.1 per cent teepol (a detergent). The detergent enhances the penetration of the disinfectant through the soil and the combination is not affected by organic matter. A high pH of about 11 or more is necessary for this disinfectant to be effective. Hnath (1983) lists a number of disinfectants that can be used in hatcheries (Table 9.3).

9.2 Major diseases of aquaculture species

As mentioned earlier in this chapter, there are several diseases that an aquaculturist may have

to deal with, particularly in semi-intensive and intensive systems of culture. Some of the major diseases are briefly described in this section and the reader should refer to treatises on fish diseases listed in the Reference section for more comprehensive descriptions of these as well as other diseases observed in aquaculture facilities. In Part II, reference will be made to diseases that are relevant to major groups of cultured species. As in the case of other aspects of aquaculture like nutrition, feed technology and reproduction, most of the existing information relates to a small number of species, such as salmon and trout, common carp, channel catfish and eel. Several diseases have been observed in species cultured in marine and brackish-water environments in recent times, but aetiological information on many of them is still incomplete. The known major diseases in aquaculture are caused by infection by pathogens of viral, bacterial, fungal or protozoan origin. Many of them have no known methods of therapy and prophylaxis is the only control measure. There are also many diseases caused by parasitic copepods and helminths.

Table 9.3 Disinfectants and their application (after Hnath, 1983).

Disinfectant	Working strength	Application	Effective against
Chlorine	1−2%	Concrete, fibreglass, butyl-lined ponds, nets, etc. footbaths	Bacteria, fungi, viruses
Sodium hydroxide	1% with 0.1% teepol; 0.5 gallon/m^2 (2.28 ℓ/m^2)	Earthen ponds, concrete, fibreglass, butyl-lined ponds, nets, footbaths, etc.	Bacteria, viruses, protozoa
Iodophors	250 ppm	Concrete, fibreglass, butyl-lined ponds, nets, etc. angling equipment, clothing, hands.	Bacteria, viruses
	100 ppm	Ova	
Quaternary ammonium compounds (Hyamine, Roccal, etc.)	As manufacturers' instructions	Nets, clothing, hands	Bacteria
Calcium oxide	As powder; 380 g/m^2	Earthen ponds, fibreglass, concrete, butyl-lined ponds.	Protozoa (whirling disease)

9.2.1 Viral diseases

Infectious pancreatic necrosis (IPN)

Infectious pancreatic necrosis is primarily a viral infection of trout and salmon, but the virus has also been isolated from a number of other fish species, including eels. IPN-like viruses have been isolated from common carp and several species of bivalve molluscs from coastal waters. It is a major fish disease problem in North America and Europe and has been found also in Japan. It occurs as an acute disease in fry and fingerlings of trout. Culture technologies and management practices can affect the severity of disease outbreaks. When the mortality rate is high, infected individuals swim in a rotating manner about their long axis. This whirling behaviour is a terminal sign and death occurs within an hour or two. Prior to this stage, the affected individuals may remain on the bottom, showing weak respiration and convulsive movements. It has been observed that very young fish or fish in poor condition may not exhibit this characteristic whirling behaviour. An overall pale pigmentation of individual fish, exophthalmia, abdominal distension and haemorrhages in the ventral areas can be noticed (fig. 9.2). Haemorrhages also occur in the pyloric caecae, and the liver and spleen are usually pale. The occurrence of clear or milky mucus in the stomach and anterior intestine is a distinctive characteristic of IPN. Necrosis and inclusion bodies are evident histologically in the pancreatic tissue. Confirmation of the diagnosis requires isolation of the virus in cell culture and identification by a serum neutralization test, using polyvalent anti-IPN virus serum.

IPN virus belongs to the birnavirus group. It grows in monolayers of fish cell cultures and induces a typical cytopathic effect. The incubation period is dependent on temperature, ranging from 6 days at 12.5°C to several weeks at 4°C. Most survivors of the infection become life-long virus carriers, intermittently shedding varying quantities of virus over a long period through urine, faeces, milt and eggs. This leads to the transmission of the virus from parents to progeny through the egg and accounts for transmission of the disease from one generation to another.

As there is no proven effective treatment for IPN, the only means of control is through

Fig. 9.2 Rainbow trout infected with IPN (below). Note the pale pigmentation and distended vent in comparison with the uninfected trout (above) and the white mucous cast from anus (courtesy of Nikola Fijan).

preventive measures, which include the incubation of virus-free eggs and the propagation of IPN-free stock in uncontaminated water supplies. Rigorous fish health inspection programmes are essential to prevent inadvertent introduction of the disease.

Infectious haematopoietic necrosis (IHN)

Infectious haematopoietic necrosis is an acute viral disease of trout and salmon fry in North America and Japan. It is caused by a bullet-shaped virus and can be transmitted from fish to fish and from parent to progeny through seminal fluids or infected eggs. The disease is generally seen in fry and fingerlings, but this depends also on the host species. For example, in chinook salmon and steelhead and rainbow trout, mortality may occur from the sac fry stage to the yearling stage. Except for rainbow trout in certain areas, older fish rarely die from IHN.

Dark coloration, weakness, abdominal swelling and pale gills are some of the external signs of the disease. The internal signs are very similar to those caused by other viral infections. In infected sockeye salmon fry, the kidney becomes translucent and speckled with pigment cells. Diagnostic confirmation requires isolation and identification of the virus by neutralization tests with anti-IHNV serum. IHN virus can reliably be detected only during the spawning season in carrier fish. Sometimes the IHN infection may be combined with that of IPN and so checks should be made for the presence of other viruses.

The primary mode of transmission is through infected eggs, but other means of transmission such as raw feedstuffs have also been recorded. Sockeye and chinook salmon and rainbow and steelhead trout appear to be the most susceptible hosts. Coho salmon and other trout species are more resistant. Differing responses have been observed and are ascribed to interactions between strains of the virus, the amount of virus present and the species, strain and age of the host. The incubation period of IHN is temperature-dependent and ranges from 5.5 days at 21°C to about 16 days at 3°C.

No drugs or chemicals are known that will control IHN outbreaks. As in other virus diseases, prevention is the only means of control.

The introduction of infected eggs and fish should be avoided. As carrier status for IHN can be reliably detected only at the spawning time and during epizootics, repeated inspections employing thorough virological samplings are necessary.

Viral haemorrhagic septicaemia (VHS)

Viral haemorrhagic septicaemia is an acute to chronic viral disease of cultured salmonids, especially hatchery-reared rainbow trout, in Europe. The disease was first recognized in Germany in 1938 and later in Denmark, where it was called Egtved disease. The new name VHS was recommended in 1966 to reduce confusion. The disease is caused by a bullet-shaped rhabdovirus, very similar in size and shape to the IHN virus. The disease causes high mortality among rainbow trout fingerlings. If exposed for the first time, older fish are subject to chronic infection. Transmission is by contact and from fish to fish through water. As the water temperature rises, losses become less and cease during spring, recurring in autumn. Stress evoked by transportation or handling of trout can cause outbursts of the disease, with high mortality.

Early clinical signs of VHS can easily be confused with those of other viral, bacterial or parasitic infections. Acutely infected rainbow trout are dark in colour, lethargic and exhibit haemorrhages in the fin sockets. Exophthalmia is common and persists throughout the course of the disease. With advance of the disease, the fish becomes nearly black and develops acute anaemia. The gills become pale in colour and bleeding occurs in gills and muscles (fig. 9.3). Signs of excitability can be seen, including erratic swimming similar to that of trout suffering from whirling disease caused by the protozoan parasite *Myxosoma cerebralis*. The diagnosis can be confirmed only by isolation and serological identification of the virus in an appropriate cell culture system. As in the case of IHN, the virus can be isolated from the carrier fish only during or soon after spawning. It can seldom be isolated from asymptomatic fish at other times.

Although rainbow trout is the main species affected by VHS, other species also can be infected. Younger fish are more susceptible to

Fig. 9.3 VHS in rainbow trout. Note bleeding in muscles and the pale gills (photograph: Pietro Ghittino).

the disease and most severe losses occur at the fingerling stage. Yearling fish generally suffer milder attacks and fish over two years old are almost completely resistant to infection. The disease is more serious at temperatures below 15−16°C. The incubation period appears to be about 6 days at 15°C and 8−11 days at about 10°C.

As there is no known cure for VHS, prevention is the best approach to control as in the case of other viral diseases. As there is strong circumstantial evidence that survivors of the epizootic become asymptomatic carriers, the transfer of such fish and the shipment of eggs contaminated by virus-laden ovarian fluids should be avoided. Healthy fish should be isolated from possible sources of infection. If contaminated water has of necessity to be used in the farm, it should be thoroughly disinfected before use. The virus survives in water for more than 24 hours at 14°C.

Infectious dropsy of carp (IDC) and spring viremia of carp (SVC)

Infectious dropsy of carp was described by Schäperclaus in 1929 as the most serious dis-

ease of common carp and several other fish. The aetiology of IDC was controversial for many years. During the past 15 years it became evident that IDC includes several aetiologically different diseases, namely spring viremia of carp, carp erythrodermatitis, swim bladder inflammation, aeromonas septicaemia and pseudomonas septicaemia. Methods of prevention and treatment of these diseases differ. Most researchers and diagnosticians have therefore discontinued use of the term IDC.

Spring viremia of carp (SVC) is an acute contagious disease caused by a rhabdovirus (fig. 9.4). The disease has so far only been recorded in Europe. It can cause high mortality in carp when the water temperature rises from 10 to 20°C in spring. Occasionally, SVC can also occur in late autumn or in winter. Mortalities cease at temperatures above 20−22°C. Silver carp and grass carp are also susceptible to SVC.

Signs of the disease include reduced movements and frequent resting, darkening of skin, bleeding in fins, skin, eyes and gills, faecal casts trailing from the protruding reddish anus, accumulation of fluid in the abdomen, protruding eyes, anaemia, inflamed intestine and

Fig. 9.4 Common carp affected by spring viremia (above), showing pale gills. Compare with the uninfected carp (below) (photograph: Nikola Fijan).

bleeding in the internal organs. Diagnosis is confirmed by the isolation of the virus in cell cultures and its identification by a serum neutralization test. Secondary bacterial infections can aggravate the course of SVC and augment the losses.

The transmission of virus from brood fish to offspring can take place through eggs or sperm. The virus is relatively stable in water and mud. It enters the fish through the gills. Infection of the fish at water temperatures above 20°C results in an asymptomatic carrier state. Such carriers may have antibodies in the blood and are resistant to subsequent infection. Infected farms may therefore have few or no losses from SVC. Experiments indicate the possibility of SVC prevention by vaccination, but more research is needed before a safe and effective vaccine can be developed.

Eradication of SVC from large farms with a

surface water supply is not possible. Small farms fed with well, spring or bore-hole water can be disinfected and kept free from SVC by the propagation of SVC-free stock.

Channel catfish virus disease (CCVD)

The most serious virus disease observed in channel catfish in culture facilities in the USA is caused by a herpes virus and occurs in fry and fingerlings less than 4 months old. It has been shown to be infective also in blue catfish (*Ictalurus punctatus*). The virus can be transmitted to fry and fingerlings through water in culture facilities. It is generally believed that channel catfish brood stock are carriers of the disease and they transmit the disease through reproductive cells and/or fluids associated with reproduction. The virus has, however, not so far been detected in alleged carrier adults.

The disease occurs about 24 hours after infection, when water temperature ranges between 25 and 30°C. Affected fish may swim erratically or hang vertically in the water column with the head uppermost. The lesions begin at the posterior part of the kidney with an increasing number of lymphoid cells and proximal renal tubular necrosis. Focal necrotic lesions also develop in the liver, spleen and the digestive tract. The spleen is generally very dark and enlarged. Oedema and necrosis of the digestive tract result in massive sloughing of the intestinal lumen. Distension of the abdomen, exopthalmia and anaemia may also occur (fig. 9.5). Haemorrhages can be found in the muscles, gills, skin and fin bases. The disease is frequently associated with a secondary bacterial infection of *Aeromonas hydrophila* or *Flexibacter columnaris*. The virus retains infectivity in pond water for about 2 days at 25°C. The incubation period is about 32–72 hours at 30°C.

Since many of the symptoms of the disease are similar to other viral and bacterial diseases, the diagnosis has to be confirmed by isolation and identification of the virus. There is no known cure for the disease and no means of immunization has been developed, even though it has been observed that some individuals which survive the disease acquire a high level of immunity. The best control measures are prophylactic, including segregation from infected stocks and the use of uncontaminated water supplies. When disease occurs, infected stock should be removed and destroyed and the facilities thoroughly disinfected with a suitable disinfectant like chlorine.

Carp pox (CP)

Carp pox is a relatively benign proliferative disease of cyprinids, known for more than 400 years in common carp in Europe. It is caused by a virus similar to a herpes virus. The disease is characterized by skin proliferation, which appears histologically as a plaque-warty hyperplasia of the epidermis. In the advanced stages of the disease, mineral metabolism may be impaired and this can result in softness of the bones. Carp pox seldom causes mortality, but the cutaneous growth reduces marketability of the fish. Certain strains and inbred lines have a genetic predisposition to this disease. The carp louse (*Argulus*) can act as virus-carrier and transmit the virus within a pond population. The occurrence of carp pox can be reduced or

Fig. 9.5 Channel catfish affected by CCVD. Note exophthalmia, extended belly and enlarged anus (photograph: Nikola Fijan).

eliminated by avoiding inbreeding or by genetic selection methods. No chemotherapy exists for the disease, but recovery can be speeded up by liming the ponds. It has been reported that infected carp recover if they are transferred to ponds supplied with large volumes of clear, oxygenated water (Ghittino, 1972).

Lymphocystis

Lymphocystis is a viral disease that occurs in several species of fresh-water, brackish-water and marine species of fish. It occurs as whitish nodules on the fins (fig. 9.6), head and, sometimes, the body of the fish. These are formed by the enlargement and encapsulation of the connective tissue cells. The disease is highly contagious and under culture conditions can spread very rapidly. It seldom causes any mortality, except when the verrucose lesions interfere with the ingestion of food. But the infected fish are difficult to sell, because of their appearance.

Like other viral diseases, lymphocystis also has no known cure and prophylaxis is the only means of control. Affected fish should be destroyed to prevent the spread of the virus to others, and the rearing facilities should be thoroughly disinfected.

9.2.2 Bacterial diseases

There are several bacterial diseases of cultured fish and shellfish and, as mentioned earlier, many bacterial infections occur in association with viral diseases, as secondary infections. Because of the association of different pathogens, the identity of diseases and description of causative organisms in the literature may appear confusing. Bacterial diseases have worldwide distribution and occur in both tropical and temperate climate aquacultures. While sanitation and prevention are the measures of choice to control the disease, there are proven chemotherapeutic agents that can be successfully used in treating many bacterial diseases when they occur.

Furunculosis

Furunculosis is a septicaemic bacterial disease occurring mainly in salmonids. The disease has almost worldwide distribution and also infects many species of cold-water and warm-water fishes, besides salmonids. The name furunculosis is derived from the furuncules or blisters that occur on the skin of infected fish (fig. 9.7). But this is not a sure sign of the disease as furuncules can occur in other types of infections

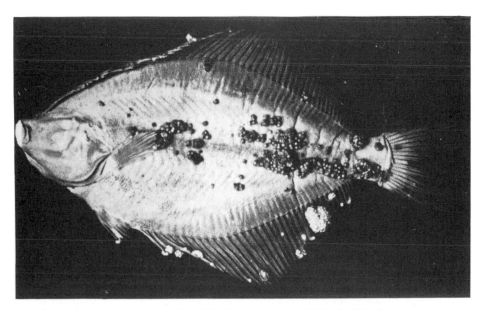

Fig. 9.6 Lymphocystis nodules on the skin of flatfish (courtesy of Nikola Fijan).

Fig. 9.7 Furuncules on an infected trout (photograph: Pietro Ghittino).

as well and also because in acute cases of furunculosis, furuncules may not be present. The gram-negative bacterium *Aeromonas salmonicida*, which is the causative agent, may survive for days or weeks in water, but cannot persist indefinitely in the absence of a carrier fish. The classic salmonid furunculosis is caused by strains of *A. salmonicida* which produce brown pigment on bacteriological media. The ulcer disease in goldfish, the carp erythrodermatitis and certain disease conditions in some other warm-water fishes are consequences of infections with strains or subspecies of *A. salmonicida* which do not produce such brown pigment.

Positive diagnosis has to be based on isolation and identification of the causative agent. The organism is typically a gram-negative, non-motile rod that ferments selected carbohydrates, produces cytochrome oxidase and yields a water-soluble brown pigment on several types of isolation agar (fig. 9.8). But special care is recommended in identifications as a number of atypical achromogenic variants have been reported. Asymptomatic carriers are very difficult to detect. To establish the absence of carriers in a brood stock, the use of serum

agglutination techniques or corticosteroid techniques might have to be employed (Bullock and Stuckey, 1975).

The incubation period for acute cases is probably 2−4 days, but in chronic cases the period may be extended by several weeks at lower temperatures. Furunculosis is usually seasonal, with peak incidence during the mid-summer months of July and August.

In carp erythrodermatitis and goldfish ulcer disease, clinical signs appear first only on the skin: small local erosions surrounded by an inflamed reddish zone develop gradually into large and sometimes deep ulcers. Seriously affected moribund fishes have also protruding eyes, fluid in the body cavity and oedematous organs. The causal agent can be isolated from the inflamed skin. Fish which recover from the disease can have scars on the skin.

Epizootics can be treated with medicated feeds. Terramycin (oxytetracycline), added to feed at the rate of 3 g per 45 kg fish, can be administered daily for 10 days. The recommended dosage of sulphamerazine is 5−10 g for 45 kg fish, fed for 10−15 consecutive days. Sulphonamides or nitrofurans can also be used, as well as antibiotics. It should be re-

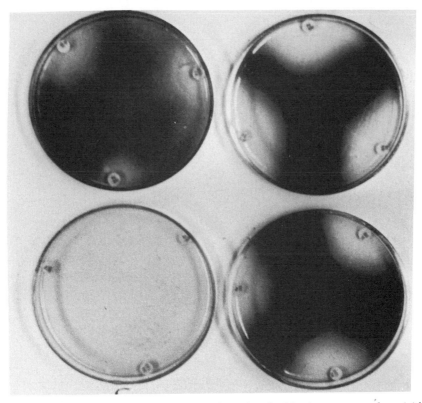

Fig. 9.8 Production of brown pigment on agar plates inoculated with *Aeromonas salmonicida*. Note the absence of pigments on the lower left-hand plate. The other plates exhibit varying zones of growth inhibition caused by different antibiotics (courtesy of Nikola Fijan).

membered, however, that in some areas the stocks have developed resistance to terramycin and sulphamerazine, probably due to the feeding of low levels of these antibiotics as a prophylactic measure.

Columnaris disease

Columnaris disease occurs in acute or chronic form in both cold-water and warm-water fishes worldwide, and is caused by the bacterium *Flexibacter columnaris* (fig. 9.9). Strains of high and low virulence have been identified. The highly virulent form attacks the gill tissue and the less virulent ones are primarily responsible for cutaneous infections.

The first sign of the disease may be the appearance of discoloured grey patches in the dorsal fin area. These lesions grow and expose the underlying muscle tissue. The lesions are prominent in the mouth and head regions and

may become yellow and cratered. The infection of virulent strains cause a 'gill rot' condition, whereas the systemic infections by less virulent strains may not show any external signs. Cutaneous infections are prevalent among most species of fish.

A preliminary diagnosis of the disease can be made based on the detection of long, slender, gram-negative rods in smears of gills or scrappings from cutaneous lesions (fig. 9.10). The diagnosis can be confirmed by isolation of the organism on cytophaga medium. Colonies of *F. columnaris* exhibit a rough, rhizoid-marginated growth that tends to extend into the agar.

Infected fish with gill or cutaneous lesions serve as a source of infection. Fish under stress due to elevated temperatures and crowding are susceptible to infection. The period between exposure and outbreak of the clinical disease varies, depending on the virulence of the bac-

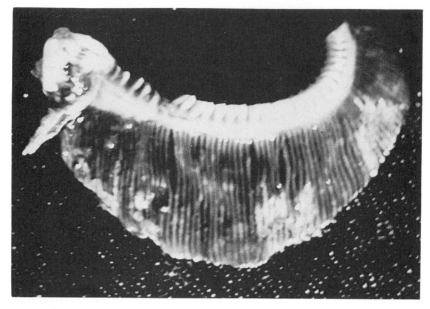

Fig. 9.9 Gill of common carp infected with virulent form of *Flexibacter columnaris*, showing advanced lesious (photograph: Nikola Fijan).

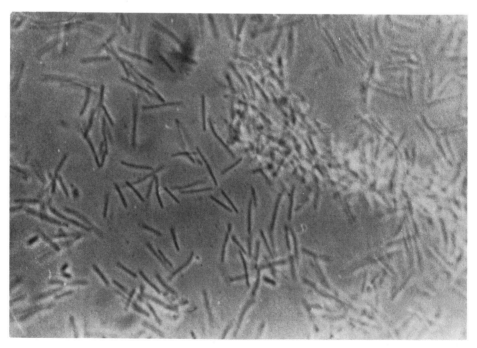

Fig. 9.10 *Flexibacter columnaris*, the causal agent of columnaris disease (photograph: Nikola Fijan).

terial strain and the ambient water temperature. High-virulence strains may induce acute disease within 24 hours, whereas less virulent forms may require from 48 hours to several weeks. The disease definitely has a pronounced seasonal occurrence, and is concentrated during the warm summer months.

Avoidance of exposure to the disease, use of disinfected water supplies, reduced population density and maintenance of lower temperature (in the case of cold-water species below 13°C) are possible means of preventing the disease. Prophylactic treatments can be adopted when other measures are not feasible. Baths of copper sulphate ($CuSO_4$) for 20 minutes at 33 ppm or of potassium permanganate ($KMnO_4$) at 2 ppm for indefinite periods have been recommended; but care should be taken in their administration, since these chemicals can be toxic to certain species in soft waters. Such baths have also been recommended in chemotherapy of the disease. Besides $CuSO_4$ and $KMnO_4$, quaternary ammonium compounds at doses of $2-3$ ppm in one-hour flow-through treatments have been reported to be effective for salmonids. Oxytetracycline (terramycin) incorporated into the food at the rate of 4 g per 45 kg fish, fed at 3 per cent body weight per day, is the usual antibiotic treatment recommended to accompany the chemical bath treatment.

Bacterial gill disease (BGD)

Bacterial gill disease is an external infection of hatchery-reared salmonids and occasionally of intensively reared warm-water species. It appears to be caused by one or more species of filamentous bacteria, including *Flavobacterium* sp. Large numbers of filamentous bacteria can be observed on the gills, accompanied by fusing and clubbing of the gill filaments. The disease generally occurs following the deterioration of environmental conditions, associated with overcrowding and accumulation of toxic metabolic products. Acute or chronic forms of the disease may occur.

Infected fish are usually lethargic and apparently lose appetite. Acute epizootics may result in high mortality of up to 50 per cent in a day. Although extensive clubbing of the gill filaments, lamellar fusion and excess mucus may be found, necrosis of the gill tissue seldom occurs. This is in contrast to columnaris disease, where extensive necrosis and erosion of the gill filaments can be observed.

The detection of large numbers of filamentous bacteria on the gills under a microscope (under a wet mount or stained with methylene blue) is the recommended diagnostic procedure. Isolation of a pure culture is not considered necessary.

The biology and survival of the aetiological organism are not fully known. Contaminated water or carrier fish are probably the source of infection, but it is almost always associated with deterioration of environmental conditions. Fingerlings are generally more susceptible to the infection and salmonids over one year in age seldom develop the disease.

Application of proper sanitation practices, avoidance of crowding and reuse of water and maintenance of an adequate flow of clean water should help to reduce the incidence of the disease. A number of compounds have been found to be effective for treatment of BGD, but most of them require multiple applications. Successful treatments are (i) potassium permanganate ($KMnO_4$) at $1-2$ ppm, (ii) Hyamine 1622 and 3500, as well as Roccal at $1-2$ ppm calculated on the basis of active ingredient, (iii) Diquat at $8.4-16.8$ ppm of the formulation, (iv) another quaternary ammonium compound Purina Four Power, at $3-4$ ppm as a one-hour flush treatment and (v) chloramine-T, in a single treatment at 10 ppm in a one-hour flush treatment.

Gill rot

Gill rot disease is reported to be one of the main infectious diseases in China and affects especially grass carp (*Ctenopharyngodon idella*) fingerlings. The causative organism has been described as *Myxococcus piscicola*. Whether it is the same as the bacterial gill disease described as affecting juvenile salmonids (Snieszko, 1958) is not clear. In juvenile salmonids, it can occur over a wide temperature range and is characterized by masses of myxobacteria on the gills and is associated with hyperplasia of the gill epithelium. Several kinds of myxobacteria can be isolated from the gills, though not from the internal organs. It is suspected that the disease is secondary to some other predisposing stress factor and that the myxobacteria are only opportunists. Chinese workers have isolated *Myxococcus piscicola* from the affected gills. The disease symptoms are described as pale coloration of the gill filaments, then sloughing and accumulation of excess slime. The gill covers of seriously infected fish are inflamed and are eroded by the bacteria forming small transparent patches. The disease is reported to be prevalent in ponds overloaded with organic matter.

Farmers are advised to treat the infected ponds with bleaching powder (containing about 30 per cent available chlorine), at the rate of about 4 kg/ha. In serious cases, a treatment dose of 1 ppm is recommended. For salmonid gill disease, the addition of antibacterial chemicals to the water is suggested, as the infection is restricted to the gills. Hyamine 1622 at 2 ppm of commercial product or Diquat at 2 or 4 ppm of active ingredient can be used as one-hour baths daily for 2−4 days.

Enteric red mouth disease (ERM)

Enteric red-mouth disease is an acute to chronic bacterial disease of intensively cultured rainbow trout and other salmonids. The causative agent is a motile, gram-negative, rod-shaped bacterium, identified as *Yersinia ruckeri*, transmitted from fish to fish by contact and through water. It causes sustained low-level losses, but in severe epizootics mortality rates can be higher, exceeding 50 per cent. Surviving fish frequently become asymptomatic carriers.

Acute cases are seldom detected. The clinical symptoms are similar to infections of *Aeromonas hydrophila* and *A. salmonicida*. The affected fish become dark and lethargic and refrain from feeding. Surviving carrier fish also show the same signs and may often have missing eyes or exophthalmia, with little avoidance reaction. During the acute stage of the disease, small bright haemorrhages occur along the gumline of the mouth and on the tongue, which together with the general inflammation give the characteristic 'red mouth' appearance (fig. 9.11). Small haemorrhages may develop on the belly and also at the base of the fins. A flaccid, fluid-filled stomach and haemorrhages of the mouth, when occurring together, give

positive evidence of the disease. An enlarged dark spleen, haemorrhagic specks on the air bladder and pyloric caecae and reddening of the posterior intestinal tract are other internal signs of the infection. Confirmatory diagnosis consists of the isolation of gram-negative motile rods of *Y. ruckeri*, with positive agglutination with rabbit anti-*Y. ruckeri* serum. The source of infection of the disease is asymptomatic carriers. No evidence of transmission from parent to progeny through eggs has been reported.

Susceptibility to the disease is obviously related to stress caused by handling or culture conditions. There does not seem to be any seasonality in its occurrence. The incubation period in young rainbow trout (7.5−10 cm) appears to be 5−19 days at about 15°C, and in young Atlantic salmon (6 cm) 9 days after exposure at 12.5°C.

As detection of the pathogen in apparently healthy carrier fish is difficult, regular health inspection and monitoring are essential to prevent the disease. Disinfection of water supplies to hatcheries and rearing facilities is another means of disease prevention. Commercial vaccines against this disease are available, and though they may not always provide total protection, they can help to control the infection and reduce losses. Therapeutics recommended are sulphamerazine and oxytetracycline (terramycin). Vaccination and/or chemotherapy should be accompanied by avoidance of adverse environmental factors and excessive handling stresses.

Edwardsiellosis

Edwardsiellosis, caused by infection by *Edwardsiella tarda* affects Ictalurids, Cyprinids and Anguillidae in the southern USA and Southeast Asia. It causes gas-filled lesions in the muscle tissues of mature fish. In the initial stages of the infection and in mild infections, small cutaneous lesions (of 3−5 mm diameter) can be observed on the flanks (fig. 9.12) and caudal peduncle of the fish. Mortality seldom exceeds 5 per cent in affected channel catfish in ponds, but may reach 50 per cent or more if the fish are transferred to holding tanks.

Colonies of *E. tarda* are dirty grey in colour and smooth. The organism is motile, gram-

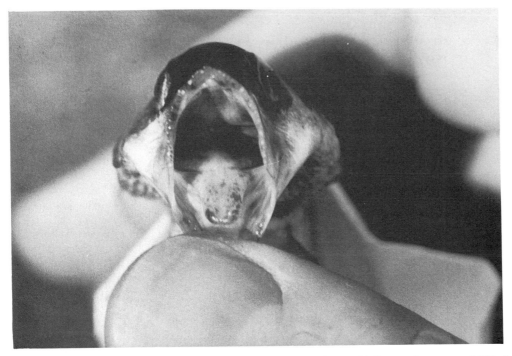

Fig. 9.11 Bleeding in the mouth cavity of trout suffering from 'red mouth' disease (courtesy of Nikola Fijan).

negative and fermentative. It occurs in organically polluted waters. Lesions develop in the infected fish 10–15 days after exposure, becoming large gas-filled cavities containing necrotic tissue. Bacteria can frequently be detected in the blood or kidney of the fish and evidently cause disfunction of the organs. Infection by *E. tarda* may often occur in association with *Aeromonas* or *Pseudomonas* infections and therefore confusion can arise in diagnosis. Recovery from the disease is rather slow. Prevention of the disease has to be attempted through avoidance of environmental stresses, improved husbandry practices and better nutrition. Treatment with antibacterials is reported to be effective. Oxytetracycline (terramycin) in the diet at the rate of 2.5 g per 45 kg fish per day for 10–12 days is the recommended dosage. Sulphonamide or furacin have also been reported to be effective.

Vibriosis

Vibriosis, caused by the bacterium *Vibrio anguillarum* which occurs in both fresh- and salt-water, has become one of the most serious diseases of cultivated marine species of fish and invertebrates. Vibrio infections are reported to have been responsible for the greatest financial loss in salmonid culture in countries like Norway. Rainbow trout, pink salmon and char can be attacked at any size or age, but Atlantic salmon are normally vulnerable only in the parr and smolt stages of development, and perhaps as they approach maturity. Sea-pen culture of Pacific salmon, particularly of coho salmon, was threatened by vibriosis until methods of control by antibiotics and immunization were developed.

Vibriosis in salmon normally occurs in smolt units, which pump water from the sea, and in sea units soon after the smolts are stocked. The vibrio usually enters the fish through surface wounds and acts mostly on the skin, where lesions are formed. The ulcers can extend deep into the muscles and internal haemorrhage, kidney damage and a swollen spleen are sometimes found in dying fish. Vibriosis is essentially a disease of spring and summer and the growth of the bacterium appears to be accelerated at

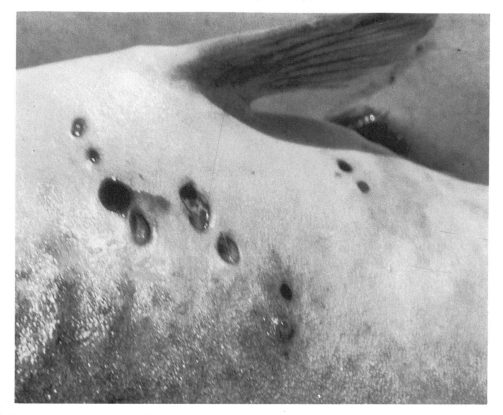

Fig. 9.12 Lesions on the skin and underlying tissues of channel catfish infected by *Edwardsiella* (photograph: Nikola Fijan).

higher temperatures. The bacterial septicaemia in shrimps appears to be mainly caused by vibrios, even though *Pseudomonas* sp. and *Aeromonas* sp. are generally associated with the disease. Disoriented swimming and increasing opaqueness of abdominal muscle in juveniles and adults are common signs of the infection. Infection occurs in larvae, juveniles and adult shrimps. Complete mortality of the stock can occur.

Recommended preventive measures for salmonids are reduction of stocking density in warm-waters and disinfection of eggs brought into the hatchery. Immunization by injection of juveniles or by hyperosmotic spray and bath immunization has been found to be useful. Oral immunization has been effective in some experiments, but is not recommended for use in commercial aquaculture. The appearance of new strains or serotypes of *V. anguillarum* has made it necessary to incorporate them also into the immunization protocol. Some cross-protection is provided by various vaccines but not enough, and not consistently (Sindermann, 1986).

Treatment of vibrios is by addition of drugs to the food, but since in the advanced stages of the infection the fish stops feeding, treatment has to be started early. The suggested dosages are 0.2 g of sulphamerazine per kg fish per day for 3 days first of all, and after an interval of a day or two another treatment of 0.2 g per kg fish for a further three days. An alternative treatment is oxytetracycline or furazolidone at the rate of 50−75 mg per kg fish for up to 10 days. The suggested treatment for shrimps is oxytetracycline at the rate of 40 mg per kg body weight per day with feed or 1 ppm baths for the larval stages (zoeae and mysis) every 48 hours. Alternatively, furanace can be used at the rate of 100 mg per feed or 1 ppm baths for the larval stages every 48 hours.

The brown spot disease of shrimps also appears to be primarily caused by vibrios, but as in the case of bacterial septicemia, other organisms like *Pseudomonas* sp. and *Beneckea* are implicated. The infection affects adult and juvenile shrimps. Brownish, eroded areas can be seen on the exoskeleton. The lesion generally results from a break on the exoskeleton, as a result of physical stress. The eroded areas may become portals of entry for secondary pathogens, causing mortality. The infection may be eliminated at moulting, except when underlying tissues are affected. Suggested treatment is administration of oxytetracycline at the rate of 450 mg per kg feed, or external treatment with 0.05−1 ppm malachite green or 20−75 ppm formalin.

Other bacterial diseases

Bacterial infections by the genus *Pseudomonas* are common in cold- and warm-water fishes. Identification of the responsible species can be done by serological methods. *P. fuorescens* infections in catfish have been controlled either by intraperitoneal injections of kanamycin (25 mg per 0.45 kg body weight) or by feeding terramycin in the daily ration (2.5 g per 45 kg fish per day). Infected trout have been successfully treated with chloramphenicol and terramycin.

There are a number of other poorly understood diseases in which bacterial infection is implicated. Although bacteria have been isolated in such cases, the actual role of the bacteria has not been determined. These include such diseases known as fin rot, fin and tail rot and peduncle disease.

The name fin rot or fin and tail rot has been given to a condition in which the margins of the fins become necrotic and slough away. Generally, the condition is associated with unfavourable environmental conditions. Under pond conditions, improvement of water quality will often result in cure. Low levels of terramycin, aureomycin and streptomycin have been found to be effective. A prophylactic treatment of potassium permanganate or 3 ppm acriflavin can also be most useful.

Pasteurella infections are sometimes found in marine species of fish, including the yellowtail, *Seriola quinqueradiata*. The symptoms of the disease are very similar to haemorrhagic septicaemia and are caused by motile gram-negative rods, showing bipolarity, especially when stained with methylene blue. The bacteria are readily isolated from the kidney, spleen and liver of affected fish. Suggested treatment is by the use of antibiotics, administered through feed.

The filamentous bacterial disease of shrimps is caused by *Leucothrix* sp. and affects larvae and post-larvae. It is generally caused by poor water quality. Growth of filamentous bacteria on the body surface and gill tips is characteristic of the disease. Heavy infestations inhibit normal swimming and cause asphyxiation. The recommended treatment is the use of cutrine plus at 0.15 ppm for 24 hours in a flow-through bath or 0.5 ppm for 4−6 hours in a static bath.

9.2.3 Fungus diseases

Growths of water moulds (Saprolegniaceae) are common in the aquatic environment. Although some forms are known to be primary invaders of fish and other aquatic organisms, most of them are saprophytic opportunists taking advantage of necrotic tissue associated with injuries, bacterial or parasitic lesions, dead and decaying eggs or generally unsanitary conditions. The most important fungal diseases are caused by members of the orders Saprolegniales, which cause a host of integumentary mycoses in most teleost fishes, and Lagenidiales, which infect eggs and early larval stages of crustaceans such as penaeids, crabs and lobsters.

Saprolegniasis

Saprolegniasis is a disease affecting the skin and gills of fresh-water fish and crustaceans caused by *Saprolegnia* and a number of non-saprolegniaceous fungii, including *Pythium* and *Leptomitus*. The three major species of *Saprolegnia* associated with the disease are *S. ferax*, *S. parasitica* and *S. diclina*. They can be identified by the characteristic profusely branched, non-septate, cotton-wool like tufts of mycelium (fig. 9.13). They reproduce asexually by means of biflagellate spores.

Environmental stress, overcrowding, poor handling and weakness caused by bacterial and

Fig. 9.13 Saprolegniasis in crucian carp. Note the cotton-like growth of the fungus hyphae on the necrotic tail and dorsal fin (photograph: Nikola Fijan).

viral infections appear to be the factors that make animals susceptible to saprolegniasis. Temperature appears to have a major role; in temperate climates epizootics occur at low temperatures, whereas in the tropics they occur in high-temperature conditions.

Lesions appear as grey-white patches on the skin, fins, eyes, mouth and gills. The colour may change to dark grey or brown as the mycelium tangle and trap debris. The mycelium invades the uppermost layer of the dermis and then ramifies laterally, eroding the epidermis. Affected eggs may be completely covered by the fungus.

Control of the disease should obviously start with avoidance of primary causes like bad handling, injuries, poor sanitation and water quality and weakness of the animals. A suggested curative treatment for fish consists of baths of (i) potassium permanganate (1 g per 100 litres water for 60–90 minutes), (ii) common salt (10 g per litre of water for 20 minutes for young fish and 25 g per litre of water for 10 minutes for older fish), (iii) copper sulphate (5 g per 10 litres of water until the fish show signs of stress) and (iv) malachite green (67 ppm dip for 10–30 seconds).

Incubating eggs may be given a daily formalin bath for 15 minutes (1–2 ml 30 per cent formol for each litre of water fed into the incubator). Malachite green can be used at concentrations of 5 mg per litre or 1 g for 200 litres of water for half to one hour. A concentration of 5 g/m^3 for 45 minutes, repeated every 5 or 6 days, has also been suggested. There are differences of opinion regarding the use of malachite green, as it is suspected that its use on trout eggs may result in genetic defects.

Branchiomycosis

Fungi of the genus *Branchiomyces* cause the gill rot which is characteristic of branchiomycosis. The disease occurs in many species, including cyprinids and channel catfish, in the summer. Two species of *Branchiomyces*, namely *B. sanguinis* and *B. demigrans*, may be involved. In the beginning of the infection, the pale gills of the fish show deep red patches (figs. 9.14 and 9.15). As it progresses, necrosis

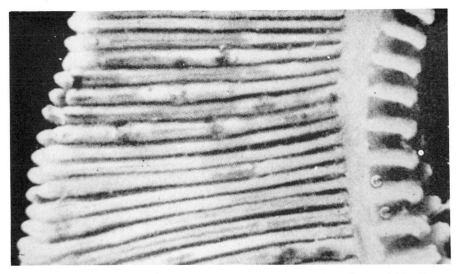

Fig. 9.14 Gills of carp affected by branchiomycosis, showing scattered swelling and bleeding, typical of early stages of the disease (courtesy of Nikola Fijan).

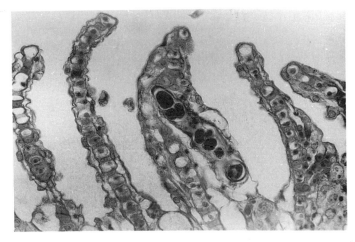

Fig. 9.15 Histological section of gill lamellae of the European catfish affected by branchiomycosis. Note the fungus in the capillaries of one of the fused lamellae in the centre (photograph: Nikola Fijan).

of the gills sets in and the colour turns yellow-brown. Secondary infection of *Saprolegnia* often occurs at this time.

To prevent the infection, dense stocking of rearing facilities should be avoided, particularly during warm weather. High concentrations of organic matter in the water supply should be avoided, and clean fresh water should be provided as often as possible. Recommended curative treatment in ponds, is 200 kg finely ground quicklime per hectare of pond area, maintaining pH below 9. Copper sulphate may be used at the rate of 8 kg per hectare of ponds of about 1 m depth. This may be applied in four monthly instalments of 2 or 3 kg each. Benzalkonium chloride also can be used in one-hour baths at a concentration of 1−4 ppm. It is reported that baths of copper sulphate (1 g in 10 litres water) for 10−30 minutes will kill all the pathogens.

9.2.4 Protozoan diseases

Numerous protozoan parasites live on fish and other aquaculture species and cause both external and internal diseases, with serious mortalities in hatcheries and rearing facilities. Even moderate numbers of these organisms on small fish may prove fatal since the infections may cause the fish to stop feeding. Except in certain cases, the infected fish may die without showing any disease symptoms other than debilitation.

Ichthyophthiriasis (Ich)

Ichthyophthiriasis, caused by the protozoan parasite *Ichthyophthirius multifiliis* (fig. 9.16), is considered to be one of the most detrimental diseases in pond culture of fresh- and brackish-water fish. All species of pond-cultured fish, including common carp, Chinese carp and trout, are susceptible to the disease. *I. multifiliis* has a round or ovoid body (0.5−1.0 mm long) with a small rounded mouth. Longitudinal rows of cilia can be found on the surface of the body, converging at the anterior end. The large macronucleus is horseshoe-shaped and has numerous contractile vacuoles. This species multiplies on fish by repeated binary fission. The mature parasites (trophonts) break the white epithelial tubercle that covers them and enter the water. They settle at the bottom and attach themselves to submerged objects. On attachment, the parasite becomes enclosed in a gelatinous cyst and multiplies. One trophont divides into as many as 2000 ciliated bodies. They emerge into the water by dissolving the cyst which encloses them, with the enzyme hyaluronidase. They swim free for 2 to 3 days and if a host is found during that period, they penetrate under the skin, grow and mature. If they do not find a host, they will die. The optimal temperature for development is 25°−26°C. Outbreaks of the disease usually occur in spring and summer at high water temperature and under conditions of over-

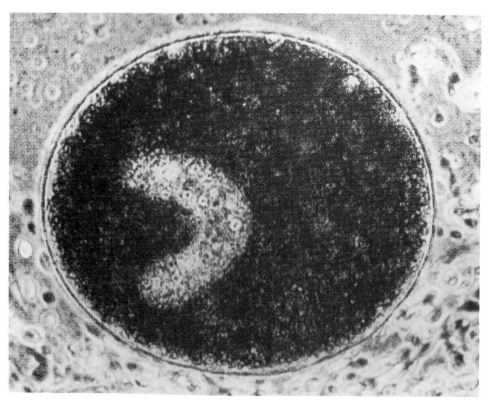

Fig. 9.16 *Ichthyopthirius multifiliis*, the causal agent of ich, under microscope (courtesy of Nikola Fijan).

crowding that facilitate the spread of the disease. In light infections, the fish show restlessness and gather in groups near the water inlet. In heavy infections in yearlings, more severe symptoms can be noticed, including acute restlessness, the fish rubbing against the bottom and sides and collecting in masses near the inlet. Small tubercles occur on the body and the fish stop feeding and cease reacting to stimuli. In advanced cases, the fish swim at the surface and rush around swallowing air. Small white tubercles cover the entire body and severe lesions of the cornea and blindness may also occur.

The fact that the non-parasitic stages of *Ichthyophthirius* are very sensitive to environmental factors make it somewhat easier to prevent infection. The destruction of carrier fish is an essential aspect of prevention. Disinfection of contaminated water or equipment is recommended. Even dilute solutions of salt (0.5 per cent) will kill encysted parasites and the ciliated bodies. The ciliated bodies can also be killed by drying the ponds or other rearing facilities. A pH below 5 and oxygen concentrations of less than 0.8 mg/ℓ are reported to be fatal. Even though vaccines are not available for immunization of fish against ich, repeated infections apparently provide relative immunity. In ponds, treatment with 0.1 ppm malachite green or 15 ppm formalin, once, twice or three times, has been reported to be effective.

Ichtyobodosis

Ichtyobodosis, previously known as costiasis, is a severe disease affecting many species of fish, including common carp and trout, especially the younger age groups. It is caused by the flagellate *Ichtyobodo necator* (*Costia necatrix*) (fig. 9.17). The parasite attaches itself to the skin of the host. Its anterior end forms finger-shaped processes at the point of attachment, which penetrate into the cell of the host and suck its contents. The parasite multiplies by longitudinal division and dies when it falls from the host. Transmission takes place through water. Under adverse conditions, the body of the parasite becomes rounded.

It can live at temperatures ranging from 2 to 30°C or higher, but multiplies rapidly at

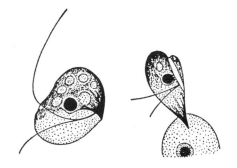

Fig. 9.17 *Costia necatrix* (from Bauer *et al.*, 1973).

20–25°C. A pH of 4.5–5.5 is very favourable for mass reproduction. In carp farms, ichtyobodosis occurs frequently in spawning ponds at higher temperatures. The incidence of the disease decreases rapidly when the young fish are transferred to rearing ponds.

A characteristic symptom of the disease is the appearance of dull spots on the sides, which eventually become fused into a continuous greyish film by the increased secretion of mucus. The fins are frequently affected, starting with erosion of the tissue between the rays, which becomes exposed. The fry becomes emaciated and therefore the head looks enlarged. The infected gills are pale and covered with mucus. The fry rise to the surface and congregate near the inlet, swallowing air.

To avoid the incidence of the disease, young fish should be given nutritionally adequate food and should not be kept too long in spawning and rearing ponds. Carp spawning ponds should be disinfected with quicklime before spawners are introduced. Carrier fish should be eliminated and the water supply should be kept free from parasites. Ichtyobodo infections can be effectively treated in ponds with 15–50 ppm formalin, 2–3 ppm potassium permanganate or 0.1 ppm malachite green. A combination of 0.1 ppm malachite green with 15 ppm formalin has also been found to be effective.

Whirling disease

Whirling disease caused by the protozoan *Myxosoma cerebralis* is one of the well known diseases of salmonids, and has been reported from many parts of the word, including the whole of Europe, North and South America,

Africa, Asia and New Zealand. It affects all species of trout and salmon, particularly the young. A common sign of the disease is rapid, tail-chasing behaviour when the fish are frightened or trying to feed. This is caused by the parasite feeding on the cartilage of young host fish. In advanced stages of the disease, skeletal deformation, including deformed heads, jaws and gill covers, as well as spinal curvature, can be observed. When exposed to the disease early in life, trout may develop 'blacktail'. Acutely infected fry reared in contaminated water may not show any symptoms before high mortality sets in. When exposed, older fish exhibit less whirling behaviour. Again, fish with light infections may not show any signs at all, but will actually be carrying spores throughout their life. Confirmation of diagnosis has to be by isolation of spores or immature forms of the parasite in histological sections.

Infected fish, contaminated water and mud are known to be the reservoirs of infection. It has been reported that the spores may survive for 10–15 years in contaminated mud (Christensen, 1972). The exact route of infection has not yet been fully determined, but it would appear that the spores released by dead or living fish (fig. 9.18) develop infectivity in mud after a period of 4–5 months (Hoffman and Putz, 1969). It has been suggested that tubificid worms may be involved in the transmission. Spores can survive as long as two months in frozen infected fish.

While rainbow trout and Atlantic and kokanee salmon may become severely diseased by the parasite, brook trout, coho salmon and lake trout seem relatively resistant to the disease. The first twelve months of the life of the fish are the most susceptible period for infection. Fish between 4 and 5 months old do not develop acute clinical signs, even when infected, and may serve as asymptomatic carriers. Spore formation in infected fish takes about 52 days at 17°C. It may take about 28 weeks after exposure for symptoms of the disease to appear.

The only means of preventing infection by *M. cerebralis* is to prevent contact of susceptible fish with the parasite. The importation of infected fish or use of contaminated water should be avoided. Since the disease has become established in certain areas, it is extremely difficult to eradicate the pathogen. The use of resistant strains has been suggested as a management alternative. Through regulation of imports of fish and fish eggs by approved certification

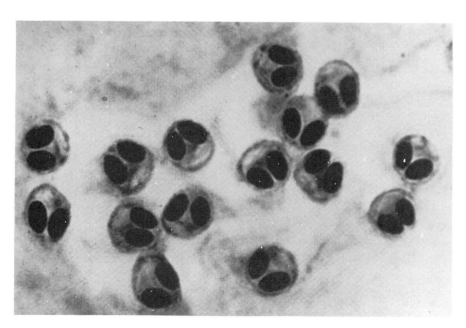

Fig. 9.18 Stained spores of *Myxosoma cerebralis* under microscope (courtesy of Nikola Fijan).

procedures, spread of the disease to uncontaminated areas can be controlled. There is no known effective therapy for the disease.

Other protozoan diseases

Of the other protozoan diseases, mention may be made of chilodonelliasis and trichodiniases. Chilodonelliasis, caused by the ciliate *Chilodonella cyprini* (fig. 9.19) affects many species of fish, including young common carp and trout, causing heavy mortality in ponds. The parasite feeds on epithelial cells, which it pierces with its protruscible pharynx. It multiplies by transverse binary division, at a rapid rate, at an optimum temperature of 5–10°C. Temperatures above 20°C are lethal to the organism. Under adverse conditions, some individuals produce resting cysts. The cysts remain viable for a long time on the bottom or in the water until they find a host.

Emaciated or undernourished fish appear to be more susceptible to infection. The physiological condition of the fish and the abundance of the parasite are obviously important factors in epizootics of the disease. Transmission seems to be through contact with carrier fish or via contaminated water.

In severe cases of infection, the body is covered with a bluish-grey film, distinctly noticeable on the dorsal side of the head. Heavily infested fish appear restless, rise to the surface, lose weight and become very lethargic. Yearling fish jump out of the water, due to impairment of cutaneous respiration.

Since the infection often takes place in wintering ponds, measures should be taken to prevent the parasites from entering these facilities. It is recommended that all fish should be given a 5 minute bath of 5 per cent sodium chloride before transfer to wintering ponds. Drying and disinfection with quicklime (2.5–4 tons/ha) would help to eliminate any cysts remaining on the bottom of the ponds from the fish wintered during the earlier season.

For curative treatment of infected fish, application of sodium chloride at a concentration of only 0.15–0.2 per cent for 1 to 2 days has been suggested. Since young salmonids may not tolerate high concentrations of sodium chloride, a bath of 0.005 per cent potassium permanganate for 10–15 minutes is recommended.

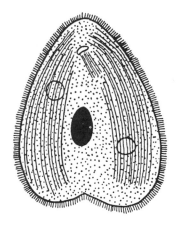

Fig. 9.19 *Chilodonella cyprini*, causative agent of chilodonelliasis (from Bauer *et al.*, 1973).

Trichodiniases caused by *Trichodina domerguei*, *T. pediculus*, *T. nigra*, *T. reticulata*, *T. epizootica* and *T. bulbosa* evoke mortality in a number of species of fish, including common carp and Chinese carps (grass carp, silver carp and bighead carp). The disease affects fry, fingerlings and yearlings of all pond fish. Adults are carriers of the disease.

The disease is transmitted either through contact with infected fish or via water. The body of the infected fish becomes dull, with a thin, whitish film of mucus, the quantity of which depends on the intensity of infection. In mild cases, the film is thin and restricted to the head and dorsum, but in severe cases the mucus covers the entire body and may flake off. At higher intensities of infection, the fish may become restless and congregate near the inlets. As the infection progresses, mortality increases rapidly. Very often trichodiniasis, chilodonelliasis and ichthyophthiriasis occur together.

Preventive and curative treatments for the disease are similar to those for chilodonelliasis.

9.2.5 Copepod infections

Copepod parasites of the families Arguilidae, Ergasilidae and Lernaeidae often infect several species of cultivated fish, such as common carp, trout and Chinese carps. The parasitic crustacea of epizootical interest are all warm-loving species with optimal temperatures above 20°C. They have a worldwide distribution. No intermediate hosts are involved in their development and all of them multiply in fish ponds if the temperature is favourable and the fish are overcrowded. The best known among them are *Argulus foliaceus* and *Lernaea cyprinacea*.

Argulosis

Argulosis caused by species of *Argulus* (fish lice), namely *A. foliaceus*, *A. japonicus* and *A. giordanii*, is one of the most common and widely distributed external infection of several species of fish including common carp, trout, grass carp, black carp, bream and eel. Outbreaks of argulosis often develop into epizootics, causing mass mortality. Argulus is a large parasite with a wide, oval, flattened greyish-green body (fig. 9.20). The organs of attachment end in curved hooks. There is a suctorial proboscis and suckers on the ventral side and four pairs of swimming legs.

The female of *A. foliaceus*, which is the most common of the three species, lays eggs on submerged objects in batches of 250–300, which are attached to substrates. Depending on the temperature, the embryonic development takes 15–55 days. The larvae that hatch out swim free in water for 2–3 hours and die unless they find a host. Eggs die rapidly if they become dry.

Younger fish appear to be more susceptible to infection. Older fish may not suffer from the infection, but would become carriers of the parasite. Because of the lack of host-specificity, *Argulus* is generally considered a greater risk in aquaculture. It attaches itself to the fish, pierces the skin with its proboscis, injecting a toxic secretion, and sucks the blood of the host. The wound develops inflammation, with profuse secretion of mucus, oedema and haemorrhages. The wound becomes necrotic and the lesion may become secondarily infected. The greater susceptibility of the young ones is ascribed to the fact that the proboscis pierces both the epidermis and dermis, whereas in adults it is able to damage only the epidermis.

Prevention of argulosis consists of isolation of susceptible young fish from older age groups and preventing contact between infected fish

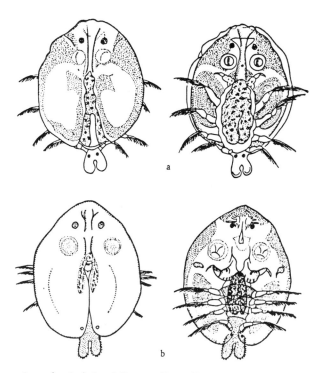

Fig. 9.20 *a. Argulus japonicus*; *b. Argulus foliaceus* (from Bauer, 1973).

and parasite-contaminated water. Ponds should be dried and disinfected after every rearing, to kill the eggs of the parasites. Sticks placed in different areas on the pond bottom can serve as traps for the eggs of the parasites – *Argulus* would attach eggs to the stick. It is reported that by daily removal of the sticks, most of the eggs can be removed and destroyed (Balarin & Hatton, 1979). Placement of wooden shields in a chess-board pattern for oviposition of *Argulus* has been suggested by Kiselev and Ivlieva (1953). They suggested removal of the shields every 15–20 days, depending on the temperature, drying them for a day and then replacing them. Effective removal of eggs can thus be achieved. For curative treatment, several chemicals have been recommended, including malathion and dipterex at 0.25 ppm and bromex at 0.12 ppm. Bromex appears to have a greater safety margin. Dip baths of lysol (1 ml lysol for 5 litres water) for 5–15 seconds or potassium permanganate (1 g in 1 litre water) for 40 seconds are also reported to be effective. Fish ponds can be treated with trichlorfon (0.5 g/m^3) for carp and eels or 0.2 g/m^3 for more sensitive species, such as trout. If necessary, the treatment should be repeated after two to three weeks.

Lernaeosis

Lernaeosis, or anchor worm disease, caused by species of the genus *Lernaea*, has been observed in a number of fresh-water fish, including common and Chinese carps. They are very sensitive to salinity and cannot survive even low concentrations of salt. The two species of epizootic importance appear to be *L. cyprinacea* (fig. 9.21) and *L. ctenopharyngodonis*. The mature females of the species have a long, unsegmented body and the head has branched processes with which the parasite can penetrate the body of the host (fig. 9.22). There are no intermediate hosts for the species and the free-living larvae parasitize the skin or gills of fish. *L. cyprinacea* will parasitize any species of fish, but *L. ctenopharyngodonis* appears to prefer the grass carp.

The metamorphosis of *L. cyprinacea* is rather complex, with three nauplii and five copepodid stages. Each female develops two egg sacs with 300–700 eggs. The optimal temperature is reported to be 23–30°C and the embryonic development takes 3 days. Hatching takes place on the fourth day. The development of nauplius stages lasts 4–5 days, followed by the copepodid stages in the next 9–10 days. The cope-

Fig. 9.21 Lernaeosis in common carp. The parasites are attached under the scales (photograph: S. Egusa).

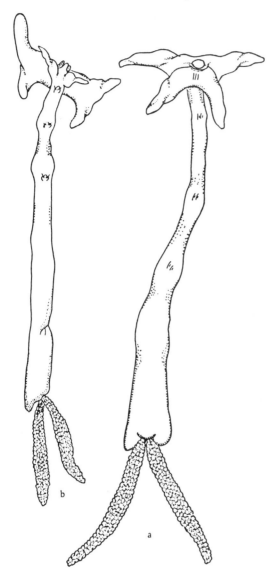

Fig. 9.22 *a. Lernaea cyprinacea* from common carp; *b. Lernaea ctenopharyngodonis* from grass carp (from Bauer *et al.*, 1973).

podid larva must find a host and attach itself for further development into the free-swimming cyclopid stage. Mating takes place during this stage, after which the males die and the females penetrate the skin of fish, settling in the muscle and remaining attached to the host. Depending on the temperature, between two and eleven generations have been observed in one year.

The parasite enters the fish-rearing facilities through water supplies. When it penetrates the skin of the host, reaching the muscles, deep ulcers, abscesses or fistulas are formed at the point of attachment. The margins of the ulcer are bright red or greyish in colour. Secondary infections may set in at this time. The inflammation at the point of penetration results in the formation of parasitic fibrous nodules. The parasite may also penetrate the liver, causing focal traumatic hepatitis.

It is difficult to eradicate adult parasites as they are very hardy and resistant to most chemicals. Potassium permanganate baths at a concentration of 2 ppm for 1−2 hours may kill most of the parasites, but it has been reported that within a short time young parasites begin to develop again. The free-living larval stages can more easily be controlled. Bromex, at a concentration of 0.12−0.15 ppm active ingredient in pond water, can kill these stages. The application has to be repeated three times, at intervals of about 7 days in summer and 12−14 days at lower temperatures.

9.2.6   Trematode infections

Among the other major causes of infection of importance in aquaculture are larvae of trematodes. Dactylogyrosis is a common disease caused by species of the genus *Dactylogyrus*, or the gill fluke, which affects common carp, Chinese carps and other fish. *Dactylogyrus* is small in size (rarely longer than 1 mm) and occurs on the gill filaments (fig. 9.23). A number of species have been identified, but the ones of epizootic interest for carp appear to be *D. vastator* and *D. extensus*. They attach themselves to the gills and, under favourable conditions, such as temperatures below 30°C, develop rapidly. They are hermaphrodites and lay eggs that fall to the bottom of the pond or other rearing facilities. The larvae which hatch out are ciliate and swim around until they attach themselves to the gills, body surface, or oral cavity of a host and begin to grow. Irrespective of the place of initial attachment, they all subsequently congregate on the gills and, when mature, start laying eggs, repeating the life cycle. *Dactylogyrus vastator* seems to prefer warm-water environments and infects mainly young carp, causing heavy mortality depending on the intensity of infection. *Dactylogyrus*

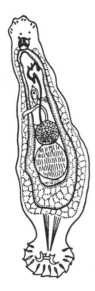

Fig. 9.23 Dactylogyrus (from Bauer *et al.*, 1973).

extensus, on the other hand, prefers temperatures below 17°C. In addition, this species occurs in both young and adult fish and is more pathogenic. The intensity of infection increases with the age of the host. In both cases, older fish are the source of infection. Infected fish become restless and collect near water outlets. The gills are damaged and become covered with mucus, inhibiting normal respiration.

The infections can be controlled by the usual sanitary measures in pond farms. Ammonia baths are recommended for treatment, at concentrations of 2 ml of 25 per cent ammonia solution per litre water for half to one minute. Bromex-50 and Dipterex have also been used successfully.

The species of *Dactylogyrus* infecting grass carp have been identified as *D. lamellatus* and *D. ctenopharyngodonis*. *Dactylogyrus hypophthalmichthys* is the common parasite of silver carp and *D. aristichthys* and *D. nobilis* of bighead and Chinese carp hybrids. They do not appear to infect other pond fish. The symptoms of the disease are generally very similar and the treatment recommended is also the same.

9.3 Public health and aquaculture

Even though the main concern of the aquaculturist is the health of the species being cul-

tured, the implications of aquaculture activities for human health and wellbeing are of equal importance, as continuation of his enterprise and public support for it will depend on its safety levels. Needless to say that his own health and that of his family and the community are directly involved. The hazards to public health are much greater in tropical areas, where there is greater prevalence of water-borne diseases. In fact in many tropical areas, particularly in Africa, conflicts between public health and fish culture interests have been one of the reasons for the abandonment of early endeavours to develop rural homestead fish farming. The large number of small, shallow, poorly managed fish ponds were considered to provide additional breeding grounds for vectors of water-borne diseases, even though their importance in the production of much-needed animal proteins was well recognized.

The major consideration relates to fish ponds becoming breeding grounds for mosquitoes which transmit malaria. The human malaria infection, brought about by four different parasite species of the genus *Plasmodium*, is transmitted through about sixty or so different species of *Anopheles* mosquitoes. The parasite is introduced into the human bloodstream as a sporozoite through the bite of an infective mosquito. In the human host the parasite develops and multiplies, first in the liver and then in the bloodstream, where it invades the red blood cells. Some of the parasites develop into sexual forms called gametocytes, which are eventually taken up by the bite of an *Anopheles* mosquito. In the mosquito the parasite undergoes various stages of development leading to the formation of sporozoites. These accumulate in the salivary glands and are injected into man during a blood meal. As all the carrier species of mosquitoes are aquatic breeders, it is necessary to ensure that pond farms do not become breeding grounds for the vectors and contribute to the spread of malaria.

Generally, pond farms are built on low-lying swampy areas that are favoured places for mosquito breeding. Properly constructed pond farms, with adequate drainage facilities, on these sites are expected to reduce the risk of mosquito breeding and thereby the prevalence of malaria. However, when shallow ponds are constructed and remain weed-infested without proper water management, they become as equally inviting for mosquito breeding as the swamps on which the ponds were built. It is generally recommended that ponds should be at least 0.61 m (2 ft) deep and preferably 0.914 m (3 ft). The inside of pond dykes should be as steep as possible to avoid too shallow margins. The ponds should be kept free from all weeds, particularly floating ones, and all emergent vegetation cut back so as to avoid mosquito breeding. Any seepage from the pond should be drained off through seepage channels. Cattle should be prevented from grazing on the embankments, as hoof prints are notorious breeding grounds for mosquitoes in humid tropics. Most fish species, particularly in the fry and fingerling stages, will feed on mosquito larvae if available and many feed on the algal vegetation that shelters them. So well-stocked ponds may be able to control the growth of algae as well as mosquito larvae. It has also been recommended that in mosquito-affected areas, the ponds may be stocked with larvicidal fish, like *Haplocheilus panchax* and *Gambusia affinis*.

Bilharzia (schistosomiasis) is another human disease transmitted through contact with water. It ranks among the most important public health problems of the tropics and subtropics and is probably only second to malaria as a parasitic disease. Human schistosomiasis is caused by at least three species of blood flukes (trematodes), namely *Schistosoma haematobium* (genito-urinary bilharziasis), *S. mansoni* and *S. japonicum* (intestinal bilharziasis). The adult forms mature in the blood of humans (or other hosts) and the eggs are laid in the plexus round the colon. The eggs escape through the bladder or intestinal wall into the urine or faeces and, on contact with water hatch, produce free-swimming larval forms called *miracidia*, which must find suitable snail hosts within a day or perish. When the larva finds a suitable intermediate host it penetrates the skin and establishes itself in the snail, passing through several generations and multiplying to form further free-living larval forms called *cercaria*. These emerge from the snail and swim about until they find a human host. They bore through the unbroken skin of the host (or through the buccal mucosa if the water is swallowed), make their way to the liver, mature and mate. The

two sexes then make their way to the terminal blood vessel near the bladder or colon, where the eggs are laid, and the cycle begins all over again.

It is thus evident that water plays a significant role in the transmission of bilharziasis. It must be contaminated by infected people and environmental conditions (water temperature, water flow, vegetation and the presence of organic foods) should be suitable for the growth of specific snail populations that serve as intermediate hosts. Evidently, the most effective and logical means of preventing pond farms and enclosures becoming a source of infection is adherence to personal hygiene by the farmhands and the neighbouring community. The use of contaminated water sources for the farm should be avoided, as far as possible. Another important measure is the control of snail populations. There are several genera of water snails that are known to be vectors. The clearance of emergent and submerged vegetation would greatly help in the control of snail infestation. There are a number of culturable fish species that feed on snails and could be used as the main or subsidiary species for culture. There are also smaller snail-feeding fish such as *Astatereochromis alluaudi* and *Haplochromus* spp., which could be introduced into ponds for the sole purpose of controlling snail populations. The system of combining duck raising with fish farming, practised in a number of countries now, could be an effective means of control, as ducks feed on snails as well as algal vegetation in ponds. The best method of avoiding snail infestation is reported to be the periodic draining and drying of ponds, followed by liming of the pond bottom. Parasitized snails are generally unable to survive prolonged desiccation and so periodic drying of the pond will be an effective means of eliminating snails carrying *Schistosoma*.

Besides the hazard of spreading water-borne diseases, possible public health problems that can be created by specific aquaculture practices also need to be considered. The most susceptible practices may be the use of large quantities of animal manures and the use of waste-water including domestic sewage in pond farming, as well as the culture of molluscs in contaminated waters. There appears to be no conclusive proof that the use of waste water, including sewage effluents, has caused human diseases. The high pH and oxygen levels in waste-water fish ponds could actually be producing quite disease-free environmental conditions, in contrast to expectations that such systems encourage fish parasites, disease and pathologies (Allen and Hepher, 1979). Bacteriological studies have shown that although a number of bacterial groups (*Aeromonas*, *Pseudomonas*, *Streptococcus* and *Klebsiella*) were isolated from the intestinal tracts of Pacific salmon and rainbow trout grown in waste water in experimental studies, none of them were isolated from other visceral organs (Allen *et al.*, 1979). There was no evidence that any of the potential pathogens had become systemic under fish-culture conditions.

Though the zoonotic aspects of fish disease are undeniably important, diseases of fish that are definitely transmissible to man are few and are limited to certain pathogenic helminths for which fish form the intermediate host. Transmission takes place through the consumption by man of raw or improperly processed fish flesh. The more important parasites are the tapeworm *Diphyllobothrium latum* and the flukes *Clonorchis sinensis*, *Opisthorchis* spp. and *Metagonimus yokagawaii*. But these have not been isolated from farmed fish.

Filter-feeding molluscs, like oysters and mussels, which are eaten raw could form greater public health risks if grown in contaminated waters. Oyster and mussel culturists should certainly avoid such areas or take the necessary action to prevent discharge of contaminants into the farming areas. The process of depuration that is adopted in organized shellfish farming and the certification of the quality of shellfish products are additional measures that prevent public health hazards caused by aquaculture.

9.4 References

Allen G.H. and Hepher B. (1979) Recycling of wastes through aquaculture and constraints to wider application. In *Advances in Aquaculture*, (Ed. by T.V.R. Pillay and W.A. Dill), pp. 478–87. Fishing News Books, Oxford.

Allen G.H., Busch R.A. and Morton A.W. (1979) Preliminary bacteriological studies on wastewater-fertilized marine fish ponds, Humboldt Bay, California. In *Advances in Aquaculture*, (Ed. by T.V.R. Pillay and W.A. Dill), pp. 492–8. Fishing News Books, Oxford.

Anderson D.P. (1974) *Fish Immunology*. T.F.H. Publications, Hong Kong.

Balarin J.D. and Hatton J.P. (1979) *Tilapia: A Guide to their Biology and Culture in Africa*. University of Stirling, Scotland.

Bauer O.N., Musselius V.A. and Strelkov Yu.A. (1973) *Diseases of Pond Fishes* (translated from Russian). Israel Program for Scientific Translation, Jerusalem.

Brock J.A. (1983) Diseases (infectious and non-infectious), metazoan parasites, predators, and public health considerations in *Macrobrachium* culture and fisheries. In *Handbook of Mariculture*, Vol. I (Ed. by J.P. McVey), pp. 329–70. C.R.C. Press, Boca Raton.

Bullock G.L. and Stuckey H.M. (1975) *Aeromonas salmonicida*: detection of asymptomatically infected trout. *Prog. Fish-Cult.*, **37**, 237–9.

Christensen N.O. (1972) Some diseases of trout in Denmark. In *Diseases of Fish*, (Ed. by L.E. Mawdesley-Thomas), pp. 83–8. Academic Press, London.

Dill W.A. (Ed.) (1973) Symposium on the major communicable fish diseases in Europe and their control. *EIFAC Tech. Paper*, **17**, suppl. 2, FAO of the UN, Rome.

Egidius E. and Anderson K. (1979) Bath immunization – a practical and non-stressing method of vaccinating sea farmed rainbow trout *Salmo gairdneri* Richardson against vibriosis. *Fish. Dis.*, **2**, 405–10.

Egidius E. *et al.* (1981) Cold-water vibriosis or 'Hitra disease' in Norwegian salmonid farming. *Fish. Dis.*, **4**, 353–4.

Egusa S. and Honma A. (1974) Fish disease diagnosis guide 1 (in Japanese). *A Diagnostic Manual of Eel Diseases Occurring Under Culture Conditions in Japan*, pp. 1–89. UNC Sea Grant Publication, UNC-SG-78-06.

Fijan N. (1968) Progress report on acute mortality of channel catfish fingerlings caused by a virus. *Bull. Off. Int. Epiz.*, **69**, 1167–8.

Fijan N. (1971) Isolation of the viral causative agent from acute form of infectious dropsy of carp. *Vet. Arh.*, **41**(5–6), 125–38.

Fijan N., Wellborn T.L. Jr and Naftel J.P. (1970) An acute viral disease of channel catfish. *US Bureau of Sport Fisheries and Wildlife, Technical Paper*, **43**.

Finlay J. (1978) Disinfectants in fish farming. *Fishery Notes*, no. 59. Ministry of Agriculture, Fisheries and Food. Lowestoft, England.

Fryer J.L., Rohovec J.S. and Garrison R.L. (1978) Immunization of salmonids for control of vibriosis.

Mar. Fish. Rev., **40**(30), 20−23.

Ghittino P. (1972) Aquaculture and associated diseases of public health importance. *J. Am. Vet. Med. Assoc.*, **161**(11), 1476−85.

Ghittino P. (1985) *Tecnologia e Patologia in Acquacoltura. Vol. 2 Patologia* (in Italian). Torino.

Hickling C.F. (1971) *Fish Culture*, 2nd edn. Faber and Faber, London.

Hnath J.G. (1983) Hatchery disinfection and disposal of infected stocks. In *A Guide to Integrated Fish Health Management in the Great Lakes Basin*, (Ed. by F.P. Meyer, J.W. Warren and T.G. Carey), pp. 121−33. Special publication, **83**(2). Great Lakes Fishery Commission, Ann Arbor.

Hoffman G.L. and Meyer F.P. (1974) *Parasites of Fresh Water Fishes*. T.F.H. Publications, Neptune City.

Hoffman G.L. and Putz R.E. (1969) Host susceptibility and effects of aging, freezing, heat and chemicals on spores of *Myxosoma cerebralis*. *Prog. Fish-Cult.*, **31**, 35−7.

Huet M. (1986) *Textbook of Fish Culture*, 2nd edn. Fishing News Books, Oxford.

Johnson S.K. (1978) *Handbook of Shrimp Diseases*. Texas Agricultural Extension Service, Texas A. & M. University, College Station.

Kinne O. and Bulnheim H.P. (Eds) (1984) Diseases of marine organisms. *Helgolander Meeresunters*, **37**(1−4).

Kiselev I.V. and Ivlieva V.K. (1953) Data on the biology of *Argulus foliaceus* and its control in fish farms. (In Russian.) *Trudy Niiporkh UkrSSR*, **9**.

Lewis D.H. and Plumb J.A. (1979) Bacterial diseases. In *Principal Diseases of Farm-raised Catfish*, Southern Cooperative Series, no. 225, pp. 19−24. Auburn University, Auburn.

Lightner D.V. (1983) Diseases of cultured Penaeid shrimp. In *Handbook of Mariculture*, Vol. I (Ed. by J.P. McVey), pp. 289−320. CRC Press, Boca Raton.

Mawdesley-Thomas L.E. (Ed.) (1972) *Diseases of Fish*. Academic Press, London.

Meyer F.P., Warren J.W. and Carey T.G. (Eds) (1983) *A Guide to Integrated Fish Health Management in the Great Lakes Basin*, special publication, **83**(2). Great Lakes Fishery Commission, Ann Arbor.

Pillay T.V.R. (Ed.) (1968) Proceedings of the FAO World Symposium on Warm-Water Pond Fish Culture. *FAO. Fish. Rep.*, **44**(5).

Plumb J.A. and Gaines J.L. Jr (1975) Channel catfish virus disease. In *The Pathology of Fishes*, (Ed. by W.E. Ribelin and G. Migaki), pp. 117−41. University of Wisconsin Press, Madison.

Reichenbach-Klinke H.H. (1966) *Krankheiten und Schadigungen der Fische*. Gustav Fischer Verlag, Stuttgart.

Richards R.H. and Roberts R.J. (1978) The bacteriology of teleosts. In *Fish Pathology*, pp. 183−204. Balliere Tindall, London.

Roberts R.J. (1978) *Fish Pathology*. Balliere Tindall, London.

Roberts R.J. and Shepherd C.J. (1974) *Handbook of Trout and Salmon Diseases*. Fishing News Books, Oxford.

Rosemark R. and Coklin D.E. (1983) Lobster pathology and treatments. In *Handbook of Mariculture*, Vol. I (Ed. by J.P. McVey), pp. 371−7. CRC Press, Boca Raton.

Sarig S. (1971) The prevention and treatment of disease of warm-water fishes under sub-tropical conditions, with special emphasis on intensive fish farming. In *Diseases of Fishes*, (Ed. by S.F. Sniezsko and H.R. Axelrod). TFH Publications, Hong Kong.

Schäperclaus W. (1929) Beitrage zue kenntnis der kiemenfaule des karpfens. *Z. Fisch.*, **27**(3).

Sindermann C.J. (1970) *Principal Diseases of Marine Fish and Shellfish*. Academic Press, New York.

Sindermann C.J. (Ed.) (1977) *Disease Diagnosis and Control in North American Marine Aquaculture*. Elsevier, New York.

Sindermann C.J. (1986) The role of pathology in aquaculture. In *Realism in Aquaculture: Achievements, Constraints, Perspectives*, (Ed. by M. Bilio, H. Rosenthal and C.J. Sindermann), pp. 395−419. European Aquaculture Society, Bredene.

Snieszko S.F. (1958) Fin rot and peduncle disease of salmonid fishes. *Fishery Leaflet* (USDI − FWS), **462**.

Snieszko S.F. (Ed.) (1970) *A Symposium on Diseases of Fishes and Shellfishes*. American Fisheries Society special publication, 5. American Fisheries Society, Washington.

The Japan Research Group of Fish Pathology (1976) Recent advances in fish pathology. *Fish Pathol.*, **10**(2), 103−259.

Thompson P.E., Dill W.A. and Moore G. (1973) *The Major Communicable Fish Diseases of Europe and North America. A Review of National and International Measures for Their Control*. Fi: EIFAC 72/Sc II. Symp. 10, FAO of the UN, Rome.

Tomasec I.I. and Fijan N. (1965) The etiology of infectious dropsy of carp. *Ann. N. Y. Acad. Sci.*, **126**, 606−15.

Warren J.W. (1983) Synthesis of a fish health management program. In *A Guide to Integrated Fish Health Management in the Great Lakes Basin*, (Ed. by F.P. Meyer, J.W. Warren and T.G. Carey), pp. 151−8 special publication, **83**(2). Great Lakes Fishery Commission, Ann Arbor.

Wolf K. (1966) Infectious pancreatic necrosis (IPN) of salmonid fishes. *US Fish. Wildl. Serv. Fish.*

Dis. Leaflet, no. 1. Washington.

Wolf K. (1972) Advances in fish virology: a review 1966−1971. In *Diseases of Fish*, (Ed. by L.E. Mawdesley-Thomas), pp. 305−31. Academic Press, London.

Wood J.W. (1974) *Diseases of Pacific Salmon: Their Prevention and Treatment*. Washington State Department of Fisheries, Olympia.

10
Control of Weeds, Pests and Predators

10.1 Weed problem in aquaculture farms

Weed infestation of aquaculture farms is a problem of varying intensity in almost all systems of aquaculture all over the world. But it assumes very severe proportions in tropical and semi-tropical pond farms, especially in 'undrainable' ponds, such as those in use in South Asia. Limited growth of aquatic plants may be useful in maintaining water quality and may serve as shelter and substrates for food organisms in ponds, but profuse and un-controlled growth affects aquaculture oper-ations in several ways. Besides restricting the movements of fish and other aquaculture species, dense growths of vegetation, particu-larly floating plants, prevent adequate light penetration into the water and thus affect their productivity. Photosynthesis and oxygen pro-duction will be reduced when pond surfaces are covered by vegetation and this may cause oxygen depletion and consequently anoxia of the cultured species. Considerable amounts of nutrients from the water and those introduced into ponds through fertilization will be used up by the weeds and consequently the growth of food organisms will be reduced, resulting in low yields of the cultured species. Blooms of algae in ponds and enclosures often lead to oxygen depletion as a result of dead and decay-ing algal mass. Mass mortality of fish can occur under these conditions. Dense growths of aquatic weeds will make fishing with nets ex-tremely difficult in ponds. Weed-infested stagnant ponds provide favourable conditions for mosquito breeding and thus become a public health hazard. In cage culture, in both fresh-and sea-water, thick growths of algae on the net cages reduce water exchange and thus affect the water quality within the cages.

Control of weed growth is not so difficult in small farms, when labour is not too expensive. As discussed in Chapter 7, in countries like China aquatic weeds are effectively used as feed or fertilizer in fish ponds. However, in large farms in most tropical countries weed control is a formidable problem. It adds sub-stantially to the operational cost as control measures have to be adopted at frequent in-tervals to prevent reinfestation. Seeds or spores may be brought in through water intake, blown in by wind or carried by birds or other animals or inadvertently by workers. If the control measures employed do not include removal and disposal of dead weeds from the pond area, the weeds decay and add to the fertility of the soil and water and thereby promote further growth of dense weeds. This problem is not unique to aquaculture farms: shallow lakes, reservoirs and irrigation channels can also be choked by dense growths of weeds, which are very difficult to control.

Several factors, individually or jointly, in-fluence the growth of particular species of weeds. Besides the geographical and climatic conditions, topography, depth of water, extent of bottom sediments, clarity and fertility of the water, access to sources of infestation, and occurrence of floods are some of the factors that are of importance. Aquaculture farms which cannot be drained and dried regularly, and where there are thick deposits of silt at the bottom, are more likely to have recurring weed problems. Persistent blooms of certain algae have been variously attributed to their ability to store nutrients for use when they are not available or to produce and liberate certain metabolites which help in the exclusion of other algae.

10.2 Common aquatic weeds

Aquatic weeds belong to various families of dicots, monocots and single-celled and filamentous algae. From the point of view of aquaculture and weed control, the macrophytic and algal weeds can be best classified according to their habits and habitat. According to Philipose (1968), they can be divided into

(1) floating weeds, which are unattached and float with their leaves above the water surface and roots under water e.g. *Eichhornia*, *Pistia*, *Azolla*;
(2) emergent weeds, which are rooted in the bottom soil but have all or some of their leaves, leaf laminae or shoots above the water surface (e.g. *Nymphaea*, *Trapa*, *Myriophyllum*);
(3) submerged weeds, which are completely submersed under water, but may be rooted in the bottom soil (e.g. *Hydrilla*, *Najas*) or free-floating (e.g. *Ceratophyllum*, *Utricularia*);
(4) marginal weeds, which fringe the shore line of the water body and are mostly rooted in water-logged soil (e.g. *Typha*, *Phragmites*);
(5) filamentous algae, which form 'mats' in the marginal area or 'scums' in the main body of water (e.g. *Spirogyra*, *Pithophora*) and
(6) algal blooms, occurring dispersed in the water body (e.g. *Microcystis*, *Anabaena*).

Water hyacinth (*Eichhornia*) is probably the most widely known floating weed, not only infesting aquaculture farms but also all other types of water bodies in Asia and many other parts of the world. The spread of the plant is truly phenomenal, as it can increase in volume by about 700 per cent within 50 days (Parija, 1934) or from a pair of plants to 1200 in four months. Two other floating weeds that deserve special mention are *Pistia* (water lettuce) and *Salvinia*, which grow very fast and cover large areas of aquaculture farms and open-water bodies. *Lemna* (duck weed), although more easy to control, can grow rapidly and cover a pond or enclosure in a short period of time.

Submerged weeds are generally more difficult to control and are therefore considered more noxious than all the others. *Hydrilla*, *Najas*, *Nitella*, *Vallisneria*, *Potamogeton*, *Ceratophyllum*, *Utricularia* and *Chara* are some of the persistent submerged weeds which require considerable efforts to eradicate.

Common emergent weeds are *Nymphaea*, *Nelumbium*, *Trapa* and *Myriophyllum*, which can more easily be controlled; some of them, like *Nelumbium* and *Trapa*, are in fact sometimes grown in association with fish.

Although marginal weeds are considered undesirable in fish ponds in many parts of the world, in some East European countries a reed belt is maintained in large fish ponds to control wave action. They are actually planted on the berm of pond dikes, by seeding as a soil root mixture, or as root or shoot cuttings. A density of at least 70 reeds/m^2 is considered necessary. *Phragmites* and *Typha* are the two common marginal weeds used in this manner.

While many of the algae are food of fish and other aquaculture species, it is the excessive growth of filamentous algae like *Spirogyra* and *Pithophora* and persistent blooms of planktonic algae such as *Microcystis* and *Anabaena* that account for their being considered sometimes as weeds in aquaculture farms.

10.3 Methods of weed control

10.3.1 Prevention of infestations and utilization of weeds

It will be logical to consider first the possibility of preventing infestation of weeds in aquaculture farms. When constructing pond farms, care could be taken to avoid very shallow marginal areas and to maintain a depth of about 0.75–0.9 m around the shoreline, to discourage growth of marginal weeds. Accumulation of silt can be reduced by preventing drainage of run-off from fertile land areas and by regular desilting of ponds. Erecting barriers or mesh filters to prevent entry of noxious weeds and their spores or seeds can be of some help. The use of netting, treated with antifouling chemicals, to make cages may reduce to some extent the growth of algal weeds in cage farms. But none of these preventive measures go far enough to eliminate entirely the weed problem in aquaculture farms. So, it is often necessary to resort to one of the four other common control methods: manual, mechanical, chemical or biological. As will be discussed later in this chapter, it will often be necessary to combine two or more of these methods to obtain satisfactory results.

The selection of the method of control has naturally to be based on the type and density of the infestation, the nature of the farm and the species that is cultured. Similarly, it is necessary to select the most appropriate time for treatment to get best results. Control measures are more effective if applied at the most vulnerable period in the life history of the weed, which is often the period of intensive production of reproductory units. For example, the best time to treat water hyacinth is during its active vegetative growth, when it is very susceptible. Plants with well-marked seasons of flowering or turion formation can best be cleared before the fruits, seeds or turions are fully formed and shed. Cutting of plants like water lilies (*Nelumbium*), *Typha* and *Phragmites* should be done when they are flowering.

Another major consideration in the selection of a method of weed control is the cost and naturally this varies very considerably between countries and locations. While estimating costs, the costs of not only the first major treatment, but also the subsequent treatments have to be taken into account. By a combination of methods, such as manual and chemical, it may sometimes be possible to bring down the costs.

The cost of weed control can be reduced if some of the weeds can be put to productive use. Several efforts have been made to convert aquatic weeds to food, fertilizer, paper, fibre and energy (biogas). For comprehensive reviews of experience in this aspect Little (1968) and NAS (1976) may be consulted. Probably the two most convenient ways in which aquaculture can make use of aquatic weeds are as fertilizer or as fodder for herbivorous species. As will be described in Chapter 29, the Chinese farmers have overcome the weed problem by their regular utilization as fodder for grass carp and other herbivores or as a fertilizer. Weeds collected from the farm and adjacent water

bodies are finely crushed and introduced into ponds (fig. 10.1). Some of the material will be directly consumed by carps and the rest fertilize the ponds for production of the planktonic and benthic organisms on which fish feed. Another option is to compost the weeds and use them as fertilizer for the ponds or for terrestrial plants grown in association with the fish farm. The methods of composting were described in Chapter 7. The weeds are harvested and spread out at the water's edge for a day or two. The wilted plants are made into a pile with some soil, ash and a little animal manure. Microbial decomposition begins spontaneously and the resultant bacteria and fungi break down the lipids, proteins, sugars, starches and cellulose fibre. The heat retained by the composting mass encourages rapid multiplication of the microorganisms. Weeds growing in nutrient-rich waters of aquaculture farms contain adequate quantities of nutrients to sustain the microbes that produce the compost. Frequent turning of the pile may be needed to maintain aerobic conditions. Weed seeds and most pathogenic bacteria are killed and the decomposed weeds become a soil-like compost.

10.3.2 Manual and mechanical methods

In small farms it is often possible to remove floating weeds and uproot marginal and emergent weeds manually, with the help of simple tools like hand scythes, wire mesh, coir nets, etc. Water hyacinth, arrow head (*Sagittaria*), water lettuce, salvinia, duck weed, *Azolla*, *Spirodella* and *Hygrorhiza* are examples of floating weeds which can thus be removed from aquaculture waters. It is often difficult to eradicate the weeds completely, and the few that remain may be enough for the water body to be recolonized. Repeated removal, combined where possible with biological or chemical methods, may be required to keep their growth under control.

Several types of mechanical equipment have been devised for weed control, but since these are generally meant for large bodies of water such as lakes, most of them can be used only in very extensive pond farms. Farms with large individual ponds measuring as much as 100 ha, found in Eastern European countries, use such equipment regularly. The most common device

Fig. 10.1 Aquatic plants being crushed for use as fodder and fertilizer in fish ponds in China.

is a weed cutter, used for cutting submerged and emergent weeds (fig. 10.2). There are several models of mechanical weed cutters; many of them consist of flat-bottomed boats fitted with cutting beams or other cutting devices which can be adjusted to cut at different depths. Many of the cutting devices consist of two cutting beams: one horizontal and the other vertical. Motor-propelled boats fitted in front with a series of circular saws (one vertical and the others horizontal) have also been used. Floating weed cutters are generally driven by paddle wheels (fig. 10.3). Amphibious boats fitted with weed-cutting devices are especially convenient for use in shallow ponds and enclosures.

Weeds have to be cut close to the pond bottom and repeated cutting will be necessary to keep the growth under check. Even in cold climates, a minimum of two cuttings per year is recommended, but in tropical climates cutting will have to be done more often. Removal of cut weeds is equally important. Large wooden rakes (4 m wide with several 30 cm long teeth) mounted on light floats can be attached to a long cable or rope and operated from the pond bank (fig. 10.4). Lange (1965) described a

sickle-bar cutter, used for controlling submerged weeds. The weeds are cut up to 1.5 m below the water surface and conveyed to a barge for disposal.

It may appear easier to remove algal growths such as of the muskgrass (*Chara* sp.), but in fact it is often difficult to remove the clumps completely by mechanical means. What remains is usually enough to recolonize the water body in a short period of time.

10.3.3 Chemical methods

Treatment with herbicide chemicals show relatively more rapid results in weed control. However, most herbicides are also lethal to cultured animals at the levels of concentration required to kill aquatic weeds, in which case they can be used in aquaculture farms only after harvests and during renovation. Another problem is the accumulation of dead weeds after herbicide application. Unless they are removed manually or by mechanical means, the weeds will decay and create oxygen depletion in the water. Further, the nutrients released by disintegrated weeds will add to the fertility of the farm and lead to further growth of weeds. It is also likely

Fig. 10.2 A weed cutter used for submerged and emergent weeds.

that herbicides will affect the development of blooms of desirable phytoplankton in the farm.

Blackburn (1968) listed several herbicides which have been used in the control of algae and floating, emergent and submerged weeds The use of inorganic fertilizers to develop a thick growth of phytoplankton that will cover the water surface and prevent the penetration of sunlight and consequently the establishment of weeds, has been recommended in some areas. But this method has not received wide acceptance, especially in the tropics, where introduction of additional fertilizers in already fertile waters often results in more luxuriant growth of macrovegetation rather than plankton. However, shading of nursery and fry ponds with dyes, and even cowdung, has been tried to control noxious algal blooms with some success.

The herbicides used show a wide range of chemical structure and their action is either by direct contact that results in the destruction of the protoplasm or by translocation to unexposed parts of the weed. They are applied on the foliage or in the water in which the weed grows (fig. 10.5). In the case of rooted vegetation, the herbicide is applied on the soil where the roots penetrate. Some of them show selective action against particular weeds, whereas others are non-selective.

The effectiveness of herbicide treatment depends on a number of factors, including the weed growth, the time and method of treatment and the nature of the water body. It is also clear that chemical treatment has to be combined with manual or mechanical removal, as well as effective farm management, to prevent reinfestation. The economics of treatment vary considerably according to country and location. More importantly, many countries prohibit the use of some of the chemicals in aquaculture waters.

Among the herbicides, 2,4−D (2, 4-dichlorophenoxyacetate) has been widely used in controlling floating and emergent weeds by foliage application. It is commercially available as an acid, sodium salts, amine salts or as esters. The amines and esters have been widely used in the control of water hyacinth, *Pistia*, *Myriophyllum*, *Inula* and *Prosopsis*, sprayed as a 1 per cent aqueous solution. Being water-soluble, Diquat (6, 7 dihydrodipyrido (1, 2−a:2′1′−c) pyrazidiinium salt) is easy to apply in pond farms and are reported to be effective against emergent, floating and submerged weeds. However, its effectiveness is greatly reduced in waters with suspended soil or organic particles, and so it can be used only in farm ponds with clear water conditions.

As mentioned earlier, the eradication of sub-

Fig. 10.3 A weed cutter used mainly for floating weeds.

Fig. 10.4 Wooden rakes used for removal of submerged weeds.

Fig. 10.5 Spraying herbicides in a fish pond in India.

merged weeds is comparatively more difficult and the application of weedicides in waters containing aquaculture animals would require the observance of greater safety margins. Some of the chemicals used for eradicating submerged weeds as well as filamentous algae, such as copper sulphate and simazine (2-chloro-4, 6 bis (ethylamine)-triazine) may affect the long-term productivity of food organisms by the persistence of the chemical in the water or by its accumulation at the bottom. Their non-selective toxicity may also be a drawback.

Copper sulphate ($CuSO_4$) is the most widely used and economical weedicide, especially in the control of algal growths. A concentration of 1–3 ppm copper sulphate pentahydrate (CSP) is reported to be very effective for several algae.

Anhydrous ammonia has been used to eradicate dense growths of submerged weeds such as *Hydrilla* and *Najas*, as well as floating weeds like *Pistia*, but the dose required to kill the weeds is lethal to fish. As ammonia acts quickly and does not leave any residual toxicity, it has been suggested that the cultured species can be saved by adopting sectional treatment.

Ammonia treatment is time-consuming and expensive, but would partly serve as a nutrient in the water. Whether such additional fertility is acceptable in the particular farm is also an important consideration in the use of ammonia as a weedicide. The recommended dose for submerged weeds is 12 ppmN and the gas is applied from a cylinder at a maximum speed of 2.75 kg/h. The weeds die within about five days. When it has to be used as a foliar spray, the use of a wetting agent is recommended to enhance its effectiveness. A foliar spray of 1 per cent aqueous ammonia solution along with a 0.25 per cent wetting agent is reported to have been most effective in controlling *Pistia* (Ramachandran *et al.*, 1975).

10.3.4 Biological control

The limitations, costs and possible side-effects of the methods of weed control have naturally led to searches for acceptable biological control measures that can be adopted in aquaculture farms. The use of several herbivorous fish and other aquatic animals has been considered and some experimental work carried out. Some of the reports are listed in the references in this chapter. Besides herbivorous fish, nutria (*Myocaster coypus*), musk-rats, manatee or seacow, ducks and geese, beetles or their larvae and snails have all been considered. Though in theory any aquatic herbivore would be useful in reducing the growth of aquatic plants, their selective feeding at different stages of life, the population density needed to exert an effective control on the plant growth and the current aquaculture techniques that also use artificial feeds make it much more difficult to use biological control methods in aquaculture farms. The situation will, of course, be different in open bodies of water, including irrigation channels, small reservoirs, lakes and swamps.

Among fish that have a high cultural potential, the grass carp, certain species of tilapia and the tawes (*Puntius javanicus*) are the ones that have proved useful in controlling dense growths of vegetation. The practice of feeding grass carp with cut grass and aquatic vegetation is described in Chapter 15. It has been estimated that 19.9 metric tons of water weed would be consumed by grass carp to produce about 195 kg fish flesh. At such a consumption rate, the grass carp can effectively keep under check weed growths in fish ponds. Depending on their size and age, fish show preferences for certain types of weeds, and consequently their effectiveness would vary to some extent in a fish farm. For example, fingerlings of 30–50 g feed on aquatic weeds like duckweed, but larger fish prefer larger vegetation. It is estimated that a population of at least 100–200 grass carp per hectare will be required to control the growth of weeds. Tawes feed on a variety of weeds, such as *Chara*, *Hydrilla*, *Nechamandra* and *Azolla*. Ponds choked with *Hydrilla* could be cleared within a month by stocking the fish at 300–375 per hectare. Stocked at levels of 125–150 fish per hectare, they consume duckweeds at a rate of 1.8 kg per fish per day. *Tilapia rendalii* and *Sarotherodon* (= Tilapia) *mossambica* have been found to be useful in controlling the growth of filamentous algae and soft submerged vegetation. Childers and Bennett (1967) reported that when more than 1000 *S. mossambica* per acre (2470 per ha) were present in a pond, they were able to eliminate algae and rooted submerged vegetation. *Sarotherodon* (= Tilapia) *nilotica* and

T. zillii are also reported to be useful in weed control. Biological control measures may involve the use of exotic species. In such cases, it will be advisable to use sterile hybrids to avoid their multiplication in aquaculture farms.

Periodic fertilization of ponds with inorganic fertilizers has been recommended to produce phytoplankton blooms to shade and kill submerged vegetation. This method is useful under certain situations, such as soon after filling a pond. But an algal screen may not be effective in preventing penetration of light, particularly in tropical areas where intense sunlight is available for long periods of time.

10.4 Control of predators, weed animals and pests

Pests and predators seldom feature in aquaculture research, but the losses caused by them are often much higher than generally recognized. It is reported that a pelican can consume between 1 and 3 tons of fish in a year. According to du Plessis (1957), ten breeding pairs of cormorants will catch about 4.5 tons of fish in a year. Herons may cause losses of up to 30–40 per cent of fry and juvenile fish in a pond farm. A heron may consume as much as 100 kg fish per year. Bird predation in shrimp ponds is reported to decrease production by about 75 per cent in Texas (USA). The losses caused by mammalian predators like otters are even greater (as much as 80 per cent of the stock), as they generally kill much more than they can eat. The measures that are presently available are only partially effective, as predatory birds and animals very soon find ways of circumventing control measures employed in aquaculture farms. Many of the pests are difficult to control in large farm areas and most control measures require continued application, involving employment of considerable labour.

10.4.1 Predators

Several species of predatory fish may gain access to aquaculture farms through water supplies or along with seed brought into the farm. Water management in farms, such as periodic draining and preparations for introduction of new stock, offers opportunities to the farmer to exercise a reasonable amount of control on predatory fish. Outdoor nursery ponds, where the post-larvae and fry are susceptible to predation not only by predatory fish, but also by insect larvae, notonectids, etc., (see Chapter 17) and amphibians like frogs, it is relatively easier to adopt control measures like the spread of oil emulsions to prevent aerial breathing of insect larvae or fencing to prevent entry of frogs. The control of avian and mammalian predators is more difficult.

Among bird predators, cormorants, fish eagles, herons and kingfishers are considered to be the worst. If adequate protective measures are not taken, large flocks of cormorants can drive fish into shallow areas by flapping their wings, and prey on them in large numbers. The shallow waters of tropical coastal ponds provide ideal conditions for birds to prey on dense stocks of cultured species. Herons and egrets are the major predatory birds in Texan shrimp ponds; grebes and shore birds (Charadri-iformes) as well as gulls (family *Laridae*) are mainly competitors for food, and prey on shrimps only when the water level is very low.

Several methods of controlling bird predation are practised with varying degrees of success. Small ponds and raceways can often be covered with nets or wire-mesh, but it is intriguing to see how some birds learn in the course of time to gain access through such protective covers. Devices like flash guns, sirens, klaxon horns, gongs, scarecrows, bamboo rattles and bells have all been tried with initial success. In small nursery ponds in Southeast Asia, farmers sometimes run lines of string on poles set in the pond and attach bright-coloured pieces of cloth or metal to the string to scare birds. An ingenious device, which consists of a windmill with mirrors that revolve and flash brilliantly, has been used in Malaysia to scare birds with some success, but it appears that if kept in the same area for a period of time its scaring effect is considerably reduced. Obviously, these devices can only be an adjunct to continued vigil by conscientious watchmen. Watching is made more difficult by the fact that some of the birds, like herons, are active primarily during the night. Some of the fish-eating birds may be protected by law, but at least the others can be

shot or caught with spring traps. They can also be poisoned or their nests destroyed.

Frogs and toads have been reported to destroy the larvae and juveniles of fish, par-ticularly of tilapia in African ponds. Some of the aquatic snakes prey on juvenile fish. Other predators are crocodiles, alligators and large lizards. All these can more easily be prevented from entering farms with proper fencing and by keeping the pond banks and surrounding areas free from dense growths of vegetation. The snapping turtle has been found to prey on catfish, but other turtles usually only compete with fish for space and food.

Otters (*Lutra* and *Aeonyx*) are probably the most destructive among the mammalian predators. They live in the immediate vicinity of water and burrow into the banks under the roots of trees. Otters are nocturnal in habit and hunt for fish mainly on clear nights. They attack relatively large fish and eat the best parts and leave the rest. The recommended control measure is to catch them with special otter traps. Large traps with sturdy solid teeth have to be used as otters can easily escape from smaller traps. The traps should be set in pass-ages generally used by the animal to enter the farm; the passages can be identified by the otters, webbed footprints and excrement. Hunting them from their holes with the help of trained otter dogs and proper fencing of the farm are other means of control.

Among the losses sustained by predation should be included poaching by man, which is extremely difficult to prevent. This problem is experienced world-wide, but its severity varies with the system of culture (e.g pond culture, cage culture, raceway culture, etc.) and the socio-cultural background of the neighbour-hood communities. The risk becomes greater when the crop is ready for harvest and the culture system makes it easy to catch large quantities in a short period of time with little effort, as in cage farms. On the other hand, in intensive farming systems using limited space, it will be possible to exercise greater vigil, unlike in large pond farms covering hundreds of hectares. Traditional anti-poaching measures include the employment of reliable watchmen, use of trained watch dogs, placing hidden ob-structions in ponds to prevent seining and fencing of farm areas. In recent times, several types of burglar alarms and even electrified fencing have been tried with varying degrees of success.

10.4.2 Weed animals and pests

Removal of weed fish, that is species of fish which compete for space and food with the cultured species, is a common practice in all types of aquaculture, and in confined areas it is often possible to achieve considerable success in this. In open-water culture systems, like those of molluscs, only very limited success can be expected. In confined aquaculture waters, weed fish gain access generally during early stages of their life history with incoming water. By the use of filtering devices at the intake, the entry of wild fish and other aquatic animals can be reduced to a large extent. There are different types of filters that can be used for the purpose. Lee (1973) describes two types of filters used in catfish farms: a saran sock filter and a box filter. Sock filters are cylindrical in shape, made by sewing together two pieces of saran screen (generally 3.7 m long and 0.9 m wide), and provided with a draw-string closing arrangement at each end. The inlet pipe is placed inside one end of the filter and the draw-string tied tightly around it. The other end is also tightly tied to prevent escape of fish or other animals. The filter is cleared regularly to remove the catch. Box filters may be made to float or fixed permanently below the inlet. They are constructed of wooden frames and screens. Common dimensions are about 2.5 m long, 0.9 m wide and 0.6 m deep. The bottom of the box is made of screen, reinforced at intervals with wooden boards. Most filters, however, would not be able to prevent the entry of small larvae and eggs of weed species. So, when necessary, some other measures like selective fishing or selective toxins will have to be used, to eradicate them from the farm.

Certain species of snails, particularly those belonging to the family Cerithidae, are major competitors for food in fresh- and brackish-waters when they occur in abundance. Large numbers of them enter farm areas as larvae and grow and multiply rapidly. They affect the growth and abundance of benthic algal complexes, which are especially important in coastal ponds of Southeast Asia. If their numbers are high, they disturb the benthic algae by loosening the sediments. The pond can become very muddy and the algal complex may break loose from the bottom on windy days and float to the surface. The wind and waves waft them to the pond bank, where they may settle and decompose, producing large amounts of hydrogen sulphide. Very dense populations (up to 34 tons/ha) of snails have been reported from coastal ponds in Indonesia. The application of 12–15 kg/ha of nicotine (commercial tobacco dust) or 15–18 kg/ha of saponin on the pond bottom after drainage of the pond is reported to be effective in controlling snails in coastal ponds (Tang, 1967). Manual or mechanical removal of snails can also be effective, if properly done after the ponds are drained. When the pond cannot be drained, application of Bayluscide (5, 2-dichloro-4-nitro-salicylicaniline-ethanolamine), at a concentration of 3 ppm in the pond water, has been suggested (Tang, 1967). But Bayluscide and other commercial preparations have residual effects for varying periods of time.

Polychaete worms are serious pests in coastal ponds. They live in burrows on the pond bottom and make the soil porous, reducing the water-holding capacity of the ponds. The growth of desirable algae is also hindered. Drainage of the pond seldom helps to eradicate the worms. Phenol has been used in the Philippines (Pillai, 1962) to control these worms after partial draining. The pond has to be flushed once or twice with fresh tidal water, before algal growth commences again and the pond becomes suitable for stocking fish. Tang (1967) reported that 2 ppm nicotine or 3 ppm Bayluscide are also effective in controlling polychaete worms after the water has been drained. Chironomid larvae which compete for food with benthic-feeding fish in coastal ponds can be controlled by repeated application of technical Y-BHC at a concentration of 0.08–0.1 ppm (Tang and Chen, 1959). But it imparts an unpleasant odour to the fish. There is also the likelihood of the larvae developing resistance to this gamma isomer.

Among the pests in coastal aquaculture

farms, probably the most noxious ones that affect the safety of the farm itself are crabs. Leakage of water from the dikes and consequent problems in maintaining the required water levels in ponds are often caused by holes made by burrowing crabs. In coastal areas, crabs are found in great abundance and so the damage that they do to the farm can be immense. Water flowing through crab holes can cause the complete collapse of dikes. Predators and weed animals can gain access to the farm through the holes. Swimming crabs (*Portunidae*) have been found to be serious predators in shrimp ponds. Net-cages in open waters are often damaged by crabs, resulting in the escape of fish from the cages.

Because of the damage that crabs do to pond dikes, farmers spend considerable time and effort in reducing crab populations in pond farms and neighbourhood areas by using chemicals or special trapping devices. Jordan (1957) reported that repeated fortnightly spraying with a 10 per cent suspension of technical BHC (containing 6.5 per cent gamma isomer) was effective in controlling *Sesarma* and *Sarmatium* species, without affecting fish. ASEAN (1978) reported the use of the insecticide 'Sevin'. It is mixed with ground-up fish and made into small balls which are then placed in crab holes above the water line. Sevin is toxic to shrimps and so if it has to be used in holes below the water line in shrimp ponds, the holes should be closed, so that the shrimps will not have access to the chemical. Another means of killing crabs is by applying calcium carbide in crab holes and pouring water into it to wet the carbide, in order to produce the lethal acetylene gas (ASEAN, 1978). Tobacco dust and several other toxic materials and insecticides have also been used as contact poisons to kill crabs.

The so-called burrowing shrimp (*Thalassina*) is another pest in certain areas, where it damages dikes by burrowing. The presence of *Thalassina* is easily detected by the occurrence of high mounds at the entrance to their holes, which are above the water line. The methods adopted for controlling crabs can be used for killing *Thalassina* as well. Special trigger-type traps have also been found to be useful in catching them.

Musk rats (*Ondatra*) and field rats dig large burrows in banks and dikes and can thus weaken the structures. Suggested ways of controlling them are capture with traps and trap-nets or by shooting.

The various fouling organisms that grow on the water control structures of ponds and cages in open-water areas can also be considered as pests in aquaculture. Regular cleaning and drying is essential to keep these in good condition. Wooden structures should be made of treated wood or painted with preservative paints that are not toxic to cultured animals, to reduce the problem. Greater use of cement concrete to build water control structures when possible will also help to reduce fouling problems.

10.4.3 Use of non-selective pesticides

Even though the selective treatments to eradicate predators, weed fish or pest animals described above are valuable in combating specific individual infestations, there are several pesticides and poisons that could be used to eradicate some or all of the predators, weed fish and pests simultaneously. The use of natural products like teaseed cake and derris powder has been very popular among aquaculturists for this purpose, because these are not harmful to man in small amounts and lose their toxicity in water in short periods. The use of chlorinated hydrocarbons (e.g. DDT, Endrin, Chlordan, gamma BHC), although effective, has to be avoided because of their long-term residual effects. On the other hand, organophosphate pesticides like Gusathion do not leave a toxic residue for more than two weeks or so after application.

Teaseed cake or saponin is widely used in fish and shrimp ponds in Asian countries to control pests and predators. Teaseed cake is the residue of the seeds of *Camellia drupisera* after extraction of oil, and generally contains 10−15 per cent saponin. A dose of 216 kg teaseed cake together with 144 kg quicklime per hectare is recommended for application in an aqueous solution on pond bottoms after reducing the water level. It is effective in killing weed and predatory fish, as well as snails and crabs. Tobacco dust, the active component of which is nicotine, can also be used as a fish toxicant to eradicate unwanted species.

Rotenone is another plant product that is widely used as a toxicant to clear aquaculture waters. It can be used as derris powder which contains 4−8 per cent of the active ingredient rotenone ($C_{23}H_{22}O_6$). Different dosages have been recommended. When applied at levels of 0.5 ppm, the toxicity disappears within about 48 hours (Hall, 1949). Lunz and Bearden (1963) found 1.5 ppm concentration of rotenone effective in controlling undesirable fish in shrimp ponds. Alikunhi (1957) reported on the safe use of concentrations of up to 20 ppm, and under tropical temperatures the toxicity continues from 8 to 12 days. Fresh derris roots are more effective than dry roots or their powder, because of higher rotenone contents. Roots should be cut into small pieces and soaked overnight in water. Soaked roots are pounded and the crushed roots are replaced in the water in which they were soaked and squeezed to press out as much of the rotenone as possible. The extract is applied in the pond at the rate of 4 g roots per m^3 water.

Many chemical fish toxicants, particularly Endrin, Dieldrin and Aldrin, have been used to clear aquaculture farms of predators and pests. Pillai (1972) has summarized the use of DDT, 2, 4−D and sodium cyanide in the eradication of predators and pests. A lethal concentration of DDT is reported to be 0.03 g/ℓ and that of 2,4−D 0.13, ml/ℓ of water. Repeated washing of the pond will be required after treatment to remove toxic effects. In the case of sodium cyanide, the lethal concentration is 1 ppm and the toxic effects are reported to disappear under pond conditions after about 96 hours. The agricultural weedicide sodium pentachlorophenate (PCP-Na) is recommended for use in shrimp ponds to kill predatory and weed fish, as the lethal concentration of 0.5 ppm of the chemical for fish is below that for shrimps. Sodium pentachlorophenate decomposes when exposed to direct sunlight and toxicity is reduced by 90 per cent after about three hours. As in all other chemical treatments, the water level in the farm should be reduced as low as possible before treatment. The aqueous solution of the weedicide should be evenly distributed and as soon as the fish are killed fresh water should be let in to dilute its concentration.

10.5 References

Alabaster J.S. and Stott B. (1967) *Grass Carp (Ctenopharyngodon idella* Val.) *for aquatic weed control*. Ministry of Agriculture, Fisheries and Food, London.

Alikunhi K.H. (1957) Fish culture in India. *Farm Bull.*, **20**, ICAR, New Delhi.

Alikunhi K.H. and Sukumaran K.K. (1964) Preliminary observations on Chinese carps in India. *Proc. Indian Acad. Sci (B)*, **60**(3), 171−88.

Allsopp W.H.L. (1960) The manatee: ecology and use for weed control. *Nature*, **188**, 762.

Anderson W.H. (1965) Search for insects in South America that feed on aquatic weeds. *Proc. 5th Weed Control Conf.*, **18**, 586−7.

ASEAN (1978) *Manual on Pond Culture of Penaeid Shrimp*. ASEAN National Coordinating Agency of the Philippines, Manila.

Avault J.W. (1965) Biological weed control with herbivorous fish. *Proc. 5th Weed Control Conf.*, **18**, 590−91.

Beynon J.L. *et al.* (1981) Nocturnal activity of birds on shrimp mariculture ponds. *J. World Maricul. Soc.*, **12**(2), 63−70.

Blackburn R.D. (1963) Evaluating herbicides against aquatic weeds. *Weeds* **11**(1), 21−4.

Blackburn, R.D. (1968) Weed control in fish ponds in the United States. In *Proceedings of the World Symposium on Warm-water Pond Fish Culture*, (Ed. by T.V.R. Pilay). *FAO Fish. Rep.*, **44**(5), 1−17.

Blackburn R.D. (1974) Chemical control. In *Aquatic Vegetation and its Use and Control*, (Ed. by D.S. Mitchell), pp. 85−98. UNESCO, Paris.

Bond C.E., Lewis R.H. and Fryer J.L. (1959) Toxicity of various herbicides to fishes. In *Transactions of Seminar of U.S. Department of Public Health, Education and Welfare*, 96−101.

Brock J.A. (1983) Pond production systems: diseases, competitors, pests, predators and public health considerations. In *Principles and Practices of Pond Aquaculture*, (Ed. by J.E. Lannan, R.O. Smitherman and G. Tchobanoglous), pp. 169−185. Oregon State University, Newport.

Bose P.K. (1945) The problem of water hyacinth in Bengal. *Sci. Cult.*, **11**(4), 167−71.

Childers W.F. and Bennett G.W. (1967) Experimental vegetation control by largemouth bass-tilapia combinations. *J. Wildl. Man.*, **31**(3), 401−7.

Crane J.H. (1963) The effects of copper sulfate on *Microcystis* and zooplankton in ponds. *Progr. Fish-Cult.*, **25**, 198−201.

du Plessis S.S. (1957) Growth and daily food intake of the white-breasted cormorant in captivity. *The Ostrich*, December, 197−201.

Frank P.A., Otto N.E. and Bartley T.R., (1961) Techniques for evaluating aquatic herbicides. *Weeds*, **9**(4), 515−21.

Hall C.B. (1949) *Ponds and Fish Culture*. Faber and Faber, London.

Hepher B. and Pruginin Y. (1981) *Commercial Fish Farming*. John Wiley and Sons, New York.

Hickling C.F. (1965) Biological control of aquatic vegetation. *Pest Articles and News Summaries (C)*, 237−44.

Hora S.L. and Pillay T.V.R. (1962) Handbook on fish culture in the Indo-Pacific region. *FAO Fish. Biol. Tech. Paper*, **14**.

Huet M. (1986) *Textbook of Fish Culture*, 2nd edn. Fishing News Books, Oxford.

Jordan H.D. (1957) Crabs as pests of rice in tidal swamps. *Emp. J. Exp. Agric.*, **25**, 99.

Klingman G.C. (1961) *Weed Control: As a Science*. John Wiley and Sons, New York.

Kessler S. (1960) Eradication of blue-green algae with copper sulphate. *Bamidgeh*, **12**(1), 17−9.

Lange S.R. (1965) The control of aquatic plants by commercial harvesting. *Proc. 5th Weed Control Conf.*, **18**, 536−7.

Lawrence J.M. (1955) Weed control in farm ponds. *Progr. Fish-Cult.*, **17**(3), 141−3.

Lawrence J.M. (1962) Aquatic herbicide data. *Agric. Handb. Agric. Res. Serv. U.S.*, **231**.

Lee J.S. (1973) *Commercial Catfish Farming*. The Interstate Printers and Publishers, Danville.

Little E.C.S. (Ed.) (1968) *Handbook of Utilization of Aquatic Plants*. FAO of the UN, Rome.

Lunz G.R. and Bearden C.M. (1963) Control of predaceous fishes in shrimp farming in South Carolina. *Contributions from Bears Bluff Laboratories*, **36**.

Mackenthun K.M. (1958) *The Chemical Control of Aquatic Nuisances*. Committee on Water Pollution, Madison.

Martyshev F.G. (1983) *Pond Fisheries*, Russian Translation Series 4. A.A. Balkema, Rotterdam.

McAtee W.L. and Piper S.E. (1937) Excluding birds from reservoirs and fish ponds. *U.S. Dept. of Agriculture Leaflet*, 120.

Mott D.F. (1978) Control of wading bird predation at fishrearing facilities. In *Wading Birds*, (Ed. by A. Sprunt IV, J.C. Ogden and S. Winckler). *Research Report* no. 7. National Audubon Society, New York.

NAS (1976) *Making Aquatic Weeds Useful*. National Academy of Sciences, Washington.

Parija P. (1934) Physiological investigations on water hyacinth (*Eichhornia crassipes*) in Orissa with notes on other aquatic weeds. *Ind. J. Agric. Sci.*, **4**, 399−429.

Philipose M.T. (1968) Present trends in the control of weeds in fish cultural waters of Asia and the Far East. In *Proceedings of the World Symposium on Warm-water Pond Fish Culture*, (Ed. by T.V.R Pillay). *FAO Fish. Rep.*, **44**(5), 27−52.

Pillai T.G. (1962) Fish farming methods in the Philippines, Indonesia and Hong Kong. *FAO Fish. Biol. Tech. Paper*, **18**.

Pillai T.G. (1972) Pests and predators in coastal aquaculture systems of the Indo-Pacific region. In *Coastal Aquaculture in the Indo-Pacific Region*, (Ed. by T.V.R. Pillay), 456−70. Fishing News Books, Oxford.

Pruginin Y. (1968) Weed control in fish ponds in the Near East. In *Proceedings of the World Symposium on Warm-water Pond Fish Culture*, (Ed. by T.V.R. Pillay). *FAO Fish. Rep.*, **44**(5), 18−25.

Ramachandran V. (1963) The method and technique of using anhydrous ammonia for aquatic weed control. *Proc. Indo-Pac. Fish. Coun.*, **10**(2), 146−53.

Ramachandran V., Ramaprabhu T. and Reddy P.V.G.K (1973) A standard field technique of clearance of water hyacinth infestations by chemical treatment. *J. Inl. Fish. Soc. Indi.*, **5**, 154−61.

Ramachandran V., Ramaprabhu T. and Reddy P.V.G.K (1975) Observations on the use of ammonia for the control of *Pistia stratiotes* Linn. *J. Inl. Fish. Soc. Ind.*, **7**, 124−30.

Saha K.C., Sen D.P. and Muhuri G.N., (1958) On the destruction of water hyacinth (*Eichhornia crassipes* Solms) by 2, 4-D. *Sci. Cult.*, **23**(10), 556.

Seaman D.E. and Porterfield W.A. (1961) Control of aquatic weeds by the snail *Marisa cornuarietis*. *Weeds*, **12**, 87−92.

Sguros P.L., Monkus T. and Phillips C. (1965) Observations and techniques of the Florida manatee − a reticent but superb weed control agent. *Proc. 5th Weed Control Conf.*, **18**, 588.

Smith E.V. and Swingle H.S. (1941) The use of fertilizer for controlling submerged aquatic plants in ponds. *Trans. Am. Fish. Soc.*, **71**, 94−101.

Spanier E. (1980) The use of distress calls to repel night herons (*Nycticorax nycticorax*) from fish ponds. *J. Appl. Ecol.*, **17**, 287−94.

Stickney R.P (1979) *Principles of Warm-water Aquaculture*. John Wiley and Sons, New York.

Stott B. and Orr L.D. (1970) Estimating the amount of aquatic weed consumed by grass carp. *Progr. Fish-Cult.*, **32**(1), 51−54.

Swingle H.S. (1957) Control of pond weeds by

the use of herbivorous fishes. *Proc. 5th Weed Control Conf.*, **10**, 11−17.

Tang Y. (1967) Improvement of milkfish culture in the Philippines. *Curr. Aff. Bull. IndoPacif. Fish. Coun.*, **49**, 14−22.

Tang Y. and Chen T.P. (1959) *Control of Chironomid Larvae in Milkfish Ponds.* Chinese American Joint Commission in Rural Reconstruction, Fish.

Series 4.

Timmons F.L. (1963) Herbicides in aquatic weed control. *Proc. 5th Weed Control Conf.*, **16**, 5−14.

Vaas K.F. (1951) Notes on water hyacinth in Indonesia and its eradication by spraying with 2, 4-D. *Contr. Gen. Agr. Res. Stn. Bogor*, **120**, 1−59.

11
Harvesting and Post-Harvest Technology

11.1 Harvesting methods

Methods of harvesting naturally depend on the culture system, the species cultured and the form in which the product is to be marketed. Properly designed fish ponds have special provisions for draining and easy harvesting, whereas in pens and similar enclosures suitable nets and other fishing devices have to be used. In open-water stocking and ranching, fishing equipment used in capture fisheries is the common choice. Harvesting is usually the most labour-intensive operation in an aquaculture farm, apart from its construction, so there have been attempts to introduce as much mechanization as possible in order to reduce labour.

11.1.1 Harvesting drainable ponds

Harvesting from drainable ponds is relatively easy, if there is a harvesting sump or similar device. In a nursery or fry-rearing pond it is almost essential to have a harvesting sump to avoid injury to fry or fingerlings during harvest. Drainage is performed at a rate suited to the size of the outlet and the drainage channels, and the fish are concentrated in the harvesting sump. From the harvesting sump, the fish can be collected by loading equipment, if necessary with the help of a net. In case the harvesting sump is considered too small for the quantity of fish, it may be necessary to combine seining and draining to harvest the fish in good condition. Some of the fish may be seined and the rest caught in the sump. When live fish are marketed, it is useful to spray fresh water or aerate the water in the sump, to avoid weakening or mortality of the fish.

In Asian brackish-water ponds, where special catching ponds or canals are provided, fish are concentrated in the catching areas by taking advantage of their habit of swimming against currents. At high tide, tidal water is allowed to flow into the pond system through the catching ponds (see fig. 6.7) and by opening the sluice gates the fish are allowed to swim from the rearing ponds to the catching ponds. From there they can easily be fished out, when the tide turns, after full or partial drainage.

In shrimp ponds in Asia, long bag nets are set in the sluice gates to catch the shrimps as they swim out from the pond with the outflow of water at low tide (see fig. 6.10). The high tide stimulates them to move around the pond and when the tide changes they move with the current and are caught in the nets. The best time for such total harvest is during the new moon or full moon periods, as at this time the shrimps are more active. But at the full moon, many of the shrimps may be moulting and the soft-shell shrimps caught will not fetch a good price in the market. Catches are best at night and a light placed over the sluice gate will serve to lure shrimps to swim towards it.

Some species of shrimps, such as *Penaeus monodon*, do not swim out of the pond easily and repeated draining is necessary in order to harvest a good percentage of the stock. Another method of harvest is by concentrating them in peripheral canals in the pond by partial draining, and then seining them from the canals.

11.1.2 Seining undrainable ponds

The so-called undrainable fresh-water or brackish-water ponds require pumping to drain. For economic reasons they are drained only very occasionally, if at all. In such cases, and when multiple harvesting and stocking are practised, it is necessary to resort to fishing with commercial fishing gear. The most common fishing equipment for pond farms is a seine net. It is well suited for harvesting most species of fish, although some species like tilapia (e.g. *Tilapia aurea*) and certain strains of common carp can escape the nets by burrowing into the bottom mud. Species like mullets, milkfish and silver carp can escape by jumping the net.

When harvesting is to be done by seining, the shape and size of ponds in farms are designed with this in mind. Usually the length of a seine net is about one and a half times the width of the pond and the depth about two to three times the pond depth. A preferred and economic length is not more than about 150 m. The mesh size of the net depends on the size of the animal to be harvested. Too small a mesh will make dragging the seine more difficult due to increased resistance in water and so a larger mesh size is preferred when the size of the animal permits it. In nursery ponds smaller-meshed seines, even ones made of mosquito nets, are used but then the smaller pond dimensions enable easy fishing. Generally, wooden brail poles are attached to both ends of the seine for convenience in pulling and keeping the net stretched vertically (fig. 11.1). The poles can be used for attaching hauling lines, when mechanized equipment is used. The head line has floats made of cork, styrofoam or other suitable floating material and the lead line has lead sinkers or lead cores. Traditional lead lines have a tendency to sink into muddy bottoms and allow the net to roll up, resulting in loss of animals caught in the net. To avoid this, the use of 'mud lines' is recommended. They consist of a number of relatively thin ropes tied loosely together, made of a material (like jute or cotton) that readily absorbs water. Mud lines skid on the bottom, without digging or lifting the mud. This reduces escape under the seine (fig. 11.2). Hepher and Pruginin (1981) described the use of a bunch of synthetic strings, bound together to form a thick band for attachment to the bottom line of seines, to prevent the line from sinking into the mud. They reported that a similar result can be obtained when polyethylene strips of about 20 cm width and about 40 cm length are attached to the bottom line. The strips should be folded over the line so that they stretch about 20 cm behind the line. To prevent the escape of fish that jump over the net, the head line is propped up using large floats.

Fig. 11.1 Seining a small-sized pond.

When dragged, the seine bows out creating a shallow bag for the catch. To increase the capacity, the central part of the seine can be made deeper so that it forms a deeper sack during operation. To increase the capacity further, and reduce escape, the seines can be equipped with special bags. Coon *et al.* (1968) described a mechanized haul seine system (fig. 11.3) which consists of two wings and a bag. The wings are composed of six sections of various length and mesh sizes. Large-sized meshes are used in the leading wing sections to reduce resistance to drag. The sections are joined together by shackles or cordage. Perpendicular breast lines at the ends of the adjoining sections are laced together. Bridles and brails attached to the edge of each wing keep the seine wings open vertically and are also used for attaching the hauling lines. Jute rope, which soaks up water, is attached to the bottom line of the wings and the bag and it keeps the bottom line down without digging into the mud during operation. The bag section has the same small mesh size as the adjoining sections of the wings.

Fig. 11.2 Harvesting a larger pond with a drag net operated from work boats. The net is provided with a 'mudline'.

The type of seine proposed by Coon *et al.* needs a pontoon-type barge for carrying and setting it, and a mechanical seine puller for hauling the net. In smaller ponds seining is generally done manually (see fig. 11.1). The water level is reduced by partial drainage to concentrate the animals, making their capture easier. Seining generally starts at the deeper end of the pond and ends at the shallow part. Dragging the net without allowing it to rise requires skill and an experienced crew. It is better to use trucks or tractors for hauling when large nets are operated. Special winches can be fitted to the vehicles for this purpose. After reaching the shallow end of the pond, the net has to be closed in and raised with the catch to the embankment (see fig. 11.2). It is comparatively easy when the seines have detachable bags, as in the mechanized seines described by Coon *et al.* In unmechanized seines, considerable labour is involved if the catch is substantial (fig. 11.4), so different types of lifting devices have been developed. Mechanized dip nets are commonly used in large farms, but need two or three men to operate. Another device used is a mechanized bucket elevator, but it can be in-

stalled only on a well-built concrete structure. Purpose-built conveyor belts, similar to ordinary grain conveyors but with soft rubber belts and covers, are efficient to use, especially when the fish have to be lifted to a high elevation (fig. 11.5). The lower hopper of the belts can be filled by hand, by mechanized dip nets or by air-lift pumps. In order to avoid any injuries to the fish and to improve product quality, specially designed fish screws for lifting fish with water have been devised. The internal spiral ribbon and the pipe housing that revolve together lift up water along with the fish, in the space between the two. Different types of centrifugal fish pumps used in marine fishing have been tried in fish farms, but because of the damage to harvested fish they are not very widely used. Special air-lift pumps can lift the fish without damage, as there are only air bubbles inside the delivery pipe and no moving parts.

11.1.3   Other methods of fishing

As mentioned earlier, the habits of certain species like the milkfish and mullets of

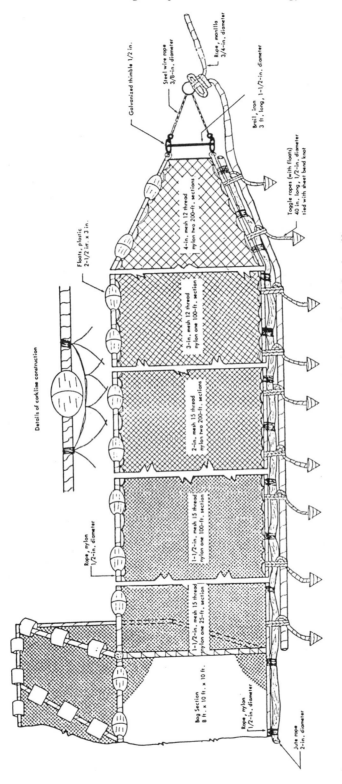

Fig. 11.3 Construction details of a mechanized pond-haul seine, showing the bag and the left wing (from *Coon et al.*, 1968).

Fig. 11.4 Manual removal of catches from a seine net.

swimming against currents is made use of in harvesting them. Even in farms without any special catching ponds, as in old traditional Indonesian tambaks, it is fairly easy to concentrate the fish near the sluice gates at high tides, from where they can be caught by means of cast nets. In large ponds, a number of fishermen may have to be employed. The main disadvantage of this type of fishing is that a large number of fish are damaged and lose

scales during removal from the net. Dip nets can also be used, even though they may not be so efficient. In milkfish ponds in Taiwan, gill nets are used for harvesting. Several pieces of netting, each of length 30 m and height about 1.6 m are joined together and towed across the pond. Before harvesting, the fish are scared in order to empty their stomachs. This is achieved by dragging a scare-line (a rope strung with a number of bamboo pieces) or by striking the

Fig. 11.5 Conveyor belts for lifting fish, used in Hungary (courtesy of Josef Kovari).

water with bamboo poles from a raft or boat. Fish with empty stomachs normally reach the markets in a better condition.

In the traditional valli culture in lagoons in Italy and other Mediterranean countries, the special trapping devices called 'lavorieri' serve the purpose of partial or sometimes even complete harvesting. Originally they consisted of screens made of reeds, installed in such a way as to lead fish into a terminal trapping chamber. In modern vallis, aluminium or other metallic screens and traps are built on concrete barrages at the inlet. Fish swim against the current at high tide and are caught in the traps. Gill nets and seines are also used for fishing from the lagoon proper.

For partial harvesting of shrimp ponds, barrier traps made of bamboo screens are used in Southeast Asia, but they have the disadvantage that many of them do not have a bottom screen, which makes removal of the shrimp from the trap inconvenient. Also it is

rather difficult to adjust the size of the mouth opening. Fyke-type shrimp traps with built-in bottoms that can be easily handled for removing shrimps are much more efficient (fig. 11.6). The mesh size of the trap, or the distance between bamboo slats (if made of bamboo), should be such that small shrimps can escape. The mouth of the trap should be adequately protected to prevent crabs from entering the trap. If operated at night with a lamp hung above to attract the shrimps, appreciable catches can be obtained. If the catches are heavy it is necessary to empty the trap periodically. Besides traps, cast nets, lift nets and seines can also be used for partial harvesting, in conjunction with lure lights at night.

In shrimp farms in Japan, harvesting is often done with pound nets. Each net consists of three or more conical net bags set in a semi-circle, with a net fence set at a right angle to the shore. The bag nets are laced together with side nets. Shrimps swim along the fence into

Fig. 11.6 Partial harvesting in shrimp ponds with traps, in Indonesia (photograph: Marcel Huet).

the pond and get caught in the bag nets. Hand-held electric equipment is reported to be used for harvesting shrimp in Taiwan (ASEAN, 1978) (See fig. 24.15). It consists of an accumulator and two bamboo poles. One of the poles is equipped with a metal tip and the other has a steel ring with a net attached. The metal tip is connected by wires to the anode of the accumulator and the steel ring to the cathode. The accumulator is carried in a back-pack or on a small raft and the operator holds one pole in each hand as he wades through the ponds. When the gear is switched on, an electric field is formed between the two poles. The shrimps jump when exposed to the electric stimulation and are caught in the net. Total harvesting of large shrimp ponds is said to be achieved by this method in Taiwan.

In multiple stocking and harvesting systems, the ponds will contain different size groups of animals and therefore special care has to be taken to return to the pond any undersized ones caught in the net. The selection of an appropriate mesh size goes a long way towards preventing the capture of undersized animals. It may, however, be necessary to perform additional sorting (fig. 11.7). Besides hand sorting, special floating graders can be used, which are usually made of wooden or metal slats set at appropriate distances apart to retain animals above a certain size and allowing smaller individuals to fall through.

11.1.4 Harvesting cage and raceway farms

It is comparatively easy to harvest stocks from intensive culture systems, particularly tank and raceway farms. They can easily be drained partially or completely as required and the animals removed by dip nets or suitable mesh.

Harvesting from cage farms can be a little more complex, depending on the location and size of the cages. Except in the case of small cages, it will not be practical or advisable to lift them for removing the stock. In many cases it may be possible to tow the cages to the shore and harvest the stock using dip nets. Where this is not feasible, boats may have to be used. When the farm has central or peripheral walkways, it is easier to raise the cages and remove the animals. But even then, some producers prefer to use seines to gather the fish and, in large farms, pumps may be used to transfer the fish to the loading area. In salmonid cage farms in Norway, feeding is stopped for a period of about a week before harvesting. This is done to empty their alimentary canals, so as to prevent quick deterioration and also to 'firm up' the flesh. During harvest the fish are removed very carefully to avoid bruising the skin and meat. Special care is taken to avoid undue stress to the fish during harvest, as the meat quality is known to be affected by stress. Generally, the fish are transported by live-hauling boats or in plastic tanks installed in boats or trucks.

Fig. 11.7 Sorting catches from a fish pond in Hungary.

11.1.5 Harvesting of molluscs

Methods of harvesting molluscs like oysters, mussels, etc., for market depend on the culture systems employed. In bottom culture, where the sea bed is planted with spat for on-growing, the most common harvesting method is dredging. From intertidal areas which are fully or partly drained at low tide, oysters or other molluscs can be easily collected by hand. But, as will be described in Chapter 26, off-bottom methods of culture are commonly practised in modern oyster and mussel farms. The oysters and mussels are grown on sticks planted in the intertidal areas or hung on long lines, special nylon bags or plastic containers. In such cases, harvesting is relatively simple and can be done by hand or from a boat when necessary. When they are grown on special platforms, as for example mussels in New Zealand (fig. 11.8), mechanical lifting devices installed on a large boat may become necessary.

In countries like France, where there is a special market for greened oysters, fully-grown flat or Portuguese oysters may be reared in small, shallow coastal ponds known as claires for up to 6 months in order to increase their weight and develop the green coloration. Claires are drained, dried and then fertilized before filling with salt water to a depth of about 25 cm. Water is exchanged at spring tides. Oysters are stocked in these ponds at a high density. They grow in weight rapidly (almost doubling in a 6 month period) and the bluish-green diatoms (*Navicula*), which grow in the claires, impart a green colour to the meat. Such large greened oysters fetch a high price in French markets.

When molluscs are dredged out from sea beds, some of the sand and silt swirled up during the operations may enter their mantle cavity. In such cases, the animals can be spread out on a cleansing plot or basin and covered with gently flowing sea water for varying

Fig. 11.8 Mechanical lifting of platform used for culturing mussels in Hauraki Gulf in New Zealand.

periods of time. This helps them to eliminate any silt or sand and also recover from the fishing stress. They are transported alive to distant markets.

11.2 Handling, preservation and transport

11.2.1 Sorting

Harvests from aquaculture farms have to be sorted according to species when polyculture is practised. When there is considerable differences in size between species mechanical graders can be used, but even then only a partial sorting is usually possible. The rest has to be done by hand (fig. 11.9). The more common need is grading according to size. Market price and consumer preference may depend very much on size; extreme examples are shrimps and tilapia. Different types of graders are in use in the fishing industry and some of them can be used in farms as well. A simple adjustable bar grader is commonly used, by adjusting the distance between the bars different size groups can be separated. Some

graders are made of revolving rollers or screens of different sizes.

Since there is usually a market preference for live animals, there will also be a need at times to sort out live fish from dead or weak ones which are not likely to survive long-distance live haulage. The live and healthy fish should be stored for transport in special holding tanks with an adequate supply of clean water. If they have to be held for longer periods, it will be advisable to aerate the water and ensure a proper water temperature. Unless storage is for unduly long periods, it is better to refrain from feeding them.

As already mentioned, certain species of fish like the American catfish and common carp, when grown in muddy ponds, may develop off-flavour caused by the compound geosmin produced by actinomycetes and a number of blue-green algae. Muddy tastes can result from the animal eating blue-green algae, such as *Oscillatoria* sp., and other microorganisms. Fish that are detected to have off-flavour should be held in flowing water for a period of 7–14 days, during which period the causative chemical is metabolized and the off-flavour removed. In higher temperatures, it takes less

Fig. 11.9 Grading and cleaning carp and tilapia before transport to markets in Israel.

time for the fish to lose the off-flavour. Off-flavours can also be caused in fish and molluscs grown in waters contaminated with chemicals from industrial wastes, such as phenols, tars and mineral oils. It is necessary to hold them in clean flowing water for longer periods to remove such flavours.

Because of the preference for live products, live hauling is practised in many areas. Oysters and mussels can fairly easily be transported alive long distances in burlap or plastic mesh sacks. Lobsters and crabs, and also to some extent the fresh-water prawn *Macrobrachium*, are transported live on ice. Live transport of shrimps is attempted only in countries like Japan where they are eaten raw. In Japan, shrimps harvested by pond nets are held in cooling tanks for periods of up to about 8 hours, depending on the climatic season, to bring the water temperature slowly down to about 12°C. The shrimps become quiescent at that temperature and the body becomes reddish

in colour. After sorting of the different sized shrimps, they are packed in cardboard boxes between layers of sawdust and transported to the markets. Ice enclosed in polyethylene or ice bags is sometimes placed on top of the layers in the packing box, especially if the transport takes a long time. When unpacked, the shrimps become active again and are displayed for sale.

Fish have to be transported in live-haul boats or trucks in water to keep them alive. Large boats with live holds are used in China for transporting carp to distant markets as far away as Hong Kong (fig. 11.10). In Norway, live-haul boats are often used to transport salmon and trout from cage farms to the packing plants. The most common means of transport is special trucks with built-in tanks or trucks suitable for carrying fish tanks. Fish tanks are usually made of fibreglass, canvas or marine plywood. Loading the fish into the tanks may be done manually with special baskets or

Fig. 11.10 Boats with live holds for transporting live fish to distant markets in China.

buckets. When large quantities are involved, a suitable lifting device has to be used. Water in the tank has to be kept in good condition during transport. As fish have to be kept under crowded conditions, it is necessary to aerate the water with mechanical agitators or airstones fed with compressed air or oxygen. Ice may be added to the tank water to reduce water temperature, particularly in summer. Some long-distance hauling trucks have more complicated designs, with facilities for filtration, oxygenation, refrigeration and recirculation of water, but these are obviously more costly to buy and maintain. A major problem in live haulage of fish is the oscillation of water in the tanks, which may cause damage to the animals and make driving of the truck difficult. So large tanks are generally provided with baffles or are divided into compartments to reduce oscillation.

11.2.2 Depuration of molluscs

As some species of molluscs, especially oysters, are eaten raw, there are strict regulations in many countries relating to bacterial concentration in the products or in the environment in which they are grown. If the median total coliform concentration exceeds 70 per 100 ml, it is mandatory to depurate them before sale. Shellfish can concentrate pathogenic organisms from sewage-polluted water. Viral pathogens, especially those causing infectious hepatitis, occur in many areas and it is often difficult to monitor them in the environment or in the shellfish. So depuration is necessary to reduce public health hazards caused by the consumption of contaminated shellfish.

Holding the animals in clean water helps them to cleanse themselves of bacteria. The three methods of depuration commonly practised consist of treatment with chlorine, ozone or ultra-violet (UV) rays. Cleansing or depuration plants are located in areas with a convenient supply of unpolluted sea water. Different systems are used to bring in the harvests and transfer them to treatment tanks. The shellfish may be spread in shallow tanks or placed in shallow baskets (about 8 cm deep) made of wood with galvanized wire mesh, plastic-coated expanded metal or plastic-coated wire mesh. The dimensions of the tanks differ very considerably according to the quantity of oysters or other shellfish to be depurated. They are made of cement concrete, with or without epoxy lining, marine plywood coated with fibreglass or of moulded fibreglass. The quality of the water and its proper circulation over the tank are very important. Cascading or splashing of water is the common practice, even though a timed air bubbling system has been incorporated in some. In order to maintain the dissolved oxygen level above 5 mg/ℓ, a flow rate of 105ℓ/ min per m^3 is generally needed. It is advisable to maintain the same salinity as in the harvest site, but it should not vary more than 20 per cent. Significant reductions in purification rates have been observed in *Crassostrea virginica* at salinities below 7 ppt.

The traditional method of cleansing the shellfish of microbes is by the use of chlorine. Initially, about 3 ppm of chlorine is added to the sea water at the time of tank filling. This excess dose makes them close their shells and enables easy cleansing of the shell. As sea water continues to flow into the tank, the concentration of chlorine decreases, resulting in a 10- to 20-fold dilution. There are reservations about the use of chlorine because of the effects of chlorine residue in the shellfish tissue. Dechlorination with sodium thiosulphate has been used in the past to remove residue, but this practice is not very common now. Many producers contend that chlorine-treated oysters retain the original flavour better than the ones treated with ozone.

Depuration with ozone treatment is commonly practised in France, Spain and Japan. Being a strong oxidizing agent, ozone can kill bacteria and viruses rapidly and therefore the time taken for purification is relatively less. It dissipates into dissolved oxygen in water and leaves no residue, unlike chlorine. In waters containing 2000−5000 coliform bacteria per litre, 1.50−2.10 g/m^3 ozone is required to sterilize the water; for a bacterial concentration of 250−1000 per litre, the requirement is about 0.75−1.15 g/m^3. It is also very effective in reducing viruses. Ozone bubbled through the water purifies the water and the shell and meat of oysters. The meat is reported to be cleansed in about 2−3 hours.

Ultra violet rays are used for disinfecting sea water for depuration in countries like the UK, USA and to a limited extent in Japan. The water can either flow through the tank or circulate. Honma (1971) described systems in which disinfected water falls in a shower on the tank and the wastes are collected at the bottom and drained out. About 300−400 oysters can be cleansed per m^3 in about 10 hours. Other designs of tanks based on different methods of delivery of disinfected water, enabling greater circulation and easy draining, have also been used. The effective penetration of UV light is limited by turbidity, as well as the depth of water. When the turbidity is high, better disinfection of viruses can be achieved by reducing the flow rate and thus increasing exposure time.

11.2.3 Preservation and processing

The care that is generally taken to preserve the quality of products during harvest, sorting and transport has already been mentioned. Since the emphasis is in marketing the product in as fresh a state as possible, only very simple preservation and processing techniques are employed for aquaculture products, except for those that are meant for export.

Salmon is probably a unique aquaculture product where extreme care is taken in handling, slaughtering and preservation. The precautions taken to reduce stress at harvest to preserve the meat quality have been described earlier. An additional means of maintaining good meat quality is by bleeding the fish when alive. Besides slowing degenerative process, it improves the appearance and taste of the meat. The usual procedure is to stab each fish with a knife behind the gills, in order to cut the major blood vessels. The carcass is then placed in tanks containing cold running water or in small floating nets, to bleed for a few minutes. Bleeding is done either on the farm itself or on arrival at the packing plant. The bled carcasses are generally gutted and cleaned and packed between layers of flake ice in polystyrene boxes. Such high-quality fresh fish fetches premium prices. Only when the market is depressed or consumers specially require it, will good Atlantic salmon be frozen. On the contrary, a good part of the trout harvested is frozen in plastic wrappings.

Where there is a demand for smoked salmonids, the fish are gutted, split and smoked for transport to markets, packed individually in styrofoam packets. Soft or hard smoking is adopted, according to market preference. The value of the product is significantly increased by this process. Smoked milkfish has a good market in the Philippines and smoked oyster meat is eaten in some countries like Japan.

Despite the preference for live or fresh products, freezing becomes necessary in certain cases because of market conditions or when the presentation of the product will improve by the process. For example, the colour of the American catfish is not very attractive to many consumers and so they are headed, gutted, skinned and then frozen for transport. Skinning is particularly difficult and labour-intensive when done by hand, but mechanical skinning devices have been developed. The same appearance problem and the general perception of it as a poor quality fish have led to the preference for raising a red-coloured hybrid variety of tilapia, which can be sold more readily in frozen fillets. They are often marketed under more attractive trade names, such as fresh-water snappers, fresh-water perch, St. Peter's fish, etc.

When long-distance transport is involved, especially for export, shrimps are preserved by freezing, following the usual shrimp processing procedures. They may be headed, shelled and deveined before freezing. Oysters, when not sold in the shell, are generally shucked by hand using an iron spatula. Mechanical shucking devices are also used in many areas. The shucked oysters are kept under refrigeration for one or two days, before packing and transport to markets.

Historically, the harvesting and processing of seaweeds have been small-scale manual operations, but with the rapid expansion of seaweed farming, mechanical devices are increasingly used. The commonly farmed seaweed *Porphyra* sp. (nori) is processed into dry sheets. The collected weeds are drained of all the water and then chopped with knives manually or with mechanical cutters that can be adjusted to cut thin or thicker slices, depending on the hardness of the thalli. A suitable quantity of water is added to the chopped weeds (usually about $13-14$ ℓ fresh water for 1 kg weeds) and mixed thoroughly. The mixture is poured on to a simple moulding device, which consists of a bamboo matting on which a frame is set up to keep the mixture in the form of a uniform sheet. The mixture spreads within the frame and the water drains through the holes in the mat, leaving the nori in a thin sheet. To reduce the labour involved, seaweed sheeting machines are widely used now. The mats with the sheets are dried in the sun or in indoor driers. When well dried, the sheets are removed from the mats and made into convenient bundles.

Canning or other types of preservation are seldom practised on any significant scale for aquaculture products, other than for mussels and oysters. Small-scale attempts have been made to find new markets for less popular species like the silver carp and tilapia by canning, with the main idea of reducing problems with fine bones. In countries like Spain, where large-scale mollusc farming is practised, it is difficult to market all the production in live, fresh or even frozen state, so a sizeable canning industry for molluscs, particularly mussel meat, has developed. Speciality products like the stuffed carp or gefilte fish in cans and oyster sauce in bottles are sometimes made from cultured species.

11.3   References

ASEAN (1978) *Manual on Pond Culture of Penaeid Shrimp.* ASEAN Coordinating Agency of the Philippines, Manila.

Billy T.J. (1974) Pond-grown catfish in the United States: present situation and future prospects. In *Fishing Products*, (Ed. by R. Kreuzer), pp. 262–6. Fishing News Books, Oxford.

Chen T.P. (1976) *Aquaculture Practices in Taiwan.* Fishing News Books, Oxford.

Coon K.L., Larsen A. and Ellis J.E. (1968) Mechanized haul seine for use in farm ponds. *Fish. Industr. Res.*, **4**(2), 91–108.

D'Ancona V. (1954) Fishing and fish culture in brackish-water lagoons. *FAO. Fish. Bull.*, **7**(4), 147–72.

Edwards D.J. (1978) *Salmon and Trout Farming in Norway*. Fishing News Books, Oxford.

FAO/WHO (1974) *Fish and Shellfish Hygiene*. FAO of the UN, Rome.

Furfari S.A. (1979) Shellfish purification: a review of current technology. In *Advances in Aquaculture*, (Ed. by T.V.R. Pillay and W.A. Dill), pp. 385–94. Fishing News Books, Oxford.

Hepher B. and Pruginin Y. (1981) *Commercial Fish Farming*. John Wiley and Sons, New York.

Honma A. (1971) *Aquaculture in Japan*. Japan FAO Association, Tokyo.

Huet M. (1986) *Textbook of Fish Culture*, 2nd edn. Fishing News Books, Oxford.

Hume A. *et al*. (1974) Studies on the acceptability of farmed fish. In *Fishery Products*, (Ed. by R. Kreuzer), pp. 253–60. Fishing News Books, Oxford.

Imai T. (1978) *Aquaculture in Shallow Seas: Progress in Shallow Sea Culture* (translated from Japanese). A.A. Balkema, Rotterdam.

Institute of Marine Fisheries (1974) *New Observations on the Use of Ozone as Sterilizing Agent of Sea Water for Cleansing of Shellfish*. France State Laboratory.

Lovell T. (1974) Environment-related off-flavours in intensively cultured fish. In *Fishery Products*, (Ed. by R. Kreuzer), pp. 259–62. Fishing News Books, Oxford.

Lovell R.T. (1979) Flavour problems in fish culture. In *Advances in Aquaculture*, (Ed. by T.V.R. Pillay and W.A. Dill) pp. 186–90. Fishing News Books, Oxford.

Pillay T.V.R. (1974) Aquaculture and fishery products development. In *Fishery Products*, (Ed. by R. Kreuzer), pp. 250–53. Fishing News Books, Oxford.

Ravagnan G. (1978) *Vallicoltura Moderna Edagricole*. G. Ravagnan, Bologna.

Stickney R.R. (1979) *Principles of Warmwater Aquaculture*. John Wiley and Sons, New York.

UNDP/FAO (1979) *Aquaculture Development in China*. ADCP/REP/79/10, FAO of the UN, Rome.

Varadi L. (1984) Mechanized harvesting in fish culture. In *Inland Aquaculture Engineering*. ADCP/REP/84/21, FAO of the UN, ROME.

Wood P.C. (1961) The principles of water sterilization by ultra-violet light, and their application in the purification of oysters. *Fish. Invest. Min. Agric. Fish. Food, G.B.* (2), **23**(6).

12
Marketing of Aquaculture Products

In previous chapters dealing with national planning and selection of species for culture, the crucial role of consumer acceptance and marketing has been mentioned. It has also been pointed out that a major strongpoint of aquaculture is that production can be market-orientated as opposed to the production-orientated marketing that has to be adopted in capture fisheries. A proper understanding of consumer demand and the consumers' attitudes and behaviour is a major asset in planning a viable aquaculture production programme. No doubt there are instances of production programmes that started without any such basic information and eventually stumbled into some success, as with tilapia in some areas, but there are many others that have experienced considerable marketing problems as in the case of silver carp, milkfish and mussels in certain countries. Chaston (1983) considered research and investment on the giant fresh-water prawn (*Macrobrachium*) as an example of efforts expended without adequate understanding of consumers' needs.

The systems and technologies of farming to be adopted will also be governed by the nature of the markets. Highly intensive systems of production which involve high production costs may be possible in certain situations only if there is a lucrative export market. If the product has to be sold in the domestic market, the possibility of using less expensive systems of production may have to be considered. It is again quite obvious that the quality and size at harvest, as well as the methods of processing and presentation, depend very much on the market.

12.1 Outlets for aquaculture products

When considering outlets for aquaculture products, one has to make a distinction between small-scale familial fish culture and large-scale commercial farming. Production from small-scale farms meant for the neighbourhood community seldom gets beyond the village market and is generally sold to consumers or fish peddlers at the farm gate in the fresh state. Except when there are a large number of farms in the area, supplies tend to be irregular and seasonal. In tropical climates, some regularity of supplies can be maintained by rearing quickly-growing species in series of ponds or other grow-out facilities and harvesting each at required intervals. Multiple stocking and harvesting procedures have also been developed in certain culture systems.

In large-scale farming, distant domestic or export markets may have to be catered to. Preservation and processing of products, long-distance transport and a variety of retail outlets may then be involved. Even in such cases, the products can be sold in live or fresh condition, as for example carp grown in mainland China sold in Hong Kong markets, yellowtail and kuruma shrimp in Tokyo markets, *Macrobrachium* grown in Martinique, in Paris markets, etc. In order to reduce the need for long transport and ensure the availability of fresh products, countries like China have encouraged the establishment of production farms in suburban locations to feed urban markets. In many countries, governmental policies have tended to promote aquaculture production in inland areas, where there is a shortage of fish supplies due to distance from the main marine fishing areas. There has also been considerable interest in locating farms near tourist centres, for the purpose of utilizing opportunities for the sale of speciality high-price products. Fish farms located near motels on highways in Yugoslavia and near fish restaurants in holiday resorts in Hungary, Greece, Italy, Taiwan and Japan have found very attractive outlets for culture products. The display of live fish or shrimps in special market ponds, pools or aquaria and the opportunity for the consumers to select the ones that they would like to eat has served as a great attraction in many areas.

While the above types of small-scale outlets serve a very useful purpose, disposal of larger production would require much greater organization. Systems that have been adopted with success are: (i) house to house delivery, (ii) sale through special fish markets or fish stalls in general markets, (iii) grocery or supermarket sales and (iv) sales to restaurants. When processing or export is involved, there is generally one or more intermediaries who process the product, pack and ship them to consuming centres.

Although one talks of market outlets and strategies in aquaculture, usually only the consumption fish are considered. Many aquaculture farms produce only fry, fingerlings, smolts or yearlings. Customers for these products are either other aquaculturists who grow them to consumption size or public or private organizations concerned with stocking open waters for recreation or commercial fishing. Producers of crustacean larvae or oyster spats also sell them to other aquaculturists, who grow them to consumption size.

12.2 Organization of marketing

As mentioned earlier, the simple system of the producer selling directly to the consumer exists only in small-scale rural aquaculture, especially in developing country situations. Medium-scale operations, raising high-valued species for urban restaurants, also undertake direct delivery of products when the farms are located in the suburbs or at short distances from the points of delivery. In other cases, marketing is usually done through intermediaries or middle men. Aquaculture products have to be distributed in as fresh a condition as possible, as consumer preference for farmed products is often based on quality and freshness. It is seldom possible for producers to undertake distribution and sales themselves, in highly dispersed distant markets. So it becomes necessary to use middle men or wholesalers, even though this will result in higher retail prices and/or lowering of the profit for the producers. Very often, marketing of fish and fishery products is dominated by middle men and market entry by individual aquaculture producers may prove very difficult. Also in areas where aquaculture products are considered to be of inferior quality or where there are prejudices against their consumption, as in the case of fish grown in sewage-fed ponds, there may be some disadvantages in making use of traditional fish marketing channels.

Another reason why many producers use the traditional fish marketing system is the opportunity to make up a deficient supply of a particular species from capture fisheries during off-seasons. As mentioned in Chapter 3, in a harmonized development of fisheries, aquaculture may be seen as a means of meeting demand for fresh products during off-seasons or filling deficits in supplies from fishing. It may be a deliberate policy to avoid any semblance of competition with capture fisheries to ensure aquaculture can develop within the fisheries sector. It is often held that the farming cost of a species that is important in capture fishery should be such that it can be sold at a price less than the current market price of the species. The reasoning is that if the availability of a species is improved as a result of successful farming, the product price will fall. If the technology does not allow it to withstand such price reduction, the farming enterprise can soon collapse.

In areas where aquaculture has developed to a significant level, the general trend now is to educate the public on the quality of farmed products and use them as selling points. Many countries have established specialized sales federations, cooperatives or similar organizations to reduce the number of intermediaries involved, harmonize marketing within the country and effectively compete in export markets. Such organizations are able to undertake useful promotional and publicity programmes and thus improve sales. They are also able to regulate production according to market demand and avoid gluts in the markets and the consequent fall in prices. Companies undertaking large-scale farming can, of course, and do organize the distribution of their products directly to major markets and to consumers. If the primary producer sells to a small number of large customers, as for example major restaurant chains, which place large orders at longer intervals, it should be possible to minimize or even eliminate the involvement of intermediaries. There will also be situations where a producer finds it not advantageous to use established distribution channels and attempts to use an alternative system. Chaston (1983) cited the example of the Scottish salmon aquaculture industry, where some processors decided to avoid the traditional wholesale/retail distribution system and started marketing their product directly to private households through mail order.

The efficiency of the physical distribution is of special importance in being able to utilize the inherent advantages of aquaculture products. As discussed in Chapter 11, the majority of aquaculture products are sold fresh or on ice. This requires delivery to the point of sale in as short a time as possible. Depending on the time required for transport and delivery, fish may be gutted or even filleted, and shrimps headed, before transport. Besides preventing deterioration of the product, considerable savings are made in shipment costs. Depending on consumer requirements, fish like salmon may be smoked and packed for transport. The success of the market depends on the speed with which handling, processing, packing and transport can be accomplished. The success of Norwegian salmon in export markets in the USA, Europe and Japan is ascribed to the incredibly short interval between harvesting and delivery at the markets. Obviously air transportation is the only means of achieving this.

When the product needs more elaborate preservation and processing, especially for export, it is generally carried out by specialized processing and exporting companies. Either the producers deliver their harvests to the processor or the processor arranges to buy the product at the farm gate and transports them to the processing plant. Some of these industrial buyers may not actually do any special processing, but may only be involved in sorting and packaging according to size and quantities preferred by customers; they then distribute the finished products to grocery stores or supermarkets, marking up prices to compensate for the services rendered.

Many of the better organized marketing arrangements referred to in this section are rare in developing countries, even though overall aquaculture production in these countries is much higher. However, conditions are gradually changing, with large-scale organized farming becoming more common. The formulation and implementation of a suitable marketing strategy would greatly assist the development of a profitable industry in these countries.

12.3 Market strategies for industrial aquaculture

12.3.1 Market research

The type of data required and the methods of evaluation vary considerably between species meant for marketing in domestic and export markets and between indigenous and exotic species. Similarly, the data required for assessing the market potential for a nationwide aquaculture programme are different from those required for determining markets for products of individual operations. The collection of any type of data and their analysis are expensive and therefore decisions have to be based on the estimated value of the expected business.

When investigating the market potential of a nationwide aquaculture development programme, detailed food consumption and dietary habits of the population, stratified geographically and socio-economically (ethnically and income-based), will be necessary. The consumption pattern of aquatic products in different groups and areas and their importance in local diets are important factors in determining the effective demand for the products. The consumer habits are affected not only by tradition and culture, but also by income level. There are species for which the demand will increase as the average income increases. On the other hand, the consumption and therefore total demand for the species may go down with increased income, if the product is not a particularly preferred one. It will also be necessary to examine whether capture fishery production of the species will be adequate to satisfy future demand, and if it is necessary to augment the supplies to meet the projected demand. The magnitude of production needed to fill the gap, if any, has to be determined.

When considering aquaculture for export purposes, additional information will be required on market trends in the importing countries and the extent of competition, besides freight and customs rates, product quality regulation and exchange rates. The ability to compete in an export market may often be dependent on the cost of production and relevant technology, availability of government incentives for export and the quality of the product.

It is much more difficult to make any realistic assessment of the markets for a non-indigenous species. Consumer attitude towards local species of comparable qualities and the price at which it can be sold in local markets may be indicative of the potential for developing a market. Test marketing may be a more useful way of assessing acceptability and demand in such cases. When markets for products of individual operations are to be evaluated, the geographic coverage can be much less, and this would make it possible to obtain much more detailed information for analysis.

Consumer habits

All the types of studies and evaluations mentioned above are ultimately based on a proper knowledge of consumer habits, within the producing country or in the importing countries. Distinction has to be made between a consumer market and an industrial market. A consumer market relates to products purchased by individuals and households for personal consumption. It is greatly influenced by traditions and social values of the community. An industrial market is concerned with the purchase of a material for the purpose of making a tangible economic return. Even though one would expect a much more rational approach in industrial buying, it is also greatly affected by the attitudes and habits of the personnel involved. It is not easy to understand the relationship between the purchasing behaviour of consumers and the main marketing variables which are described as 'product quality and price, market location and market promotion'.

When the required information on consumer habits is not available from previous market research, it will be necessary to carry out special surveys to obtain data on consumer preferences, attitudes and consumption patterns. Shang (1981) grouped consumers broadly into three categories: householders, restaurants and institutions. To these should be added industrial consumers. For each group, stratified samples will have to be studied. These comprise households on socio-economic characteristics; restaurants on the basis of their customers and classification of the standards and nature of the cuisine; and institutions according to their basic functions (hospitals, schools, children's and old people's homes, etc.). The socio-economic characteristics of those who consume the product, and of industrial consumers on the basis of their business and value added to the product will have to be considered. The quantity of the product or products purchased during a given time in the recent past, buying and eating habits and the consumers' view of prices, quality and substitutes are important pieces of information to be collected. The frequency of repurchase of the product and intentions for future purchasing would be necessary to assess levels of satisfaction with the product.

There are several methods of collecting consumer data and each has its drawbacks and merits. The most commonly employed method is through the medium of questionnaires. To obtain useful answers, questionnaires have to be prepared with great skill and understanding of a respondent's psychology. Some appropriate questions are shown in the sample questionnaire reproduced from Shang (1981). Open-ended questions usually elicit very general answers, which may not be of much use in evaluations. The selection of the consumers, the number to be surveyed and the methods of contacting them are very important. Though a large random sample would be the best, cost considerations may make it necessary to limit the number; but it should be as representative as possible of the individuals who purchase or influence the purchase of the product. The method of contacting respondents will depend largely on their behavioural patterns and communication facilities. Response rates to mailed questionnaires and enquiries over the telephone are not always satisfactory. Personal interviews are probably the best method, as the interviewer can try to influence the respondent to provide all available information through supplementary questions and also make some judgements on the accuracy of the answers.

When cross-tabulated and analysed, the data can give useful estimates of the relationship between a number of factors under study, such as acceptability of the product or quantity consumed as a function of income, race, religion, social status and life style.

When a product is unknown to the potential consumers surveyed, as for example a non-indigenous species under consideration for introduction, any amount of interviews and questionnaires would give only hypothetical answers, which can be only of very limited value. In such cases, test marketing is probably the best means of obtaining useful information.

Income and price elasticities

Using time-series data obtained through the survey on per capita consumption, per capita income, prices of species concerned and prices of competing products, the elasticities of demand (price, income and cross-elasticities) can be calculated by the regression method. These

---

Sample questionnaire for a consumer survey

Name of respondent ⎯⎯⎯⎯⎯⎯⎯⎯⎯⎯⎯⎯⎯⎯⎯⎯⎯ Date of enumeration ⎯⎯⎯⎯⎯⎯
Address of respondent ⎯⎯⎯⎯⎯⎯⎯⎯⎯⎯⎯⎯⎯⎯⎯ Enumerator ⎯⎯⎯⎯⎯⎯⎯⎯⎯

(1) Number of people in family ⎯⎯⎯⎯⎯⎯⎯⎯⎯⎯⎯ Ages ⎯⎯⎯⎯⎯⎯⎯⎯⎯⎯⎯
(2) Profession of the head of family ⎯⎯⎯⎯⎯⎯⎯⎯ Level of education (yrs) ⎯⎯⎯⎯
(3) Number of working family members ⎯⎯⎯⎯⎯⎯ Types of jobs ⎯⎯⎯⎯⎯⎯⎯
(4) Monthly family income (range) ⎯⎯⎯⎯⎯⎯⎯⎯⎯
(5) When did you last eat fish ⎯⎯⎯⎯⎯⎯ Type of fish ⎯⎯⎯⎯⎯⎯ Amount⎯⎯⎯⎯kg
(6) How often do you eat fish ⎯⎯⎯⎯⎯⎯⎯⎯⎯⎯⎯⎯⎯
(7) Have you tried the species (species concerned)? Yes ⎯⎯⎯⎯⎯⎯ No ⎯⎯⎯⎯⎯⎯
  If no, ask why ⎯⎯⎯⎯⎯⎯⎯⎯⎯⎯⎯⎯⎯⎯⎯⎯⎯⎯⎯⎯⎯⎯⎯⎯⎯⎯⎯⎯⎯⎯⎯⎯
  Too expensive ⎯⎯⎯⎯⎯⎯⎯⎯⎯⎯⎯⎯⎯⎯⎯⎯⎯⎯⎯⎯⎯⎯⎯⎯⎯⎯⎯⎯⎯⎯⎯⎯
  Not available all the time ⎯⎯⎯⎯⎯⎯⎯⎯⎯⎯⎯⎯⎯⎯⎯⎯⎯⎯⎯⎯⎯⎯⎯⎯⎯⎯⎯
  Not familiar with the species ⎯⎯⎯⎯⎯⎯⎯⎯⎯⎯⎯⎯⎯⎯⎯⎯⎯⎯⎯⎯⎯⎯⎯⎯⎯
  Other (specify) ⎯⎯⎯⎯⎯⎯⎯⎯⎯⎯⎯⎯⎯⎯⎯⎯⎯⎯⎯⎯⎯⎯⎯⎯⎯⎯⎯⎯⎯
  If yes: When did you buy it last? ⎯⎯⎯⎯⎯⎯⎯⎯⎯ Amount bought ⎯⎯⎯⎯⎯⎯⎯ kg
  In which form: (a) live, fresh, frozen, salted, dried, etc.
                 (b) whole, gutted, fillet, etc.
Where did you buy? Supermarket, fish market, fish stalls, fish peddlers, etc.
  How often do you buy it? ⎯⎯⎯⎯⎯⎯⎯⎯⎯⎯⎯⎯⎯⎯⎯⎯⎯⎯⎯⎯⎯⎯⎯⎯⎯
  How does it taste? Excellent ⎯⎯⎯⎯⎯⎯ Good ⎯⎯⎯⎯⎯⎯ Poor ⎯⎯⎯⎯⎯
  Do you intend to consume more of this species in the future? ⎯⎯⎯⎯⎯⎯⎯⎯⎯⎯
  If no, ask why:
  Do not like taste ⎯⎯⎯⎯⎯⎯ Too expensive ⎯⎯⎯⎯⎯⎯ Too bony ⎯⎯⎯⎯
  Not always available ⎯⎯⎯⎯⎯⎯ Low quality ⎯⎯⎯⎯⎯⎯ Others (specify) ⎯⎯⎯⎯⎯
  If yes, ask
  The size of fish preferred
  The form of fish prefered
  How often do you intend to buy it?
      Once per week, or more ⎯⎯⎯⎯⎯⎯⎯⎯⎯⎯
      Once per month ⎯⎯⎯⎯⎯⎯⎯ Less than once a month ⎯⎯⎯⎯⎯⎯⎯⎯⎯⎯
  Other species you prefer (in order of preference) ⎯⎯⎯⎯⎯⎯⎯⎯⎯⎯⎯⎯⎯⎯⎯
  In your opinion what is the closest substitute to the species ⎯⎯⎯⎯⎯⎯⎯⎯⎯⎯

---

elasticities help in assessing the market potential for an aquaculture product.

Shang (1981) described the ways in which price and income elasticity values can be used to estimate market potentials. The price elasticity of demand is defined as the percentage change in demand created by a one per cent change in price in either direction, while other factors (such as income and prices of substitutes) remain unchanged. The demand for the product is considered to be elastic when the absolute value of the price elasticity is greater than one. This would show that increased production can be absorbed by the market as a result of a relatively small decline in price. Producers will realize higher incomes and consumers will pay lower prices.

The demand for the product is considered inelastic when the absolute value of the price elasticity is less than one. An increase in production will lead to substantially reduced prices. Consumers will be benefited, but producers will have to sell at lower prices.

Income elasticity of demand measures the percentage change in the quantity of demand as a result of a one per cent change in income, when other factors remain unchanged. An increase in income will induce proportionately more demand if the income elasticity is greater than one. On the other hand, if the income elasticity is less than one, an increase in income will result in a less than proportional increase in the quantity of demand. In cases where the income elasticities are negative, the demand will decline as income increases. Obviously species with negative income elasticity have only limited potential for culture. By multiplying the projected rate of growth of real per capita income by the estimated income elasticity of demand for a species, the rate of growth which may result from increased income can be projected, if other factors remain constant.

The total effective demand for a given species, resulting from increases in population and income, can be estimated using the following formula:

$$D_t = D_0 [1 + N + (Ee)]$$

where

D_t = total demand at year t

D_0 = total apparent consumption at the base year (total apparent consumption is defined as total local production plus imports minus exports)

N = rate of growth of population between base year and year t

E = rate of growth of income between base year and year t

e = income elasticity of demand

t = number of years after the base year

It will be advisable to make separate estimates for urban and rural areas, because of the differences in their population and income growth patterns.

Test marketing

The need for test marketing or sales experiments to assess consumer acceptance and demand for new species has already been referred to. Besides determining the acceptability of newly introduced and unfamiliar species, this method can also be used to forecast consumer response to changes in product specifications and presentation, price changes, etc. Properly designed test marketing permits actual observation and recording of consumer behaviour and reaction, rather than depending on answers to questionnaires or interviews with their inherent limitations. It has, however, to be recognized that it is costly and time-consuming. Control of all the factors that affect sales will also be difficult in such tests.

Tests should ideally be designed to evaluate (i) the actual product trial rate, (ii) the level and frequency of repeat purchase, (iii) the relative effectiveness of various marketing plans (e.g. a high- versus a low-weight promotional programme), (iv) consumer acceptance of product benefit claims, (v) reaction of the trade to the new product and (vi) potential problems of establishing an effective distribution channel (Chaston, 1983). The retail stores or markets selected for testing should be located in areas that have a population structure similar to that of the nation as a whole. Different sizes of the species likely to be produced and the product forms (e.g. fresh, on ice, fillets, frozen, smoked, etc.) that can be sold at different price ranges should be introduced. Detailed data on the buyers and their purchases should be recorded for evaluation. It has been suggested that only a few variables should be tried at a time, but this may result in prolongation of the test period. In fact, a critical problem of test marketing is the duration of the tests. It is essential to continue until an evaluation of the trial and repeat purchasing can be made. The longer the duration of the experiment, the greater will be the accuracy of the results. However, it will be necessary to limit the duration for reasons of cost and the need for early decisions on the introduction of the species or product. If the first-time purchase or repeat purchase frequency is low, it may be necessary to continue the trials.

Ready acceptability of the product will be characterized by high rates of trial and repeat sales. If the trial rate is high, but the repeat sales are poor, it may show lack of customer satisfaction. The opposite situation of low trial but high repeat sales would indicate that, with better promotional work, the product can become a success. Low trial and repeat sale rates can be taken as proof of the unattractiveness of the product in the markets.

If the results of the test marketing are positive, an estimate of likely market demand can be attempted, using the sales data. The ratio of the number of buyers to the household population of the test area and to the quantity of sales can be used as an indicator of the approximate market demand in the area. The national market demand can be roughly estimated, using the ratio of the household population in the testing area to the entire country.

If the purpose of the trial marketing is to assess the effect of selected stimuli, such as changes in the quality or presentation of products, it will be necessary to select an adequate number of representative stores or markets to serve as controls. The data relating to sales from the test stores and markets should be compared with those in the control markets. An appreciable improvement of sales in the test marketing would clearly show the beneficial effects of the changes.

12.3.2 Formulation of market strategies

The present organization of marketing available for aquaculture products has been described earlier in this chapter, and it appears to meet the needs of a relatively small industry. However, the expected future expansion of the industry may warrant a reappraisal of existing market arrangements and the formulation of appropriate strategies. Cracknell (1979) suggested that the structure and philosophy of hunted fish channels of distribution should not be adopted in the marketing of aquaculture products, and a suitable model might be the organization which undertakes coordinated distribution and marketing of farmed meats. The strategies selected could relate to individual enterprises or farms or to the whole national aquaculture development plan. In either case, the main thrust of the strategy should be the full utilization of the advantages that aquaculture offers for satisfying consumer needs and tastes. Such a strategy can be expected to improve the economic viability of farming operations and help in meeting the objectives and targets of aquaculture.

Market segmentation and channels

The objectives of most aquaculture enterprises allow the identification of defined groups of customers or consumers for their products. Depending on the nature and price of the product, it can be low-income rural populations in need of a low-cost protein food or the discriminating high- and medium-income consumers who prefer farm-fresh products of high quality. Marketing may be designed to reach these selected groups, through what marketing experts call marketing segmentation. Restaurants and institutions are prime customers for aquaculture products. Besides product quality, restaurants need regularity of supplies to include the product in their menus. Regularity of supplies will also be needed for institutional customers, but it may be possible for them to use substitute products occasionally if the price and quality are acceptable. Supermarkets and grocery stores also look for regularity of supplies. Farming operations will therefore have to be adjusted to facilitate round the year supplies, when possible. Decisions on the location of the outlets are important and consumer survey data should indicate the most attractive locations.

For policy and business reasons it may become necessary to participate in an undifferentiated market and sell through capture fishery outlets. Even in such cases, aquaculture enterprises can benefit by restricting sales of their products to off-seasons of fishing, and thus avail themselves of opportunities to raise their prices.

In developing country situations, where a suitable infrastructure for fish marketing does not exist, the establishment of adequate facilities for wholesale and retail marketing, as well as storage and transport, should get special attention. The establishment of fish markets has traditionally been the responsibility of governments and public bodies in many countries.

Product quality and presentation

In a strategy for marketing aquaculture products, greater attention has to be devoted to product quality and presentation. Experience has already shown the value of these in establishing lucrative markets. The production of pan-size salmon and trout has helped to expand markets for the species and improve the economics of operations. The presentation of uniform-sized fish, fresh or frozen, could attract more customers. When there is a consumer demand for animals which are too small to be fished from natural stocks, as in the case of abalone, aquaculture could be directed to meet this demand. The potential for producing fish of the required colour and quality by suitable feeding during grow-out, and by appropriate processing, has been well demonstrated in sea farming of salmon and trout. The data collected through market research can form the basis for decisions on product lines that should be developed. In the case of aquaculture products like oysters and mussels, which may be eaten raw, there is greater public concern on possible contamination, when grown in waters that are likely to be polluted. Cleansing and purification procedures needed to ensure product quality have been discussed in Chapter 11. Marketing of such products in packages containing certificates of inspection by competent authorities would greatly enhance customer confidence in the quality of the product.

Packaging of the product is equally important in building up consumer acceptance. Besides protecting the product and improving its shelf life, packaging fulfils a promotional function as well. This is especially important when sales are through self-service supermarkets and grocery stores. The package should gain attention, describe the product features, provide consumer confidence and sustain the product image already established in the consumer's mind through advertising (Chaston, 1983). Even a brand name for the product, denoting the image that is intended to be created in the customer's mind, can go a long way in promoting the product. In many developing countries, where the major outlets for aquaculture products are open markets, the priority will be in providing fish stalls and sanitary conditions. Such markets offer opportunities for keeping and selling live fish and shellfish. This will be particularly attractive to customers who eat them raw and need them in an absolutely fresh condition.

The export of aquaculture products involves a more complex organization and often includes processing, quality control, long-distance shipping, import formalities and distribution in the importing country. If there is an established export trade of sea food, it would be simpler to utilize it, unless the magnitude of export warrants the expenditure and efforts involved in entering a foreign market independently and managing it to the best advantage of the exporters. Very often it will be necessary and possible to obtain governmental support for export trade, and this should be made use of, particularly by small producers.

Market promotion

The importance of market promotion in popularizing new products and non-indigenous species has already been referred to. The consumers have to be made aware of the comparative advantages of these products through promotional activities. Of the various promotional activities directed towards customer communication, advertising and personal selling are considered to be the ones that have long-term effect, in so far as sales are concerned. Tasting sessions, free samples, low-price introductory sales, etc., can have a short-term impact on sales. Public media can be used in educating potential consumers on the value of aquatic products in general as food, and the advantages of farmed products in particular. It is also an appropriate means of removing misconceptions about the food values, flavour or health hazards associated with farmed products. Publicity by producers will be less convincing than reports by public media. Newspapers, magazines, radio and television could all be utilized when possible, for educational purposes. Selection of the media should naturally depend on the section of the population to be reached.

The most effective means of marketing a product may be through personal calls to every potential customer. This will enable personalized communication and provision of additional detailed information that the customer may need. When large groups of potential customers have to be reached, personal calls will become impractical. The alternative then will be advertising, even though it has the inherent disadvantage of being only a one-way communication process. Initially, the purpose of advertising is to generate awareness among potential customers. Even after a product has established itself in the market, advertising will have to continue to sustain consumer awareness of particular species and product lines.

Advertising is expensive and should therefore be properly planned to reach the audience to whom the message is directed. The message itself should be based on the results of market research on consumer behaviour and the most important benefit factor that will influence their purchase decision. It has, however, to be recognized that advertising alone, without the other components of the so-called marketing mix (product, price and location) will not lead to increased sales.

12.4 References

Anderson A.M. (1973) Developing markets for unfamiliar species. *J. Fish. Res. Board Can.* **30**(12) Part 2, 2166−9.

Chaston I. (1983) *Marketing in Fisheries and Aquaculture*. Fishing News Books, Oxford.

Cracknell T.J. (1979) Development of vertically-integrated fish farming in Europe. In *Advances in Aquaculture*, (Ed. by T.V.R. Pillay and W.A. Dill), pp. 34−40. Fishing News Books, Oxford.

Pillay T.V.R. (1981) Some experiences in aquaculture development. *Proc. Gulf Carib. Comm.*, 64−72.

Pillay T.V.R. (1985) Some recent trends in aquaculture developments. In *Conference Proceedings AQUANOR 85, Trondheim, Norway*, 61−4.

Shang Y.C. (1981) *Aquaculture Economics: Basic Concepts and Methods of Analysis*. Westview Press, Boulder Croom Helm, London.

13
Economics and financing of aquaculture

13.1 Economic viability

In Chapter 3, on national planning of aquaculture development, the characteristics of programmes for social benefits and commercial profit have been outlined. The distinctions between the two become less clear when the economics of the operations are considered, as on a long-term basis both types of activities have to be economically profitable to be sustainable. Socially-oriented aquaculture can expect governmental support in the form of easy loans and grants, subsidies and free technical advice and assistance. Such support is generally time-bound and intended to improve socio-economic and nutritional conditions of communities. Although under certain circumstances it may be maintained on a prolonged basis, it is expected that the targeted improvements will be eventually achieved and the assistance can be phased out. On a somewhat similar basis, incentives may be offered to commercial enterprises for initial periods in the form of tax rebates and exemption, concessional loans, etc. In both cases, economic viability will be essential for continued operation and future expansion. Whether it is a small-scale farmer, or an entrepreneur involved in large-scale production, the attractiveness of aquaculture depends to a very great extent on the economic benefits. The social benefits are often closely intermingled with economic benefits. The decline of the numerous homestead fish ponds established for improved nutrition of rural people in parts of Africa is adequate proof of this.

Despite the basic importance of economic viability, very little attention has been paid to this aspect, and promotion of aquaculture has suffered considerably for lack of appropriate data and documentation on relevant evaluations. Costs and earnings are undoubtedly site-specific, and what is economic in one area may not be so in another. However, rough evaluations of the feasibility of an enterprise or project can be done on the basis of similar ones operating under as comparable a situation as possible. It is also necessary to compare the costs and earnings of other similar activities for which opportunities exist in the area, before making decisions on investment in aquaculture.

Problems of obtaining reliable data from commercial operations are not peculiar to aquaculture. But, unlike in many other sectors, aquaculture has until very recently been carried out almost entirely by individual farmers or small groups that seldom maintained the type of data required for proper economic analysis. So, the paucity of adequate and appropriate data has been a major constraint. Most of the presently available data on commercial operations are based on special surveys undertaken by investigators. The limitations of laboratory and experimental data for economic evaluations are well known. Pilot productions, which could be of value for this purpose, are too few to be of much help. It has yet to become an accepted practice for aquaculture scientists to incorporate economics as an important variable in experimental designs. In the absence of this, many experimental findings remain of only technical significance and do not find application in production programmes.

The lack of economic data also seriously affects access to suitable financing that is badly needed for the development of the aquaculture industry. Not only does it constrain investment financing, but it also makes the organization of a proper risk insurance extremely difficult.

13.2 Data requirements

The nature of data required will naturally depend on the purpose and type of evaluation to be carried out. Since most types of technological studies in aquaculture should include economic aspects, it may not be practical to list all items of economic data that would be useful. However, a review of the basic data that are essential for proper development planning and farm level management can be attempted here.

13.2.1 Assets and liabilities

In most types of aquaculture, the capital costs of establishing the farm are generally the highest. A precise and detailed record of the established assets will therefore be required. Separate values of the costs of the land, facilities such as hatchery systems, rearing and grow-out facilities, the water supply system and buildings (residential, storage, office, laboratory, workshop, garage, processing and packing, feed storage, etc.) will be needed. Data on the cost of various types of equipment acquired for the farm will also be necessary. These will include farm equipment like pumps, generators, refrigerators and cold stores, aerators, feed mill, feed dispensers, vehicles, etc.

It is necessary to have estimates of the years of economic life of these major farm assets, in order to calculate their depreciation, which will be a fixed cost to be considered for most economic evaluations. The depreciation is derived by dividing the cost of the asset by the estimated years of economic life. If there is a salvage value for it after the estimated period, it is subtracted from the initial cost before the annual depreciation is computed. In the case of assets constructed or acquired a long time ago, the current replacement cost should be the basis for calculating depreciation. In the event that any of the fixed assets are shared with other activities, it will be necessary to determine the share of the assets that can be claimed for aquaculture, for purposes of calculating depreciation. Depreciation would not be applicable in the case of land or for regularly maintained fish ponds, as generally their value only appreciates and all maintenance costs are accounted for under operating costs.

Information on liabilities of a long-term nature, such as loans and mortgages, and terms and amounts of payment of interest and instalments will be required.

For assessment of productivity, data based on either the area of the farm or the volume of water, as appropriate, will be needed.

13.2.2 Variable and fixed costs

Variable and fixed costs constitute the main inputs in an aquaculture enterprise. Variable costs, as the name implies, vary with the level of production, whereas the fixed costs are not affected by it. The variable costs can be divided into two: production and labour. Similarly, the fixed costs can be classified into indirect operational costs and administrative costs and salaries. The main types of data required for each of these costs are listed below.

Variable production costs:
Water fees
Farm preparation and maintenance
Purchase of fry, fingerlings or brood animals
Purchase of feeds or feed ingredients
Purchase of fertilizers, chemicals and drugs
Purchase of electricity and fuel
Purchase of product containers and packing materials
Freight and transportation costs
Adjustments for inventory changes in feeds and other major production materials
Variable labour costs:
Number of man-hours per day and number of man-days per month/year
Wages in cash and kind
Board and lodging provided
Indirect operational costs:
Running, maintenance and service of tools and equipment
Fixed costs:
By definition, fixed costs are those that do not vary with production. This would include interest on both own and borrowed capital (debt and equity) and depreciation of assets. It would also include the following administrative costs:

Salaries and wages to administrative

personnel
Imputed salary of owner(s)
Telephone, postage and office accessories
Travelling expenses
Commissions and contingencies
Auditing, legal and technical assistance
Insurance

It is obvious that the administrative costs are applicable only in the case of large-scale enterprises.

In the absence of a system of detailed book-keeping at the farm level, the above types of data can be obtained only by detailed study of representative sample farms, by trained enumerators. An alternative is to base evaluations on data from pilot farms. Besides the normal economic analysis, pilot farms will also make it possible to test under practical farm conditions different management strategies, such as farm mechanization, modified production cycles, changes in product size, etc., and assess their economic implications.

13.2.3 Operating income

Details of the farm output (species, quantity) and their unit price and the sales in cash or credit should be available. The imputed values of products consumed on the farm and given away as gifts or as payments in-kind, should be calculated. In cases where multiple stocking and harvesting are practised, inventory adjustments of the farm stock (increase or decrease from the last inventory) will be necessary. The farm stock may be expressed in number or weight of each species in the farm. In farms that produce subsidiary crops in association with fish, shellfish, etc., details of the products and income from their sale should be collected in the same manner as for the primary products.

It will be appropriate to consider subsidies or other tangible government support as part of the operating income and so the actual or imputed value of this should also be taken into account.

13.3 Analysis of data

Economic analysis of farm performance can be used for a number of purposes. In the allied field of agriculture, the main aims of such studies have been described by Yang (1965) as: (i) determination of the relative profitability of various farm enterprises; (ii) assessment of causes or reasons for variation in the unit costs of production; (iii) establishment of efficiency and management standards; (iv) description of the most efficient practices and techniques of farm operation and (v) determination of the optimum input requirements for each farm enterprise. Taking into account the present state of fish farm economics, Berge (1979) suggested that analysis of data on costs and earnings serve the purpose of (i) helping managers of farms in systematically and realistically studying their own operations, (ii) facilitating comparisons between farms, forming the basis for entrepreneurial decisions on matters such as improvements in the efficiency of the enterprise and (iii) providing the basis for policy decisions relating to fish farming and facilitating cooperative action with regard to marketing, supply of feed and stocking material, etc. Shang (1981) emphasized that cost and earnings analysis would provide the information necessary to determine the relative profitability of various production techniques or systems; compare the productivity of major inputs, such as land, labour and capital, with that of alternative production activities and improve the efficiency of the farm operations. He also underlined that the results of economic analysis are not only for fish farmers or aquaculturists, 'but also for economists and policy makers, who make comparisons among different farm groups classified by size, ownership and so on'. Since the normal cost-return analysis is static, he suggested that variations and interactions of factors affecting production and profit should also be considered. 'Cross-section data collected from a survey can be analysed by regression methods.' For example, the yield per hectare can be a function of the capital input, man-days employed, amount of feed or fertilizer used, farm size, level of technology, etc. Useful information on how the various factors affect production levels can be obtained from the magnitude of the regression coefficients.

Despite the usefulness of economic evaluation in the development of suitable aquaculture technologies and production programmes, investment appraisals and farm management, its potential has not yet been utilized to any great extent for reasons already indicated. Recognizing this, some attention is now being paid to this neglected aspect of aquaculture science. Documented data on the operations of individual farms or production systems are now becoming available for certain areas, and these have been used in determining profitability and rates of return on investments and sensitivity studies. Some of the available information will be referred to in Chapters 15 to 29 in descriptions of specific farming techniques. A review of available reports would show that, even now, evaluations are frequently based on hypothetical and assumed values which have not been validated by actual case studies or farm surveys.

The methods of economic analyses are described in standard textbooks, but those that have been considered as applicable to aquaculture are summarized here.

13.3.1 Evaluation of farm performance

The basic data required for farm performance analysis, as proposed by Shang (1981), are capital costs, annual operating costs and gross revenue (income). The indicators of performance are as follows:

(1) Profit, which is defined as the difference between the gross revenue and the total annual operating cost of production.

(2) Rate of return on investment, which can be derived by dividing the profit by the total investment on fixed assets.

(3) Rate of return on annual operating cost, which is obtained by dividing the profit by the total operating cost.

(4) Value of production per unit of major input. This can be expressed as the weight of production per unit area of water surface or volume of water (kg/ha or kg/m^3), weight of production per man-month, weight of production per unit of feed, or unit of capital, etc. These values may partially indicate the operational efficiency.

(5) Cost of input per unit of production; for example cost per kg, man-hour per kg, feed per kg, etc.

(6) Pay-back period, which is the number of years required to recover the investment.

(7) Break-even point, which can be defined as the amount of income where the income (minus variable costs) is sufficient to cover the fixed costs, and there will be no profit and no loss. The break-even price will then be the total operating cost divided by the quantity of production, and the break-even production, the total operating cost divided by the unit price of the product.

A simplified example of the cost-return analysis of a 1 ha shrimp farm partly based on Shang (unpublished) is given below:

(1) Initial costs:

Item	Cost ($)	Economic life	Salvage value	Annual depreciation ($)
Pond construction	10 000	na	na	—
Water supply system	4 000	10	0	400
Stores, workshop	1 000	5	0	200
Nets	200	5	0	40
Pumps	500	5	0	100
Others	500	5	0	100
Total	16 200			840

(2) Annual operating costs:

	Quantity	Unit price ($)	Total cost ($)
Variable costs			
Fry	20 000	10 per 1000	200
Feed	2 000 kg	0.5	1000
Electricity			200
Hired labour	2 man-months	1000 per man	2000
Others			200
Subtotal			3600
Fixed costs			
Land lease			200
Interest on loan (10%)			1620
Depreciation			840
Maintenance			200
Operator's labour	1 man-month	1500/per man	1500
Tax			200
Subtotal			4560
Total cost			8160

(3) Income:

	Production (kg)	Unit price ($)	Income ($)
Shrimp	2000	5	10 000

(4) Indicators:

(a) profit = 10 000 − 8160 $\quad = \$1840$

(b) rate of return on initial cost $\quad = \dfrac{1840}{16 200} \quad = 11.36\%$

(c) rate of return on operating cost $\quad = \dfrac{1840}{8160} \quad = 22.6\%$

(d) production $\qquad$ kg/ha $\quad = 2000$

$\qquad\qquad$ kg/man-months $\quad = \dfrac{2000}{3} \quad = 667$

$\qquad\qquad$ kg per unit of feed $\quad = \dfrac{2000}{2000} \quad = 1$

(e) pay-back period $\quad = \dfrac{16 200}{(1840 + 840)} = \dfrac{16 200}{2 680} = 6$ years

(f) break-even price $\quad = \dfrac{8160}{2000} \quad = \$4.08$

(g) break-even production $\quad = \dfrac{8160}{5} \quad = 1632$ kg

Such evaluations to determine the economic performance of a farm or to compare the economic aspects of different farm management systems, as for example extensive versus semi-intensive or intensive, or mono-culture versus polyculture, have been used in many instances.

Berge (1979) proposed the use of 'contribution' for fish farm accounting, i.e. the operating income minus variable costs, which indicates what remains of the income for fixed costs and net income. This method distinguishes fixed costs from direct variable costs throughout the analysis. Three key values can be calculated by this method:

(1) degree of contribution, which is the contribution as a percentage of the income:

$$\frac{\text{contribution}}{\text{income}} \times 100$$

(2) break-even point:

$$\frac{\text{fixed cost}}{\text{degree of contribution}} \times 100$$

(3) security margin:

$$\frac{\text{income break-even point}}{\text{income}} \times 100$$

Berg recommended this method of accounting for comparison of different aquaculture enterprises and for working out key indicators for rentability, liquidity, solvency, etc. Examples of calculations of contribution and rentability budget, taken from Berg (1979), are reproduced in Tables 13.1 and 13.2 respectively.

There are also other methods of economic analysis. Gerhardsen (1979) proposed a simple model which can be used for analysing a specific type of production, the total production of an enterprise or the total production of a region. While return on investment is the most common measure of profitability, another useful measure under certain situations is the ability to pay wages. Both measure the same performance, but from different points of view. The suggested model, based on fish production of a hypothetical aquaculture enterprise in Norway, is reproduced in Table 13.3. Other useful economic indicators which can be calculated from this model are (i) investment per annual man-hours and per kg produced, which indicate the capital intensiveness of the enterprise, (ii) production per annual man-hours, which measures its productivity and (iii) production per volume or surface unit of water, which indicates the level of intensity of the operation.

13.3.2 Sensitivity analysis

The analysis referred to earlier is static and describes a given situation. It will often be necessary to examine the effect of variations and interactions of factors influencing income and variable costs. In planning and managing an aquaculture enterprise, there will be a continuing need to evaluate the sensitivity of return on investment to changes in major production costs, as well as price of products. Variations in costs may arise from changes in the costs of inputs or the adoption of new technologies. Before introducing new techniques, as realistic an analysis as possible has to be carried out to assess their impact on operating costs. Variations in price can be expected as a result of changes in demand, increased supplies and competition. In the case of export products, changes in foreign market conditions, including currency exchange rates and import regulations, will have a considerable impact on the income. It will also be necessary to examine whether the farming of species meant for export can be maintained, if the export market ceases to exist and the product has to be sold in the domestic market at lower prices. Fall-back positions in terms of technology, diversification of species and product development and their effects on profitability will have to be worked out.

As an example of sensitivity analysis, evaluations of catfish production costs and prices are reproduced in Tables 13.4 and 13.5 from Greenfield (1970).

When there is only a minor change in a production system, which results in a partial change in cost and returns, complete recalculation of economic viability can be avoided by using the partial budgeting method. In this method the benefits are first estimated by the increase in income due to the change, ignoring incomes that will not change. Then the reduction in cost is estimated, if one proceeds with the venture with the changes. The next step is to estimate the added cost due to the

Table 13.1 *Contribution Calculation* Product − Rainbow Trout 3 years, 2.5 kg/Fish, Unit − 100 kg live Weight (from Berg, 1979).

Income

Production	Saleable production	Prices (Nkr)	Income (Nkr)
Rainbow trout			
Grade I	80	15	1200
Grade II	20	11	220
Total	100		1420

Variable costs

Feed	Quantity (kg)	Variable cost	
		Per unit (Nkr)	Total (Nkr)
Dry feed, pelleted	40	2.20	88
Trash fish	400	1.00	400
Shrimp waste	50	1.00	50
Total			538

Others	Quantity	Variable cost		Income (Nkr)
		Per unit (Nkr)	Total (Nkr)	
Fingerlings	50 fish	2.00	100	
Veterinary assistance			20	
Insurance			100	
Packing			40	
Freight			20	
Ice			5	
Miscellaneous			10	
Total:			295	833
Contribution				587

Conditions:
(1) period of growth: 3 years
(2) labour: 11.5 hours
(3) volume of water: 13 kg/m^3
(4) cost: Nkr 25/hour (US\$ = Nkr 5.5)

change, again ignoring the costs that will not be affected. Then the income foregone due to the change has to be estimated. Finally, the sum of the increased cost should be subtracted from the sum of increased benefits. A positive result would show the profitability of the change and the negative result the loss in profitability due to the change.

13.3.3 Minimum farm size

An important decision in a commercial aquaculture venture is the minimum economic size of the farm. Economies of scale are associated with mass production and large-scale enterprises. In many instances, when all inputs are doubled, the output may be more than doubled,

Table 13.2 Rentability budget (from Berg, 1979).

Type of product	Quantity (kg)	Nkr/kg	Sales	Variable production cost	Variable labour cost	Total variable cost	Contribution	Labour hours	Maximum volume of water (m³)	Contribution per hour	Contribution per m³
Rainbow trout	100 000	14.20	1 420 000	833 000	287 500	1 120 500	299 500	11 500	7 700	26.00	3.90
Salmon	60 000	23.00	1 380 000	567 000	312 500	879 500	500 500	12 500	6 000	40.00	8.34
Total	160 000	17.50	2 800 000	1 400 000	600 000	2 000 000	800 000	24 000	13 700	33.30	5.84

Fixed cost: Nkr 450 000
Profit: Nkr 350 000
US$ 1.00 = Nkr 5.5

Degree of contribution:

$$\frac{800\,000 \times 100}{2\,800\,000} = 28.8\%$$

Break-even point:

$$\frac{450\,000 \times 100}{28.8} = 1\,573\,000$$

Security margin:

$$\frac{(2\,800\,000 - 1\,573\,000) \times 100}{2\,800\,000} = 43\%$$

Table 13.3 Model for economic analysis in aquaculture (from Gerhardsen, 1979).

Volume of water		20 000 m^3
Fish produced		160 000 kg
Maximum production capacity		200 000 kg
Gross operating income (sales)		Nkr[†] 2 500 000
variable production costs	Nkr 1 375 000	
variable labour costs	500 000	
		Nkr 1 875 000
Gross profit:		Nkr 625 000
salaries	Nkr 65 000	
indirect costs	200 000	
interest on debts	100 000	
depreciation	50 000	
		Nkr 415 000
Net profit or operating income:		Nkr 210 000
Total man-hours[‡]		26 000
Total investment		1 805 000
Return on investment		17.2%
Ability to pay wages		Nkr 29.81/hour
Investment per employee per year[‡]		Nkr 138 000
Investment per kilogramme produced		Nkr 11.28/kg
Production per employee per year[‡]		12 300 kg
Production per cubic metre of water		8.00 kg/m^3
Utilization of capacity[§]		80%

* 'Production' is defined as quantity of fish sold plus increase or minus decrease in stocks of fish. This is the quantity from which the operating income is calculated.
† US$ 1.00 = Nkr 5.5.
‡ Yearly man-hours per employee = 2000 hours.
§ Utilization of capacity means the calculated maximum production in kg in relation to the achieved production in kg gs for the period under review.

showing what is known as increasing return to scale. Economics of scale are generally characterized by the use of higher levels of capital and technology per man-year and a low number of working hours per unit produced. Whether to use the economies of scale or not will depend on the objectives or targets of the project. But in most cases, it will be necessary to determine the minimum size of operation to ensure its economic viability. An investor or a farmer will have to evaluate various farm size combinations to determine the effect of size on production costs and profit. As an example, the study of Mitchell and Usry (1967) on catfish farming in Mississippi (USA) can be cited. Based on the comparison of the prevailing production costs and profits they came to the conclusion that a new farming operation as small as 30 acres (12.14 ha) with a total investment of about $40 000 can make a profit in a few years. The summary of results given in Table 13.6 and fig. 13.1 show the decrease in production costs for a farm up to 500 acre in size.

The profit or loss compared to farm size and sale price (fig. 13.2) show that a 500 acre (202.34 ha) farm (requiring $490 000 total investment) will have very competitive costs compared to farms of even 5000 acres (2023.4 ha).

Table 13.4 Sensitivity of return on investment to major production cost variables (before tax and not including interest on fixed investment) (after Greenfield, 1970).

	4 Pond 160 acre unit* (%)	10 Pond 120 acre unit[†] (%)
Land value		
$150 per acre (0.40 ha)	18	8
250[‡]	14	7
350	12	6
Growing period		
1 year	23	13
$1\frac{1}{3}$ years*	14	7
2 years	6	2
Cost of fingerlings		
1 c each	20	11
4 c[‡]	14	7
8 c	7	0
Harvesting costs		
1.0 c per pound (0.45 kg)	17	(2.0 c)[§] 11
2.1 c[‡]	14	(4.0 c) 7
5.0 c	8	(7.0 c) 1
Stocking rate		
1600 per acre (0.4 ha)	28	19
1200[‡]	14	7
1000	7	1

* 64.7 ha.
[†] 48.6 ha.
[‡] Typical value under average management.
[§] Higher minimum harvesting costs for a 10 pond, 120 acre unit.

Table 13.5 Price sensitivity of channel catfish (after Greenfield, 1970).

Price per pound* (US $)	Profit per acre[†] (US $)	Return on fixed investment (%)
0.44	1.31	28
0.42	1.10	23
0.40[‡]	0.89	19
0.38[§]	0.69	14
0.36	0.48	10
0.34	0.27	6
0.32	0.06	1
0.30	−0.15	− 3
0.28	−0.36	− 8
0.26	−0.57	−12
0.24	−0.78	−16

* 1 lb = 0.45 kg
[†] 1 acre = 0.4 ha
[‡] 1969 Average price to producers
[§] 1968 Average price to producers

The production costs fall rapidly as the farm size and consequently the investment increase. A $25 000 investment was found to produce at a cost of approximately 34 c a pound and a $100 000 investment would bring production costs down to about 17 c per pound. To get production costs down to 15 c per pound, an investment of $250 000 is required.

13.4 Financial and economic feasibility of investment

Being a new industry, it has not always been easy to obtain financing for commercial-type aquaculture. Financing has become particularly difficult because of the failure of a number of badly planned projects. The longer gestation periods of aquaculture projects have added to these problems, even though the export potential of some types of farming has created considerable private and public sector interest.

Table 13.6 The effect of farm size on production costs and profits (from Mitchell and Usry, 1967).

Farm size (acres)	Capital (in 000 $)		Production costs (cents per pound) ($)	Return on equity capital	
	Total	Estimated equity		Sales price 35 c/pound (%)	Sales price 20 c/pound (%)
30	42	17	0.23	42	($2000 loss)
100	100	42	0.17	56	13
500	490	200	0.15	89	22
5000	4699	2021	0.15	90	22

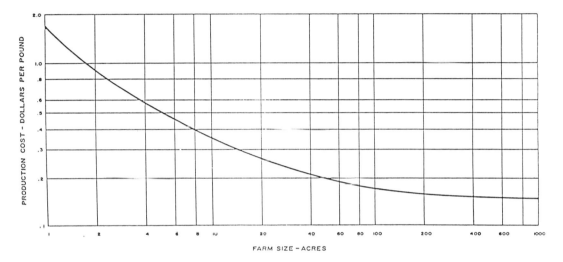

Fig. 13.1 The effect of farm size on production cost (from Mitchell and Usry, 1967).

Experience so far clearly points to the need for appropriate feasibility studies of investments to enable informed decision-making and avoidance of unforeseen failures.

The major sources of financing for large-scale aquaculture projects presently appear to be owner's equity and institutional financing, or a combination of both. Socially-oriented small-scale projects may often be financed partially or wholly from public-sector resources. Criteria for decision-making for investments will naturally depend on the source and terms of financing, but the need for financial or economic feasibility analysis is equally applicable for all projects.

Among the many methods available for feasibility studies, the most widely used ones for appraisal and comparison in aquaculture are the pay-back period, the average rate-of-return and discounting or present value (Shang, 1981). The last method (which includes the net present value, the internal rate-of-return and the benefit-cost ratio) is considered the most useful means of assessing the economic feasibility of an investment.

The pay-back period is sometimes referred to as simply an estimate of the time required to recover the initial investment out of the expected earnings from the investment, before any allowance for depreciation.

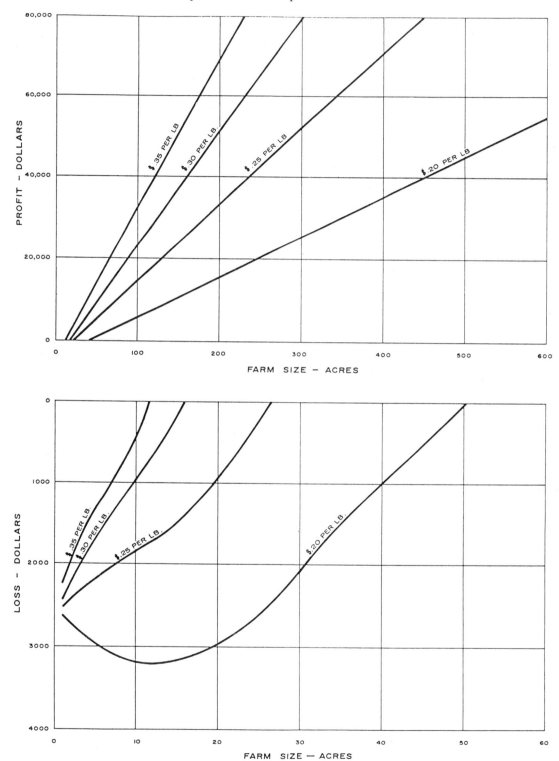

Fig. 13.2 Profit or loss compared to farm size and sales price (from Mitchell and Usry, 1967).

It does not take into account profits realized after the recovery of the initial investment recovery period and the timing of expenditures and income. However, it helps an investor to have some assurance of the recovery of his capital in a risky venture, though it will not help him to assess the merits of the investment in relation to other investment opportunities that may exist.

The average rate-of-return is the average annual profits expected after depreciation divided by the project's initial investment:

$$\frac{\text{Average}}{\text{annual rate}} = \frac{\text{Average annual profit expected after depreciation}}{\text{Initial investment}}$$

This method permits comparison of return on investment with other possible investments in projects having similar expenditure and earning profiles. But it does not take into account the timings of expenditure and earnings. A good part of the capital outlays is made in the beginning of an aquaculture project and incomes may take some time to build up, but will continue for long periods. These timings are of crucial importance in judging the feasibility of the venture.

The discounting method or the present-value method takes this aspect into full account and provides a better indication of the advisability of the investment. This method is based on the concept that the value of an investment today is much greater than its future value. In the normal course of events, a sum of money, if invested or loaned today, will start earning interest and with reinvestment can be expected to increase as a function of time. If a sum P_0, is loaned at an interest rate of r for n years, at the end of the nth year it can be expected to have the value $P_n = P_0 (1 + r)^n$. Today's value of P_n can be found by solving:

$$P_0 = \frac{P_n}{(1 + r)^n}$$

or

$$P_0 = \frac{P_i}{(1 + r)^i}$$

where i is the number of years in operation.

The present value of any series of future cash outflows and inflows can be determined by the above formulae. Pre-calculated tables are available for the present value of a dollar received or spent at one point in time for 50 years, $(1 + r)^n$, and the present value of a dollar received annually at the end of each year at various discount rates $[1 - (1+r)^n]/r$. So it will not be necessary to calculate the value every time.

13.4.1 Financial analysis

The financial analysis of commercial aquaculture projects, as well as the economic analysis of socially-oriented aquaculture projects, can be carried out by using the discounting method. The data required are very similar to the ones used in the evaluation of farm performance. These data should be as realistic as possible and preferably based on actual experience in the area. Inflation and relative price changes in the future should be considered. If the total capital investment (including owner's equity and borrowed capital) is treated as an initial cost of the project, depreciation and interest on borrowed capital should not be included in the annual operating cost, to avoid double counting. In case the owner's equity is treated as an initial cost, the amount of borrowed credit should be treated as an income at the time of borrowing. The payment of the principal and the interest on borrowed capital should then be included as operating costs. The annual cost of the loan (principal and interest) can be calculated using the amortization formula:

$$\text{annual amortization} = A \left[\frac{r}{1 - \dfrac{1}{(1 + r)^n}} \right]$$

where
A = the amount of the loan
r = the interest rate
n = duration of the loan (in years)

Net present-value method

One way to assess the feasibility of an investment project by the discounting method is by the use of the net present value. This is done by subtracting the costs from the returns (benefits) on a year-to-year basis to arrive at the net profit (or benefit) stream, which is then discounted into a present value estimate as shown in the equation below:

$$\text{NPV} = \frac{A_1}{(1+r)^1} + \frac{A_2}{(1+r)^2}$$

$$\ldots + \frac{A_n}{(1+r)^n} + \frac{S}{(1+r)^n}$$

Where

NPV   = net present value
$A_1, A_2, \ldots A_n$ = net profit of individual year (difference between total profit or benefit and total cost)
r = discount rate
n = the number of years in operation
S = salvage value of the asset in year n

As an example, the net present value of net benefit (income) of a hypothetical project is given in Table 13.7. The net present value can also be calculated by discounting independently the stream of annual cash flow (B) and cash outflow (C) and subtracting the sum of the latter from that of the former as shown below:

$$\text{NPV} = \sum_{i=1}^{n} \frac{B_i}{(1+r)^1} - \sum_{i=1}^{n} \frac{C_i}{(1+r)^1}$$

If the NPV is more than 0, the investment would be profitable; if it is less than 0, it will not be profitable. If the NPV is equal to 0, it would be a break-even situation.

The most appropriate rate of discount for calculations is the opportunity cost of the capital, and there are practical problems in establishing this rate. As it is not possible to determine the appropriate rate at the project level, it will be necessary to obtain it from a central planning agency.

Benefit-cost ratio

The benefit-cost ratio is defined as the ratio of the present total value of the benefits to the

Table 13.7 Calculation of present value (in 000$).

Year	Net benefit	Discount rate (10%)	Net present value
1	−90	.909	−82
2	−19	.826	−16
3	10	.751	8
4	15	.683	10
5	28	.621	17
6	28	.565	16
7	30	.513	15
8	28	.467	13
9	25	.424	11
10	68*	.386	27
Total			19

* Including a salvage value of $38 000

cost. This method is more appropriate to determine the social and economic worth of a project. The net and gross benefit-cost ratios can be estimated as follows:

$$\text{Net ratio} = \frac{\sum_{i=1}^{n} \dfrac{B_i}{(1+r)^1}}{\sum_{i=1}^{n} \dfrac{K_i}{(1+r)^1}}$$

$$\text{Gross ratio} = \frac{\sum_{i=1}^{n} \dfrac{R_i}{(1+r)^1}}{\sum_{i=1}^{n} \dfrac{K_i + C_i}{(1+r)^1}}$$

where

B = the net annual benefit (income)
K = the capital outlay for assets
R = the gross annual benefit or income
C = the annual operating cost

An example of the benefit-cost ratio calculation is shown in Table 13.8. The gross benefit ratio at 10% discount will be $\frac{375}{355} = 1.06$. An investment with a benefit-cost ratio greater than one can be considered as feasible. The problem of choosing an appropriate rate of discount is encountered in this type of analysis as well.

Table 13.8 Benefit-cost ratio calculation (in 000$).

Year	Capital cost	Equipment and operating costs	Total costs	Total benefits	Discount rate 10%	Discounted costs	Discounted benefits
1	90	0	90	0	0.909	82	0
2	0	59	59	40	0.826	49	33
3	0	50	50	60	0.751	37	45
4	0	50	50	65	0.683	34	44
5	0	52	52	80	0.621	32	50
6	0	52	52	80	0.565	29	45
7	0	50	50	80	0.513	26	41
8	0	52	52	80	0.467	24	37
9	0	55	55	80	0.424	23	34
10	0	50	50	118*	0.386	19	46
						355	375

* Including a salvage value of $38 000

Internal rate-of-return

The internal rate-of-return of an investment (K) is the discount rate that makes the present value of the net cash inflow equal to zero. The difficulties of choosing an appropriate discount rate encountered in the earlier two methods (net present-value and benefit-cost ratio) can be avoided to a considerable extent. If the benefit (or incomes) and costs are known, it can be solved by setting the left-hand side of the net present-value formula (see previous page) equal to zero.

$$\sum_{i=1}^{n} \frac{A_i}{(1 + K)^i} = 0$$

It represents the average earning power of money used in the project over the project's life. It can be calculated by computing the rate of return on investment or the return on equity. If the total capital outlay is treated as investment, K is the internal rate on investment. On the other hand, if the owner's equity is treated as the initial investment, K is the internal rate of return on equity. This K can be a measure for comparing returns on other investment opportunities.

The appropriate rate of discount has to be found through a process of trial and error. By trying several rates, a close approximation can be reached. When the calculated discount rates are positive, a higher discount rate has to be used. If the result is negative, a lower discount rate should be used. To shorten the trial-and-error computations and find an approximate discount rate, the following method can be used. Divide the capital cost by the average expected annual net income (if relatively stable over time). From the precalculated tables of present values (Shang, 1981) it will be found that the result of the division lies in a range of discount rates at a given analysis period. The internal rate of return (IRR) can then be interpolated by the following method (Gittinger, 1972; Shang, 1981)

IRR = Lower discount rate + Difference between the two discount rates

$$\times \left\{ \frac{\text{Present worth of the net income at the lower rate}}{\text{The sum of the absolute value of the present worth of net income at the two discount rates}} \right\}$$

As an example, Table 13.9 sets forth the internal rate of return on a $110 000 investment, with an expected annual average profit of about $23 200 for ten years (after the first two years). To find the first approximate discount rate, the capital cost is divided by the average expected annual income: $\frac{110\,000}{23\,200} = 4.74$. This discount

Table 13.9 Internal rate of return calculation (in 000 $) (courtesy of Y. C. Shang).

Year	Capital costs	Equipment and replacement costs	Operating costs	Total costs	Total income (benefit)	Net income (benefit)	Discount rate 16%	Net present value	Discount rate 11%	Net present value
1	90	0	0	90	0	−90	0.862	−78	0.90	−81
2	0	9	50	59	40	−19	0.743	−14	0.812	−15
3	0	0	50	50	60	10	0.641	6	0.731	7
4	0	0	50	50	65	15	0.552	8	0.659	10
5	0	2	50	52	80	28	0.476	13	0.593	17
6	0	2	50	50	80	28	0.410	12	0.534	15
7	0	0	50	50	80	30	0.354	11	0.482	15
8	0	2	50	52	80	28	0.305	9	0.434	12
9	0	5	50	55	80	25	0.263	7	0.391	10
10	0	0	50	50	80	68*	0.227	15	0.352	24
Total of net present value								−11		14

* Including a salvage value of $38 000.

factor lies between 16 and 17 per cent at year 10, according to the table of precalculated present values. Taking this as a starting point for trial-and-error computation, the present value of net income (benefit) is −11, at 16 per cent, which indicates that the discount rate used is too high, and it is 14 at 11 per cent, which indicates the discount rate used is too low. The true internal rate of return is interpolated by following the method suggested by Gittinger (1972).

The lower discount rate is 11 per cent and the differences between the two discount rates (11 and 16 per cent) is 5 per cent. The present value of the net income (benefit) at the lower discount rate of 11 per cent is 14. The sum of the absolute value of the present worth of the net income at the two discount rates is 14 + 11 = 25. Therefore

$$\text{IRR} = 11 + 5 \left(\frac{14}{25} \right) = 11 + 2.8 = 13.8\%$$

It has been suggested that the interval between discount rates should not be more than 5 per cent. The interpolated rate of return usually overstates the true return (Shang, 1979). In financial analysis, the internal rate of return is often calculated on equity. In this case, payments of interest and amortization are treated as cash outlays, and the borrowed capital as receipts.

13.4.2 Cash flow analysis

In any investment project, it is important to examine the interaction between the average rate of return or rentability and liquidity. Good rentability and high profits do not necessarily involve liquidity. This is particularly important in aquaculture, because of the relatively long gestation period of the projects and the scarcity of venture capital. All available equity and credits may be exhausted before the flow of income becomes large enough to balance the outflow of expenses. If such a situation arises and the project is left without funds to meet operating costs, it is bound to fail, even though it is potentially viable. It is therefore very important to project realistic annual cash flows, to ensure appropriate financing according to an operational schedule and to implement a proper balance of expenditure and income to meet business conditions.

The projected cash flow of an investment project for shrimp farming is given in Table 13.10 as an example. Such estimates help not only in arrangements for financing projects, but also in the selection of farm management systems and technologies most suited from the point of view of project financing.

13.4.3 Socio-economic analysis

While the evaluation of economic performance of aquaculture projects is relatively easy, the

Table 13.10 Cash flow projection for a 50 ha shrimp farm operation, for projected years 1–10, in 000 US$ (courtesy of P. Kungvankig).

	1	2	3	4	5	6	7	8	9	10
Cash inflow										
Gross income (1)	–	1392	1392	1392	1392	1392	1392	1392	1392	1392
Loans	800	–	–	–	–	–	–	–	–	–
Shareholders' contribution	200	–	–	–	–	–	–	–	–	–
Working capital	500	500	–	–	–	–	–	–	–	–
Total cash inflow (2)	1500	1892	1392	1392	1392	1392	1392	1392	1392	1392
Cash outflow										
Investment cost (3)	849	–	–	–	–	–	–	–	–	–
Operating cost (4)	603.4	603.4	603.4	603.4	603.4	603.4	603.4	603.4	603.4	603.4
Debt service										
Amortization of loan and interest (20%)	160	160	160	360	320	280	240	–	–	–
Repayment of working capital	–	500	500	–	–	–	–	–	–	–
Interest on working capital (20%) (5)	100	200	100	–	–	–	–	–	–	–
Total cash outflow (6)	1710.4	1463.4	1363.4	963.4	923.4	883.4	843.4	603.4	603.4	603.4
Net cash flow (2–6)	(210.4)	428.6	28.6	428.6	468.6	508.6	548.6	788.6	788.6	788.6
Cost/benefit										
Financial benefit (1)	–	1392	1392	1392	1392	1392	1392	1392	1392	1392
Cost (3+4+5) = (7)	1452.4	803.4	703.4	603.4	603.4	603.4	603.4	603.4	603.4	603.4
Net benefit	(1452.4)	588.6	688.6	788.6	788.6	788.6	788.6	788.6	788.6	788.6

estimation of social benefits is much more difficult. Many intangible and unquantifiable benefits are involved. It will often be impractical to separate out aquaculture-derived benefits in communities served by integrated rural development projects. For example, improvement in the nutritional status, employment and income may often be combined with other social development programmes and it is extremely difficult to apportion with any amount of preciseness, the contribution of aquaculture to the combined benefits derived from community activities. Even when it is possible, numerous field studies and collection of data over long periods of time may be needed. But, as pointed out in Chapter 3, when an aquaculture project is designed as the main instrument to meet a specific need of the population it will be easier to evaluate in quantifiable terms the project's performance.

For example, if a project is designed to upgrade the economic status of a community of twenty families living below the poverty line, the immediate target may be to bring their income to the national average. This may involve, let us say, doubling of their present

annual income of $600. So, the project may be designed to bring in an annual net income of $1200, which would probably need a pond farm with the capacity to produce about 3 tons of fish annually. In such a project, it should be possible to determine how far the targeted increase in family income has been achieved and what the spill-over effects of these increases have been.

Many of the government-sponsored development programmes often have broader objectives and targets. Besides social benefits, they may be aimed at the overall development of the aquaculture sector. Analysis of such projects cannot therefore be based on purely economic or business criteria. Shang (1981) discussed possible ways of estimating direct benefits and costs of public programmes of aquaculture development and also their secondary benefits and costs.

The direct benefits that can be estimated are the increase in value of outputs or the reduction in production and marketing costs. An increase in value of outputs generally takes place as a result of expansion of production areas, adoption of new technologies, introduction or improvement of credit facilities to farmers or assistance in improving the handling, preservation and marketing of produce. A reduction in production and marketing costs may be achieved by the use of new technologies and assistance in preservation and marketing, as well as by improvement of supporting facilities like hatcheries, feed mills and fertilizer distribution. Projects that are directed towards preventing a predicted reduction in production due to identified reasons have to be evaluated on the basis of the magnitude of production decline that has been prevented.

In the case of aquaculture projects, aimed at import substitution or for export to earn foreign exchange, the net benefits will be in terms of net savings or earnings of foreign exchange. If conditions are such that the farm-level prices would not represent the proper prices of the benefits, 'shadow prices' will have to be selected. Examples of such situations are (i) when the government keeps down the domestic price of products through taxation or other means to promote export and (ii) when the government provides price support to encourage local production. The 'shadow price' is the value that reflects the true value of the product to society. In a situation as in (i), the shadow price will be the world market price and in (ii) a price lower than the existing one, approximating the actual price which would be obtainable under local conditions if there were no government price support.

Estimates of direct costs are relatively easy to make, as they are more readily available. An opportunity cost has often to be included in estimating the cost of production elements, as it reflects the true cost to society. The actual price paid for the project inputs may not be the same as their opportunity cost, and therefore adjustments have to be made in the costs for analysis. An example is the opportunity cost of unskilled labour, who could find other seasonal employment, and labour that needs training and relocation to work on the project. The treatments of costs in socially-oriented development and private commercial projects differ in certain respects. For example, a government subsidy is considered a cost to society, but it is treated as a return in private projects. Interest paid on capital borrowed is a cost in private-sector projects, but is not a cost to society, as it

forms a part of the capital returned, which becomes available to the society as a whole. Similarly, taxes and duties are costs to private projects, but these are incomes to society.

Estimation of indirect effects should include benefits and costs to other segments of the same sector or to other sectors of the economy. If the spill-over effects of public investment projects affect the output of other producers this should be taken into account in socio-economic analysis. Examples of adverse effects of aquaculture projects are disadvantages to recreational facilities and pollution of water sources due to farm discharges. Examples of beneficial effects are improvement of nutritional status of communities, reduction in water-borne diseases, stabilization of fish supplies, etc.

Estimation of 'value added' by aquaculture in a region can be performed using the benefit and cost data mentioned earlier. Value added consists of payments made in the form of wages, salaries, rent, interest, depreciation and profit. The sum of these payments is useful in assessing the first impact of the project on the local economy. Additional spending and incomes may then be generated, and if the spending pattern is known, the total income generated within the local economy can be calculated by what is known as the 'income-impact multiplier'. This is a more complex analysis and may not be justified in the case of modest aquaculture projects.

There are differences of opinion regarding the choice of an appropriate social rate of discount for benefits and costs of public investments, but the before-tax marginal rate of return (i.e. the additional earning that could have been made elsewhere or in the private sector) is considered suitable for all practical purposes. This has to be obtained from a central planning agency, as it is difficult to estimate it at the project level.

13.5   Risk and insurance

Even though comparative figures are not readily available, it is generally held that the risk in aquaculture is substantially greater than in any other form of animal husbandry (Gerhardsen, 1979) and this is mainly due to the fact that production takes place in water, which is not easily observed and controlled by man. It is said that there are very few other stock-rearing industries that are so exposed to such a rapid and extensive loss of stock from so many varied causes (Secretan, 1979). Risks of loss or loss of value have been broadly listed as: pollution, disease, food poisoning, failure of water supply, break-down of equipment and machinery, net and cage failure, predation, extreme (hot or cold) weather conditions, power failure, poaching, negligence, floods and other natural disasters like cyclones, typhoons and hurricanes and malicious damage. Of these, which ones account for maximum losses is difficult to say. Gerhardsen (1979) published the result of an investigation of trout culture in Norway (Table 13.11) which showed that disease was the most frequent reason for losses.

Secretan (1986) reported that over 20 per cent of losses handled by the Aquaculture Insurance Service of England were due to diseases. Pollution, which was expected to be a major cause of loss, accounted for only 3.65 per cent of the losses. When it is remembered that many of the serious diseases have no known cures, and infected fish have to be destroyed, the magnitude of the risk involved will become evident. To the risks mentioned above should be added other business risks like price risk and other sundry risks like claims on customers and advances to suppliers.

This high-risk status seriously affects the availability of venture capital for aquaculture. In evaluating investments, cash flows are discounted at a high-risk rate and this may affect the attractiveness of the project to investors.

A means of limiting the risk of an owner's capital is the formation of companies, for example a joint stock company or limited partnership, with limited liabilities (Gerhardsen, 1979). However, this only helps to limit individual risk. Insurance, when possible, is probably a better way of covering risk and represents security of the interests of all those who are financially interested in the venture, including investors, shareholders, bankers and suppliers of equipment and supplies. An insurance on the important insurable interests of an operation will probably make it easier for even a small operator to obtain the necessary bank credit. Aquaculture insurance is a new and developing industry and presently serves mostly

Table 13.11 Reasons for losses in some Norwegian trout culture enterprises in 1970 (from Gerhardsen, 1979).

Cause of loss	Percentage of total losses
Disease	28
Climatic conditions	19
Faulty construction	16
Pollution	9
Failure of water supply	9
Employee error	7
Predation	5
Transport	5
Silting	2
Total	100

industrially advanced countries only.

Insurance is concerned with the spreading of risks and hazards of the industry among policy holders. Being a new industry dealing with a high-risk activity, both the underwriters and insurers face problems in choosing the type of risks to be covered. Premium rates remain high because of the absence of any track record for the insured and the large number of claims that the insurers have to settle.

The most important asset to be insured in an aquaculture enterprise is the stock of species raised. As most of the risks listed earlier are inter-related and not adequately defined for legal purposes, an 'all risks' coverage may be the best guarantee. It may be possible to exclude from this selected individual risks (i.e. risks considered not applicable or important) and thus reduce the premium to be paid.

The main areas of liability that are important in aquaculture are employers' liability, public liability and products liability. Depending on the legal system in the country, provision can be made to meet these through insurance coverage. Employer's liability or workmen's compensation is fairly easily determined, based on the nature of the work carried out by them. Public liability can involve somewhat more complicated situations. This could include third-party personal injury or property damage, especially in open-water cage or raft culture. In countries where products liability laws exist, it is important to have insurance coverage for this. Death or disability caused by the consumption of contaminated aquaculture products or damage caused by the supply of infected fry or fingerlings can result in payment of considerable compensations.

Risk management that can be effected through cooperation with the insurance industry relates to pure risks as opposed to business risks. These pure risks include (i) natural disasters, (ii) technical (breakdown and failure of equipment and plants), (iii) theft, poaching, negligence and (iv) personal risks. It is believed that losses due to all these can be reduced by proper management practices, making use of the experience of insurers (Secretan, 1979).

13.6 References

Aplin R.D. and Casler G.L. (1968) *Evaluating Proposed Capital Investments with Discounted Cash Flow Method*. Department of Agricultural Economics, Cornell University, Ithaca.

Berg L. (1979) A proposal for economic investigations of fish farms with special reference to book keeping and financial analysis. In *Advances in Aquaculture*, (Ed. by T.V.R. Pillay and W.A. Dill), pp. 239–46. Fishing News Books, Oxford.

Brown E.E. (1979) Fish production costs using alternative systems and economic advantages of double-cropping. In *Advances in Aquaculture*, (Ed. by T.V.R. Pillay and W.A. Dill), pp. 235–9. Fishing News Books, Oxford.

Collins R.A. and Delmendo M.N. (1979) Comparative economics of aquaculture in cages, raceways and enclosures. In *Advances in Aquaculture*, (Ed. by T.V.R. Pillay and W.A. Dill), pp. 472–7. Fishing News Books, Oxford.

Gerhardsen G.M. and Berg L. (1978) Notes on the economics of aquaculture. *Papers on Fisheries Economics*, **18**. The Norwegian School of Economics and Business Administration, Bergen.

Gerhardsen G.M. (1979) Aquaculture and integrated rural development, with special reference to economic factors. In *Advances in Aquaculture*, (Ed. by T.V.R. Pillay and W.A. Dill), pp. 10–22. Fishing News Books, Oxford.

Gittinger J.P. (1972) *Economic Analysis of Agriculture Projects*. John Hopkins University Press, Baltimore.

Greenfield J.E. (1970) *Economics and Business Dimensions of the Catfish Farming Industry*. Bur. Comm. Fish., U.S. Dept. of Int., Ann Arbor.

Leopold M. (1978) *Main Problems of Fish Culture Economics*, working document. EIFAC, FAO of the UN, Rome.

MacFarlane I.S. and Varley R.L. (1979) Risks, mortality and insurance of European trout farms. In *Advances in Aquaculture*, (Ed. by T.V.R. Pillay and W.A. Dill), pp. 70–74. Fishing News Books, Oxford.

Mitchell T.E. and Usry M.J. (1967) *Catfish Farming – A Profit Opportunity for Mississippians*. Mississippi Research and Development Center, Jackson, Mississippi.

Pillay T.V.R. (1977) *Planning of Aquaculture Development – An Introductory Guide*. Fishing News Books, Oxford.

Secretan P.A.D. (1979) Insurance and risk management for aquaculture industry. In *Advances in Aquaculture*, (Ed. by T.V.R. Pillay & W.A. Dill), pp. 63–70. Fishing News Books, Oxford.

Secretan P.A.D. (1986) Risk insurance in aquaculture. In *Realism in Aquaculture: Achievements, Constraints, Perspectives*, (Ed. by M. Bilio, H. Rosenthal and C.J. Sindermann), pp. 535–41. European Aquaculture Society, Bredene.

Shang Y.C. (1972) *Economic Feasibility of Fresh Water Prawn Farming in Hawaii*. University of Hawaii, Honolulu.

Shang Y.C. (1979) *Guidelines for Identification and Preparation of Aquaculture Investment Projects* (unpublished).

Shang Y.C. (1981) *Aquaculture Economics: Basic Concepts and Methods of Analysis*. Westview Press, Boulder/Croom Helm, London.

Webber H.H. (1973) Risks to the aquaculture enterprise. *Aquaculture*, **2**(2), 157–72.

Yang W.Y. (1965) Methods of farm management investigation. *FAO Agric. Dev. Paper*, 80.

14
Farm Management

Experience in the development of large-scale aquaculture ventures during the last couple of decades has given rise to the view that the key to success is not just adequate technology, but also efficient farm management. According to Huguenin and Colt (1986) the ability to organize and implement an aquaculture technology, which is a complex combination of technical, economic, marketing, social and political elements towards some specific goals, is a managerial process. There is a general belief that the importance of management relates only to large-scale enterprises and not to small-scale aquaculture. Webber and Riodan (1979) pointed out that 'new problem areas are engendered and many of the old problems become more critically significant' as small-scale fish farms, owned and operated by single-family units primarily for subsistence or at best for a small cash crop, evolve into large-scale agri-business enterprises incorporated and conducted for economic profit. While, indeed, there will be differences in management problems between these two types of farming, it would seem clear that management plays an important role in small-scale aquaculture as well. This is evidenced by notable differences that can be observed between the performance of small-scale farms in the same area, operated under similar conditions, using the same technology. At least a part of this difference in performance can be ascribed to differences in farm management practices. The ability of the farmer to manage his resources, including the know-how, land, water, labour, capital and time, to the best advantage for achieving his goals will to a large extent determine the performance of his farm. The role of efficient farm

management has been well accepted in the allied fields of agriculture and animal husbandry, irrespective of whether they are large-scale or small-scale operations.

14.1 Concepts and economic principles of farm management

The term 'farm management' is used to convey different concepts by different people. Aquaculturists often tend to consider it as the overall technical operation of the farm and supervision of day-to-day activities. Good farm management expertise is often considered to be the same as practical experience in the application of aquaculture technologies in the field. Proper and timely maintenance of the farm and its installations, successful methods of brood stock manipulation, breeding, seed production, stocking, feeding, disease and pest control, proper water management, including the maintenance of water quality, protection of the stock from poaching, harvesting and marketing are the major elements of this concept of management.

The science of farm management, which is relatively new and developed in agriculture and animal production, is based on the concept of a farm as a business and consists of the application of scientific laws and principles to the conduct of farm activities. Originating in production or agricultural economics, it is now accepted as a multidisciplinary science (Dillon and Hardaker, 1984). Yang (1965) defined it as 'a science which deals with the proper combination and operation of production factors, including land, labour and capital, and the choice of crop and livestock enterprises to bring about a maximum and continuous return to the most elementary operation units of farming'. He considered it a pure science because it deals with the collection, analysis and explanation of facts and the discovery of principles; and at the same time an applied science because the ascertainment and solution of farm problems are within its scope.

Farm management involves a continuous process of economizing and therefore the relevant basic theory of farm management is economics. However, it has to draw heavily on biology, technology, meteorology, sociology, psychology and related disciplines to optimize the use of scarce resources. While scientific research to develop technologies is performed in laboratories and experimental stations, farm management research is done in the field by collecting and analysing information from individual farms, to discover or verify successful farming practices under specified circumstances. Its aim is to plan optimum farm organization and management practices for higher production efficiency and maximum farm earnings. Although field experimentation can be a useful means of generating the necessary information, cost considerations often militate against its use in developing farm management methods. Experiments have admittedly the advantage of elucidating clearly input/output relationships, by varying the level of selected inputs. By replication and statistical methods, the reliability of the results and significance of the differences can be measured. However, it usually fails to discover the interactive effects between factors. Because of this, much of the data required for management research are obtained through farm management surveys, financial book keeping and the study of farm practices, including costs, use of land, water, labour and other material requirements. The results of management research can be used by farm managers in planning their activities or by governments in formulating farm policies.

The main elements of economic principles considered in agricultural farm management are comparative advantage, diminishing returns, substitution, cost analysis, opportunity cost, enterprise choice and goal trade-off.

Comparative advantage relates to the determination of the most economically suitable crops for a farm or area, from among the different crops that could be grown there. Since the comparative advantage can change as a result of changes in technology, input and transportation costs, farm product prices, etc., it will be necessary to evaluate the advantages on a continuing basis.

The principle of diminishing physical and economic return determines the best level for any production practice. It helps in considering the level of output produced from a set of fixed resources, taking into account the variable factors. Diminishing economic returns appear when diminishing physical returns are converted into value, generally measured in money terms. For example, in considering the use of weedicides in a farm, the farmer has to balance the money cost involved against the expected money value of the increased yield, or losses prevented, in order to decide whether it pays from the financial stand-point. It may be that he should use the weedicide up to the point where the last unit of application is expected to pay for itself.

The principle of substitution refers to the selection of the most economical method, measured in the most appropriate terms (e.g. physical labour, time or money) to suit his conditions. For example, the farmer has the option to use manual labour, mechanical equipment or chemical means to control weeds in his fish pond. He has to decide which of the methods he should use, taking into account the performance and cost of each. In substituting one method for another, he has to ensure that the saving is greater than the cost of the technique added. This principle is of special importance when decisions have to be made on the adoption of new practices.

The principles of cost analysis have been dealt with in Chapter 13. Even though the farmer may have some control over the costs of production on his farm, he has little control over the prices he receives for his produce. It is obvious that, under normal circumstances, a farmer must reduce his costs per unit of output if he is to increase his net farm income. While the fixed costs remain the same regardless of how much he produces, variable costs change as the size of operation changes. The classification of a particular cost as fixed or variable depends partly on the nature and timing of the management decision considered. For example, land rent becomes a variable cost in relation to a decision to lease more land; but for land already leased and being used, the rent is a fixed cost.

In Chapter 3, the importance of opportunity cost in farm planning and decision-making are indicated. This concept relates to the cost of any choice in relation to the value of the best alternative foregone. For example, if a farmer can earn a profit of $1500 from a farm growing milkfish and $3000 by growing shrimps, the opportunity cost of growing milkfish is $3000. If he persists in growing milkfish he should recognize that he is earning $1500 less profit than he could have earned. Although for certain reasons he may continue with milkfish, the general principle is that the land, water, labour and capital should be used where they will add most to the income. The income may be measured directly as money, or in some broader terms such as satisfaction or utility.

Enterprise choice is made by a farmer, making allowances for the relationship with other activities or enterprises on his farm. Enterprises can be supplementary or complementary, as in aquaculture integrated with crop and animal farming. As far as operation of his farm is concerned, the overall goal of the farmer is to make the most efficient use of whatever resources he has.

All the considerations summarized above relate largely to internal allocation of resources to enterprises and activities that will maximize the net return. The principle of goal trade-off implies the existence of multiple goals which will often compete with one another, such as cash income, utilization of unproductive land, export earnings, etc. The farm may be managed to achieve that mix of goal attainments which gives the farmer the best level of overall satisfaction across his multiple goals. There may have to be some trade-off, ensuring that the gain in satisfaction from the relatively more important goal is greater than the decreased satisfaction incurred on the other goals.

Application of the above economic principles in farm management is very much influenced by two factors that are somewhat unique to farming of animals and plants, whether aquatic or terrestrial. One is the varying degree of uncertainty under which annual operations have to be planned. The uncertainty may refer to the climate that will prevail, natural disasters that can occur, incidence of pests and diseases, the performance of new technologies adopted, the prices and competition that may be faced in the markets and the political environment in which the enterprises has to operate. Decisions are made under such uncertainties and therefore call for the exercise of personal judgement by the manager about the risks that he faces in the application of the various economic principles. Conclusions based on historic data can be of only partial assistance and decisions have to be on the basis of estimated future possible yields, costs, prices and technology. The other important factor is the orientation of the farm: whether it is completely market-oriented and operating commercially in a money economy or whether it is subsistence or semi-subsistence farming. A good majority of small-scale farmers, and almost all large-scale farmers, have contact with markets through which they receive money as total or partial income.

It may appear that the economic theories mentioned above do not apply to small-scale farmers who operate outside the cash economy. But in point of fact, they are very pertinent to their operations and can be used to assess the gains and losses, irrespective of whether money value or some other measure (such as utility or satisfaction) is used. When applicable, money is a very convenient measure, as it enables comparisons between farms and the aggregation of individual farm performance to regional and national aggregates. When gains and losses involve both cash and non-cash elements, the trade-off or exchange rates between them will be specific to each farm. So it has to be recognized that money is a compromise measure. While it may be the best basis for analysis, it is less than an adequate approximation, depending on the extent to which trading guides are available on the money value of non-cash gains and losses.

14.2 Application of farm management principles in aquaculture

As indicated earlier, the concepts and principles of farm management have been developed on the basis of agricultural and livestock farm operations. In the absence of relevant data and appropriate research, the applicability of these principles has not been explicitly tested in aquaculture. The nature of aquaculture research that is being promoted presently does not include the type of applied research considered necessary for developing farm management procedures. The reliance on the green thumb for successful farming is admittedly a risky approach, as has already been shown in many instances. To some extent, the present state of the aquaculture industry accounts for some of the problems relating to the development of an applied science of farm management. It is seldom possible to carry out farm management analysis, if records and accounts of operations are not available. Unfortunately, reliable farm data are very scarce and this is a major handicap in the development of aquaculture management procedures.

Managing a farm business often involves (i) organizing the farm, (ii) planning and directing its operation from day to day, (iii) planning or conducting the buying and selling and (iv) arranging financing and credit. Planning and organizing the farm operation is not a once-only task, and at least some of it has to be done on a continuing basis at the beginning of each rearing season.

Most of the present-day aquaculture farms are too small to afford a manager who can devote his time entirely to managing. Very often they are family farms, where the owner is also the manager, and he along with his family members and hired labour undertake all the work involved. He does all the organizing and reorganizing and he also does the buying and selling during his spare time. Larger farms are likely to present more varied management problems than family farms, but most tasks of management are the same for both. It is, however, true that on larger enterprises particular functions can be organized more carefully. It is frequently possible to obtain the services of a specialist. However, on larger farms there is the added task of supervising the work of labourers spread over a large area. Because of the larger number of workers employed, management/labour relations take on new forms. Another distinction is that larger farms are likely to employ more equipment, which introduces the problem of maintaining it in good working condition.

14.2.1   Farm business analysis

In Chapter 13 methods of estimating the economics of aquaculture farms have been described. Farm business analysis based on data collected through farm surveys is intended to ascertain the relationship between management factors and income. It elucidates factors which affect farm success and failure, so that an individual farmer may recognize his weak points by comparing his performance with relevant standards and take the necessary steps for improvement. There is little doubt that management studies would prove most useful in the development of aquatic farming. The methods of data collection and analysis for agricultural farm management analysis are described in standard text books and could possibly be used in aquaculture farm studies, depending on local situations. In many developing country situations, the simpler methods of tabular and budget analysis may be adequate and more appropriate. In others, more complicated techniques such as linear programming and production function analysis can be employed to provide more direct guides to relevant research conclusions. The need for the preparation of a large number of alternative budgets to determine a near optimum plan can thus be avoided. However, there are grave doubts expressed by many experienced farm planners about the practical value of a precision tool like linear programming at the farm level, when the precision of the input/output data is in doubt. The increasing availability of electronic computers makes it easier to carry out the complicated computations required, but without reliable farm records and accounts not much progress can be made.

Stamp (1978) has described a pioneering attempt to use a computer as a management aid for planning, budgeting, keeping records and accounting at the farm level in shrimp farming in the USA. But because of the need for a computer model that a farm manager can understand and use for decision-making, he considers the sophisticated methods of value only in research, rather than in farm-level management. Nevertheless, it is clear that these methods will find greater application in aquaculture when farm management research based on farm business surveys becomes a reality.

14.2.2 Planning and organization of farm business

Sound planning of the organization, financing and operation forms the basis of increased productivity of an existing farm business or of new ones proposed to be established. Several questions have to be answered in deciding on the organization of a farm. These would primarily relate to the species and systems of culture to be adopted, the design and construction in the case of new farms, levels or intensity of production to be selected, equipment and labour to be used, etc. These decisions have to be based not only on technological information, but also on the analysis of farm performance data when possible and on the economic principles of comparative advantage, diminishing returns, substitution, cost analysis, opportunity cost, enterprise choice and goal trade-off, mentioned earlier. Each one of these principles is applicable to small- and large-scale aquaculture enterprises, and it is only the scarcity of farm performance data that restricts its use.

It is essential to assess the comparative advantage of the species which can be cultured economically in a certain area and the advantages of monoculture versus polyculture, based on existing technology, production costs and markets. If adequate farm data are available, it should be possible to determine the intensity of production, e.g. extensive, intensive or semi-intensive, that would yield the maximum economic return. Feeding and fertilization rates and stocking levels are obviously susceptible to the principles of diminishing returns. The question of whether to substitute labour with mechanization, use artificial feeds instead of fertilizing pond farms, produce fry and fingerlings on the farm instead of buying them from outside producers, is the type of substitution that an aquaculturist has to consider. When deciding on mechanization or the greater use of equipment, the manager has to take into account not only the cost of initial investment but also the cost of maintenance and replacement. He would also have to consider possible joint ownership of equipment or hiring equipment for specific uses. Cost analysis to reduce production costs and increase profits is an obvious requirement of any aquatic farm. In making decisions on the establishment of an aquaculture farm, a farmer

has necessarily to take full account of the opportunity costs. It will be necessary for him also to consider relations with other enterprises or activities, as for example crop cultivation or livestock farming. As mentioned earlier, the systems of fish/rice farming and integration of pig and duck raising with fish culture are examples of enterprises combined to make the most economical and efficient use of farm resources. Decisions on farming are often based on certain goals, such as increasing the family income, entering the export trade, competing in the domestic fish trade, etc. These goals may change with time and circumstances, but the farmer has to take decisions to achieve gains in his primary goals, by making any trade-off that may become necessary.

14.2.3 Operation and financing

There is considerable interaction and overlap between the organization and operation of a farming business. As in the case of agriculture, some of the planned operations may have to be changed due to unexpected climatic conditions, including natural disasters. Decisions have to be made on operational details on a day-to-day or week-to-week basis. Whether it is a small farm of a few ponds or a large farming enterprise, there will be the need to implement the various operations according to an appropriate plan. Often, planning would make all the difference between work completed on time and always being behind. In the absence of suitable operational schedules, it will be very difficult for a farmer or farm manager to make the best use of the hired labour.

Despite efficient organization and well-planned day-to-day operation, a manager may not be able to achieve a good balance sheet unless he is able to handle the commercial aspects of his business properly. He has to use good judgement in buying the inputs and equipment needed and also in hiring labour. He has to follow the markets closely and buy supplies in bulk when possible, to economize on his costs. Participation in cooperatives may, in certain circumstances, help him in buying farm requirements at reasonable wholesale prices.

As mentioned in Chapter 12, the production has to be scheduled in such a way that it can be sold when the prices are good. Obviously, if the farm has to sell most of its production when the prices are low, and has little to sell when the price is high, the profitability of the operation will be greatly affected. At the same time, it is important to ensure that the product is sold at the time of the year when it sells at the highest margin over costs, which need not necessarily coincide with the time when the prices are the highest.

Timely decisions have to be taken for adequate financing of the business, and a small-scale farmer is at a great disadvantage in getting the right type of advice in this regard. Some are persuaded to borrow too much and too often, whereas others are too timid to borrow. It is a sound policy to take advantage of opportunities that may occur for expanding the farm or intensifying its production, even when funds have to be borrowed. However, the benefits and liabilities should be carefully analysed and weighed against each other, before decisions are taken. The magnitude of the increased income that is likely to be made, the time it will take to achieve the increase and the risks and uncertainties that are involved have all to be carefully considered. These factors are also relevant to decisions on investments in starting a new farm business. The best time to borrow is as important as selection of the best time to repay the loans. As indicated in Chapter 13, the maintenance of a satisfactory cash flow is a basic requirement for the stability of the farm business.

14.2.4 Labour management

Although labour relations in a large farm take on a character similar to that in factories, labour management is not a problem restricted only to large farms. It is also important in small-scale and family farms. In a family farm, the farmer-manager directs his own labour along with his family and such hired labour as he may employ. In larger farms, the manager often has a number of alternatives for using labour, such as increasing the labour force to implement more intensive farming, combining mechanization with manual labour to increase production or reducing the labour force and substituting it with machinery, etc. Decisions on such alternatives depend very much on the skills and judgement of the manager. They are also strongly influenced by the special characteristics of farm work. Unlike in factories, there are relatively few repetitive jobs in aquaculture farms. Even though chores may be repeated each day, frequent changes are required to meet the constantly changing water and weather conditions and the increasing magnitude of the biomass in the farm. Most workers find themselves doing a wide range of tasks over the year, with very little opportunities for specialization.

The main difference between family labour and operator labour in an aquaculture farm would appear to be personal motivation and appreciation of the importance of the job being carried out. Programming of family labour should take this into account. Usually hired labour is interested in wages, working conditions, security and the progress they can make. From the point of view of the manager, the greater the work he can get out of his labour hired on a time-rate basis in a given period of time, the lower his labour cost. Good management involves forestalling successfully any conflicts that may arise due to differing motivations.

In large farms, the operator seldom works with his men and he may employ a manager or a foreman to supervise the work force. The labour often works in groups and generally there is a division of labour. The employer/employee relationship becomes an important factor in the successful operation of such farms. Many of the basic principles of personnel management are applicable in the management of labour under such situations.

14.2.5 Decision-making

Throughout the above discussion on farm management, the need for decision-making on various aspects of organization and operation has been recognized. All the data analysis and application of economic principles cannot completely replace the manager's task of decision-making. They can only help him to make rational decisions, but at the end of the day these have to be based on his personal judgement. In actual practice, most decisions are based on risky choices. A decision-maker has to choose between different alternatives, some of which have consequences that are uncertain.

As a result of theoretical studies by economists, statisticians and management specialists, decision theories or decision analysis approaches have been developed to indicate which alternative he ought to take in line with his goals. But still the decision will be based on the personal belief of the decision-maker about the occurrence of uncertain events and his personal evaluation of potential consequences. Most managerial decisions have elements of uncertainty and are therefore risky. In a risky business like aquaculture decisions naturally become additionally risky. Even though it is not always possible to rationalize risky choices, procedures have been developed to systematize the process through decision analysis. It is a logical procedure for bringing together all the pertinent aspects of a decision environment. Still the personal element, including perceptions of the risks involved and personal attitude to consequences, plays a major role in the final decision, even though it is preferable to handling complex decisions by mere intuition.

How much effort and time should be spent on decision analysis should depend on the importance and nature of the decision and the time and cost involved in the analysis. Even an analysis at the simplest level, consisting of evaluating the various choices available and asking questions about the consequences of each and the chances of their success, can lead to better management decisions. When risky decisions are taken it will also be necessary to consider ways of covering major losses sustained as a consequence of the decision, by appropriate risk insurances as described in Chapter 13.

14.3 References

Black J.D. *et al.* (1947) *Farm Management*. Macmillan, New York.

Dillon J.I. and Hardaker J.B. (1984) Farm management research for small farmer development. *FAO Agric. Serv. Bull.*, **41**.

Hilton N. (1983) Farm planning in the early stages of development. *FAO Agric. Serv. Bull.*, **1**.

Huguenin J.E. and Colt J. (1986) Application of aquaculture technology. In *Realism in Aquaculture: Achievements, Constraints, Perspectives*, (Ed. by M. Bilio H. Rosenthal and C.J. Sindermann), pp. 495–516. European Aquaculture Society, Bredene.

Ministry of Agriculture, Fisheries and Food (1963) *The Farm as a Business. Vol. 1 Introduction to Management*. HM Stationery Office, London.

Spivey W.A. (1973) Optimization in complex management systems. *Trans Am. Fish. Soc.*, **102**(2), 492–9.

Stamp N.H.E. (1978) Computer technology and farm management economics. *Proc. World Maricul. Soc*, **9**, 383–92.

Webber H.H. and Riodan P.F. (1979) Problems of large-scale vertically integrated aquaculture. In *Advances in Aquaculture*, (Ed. by T.V.R. Pillay and W.A. Dill), pp. 27–34. Fishing News Books, Oxford.

Yang W.Y. (1965) Methods of farm management investigations. *FAO Agric. Dev. Paper*, **80**.

Yang W.Y. (1984) Planning for action in agricultural development. *FAO Agric. Serv. Bull.*, **2**.

Part II

Aquaculture Practices

The succeeding chapters deal with culture technologies as presently practised for large-scale production the world over. Chapters 15 to 28 refer to the culture of groups of major aquatic species, and Chapters 29 and 30 to specialised culture systems using different species, *viz* integrated and open-water aquaculture. Because of space limitations, it has not been possible to describe separately and in detail the practices relating to all the species listed in Chapter 5. However, efforts have been made to cover all the important ones that contribute to the present world aquaculture production.

Although there are some differences in culture technologies between various species in each group, the similarities are greater, and these are brought out in the descriptions that follow. The value of concentrating research and technology development to a lesser number of species to achieve rapid progress in aquaculture has been mentioned earlier, and an attempt here to concentrate on the most important species and groups is in line with this approach.

The references at the end of each chapter, particularly some of the manuals listed, should be consulted for more detailed information on specific culture technologies. As many of the general aspects of aquaculture have been considered in other chapters, greater attention will be focused here on spawning, production of young and grow-out technologies.

15
Carps

15.1 Main species of carps (family Cyprinidae) used for culture

15.1.1 Common carp

Of all the species of finfish or shellfish used for aquaculture, carps undoubtedly have the oldest history. The common carp (*Cyprinus carpio*) is probably one of the few aquaculture species that can be considered to have been domesticated.

The common carp is presently cultured all over Asia, in most parts of Europe including the USSR, and on a small scale in some countries of Africa and Latin America (particularly Brazil). It has also been introduced in North America and Australia. Though considered to be a sport fish as well, its main importance is as a food fish. In fact the prejudice towards common carp in some countries has been created by anglers who consider it a pest in sport waters, because it muddies the water when routing around for food on the margins and the bottom of water bodies.

There are three recognized varieties of common carp: the orange-coloured scale carp (*C. carpio* var. *flavipinnis*), the partially-scaled mirror carp (*C. carpio* var. *specularis*) and the virtually scaleless leather carp (*C. carpio* var. *nudus*). There is also a variety with only one row of big scales on the lateral sides. The normally-coloured or orange-coloured scale carp and the mirror carp are the varieties preferred for culture, mainly because of their faster growth rates (figs 15.1 and 15.2). Several races and strains of common carp have evolved or have been created through breeding programmes. A well-known race is the 'big-belly'

carp of China, which is a hardy fish that starts breeding at the early age of about 6 months and has relatively large gonads, which accounts for its name 'big-belly'. The Japanese Yamato carp appears to be related to the Chinese big-belly. The Punten carp of Indonesia is a quick-growing race.

There are a number of other geographical races such as the Galician and Franconian races of the mirror carp and the Aischgrund carp of Germany and Royale of France. They are distinguished by their body form, particularly by the length/height ratio and body thickness. The main varieties cultured in the USSR are the scale carp, the mirror carp (with scales scattered all over the body) and the Ukrainian frame carp (with scales framing the sides of the body). A number of breeds or strains of the common carp have been developed by genetic breeding in East European countries, especially in the USSR, Hungary and in Israel. Kirpichnikov *et al.* (1979) referred to local strains of the Ropsha scaly carp based on selection of hybrids between the Amur wild carp and Galician cultured carp and also the Ukrainian-Ropsha hybrid scaly carp, obtained through crossing the Ropsha carp with the Ukrainian 'frame-scaly' carp (fig. 15.3). Most of these strains have been developed for the purpose of improving the growth rate, suitability for the climatic conditions, time of maturation and spawning. Hungarian races of common carp have been cross-bred to produce hybrids with improved egg fertility and increased growth with lower feed consumption (Bakos, 1979) (figs 15.4 and 15.5). Recently, a strain of common carp known as the Heyuan carp has been developed in China by crossing the female 'purse carp' with the male 'yuanjiang

Fig. 15.1 Orange-coloured scale carp.

Fig. 15.2 Mirror carp.

Fig. 15.3 The Ropsha strain of scaly carp.

Fig. 15.4 Selected Hungarian strain of mirror carp.

carp'. This hybrid strain is reported to grow faster (by over 30 per cent) than wild carp, is easy to catch, has a larger body weight and apparently greater disease resistance.

The common carp is an omnivore and in nature and in culture ponds it feeds on a wide variety of plant and animal matter. The young carp, up to a length of about 10 cm, feed on protozoa and zooplanktonic organisms such as copepods and cladocerans. Above that size, they start feeding on benthic organisms, such as insect larvae (especially chironomid larvae),

Fig. 15.5 Selected Hungarian strain of scale carp.

worms and molluscs, together with large quantities of vegetable matter and epiphytic organisms. The carp's habit of sucking food organisms in the mud on the pond bottom and margins makes the water muddy and weakens the base of pond dikes. However, when fed on artificial feedstuffs or processed feeds, this habit is greatly curtailed.

15.1.2 Chinese carps

A group of carp that has become equally or more important in aquaculture consists of the five species popularly known collectively as the Chinese carps: the grass carp (*Ctenopharyngodon idella*) (fig. 15.6), the silver carp (*Hypophthalmichthys molitrix*) (fig. 15.7), the bighead (*Aristichthys nobilis*) (fig. 15.8), the black carp (*Mylopharyngodon piceus*) (fig. 15.9) and the mud carp (*Cirrhina molitorella*). The historical reason for the origin of their

culture in China has been mentioned in Chapter 2.

It would appear that the very popular polyculture system also originated with the culture of these species. Since they do not spawn naturally in ponds, larvae and fry had to be collected from the natural spawning grounds in the rivers. It was not easy to sort young larvae completely, and so all the species had to be cultured together. Experience showed the compatibility of the species in ponds and the higher production that the farmers could obtain by combined culture.

The grass carp is herbivorous and feeds on macrovegetation, including grass and aquatic plants; the silver carp feeds on plankton, mainly phytoplankton; the bighead consumes the macroplankton and the black carp feeds on snails and other molluscs at the bottom. The mud carp feeds primarily on detritus. To this combination is added the omnivorous common carp, which the farmers consider a scavenger in the pond. The grass carp starts feeding on macrovegetation when about 2.5−3 cm in length, and is reported to ingest up to 50 per cent of its weight in the form of land plants (Martyshev, 1983). It stops feeding at a temperature of 10−12°C; at temperatures above 20°C, it eats large amounts of grass. The silver carp starts feeding on algae at a length of about 1.5 cm. It can feed on microscopic phytoplankton, as small as 30−40 µm, by filtering it through its very efficient gill rakers. The larvae of bighead feed on unicellular planktonic organisms. The fry and adults feed on both phyto- and zooplankton and thus compete to some extent with silver and common carps; however, the bighead generally prefers larger organisms. The adult black carp feeds on benthic animals, but shows a special preference for molluscs. Though it is not considered a high-value fish outside China, it serves a specific role in controlling the growth of molluscs in polyculture ponds. Mud carp does not grow to a large size and is essentially a subtropical species, and so its use in polyculture is somewhat restricted, even though it is greatly esteemed by consumers.

Despite considerable interest in Chinese carps, their culture was restricted initially to only a few neighbouring countries with large Chinese populations, until methods of induced

Fig. 15.6 Grass carp, *Ctenopharyngodon idella*.

Fig. 15.7 Silver carp, *Hypophthalmichthys molitrix*.

Fig. 15.8 Bighead, *Aristichthys nobilis* (photograph: Paul Osborn).

Fig. 15.9 Black carp, *Mylopharyngodon piceus*.

breeding were developed in China, the USSR and India during the period 1960−62. Starting with the USSR, most of the East European countries have made progress in the culture of these carps, known in these countries as phytophagous or herbivorous carps. The more popular species have been the grass carp, because of its ability to control macrovegetation, and the silver carp, as it can utilize the dense blooms of phytoplankton. But the consumer preference continues to be for the common carp and countries like Hungary have depended on export markets in the Middle East for their production of silver carp. Many countries in Asia, some in the Middle East (particularly Israel and Egypt) and some countries in South America (especially Mexico) have introduced Chinese carps for pond culture. The main interest in Western Europe and the USA has been in using the grass carp as a biological weed control agent, and the species has been introduced for this purpose. To prevent breeding of the species in weed-infested water bodies, efforts have been made to produce all-male grass carp by administration of methyltestosterone, with limited success (Stanley and Thomas, 1978).

15.1.3   Indian carps

A third group of carps is the Indian carps: catla (*Catla catla*) (fig. 15.10) rohu (*Labeo rohita*) (fig. 15.11), mrigal (*Cirrhina mrigala*) (fig. 15.12) and calbasu (*Labeo calbasu*). In India these species are referred to as the major carps, to distinguish them from a number of other cyprinids that grow only to smaller sizes, known as minor carps. For more or less the same reasons as the Chinese carps, these species have also been cultured together in traditional pond culture. Even though the farmers were able to spawn them in special 'bundh' types of ponds, most of the larvae and fry required for culture had to be collected from their natural spawning grounds in the rivers. It was extremely difficult to sort out the species at the early stages, and was easier to rear them together rather than separately. Addition of Chinese carps, such as grass carp, silver carp and common carp, to this combination resulted in higher yields than from polyculture of Indian carps only. This combination of Indian and Chinese carps is referred to as composite carp culture in India.

Catla is considered to be a surface and column feeder. Larvae and young fry feed on planktonic unicellular algae. After reaching a length of about 2 cm, the fry start feeding on zooplanktonic organisms, showing a preference for protozoans and crustaceans. The adults feed on different types of algae, planktonic protozoa, rotifers, crustaceans, molluscs and decayed macrovegetation. They show a specific preference for planktonic organisms. Rohu is a column feeder in ponds. Larvae and fry feed on unicellular algae and zooplanktonic organisms. Adults feed on various types of vegetable matter including decaying aquatic plants, algae, etc. The adult mrigal is a bottom feeder. Larvae and fry have about the same feeding habits as the other Indian carps, but the adult fish feed

Fig. 15.10 Catla, *Catla catla*.

Fig. 15.11 Rohu, *Labeo rohita*.

Fig. 15.12 Mrigal, *Cirrhina mrigala*.

on algae, diatoms, higher plants and detritus. Calbasu is also a bottom feeder, feeding on benthic and epiphytic organisms and organic debris from the pond bottom. Larvae and fry feed on unicellular algae until they reach about 2 cm, after which they prefer phyto- and zooplankton. The combination of surface, column and bottom feeding fish has been the basis of Indian carp culture. As indicated earlier, the Chinese silver carp and grass carp and the common carp have been included to improve production. Several ratios of these species have been tried in experimental stations and it was reported that a combination of grass carp, catla, rohu and common carp gave the best results.

15.1.4　Species combinations

In both China and India, there is a trend towards increasing the number of species in pond polyculture for the purpose of higher production. While the value of combined culture of certain species in ponds under specific culture conditions is accepted, the general concept of exclusive ecological niches for each species of carp does not seem to hold, particularly in the context of the increasing number of species added to the combinations. There is considerable overlap in the feeding habits and spatial use of the ponds by the different species. However, some of the species in the combinations can have specific roles in the maintenance of the oxygen regime and sanitary conditions in the ponds. For example, the silver carp can be of considerable value in controlling algal blooms and reducing oxygen consumption. Though not very clearly proved, it is generally believed that some species feed on the excreta of others and thus reduce the accumulation of organic matter on the pond bottom. As mentioned earlier, the common carp and mrigal are considered as scavengers in polyculture ponds as they may consume the faeces of the grass carp and the silver carp, which contain large amounts of undigested plant matter.

The above possible advantages have to be weighed against some of the requisites of polyculture. The need to produce or purchase the required number of fry or fingerlings of a selected number of species and to maintain the balance of species in order to avoid competition calls for special care, skill and effort on the part of the farmer. It is not very easy to adopt supplemental feeding in an economical way. Additional labour would be involved in sorting out the different species after harvest. Consumer acceptance of the different species varies in most areas, and in fact difficulties have been experienced in finding markets for certain species like the silver carp in some countries.

A critical analysis of the biological and economic aspects shows that the value of polyculture depends very much on the situation in a particular area, and is probably not so widely applicable as generally considered. It is necessary to compare the productivity and economics of monoculture of the more important species, before making decisions on extending species combinations.

Another cyprinid species that needs to be mentioned is the tench (*Tinca tinca*), which has been associated with common carp culture in Europe and introduced into upland areas of countries like India many years ago. However, there is presently very little interest in its culture and it does not appear to figure prominently in aquaculture production.

15.2　Culture systems

As a group, carp appear to be especially suited for pond culture, although other systems of culture, in cages and rice fields and stocking in open waters, are carried out experimentally or commercially on a small scale. Small ponds or pools with running water have been used in countries like Japan and Indonesia, for intensive culture of common carp, but most of present-day aquaculture is carried out in stagnant or semi-stagnant ponds. As carps feed on low trophic levels, it is possible to produce most of the food needed through fertilizing ponds and this greatly reduces feed costs. Further, different types of organic manures and farm wastes can be used as fertilizers, and this contributes to the farm economy. Species like the grass carp that feeds on macrovegetation can be fed at little cost with land grass and most types of aquatic vegetation, in a fresh or decayed state. Because of these advantages and the comparative ease with which they can be reared in small and large farms, including village ponds, carps are generally perceived as low-valued fish, suited for large-scale production to feed poor communities in rural areas. This is not the case in all areas. In countries or regions where there is a distinct consumer preference for carps, some of the species sell at comparatively high prices − for example the grass carp in China, rohu in Eastern India and the common carp in Israel, Eastern Europe and some West European countries.

Though all the carps used in aquaculture are fresh-water species, they can tolerate salinities up to 10−11 ppt, and sometimes even grow better at salinities of about 5 ppt.

As will be discussed in Chapter 30 the carps, particularly the Chinese and Indian carps and common carp, have been widely used in their native countries for open-water stocking. Some of the other Asian countries have also transplanted them in open waters. Common carp has been the main species for rice-field fish culture, but other species like tilapia have replaced them in recent years as the major species in some countries.

Most of the recorded cases of cage culture of carps are experimental in nature, but there is a common traditional practice of growing them in bamboo cages in Indonesia and small-scale cage farming is practised in Chinese lakes. In Indonesia, cages are installed on the bottom of canals conveying large amounts of domestic wastes (fig. 15.13). The fish obtain their food mainly from the bottom of the canals, where chironomid larvae and other benthic organisms abound. The farmer may introduce limited quantities of supplemental feeds. Obviously this type of culture differs very considerably from the present-day culture of fish in floating cages, based mainly on artificial feeding. Such practice appears to be adopted only in Japan for growing carp in lakes, reservoirs and large irrigation or farm ponds.

Carps are grown in a wide range of pond facilities, ranging from small undrained village ponds to large, well laid out pond farms several hectares in area. Most such farms have a series of nursery and rearing ponds of different sizes. Rearing ponds are comparatively much larger in Eastern Europe than in Asia. If breeding is done on the farm, brood ponds (and in the case of common carp, spawning ponds) may be included in the farm layout. Many farmers purchase their fry from commercial fry producers and so require only nursery ponds in their farms (fig. 15.14).

Although monoculture of the common carp is practised by many farmers, the Chinese and Indian major carps are almost always grown in polyculture systems. As mentioned earlier, several combinations of Chinese and/or Indian carps and the common carp are used very widely. The crucian carp (*Carassius auratus*) and the wuchang fish (*Megalobrama amblycephala*) are often added to this combination in China. However, each combination is based on

Fig. 15.13 Cage culture of common carp in canals in Java, Indonesia (photograph: Marcel Huet).

Fig. 15.14 A fry market in Indonesia, where farmers buy their requirements of fry.

one or two species (grass, black or silver carp) as the major elements and the others are only complements to the major species. Polyculture of common carp, silver carp, tilapia and the grey mullet (*Mugil cephalus*) in fresh-water ponds has become a common practice in Israel. This combination is valuable because of the control of filamentous algae by the common carps the consumption of planktonic algal blooms by the silver carp and consequent improvement of oxygen regime, and the feeding on the organic ooze at the pond bottom by the tilapia. More importantly, such polyculture has helped considerably to overcome problems created by limits on carp production and to diversify and increase overall pond fish production.

15.3 Spawning and fry production

Among the important carp species used in aqua-culture, it is only the common carp that spawns naturally in ordinary fish ponds. Until about

two decades ago, the culture of Chinese carps and most Indian major carps was dependent on eggs, larvae or fry collected from riverine spawning areas. With the widespread practice of induced breeding, the use of wild seed has completely disappeared in China. In India and Bangladesh, some of the seed required are still collected from the wild, with different types of collecting nets. A very common spawn-collection tool consists of a funnel-shaped, close-meshed net with a cloth receptacle attached at the cod-end to hold the catches (fig. 15.15). The net is set in the shallow margins of flooded rivers with the mouth of the net facing the current. Fertilized eggs and larvae brought in with the current collect in the cod-end receptacle, from where they are periodically scooped out.

Some of the seed used for culture in India are produced by breeding the major carps in special types of ponds known as 'bundhs'. These are seasonal or perennial ponds or impound-ments, where riverine conditions are simulated

Fig. 15.15 Collection of Indian carp eggs and larvae from a river in India.

during the monsoon season. Large quantities of rainwater from the catchment area flow into the bundh after a heavy shower and create extensive shallow areas along the margin which serve as suitable spawning grounds for the carp. The perennial type of bundh is used as a brood pond and will therefore already have a stock of breeders when the marginal areas are inundated by rainwater and conditions become suitable for spawning. In the seasonal bundhs, a selected number of brood fish in the ratio of two or three males to a female is introduced. The mature fish move to the shallow areas near the margin and adjoining areas, and spawn after a short period of courting. The fertilized eggs settle on the shallow areas and can easily be collected with small pieces of close-meshed netting for incubation and hatching in improvized hatching pits, double-walled hatching hapas or cement cisterns. There is, as yet, no consensus of opinion regarding the factors which prevent these carps from spawning in ordinary ponds or the exact conditions which facilitate spawning in bundhs. It is believed that the cumulative effect of the conditions created by heavy rains and flooding of shallow areas induces gonadal hydration, contributing to final maturation and spawning. It would appear that the change of environmental conditions caused by heavy dilution with fresh rainwater triggers spawning.

Some of the recent developments in the use of bundhs for carp breeding are aimed at increasing the percentage of successful spawning. By storing rainwater in a reservoir located at a higher elevation, the bundhs below can be filled whenever needed, making it possible to conduct breeding operations without having to wait for a satisfactory rainfall. Injection of about 10–20 per cent of the brood stock with pituitary hormones is reported to have the effect of complete spawning of the entire brood stock. Sinha *et al.* (1979) were able to obtain the spontaneous spawning of silver and grass carps in bundhs, without having to resort to stripping.

15.3.1 Breeding of common carp

As the common carp breeds naturally in confined waters, several methods of propagating the species have been developed in different areas. The simplest allows uncontrolled breeding in communal ponds, with shallow marginal areas covered with grass or aquatic vegetation which serve as substrates for their adhesive eggs. A more advanced method uses special spawning ponds for spawning, hatching and larval rearing. The most familiar type of carp spawning pond is probably the Dubisch pond, named after the Silesian fish farmer who developed it. It is a square or rectangular-shaped shallow pond (8−10 m^2), generally surrounded by a reed fence for protection from chill in temperate climates. It has a peripheral 40−50 cm deep ditch the rest of the pond being only 20−30 cm deep. In the centre of the pond is a sloping spawning area covered with meadow grass. The Hofer type of pond is a variation of this, without a peripheral ditch but with a harvesting ditch near the monk. In a carp farm, a number of such spawning ponds may be built to spawn an adequate number of fish, when the temperature conditions are suitable. Before the spawning season, the ponds are dried and, if necessary, treated with lime to eradicate unwanted organisms. The ponds are filled when the water is sufficiently warm (above 18°C) and selected brood fish are introduced at the ratio of up to six males to three females. They usually spawn within 24−48 hours. The brood fish are removed after spawning and the eggs are left to hatch in the pond. Within a week after hatching, the fry are removed to nursery ponds for further rearing. In present-day carp culture, efforts are made to exercise greater control, so as to achieve a higher percentage of spawning success and hatching rates.

The spawning season for common carp in temperate climates is in the spring, when water temperature rises above 18°C. By the manipulation of environmental conditions and the use of selected races or strains, farmers have succeeded in extending the breeding season to suit particular culture requirements. In tropical climates, it has been possible to breed common carp at any time of the year.

15.3.2   Selection and segregation of brood stock

Proper selection of brood fish is very important in obtaining best results in breeding as well as in later grow-out. Many farmers select the largest, fast-growing fish, or the ones with the desired body shape, on the assumption that these characteristics will be inherited by the progeny. It is not advisable to select from the same brood stock or their offspring, as this may lead to inbreeding depression of growth rate and a predominance of deformed fry. As far as possible, brood stock should be selected from divergent sources.

Generally, two-year-old fish weighing about 2−3 kg are used for breeding, and they can be bred every year for several years. Larger fish may be more difficult to handle, but large females spawn more and larger eggs, and the hatchlings are also larger and survive better. About 100 000−150 000 eggs are produced per kg body weight.

At least three to four months before the breeding season, the brood stock are removed and stocked in segregation ponds. Males and females can be distinguished by external features during the spawning season. The female has a swollen abdomen due to the developing ovaries and in the males the milt runs freely when the abdomen is gently pressed. Chinese farmers identify older males from the tubercles on the sides of the head and on the pectoral and ventral fins. It is desirable to segregate the males and females into separate ponds, to avoid unwanted spawning. In segregation ponds, the stock is maintained under uncrowded conditions and fed on protein-rich natural food and supplemental feeds to assist faster gonadal development. High carbohydrate feeds have to be avoided, in order to prevent the accumulation of fat. During the period of segregation, it is necessary to prevent stress through netting or stimulation of spawning through the flow of fresh water into the pond. The pond is cleared of any substrates for eggs, such as weeds and grass, to avoid stimulating wild spawning.

15.3.3   Spawning and larval rearing

Spawning of common carp can be carried out in spawning ponds or in a hatchery, using induced spawning methods. Obviously pond spawning is easier for farmers, especially when they do not have the necessary hatchery facilities. However, induced spawning and hatchery rearing of larvae provide greater control over the propagation procedures and greater survival rates of larvae when reared in indoor tanks. Spawning can be carried out in small or large ponds, although smaller ponds are easier to manage. It is possible to spawn common carp in cement cisterns or even in cloth tanks (hapas), but then the fertilized eggs or early larvae will have to be transferred to nursery tanks or ponds.

There are different types of substrates or egg collectors used in carp spawning ponds. Mats made of fibres of the indjuk (*Arenga* spp.), known as 'kakabans' in Indonesia, are commonly used now by many farmers. Half to one metre long fibres are arranged in the form of a 40−70 cm mat, pressed longitudinally between two bamboo lathes. They are placed on long bamboo poles held in place between two pairs of shorter poles driven into the bottom (fig. 15.16). The mat remains slightly submerged, but floats with the bamboo pole and adjusts itself to changes in the water level. Water weeds such as *Ceratophyllum* and *Myriophyllum*, kept in place within a bamboo frame, are also used by some farmers in Asia. Israeli farmers use branches of pine, casuarina or cypress as egg collectors.

Artificial spawning mats made of synthetic material can also be used. The mat area needed as substrate is about 10 m² for every 2−3 kg females. The spawning mats are easy to handle and it is easy to estimate the number of eggs spawned. If only the underside of the mat has been used by the fish for depositing eggs, the mat can be turned over for deposition of more eggs. Such spawning mats can be placed in cement cisterns or cloth tanks (hapas) installed in ponds, filled with fresh clean water from a natural source, for incubation and hatching of eggs.

It is essential to ensure that the water supply to the spawning pond is from a natural source and not from another fish pond where large numbers of fish are held. The importance of fresh water in triggering spawning has already been mentioned. Ripe spawners from the segregation ponds are released into the spawning pond after it is filled. The number of spawning females (2−5 kg weight) introduced per hectare of pond area is about ten, and the ratio of females to males is about 2:3. Weight-wise the ratio may often be 1:1, as the males are smaller in size and weigh less. The ripe fish generally spawn in the morning after they are introduced into spawning ponds or tanks.

In Israel, where fish culturists often spawn fish late in the season (by August), to prevent the progeny from maturing and spawning before reaching market size during the succeeding spring, the brood fish are injected with pituitary extract to stimulate spawning. Females are injected with one pituitary per kg body weight and the males with half that dose. The number of eggs in late season is much less and so a larger number of females is introduced per hectare of spawning pond − as much as five to seven times the usual number.

The incubation period of eggs depends on temperature conditions, varying from 7 to 7.5 days at 16−17°C to 2 days at 30°C. The temperature most suited for hatching is considered to be 20−22°C or 60−70 day-degree (Woynarovich and Horvath, 1980).

Fig. 15.16 Kakabans used for carp spawning in Indonesia.

15.3.4 Hypophysation

Common carp

When conditions are not favourable for successful natural spawning in ponds, or when large quantities of fry have to be produced, it may be advantageous to resort to hypophysation techniques to induce spawning, and use hatcheries for incubation and larval rearing. This helps to improve the survival of hatchlings, by reducing predation by insects and other enemies.

Ripe brood fish kept in segregation ponds can be used directly for hypophysation. During the normal breeding season, a good common carp female produces about 150 000 eggs per kg. The eggs measure 0.9−1.6 mm in diameter. In practice, not all females respond to hypophysation and so it is considered advisable to inject double the number required. The number of males to be injected will be about two-thirds the number of females. Based on these factors, the total number of females to be hypophysed has to be determined.

It is considered advisable to disinfect the brood stock brought into a hatchery for hypophysation. Giving a 40 ppm formalin bath for about 2 hours is a common practice in Israel. The methods of preparing hypophysis extracts have been described in Chapter 8. The doses and sequence of injections are not standardized and practices vary considerably. Hepher and Pruginin (1981) described a tested procedure of injecting half a pituitary per kg female spawner one day after transfer into indoor hatchery tanks, by which time the fish has acclimatized itself to hatchery conditions. A second injection is given 8 hours later with 0.8 pituitary per kg female. The males are injected only once, 24 hours after transfer to the hatchery, with 0.5−0.6 pituitary per kg. Woynarovich (1975) recommended a single injection of one pituitary, at about 2.5−3.7 mg dried pituitary per kg body weight of the recipient. Dried pituitary is pulverized and dissolved in a solution of 0.6−0.7 per cent common salt (NaCl) and pure glycerine in the proportion of 70 parts salt to 30 parts glycerine. This solution is injected intramuscularly between the base of the dorsal fin and the lateral line. The injected males and females are kept in separate tanks.

As indicated earlier, the time taken for the eggs to ripen for spawning or stripping depends on the water temperature. Many fish culturists anaesthetize the brood fish before stripping. The most common anaesthetic used is a solution of ethyl-M-aminobenzoate at a concentration of 100 ppm. After a bath in the solution for 3−5 minutes, the fish are completely anaesthetized. First, the female is stripped by gently squeezing the abdomen towards the tail and the eggs that flow easily are collected in bowls or basins. Then the males are similarly stripped for milt, which is collected over the eggs in the same container, and the contents are mixed immediately with a feather or a plastic spoon. For every litre of eggs, two to three litres of milt will be needed to ensure proper fertilization. The adhesive nature of the eggs makes them clump together and hamper proper fertilization. The stickiness can be eliminated by treating the eggs with a solution of sodium chloride and carbamide (urea $(CO(NH_2)_2)$) (40 g sodium chloride and 30 g carbamide dissolved in 10 ℓ clean water), equivalent in quantity to the eggs. The solution is first poured over the mixture of eggs and milt and stirred with a plastic spoon or feather for about 5−10 minutes. As the eggs begin to swell, small quantities of the solution may be added at intervals, as required. Within about one to one and a half hours, swelling of the eggs will have stopped and the first cleavage will have occurred. The sticky layer will have dissolved, but to remove it completely the eggs have to be washed with a 0.05−0.07 per cent tannic acid solution for about 20 seconds. The washing has to be repeated up to five times, the solution being diluted each time by about 0.01 per cent by the addition of water to the stock solution. Finally, the eggs are washed in fresh water for about 5 minutes. The water-hardened eggs, which measure 2−2.1 mm in diameter, are ready for incubation in hatchery jars or other suitable containers.

There are different types of incubators used for hatching common carp eggs, ranging from simple double hapas (fig. 15.17) to Zoug jars with temperature-controlled water supplies. Essentially, a hatching hapa consists of a fine mesh (0.5 mm) sieve-cloth tank about 2 × 1 × 1 m in dimension, with an inner hapa or chamber made of the same material with a mesh of 2−2.5 mm. The whole device is placed

Fig. 15.17 A number of double hapas being used for hatching Indian carp eggs.

in a protected water body where the water is well-oxygenated. The fertilized eggs are spread in the inner hapa. The hatched larvae fall through the larger meshes of the inner hapa and are retained by the outer hapa. After the hatching is over, the inner hapa is removed together with the dead eggs, egg shells and other debris to avoid deterioration of the water quality in the hapa.

The Zoug jars (see fig. 6.42), named after Lake Zoug (Switzerland) where their use originated, are large (60–70 mm in height and 15–20 cm in diameter) inverted bottles, with open tops and narrow bottoms. In a hatchery they are installed vertically in a series, with the narrow neck directed downwards attached to the water lines with suitable taps. Water is supplied from below and the outlet is at the top. The capacity of a jar is usually 6–8 ℓ and it can carry about 1–5 ℓ eggs. The flow of water varies between 2 and 6 ℓ/s, with an average of about 4 ℓ/s. The MacDonald or chase jar, which is similar to the Zoug jar, is cylindrical in shape (40–50 cm in height) with a circular bottom on which it can stand. A tube is fitted inside at the bottom for the water

supply and the flow of water in the jar keeps the eggs rotating slowly.

Farm-fabricated incubation funnels are also used very widely. They can be made of plexiglass or fibreglass in any dimension required. Israeli farmers use plexiglass incubation funnels 80 cm long and 60 cm in diameter at the upper rim. Water is supplied from a bottom inlet, which is about 1.23 cm in diameter. Water overflows from an outlet at the top and this overflow can also be used for collecting the hatchlings. Each such funnel can be used to hatch about 175 000 carp eggs. Woynarovich and Horvath (1980) illustrated a number of designs of incubation funnels that can easily be made with plastic and sieve-cloth material, with simple sprinkler-type water inlets.

The eggs hatch out in the incubators within 2–7.5 days, depending on the temperature. At a temperature of about 25–26°C, the hatching takes place in 2–2.5 days, whereas at temperatures of 16–17°C, it may take up to 7.5 days. There will usually be some unfertilized eggs and these eggs are the focus of fungal (*Saprolegnia*) infection. A common method of controlling fungus in the incubator is by the application of malachite green at a concentration of 0.02 g/ℓ water for about 20−25 minutes, after stopping the flow of water. When the flow is resumed the chemical is washed out from the incubator.

The hatchlings are removed to indoor nursery tanks or similar containers. It is desirable to rear the hatchlings under controlled conditions, up to the fry stage. Nursery tanks are usually shallow, with a depth of about 0.5 m. About a million fry can be reared in a 20 m^2 tank with a water exchange rate of one litre per minute per square metre. Best results are obtained when the larvae are fed on natural food organisms collected from the ponds or other water bodies or cultured in the farm. Some farmers use manufactured starter feeds and, where feasible, brine shrimp produced on the farm.

Chinese carps

Although there are records of Chinese carps breeding in rivers in Japan and, as mentioned earlier, Indian carps have been bred in special bundh-type ponds in India, the main means of propagation of these species is through hypophysation. Though some interspecific hybridization of Chinese carps has been done, there do not appear to be any races or hybrid strains with well-defined cultural characteristics that a farmer can use for propagation. Chinese farmers get their brood fish from fish farms, rivers, reservoirs or lakes. Females of silver carp over 3 years, bighead 4 years, grass carp 5 years and mud carp 3 years are selected for propagation. As the males mature earlier, one year younger males of all these species can be used. Larger individuals are usually preferred, because of the larger quantity of eggs and milt that they have.

Although many farmers keep all the species together in the same brood ponds, it is preferable to maintain them separately as there are some differences in the requirements of each species. Grass carp prefer clean water and so there should be a regular renewal of water. They are fed on cut grass, sprouted grains, rice bran, corn meal and oil cakes. The feeding rate for grass is about 15−20 per cent of the body weight and for other foodstuffs 2−2.5 per cent of the body weight. About 20−30 brood fish can be kept in a 1000 m^2 brood stock pond.

Silver carp brood fish prefer small fertilized ponds with a good growth of phytoplankton. The fish are fed with soybean flour or rice bran. In a 1000 m^2 pond about 30−50 brood silver carp can be kept. The bighead needs water with a rich growth of zooplankton, which can be produced through regular manuring. The stocking rate in a 1000 m^2 pond is about 20−25 fish. The mud carp and black carp can be stocked at higher rates, even up to 100−150 in this size pond.

Spawning and larval rearing

The brood stock for spawning are selected largely on the basis of their size and the extent of maturation. Sometimes a sample of eggs is removed and examined to determine the stage of maturity. The maturity of the males is easily determined. If the milt oozes out on gentle pressing of the abdomen towards the genital pore, they are ready for spawning.

Actual propagation techniques differ very considerably between countries. Induced breeding by hormone injections is the most common technique, except for bundh breeding practised on a limited scale in India. Countries outside China, including East European countries and Israel, use indoor hatcheries, with stripping, fertilization and hatching, on lines very similar to those described above for common carp. In China itself, most farms now use special spawning ponds for spawning and hatching of eggs, as these are considered to be more efficient and cost-effective.

The inducing agents presently used are pituitary extracts of the common carp, silver carp or bighead and human chorionic gonadotropin (HCG), as well as luteinizing release hormone (LRH) or luteinizing release hormone-analogue (LRH-A). There is no species-specificity in the response to pituitary extract, but the common carp pituitary continues to be a widely used agent. Chinese fish culturists report that the hypophyses of silver carp and bighead are fatal to grass carp due to heterotypic protein. HCG is more effective on silver carp and bighead and its effect on grass carp and mud carp is rather weak. The grass carp, bighead and black carp respond well to LRH-A, but the silver carp appears to be less responsive to it.

In East European countries, pituitary extracts are generally administered in two doses. The first preparatory dose is about one-tenth of the total dose, i.e. about one dry gland (3–3.5 mg) for each female at the beginning of the spawning season. If, however, the water temperature is about 22–28°C, which is the optimum for maturation, half that dose will be enough as a preparatory dose, except in the case of the bighead, which needs one whole gland at any time during the propagation season. The most common practice in Israel is an initial injection of 0.2 pituitary per kg body weight, followed by a second injection of 0.5 pituitary per kg, 24 hours after the first. Eight hours later a third injection of 0.7 pituitary is given. When only two injections are given, 0.3 pituitary is administered first followed 24 hours later by 1 pituitary per kg female body weight. The interval between the preparatory injections varies according to the temperature conditions, and consequently the rate of maturity. During the beginning of the propagation season it may vary between 6 and 24 hours, but less later. In the case of mud carp, an interval of only 3 to 5 hours will be needed. However, the common dose is one gland or 3–3.5 mg pituitary in a 0.65 per cent salt solution per kg body weight of medium and small females. The males need only one injection of 0.6 pituitary. The injection is given, taking the necessary precautions and care as described for the common carp. It is usually given in the dorsal part of the body. In China, both intramuscular and coelomic injections are practised and both are considered to be equally effective. In India, the females are normally given two intramuscular injections of gland extracts at the rate of 10–14 mg per kg body weight, the first one being 3–4 mg per kg body weight. The second injection is given after a 6-hour interval. Males are given only one injection.

At temperatures between 20 and 22°C, the brood fish become ripe for stripping in 12–10 hours and at temperatures between 26 and 28°C, in 9–7 hours. Stripped eggs should be fertilized immediately with adequate quantities of milt, as in the case of common carp. For one litre of eggs, about 10 ml milt will be required. After mixing the eggs and milt for a minute or two, clean water is poured over the mixture, at the rate of about 100 ml water for one litre eggs. Stirring of the eggs is continued for 2 or 3 minutes. The eggs begin to swell on contact with water. More water is then added so that 3–5 cm of water is maintained above the eggs. About 30 minutes after fertilization, the eggs should be washed with water three or four times and then they are ready for incubation.

Any of the types of incubators used for common carp can be used for Chinese carp as well. Plastic funnel or sieve-cloth funnel incubators are more commonly used (see figs 6.43 and 6.44). Because fertilized Chinese carp eggs swell to 50–60 times their original volume, fewer eggs can be incubated in a funnel as compared to common carp. In a funnel of 8–10 ℓ capacity, about 2–3 ℓ eggs can be incubated. Since the eggs are more fragile and the egg shells much thinner, they are handled more carefully and the water circulation in the incubator is slower. In hatcheries in Eastern Europe, formaldehyde is added at the rate of 1 part in 20 000–30 000 parts hatchery water, to prevent bacterial and fungal infections. Woynarovich (1975) has devised a simpler and less risky method of treating developing fertilized eggs, with a solution of 5–10 g tannin in 10 ℓ clean water, two or three times. The first treatment is about 6–7 hours after fertilization and the last 6–8 hours before expected hatching. At temperatures of 26–28°C, hatching is completed in 24–28 hours, but if the temperature is lower (about 22–25°C) it will take 32–35 hours. The hatchlings swim upwards and eventually out of the jars or funnels into suitable tanks for further rearing.

The hatchlings can be reared indoors in larval tanks for about 10–11 days and then transferred to nursery ponds. If necessary facilities are not available for indoor rearing, the hatchlings can be reared in sieve-cloth tanks or hapas outdoors (fig. 15.18). After the yolk sac is absorbed, they must be fed with planktonic organisms like rotifers, ciliates, etc. Israeli farmers feed the larvae with trout starter feeds or soybean meal.

Specially designed spawning and hatching pools are now in common use for fry production of Chinese carps in China (see fig. 6.37). It has the advantage that large numbers of brood fish can be spawned, with minimum handling. Spawning, incubation and hatching can be carried out in separate pools or in interconnected pools without the need for transferring eggs. The operation of these facilities requires minimum skill on the part of the

Fig. 15.18 Hatchlings being reared in outdoor hapas in China.

workers. Generally, better hatching rates are obtained by this method, although visual monitoring of the state of fertilization and development of eggs is more difficult.

The spawning pools are generally located near the brood stock ponds and are usually circular cement tanks, (see fig. 6.37) 8−9 m in diameter and 1.2−1.5 m deep, with a capacity of 50−60 m^3. The bottom of the pool slopes towards the centre, where there is an outlet tube leading to the egg collection chamber (fig. 15.19). Sometimes the pools are elliptical in shape, with a maximum width of 5 m and a length of 15−18 m. Two inlets, one at the broader end and the other at the narrow end, are connected to the egg collecting chamber. The water inlets are placed on the walls of the tank in a tangential position so that the incoming water will create a circular flow. A flow of clean fresh water is maintained in the pool at a rate of about 200−400 ℓ/s. There is an egg collection chamber for each pool and this is to facilitate collection of eggs, estimation of quantity and transfer to the incubation pools. The outlet pipe of the spawning pool is fitted with a net to collect the eggs. The collection net made of bolting silk is long (260 cm) and conical with

a diameter of 45 cm at the rim and a collecting box (60 × 40 × 30 cm) at the cod-end. The floating eggs are carried by the water flowing from the spawning pool into the collecting box, from where they can be removed at regular intervals.

The incubation and hatching pools are also circular in shape, 3.5−4.0 m in diameter and 1 m deep, and have one or more incubation chambers (fig. 15.20). When there are a number of incubation chambers they are built concentrically, with cement walls separating them. It is important to maintain proper water circulation, for which several faucets are installed tangentially at the bottom of the pond. If the water flow is not adequate to create the velocity of 0.2−0.3 m/s, paddle wheels may be used to increase the circulation.

The brood fish are introduced in the spawning pond after hormonal treatment in the proportion of three males to two females. They spawn after the second injection of HCG or pituitary extract, generally at dawn. The effective dose of HCG is about 1100 IU/kg, but it is less effective in grass and mud carp. The optimum dose of LRH-A is 10−13 mg/kg for females and half of that for males. For silver and

Fig. 15.19 A spawning pool with the egg-collecting chamber at the centre (photograph: K.G. Rajbanshi).

Fig. 15.20 Incubation and hatching pools: note the concentric multiple chambers in the pools in the centre.

bighead carp the dose is 200 IU HCG/kg. After spawning, the brood fish are removed to the brood stock ponds. The floating eggs can either be collected and removed to a hatching pool or the spawning ponds can be connected to a hatching pool, to which the fertilized eggs will be carried with the water flow. The size of the incubation and hatching pools varies, but many of them have a diameter of about 2.9 m, depth of about 0.8 m and a capacity of 4.6 m³. Such pools can incubate about 700 000−800 000 eggs per m³, with an average hatching rate of 80 per cent. In some small farms in China, incubation is done in large clay jars (150 ℓ capacity, containing 150 000−200 000 eggs) with a vertical flow of water. Larger farms use a battery of jars, producing as many as 8 million larvae per year. Some farms use portable, funnel-type incubators. The larvae are removed to nursery ponds after the absorption of the yolk sac, which may take 4−5 days, depending on the temperature.

Indian carps

As indicated earlier, a considerable proportion of the fry and fingerlings of Indian carps used in India is raised from eggs and larvae collected from natural spawning grounds in rivers. Some are also bred in special 'bundh' type ponds as described in Section 15.3. Induced breeding of Indian carps by hypophysation is now increasingly used in India, as well as in other countries of Asia.

For successful hypophysation, an adequate stock of selected brood fish is reared with special care for a period of 4−5 months before spawning. The brood stock ponds are generally 0.2−0.5 ha in area, with a minimum depth of water of about 1.5 m. The selected brood fish are generally two to three years old and 2−4 kg in weight. They are stocked at the rate of 1000−2000 kg/ha. The ponds are periodically fertilized to produce enough planktonic organisms, taking care not to produce any algal blooms. Supplementary feed of groundnut oil cake and rice bran in equal quantities may be given at the rate of 1−3 per cent of body weight of the brood stock. It is advisable to change the water or add fresh water to hasten the maturation of brood fish. Generally, all the major carp species are kept together, but when enough ponds are available the species and sexes may be segregated. The males can be distinguished by the roughness of the pectoral fin, whereas the pectoral fins of the females are smooth.

Induced breeding is normally carried out at the onset of the monsoons, when there is an accumulation of rainwater in the ponds and the water temperature has somewhat decreased. The mature female has a bulging abdomen and swollen reddish vent. The extent of maturation can be determined by examining cathetered eggs from a female. Mature males eject milt when the belly is gently pressed near the vent.

Pituitaries of the common carp are considered the best for induced breeding of Indian carps, but in most cases the glands from the same species are used, largely because of easy availability. Pituitaries of many other species and HCG have been used for induced breeding, but there is as yet no consensus of opinion about the effectiveness of many of them. Pituitaries of several species of catfish and other species from fresh- and brackish-water sources and partially purified salmon pituitary gonadotropins have been tried, and a summary of the results are given by Jhingran (1982).

The dosage of pituitary gland extract to be administered depends on the stage of maturity of the fish and environmental conditions such as rain and water temperature. A priming dose of 1−2 mg per kg brood fish, followed by a resolving dose of 6−8 mg per kg 6 hours later, is suggested for rohu, catla and mrigal. Natural spawning in hapas or other containers can be expected 4−6 hours after the second injection, in all three species. The optimum temperature for spawning for all the species appears to be around 27°C, though they will spawn in a temperature range of 24−31°C, except catla, which prefers cooler water.

After the first injection, the brood fish are released in breeding hapas in the proportion of one male to two females. Breeding normally takes place within 3−6 hours after the second and final injection. Stripping is not required as the treated fish spawn and fertilize the eggs naturally. Though Indian major carps spawn only once in a year in their riverine habitats, during the monsoon, it has been possible to spawn rohu and mrigal twice during a season at intervals of 30−60 days.

The water-hardened eggs are hatched in hatching hapas (see end of Section 6.53, Hatchery equipment). About 50 000−100 000 eggs can be hatched in a hapa 2 × 1 × 1 m size (fig. 15.21). At an optimum water temperature of 26−31°C, the hatching time is 16−18 hours and the hatchlings fall into the outer hapa. When hatching occurs in hatchery jars, 50 000 eggs

Fig. 15.21 Water-hardened eggs being transferred to hatching hapas.

are kept in each jar of about 6.35 ℓ capacity, with a water flow of 600–800 ℓ/min. The hatching time in the jars is about 2–3 hours less than in hapas. Though the Indian major carps spawn in a wide range of water temperatures under hypophysation, percentages of fertilization and hatching rate are better when the temperature is about 30°C. After all the eggs hatch out, the inner hapa with the egg shells and debris can be removed. The larvae can be reared in the outer hapa for two to three days, after which they should be removed to nursery ponds for further rearing.

15.3.5 Nursing of carp fry

Nursery practices for all species of carp are based on the need to provide post-larvae and fry with the right type of food, environmental conditions and protection, for high survival rates and growth. The critical period is when the egg yolk is fully absorbed and the larvae start feeding. Zooplankton of the right size are the most efficient food at that time. By the use of nursery tanks or small ponds, greater control of water quality and predation by insects and their larvae, as well as predatory fishes, etc., can be achieved. One of the essential requirements in nursery farms is the eradication of the many pests and predators that infest fish ponds (see Chapter 10).

Although several types of nursery structures are used, including troughs and cement cisterns, the most commonly used in carp farms are specially constructed nursery tanks measuring over 100 m^2, and nursery ponds measuring up to 2000 m^2. In such structures the post-larvae or fry can be raised to the fingerling stage within a period of about one month.

Before transfer of the post-larvae or early fry for nursing, the tanks or ponds have to be properly prepared. Woynarovich and Horvath (1980) describe the use of soaked and decaying hay as a substrate under a layer of 5 cm of water in the tanks to stimulate the growth of rotifers and *Paramoecium*, which will serve as food for the fry. Stocking is done at the rate of 1000−2000 per m^2. From the third or fourth day after the release of fry, the tanks are regularly manured during early morning hours, at the rate of 1 kg of fresh manure for every 10−15 m^3 water surface. The water level is raised daily by 2−3 cm. If available, artificial starter feeds can be given from the seventh day onwards. If not, the manuring is continued to provide natural food. A crop of advanced fry can be raised in such tanks in 10−15 days. If they have to be grown to a fingerling stage, the density is reduced by about 25 per cent.

When earthen ponds are used, the preparatory treatment consists of draining the pond bottom and drying it. After refilling the ponds, a suitable insecticide is applied to eradicate aquatic insects, etc. To obtain a good standing crop of rotifers, application of chemical fertilizers (1 kg superphosphate, 1.5 kg ammonium nitrate and 1.5 kg carbomide per 100 m^2 pond surface) is recommended. When a dense growth of rotifers has appeared, the post-larvae or early fry are stocked in the pond. As *Cyclops* form a major enemy of the post-larvae, it is necessary to ensure their absence in the pond. Cladocerans like *Moina* and *Daphnia* may be introduced into the pond to multiply and be available for the mature fry to feed on.

In the rearing of Chinese carp fry to a size of 3 cm, monoculture is preferred. The fry ponds are generally 1000−2000 m^2 in area and 1.2 m deep. They are prepared by draining and application of quicklime at the rate of 750−1125 kg/ha, depending on the amount of silt at the bottom. Tea seed cake, derris powder or bleaching powder can also be used to eradicate pests and predators. The ponds are fertilized with green manure or organic manure supplemented with inorganic fertilizers. The optimal stocking rate is 1.5−2.25 million larvae, about 70 mm in length, per ha for a culture period of 15−20 days. The depth of water is raised from 50−70 cm at the beginning of the culture period by 10−15 cm every three to five days.

It is advisable to provide artificial fry feeds to obtain a rapid growth of the fry. A common fry feed is made of yeast (40 per cent), blood meal (25 per cent), fermented and pre-digested soya (20 per cent), fine quality fish meal (10 per cent) and soyoil (5 per cent), all finely ground and sieved through 100−150 μm mesh and fed at the rate of 0.5−1.0 kg food per 100 000 fry per day. After 10 days, the size of the feed particles can be increased to 400−500 μm. Green manuring is often adopted to increase the availability of natural food in the pond. In Israeli fish farms, the fry are fed with ground cereal grains (sorghum, wheat, etc.) and at a later stage, after they reach about 10 g in weight, with whole cereal grains.

The duration of fry rearing and the size to which they are grown before stocking in rearing ponds vary considerably. The most common practice appears to be to grow them in nurseries for about a month and then in grow-out ponds to market size. Some farmers transfer the fry to well-fertilized larger ponds (1−2 ha) and grow them to advanced fingerling size at lower population densities. Some farmers hold the fingerlings in holding ponds at high densities (a standing crop of 10 tons/ha or more) to become available for late stocking. In such cases the fingerlings become stunted, but when introduced in rearing ponds grow rapidly under favourable conditions.

15.4 Grow-out and polyculture

Of all the important carps used in aquaculture, it is probably only the common carp that is produced in monoculture, whereas both Chinese and Indian major carps are almost always grown in polyculture. Even common carp is now increasingly used in polyculture with some of the species of Chinese or Indian carps, especially with (i) grass carp and silver carp, (ii) catla and rohu or (iii) tilapia, the grey mullet and silver carp.

15.4.1 Stocking rates

In the grow-out of carp, either in monoculture or in polyculture, the basic objective is the production of an optimum quantity of the desired size of fish, at minimum cost. There are a number of interdependent factors that affect productivity and cost. Stocking rate or the density of fish in the pond, the quality and quantity of food produced by fertilization or artificial feeds supplied to the pond, water temperatures, availability of oxygen and build-up of metabolites in the pond, are all factors that influence growth rate and production. The size of fish at stocking, the duration of culture and the size at which the fish are harvested will also influence the total yield. The growth potential of the genetic strain used is another important factor to be considered. To this should be added the influence of natural productivity of fish food in the ponds, even when fertilization and feeding are adopted.

Many fish farmers adopt the system of multi-size stocking, which involves stocking fry, fingerlings and young adults belonging to different size-groups in the same pond, in order to utilize the food resources more efficiently. This practice involves periodic harvesting of the marketable fish and in some cases even additional stocking. There is also the practice of multi-stage stocking which consists of stocking fish in progressively larger ponds as they grow in size, reducing the stocking rates as required.

From the above it would be clear that the formula for determining the number of fish to be stocked, obtained by dividing the expected total production by the expected individual growth and adding it to the expected loss due to mortality, can only give a general indication. As many of the influencing factors are site-specific, it is necessary to decide on the stocking rate or population density in a pond according to the culture practices adopted, environmental conditions and market requirements.

Several stocking rates and grow-out practices are in use in different areas. In Europe, because of climatic conditions, it generally takes three years to grow the fish to the preferred market size of 1000–2000 g weight, except in southern Europe where marketable fish can be produced in two summers. This involves keeping fingerlings and yearlings in wintering ponds. In tropical and semi-tropical regions, fish of 600–1000 g or more can be grown in one year or less. In monoculture, under normal management, a stocking rate of 4000–5000 fingerlings, of 2.5–5 cm length, per ha or 2000–3000 fingerlings, of 5–10 cm length, per ha is recommended. With intensive feeding and aeration of ponds, higher rates of stocking can be adopted. If the fish are to be raised to market size in a shorter time, a lower stocking density of 3000–5000 per ha may be necessary.

15.4.2 Polyculture

Polyculture is now the most common practice of carp culture and several species combinations and stocking rates have been developed. These combinations are not always targeted to obtain the maximum biomass of fish from a unit area, but are often based on one or two species as the main crop for which there is the highest market demand, and the others as subsidiary compatible species that will utilize parts of the food resource that may be wasted otherwise.

Polyculture is adopted even in the rearing of fingerlings in Chinese carp culture. However, because of the overlap in feeding habits, when bighead is a major species, silver carp is excluded; similarly grass carp and black carp are seldom raised together. Some typical stocking rates are summarized in Table 15.1. The duration of fingerling production from 30–40 days old fry ranges from 120 to 270 days, extending from about early July to January, and in some cases up to March or April. Adult fish for the market are raised from these fingerlings in one to two years. Bighead, silver and mud carp reach market size in one year and grass and black carp in two years. The rate of growth and the size reached vary considerably between different parts of China, depending on climatic conditions.

Besides carp, a number of other species are stocked in polyculture, the more common ones being tilapia (*Tilapia mossambica*), Wuchang fish (*Megalobrama amblycephala*) crucian carp (*Carassius auratus*), red eye (*Squaliobarbus curriculus*) and white croaker or white amur bream (*Parabramis pekinensis*). Snakehead (*Ophicephalus (= Channa) argus*) and mandarin fish (*Siniperca chautsi*) are sometimes added to feed on weed fish and other unused

Table 15.1 Stocking rate in polyculture for fingerling production (after Shan-Jian, 1983).

Stocking rate (10^3/ha)						Size of fingerlings at harvest (cm)					
Grass carp	Black carp	Silver carp	Bighead	Common carp	Wuchang fish	Grass carp	Black carp	Silver carp	Bighead	Common carp	Wuchang fish
2–4	–	8–12	–	–	–	16–20	–	11–13	–	–	–
4–6	–	20–25	–	–	–	13–15	–	8–10	–	–	–
10–25	–	4–5	–	–	–	8–13	–	11–13	–	–	–
2–4	–	–	8–12	–	–	16–20	–	–	11–13	–	–
4–6	–	–	15–20	–	–	13–15	–	–	8–10	–	–
10–25	–	–	4–5	–	–	8–13	–	–	11–13	–	–
–	5–6	4–5	–	–	–	–	13–15	13	–	–	–
–	5–6	–	4–5	–	–	–	13–15	–	11–13	–	–
1	–	–	4	5–6	–	13	–	–	13	8–10	–
1	–	–	0.08–0.1	–	15–20	16–20	–	–	0.25–0.5 kg	–	7

aquatic organisms. Many farms adopt the system of multiple stocking and harvesting in rotation. Ponds are stocked with fingerlings in high densities and as they grow in size the bigger fish are harvested and additional fingerlings added. In this way, grass carp can be harvested up to six times a year and mud, black and common carp up to twice a year. As is to be expected, annual yield per ha varies between farms and regions within China, from 310 kg/ha in extensive farming to about 3000 kg/ha in semi-intensive systems and about 8700 kg/ha in intensive systems. Much higher production has been shown to be possible under experimental conditions.

Hepher and Pruginin (1981) reported on the results of polyculture with comparatively fewer species in a sub-tropical climate in Israel. Higher weights of the stocking material, intensive fertilization and supplementary feeding and temperature conditions contributed to the higher yields of up to 10.5 tons/ha.

In polyculture of Chinese carp in fresh-water ponds in Taiwan, tilapia and some of the coastal species such as milkfish and mullet are included. A small number of the carnivorous sea perch (*Lateolabrax japonicus*) are also added to control small weed-fish and young tilapia that may be produced by wild spawning in the ponds.

The traditional Indian major carp culture in India is an extensive production system with only limited fertilization. The different species are stocked in varying ratios, the most common being; Catla 30 per cent, rohu 60 per cent, mrigal 10 per cent. When calbasu is included, the percentage of rohu is reduced to 50 to provide for 10 per cent calbasu. By altering these ratios according to the primary production in the ponds, and more intensive stocking and supplemental feeding with locally available feedstuffs like oilcake and rice bran, higher rates of production have been obtained. In recent studies in polyculture or composite culture in India, the common carp and some of the Chinese carps (mainly silver and grass carp) have been added.

In later studies, this combination has been further enlarged by adding a small number of carp hybrids (calbasu male × catla female), grey mullet and a carnivore, chital (*Notopterus chitala*), to control weed-fish. With a high stocking density and greater use of fertilizers, higher rates of production of marketable fish have been achieved in culture periods of one year or less. It is believed that the maximum yield that can be obtained in polyculture under Indian conditions is around 7 tons/ha in 8 months of rearing and 10 tons/ha in one year (Tripathi, 1983).

Polyculture of common carp together with tilapia and other species like nilem (*Osteochilus hasselti*), tawes (*Puntius javanicus*), kissing gouramy (*Helostoma temmincki*) and gouramy (*Osphronemus goramy*) has been practised in Indonesia. In polyculture of common carp, generally 80 per cent would be common carp and the rest tilapia. Two other common combinations are with tilapia as the major species and kissing gouramy or tawes as the major species.

Rainbow trout (*Salmo gairdneri*) is sometimes grown in combination with common carp in countries like Poland and Czechoslovakia. One-year-old trout are stocked at the rate of 1200–1500/ha, forming about 15 per cent of the stock in the pond. The trout feed on the abundant weed-fish and carp fry produced by wild spawning. Under favourable water conditions (mainly oxygen levels and temperature) an additional production of about 40 kg/ha of trout is obtained, in addition to carp (Lavrovsky, 1968). Other polyculture systems which include carp are the stocking of the European catfish (*Silurus glanis*) in carp ponds to control weedfish, and of carp and the grey mullet in eel ponds to utilize the large amounts of feedstuffs that eels leave uneaten.

15.4.3 Pond fertilization and feeding

Pond culture of carp is in most cases based on fertilization and supplemental feeding. Organic and inorganic fertilizers and their use in fish ponds have been described in Chapter 7. In most countries, particularly in Asia, there is a distinct preference for the use of organic manures including green manure and compost in carp ponds. It has been demonstrated that manure increases zooplankton and chironomid production in carp ponds, probably as a result of the high production of bacteria and protozoa developing on the organic matter of the manure. The decomposing organic matter or detritus has a high protein content, possibly due to the growth of bacteria, protozoa and microalgae. The rate of application of manure has necessarily to be based on the environmental conditions and stocking densities. Besides the dosage, the mode of application is also important. Application of the entire quantity of manure required in one lot, or sporadically, sometimes results in the development of algal blooms, especially under tropical conditions. It also affects the oxygen concentration in the ponds. A dose of 100–120 kg dry matter per day can be used safely under most situations. The quantity of manure to be applied has to be increased with increasing standing crop of fish in the ponds. Based on East European practices, Woynarovich (1975) recommended a lower dosage of 20–35 kg per day per ha. The rate of organic manure application in India is 10 000–20 000 kg/ha (wet weight) per year. Often inorganic and organic fertilizers are used

alternately: inorganics at fortnightly intervals and organics at monthly intervals.

One of the major advantages in the use of chemical fertilizers is that since the nutrient contents are generally standard, the dosages required can be easily determined and followed with some confidence. In intensive methods of culture with higher densities of fish in the ponds (up to 3000 fish per ha), fertilization with a standard dose of 60 kg/ha single superphosphate and 60 kg/ha ammonium sulphate or liquid ammonia every two weeks is considered necessary in Israeli carp ponds when the temperature is over 18–20°C. In medium to highly productive polyculture carp ponds in India, where supplementary feeding is adopted, NPK fertilizers are applied in the ratio of 18:8:4, at the rate of 500 kg/ha per year. In less productive ponds, low in nitrogen and phosphorus, the fertilization dose is increased to 120 kg N and 90 kg P_2O_5 per hectare.

The feeding habits of the different species of carp have been described earlier. The nutritional needs of common carp have been studied in some detail (see Chapter 7). The major advantages of the common carp are that it can digest carbohydrates and accepts several feedstuffs such as cereals, legumes, oil cakes, slaughterhouse refuse, ground trash fish, etc. The most common feed used in Israel is sorghum, although broken or poor quality wheat and bitter lupine have also been used. Out of a wide variety of feedstuffs (see Chapter 7), rice bran is most popular in both nursery and rearing ponds for carp in Asian countries. Rye and barley are the main feedstuffs used in European carp farms.

Several countries now use processed feed mixtures or pellets in carp farms, but usually as supplementary feed along with fertilization. Hepher and Pruginin (1981) give the following two formulations of carp feeds used in Israel:

a diet containing 18 per cent protein:

80–90 per cent finely ground wheat
 5–10 per cent fish meal
 5–10 per cent soybean meal

a diet containing 25 per cent protein:

60–70 per cent wheat (partly replaced sometimes by other cereals)

15 per cent fish meal containing 65 per cent protein

15–25 per cent soybean meal

3–4 per cent soapstock oil.

The diets formulated on the basis of the nutritional needs of common carp appear suitable for Indian and Chinese carps as well, because the dietary protein requirements are very similar. The crude protein levels required for optimum growth at 30°C have been estimated as 450 g/kg for common carp, rohu, mrigal and grass carp and 410–430 g/kg for the grass carp (Sen *et al.*, 1978; Singh, 1983).

15.5   Diseases and mortality

A number of diseases that affect carps have been described in Chapter 9 and consideration was given to the role of the environment in the prevalence of diseases and mortality of stocks. The culture system and culture technologies also play an important role in the occurrence of diseases. Diseases are generally less rampant in extensive types of carp culture in comparison to intensive systems like cage rearing. Intensively stocked polyculture carp ponds in China are heavily loaded with organic manure and bacterial, fungal and viral diseases are comparatively common in these. Haemorrhagic septicaemia, gill-rot, enteritis, Myxosporidiasis, Ichthyophthiriasis, Dactylogyrosis, Trichodiniasis, Saprolegniasis, Synergasilosis and Lernaeosis are common in the carp ponds.

Enteritis in Chinese carp ponds is caused by *Aeromonas punctata* and can be recognized by erythema on the abdomen, decayed fin rays and swollen and reddish anal openings. The intestine becomes purplish in colour and the blood capillaries are inflamed. The intestine is filled with pus-like slime and ulcerated mucus membrane. Grass carp is the main species affected and mortalities of up to 70–90 per cent can occur, particularly during the period May/June. Poor feeds are believed to be responsible for weakening resistance to the pathogen. Disinfection with bleaching powder and provision of better feeds and feeding methods are recommended as preventive measures. As a cure, treatment with medicated feeds containing sulphaguanidine, at a dose of 10 g per 100 kg body weight of fish on the first day, and half that from the second to the sixth day, is reported to be effective.

Myxobolusis is one of the common Myxosporidian diseases of grass carp, particularly dangerous to fry when the ponds are heavily stocked and food availability is low. The spore of the parasite attaches to the intestine of the fish. The plasmodium that emerges from the spore sucks nutrition from the host and develops into spores on the intestinal wall. As a result, heavy mortality of fry can occur. The control measure is thinning the stock to enable better growth and resistance to infection.

Infestation by the tape worm *Bothriocephalus gowkongensis* is also reported to occur in young grass carp below 10 cm in length, but not in the adult. The diseased fry is milk-white in colour, thin and feeble, and suffers from pernicious anaemia. The intermediate host is a cyclops, in which the coracidium of the worm undergoes its development into procercoid. When the grass carp fry eats the infected cyclops, the procercoid develops into plerocercoid and adult stages. Mortality caused by the infection can be as high as 90 per cent during winter seasons. The only means of eradicating the parasite is by disinfecting the pond with quicklime at a concentration of 0.5 ppt.

In polyculture ponds in India Ichthyophthiriasis, fin and tail rot, Saprolegniasis and infestation by *Argulus, Lernaea* and *Ergasilus* are common. The major diseases in carp ponds in Israel are also similar. Carp ponds in Eastern and Western Europe appear to be more prone to disease problems. It is believed that this is related to the greater stress that the fish are exposed to during severe winters. The problem of infectious dropsy of carp has already been discussed in Chapter 9. Gill-rot or branchiomycosis, costiasis, ichthyophthiriasis and dactylogyrosis are the other more common diseases affecting carp in Europe.

One of the major causes of large-scale mortality in carp ponds, particularly in tropical and sub-tropical regions, is oxygen deficiency and resultant anoxia of fish. As carp culture is mainly done in highly manured ponds with dense stocking of fish, there is a greater chance of oxygen deficiency occurring. The rate of oxygen production through photosynthesis is very high during day time in highly fertilized ponds with a rich growth of phytoplankton.

During the night, photosynthesis and oxygen production cease and the only source of dissolved oxygen is the atmospheric oxygen dissolved in the water due to wind and water movements. The consumption of oxygen by the organisms in the pond continues throughout day and night and in highly productive fish ponds the balance may be greatly affected. Oxygen deficiency and anoxia can occur if the conditions are not properly monitored and the necessary action taken to rectify it. Though dense growths of algae produce oxygen, they also reduce light penetration in ponds and reduce oxygen production through photosynthesis in the lower layers. Intensive manuring, which increases the organic loading of pond waters, results in high rates of production of bacteria, protozoa, zooplankton, etc., thus adding to oxygen expenditure. High temperatures lead to a lower dissolved oxygen concentration in water and stimulate higher consumption of oxygen through increased respiration. Under cloudy atmospheric conditions, photosynthesis and dissolved oxygen production are reduced as a result of lower light irradiation and this often causes large-scale mortality due to anoxia under monsoon weather conditions in Asian countries.

Fish that feed on microscopic algae, like the silver carp, and those that feed on zooplankton, like the bighead, can help in controlling the occurrence of anoxia. The presence of detritus-feeding fish helps to reduce the accumulation of organic matter. Even when special care is taken to stock ponds with proper ratios of these species, unfavourable conditions can arise. In such circumstances, the two most common measures adopted are addition or exchange of water and aeration. The addition of fresh water, if possible by cascading or spraying, increases the dissolved oxygen content of the ponds. Where possible, it is better to drain some of the pond water and replace it with fresh, well-oxygenated water from an outside source, or even pump back the same water in a spray so as to increase the dissolved oxygen content to about 3 mg/ℓ. Aeration is a more efficient means of controlling the depletion of oxygen.

Different types of aerators are now in use in fish ponds (figs 15.22 and 15.23). Kepenyes and Varadi (1984) divided pond aeration into two types: hydraulic and air diffusion. A water pump is the basic device for hydraulic aeration, whereby the energy content of the water is increased. The air intake device can be a cascade, a sprinkler, an ejector or an air-intake head. Surface aeration can be provided by a simple open impeller or a centrifugal pump. The basic unit of an air diffusion aeration device is the air compressor, which can be a root-type blower, a ventilator, a compressor or a membrane pump. The air penetrates the water through a porous material or perforated tubes installed in the pond. When the air is diffused through a perforated pipe, large bubbles of up to 10 mm in diameter are formed. If the diffusor is a porous material, fine bubbles of 2−5 mm diameter can be produced. The smaller the bubbles, the larger is the total surface for a given volume of air and the lower their buoyancy, which improves oxygen transfer. As the bubbles pass through water, part of their oxygen content is dissolved in the water. An upward water movement also results, which creates a mixing effect.

When a sprinkler is used, the water jet comes from the nozzle with high velocity and falls into the water in drops. The turbulence caused when the water drops hit the water surface increases the diffusion of air into the water. The water jet can be directed down into the water surface, forcing air bubbles into the water. When aeration is performed with an ejector, the water is passed through a venturi-type diffusor where its pressure is reduced below the atmospheric pressure, thus allowing air to penetrate into the water. Special air-intake heads consist of a propeller inside a pipe through which water passes. As the propeller is driven by the water current, the pressure decreases on the surface of the blades and the air is sucked in. Cascades can be used for aeration and break the water into small drops, increasing the interfacial area. Surface aerators are placed directly in the fish pond and, by rotation of a paddle-wheel device with a vertical or horizontal axis, water is discharged into the air. These aerators generate a vertical or horizontal water current in the fish pond.

In carp ponds in China, aerators have become standard equipment. Frequent operation, often once a day for an hour or two, has been found to increase yields by as much as 14−28 per

Fig. 15.22 A type of aerator commonly used in Chinese ponds.

Fig. 15.23 Aerating carp ponds in Israel through a system of perforated pipes.

cent, as a result of better circulation of nutrients and consequent plankton growth. At times of oxygen deficiency, aerators are operated for two or three hours in the morning or around noon, particularly on overcast days.

15.6 Harvesting and marketing

The harvesting size for carp varies considerably, depending on consumer preference, climatic conditions and culture techniques. In the majority of countries where carp culture is practised, consumer preference is for a fish of at least 0.5 and preferably 1–1.5 kg weight, except in some Southeast Asian countries where smaller-sized fish are acceptable. In India, in the new improved systems of polyculture the required marketable size is reached in one year. In Israel, the marketable size of 0.5–0.6 kg is achieved in about four to six months. In Europe, it usually takes two to three years to grow them to the preferred weight of 1–1.5 kg.

The most common means of harvesting carp is by seining, as described in Chapter 11. The problems of seining common carp and silver carp have also been mentioned. In ponds with proper harvesting sumps, it is always preferable to harvest by draining, especially when the fish are to be sold in live condition. In most areas, the catches are sorted out before marketing. As described in Chapter 11, wherever possible, carp are sold in live condition or fresh on ice. In many countries, live fish transport trucks or boats are used for long- and short-distance transport of carp.

15.1.7 Economics of carp culture

Even though comparable recent data on commercial culture are not readily available, the economic viability of carp culture has never been in doubt, in areas where there is a market for carp and appropriate technologies are used. The very fact that the food requirements for carp can be reached fully or partly by pond fertilization, and that inexpensive feedstuffs can be used as supplemental feeds, helps to bring down considerably one of the major costs of production. The cost of producing fry or fingerlings, another major production cost, can be kept at a desired level by the adoption of less expensive techniques. The established systems of polyculture also contributes to the profitability of carp culture. Though capital investments for farm construction and equipment can be high, with proper maintenance ponds can be used almost indefinitely. So it would appear that consumer acceptance and price levels in the market place, are two of the major factors that determine economic viability in many situations.

References

Alikunhi K.H., Sukumaran K.K. and Parameswaran S. (1971) Studies on composite fish culture: Production by compatible combinations of Indian and Chinese carps. *J. Ind. Fish. Assoc.*, **1**(1), 26–57.

Bakos J. (1979) Crossbreeding Hungarian races of common carp to develop more productive hybrids. In *Advances in Aquaculture*, (Ed. by T.V.R. Pillay and W.A. Dill), pp. 633–5. Fishing News Books, Oxford.

Chakrabarty R.D. *et al.* (1979) Intensive culture of Indian major carps. In *Advances in Aquaculture*, (Ed. by T.V.R. Pillay and W.A. Dill), pp. 153–7. Fishing News Books, Oxford.

Chaudhuri H. *et al.* (1975) A new high in fish production in India with record yields by composite fish culture in freshwater ponds. *Aquaculture*, **6**, 343–55.

Hepher B. and Pruginin Y. (1981) Commercial fish farming. John Wiley & Sons, New York.

Jhingran V.G. (1982) Fish and fisheries of India. Hindustan Publishing Corporation (India), Delhi.

Jhingran V.G. and Pullin R.S.V. (1985) *A Hatchery Manual for the Common, Chinese and Indian Major Carps*. Asian Development Bank, Manila and ICLARM, Manila.

Kepenyes J. and Varadi L. (1984) Aeration and oxygenation in aquaculture. In *Inland Aquaculture Engineering*, pp. 473–507. ADCP/REP/84/21 FAO of the UN, Rome.

Kirpichnikov V.S. *et al.* (1979) Selection of common carp (*Cyprinus carpio*) for resistance to dropsy. In *Advances in Aquaculture*, (Ed. by T.V.R. Pillay and W.A. Dill), pp. 628–33. Fishing News Books, Oxford.

Lavrovsky V.V. (1968) Raising of rainbow trout (*Salmo gairdneri*) together with carp (*Cyprinus carpio*) and other fishes. In *Proceedings of the World Symposium on Warm-water Pond Fish Culture*, (Ed. by T.V.R. Pillay). *FAO Fish. Rep.*, **44**(5), 213–17.

Murty D.S. *et al.* (1978) Studies on increased fish production in composite fish culture through nitrogenous fertilization with and without supplementary feeding. *J. Inl. Fish. Soc. India*, **10**, 39–45.

Pearl River Fisheries Research Institute (1980) *Pond Fish Culture in China*. Pearl River Research Institute, China National Bureau of Aquatic Products, Guangzhou.

Rabanal H.R. (1968) Stock manipulation and other biological methods of increasing production of fish through pond fish culture in Asia and the Far East. In *Proceedings of the World Symposium on Warm-water Pond Fish Culture*, (Ed. by T.V.R. Pillay). *FAO Fish. Rep.*, **44**(4), 274–88.

Ranadhir M. (1983) Economic analysis of composite fish culture. In *Lectures on Composite Fish Culture and its Extension in India*. NACA/TR/83/7.

Sen P.R. *et al.* (1978) Observations on the protein and carbohydrate requirements of carps. *Aquaculture*, **13**, 245–55.

Shan-Jian *et al.* (1983) *Lectures on Introduction to Freshwater Fish Farming in China*, (Ed. by T.E. Chua). Network of Aquaculture Centres in Asia, NACA/TR/83/9.

Singh B.N. (1983) Nutritional requirements of carps. In *Lectures on Composite Fish Culture and its Extension in India*. NACA/TR/83/7.

Sinha V.R.P. (1979) Breeding of silver carp *Hypophthalmichthys molitrix* (C. and V.) and grass carp *Ctenopharyngodon idella* (Val.) in a bundh type tank in West Bengal. *Curr. Sci.*, **44**(7), 230–31.

Sinha V.R.P., Jhingran V.G. and Ganapati S.V. (1979) A review on the spawning of Indian major carps. *Arch. Hydrobiol.*, **73**(4), 518–36.

Stanley J.G. and Thomas A.E. (1978) Absence of sex reversal in unisex grass carp fed methyltestosterone. In *Symposium on Culture of Exotic Fishes*, pp. 194–9. Fish Culture Section, American Fisheries Society, Auburn.

Sukumaran K.K. (1983) Induced breeding of carp and its constraints. In *Lectures on Composite Fish Culture and its Extension in India*. NACA/TR/83/7.

Tripathi S.D. (1983) Recent advances in composite fish culture. In *Lectures on Composite Fish Culture and its Extension in India*. NACA/TR/83/7.

Wohlfarth G., Lahman M. and Moav R. (1962) Genetic improvement of carp IV. Leather and line carp in fishponds of Israel. *Bamidgeh*, **15**(1), 3–8.

Woynarovich E. (1975) *Elementary guide to fish culture in Nepal*. FAO, Rome.

Woynarovich E. and Horvath L. (1980) The artificial propagation of warm-water fin fishes – a manual for extension. *FAO Fish. Tech. Paper*, **201**.

16
Trouts and Salmons

Culture of trouts and salmons (Salmonidae) originated much later than culture of carp, but greater scientific effort has been concentrated on these groups. A considerable part of basic information relevant to fish culture (see Chapters 7 to 9) has been derived from laboratory and field investigations, especially in the case of trout. Salmonid culture has a relatively long history in Europe and North America. The main interest of early salmonid culturists was hatchery production of young ones for introduction in new areas or to enhance existing populations in the native habitats of the species, mainly to improve or maintain sport fisheries. It is only in the last few decades that the feasibility of applying the available technical know-how to commercial production of this high-valued group of fish has been fully appreciated. The general concept of salmonids as highly expensive fish to raise in farm and as a luxury food beyond the reach of the common man has been brought into question by the rapid expansion of trout farming in Europe. Here production has increased steeply and prices have come down to a level comparable to the less expensive species without affecting the producer's profit to any appreciable extent. The recent upsurge in the production and profitability of the Atlantic salmon and trout in sea-water cages in Norway and Scotland and its distribution to far-off markets around the world have served to demonstrate the potential of salmonid culture. There are indications that salmon might also become a less expensive and fairly common product in many areas, if the present rate of expansion of salmon culture continues.

A major constraint to expansion of salmonid culture is the availability of adequate quantities of water of the required quality. Water quantity requirements depend on temperature conditions and the type and intensity of culture. It has been suggested that a fresh-water rainbow trout farm using surface water in a temperate climate should have available a supply of about 5 ℓ/s for every ton of fish produced, although a lower level may be sufficient when temperatures decrease. Because of the need to have clean water of the appropriate temperature, water from springs, bore-wells or clean flowing streams has been used for culture. Spring water is considered essential for a trout hatchery and is recommended for use in rearing up to swim-up stage. However, such water sources are limited and the water available may not be adequate. The quantity of water naturally limits the number of ponds or other culture facilities, even when methods of aeration are adopted. Recirculation of water could improve water availability but, as pointed out in Chapter 6, the high cost involved restricts its wider use. It is to overcome this limitation that salmonid culturists have turned increasingly to cage and pen farming in fresh or sea water.

Among water quality requirements, the most important are temperature and oxygen concentration. Temperatures around 10–18°C are considered optimal for the growth of rainbow trout and are not allowed to exceed 21°C. Slightly lower temperatures are preferred for other salmonids. Water for salmonid hatcheries usually has 100 per cent oxygen saturation. A pH of 7–8 is preferred, and must be maintained when surface water is used during periods of rain. Spring water may sometimes have a high dissolved iron content and in a hatchery it can precipitate as a result of bacterial action and settle on eggs or the gills of fry. Such water should be avoided or treated to remove the iron before use. Water from bore-wells or cooling water from power-stations (that is often used to supplement water supplies and to raise water temperature for better growth in cold climates) may be supersaturated with nitrogen. Supersaturation of around 107 per cent can cause gas bubble disease and mortality of fish. Nitrogen absorbed into the blood at a super-saturated level begins to fall to the normal saturation level and during this process the gaseous nitrogen comes out of solution in the blood vessels, causing gas bubbles to form and eventual mortality of the fish. So, such water has to be 'degassed' by exposure to air by suitable means, such as allowing it to fall through a stack of perforated aluminium plates. The concentration of CO_2 has to be maintained below 10 mg/ℓ.

16.1   Trouts

The trout species of the greatest importance in aquaculture is undoubtedly the rainbow trout (*Salmo gairdneri*) (fig. 16.1). Native to the Pacific Coast drainages of North America, the rainbow trout has been introduced since 1874 to waters on all continents except Antartica. Its range extends into low latitudes, at higher el-evations. Trout waters are maintained in the upland areas of many tropical and sub-tropical countries of Asia, East Africa and South America, and commercial trout farming has developed in Central and South America and to a limited extent in some Asian and African countries like India and Kenya. As can be expected in a widely distributed and adaptable species like the rainbow trout, several local forms have developed, some of them described as distinct separate species or subspecies (see MacCrimmon, 1971, for a list of some of them). Several strains have also been developed through mass selection and cross-breeding for improved cultural qualities.

The brown trout (*S. trutta*) is the indigenous trout of Central and Western Europe, and was the first trout to be artificially propagated. It has also been introduced into several countries around the world for developing sport fisheries, by stocking natural water bodies. Because of their slower growth rate, poor utilization of artificial feeds and stringent water quality requirements, commercial cultivation of this species has not developed to any appreciable extent, as compared to the rainbow trout.

Another trout which has received some at-tention is the brook trout (*Salvelinus fonti-nalis*), native to north-eastern North America. The species was introduced into Europe to-wards the end of the 18th century, and to a number of other areas where the water tem-perature is between 12 and 14°C (not higher than 18°C). Because of the rather demanding environmental requirements and susceptibility to infectious diseases and water pollution, its culture is not very common now in spite of its fast growth rate under favourable con-ditions and the high level of acceptability by consumers.

As indicated earlier, it is the rainbow trout that has now become the mainstay of large-scale salmonid aquaculture on a world-wide basis. The two main varieties of importance are the sea-going form known as the steelhead and a land-locked fresh-water form. The steelhead grows very rapidly in salt water, reaching 6−9 kg in the sea in about three years. The fresh-water form, which is the one usually used in commercial aquaculture today, attains a weight of 4.5 kg or more under favourable conditions. Sedgwick (1985) reported that this fresh-water form attains weights of up to 9 kg in the South American Andes, indicating the potential of this species under suitable con-ditions. Even under normal conditions, the rainbow grows faster than all other trouts and is more adaptable to higher temperatures than the others, which enables its cultivation under a wider range of climatic conditions and the utilization of higher water temperatures for rapid growth rates.

The optimum water temperature in a rainbow trout farm is below 21°C, although the lethal limit is in the region of 25−27°C, in which the animal may survive for short periods but may not grow and be active. Fish culturists in Europe prefer to maintain temperatures ranging over 10−15°C for as long as possible in their farms. While higher temperatures would assist higher levels of metabolism and growth, it has to be remembered that the dissolved oxygen content decreases as water temperature increases. This situation can, however, be overcome by re-ducing the stock density or by using special measures to increase the oxygen concentration

Fig. 16.1 The rainbow trout, *Salmo gairdneri.*

of the water. The minimum concentration needed for both rainbow and brown trout is 6 ppm, but it is not allowed to fall to that level and is maintained in a fully saturated state.

The potential for expansion of trout farming, utilizing the advances of culture technology and product development, is much greater than is generally appreciated. In the vast areas of the South American Andes, where there are no other comparable cold-water species available for aquaculture, there appears to be considerable possibilities for developing large- as well as small-scale trout culture. The experience in trout farm development in the upland areas of Mexico clearly indicates its potential in creating rural self-employment and commercial production, in areas with adequate supplies of good quality water. In countries where commercial trout farming is already well developed, as in Europe, increases in harvesting size from portion size (170–230 g) to 350–450 g for the fresh fish market and to 1.5–3 kg for fillets and smoked trout are expected to stimulate market demand. This can be expected to lead to more intensive production in existing farms and wider utilization of new technologies such as cage farming.

16.1.1 Culture systems

As mentioned earlier, the original system of trout culture consisted of hatchery propagation and rearing of young for stocking streams, lakes and other water bodies. This system will be discussed further in Chapter 30. The techniques of brood stock development, stripping, fertilization, incubation of eggs and larval rearing are very similar for rainbow and brown trouts. Although the natural methods of spawning, fertilization and incubation of eggs in redds (depressions or nests) on the stream bed are used for some salmons, they are hardly used in the propagation of trouts.

Commercial trout farming in ponds originated in Denmark some 30 years ago. The farms consist of earth ponds excavated on more or less level ground, with water from a nearby river or other suitable water source, ideally supplied by gravity. When energy is not very expensive, water may be supplied by pumping. Figure 16.2 shows the typical layout of a Danish trout farm. The general principles followed are the same as described for pond farms in Chapter 6. The ponds are rectangular in shape, usually about 30 m × 10 m, with the bottom sloping

Fig. 16.2 A typical Danish pond farm in Brons, Denmark.

towards the outlet and a depth of approximately 1 m at the upper end and 1.7 m at the lower end. The capacity of such a pond is about 1500 kg rainbow trout. Land-based pond farms, supplied with pumped sea water and tidal enclosures are also used on a limited scale for trout culture.

Trout (rainbow, brook and brown trout) are grown in farm and ranch ponds in many parts of the USA. The ponds are meant mainly for recreational fishing and to provide an additional crop of highly relished food to the farmer and his workers. The farmer purchases fry or fingerlings from hatcheries in the spring and harvests them before the winter freeze-up. If the farmer is able to over-winter the stock, he grows at least a part of his stock for a second year, when he can expect reasonably large fish for angling. The best results are obtained with rainbow trout.

Probably the most widely used system of trout culture is in raceways, originally used in North America, for rearing trout for restocking sport waters. As described in Chapter 6, raceways consist of long continuous channels or a series of channels divided by cross-walls. They are generally constructed of earth, brick or cement concrete, sunk in the ground or built above ground level (fig. 16.3). The channels are narrow, with a width of 2−4 m and a depth of less than 1 m. The most important requirement of raceway culture is a plentiful supply of clean water, flowing through the channels. Most raceway farms depend on spring water with a constant temperature and flow velocity (fig. 16.4). Stocking density depends on water temperature, but is around 4−5 kg/m^3 when there is a water exchange of 2.5 ℓ/min per m^3.

Another system in use, especially in Europe including the UK, is tank culture. Typical trout tanks are 4−10 m in diameter and 1.6 m deep (see fig. 6.17). Tanks are generally sunk in the ground, leaving about 30 cm above ground. The drainpipes are laid in trenches and connected to prefabricated outlet sumps. Many new trout tank farms have a central fish grading arrangement. Each tank has a separate outlet pipe connected to a separate main, leading to a sump where the fish can be graded. The fish, along with the water, are pumped out through the grading sump.

Cage farming of trout is a rapidly expanding

Fig. 16.3 A concrete raceway farm for trout in South Korea.

system and is increasingly being used in both fresh- and salt-water environments. Though presently restricted in its use to sheltered areas of the sea and inland impoundments (fig. 16.5 see also figs. 6.25 and 6.26), considerable development is taking place in the design and construction of cage farm units which can be operated in more exposed areas and can withstand rough weather conditions. Thus unforeseen opportunities for extending cage-culture systems of trout and other salmonids are opening up. As mentioned in Chapter 6, there are several designs of cages presently in use.

A trout culturist may produce only eyed ova for shipment and sale to producers elsewhere, or may hatch them and rear the hatchlings to fry or fingerling stages for restocking. Many rear them to market size for sale to consumers either fresh or after processing.

A system of trout production somewhat similar to restocking is what is often called put-and-take fishing, or put-and-take stocking for the benefit of sport fishermen. This has been in practice in North America for a long time, and

is becoming common in Western Europe. Trout are raised to market size and then stocked in sport waters, where anglers are allowed to fish for a fee. The producer can combine production for the market with sale for stocking, and can also combine trout farming with a put-and take fishery of his own. The main difference from normal restocking is the need to grow the trout to an adult size preferred by anglers. Rainbow trout are considered to be an excellent fish to use, as they grow rapidly and have high sporting potential.

Monoculture is the most common practice in trout culture, and intensive systems are considered necessary in most situations to make the operation economically attractive. However, in areas where climatic conditions are conducive, double-cropping systems have been introduced (Brown *et al.*, 1974; Brown, 1979). The channel catfish (*Ictalurus punctatus*) is grown in ponds or raceways for seven months from April to October, and the trout from middle of November to March. The ponds are thus under production throughout the year,

Fig. 16.4 An earthen raceway farm for trout in northern Greece.

instead of only part of the year as necessitated by temperature conditions. This is reported to bring down the cost of production and to increase return on investments. The practice of stocking trout in carp ponds to feed on the hatchlings and carp fry produced by wild spawning in some East European countries is a form of polyculture. They are stocked at the rate of 10–15 per cent of the carp stock and harvested during the second year, producing about 20–50 kg/ha of marketable trout.

16.1.2 Development of brood stock

Although brood stock for propagation can be taken from natural open waters or from culture installations, most of the rainbow trout brood stock is obtained from farms. In Europe, brown trout brood fish are caught from open waters when they start migrating upstream for spawning. Many trout farms have developed their own strains of trout with specific characteristics for rapid growth, better food conversion, early or late maturity, larger egg size, etc. Scientifically planned selective breeding of rainbow trout has been carried out in a number of

areas, as a result of which strains have been developed through several generations, with high percentages of spawning at an age of two years, and increased egg and fingerling production. Fast growing strains with large eggs are now available for culture. The progeny of crossbred salmonids are often sterile, but in some cases the males or females only are infertile. Fully fertile cross-breds are usually obtained when crossed with closely related species. With the development of trout farming in sea water, interest in late-maturing, fast growing strains or sterile hybrids has increased, and considerable research is now under way in this direction. A recent development is the use of sex-reversed all-female brood stock, in order to produce all-female progenies that grow faster. Functional males are produced by oral administration of the male hormone 17-methyl testosterone through starter feeds containing 3 mg per kg of feed, at the fry stage. Higher levels of hormone may cause sterility, and lower levels may result in low percentage masculanisation.

A farmer has the option to carry out his own selective breeding or depend on specialized

Fig. 16.5 A cage farm for trout in an irrigation reservoir in northern Portugal.

breeding centres to obtain eyed ova of the desired strains for incubation and rearing. Eggs of early or late spawning fish can be obtained, in order to spread the hatching time, grow-out period and attainment of market size over as long a period as possible. Trout hatcheries in the southern hemisphere, where climatic patterns are reversed and consequently also the breeding seasons, are a major source of egg supply to hatcheries in the northern hemisphere during the off-season. When eggs are purchased, strict regulations pertaining to infectious and communicable diseases are observed, to prevent introduction of diseases.

If the fish culturist decides to carry out propagation in his own farm, a suitable stock of brood fish is reared in special brood ponds. A density of about 8000/ha is recommended in small ponds with a current of water. It is considered best to feed them on natural food, but if artificial feeds have to be given, the quantity of feed is gradually reduced before spawning and the fish transferred to holding tanks before propagation.

Although two-year-old trout start spawning, females are seldom used for propagation before they are three or four years old. Males of two to four years are considered to be the best for breeding. The quantity of eggs or milt increases with increase in size of the brood fish. Larger females have larger eggs and hatch into larger alevins (hatchlings, sac-fry).

The number of brood stock required naturally depends on the number of fry or fingerlings needed. The number can be calculated by extrapolation backwards, based on the expected survival rates of alevins to fry or fingerling stage, fertilization and hatching rates and the fecundity of the parents available. There are differences of opinion about the proportion of males to females required, but one male to three females is generally considered satisfactory. The males and females are usually held in separate ponds or tanks. The state of maturity is examined at regular intervals, so that as soon as a fish is ripe it can be removed for stripping.

In a commercial fish farm it is not easy to prevent inbreeding while building up a brood stock. As will be described later, when the eggs are fertilized by the dry method the largest number are fertilized with the milt from the first two or three males (in descending order, the first one fertilizing the largest number) and the later ones have lesser chance of finding any unfertilized eggs. To avoid this and to reduce inbreeding, it has been suggested that a mixture of milt from a number of fish should be used for fertilization. Milt from a number of male fish is collected, avoiding any admixture of water, mixed together, and kept in a cool dark place. The mixture is viable for several hours. The required amounts are removed by pipette for fertilization after the eggs have been stripped from the females.

Another means of reducing inbreeding is by using cryopreserved sperm of fish from selected sources. If such preserved sperm is available from dependable sources, the need to maintain male brood stock will be minimized. There is also the possibility of cryopreservation being carried out in the farm itself and of using a mixture of cryopreserved milt for fertilization.

16.1.3 Techniques of propagation

Stripping and fertilization

The methods of propagation of different species of trout are very similar; in fact the basic procedures are the same for almost all finfishes. Brood fish are removed from brood or holding ponds/tanks for propagation as soon as they are ripe. The dry method of fertilization without admixture of water is most commonly followed. The female brood fish are taken out and wiped gently with a towel to remove all adhering water, after which the eggs are stripped and collected in a dry bowl. Depending on the size of the fish and consequently the quantity of eggs, two to four females may be stripped at a time. Afterwards, the males are stripped and the milt poured over the eggs evenly so as to cover as many eggs as possible. Then they are mixed well with a feather or spoon and allowed to stand for a minute or two. Then water is poured in slowly to fill the bowl and allowed to stand for about 10 minutes. After successive changes of water to remove surplus milt, the eggs are transferred to a bigger container, such as a bucket about two-thirds full of water.

For mono-sex culture of females, all-female eggs are produced by fertilizing normal female eggs (XX chromosomes) with milt from sex-reversed masculanized females (XX chromosomes). The mature testes of sex-reversed fish are large and rounded with several lobes, carrying as much milt as a normal male. The testes are removed by cutting open the abdomen, placed in a dry container cooled on ice, and carved criss-cross for the fluid milt to drain out into containers. An equal volume of a standard extension fluid is added. The spermatazoa then become motile, and can be used for fertilizing normal ova. One advantage of the technique is that only the brood stocks are sex-reversed, and they can be grown separately, while the marketed fish are not exposed to any hormonal treatment.

Incubation and rearing of hatchlings

Different types of incubators are used for hatching salmonid eggs, but according to many trout culturists the best is the system of troughs and California 'baskets' (fig. 16.6). These have the advantage that the eggs and alevins can be easily observed and the whole process of incubation can be properly monitored. The troughs are usually 40−50 cm wide and 20 cm deep. The length varies, but a convenient length is about 3−4 m. Rectangular baskets are placed in these troughs, with their top edges sitting on the sides of the trough. The perforated bottom of the basket rests about 3 cm above the bottom of the trough. The perforations are such that they retain the spherical eggs but allow the hatchlings or alevins to fall through to the water below. The baskets are arranged in such a way that water will be forced through the mesh to aerate the eggs. The eggs are spread one to two layers thick. Water flows from the top end of the trough, under the basket, and passes up through the perforated base through the eggs and passes to the next basket, until it reaches the end of the trough. An average-sized trough would need a continuous water flow of 5000 ℓ/day for every 10 000 eggs.

When large numbers of eggs have to be incubated and when there are space limitations in the hatchery, it may be necessary to use battery incubators (see Chapter 6). These consist of vertically arranged series of trays held

Fig. 16.6 A trough and basket incubator (From Stevenson, 1980).

in guides, one below the other. The inner part of the tray, which is similar to the egg basket described above, though generally larger in size, has a perforated base and the eggs are spread on it. Water flows into the outer part of the top tray, under its perforated base and up through the eggs. It passes through the edge of the inner tray or an outlet arrangement to a pipe and into the side of the next tray. The water flow required depends on the size of the system and temperature conditions. Provision is made for a continuous flow of 1 ℓ/min per 1500 eggs, although only half of it may be necessary under normal conditions. Dead eggs are removed regularly from the incubators. Some culturists treat the eggs with malachite green twice weekly (3.75 g dissolved in 3 ℓ water), but others consider it unnecessary if the hatchery has a good quality water supply and is maintained in sanitary conditions. Chemical baths are believed to affect hatching rates and the condition of the hatchlings.

The time taken for hatching varies mainly with the temperature of the water and ranges roughly between 100 days at 3.9°C and 21 days at 14.4°C. From the 'eyed' stage (when the eyes can be seen through the egg shell) until hatching, the eggs are quite tough and can withstand handling and transport. Rainbow trout eggs take about 370 day-degrees (number of days after fertilization multiplied by water temperature (°C) over the period) for hatching. If heated water can be used, the hatching can be speeded up; for example, eggs fertilized in January can be hatched at the end of February or earlier. The hatchlings remain in the hatchery baskets until they reach the swim-up stage and all the yolk has been absorbed. It is advisable to remove the egg shells from the basket with a suction device.

Fry rearing

The swim-up fry can be reared in the hatchery tray itself for some days, after removal of the egg baskets. In the case of battery incubators, it is essential to transfer them to a rearing tank. Indoor concrete or fibreglass tanks are considered most suitable for fry rearing. It is easier to maintain a regular current of water in such tanks and this helps maintain a uniform distribution of the fry. Circular tanks, where fry distribute themselves freely, can carry larger numbers of fry than traditional long troughs. Both circular and square tanks with rounded corners are in common use for trout and salmon fry rearing. Usually they are 2 m in diameter, or 2 m × 2 m square, with depths of 50−60 cm. An elbow pipe delivers the water on the side of the tank, below the water surface, in such a way as to create a water circulation. The drain is in the centre of the tank, with a flat screen leading into a sump below the tank or a vertical cylindrical screen round a central drain pipe. The sump or drainpipe is connected to an elbow pipe on the side under the tank, which can be used to regulate the water level in the tank.

Instead of circular or square tanks, traditional rectangular tanks made of concrete or fibreglass are also in use. Generally they are 3–4 m long, 70–80 cm wide and 50–60 cm deep. The water flows in at one end and out at the other, through fine-mesh screens. Italian trout farms use a device known as an 'embrionatori' for incubating large numbers of eggs. It has a capacity of 40–50 ℓ and is supplied with a water flow of 6 ℓ/min. It can incubate half a million trout eggs, and hatching is completed in about 3 weeks at temperatures of 13–14°C.

In modern fish hatcheries, specially prepared starter feeds are given to the fry, using automatic feed dispensers. The quantity of feed and the frequency of feeding are programmed according to the size of the fry and temperature. When they start feeding for the first time, it is customary to feed five or six times a day, and as they grow older feeding is reduced to twice a day. In areas where prepared dry feeds for fry are not available, the farmer may have to fall back on conventional feeds consisting of meat, liver and fish finely ground to a size that can be swallowed by the fry. They are forced through the finest plate of a grinder several times, after all fat, skin and connective tissues have been removed. Some farmers use 1–2 per cent salt to improve the binding quality of the mixture. Unsalted hog liver with up to 50 per cent dry meal makes a suitable mixture. Mechanical mixers are used to mix the feed ingredients, to achieve the right consistency. The mixed feed is fed to the fish as soon as possible after it is prepared, and distributed over a wide area so that every fry will have the opportunity to feed.

Besides careful feeding, successful rearing involves constant care of the tanks and the fry. Though the period of fry rearing depends on culture requirements, it is normally 10–12 weeks under controlled conditions, so that the fry become immune to infections like whirling disease or myxosomiasis. At the end of the first summer feeding period, the fry are graded and transferred for further rearing to fingerlings and adult fish.

16.1.4 Grow-out

Grow-out in tanks, raceways and ponds

The fry continue to feed and grow throughout the year, if sufficient quantities of warm water are available. If the water temperature falls below about 5°C, feeding may cease and overwintering procedures will be needed. Though trout can be raised on natural foods, most farms at present use more or less complete artificial feeds.

There is a greater use of concrete or fibreglass tanks for growing yearling or two-year-old trout. The same tanks, particularly the circular tanks used for fry rearing, can also be used for grow-out of yearling trout. Tanks with a good circulation of water carry 25–35 kg fish per m³ water. Long raceway type tanks, 30 m × 3 m and about 1.2 m deep are commonly used. Circular tanks of about 5–12 m diameter and 0.75–1 m depth are also in use. In circular tanks with a flow of 10 ℓ/min of water, 1000 fry are reared per m². The silo type of tanks, briefly described in Chapter 6, have been used experimentally for high density culture of rainbow trout. A unit 5 m high and 2.29 m in diameter with a flow rate of about 28.4 ℓ/s is reported to carry 2820 kg trout without any problems. This is a stocking density of 136 kg/m³ or 27.5 kg/m³ per second of water flow. Raceways are widely used for grow-out as well as fry and fingerling production in North and Central America and Europe. Each concrete raceway system may be up to 500 m long (10 m wide and 1–2 m deep), divided into several segments, with arrangements for aeration. During the grow-out period of about 4.5–5 months, fish are graded and sorted by automatic equipment and transferred by pumping through water pipes to different raceway segments.

In the Danish type of freshwater trout farm, with sufficient fresh water supplies, 25–50 fry are stocked per m². It is possible to produce up to 30 kg/m² with proper feeding and abundant water supply. In ponds with a high flow-through of water, higher stocking densities are adopted. It is believed that ponds with a high degree of aeration can support a stocking density up to five times greater than unaerated ones. A safe high density of stocking is around 60 kg/m^3, with a water replacement of at least four to six times per day.

Grow-out of trout in land-based salt-water ponds seems to have originated in Denmark and Japan. One of the main reasons for utilizing sea water is to take advantage of its higher temperatures in colder northern countries in the winter season. Also, sites with abundant supplies of clean sea water can be found more easily in many areas for land-based farming as well as cage and enclosure farming. Young trout and salmon acclimatize to sea water conditions, but it is advisable to acclimatize them over a period of time by gradually increasing the salinity. When the fish is transferred to sea water, water from its body fluid is drawn out by osmosis because the salt concentration in the body is slight. Water loss is countered by drinking sea water, and this results in accumulation of excess salt in the body, which has to be excreted through special cells in the gills. The migratory steelhead acclimatize more readily, because of the increase in the number of salt-excreting cells in the gills during the period when they change into smolt. If other trout races are fed with a high-salt diet while in fresh water, their ability to adapt to sea water can be increased. Rainbow trout are ordinarily transferred to sea water when they are about 70–100 g in weight, but 150–200 g fish would acclimatize with less mortality. The growth rate of rainbow trout in sea water is reported to be double that in fresh water.

Shore-based farms consist of tanks of different sizes, with a pumped water supply for rearing fry to market size. The majority of the farms have tanks for rearing fry and ponds of different sizes for fingerlings, market fish and brood fish. If all sizes of fish have to be reared, there should be access to both fresh and sea water. The acclimatization of fingerlings to sea water can best be done gradually over a long period of time, by very gradual dilution with sea water. Pumping will be a major additional cost, but the ease with which the farm can be serviced and the stock monitored has made this system acceptable for growing fingerlings and smolts. Market fish are grown in these when the price is high enough to make it economical.

Instead of shore-based farms, tidal enclosures (see Chapter 6) have been used on a limited scale, where free exchange of water is possible during the course of low and high waters. An enclosure can hold about 8–10 kg rainbow trout per m^3 water. The fish are grown to sizes of 3–4 kg.

Grow-out in cages

The trout culture system that has received considerable attention in recent times is cage culture in fresh- and sea-water environments. The design features of cages have been described in Chapter 6. There is very little difference between the designs of cages used in fresh water and in protected areas of the sea, except that the marine cages are made stronger to withstand rough weather conditions. Floating cage culture has several advantages, among which the more important are the relatively high level of control the fish culturists can have on the stock and the possibility of starting operations on a small scale with a small investment (for example, starting with only two or three cages) and building on in the course of time to the maximum size that the site and local regulations permit. Rainbow trout is the most common species used in cage culture and fingerlings are stocked in cages in spring and are harvested in the autumn after a culture of 1.5 years or stocked in the autumn for harvest after one year. Fingerlings of about 70 g weight can attain a size of about 3 kg in less than 1.5 years. As male rainbow trout has a tendency to mature under two years of age, many farmers prefer to grow all-female or sterile stock for sea water grow-out. An innovation, started by some production units in Norway, is to hold the brood fish in spring water, with a temperature of about 7°C, through winter and strip them in January or early February. By using heated water at about 10°C, the incubation and fry growth are accelerated, so that by the autumn of the first year itself, smolts are produced which can withstand transfer to sea water. The duration of cage culture is restricted to one year, as otherwise the fish will start maturing in the cages. Even though the fish are smaller, the whole cycle of production is reduced to 1.5 years and all the cages can be kept under production throughout the year.

The stocking density of fish in the cage is 10–20 kg/m^3 at harvest, depending on water circulation. Generally, larger quantities of fingerlings than are required to obtain the above density are stocked in each cage and as they grow larger they are graded out to other cages.

The double cropping system of trout and channel catfish, in the southern USA, is an attempt to improve the income from raceway trout culture. Rainbow trout are cultured for about 132 days when the temperature is below 21°C and after the fish are harvested and when the temperature goes above this level, channel catfish (*Ictalurus punctatus*) are reared for over 200 days.

16.2   Salmons

As will be discussed in further detail in Chapter 30, typically anadromous salmons have been the focus of ranching programmes, and the Pacific salmon (*Oncorhynchus*) is the most important in this respect. The largest species is the chinook or king salmon (*O. tshawytscha*) which occasionally reaches a weight of 45 kg. Maturity and spawning usually occur after 4 years, but can occur earlier or later (between 2 and 7 years). The fresh-water nursery period also varies from 3 months to 2 years, but the usual period is 6 months. Coho salmon (*O. kisutch*) is a hardy species and has been widely transplanted. It reaches a weight of 15 kg, maturing at an age of 2 to 4 years and migrating

upstream in late fall or winter, after other salmons have completed spawning. The fresh-water nursery period is between 1 and 2 years. The sockeye salmon (*O. nerka*) spawns in streams connected to lakes. Spawning takes place below the lake and the young ones spend 2 to 3 years in the lake before migrating to the sea, where they live for a year or two. Though they occasionally reach 100 cm in length, the average size is much lower. The chum or dog salmon (*O. keta*) reaches a length of 1 m and a weight of 20 kg, whereas the pink salmon or humpback salmon (*O. gorbuscha*) reaches only about 4–5 kg. Both species spawn in streams or rivers and migrate to the sea within a year of hatching, the pink salmon returning as 2-year olds, and the chum salmon as 2- to 6-year olds.

Though restocking and ranching of the Atlantic salmon (*Salmo salar*) have been undertaken in the past and continue to be done, this is the main species presently used for large-scale sea farming (fig. 16.7). Some of the Pacific salmons, mainly coho and chinook salmons, have also been cultured in floating cages. As the general culture technologies are similar, the methods relating to the Atlantic salmon shall be described here in some detail. However, there is one major difference between the Pacific and Atlantic salmons: the Pacific salmon normally dies after spawning, whereas some of the spent fish (or kelts) of the Atlantic salmon survive and return to the sea, ascending

Fig. 16.7 The Atlantic salmon, *Salmo salar* (photograph: Ola Sveen).

streams two years later to spawn again. This makes it possible to use brood fish of this species for repeated spawning, if necessary. The size attained by the Atlantic salmon in nature varies greatly, depending on the length of its ocean residence, but weights of up to 36 kg have been recorded. The usual weight is 2–10 kg, averaging about 4.5 kg.

16.2.1 Culture systems

As in the case of trout, much of the earlier efforts at salmon culture were directed towards transplantation and hatchery production of young ones for stock enhancement or ranching. It is only in recent years, in the latter part of the 1960s, that the possibility of growing salmon received commercial attention. The decreasing price of trout and the high market value of salmon served as major incentives. Pioneering efforts in Scotland and Norway led to a modern cage culture industry of the Atlantic salmon, with specialization in egg and smolt production, grow-out for market fish, processing and marketing of smoked and frozen fish, feed and equipment manufacture, sales organizations, risk insurance arrangements, etc. Cage farming has spread to a number of other areas including Sweden, Iceland, Ireland, North America Japan, New Zealand and Chile. The types of cages used for salmon farming have been described in Chapter 6, and are very similar to those used for trout in sea water.

Another system of salmon culture is to grow them in impoundments of the type described in Chapter 6 (see fig. 6.32). Because of the high capital involved in such large-scale operations and the scarcity of suitable sites, this system is not commonly used. However, floating pen culture is not uncommon in Norway. Pens are built to enclose about 300–700 m² of the seashore (see Chapter 6). The management procedures for these pens are very similar to those for floating cages, except that for harvesting fish, repeated seining is necessary. Each enclosure can hold about 20 tons of fish, and if at any time the oxygen levels in the enclosure fall too low, a floating aerator is used to increase oxygen concentration.

Land-based pond or tank systems with supplies of pumped sea and fresh water are used only for growing smolts of salmon.

16.2.2 Techniques of propagation

As the basic techniques of propagating trout and salmon are similar in many respects, this section will focus mainly on the differences. The brood stock for artificial propagation are obtained from natural sources or from farm-raised stocks. When the fry or smolts are to be released into open waters, as in the case of Pacific salmons, it is preferable to have brood fish from the rivers or sea. On the other hand, when the fish are to be raised from egg to marketable adults in captivity, selected captive stocks with the desired genetic qualities are more valuable. When necessary, brood fish from natural stocks are also used and, as mentioned earlier, a good percentage (frequently over 70 per cent) of these fish can be used to provide eggs and milt a second time two years later, if necessary. Brood fish are collected before the onset of maturity (before they become grilse) and many stocks reach this stage in one year in the sea. As described in Chapter 30, artificial spawning channels are used to supplement natural breeding of Pacific salmon, and it is reported that the best results in ranching are obtained when native stocks are used in hatchery production of fry.

Stripping, fertilization and hatching

Brood fish can be held in sea or fresh water for stripping, but sea-grown fish are slowly acclimatized if they have to be transferred to fresh water. Mature male salmon can be recognized by their hooked lower jaw. When milt and eggs ooze out on gentle pressure on the belly the brood fish are suitable for stripping (figs 16.8 and 16.9). Brood fish are often anaesthetized with MS 222 or chlorobutanol ($C_4H_7Cl_3O$). The dry method is used for fertilization and care is taken to ensure that the eggs and milt do not come into contact with water before they are properly mixed together. As the stripped eggs and milt are sensitive to light, they are kept away from direct sunlight. The average sex ratio for propagation is four to five females to one male.

Since the Pacific salmons die after spawning, it is not necessary to strip them. The eggs are taken by incision, after killing the female by a blow or using a mechanical device that pierces

Fig. 16.8 Stripping a female salmon (photograph: Ola Sveen).

its head. An incision is made on the side of the fish with a blunt-tipped knife from behind the pectoral fins and slightly to one side of the median ventral line up to the genital papilla. Eggs start falling into a spawning pan and the body cavity is kept open by hand to let all the ripe eggs come loose. The milt is generally stripped from the males, which may or may not be killed.

The water-hardened eggs can be transported,

if necessary, for incubation at a distant hatchery. If the hatchery is nearby, it is possible to carry the eggs and milt in plastic containers and fertilize there. The trough and battery type incubators are commonly used. Some enterprises in Norway use a silo-type incubator for hatching large numbers of eggs. This is an upright plastic cylinder containing egg baskets. Water enters from the bottom and flows up past the eggs and leaves at the top of the

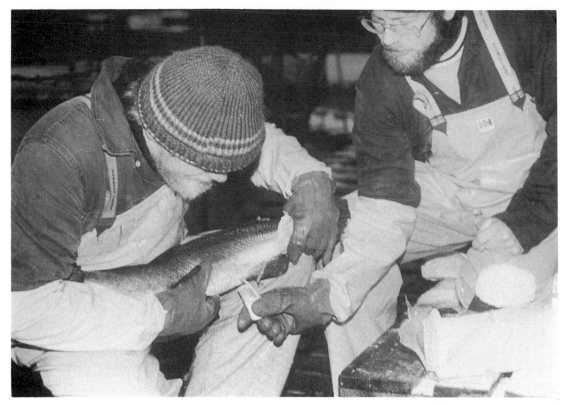

Fig. 16.9 Milt being collected from a male salmon (photograph: Ola Sveen).

container. Each silo can hold about 25 ℓ eggs and requires only 8 ℓ/min water. Since only such a small quantity of water is required it is not expensive to supply heated water for accelerating hatching.

The Atlantic salmon eggs become eyed after about 245 day-degrees, and may hatch after about 510 day-degrees. By controlling the water temperature in the hatchery, the rate of development is manipulated to obtain hatchlings at the most convenient time.

Yolk fry of Atlantic salmon take about 290 day-degrees, that is about 4−5 weeks at 8°C, to absorb the yolk completely and start taking external food. This is the most critical period and mortality at this stage is greater in salmon than in rainbow trout. A water temperature of 8−9°C is necessary for salmon fry to start feeding. Fry are reared indoors in fibreglass tanks holding water to a depth of about 25 cm. Often the same tank is used for rearing young salmon throughout their fresh-water life, until they are ready to smolt and acclimatize to sea water.

Salmon fry can be reared in high densities, up to about 10 000 fry per m² tank surface being usual (fig. 16.10). After they start growing, they are thinned out to give more space for growth. The density is regulated to about 25 kg/m³ and the fry graded to ensure uniform size in each tank. Water flow in the tank is about 1 ℓ/min per kg fish.

Smolt production

Rearing of fry to smolt stage can be in indoor tanks up to the first summer's growth. Larger outdoor tanks are also used for this purpose, although not so commonly as for rainbow trout. There are a wide variety of tanks in use, but many farmers seem to prefer circular or square ones, with circulating water for growing salmon parr. Their size does vary considerably, but large ones may have a diameter of about 10 m. While small tanks made of fibreglass are com-

Fig. 16.10 Salmon fry rearing in fibreglass tanks (photograph: Ola Sveen).

mercially available, larger ones are constructed above ground on the site. A circular tank of 10 m diameter and 1 m depth can hold about 20 000 salmon up to the smolt stage.

Feeding with a special salmon diet is performed manually or using timed automatic feeders. Besides regular cleaning of tanks to remove algal growth, young fish are usually graded by size several times during their freshwater growth, using special grading devices or sorting machines.

The smoltification of Atlantic salmon occurs in spring, when it undergoes physiological pre-adaptation for life in sea water. By the use of water at a higher temperature and with adequate feeding and care, many farmers are able to induce 50–80 per cent of the fish to smolt at the age of one year, during the spring of the second year after hatching. The rest smolt when two years old, i.e. during the spring of the third year after hatching. The minimum size for smolting is about 15 g, but the desired size at the end of the first year is about 30–50 g. Larger size smolts are preferred for grow-out and therefore fetch a better price. Because of this, many smolt producers use heated water with recirculation.

16.2.3 Grow-out

As indicated earlier, the most common method of culturing salmon to market size is in floating sea cages or pens. Before transfer of the smolts to cages or pens, it is the general practice to acclimatize them gradually through intermediate salinities, over several months. Such acclimatization helps to reduce mortalities during transfer to sea farms, and prevents the sudden shock of transfer to sea water that otherwise affects feeding for weeks, and therefore their growth during the first year at sea. Smolts are transported in boats or by trucks to the site of grow-out farms, as convenient. Smolts transported by sea are generally acclimatized before transport in sea water. When transport is by truck through inland areas, smolts are carried in fresh water and the acclimatization carried out at the grow-out site. Where feasible and economical, transport by air is carried out in special tanks or plastic bags filled with water and oxygen.

Stocking of cages and enclosures is undertaken in spring and harvests taken after two or three years, depending on market requirements. The stocking density of fish in each cage may vary between 10 and 35 kg fish per m³ cage at harvest. As in the case of rainbow trout, higher numbers are stocked initially and are thinned out later as they grow in size. Besides adequate feeding and frequent monitoring of the oxygen concentration of the water, the health of the fish is regularly checked. Harvesting from cages is easy. The net is pulled up until the fish are confined in a small volume of water, from which they are removed by dip nets. The use of low grilsing strains and improved rearing strategies have eliminated the need for grilse harvesting.

16.3 Feeds and feeding of salmonids

Much of our present knowledge of fish nutrition and feed technology is based on work carried out on salmonids (refer to Chapter 7). Although there are many gaps in the basic information, a sizeable fish feed manufacturing industry has developed as a result of this work and there is an expanding demand for feed in trout and salmon farming, especially in Europe and North America. Manufacturers do make separate feeds for salmon and trout, but experience seems to suggest that they are interchangeable, although salmon diets normally contain a higher percentage of animal protein.

Large-scale farming of trout started in Denmark with the use of trash or industrial fish that was available at low prices as feed. It continues to be used for both trout and salmon farming in Scandinavian countries, even though there is now a greater use of processed commercial feeds. The whole fish or waste left after filleting in processing industries and fish silage are also used. Other fresh feeds, like slaughterhouse offals, have been used as feed in small-scale farming in other parts of the world, but such material is not available in sufficient

quantities to sustain any large operations.

Species of white (non-oily) fish are preferred as salmonid feed, because fish with a high fat content are more difficult to store and the fat soon becomes rancid. Wet diets prepared with white fish have a low fat and high protein content (approximately 5 per cent fat and over 80 per cent protein, dry weight). It may therefore be necessary to add extra fat to the mixture, so that part of the energy requirements of the fish can be met and more of the proteins become available for growth. When species with somewhat higher fat contents are used, it is not necessary to add extra fat. The most common industrial fish used for salmon and trout feeding in Norway is the capelin (*Mallotus villosus*). Other fatty fish used are sprats (*Clupea*) and sand eels (*Ammodytes*). Fish of the herring family are not used because of the presence of thiaminase which destroys vitamin B_1. It is necessary to add thiamin to the diet, if the fish are fed with raw herring or a diet containing herring meal.

Although some farmers in Norway feed the fish with whole capelin, the general practice is to mince the fish with a binder to improve the consistency and add vitamin and mineral mixes to ensure that there is no deficiency in the essential trace elements. The protein content of whole fish is generally around 17–18 per cent of the wet weight. The commercially available binding meals, which are mostly carbohydrate, contain 10 per cent or less protein and about 3 per cent fat. Meals containing 35–40 per cent protein are also available commercially for adding to fresh food diets, which contain about 50 per cent animal protein and about 7–10 per cent fat. Some formulations may contain the necessary vitamins and minerals, in which case 5–10 per cent of this meal is added; if not, only about 1 per cent is added to keep the food together and reduce wastage. Shrimp waste is often added to give a distinctive pink colour to the flesh of salmon and trout at the rate of 10 per cent of the diet, which gives a concentration of about 5–6 ppm of the pigment astaxanthin in the prepared feed.

Special wet feed dispensers are available, but it is more common to feed cage fish with wet feed by hand (fig. 16.11). If mechanical dispensers are used, the feed should have a smoother consistency, suitable for extrusion

Fig. 16.11 Hand feeding salmon in a cage farm (photograph: Ola Sveen).

through a dice. Most farmers feed their fish as often as possible, as they believe that frequent feeding with small quantities, rather than occasional feeding of large quantities, gives better growth rates. However, feeding once or twice a day has been found to be equally satisfactory. Feed dispensers used for wet feeds have been described in Chapter 7. A type of feeder especially suited for salmon and trout grown in tanks consists of a canister containing the food mixture travelling on a track over the fish tanks. As it passes over each tank, a piston expels a measured amount from a nozzle. Mobile dispensers have been made from modified slurry tankers from which feed is forced out in a jet by compressed air. Wet feed is also sometimes pumped along pipelines directly to the raceways or cages in large farm units. Very good wet feed mixtures give conversion ratios of up to 5:1, but poorer qualities give only about 8:1.

Attempts have been made to reduce the percentage of animal protein in salmonid diet. Soybean meal now forms 10—30 per cent of commercial diets, but the quantity of fish meal

is not reduced to less than 20 per cent. A choice of pigmented or unpigmented feed is commercially available in larger pellet sizes. Around 40—60 ppm of the artificial carotenoid canthaxanthin or astaxanthin is added to dry feed to give a darker red colour to salmonid flesh. Pigmented feed is given for only 3—6 months, depending on temperature, before the fish are harvested for slaughter. Many salmonid farmers prefer to feed their fish with moist pellets that contain 20—50 per cent moisture, as against 12 per cent moisture of dry pellets.

16.4 Salmonid diseases

The high-density rearing of trout and salmon and their large-scale transplantation in the egg, fry or smolt stages give rise to conditions suited for the transmission and spread of various diseases and environmental hazards, like the toxic algal blooms of *Chrysochromulina polylepis*, that developed on the Norwegian coast in 1988. Many of the diseases described in

Chapter 9 are those diagnosed from trouts and salmons. The most common virus diseases are infectious pancreatic necrosis (IPN), infectious haematopoietic necrosis (IHN) and viral haemorrhagic septicaemia (VHS). Among the bacterial diseases are furunculosis, bacterial gill disease, enteric red mouth disease and vibriosis. The whirling disease caused by the protozoan parasite *Myxosoma cerbralis* infects almost all species of salmonids, particularly the young. Costiasis and icthyophthiriasis (ich or white spot disease) are also quite common. Infections by the copepod *Argulus* and the trematode *Gyrodactylus* do not cause direct mortality, but adversely affect the growth rate and marketability of the fish.

The diagnosis of these diseases and preventive and curative measures are discussed in Chapter 9. The absence of any feasible curative measures for some of these diseases, particularly the viral infections, emphasizes the need for scrupulous observance of high levels of sanitation in salmonid culture establishments. Many of the diseases cause large-scale mortalities in fry and fingerlings and this clearly indicates the need for special care during hatchery and nursery operations. The high value of salmonids makes it possible to adopt immunological techniques, especially in the case of viral diseases, for which there are no known cures. Regulatory measures that are implemented in some countries to control the spread of infectious diseases of salmonids should serve as a major means of protecting the very important salmonid aquaculture industries that are developing in many parts of the world, especially in the northern hemisphere.

16.5 Harvesting and processing

In the type of high-density culture adopted for trout and salmon, harvesting does not pose any special problems. In land-based, large pond farms and in enclosures seining is adopted. In most cases, ponds are drained and the fish gathered in a sump. Methods of harvesting from cages and raceways have been briefly described in Chapter 11. As the marketability of the high-valued salmonids depends very much on quality, it is necessary to ensure that the fish are handled with the utmost care and

speed during harvest, sorting and transport. The best way for long distance transport of trout is in live tanks (see Chapter 11).

Portion-size fish are usually allowed to die by suffocation, but larger fish have to be slaughtered as quickly as possible. If allowed to die slowly, lactic acids are released into the tissues and these will accelerate the autolytic process and deterioration of the flesh. Where possible the fish are kept in a suitably-sized tank (for example $2 \times 1.5 \times 1.5$ m) with water and killed instantly by passing electricity (about 600 volts) into the water. The practice of bleeding carcasses in Norway has been described in Chapter 11).

Portion-size trout and salmon are preferred in many markets in the live or fresh condition. Both small and large fish are also marketed in the frozen, smoked (hot- or cold-smoked) filleted and packaged forms. The product form naturally depends on the consumer preferences, but the demand for ready-to-cook products is increasing in most urban areas. Much of the success of Atlantic salmon farming appears to have been due to the high quality of smoked salmon exported to lucrative markets. Hot-smoking is performed mainly for smaller fish, which are partially cooked in the process. Larger fish of around 2.5 kg are split and, after removal of the backbone, are smoked slightly above the ambient temperature. For hot-smoking, fish are brined by immersion in a brine solution for 2−5 hours depending on the size of the fish and then smoked at different temperatures, starting with 30°C for half an hour, followed by 50°C for another half hour and then 80°C for three-quarters to one hour. The smoked fish are cooled down to 4−10°C before being packed. If quick-frozen, they can be stored for about a year, otherwise they will stay fresh for only a week at low temperatures of 0−4°C. For cold-smoking, a fish with a fat content of at least 15 per cent of the body weight is preferred. Large fish are gutted and cleaned and the sides cut back from the neck to the tail or the vent. The sides which are carved across are salted by placing them on a layer of salt. After 12−14 hours they are washed to remove the surface salt. Another method is to immerse the fish in 8−10 per cent brine solution. Dry-salted or brined fish are dried in cool air and smoked at ambient temperature, not exceeding 28°C, in natural or forced-draught kilns.

16.6 Economics

Investment costs for the establishment of salmonid farms are comparatively high, and because of the need for artificial feeds rich in animal proteins, a supply of high-quality water and the employment of skilled labour, production costs also tend to be on the high side. In most cases, these are compensated for by the high volume of production and reasonably high prices of the diversified products that can be marketed. Obviously profitability is dependent on a number of factors and it is therefore difficult to generalize on the levels of return that can be expected. According to Sedgwick (1985), a well-sited and well-run unit for trout with good marketing should return between 35 and 40 per cent profit on turnover before tax, and pay off the invested capital in less than three years.

Compared to one-year cycle of production for portion-size fish, the rearing of larger fish gives a higher return per ton of fish produced, in areas where there is a demand for them and they fetch a premium price. However, to grow the fish to a larger size, more equipment and labour are needed. Edwards (1978) showed that if the farmer decides to use the required additional resources, he would achieve a comparable return.

Experience in growing Atlantic salmon shows that it is more profitable than growing trout. Berge (1978), who compared the economics of salmon and trout farming, concluded that the higher profitability of salmon farming is due to the higher price of the product. He found that the cost of production of salmon is about 19 per cent more than that of trout, but because of their later maturing, salmon can be grown to a larger size and this enhances their market price. From a comparison of different types of farms in Norway, he suggested that farms which undertake the complete process of rearing from egg to market fish are far more profitable than those which undertake only one part of the operation. Similarly, larger enterprises, producing over 20 tons, show higher profits. The returns on investment are negative in the case of some smaller farms.

16.7 References

Berge L. (1978) A study of costs and earnings in Norwegian pondfish farming for 1974. In *Notes on the Economics of Aquaculture*, pp. 19–36. Papers on Fisheries Economics, The Norwegian School of Economics and Business Administration, Bergen, No. 18.

Brown E.E. (1979) Fish production costs using alternative systems and the economic advantages of double-cropping. In *Advances in Aquaculture*, (Ed. by T.V.R. Pillay and W.A. Dill) pp. 235–9. Fishing News Books, Oxford.

Brown E.E., Hill T.K. and Chesness J.L. (1974) *Rainbow Trout and Channel Catfish − A Double Cropping System*. Research Report 196. Agricultural Experiment Station, University of Georgia, Athens.

Buss K., Graff D.R. and Miller E.R. (1970) Trout culture in vertical units. *Progr. Fish-Cult.*, 32(4), 187–91.

Edwards D.J. (1978) *Salmon and Trout Farming in Norway*. Fishing News Books, Oxford.

Ghittino P. (1973) Note di moderna tecnologia in troticoltura. In *La Troticoltura*, pp. 21–6. Associazione Piscicoltori Italiani, Treviso.

Ghittino P. (1983) *Tecnologia e Patologia in Aquacoltura*, Vol. 1. Tecnologia, Torino.

Huisman E.A. (1976) Hatchery and nursery operations in fish culture management. In *Aspects of Fish Culture and Fish Breeding*, (Ed. by E.A. Huisman), pp. 29–50. Miscellaneous papers, Landbouwhogeschool Wageningen, The Netherlands, No. 13.

Laird L.M. and Needham T. (Ed.) (1988) *Salmon and Trout Farming*. Ellis Horwood Ltd., Chichester.

Leitritz E. and Lewis R.C. (1976) Trout and salmon culture. *Fish. Bull.*, **164**.

MacCrimmon H.R. (1971) World distribution of rainbow trout (*Salmo gairdneri*). *J. Fish. Res. Bd Can.*, 28(5), 663–704.

O'Malley H. (1920) *Artificial Propagation of the Salmons of the Pacific Coast*. Bureau of Fisheries Document, No. 879, Washington.

Pillay T.V.R. (1973) The role of aquaculture in fishery development and management. *J. Fish. Res. Bd Can.*, 30(12), 2202–17.

Rabanal H.R. and Shang Y.C. (1979) Economics of various management techniques for pond culture of fin fish. In *Advances in Aquaculture*, pp. 224–35. Fishing News Books, Oxford.

Sedgwick S.D. (1985) *Trout Farming Handbook*, 4th edn. Fishing News Books, Oxford.

Stevenson J.P. (1980) *Trout Farming Manual*. Fishing News Books, Oxford.

17
Catfishes

Catfishes belonging to Ictaluridae, Claridae, Pangasidae and Siluridae are widely distributed in different parts of the world, and their culture has been a traditional practice in some parts of Southeast and South Asia. Like many other warm-water species, catfish are valued as high-quality fish in certain areas, whereas in others they are considered as medium- or low-quality fish. Their hardy nature and ability to remain alive out of water for long periods have been of special value in tropical countries, and there is a specialized trade in 'live fish' (a collective name for species that have accessory respiratory organs and can live out of water for long periods) in some areas, as in eastern India. Catfish are also valued for recreational fishing in the southern parts of the USA. Nevertheless, the recent widespread interest in commercial catfish farming was largely generated by the development of a multimillion dollar catfish farming industry in the southern USA. Since the 1900s, considerable research and development efforts have been directed towards the formulation of suitable methods of farming and the processing and promotion of catfish products. Channel catfish, the main species used in farming in the USA, have been transplanted to a number of countries in southern Europe, Africa and Central America, but no comparable enterprises have developed in these regions. These introductions have, however, resulted in greater attention being paid to improving techniques of farming local species of catfish in many countries.

17.1 Channel catfish

The most important aquaculture species of catfish in the USA is the channel catfish (*Ictalurus punctatus*, family Ictaluridae) (fig. 17.1), although the white catfish (*I. catus*), which are more tolerant to crowding, higher temperature and low oxygen levels and the blue catfish (*I. furcatus*), which grow more uniformly and dress-out better also have farming potential. Among the many catfish cultured in Asia, the most important is *Clarias batrachus*, although the slower-growing allied species, *C. macrocephalus*, has a higher consumer preference. The species of catfish that has received greater attention in Africa is the so-called African catfish *C. lazera*. This species is now considered synonymous with sharp tooth catfish, *C. gariepinus*. Recently, efforts have also been directed towards developing a technology for the culture of the brackish-water catfish *Chrysicthys* spp. Catfish farming in southeastern Europe is based on the sheatfish or wels, *Silurus glanis*.

17.1.1 Culture systems

Pond culture is the most common culture system of catfish in all parts of the world, and most of the present-day production comes from small- and large-scale pond farms. Though more than one species of catfish may be reared in the same pond, polyculture of the type employed in carp farming is seldom practised. Exceptions are the small-scale rearing of *Pangasius* with tawes (*Puntius gonionotus*) and sepat siam (*Trichogaster pectoralis*) in Thailand, and the grow-out of yearlings of sheatfish in carp ponds. The double-cropping system of channel catfish with rainbow trout in the southern USA has already been described in the previous section.

In predominantly catfish farming areas, preference is usually for intensive farming systems. The hardy nature of the species makes high-

Fig. 17.1 Channel catfish, *Ictalurus punctatus* (from Huet, 1986).

density culture possible. Nevertheless, as will be discussed later, excessive stocking densities and very intensive feeding have created several management problems in some areas, such as Thailand.

Raceway culture is one of the intensive systems of catfish farming employed in the USA. Raceways constructed of concrete, asphalt, concrete blocks or earth are used. The degree of production intensity depends largely on the abundance of the water supply. Smaller raceways with water supplies of high volume and high velocity are used for highly intensive production, whereas larger raceways with a lower water flow are utilized for semi-intensive systems of production. A recent development is the use of circular and linear tanks for growing fish to market size.

The catfish *Pangasius* has been cultured from ancient times in floating bamboo cages in Kampuchea and Thailand. This system of culture is now carried out in Europe and the USA as well, using new and improved types of cages and feeds. However, cage culture production of catfish in any of these areas is only a small percentage of that of ponds and raceways.

A culture system of some importance in the USA consists of the development and management of what are called fee-lakes, pay-lakes or put-and-take fishing. Operators of such establishments produce their own fingerlings for stocking lakes or ponds, or buy fingerlings from other producers. Recreational fishermen are allowed to fish in these waters for a fee, based on the quantity of fish caught or the duration of recreational fishing.

17.1.2 Propagation and grow-out

Brood stock

Normally brood fish of channel catfish are about 3 years old and weigh at least 1.36 kg and a female of that size can be expected to spawn 6000–9000 eggs. Larger brood fish of up to 4.5 kg are also used, but the smaller ones are easier to handle. Though brood fish can be obtained from streams or lakes, farmers prefer cultured brood fish grown on a diet rich in animal protein in a clean and healthy environment. In the absence of strains developed through selective breeding, farmers generally choose large individuals that look healthy. To reduce inbreeding depression, it is considered advisable to introduce some brood fish from outside sources every year. New brood stock, as well as those that have already been spawned, are reared in brood ponds.

The size of the stocking ponds and the stocking density depend on the size of individual fish and the rate of growth required. Smaller fish of about 1–1.4 kg are stocked at the rate of 340–450 kg/ha and larger fish at a rate of about 900 kg/ha. The natural food produced in a well-fertilized pond (including minnows, tadpoles and other forage organisms) can sustain about 340 kg/ha, but supplemental feeding is needed at higher stocking rates. Daily feeding at about 3.4 per cent of body weight is recommended, depending on the availability of natural food. It is a common practice to feed maturing fish once a week, during late winter and early spring. Additional food consisting of fresh or frozen meat, fish or beef liver, beef heart or other low-priced meat products is given at the rate of 10 to 15 per cent of the body weight. These items of diet are believed to meet the additional needs of minerals and vitamins of brood fish during gonadal development. Feeding is generally stopped if the water temperature goes below 70°C, though some culturists recommend a low-level feeding of 0.5 per cent of the body weight every 4−5 days during this period. The required sex ratio of brood stock depends on the method of spawning, but generally a 1:1 ratio is considered suitable.

The sexes can be differentiated by the secondary sexual characteristics developed during the spawning season. The female develops a well-rounded abdomen and the genital pore becomes raised and inflamed. The head of the male is wider than the body, with darker pigments under the lower jaw and on the abdomen. A large protruded genital papilla is another distinguishing characteristic.

Spawning and fry production

Channel catfish can be spawned in ponds, special pens and in aquaria or similar containers. Provision of a suitable nest is the major requirement for pond spawning as the natural nesting sites, such as holes in banks or submerged stumps in the natural habitats may not be available in ponds. Shallow ponds of about 0.4 ha usually serve as spawning ponds, but bigger ponds of up to 2 ha are also sometimes used. Spawning nests may be made of nail kegs, wooden boxes, hollow logs, large milk cans, concrete tiles, metal drums, etc. The number of nests needed is dependent on whether the fertilized eggs are allowed to hatch in the spawning pond itself or are to be removed and hatched indoors in hatchery troughs. It is possible to use fewer nests than pairs of fish, as

not all fish spawn at the same time, and so one nest for each two pairs of fish is considered a good rule of thumb to follow. The nests are placed around the edge of the pond at depths varying from 15 to 150 cm water, with the open end facing the centre of the pond. The number of brood fish to be introduced depends on whether the eggs are to hatch out in the pond or will be removed for hatching indoors. If the hatching takes place in the pond, a stocking density of 50 females per ha is recommended, whereas if the spawn is to be removed for hatching the density can be raised to 125 females per ha.

Channel catfish spawn in late spring and summer, depending on the strain and the geographic region, at a temperature between 21 and 29°C. After introducing the brood fish in the ponds, the nests are checked regularly to see whether spawning has taken place. As too frequent checking may disturb the spawners, checking only every three days is recommended. Following spawning, the male catfish guards the eggs during incubation and fans them with his fins to keep a current of well oxygenated water flowing over them.

Pen spawning is a relatively more controlled method of propagation (fig. 17.2), where it is possible to ensure mating between a selected pair of brood fish, which is not possible in ponds. Pens are usually placed in ponds, but flowing streams are also suitable. Different sizes of pens are used, but generally they are not larger than 2 m × 3 m, the depth being less than 1 m. They are constructed of wood and welded steel wire mesh, (2.5 cm by 5 cm mesh) and are embedded in the pond bottom with about 0.3−0.6 m of the pen above the water. A spawning nest is placed in each pen which is stocked with a selected pair of brood fish. The female should be slightly smaller than the male, as the male guards the nest after spawning and fighting between the parents often occurs during this period. If larger, the female may chase the male away or even kill it, with the result that the eggs would not receive parental care and might even be eaten by the female. After introduction of the brood fish, the pen is checked on at least alternate days. If spawning has occurred, the female is removed from the pen and returned to the brood pond. Either the male fish is left to hatch the eggs or the

Fig. 17.2 Channel catfish spawning pens in a pond (from Stickney, 1979). By permission of John Wiley & Sons Inc.

egg mass is collected and hatched indoors. If hatched in the pen, the fry swim out of the nest through the wire mesh into the pond. If the eggs are removed for hatching, each pen is stocked with a second female or a new male and female pair after removing the male that has spawned once.

The aquarium method of spawning is used only when a small number of fish are to be spawned at a time and greater control of spawning is required. Aquaria of 100−220 ℓ capacity with a constant flow of water are used. Fish can be spawned at any season of the year, one pair at a time, by injecting the female with pituitary extract or HCG. The dose required is about 13 mg pituitary extract per kg body weight, or an average of 1760 mg/kg of HCG. If spawning does not occur within 24 hours, a subsequent injection may be needed. If the fish does not respond to three injections, it is replaced with another female. In the aquarium method, the eggs are hatched artificially.

If the eggs hatch in the spawning pond itself, or in the pen, the fry are collected (after driving off the male or removing it as appropriate) into suitable containers and transferred to fry ponds for rearing. Artificial hatching techniques are preferable; especially in pen spawning, as the spawning nests can be re-utilized quickly and the culturist will know more accurately the number of eggs hatched. It reduces losses to predacious insects and other fish, including the

parent females. Though channel catfish eggs can be incubated in hatchery jars, troughs are more commonly used in commercial farms.

Hatching troughs are usually made of sheets of aluminium or stainless steel; their dimensions vary, but they are segmented by partitions across the middle, each segment having a drain and an inlet pipe. The segments make it possible to separate eggs in different stages of development. The eggs are kept in 7.5 cm deep baskets, made of 0.6 cm meshed hardware cloth and suspended by wire from the side of the trough, so that the water line is below the top of the basket. Paddles attached to a revolving shaft cause the movement of water on the eggs, in a manner similar to that made by male fish when they guard the nests in ponds.

Six to ten days are required for catfish eggs to hatch at temperatures in the range 21−24°C. To combat the growth of fungi, troughs are flush-treated once or twice each day with malachite green at the rate of 2 ppm in the water. The treatment is stopped 24 hours prior to hatching, as the chemical is highly toxic to fry. As the eggs hatch, the fry swim out of the baskets through the hardware cloth into the trough.

Fry can be reared in troughs or ponds, but trough rearing is preferred since it allows greater control and the fry are less exposed to predators. Rearing troughs are generally 2.5−3.7 m long, 30−38 cm wide and 25−38 cm deep, made of wood, metal, plastic or fibre-glass. A water flow of about 0.06−0.3 ℓ/min and a temperature range between 24 and 29.5°C, are maintained. Fry reared in troughs have to be given complete feeds after the yolk sac is absorbed, which takes about 8 days after hatching. Floating and non-floating types of commercial feeds, containing 28−32 per cent protein and other nutrients, are fed five or six times each day at the rate of 4−5 per cent of the fry weight. Many culturists feed fry initially with fine particle trout starter rations, that contain about 50 per cent protein. The protein percentage is reduced and the pellet size increased as the fry grow. In a few weeks, a typical catfish ration containing about 30 per cent protein is given. Fry can be raised to fingerling stage in troughs or moved to rearing ponds after they have grown to 2−4 cm.

Fingerling ponds are frequently about half a hectare in size, and when stocked at the rate of 90 000−125 000/ha, fingerlings of 15−20 cm can be produced in about 4 months. Predatory aquatic insects can be controlled with a layer of oil on the pond surface to prevent the insects from breathing atmospheric air. Fry reared in ponds may not need complete feeds as for those in troughs, but it will be safer to use them as the quantity of natural food in the pond is not always predictable. The ration of artificial feed is gradually increased from about 0.5 kg per 0.8 ha pond to about 4−5 per cent of the fry body weight. Starter feeds in meal or ground mash form may be changed to small pellets when the fish reach about 2.5 cm in length. Many farmers produce only fingerlings, for sale to those who grow them to market size or stock recreational waters. There is greater demand for larger fingerlings of 15−20 cm, rather than smaller 5−10 cm ones. Raising of fingerlings of this size covers the first growing season.

Grow-out

Grow-out of channel catfish to market size takes a little less than two years after hatching, or one year from the fingerling stage. The usual market size is 500 g to 1.4 kg, though many are harvested at 450 to 600 g size. Fingerlings are generally stocked in grow-out facilities in the spring, and harvested in about 7 months in October or November.

The most common grow-out facilities are pond farms and they generally seem to be more cost-effective than other systems. The ponds are prepared for stocking by eradication of weed fish by application of rotenone. New ponds are fertilized with 16−20−4 or 16−2−0 NPK fertilizer at the rate of 56 kg/ha. The stocking density depends mainly on the quantity and quality of the water supply and the desired size of the market fish. In ponds with a depend-able water supply, a stocking density of about 3700−4900/ha is common. At the lower range of this density, the fish would weigh 500−600 g at the end of the growing season. To obtain a fish of about 1.2 kg weight, a third year of growth is needed. A stocking density of 2000−2500/ha is recommended for this purpose. Producers very often thin out the stocks from their stocking ponds after the second year of growth and maintain the lower density required for a third year of growth.

The use of commercially produced formu-lated feeds is a common practice in the catfish culture in the USA. In ponds where the fish have access to natural food, a feed containing 25 per cent protein may be enough, whereas in others a complete feed containing 30 per cent protein is required. Commercial catfish diets are made in different forms, the most common one being the extruded or hard pellet. Other forms used are dry meals and crumbles for feeding fry, floating pellets, semi-moist pellets containing 25−30 per cent water, and agglom-erates prepared by rolling finely ground dry formulated feed into balls for fingerling and adult feeding.

The most common methods of feeding are hand feeding and self or demand feeding. Many farmers prefer the former practice, mainly because they can regulate the feeding more easily and closely observe the feeding activity. Self or demand feeders permit the fish to obtain feed when they want and reduce over-feeding, which appears to be a common problem in channel catfish culture. Blow feeders, or low-flying aeroplanes that dispense feed, are made use of in extensive farms. Feed conversion in commercial grow-out facilities during the second year of growth is generally 2 kg feed for 1 kg fish, although better conversion has been achieved with improved feeding and manage-ment. An average production in commercial pond farms is around 1500 kg/ha, although production of up to 3000 kg/ha has been reported.

Raceways with an adequate flow of water that allows an exchange twice every hour in each segment can produce much greater quantities of fish than ponds. The stocking density depends on the size to which the fish are to be grown. The usual rate is about 2000 fingerlings per 120 m^3 raceway, where with proper feeding and exchange of water about 1 ton of fish can be produced in 180–210 days. In raceways with a lower water flow, only lower densities of fish can be raised. For example, with a flow of 9.5 ℓ/s, only 3500–5000 fingerlings can be stocked. Stocking is carried out in spring and early summer, with advanced fingerlings of 15–20 cm size. A complete feed, nutritionally richer than the feed used in ponds, is provided in raceways. The rate of feeding is usually 4–5 per cent of body weight twice a day, for two months after stocking, after which it is gradually reduced to about 3 per cent.

Tank culture of catfish, though only introduced recently, is reported to give very satisfactory results. Farmers who have adopted this technique report that a circular tank of 6 m diameter and 0.6 m depth gives the same production as a 0.4 ha pond. High rates of stocking are possible if the tanks are provided with aerators. Complete feeds at the daily rate of 3 per cent of the body weight are fed twice a day. Protection from bright light appears to give better growth rates and so the tanks are either covered or housed indoors.

Cage culture of catfish is of greater value when the cages are installed in open waters rather than in ponds. The stocking rate is about 65 fingerlings per m^3. Complete floating feeds are usually given at the rate of 3.5 per cent of body weight per day. This is gradually reduced to 2 per cent, depending on the quantity consumed.

Diseases

The major diseases that afflict channel catfish have been described in Chapter 9. The only virus disease diagnosed is the channel catfish virus disease (CCVD), which may cause large losses of fingerlings in a short time. Like most viral diseases, the only means known of elim-

inating CCVD is by the destruction of all brood stock associated with the epizootic. Haemorrhagic septicaemia and columnaris disease are important bacterial diseases of catfish that cause considerable mortality. A variety of protozoan diseases also infect the species, of which ichthyophthiriasis or 'ich' is the most important. Eradication of the parasite is possible only during its free-swimming stages (see Chapter 9) and repeated treatments over a period of days or weeks are needed to eradicate all. Costiasis is another common protozoan infection, which causes high levels of mortality, especially among fingerlings.

Various species of the external parasite *Trichodina* affect channel catfish. They occur on the body, fins and gills. Trichodiniasis is characterized by the appearance of irregular white blotches on the head and dorsal surface of infected fish. There may also be fraying of the fins and loss of appetite. Epidermal necrosis and excessive production of mucus may occur. Dips in 30 ppt salt water, a 1:500 solution of acetic acid or a 1:4000 solution of formalin form the usual treatment.

Myxosporidian parasites of the genus *Henneguya*, monogenetic trematode *Gyrodactylus* and the copepod parasites *Ergasilus*, *Argulus* and *Lernaea* can cause mortality among catfish.

Harvesting and marketing

Seining and draining are the two main methods of harvesting catfish from pond farms. Draining is an effective method as it allows better management of pond soils, but in areas where refilling the ponds requires pumping it involves an additional major cost. Also, there appear to be some practical difficulties in synchronizing the completion of draining with the delivery time of the fish to live haulers. Seining has an advantage in this respect and it also permits partial harvesting. If the mechanized seining equipment described in Chapter 11 is used, labour requirements can be reduced. However, larger investments for equipment are required and complete harvesting will not be possible. In recreational waters, channel catfish are caught by hook and line, using an ordinary fish hook on a pole, trotline or spinning rod.

After harvest, the fish may have to be held for several days before marketing. Vats, tanks or small ponds are used for holding. The fish have to be fed at maintenance levels if the holding period is longer than a day or two. If large harvests are involved, some culturists use 'live cars' (rectangular enclosures of netting buoyed by a series of floats) into which the catch from a seine can be transferred through a framed opening. The live car itself can be opened or closed by means of a draw-string. The problem of off-flavours developed by catfish in ponds and methods of eliminating them have been referred to in Chapter 11.

Catfish are usually hauled in tanks made of wood, fibreglass or aluminium. About 1 kg fish can be hauled in every 5 ℓ water, with a change of water every 24 hours. Devices for aeration are installed in the tank. In most cases, agitators that stir the surface water suffice. Deep tanks may, however, need aeration from the bottom.

Catfish are marketed as whole fish, dressed fish or as steaks and fillets. The dressed or pan-dressed fish is the most popular product, with the viscera, skin, head and some of the fins removed. Fish of 500−650 g are well-suited for this type of product. For steaks, larger fish of about 900 g or more are used. Fillets are made from small or large fish. Usually such processing and packaging is done in either farmer-operated facilities or in large processing plants.

The product is packaged for quick-freezing or for marketing fresh, wrapped in polyethylene film or bags. Frozen fish are stored at −1° to 1.7°C before sale.

Economics

As in all types of aquaculture, the economics of catfish culture show considerable variations between farms, depending on the location and culture system employed. Available data indicate that under prevailing conditions, pond farming is probably the most profitable system, although production up to fingerling stage only may sometimes give a better profit. Larger farms are more profitable than smaller ones. Production has to be at least 1500−2200 kg/ha to make a reasonable profit. The benefits of double-cropping with trout have been discussed in the earlier chapter on trout farming. Some farmers rotate rice, catfish and soybeans to yield better returns. The agronomic crops benefit by the improved nutrient level of the soil, caused by the accumulation of excrement and unconsumed feeds during fish culture.

17.2 Asian catfishes

The most important aquaculture species of Asian catfish is *Clarias batrachus* (family Claridae) (fig. 17.3). Commercial culture of this species is undertaken on a large scale in Thailand. In South Asian countries, it forms a major species in the so-called 'live fish' culture in swampy areas. Another important *Clarias* species is *C. macrocephalus*, which is preferred by the consumer for its appearance and eating quality. However, because its growth rate under culture is comparatively slow and there is a scarcity of fry, its culture has received less attention.

The ability to adapt to fresh and brackish waters with a very low oxygen content and to grow under generally poor environmental conditions makes *Clarias* extremely valuable for small- and large-scale rural fish farming. They are usually cultured to market size in pond farms, but fry are sometimes grown in floating baskets. In eastern India and Bangladesh, partly improved swamps are used for growing these species, along with another catfish, *Heteropneustes fossilis*, the climbing perch *Anabas testudineus* and the snakehead or murrel, *Channa* spp. High-density tank culture in recirculating water systems has been tried on an experimental scale and found to be profitable (Tarnchalanukit *et al.*, 1982), but has not yet been adopted for commercial production.

As mentioned earlier, another important Asian catfish *Pangasius sutchi* (family Pangasidae) has been cultured in ponds and cages for many years in Thailand and Kampuchea. Two other species of the genus, *P. larnaudi* and *P. pangasius*, are also cultured in ponds on a limited scale.

17.2.1 Spawning and fry production of *Clarias* spp.

Clarias batrachus will readily spawn in ponds and other confined waters, if the necessary environmental conditions are available. Even though methods of pond breeding have been developed, farmers still depend to a large ex-

Fig. 17.3 Asian catfish, *Clarias batrachus*.

tent on fry collected from natural waters, irrigation canals, rice fields, etc. Fry collection is performed during May to October. The fry are found in nests on the margins of water bodies, about 50 cm below the water surface. About 2000–15 000 fry can be found in a nest. Small, fine-meshed hand nets are used to transfer the fry from the nests, which are then transported to nursery ponds for rearing.

Natural spawning of *C. macrocephalus* is very similar to that of *C. batrachus*. They spawn in rice fields during the rainy season. The females make small, round hollow nests, about 30 cm in diameter and 5–8 cm deep, in the grassy bottom in shallow waters. The eggs are deposited in the nest and, being adhesive, stick to the surrounding grass. The male guards the nest and the female stays nearby.

Pond breeding starts with the procurement of brood fish from wild or captive stocks. They are held in small brood ponds or in holding sections of spawning ponds until they reach maturity. The brood stock is given a high protein diet every day, consisting of a mixture of 90 per cent ground trash fish and 10 per cent rice bran, at the rate of about 10 per cent of body weight. The fish attain maturity in about one year, when they weigh around 200–400 g.

Spawning ponds are usually 8000–16 000 m² in area. Nests in the form of 30 cm deep hollows (20–30 cm in diameter) are made on the banks about 20–30 cm below the water line, to resemble nests built by the species in natural waters for spawning. If the brood stock has to be held in the spawning pond, a section of about 20–30 per cent of the pond is dug deeper by about 1 m.

The spawning season is generally during the rainy months and extends from March or April through to September or October. But it has been observed that the species can spawn all the year round, if the pond water is changed with fresh water from outside sources. Brood fish are stocked at the rate of about one pair per 4 m² of the holding area of the pond, prior to commencement of breeding. The sexes can be distinguished easily when the fish have grown to a size of 20 cm: the anal papilla of the male is pointed, while that of the female is oval in shape. During the spawning season, the abdomen of the female is comparatively more distended.

Initially only the holding area of the pond is filled. Feeding is stopped and after one or two days the pond is filled to the maximum level with fresh water from an outside source. This stimulates the fish to spawn within a day or two, in the nests on the pond margins. If the spawning pond has no holding area, brood stock held in brood ponds can be introduced into the spawning ponds directly after filling with fresh water. The eggs are round, yellowish-brown in colour and 1.3–1.6 mm in diameter. They adhere to the soil or grass and are guarded by the male, as in the case of channel catfish. Hatching takes place in the nest within 18–20 hours at a temperature of 25–32°C. The hatched fry remain in a school in the nest and are removed with small scoop nets within 6–9 days after spawning. Each female produces about 2000–5000 fry. The fry are held in net enclosures for transfer to nursery ponds or for sale to nursery pond operators.

The same brood stock can spawn again after a ten-day period. In spawning ponds with holding areas, the brood fish can be made to return to the deeper areas by reducing the water level. Feeding is resumed for a ten-day period, after which the water level is raised again and feeding stopped. The fish can spawn again and the cycle can be repeated several times. Up to thirteen crops of fry from a brood stock have been recorded (Kloke and Potaros, 1975).

Clarias macrocephalus do not seem to spawn in confined waters of fish ponds and so hypophysation has to be resorted to. Intramuscular injections of common carp pituitary at the rate of 26−39 mg per fish of 23−30 cm length are reported to have been successful in inducing spawning in aquaria or hapas (cloth tanks), between June and July. The females can be stripped 10−12 hours after the injection, at temperatures of 29−32°C. After fertilization of the eggs with milt stripped from males, incubation is carried out in cloth tanks installed in ponds with flowing water. The eggs hatch out within 24−30 hours at the temperature range mentioned above. Yolk is fully absorbed within two days after hatching.

Nursery ponds for *Clarias* spp. are 400−1000 m^2 in area, with depths of 0.8−1 m. Before releasing the hatchlings for rearing, the ponds are fertilized with chicken manure and rice bran to produce an adequate quantity of food organisms. Initially the ponds are filled to only about 50 cm depth. Stocking is generally performed at the rate of 1000−3000 fry/m^2, but if it is planned to grow the fry for a longer time a lower stocking density has to be adopted. The rearing period varies from 15 to 35 days, depending on the size of fry required. Larger fry are preferred for stocking, as survival and production will then be higher. Steamed poultry eggs are considered to be the best feed for fry soon after the yolk sac is absorbed. For the first day or two, they are fed twice a day at the rate of 10 eggs per 100 000 fry. After that they are fed on ground trash fish twice a day, at the rate of 1 kg per 100 000 fry. Every two or three days the quantity is increased by about 1 kg. The ponds are filled to their normal depth of about 1 m a few days after stocking. No addition or exchange of water is generally practised. The fry are harvested with scoop nets after the ponds have been drained and the fry concentrated in a catching sump. The fry are usually grown to only about 3−5 cm length, even though there is a greater demand for larger fry, because of lower survival rates during prolonged rearing.

17.2.2 Propagation of *Pangasius*

Pangasius sutchi does not spawn in captivity and so hypophysation techniques have to be adopted for its propagation. Brood fish can be obtained from wild stocks or from culture ponds. Usually, 3-year-old fish that weigh about 4−5 kg are selected. The brood stock can be held in ponds or in floating cages. Brood ponds of about 1000 m^2 surface area are stocked at the rate of 1 kg fish per m^2 surface. The males and female spawners are separated well ahead of the spawning season, which generally starts in June and continues through to September. The brood stock is fed with a high protein diet (about 35 per cent protein), similar to that given to *C. batrachus*, consisting of ground trash fish and rice bran. An alternative feed is a mixture of fish meal (35 per cent), peanut meal (35 per cent), rice bran (25 per cent) and broken rice (5 per cent). Addition of a vitamin premix (1 per cent) or ground fish, once a week, is recommended three or four months before the spawning season. Occasional exchange of pond water is also recommended.

Floating cages used for brood stock rearing measure about 5 × 3 × 1.5 m and are stocked at the rate of two fish for every 1−2 m^3 of the cage. The cage is installed in running water or in a water body where there is sufficient current to remove waste products from the cages. Covering the cages with aquatic weeds or similar material offers an additional protection for the fish. The feed given in cages is the same as in ponds.

Pituitaries of the same species or of *C. batrachus* have been used for hypophysation. The sexes of mature fish are distinguished by the distended abdomen of the female and the easy emission of milt by the male on gentle pressure near the genital pore. Before injection, selected brood fish are removed and males and females held separately in tanks or cloth hapas. Pituitary extract is injected between the dorsal fin and the pectoral fin or at the base of the pectoral fin. The male is given one injection and the female two injections. The first dose for the female is the extract of one gland of a fish of about equivalent size. The second dose, given after about a twelve-hour interval, is about one and a half to three times that of the first. The dose for the male is about a quarter of the dose for the second injection of the female, and is administered at the same time that the female is given the second. Ovulation takes about 8−12 hours after the second injection, at a temperature of 28−32°C. The eggs are gently stripped and fertilized by the dry method. After a couple of minutes, the fertilized adhesive eggs are spread on special egg collectors in the form of mats of palm fibres, roots of aquatic plants or fine meshed netting. One litre of eggs would need about 10 m² of egg collectors, which are transferred to hapas made of fine-mesh cloth and held in flowing water for hatching. Spraying of water into the hatching hapa improves the oxygen supply.

Most of the fertilized eggs hatch out in 24−26 hours at the temperature range of 28−32°C. After the eggs have hatched out, the collectors are removed. The swim-up larvae appear in 10−12 hours and the yolk sacs are absorbed in about two days after hatching. It is important to provide adequate quantities of live food like *Moina* or other zooplankton, as otherwise they become cannibalistic. From the third day after hatching, the larvae can be fed on ground boiled egg yolk and waterfleas or other zooplankton, in small quantities, several times a day. Five-day old larvae will eat ground liver or cooked feeds. They can then be removed to rearing ponds. Methods of fry rearing are very similar to those for *C. batrachus*.

17.2.3 Grow-out of Asian catfishes

For grow-out of *C. batrachus*, small ponds of about 200−1000 m² area are used in Thailand (fig. 17.4). Normally the ponds are not fertilized, but between crops the ponds are dried and occasionally treated with light doses of lime. They are filled to a depth of 50−80 cm. Stocking is generally performed in March or April and the duration of culture is usually 3−4 months. After the first harvest in July or August, the ponds are stocked again between July and September for a second crop. In order to avoid the period when wild-caught catfish are available in the market, the second harvesting may be delayed until the following February or March.

To offset high mortality in commercial ponds, most farmers stock about 200 fingerlings/m², although the recommended stocking rate is less than half of that. The main feed is ground trash fish with rice bran (in a ratio of 10:1 by weight), which forms a sticky paste. By the fourth month the proportion of rice bran is generally increased to two parts, sometimes with one part of cooked broken rice added. The feeding rates are by no means standardized, and each farmer depends on his own experience. Soon after stocking, 2 kg feed is given every day in two equal quantities for every 10 000 fingerlings. Every two weeks the quantity is doubled, through gradual increases. Many of the problems of poor water quality, diseases and mortality in catfish ponds are due to the heavy rates of feeding. Water in many ponds gives the appearance of a pea soup, with quantities of uneaten feed and blooms of algae. Feeding with more than 0.5 kg feed per m² has been found to result in water pollution and fish mortality, in spite of the hardy nature of catfish and their ability to breathe atmospheric air. Frequent water exchanges, when possible, are an efficient means of pond sanitation and maintenance of water quality standards. The average yield of *Clarias* in Thailand is 29−32.6 tons per year in a 1600 m² pond. A production of up to 100 tons/ha per 4-month period has been reported.

The first harvesting of *Clarias* is done within 3−4 months, around June/July, when the fish have grown to 25−30 cm, weighing 200−300 g. The second harvest between February and May, after 5−6 months' rearing, produces fish of 35−40 cm length, weighing 400−450 g. The fish are transported alive to markets in metal boxes (180 ℓ capacity) with very little water.

Pangasius ponds are usually somewhat larger than *Clarias* ponds, but the methods of grow-out are generally similar. *Pangasius* is stocked at the rate of two fingerlings of 10−15 cm size per m² pond surface. The duration of culture is 12−15 months.

Cage culture of *Pangasius* is common in Kampuchea, Thailand and in recent years in Vietnam (Pantulu, 1979). In Kampuchea, cages are made of bamboo poles and splints. They are box-shaped when installed separately but when trailed behind a fisherman's boat, as is often done, they are arranged to fit the shape of the boat. The sizes vary considerably, but the larger cages are 40−50 m long, 4−5 m wide

Fig. 17.4 Catfish harvest from a pond in Thailand.

and 2.5−3.0 m high. Small cages are 4 m × 4 m with a depth of 2.5 m. A number of small cages may be lashed together to form a floating cage farm, buoyed with air-tight metal drums and walkways for feeding, harvesting, etc. In Vietnam, box-shaped cages are made of wooden planks with mesh-wire panels on the sides through which a free flow of water is possible. There is a floating cabin on the cage farm for the owner or caretaker to live in. The whole installation is moored in the river near the shore, or secured directly to the shore.

The cages are usually stocked with wild fry or fingerlings. In Kampuchea, a large cage of the dimensions mentioned above may be stocked with 6000−10 000 fry during the period June to August. They are fed with cooked vegetables like pumpkin, banana and a combination of cooked rice and rice bran. As the fish grow in size, they are fed on live and dead weed-fish and kitchen refuse. Harvesting is carried out during the months February to May, when the fish have grown to 1.5−2.5 kg each.

In Vietnam, cages are stocked with fry 38−63 mm in length from August to October. Sometimes advanced fingerlings, obtained from rice fields or rivers, are stocked. The stocking density of fry in cages is about 93 fry/m^3. *Pangasius* are fed on vegetable matter, such as chopped leaves, rice bran and forage fish. This may be supplemented with cooked or uncooked meat of mussels, snails, etc. Harvesting is usually carried out in March to August, after a culture period of over 10 months on average. The production ranges from 3000−25 000 kg/year per cage of capacity 1600 m^3.

17.2.4 Diseases

As mentioned earlier, the environmental conditions in high-density culture ponds of *Clarias* are conducive to a heavy incidence of disease and mortality, which often decimates almost half the fish stocked. The three most common diseases of cultured catfish are *Trichodina* infection of the gills, bacterial infection of the kidney and *Gyrodactylus* infection. Infections by *Aeromonas* spp., *Flavobacterium* spp., *Flexibacter columnaris*, *Pseudamonas* spp. and *Edwardsiella tarda*, have been identified in *C. batrachus* and *C. macrocephalus*. Most of these infections are believed to be brought in with the fry or fingerlings. Treatment with 25–50 ppm formalin in the pond or a one-hour bath of 250 ppm formalin in tanks is recommended, before the fry or fingerlings are stocked.

Accumulation of H_2S is another cause of mortality in *Clarias* ponds (Colman *et al.*, 1982). Dissolved oxygen levels do not appear to be so critical for the survival and growth of the species.

17.2.5 Economics

Studies of the economics of *Clarias* farming in Thailand have shown that, in spite of risks involved, most farms make substantial profits (Kloke and Potaros, 1975). For a single crop, the net income to gross returns averaged 37.7 per cent, the net income to the total cost ratio averaged 71.4 per cent and on an annual basis the return of the total capital averaged 108.1 per cent (Kloke and Potaros, 1975).

17.3 African catfish

Clarias lazera (= gariepinus), known as the African catfish, sharp tooth catfish or the Nile catfish (fig. 17.5), is a recent addition to aquaculture in Africa, which has been largely dominated by tilapia. Though its potential for farming has been demonstrated, its culture presently seems to be restricted to the Central African countries, the Ivory Coast and, on an experimental scale, to Egypt. Other catfish, such as *Chrysicthys* spp., are also being experimented with for pond and cage culture.

Clarias lazera can best be described as an omnivore, often feeding on vegetable matter, aquatic invertebrates, small fish, detritus, etc. Though it normally survives on dissolved oxygen, it comes to the surface and breathes atmospheric air when the oxygen concentration of water becomes low. The fish have been observed to reach over 130 cm in length and 12.8 kg in weight. A high degree of hardiness,

the ability to feed on a variety of feedstuffs and good growth and survival in poorly oxygenated waters have made it an attractive fish for rural aquaculture. The species can grow in brackish water in salinities of 10 ppt and survive in salinities up to 29 ppt.

The most common system of culture for this catfish is in pond farms, either in monoculture or in combination with tilapia, which has been shown to be a compatible species under pond conditions. Experiments have shown that the species is highly suitable for high-density tank culture (Hogendoorn *et al.*, 1983), but such systems have not yet been adopted on a commercial scale.

17.3.1 Spawning and fry production

Clarias lazera are reported to become mature under natural conditions at the size of about 32 cm (two or three years old) and spawn in the flooded rivers. Under pond conditions they mature in about 7 months, when they have attained a weight of 200–300 g. The spawning season varies between regions. In Egypt and Central Africa it is between July and September and in West Africa in April and May. They seem to spawn only once with the onset of the rainy season, under natural conditions, but can be bred throughout the year in captivity.

Observations confirm that spawning is stimulated by floods or increased levels of water in ponds due to rain or exchange with fresh water, as in the case of *C. batrachus*. Eggs are ejected in several batches (15–50 batches) during the extended mating and spawning, at temperatures above 17°C. The eggs adhere to sedges and grass.

While it is fairly easy to spawn mature fish in ponds by simulating changes in water levels, as happens during floods in nature (by draining the ponds partly and filling them suddenly with fresh water), the survival of offspring is generally very poor. Unlike some of the other catfish, *C. lazera* does not seem to show any parental care, and under pond conditions the larvae and fry seem to become cannibalistic. Additionally, there is a fair amount of predation by frogs and other aquatic animals in ponds. Because of these limitations, methods of induced spawning are adopted for the production of fry.

For induced spawning, brood stock from

Fig. 17.5 The African catfish, *Clarias lazera*.

natural habitats or culture ponds can be used. Ripe females in captivity range in size from 28–65 cm, weighing 175–1600 g. Females can be identified by the rounded vent with a longitudinal cleft, and the males by the elongated urogenital papilla. From experimental studies it is concluded that the best means of stimulating ovulation in females is by injection of the hormone product desoxycorticosterone (DOCA). The suggested dose is a single intraperitoneal injection of 5 mg DOCA per 100 g weight of fish. Injected females and mature males (which may not require any injection) are kept in separate tanks for about 10 hours, after which they are placed together in a tank or cement cistern, usually in the evening. Spawning occurs during the night, about 10–16 hours after injection of the female. The eggs can easily be collected in the morning and hatched in separate containers. The main problem with this method is that the couples often inflict fatal injuries on each other. To avoid this, the injected females can be stripped and the eggs fertilized artificially. First the females are injected with the above dose of DOCA, usually in the morning, and the males in the evening. The female is stripped about 10 hours after the injection. Males cannot be hand stripped because of the structural peculiarities of the seminal vesicle. If pressure is exerted on the abdomen, the milt will pass to the dorsolateral lobes of the vesicle and not to the genital opening. So the males have to be killed and the sperm collected directly from the vesicles. Injection helps to increase the yield of milt three to five times. Embryonic development is completed in about 24 hours after fertilization at temperatures around 26°C. The yolk sac is absorbed in 6 days and the larvae start feeding when about 3 days old.

Clarias lazera can be spawned by hypophysation as well, like many other species (Hogendoorn, 1979). Acetone-dried carp pituitary at a dose of 4 mg per kg body weight is adequate to ripen females. As mentioned earlier, injection of males does not seem to help in stripping them, and they have to be sacrificed to obtain milt. The females can be stripped 11−16 hours after injection. At 20°C, hatching of fertilized eggs occurs in about 48 hours, when the hatchlings can be transferred from incubators to a trough for rearing.

The stocking rate of fry is generally 10 000−20 000 per ha. Fed on natural food, they reach a weight of around 10 g in about three weeks. Thereafter, the fry start feeding on larvae of aquatic insects. Artificial feeds can be given to the fry at this stage. Though amphibians, aquatic insects and occasionally wild fish prey on the fry, the main reason for low survival in ponds appears to be the lack of appropriate feed.

17.3.2 Grow-out and feeds

Since most of the research efforts on *C. lazera* have been directed towards developing methods of spawning and rearing of fry, there are only limited data available on commercial grow-out procedures. In rural fish culture, the fish normally live on the natural productivity of the ponds, which may be enhanced by the addition of small quantities of fertilizers. In recent years, attempts have been made to grow pigs or ducks in association with fish ponds, as pig manure and duck droppings add considerably to the productivity of the ponds.

In Central African countries, the fish have been reared in ponds heavily fertilized with pig manure and fed on different locally available feedstuffs. Brewery waste and peanut cake are used as the main supplementary feed in some centres. Table 17.1 shows the composition of a diet used to determine the growth rate of fingerlings (9 g weight), stocked at densities in the range 1−10 per m^2 (Richter, 1976). The results showed that growth and feed conversion with pelleted feeds containing vegetable products were very satisfactory at stocking densities of two fish per m^2. The fish grew to an average weight of 600 g in about 200 days. The estimated production was in excess of 12 tons/ha per year.

It has been found that the African catfish seldom make use of demand feeders, and hand feeding is the most practical feeding method. It has also been observed that the males grow much faster than females. De Kimpe and Micha (1974) found that in four months' rearing, the males attained a weight of 427 g, while the females reached a mean weight of only 292 g.

In integrated farming with pigs, catfish of 95 g mean weight, stocked at the rate of one fish per m^2, grew to over 380 g in 4.5 months' rearing, giving an annual yield of 7 tons/ha.

Although a number of parasites have been identified from *C. lazera*, mortality due to major infections has not been reported. *Trichodina* infestation of fry has been observed in ponds and this can be controlled by treatment with 50 ppm formalin for about an hour.

Table 17.1 Composition of feed for *C. lazera* fingerlings based on vegetal and animal byproducts in Bangui (Central African Republic) (from Richter, 1976).

Product	Total dry matter (%)	Chemical composition of dry matter					In feed (%)
		N-free extract (%)	Lipids (%)	Total nitrogen (%)	Cellulose (%)	Minerals (%)	
Brewery waste	50.0	46.4	7.8	22.8	18.8	4.2	15
Corn bran	87.7	59.7	3.8	14.4	14.5	7.5	15
Cotton cake	93.1	28.5	7.4	47.3	9.6	5.4	45
Sesame cake	94.4	14.4	53.0	22.1	4.8	5.7	7.75
Bonemeal	91.9	−	7.1	52.1	−	40.8	2
Rice bran	88.4	56.9	3.8	8.7	22.6	8.0	15
Concentrated vitamins							0.25

17.4 European catfish (family Siluridae)

The European catfish, *Silurus glanis*, known also as the sheatfish or wels (fig. 17.6), is a highly relished fish in many parts of Europe, particularly in the east and central regions. Besides the taste, the high dressed weight (66 per cent) and the absence of intramuscular bones make it an especially valuable species for filleting and processing. It is generally described as a highly voracious species, feeding during adult stages on fish and other aquatic animals. In fact, its culture appears to have been largely as a predator in carp ponds to control weed-fish, or for stocking recreational waters. In recent years, efforts have also been directed towards monoculture of the species in ponds and cages.

17.4.1 Spawning and rearing of fry

Pond spawning

Brood fish for spawning are generally selected from captive stocks. Fish less than 4 kg in weight are normally not selected as they are seldom mature below that weight (fig. 17.7). Fish weighing 6−10 kg are preferred, but sometimes even larger ones weighing up to 25 kg are used. As secondary sexual characteristics are not very pronounced, it is somewhat difficult to distinguish the sex of brood fish. Females have a comparatively more oval, convex and blunt genital papilla, a swollen periphery for the anal opening, a less pigmented and more round abdomen and a more oval head. The brood stock are selected and sexed about two months before the spawning season, which is towards the end of spring when the water temperature reaches 20−22°C. The sexes are separated when the water temperature reaches 12−15°C, otherwise the males may injure females by biting. They are held in special holding ponds and fed on trash fish, frogs, tadpoles, etc., at the rate of 2.5 to 3.0 times the total weight of the fish. They readily accept pelleted feeds as well.

Pond spawning is carried out in small ponds, of area 100−200 m², with depths of about 1 m. A number of nests are placed in each pond to

Fig. 17.6 The European catfish, *Silurus glanis*, grown in a cage farm in Hungary.

induce the fish to spawn. The nests are made of roots of willows, pine branches or similar material, in different shapes. They are usually tent-like or pyramidal in shape (fig. 17.8). A male and female pair are introduced in every nest as the ponds are filled. Occasionally the spawners are subjected to hypophysation with carp pituitary (at the rate of 4 mg/kg) at the time of transfer to the spawning pond, and this has been reported to give better results. The male fish cleans the nest and after a period of courtship spawning takes place, generally in the morning. The fertilized eggs which adhere to the nest are guarded by the male. The nests are subsequently dismantled and the roots or

Fig. 17.7 A brood European catfish selected for spawning from a brood pond in Hungary.

branches are suspended in a box made of 1.0–1.5 mm mesh net, near the water inlet of the pond, for hatching. Hatching can also be done in tanks, troughs or hatching jars.

Hatching occurs 2 to 3 days after ovulation, at 45–69 degree-days. The unpigmented yolk-sac fry school together and avoid direct light, hiding under grass or other materials in the pond. Within 4 to 5 days after hatching (140 degree-days), the pigmented fry start feeding. In hatching boxes, tanks, troughs, etc., the fry can be fed on plankton or specially prepared doughlike mixtures such as boiled eggs, fish meat and flour. Trout starter-feeds have been found to be suitable for feeding the fry up to one month. Feeding is initially done five or six times daily, and later the frequency is reduced to three or four times.

Hypophysation

As mentioned above, the sheatfish can be spawned by the administration of carp pituitary. The dose required varies between 3.0 and 4.5 mg/kg, depending on whether it is the beginning, middle or end of the spawning season. Higher doses are indicated during the beginning than at the end. Usually two injections are given intramuscularly, the first one of about one-tenth the total dose and then the rest after 24 hours. Males are given only half the dose given to the female. Ovulation takes place at 430–460 hour grade. A fish weighing 16 kg was reported to yield 1845 g or about 300 000 eggs (Fijan, 1975). Stripped eggs are fertilized with milt in a 0.3 per cent common salt solution. Because of the structure of the seminal vesicle, there may be difficulty in obtaining enough milt by stripping, in which case half the testes may be removed by dissection and crushed to obtain the milt needed for fertilization. The eggs are fertilized in about 2–5 minutes, after which they are incubated for hatching.

Because of the sticky nature of the eggs, they tend to clump together. After 10–12 hours, the water flow in the incubator is cut off and the clumped mass treated with a 0.3–0.5 per cent solution of alkaline protease enzyme for 2–3 minutes, to dissolve the sticky layer and to separate the eggs. The eggs can be treated with malachite green solution (5 ppm for 30–60 minutes) to prevent fungal infections.

The hatched larvae are reared in fine-meshed (0.8–1.0 mm mesh) sieve-cloth tanks. They are provided with the type of feeds mentioned above and also shaded spots within the tanks into which they can withdraw from the light, when necessary. Two- or three-week-old fry can be used for stocking ponds.

17.4.2 Grow-out

Grow-out of fry to fingerling and market size is generally carried out as monoculture in ponds. Fingerlings have been raised very successfully up to 6 months, fed on trout starter-feeds and pellets. Small ponds rich in plankton, are stocked at the rate of 10–15 fry per m². They are fed with pelleted feeds or with ground fresh meat or fish, several times a day.

Yearlings can also be grown in polyculture in

Fig. 17.8 Artificial nests used for spawning European catfish in Hungary (from M. Huet, 1980).

carp ponds. Fingerlings of both carp and sheat-fish, weighing 25–30 g, are reared together, at stocking rates of 3000–5000 per ha. By autumn of the second year they attain a weight of 900–1100 g.

Losses due to predation are more pronounced in the larval and fry stages. Among the diseases reported, the most common one that results in high losses is ichthyophthiriasis. Larger fish may suffer from branchyomycosis.

17.5 References

Aprieto V.L. (1974) *Early Development of* Clarias macrocephalus *Günther Reared in the Laboratory (Pisces: Claridae)*. UPMF Technical Report, 3, University of the Philippines, Quezon City.

Brown E.E., LaPlante M.G. and Covey L.H. (1969) *A Synopsis of Catfish Farming*. University of Georgia, Athens.

Carreon J.A., Ventura R.F and Almazan G.F. (1973) Notes on the induced breeding of *Clarias macrocephalus* Günther. *Aquaculture*, **2**, 5–16.

Chuapochuk W.T.W and Na Nakorn P.S.U. (1982) *Pla Duk Culture in Circular Concrete Ponds with Water Recirculating System*. Kasetsart University Fishery Research Bulletin, No. 13.

Colman J.A. *et al.* (1982) *Pond Management, Water Environment and Fish Grow-out Performance Relationships in* Clarias *Culture Trials*. National Inland Fisheries Institute, Bangkok.

De Kimpe P. and Micha J.C. (1974) First guidelines for the culture of *Clarias lazera* in Central Africa. *Aquaculture*, **4**, 227–47.

El Bolock A.R. (1976) Rearing of the Nile Catfish, *Clarias lazera*, to marketable size in Egyptian experimental ponds. In *Symposium on Aquaculture in Africa, CIFA Technical Paper*, **4**, Suppl. 1, 613–20.

Fijan N. (1975) Induced spawning, larval rearing and nursery operations (*Silurus glanis*). In *Workshop on Controlled Reproduction of Cultivated Fishes, EIFAC Technical Paper*, **25**, 130–38.

Grizzell R.A., Dillon O.W. and Sullivan E.G. (1969) *Catfish Farming – A New Farm Crop*. Farmers Bulletin No. 2244, US Department of Agriculture.

Hogendoorn H. (1979) Controlled propagation of the African Catfish, *Clarias lazera* (C & V) I. Reproductive biology and field experiments. *Aquaculture*, **17**, 323–33.

Hogendoorn H. (1983) *The African catfish, (*Clarias lazera C & V, 1840*) – a new species for aquaculture*. Dissertation, Agriculture University, Wageningen.

Hogendoorn H. *et al.* (1983) Growth and production of the African Catfish, *Clarias lazera* (C & V) II.

Effects of body weight, temperature and feeding level in intensive tank culture. *Aquaculture*, **34**, 265–85.

Huet M. (1986) *Textbook of Fish Culture*, 2nd Ed., Fishing News Books, Oxford.

Inoue K. and Swegwan S. (1970) Economic survey on catfish culture in Suphanburi Province, Thailand. *Thai Fisheries Gazette*, **23**(2).

Kellehar M.K. and Vincke M. (1976) Preliminary results of studies on the survival of *Clarias lazera* fry in Africa. In *Symposium on Aquaculture in Africa, CIFA Technical Paper*, **4**, Suppl. 1, 487–96.

Kloke C.W. and Potaros M. (1975) The technology and economics of catfish (*Clarias* spp.) farming in Thailand. *IPFC Occasional Paper*, 1975/2.

Lee J.S (1973) *Commercial Catfish Farming*. The Interstate Printers and Publishers, Danville.

Martyshev F.G. (1983) *Pond Fisheries*. Russian Translation Series, 4, A.A. Balkema, Rotterdam.

Micha J.C (1976) Synthese des essais de reproduction, d'alevinage et de production chez un Africain: *Clarias lazera* Val. In *Symposium on Aquaculture in Africa, CIFA Technical Paper*, **4**, Suppl. 1, 450–73.

Pantulu V.R. (1979) Floating cage culture of fish in the Lower Mekong Basin. In *Advances in Aquaculture*, (Ed. by T.V.R. Pillay and W.A. Dill), pp. 423–7. Fishing News Books, Oxford.

Potaros M. and Sitasit P. (1976) Induced Spawning of *Pangasius sutchi* (Fowler) *Technical Paper* No. 15, Freshwater Fisheries Division, Department of Fisheries, Bangkok.

Richter C.J.J (1976) The African catfish, *Clarias lazera* (C & V). A new possibility for fish culture in tropical regions. In *Aspects of Fish Culture and Fish Breeding*, (Ed. by E.A. Huisman), pp. 51–71. Miscellaneous Papers 13, Landbouwhogeschool, Wageningen.

Rogers B.D. and Madewell C.E. (1971) *Catfish Farming – Cost of Producing in the Tennessee Valley*. Circular Z-22, TVA, Alabama.

Sidthimunka A. (1972) *The Culture of Pla Duk* (*Clarias spp*.). Department of Fisheries, Bangkok.

Sidthimunka A, Sanglert J. and Pawapootanon O. (1968) The culture of catfish (*Clarias* spp.) in Thailand. In *Proceedings of the World Symposium on Warm-water Pond Fish Culture*, (Ed. by T.V.R. Pillay). *FAO Fish Rep.*, **44**(5), 196–204.

Tarnchalanukit W. *et al.* (1982) *Pla Duk Dan Culture in Circular Concrete Ponds with Water Recirculating System*. Kasetsart University Fishery Research Bulletin, 13.

Tiemeir O.W. and Deyoe C.W. (1973) *Producing Channel Catfish*. Bulletin 576, Kansas State University of Agriculture and Applied Science, Manhattan.

Tongsanga S., Sidthimunka A. and Menasveta D. (1963) Induced spawning of catfish (*Clarias macrocephalus* Günther) by pituitary hormone injection. *Proc. Indo-Pac. Fish. Coun.*, **10**(2), 205–13.

Woynarovich E. and Horvath L. (1980) The artificial propagation of warm-water finfishes – a manual for extension. *FAO Fish. Technical Paper*, **201**.

18
Eels

Eels (family: Anguillidae) are considered a delicacy in some countries, while in others they are not eaten at all or have only limited demand. Traditionally, Western Europe and Japan have been the main areas where there is high demand for eels. Probably the earliest form of eel culture, as distinct from the stew or holding ponds in Roman times, is the rather extensive system of lagoon farming along the Mediterranean coast. In Italian lagoons, eels form an important polyculture species, with grey mullets, seabream and seabass. The rapid expansion of eel farming in Japan from about the middle of the 19th century, aroused considerable interest in intensive farming of this group and eel culture enterprises have developed in a number of countries in Europe, especially Italy, Germany and France. Taiwan has become a major exporter of cultured eels to Japan. As will be discussed later, aquaculture of eels continues to be based on seed eels collected from the rivers. Although some laboratory-scale progress has been made in maturing and fertilizing the eggs of some species of eels, it has not yet been possible to develop a system of artificial propagation. Reliance on natural supplies has led to periods of scarcity of elvers, restricting the expansion of culture enterprises. For instance, for almost a decade from 1952 Japan had to depend on the import of elvers from abroad, even to maintain the existing farms. As a consequence of this, the collection and export of elvers to Japan became an industry of some magnitude in a number of countries.

Although there are some 16 species of eels, the most important ones from the point of view of large-scale aquaculture are *Anguilla anguilla* (= *vulgaris*) in Europe (fig. 18.1) and *A.*

japonica in Japan and Taiwan. They are known to be catadromous species and migrate from rivers and other inland water bodies into the sea for breeding, and the glass eel or leptocephali return to inshore waters and eventually migrate up the rivers. The Japanese eel spawns not very far from the coast, but the European eel migrates far out to the Sargasso Sea area of the Atlantic to spawn. The leptocephali of the European eel (*A. anguilla*) reach the continent three years after hatching and enter the rivers after they have metamorphosed into elvers; whereas the Japanese eels (*A. japonica*) enter the rivers as elvers within a year of hatching. The upriver migration starts when the temperature of the river has risen, in about May, and continues until the end of August. Once they start the upriver migration, they show remarkable endurance and ability to overcome barriers and obstacles and form thick shoals.

18.1 Culture systems

The most common method of eel culture is in pond farms (fig. 18.2), the ponds being comparatively small. The elver ponds are about $100-350$ m^2 and the stocking ponds about $1000-1500$ m^2. They may be of the still-water type with occasional exchange of water or running water ponds with a flow-through of varying velocity. The latter type permits intensive stocking and feeding for high production.

Another system practised in areas with abundant supplies of warm water from springs is the tunnel method. Rearing is carried out in concrete tanks of 1 m^2 surface area and a depth of 1 m, with an inlet tank and an outlet tank. Water enters the main tank from the inlet tank through a 23 cm diameter pipe and drains out

Fig. 18.1 The European eel, *Anguilla anguilla*, from a farm in Italy.

Fig. 18.2 A pond farm for raising eels in Taiwan (from *Fish Culture Bulletin*, **7**, (3−4)).

through the outlet tank. This system is suited for highly intensive production.

The recirculation system has also been adopted on a limited scale for eel culture, to enable highly intensive tank culture in areas with a limited water supply.

The systems mentioned above are based on the use of fresh water. Making use of the adaptability of eels, systems of culture in brackish and sea water have also been developed. The extensive culture in the Mediterranean lagoons (valli) is based on elvers which are allowed to enter the impounded areas through the manipulation of tidal flows and which are grown in the fertile lagoon waters. In ponds supplied with sea water, eels are reported to grow faster, even though the higher rate of H_2S formation in salt-water ponds creates problems when the oxygen concentration or pH is low.

The culture of eels in net enclosures in sheltered bays, back waters, etc., has been tried to reduce the capital costs of pond construction and avoid problems of water supply. As eels grow better in warm-water environments, wherever possible heated water from thermal stations or industrial sources, as well as from hot springs, has been used.

18.2 Collection and rearing of seed eels

As indicated earlier, eel culture is based on seed eel collected from the wild. Several attempts have been made to propagate *A. anguilla* and *A. japonica* artificially, starting as early as the 1930s (Boucher *et al.*, 1934). Spontaneous release of eggs was obtained by Fontaine *et al.* (1964) in the European eel. Boëtius and Boëtius (1967) were able to mature males of the species by weekly injections with carp pituitary and maintaining them in sea water at a temperature of about 14°C. Several other workers have subsequently succeeded in maturing the males, as well as obtaining the release of mature eggs from females, but artificial fertilization of the European eel has not been a success. The Japanese eel have been stimulated to spawn by hormone injections and by keeping the brood fish in sea water at a temperature of about 23°C (Yamamoto *et al.*, 1975), but the larvae could be reared only until the sixth day.

In Europe, the collection of elvers is done either during winter and spring or in the beginning of summer in June and July, when they ascend the rivers. They seem to be able to migrate up the rivers at lower temperatures of 2–10°C, unlike the Japanese eels. The early migrants are smaller in size (about 7 cm in length), but the later ones are larger (15–20 cm in length). These are probably the elvers that hatched out the previous year and they ascend further up the river than the smaller ones. The elvers for restocking or rearing are collected during the earlier migration which starts around December. The best catches are generally obtained from February to May. They are captured with large wire-meshed sieves or large nets similar to a plankton net dragged from a powered boat. The catches are stored in aerated tanks for transport ashore.

In Japan, the elvers enter the rivers from October through to late May, when the water temperature reaches 8–10°C. They are caught in scoop nets at night, using bright lights as attractants, or by using fine-meshed bag nets set across the river. Special elver traps may also be set near obstructions across the rivers, where the elvers are likely to congregate. In Taiwan, the elver catching season is from October to March. They are caught with scoop nets, drag nets or eel traps.

Elvers need careful handling after capture and during rearing. It is a common practice to condition them after capture for a day, in special bamboo baskets or tanks. They can be transported to distant farms, packed in wooden boxes. Intercontinental shipments of elvers have also been made in polythene bags, after conditioning at low temperatures of 4–7°C.

Before release of elvers into nursery ponds, many farmers give them a bath of malachite green to prevent infection. To protect the elvers from cold winds, the ponds may be covered with vinyl sheets. Some farmers use electric heaters to maintain the water temperature at about 10°C. A common system of rearing elvers in Japan is to stock them initially at higher densities in a series of ponds, and when they grow larger to transfer them to a series of progressively larger ponds at lower densities. The first series is about 165 m^2 in size, with a depth of about 40 cm, and the stocking rate is about 500–600 g elvers per m^2. The larger ponds are about 200 m^2 in area. Most eel ponds have vertical concrete or brick walls and sandy soil bottoms, although there are some with steep mud walls and mud bottoms. The seed eels can climb considerable distances up the wall, especially during heavy rains, and so it is desirable to have some protective devices at the top of the walls to prevent their escape.

Before the elvers are released, the nursery ponds are disinfected with lime. As mentioned earlier, the stocking density in Japan is 500–600 g per m^2, but in Taiwan it may be almost ten times that. European elvers are stocked at the rate of 3500–10 000 g per ha. Feeding is started when the water temperature is about 15°C. As they are nocturnal in feeding habits, the feeding spot is covered with boards or other suitable material, to make it as dark as possible. Small worms are considered a suitable first feed for elvers, and after two or three days fish flesh is added in progressively increasing quantities until about the tenth day, when a paste of minced fish is given. Formulated eel diets are now in use in many farms. The recommended protein level in practical diets for elvers is 50–60 per cent and for sub-adults, 40–45 per cent. Even though the amount of feed consumed by elvers depends on the condition of the water and the temperature, the recommended normal daily ration is about 30 per cent of the total weight of the released elvers, fed in several lots. Adequate feeding is important to reduce cannibalism. When the elvers overgrow the capacity of the pond (indicated when they congregate near the surface and breathe atmospheric air), the stock is thinned out. Eels can be scooped out easily from near the feeding place and transferred to larger ponds. The stocking rate then is about 150–200 g per m^2. In about 4 months the elvers grow to around 7 g, and in another 4 months they reach about 100 times their size at stocking. It is essential to ensure that no filamentous algae develop in the ponds, as elvers are likely to hide under them and refrain from feeding. Usually elver farmers start catching and selling some of the stock as they grow in size, according to market demand.

18.3   Feeds and grow-out of adult eels

In the brackish-water vallis of the Mediterranean, the elvers enter the lagoons and estuaries and grow in the sheltered areas, feeding on the natural food available. In modern vallis, special eel ponds are built for monoculture and over-wintering. More intensive stocking and artificial feeding are now practised to improve production. A stocking rate of 50/ha is common and it takes 4–7 years for these elvers to grow to market size. They are caught in 'lavorieri' (traps) when they swim against the tidal inflow at the sluice gates.

In Japanese eel farms (fig. 18.3), raising of adult eels starts with seed eels weighing about 20 g. They are grown to about 150 g in ponds ranging in size from 3300 to 10 000 m^2. The most convenient size of pond is usually about 3300 m^2 (depth about 1 m). The best results are obtained in areas with a plentiful supply of warm water. Generally, the dikes are lined with wooden planks or stonemasonry, to prevent eels from burrowing into the dike. The average stocking rate is 500–700 g per m^2. The feeding rate is about 10 per cent of the total weight of the stock, if fresh fish are used as feed. The compound feeds used generally contain 46–52 per cent protein and 3–5 per cent fat. During the early years of eel culture in Japan and Taiwan, silkworm pupae formed a major item of feed, but in recent years compound feeds are more commonly used.

The feed, usually in a powder form, is mixed with water to make a paste. The paste is placed on a mesh tray in the feeding area and the eels go through the net mesh or climb on to the tray to feed (fig. 18.4). The uneaten feed can easily be removed. Eels feed avidly at warmer temperatures on clear, windy and dry days.

Most eel ponds are provided with motorized paddle wheels to aerate the water, particularly at night and in the early mornings. Some farms in Japan and Taiwan have a corner of the pond partitioned off with wooden boards into a pool at the inlet, with a suitable entrance for the eels. Water wheels are installed in these pools. During the night, when the oxygen concentration is reduced, the eels congregate in these pools. Some farms in Taiwan install compressed air blowers at the pond bottom to increase the oxygen supply.

As the growth rates of individual eels differ considerably, it is essential to sort out the stock and restock them in separate ponds, to obtain eels of marketable size. Japanese consumers prefer 100–200 g eels, and this size is reached in one year after stocking. In Taiwan and Europe, larger eels are preferred and this re-

Fig. 18.3 An eel farm in Japan. Note the intensive aeration and water circulation in the tanks.

quires a second year of growth. Japanese eel growers export large eels to Europe.

Very high production rates of up to 26 tons/ha have been achieved in recent years in intensive farming systems, particularly with the use of heated water or in running water ponds.

18.4 Diseases and mortality

Eels appear to be comparatively more prone to diseases and resultant mortality than many other aquaculture species. Unstable temperature conditions, accumulation of uneaten feed and decayed algal blooms are direct or indirect causes of mortality of elvers and adult eels. Despite their ability to breathe atmospheric air, they seem to be very susceptible to low dissolved oxygen concentrations in stagnant ponds. Fluctuations in temperature affect their feeding activity very considerably and reduce resistance to disease. Susceptibility to disease also seems to be accelerated by over-wintering practices. The above factors contribute to the incidence of high mortality in eel farms.

Fungal infection or the cotton cap disease, as it is called in Japan, is a common cause of mortality in Japanese ponds. It has been demonstrated that the fungal infection is only a secondary condition and the primary cause of the disease is a pathogenic bacterium. The disease is recognized in spring and autumn, at temperatures between 15 and 20°C. White patches of *Saprolegnia* develop and spread on the bodies, and within a week or two large numbers of eels die. This disease usually appears several days after the first feeding in spring, when the weather conditions are unstable, or when some of the stock were already infected during winter. Increased stock density (due to stocking of additional eels in the late autumn) and low or very high pH of the water (below 6 or above 10) appear to be conducive to infection and high mortalities. According to Honma (1971), feeding with medicated feeds containing furazolidon or thiazine for about a month after the start of feeding reduces the mortality rate to a considerable extent.

The red disease of pond-cultured eels affects elvers as well as adult eels. In the Japanese eel it occurs mostly in temperatures of about 28°C

Fig. 18.4 Feeding eels in a pond in Taiwan. Feed in the form of a paste or dough is placed in a tray for feeding.

in summer, but can occur also in spring. In the past the disease was thought to be caused by *Aeromonas punctata* (= *hydrophila*) or *Paracolobactrum anguillimortiferum* or a combination of the two (Hoshina, 1962). From later investigations (Egusa, 1976), however, it was shown that the disease is actually caused by *Edwardsiella tarda* and that *P. anguillimortiferum* was a misidentification. However, because of nomenclatural priorities the valid name of the pathogen now is *E. anguillimortiferum*.

The red disease is characterized by macroscopic putrefactive lesions in the kidney or liver, frequently causing high mortalities, mainly in the summer months. Most of the seriously affected eels have a swollen anal region and marked reddening of the anal and urinogenital apertures. The fish that survive the attack are reported to develop strong immunity against further infection. Medicated feeds containing chloromycetin or sulphadiazine have been found to cure the disease

effectively. Proper disinfection of the infected ponds is carried out prior to the release of new stock.

The causative agent of the red disease infecting European eels has been identified as the bacterium *Vibrio anguillarum*, which is a common pathogen of many marine fishes. The red spot disease caused by *Pseudomonas anguilliseptica* has been observed in Japanese eel ponds. It is characterized by sub-epidermal petechiae on the body surface. The occurrence of large numbers of relatively long rods in the bloodstream in advanced stages is another characteristic. The optimum temperature for the growth of the bacterium infecting European eels is 15–25°C. The disease generally occurs from spring to early summer and secondarily in the autumn, when the pond water has a temperature of about 20°C, and ceases when the temperature exceeds 25°C. The wall tissues of the circulatory organs are inflamed as a result of the infection and this leads to systemic haemorrhage. Recent studies seem to show that this disease is confined to brackish-water ponds and outbreaks occur when the water temperature is less than 26°C.

The white spot disease caused by parasitic sporozoa-like species of *Pleistophora*, *Myxidium* and *Myxobolus* often causes much damage to elvers and sometimes to adult eels. The affected eels become thin, and in elvers the body becomes black and the pigmentation disappears in patches. They swim vertically up and down in the surface layer of the pond. The maximum infection occurs in the kidneys and muscles. As no suitable cure has been developed, it is important to remove infected fish and destroy them to prevent the spread of the infection.

The bubble disease or gas embolism is a common disease of elvers. Bubble-like tumours occur, especially in the head region, due to excessive oxygen or nitrogen in the water. Sometimes the gas may accumulate in the muscles or even in the blood vessels, blocking blood circulation and causing death. Usually the disease is controlled by the introduction of clean water with a lower gas content and lower temperature. If the water supply is high in nitrogen, as in the case of ground water, aeration can be of help.

The crustacean parasite *Argulus giordani* has

caused large-scale mortality of eels in Italian vallis, which, in some areas has even led to a complete cessation of eel culture. Infection by the anchor worm *Lernaea cyprinacea* has been an important cause of mortality in eel ponds in Japan, but due to preventive and curative measures now available it is no longer a major problem.

Branchionephritis or branchial kidney disease has been identified in recent years, the cause of which has not yet been determined. The disease has been responsible for considerable losses in eel farms in Japan. The skin of the gill lamella swells, causing adhesion, and the inflamed kidney shows signs of bleeding. It would appear that the salt metabolism is impeded as a result of dehydration caused by the histological hindrance and consequent high density of the blood and drop in salt concentration. Release of the diseased fish in salt water containing 0.4–1.0 per cent sodium chloride has been found to reduce mortality.

Among other diseases reported to occur in eel farms are gill erosion, caused by bacteria, and ich, caused by *Ichthyophthirius multifiliis*.

18.5   Harvesting and marketing

Eel culture, as practised in most areas, involves partial harvesting and stocking at regular intervals. For thinning of stocks or capture of marketable eels, a scoop net is generally used in the feeding area, where the fish congregate at the usual feeding time. The day before harvesting feeding is stopped, so that on the day of harvesting the eels readily gather at the feeding spot. During fishing, fresh water is let in to avoid a reduction in oxygen levels. At times when eels fail to assemble in large numbers due to poor water conditions, a seine net with a fine-meshed bag at the centre is used. The seine is dragged from the deeper part of the pond towards the inlet, and finally the catch is removed by small dip nets. Seining is repeated two or three times and this helps to aerate the pond. Often fresh water is added to the pond through a hose, or a water mill is operated to replenish oxygen. This type of harvesting is performed in Japan in the summer months.

Complete harvesting in winter months is carried out by draining the ponds. Draining is done on warm and windless days. If the draining is done in the morning, most of the eels will not burrow into the bottom soil and will swim out with the water flow. A few eels buried in the mud can be stirred out using a T-shaped wooden scraper. If many eels still remain, the ponds are partially refilled and drained again during the night, when the eels will swim out with the drained water. After all the eels are caught, the pond bottom is treated with lime and stirred several times before allowing it to dry.

Sorting the catches is done soon after harvest. Smaller ones (below about 120 g weight) are generally used for further rearing if caught in summer or, if caught in winter, are kept for over-wintering and rearing during the following year. Larger sizes are sorted according to the size preferred in the markets (fig. 18.5). European markets prefer larger fish, above 250 g, but eels of this size are generally not eaten in Japan and so are exported to Europe. Eels of 120–150 g and 160–250 g are the preferred sizes in Japan, but the size preference varies according to the region. Taiwanese producers reserve these size groups for the Japanese market.

Before shipment of live eels, they are starved for a few days and kept confined in a limited space. This is meant to reduce the accumulation of fat and destroy any off-flavour they may have acquired. Besides this, the eels become conditioned to the lower levels of oxygen required for safe shipment. The starvation period ensures more sanitary conditions in the containers in which the eels are transported.

The most common method of conditioning used now is to stack up a number of baskets containing about 3 kg eels each and provide a shower of water from above the stack for three to four days. Elvers for transport are also acclimatized in the same manner, but the containers are made of fine-mesh screens and can hold a higher weight of about 5 kg in each container. The conditioning of elvers lasts only for half to one day.

Eels are shipped to markets alive in ordinary vinyl bags, each containing about 10 kg with about 1 kg ice, for short journeys. For longer journeys by train or truck, the eels can be packed in double-polythene boxes filled with ice and oxygen. Aerated tanker lorries are also used for the transport of live eels taking up to one week. Many of these lorries can carry 15

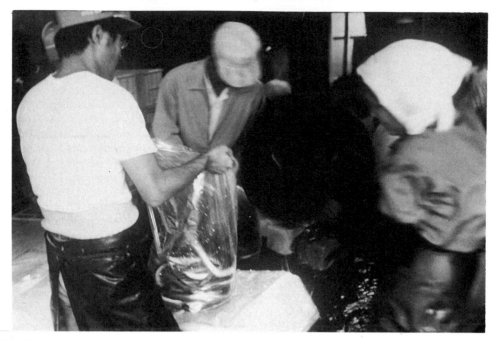

Fig. 18.5 Eels being sorted and packed according to size for transport to markets in a Japanese farm.

tons of live eels in about 15 tons of water, with continuous aeration by a compressor. If the water is changed every four days, they can be transported this way for up to two weeks.

Dead eels are generally quick frozen and glazed for transport. Eels are frozen either whole or after gutting and cleaning, in blocks in a plate-freezer or in an air-blast. Storing is done at −20°C. Each eel or block of frozen eels is glazed to prevent oxidation, which causes rancidity in eels with a high fat content. The glazed eels or blocks are wrapped in polyethylene and sealed. At −20°C, frozen eels can be kept in good condition for about six months.

There are speciality markets for smoked eels in many countries. Eels are generally hot-smoked after gutting, cleaning and brining for about 10 minutes. First they are smoked in smoking kilns for about an hour at 35°C, then for about half an hour at 50°C and finally for one hour at 73°C. This gradual increase in temperature is to enable uniform drying. The smoked product is allowed to cool before packing, to prevent the formation of moulds. Since the eels are cooked during the smoking process, they are ready to eat and the shelf life is about

3−4 days. The smoked eels can also be canned in vegetable oils, processed at a temperature of about 102°C.

18.6   Economics

Eel farming is considered a highly profitable venture in all the countries where there is a dependable supply of elvers. Besides irregularity in the availability of elvers, the susceptibility of eels to environmental changes affects production and consequently profitability. Shang (1973) estimated the rate of return on investments in an eel farm in Taiwan to be about 65 per cent. Based on experience in eel farming using heated water effluents in France, Lemercier and Serene (1981) estimated the internal rate of return to be 17.5 per cent and the pay-back period as between 9 and 10 years, using an actualization factor of 17.5 per cent.

18.7   References

Bardach J.E., Ryder J.H. and Mc Larney W.O. (1972) *Aquaculture*. John Wiley & Sons, New York.

Boëtius I. and Boëtius J. (1967) Studies in the European eel *Anguilla anguilla* (L). Experimental induction of the male sexual cycle, its relation to temperature and other factors. *Medd. Dan. Fisker — Havunders.* (Ny. Ser.), **4**(11), 339–405.

Boucher S., Boucher R. and Fontaine, M. (1934) Sur la maturation provoquée des organes génitaux de l'anguille. *C.R. Séances Soc. Biol. Paris*, **116**, 1284–6.

Chen T.P. (1976) *Aquaculture Practices in Taiwan*, pp. 17–28. Fishing News Books, Oxford.

Egusa S. (1976) Some bacterial diseases of freshwater fishes in Japan. *Fish. Pathology*, **10**(2) pp. 103–14.

Fontaine M. *et al.* (1964) Sur la maturation des organes génitaux de l'anguille female (*Anguilla anguilla*, L.) et l'émission spontanée des oeufs en aquarium. *C.R. Hebd. Séances Acad. Sci., Paris (D)*, **259**(17), 2907–10.

Forrest D.M. (1976) *Eel Capture, Culture, Processing and Marketing*. Fishing News Books, Oxford.

Ghittino P. (1983) *Tecnologia e Patologia in Acquacoltura. Vol. 1. Tecnologia*. Torino, Italy.

Honma A. (1971) *Aquaculture in Japan*. Japan FAO Association, Tokyo.

Hoshina T. (1962) Studies on red disease of eel. (In Japanese.) *J. Tokyo Univ. Spec. Edn.*, **6**(1).

Huet M. (1986) *Textbook of Fish Culture*, 2nd edn., pp. 217–21. Fishing News Books, Oxford.

Lemercier P. and Serene P. (1981) Commercial eel farming using heated effluent in France. In *Aquaculture in Heated Effluents and Recirculating Systems*, Vol. 2 (Ed. by K. Tiews), pp. 587–92. Schriften der Bundesforschungsanstalt für Fischerei, Berlin.

Matsui I. (1979) *Theory and Practice of Eel Culture*. (Translated from Japanese.) Amerind Publishing, New Delhi.

Ravagnan G. (1978) *Vallicoltura Moderna Edagricole*. Bologna, Italy.

Shang Y.C. (1973) *Economic aspects of eel farming in Taiwan*. Taiwan China Joint Commission on Rural Reconstruction.

Usui A. (1984) *Eel Culture*. Fishing News Books, Oxford.

Yamamoto K., Yamaguchi K. and Moricka T. (1975) Pre-leptocephalic larvae of the Japanese eel. *Bull. Jap. Soc. Sci. Fish.*, **41**(1), 29–34.

19
Tilapias

Tilapias (family Cichlidae) are natives of Africa. They have been introduced into a large number of tropical and sub-tropical countries around the world in the last four or five decades, either accidentally or deliberately. Aquaculturally this group of species has had a rather chequered history. Tropical aquaculturists who experienced considerable problems in controlled spawning of fishes were initially excited by the availability of a species that could breed in almost any type of water body. Being herbivorous or omnivorous, it was comparatively easy to feed the species of this group. They were found to be hardy and could be reared in fresh, brackish and even sea water. Even though the darkish coloration of the fish was not very attractive to some, large fish were well-relished when presented under different names such as 'lake fish' or 'bream'.

Because of these favourable characteristics, tilapias were considered ideal species for rural fish farming. In early efforts to develop fish culture at a subsistence level in Africa, oriented to improving the nutrition of rural populations, tilapias were the natural choice, especially in the Belgian Congo (now Zaire). But some of the perceived advantages proved to be real constraints to profitable fish farming, particularly the prolific breeding, which resulted in overpopulation of small stagnant ponds. Experience in Asia in the early days of tilapia farming, after the accidental introduction of *Tilapia mossambica* in Indonesia, created opposing reactions in countries of that region. While governments of some countries like Thailand gave active support to stocking every water body with this species, others, like India, tried to restrict its culture to prevent possible escape of the species into the major river systems.

Despite some of the advantages of tilapia as candidate species for small- or large-scale culture, it was soon realized that the culture technique for producing marketable fish was not as easy as originally believed. Their early maturation and frequent breeding, especially in tropical climates, affected growth rates to such an extent that special measures of stock management and feeding were found necessary to produce fish suitable for human consumption. Even in areas where small fish were acceptable as table fish, stunted tilapia could be used only as livestock feed. There were very few profitable tilapia farms at that time, and naturally interest in tilapia culture dwindled for some time.

In recent years, the status of tilapia as a culture species has risen again, as a result of efforts by enterprising farmers and aquaculture technicians. Enthusiasm for the species has become so high that some have begun to describe tilapia as the future 'aquatic chicken'. Irrespective of whether this is an exaggeration or not, it is clear that workable technologies are now available for raising some of the species or hybrids of tilapia on a profitable basis, even though many problems still remain to be solved.

19.1 Cultivated species of tilapias

Thys (1969) described at least 77 species (besides a number of sub-species) of *Tilapia*, and Jhingran and Gopalakrishnan (1974) listed 22 species that have been used in experimental or production-scale fish culture. Considerable confusion exists on the taxonomic status of many of them. Because of the overlap of morphological characteristics, taxonomists have tried to split the genus of *Tilapia*, based largely on their breeding behaviour which also coincides roughly with macrophytophagous, microphagous or omnivorous feeding habits. The substrate spawners, which make nests on the bottom of water bodies and spawn in them, retained the name *Tilapia* and the mouthbrooders, which incubate the fertilized eggs in the mouth of the female or male parent, came under a new genus *Sarotherodon* (meaning 'brush toothed') (Trewavas, 1982). Later, a new genus *Oreochromis* was constructed to accommodate species which spawn in nests on the bottom of water bodies, but brood the eggs in the mother's mouth. Though in most cases *Tilapia* spp. have coarse teeth and feed on macrophytes, *Sarotherodon* spp. have fine teeth and feed on unicellular and filamentous algae. The feeding habits are, however, highly flexible and are not a strong diagnostic characteristic. Recently, two alternative classifications were proposed; one includes five genera, *Tilapia*, *Sarotherodon*, *Oreochromis*, *Tristromella* and *Danakilia*, and the other only one genus, *Tilapia*, with seven sub-genera: *Heterotilapia*, *Pelmatilapia*, *Sarotherodon*, *Oreochromis*, *Nyasalapia*, *Alcolapia* and *Neotilapia* (Fishelson and Yaron, 1983). These revisions of the classification have not eliminated the confusion and other taxonomists prefer the continued use of the broad genus *Tilapia* for all the species. As aquaculturists may find it difficult to keep up with the frequently changing nomenclature, in this chapter it is proposed to use the generic name *Tilapia* for all the species, with no reference to the sub-genera.

For commercial aquaculture, the more important species of tilapia are: *T. rendalli*, *T. zillii*, *T. mossambica*, *T. hornorum*, *T. nilotica* (fig. 19.1), *T. aurea* and *T. melanotheron*. Two other species, *T. andersonii* and *T. spilurus*, also seem to be of some importance. It is believed that very few pure strains of these species are used in fish farms, and natural cross-breeding has occurred in many areas. As will be discussed later in this section, interspecific hybrids including red-coloured hybrids have been cultured commercially on a limited scale.

Fig. 19.1 Nile tilapia, *Tilapia nilotica*, a widely cultivated species.

Though essentially a tropical species which cannot survive temperatures below 10°C, tilapias have been introduced for commercial culture in sub-tropical areas and even in temperate areas for indoor culture under controlled temperature conditions.

19.2 Culture systems

Tilapias are euryhaline and grow well in brackish and salt waters. *T. mossambica* and *T. zillii* can grow even in hypersaline waters above 42 ppt. The hybrid red tilapia seem to grow best in brackish- and sea-water environments. Species like *T. aurea* and *T. zillii* do not appear to breed in high salinities, but *T. mossambica* reproduce at as high a salinity as 49 ppt (Popper and Lichatovich, 1975).

The most common and widely practised system of culture of tilapia is in earthen ponds and similar impoundments (fig. 19.2). As the species can survive in a restricted space, all sizes of ponds have been used, including those measuring less than 100 m². In pond culture, attempts have been made to control overpopu-

lation by stocking a certain number of predators (2–10 per cent of the stock), like *Hemichromis fasciatus*, *Lates niloticus*, *Clarias lazera*, *Micropterus salmoides*, *Channa striata* and *Cichla ocellaris*. In brackish- and salt-water ponds, *Elops hawaiensis* and *Dicentrarchus* spp. have been used as predators. In order to reduce breeding and increase production, mono-sex culture of males is carried out in a number of tilapia farms. The techniques of separating the sexes, or producing mono-sex stocks by hybridization or by sex-reversal, have not been perfected to the extent necessary to ensure the complete absence of female fish. A few female fish in the ponds can cause uncontrolled breeding.

In many areas, tilapias are produced mainly by polyculture. They have been used as a compatible species with a number of freshwater fish, including carps, grey mullets, *Clarias lazera*, *Heterotis niloticus* and the Amazonian characid tambaqui (*Colossoma macropomum*).

Intensive monoculture of tilapia in indoor tanks is carried out in colder climates, as in

Fig. 19.2 A pond farm growing tilapia in Costa Rica.

China, using warm water during the winter season. Over-wintering of fry during the cold season and stocking in open ponds during the spring is also a common practice in temperate regions. The economic viability of these systems depends very much on the local market value of the species.

Cage culture of tilapia in both fresh and salt water has received considerable attention, not only for more intensive production, but also as a means of controlling wild spawning and over-population. Although in many areas it is still in an experimental or pilot scale only, there are some successful commercial operations, as for example in the Philippines and Costa Rica (fig. 19.3). Pen culture of tilapia in open waters of lakes is practised in the Philippines. Tank and raceway culture are also done on a very limited scale, for producing marketable fish (fig. 19.4) or bait fish.

In some of the countries of Southeast Asia, especially in the Philippines and Thailand, rice-field culture of tilapia is practised on an appreciable scale. To obtain marketable fish within the short period of rice cultivation or between crops, mono-sex culture has to be adopted. Tilapias have also figured as important species in integrated animal and fish farming systems in several Asian and some African countries.

Stocking in open waters has been carried out in a few countries to enhance or develop commercial fisheries. A notable success is the development of self-sustaining stocks in lakes and reservoirs in Sri Lanka and in Lake Kinneret (Tiberius) in Israel. Stocking has also been undertaken in lakes or reservoirs in East Africa (Kenya, Tanzania, Uganda, Rwanda, Zambia, Zimbabwe, etc.) and in central Florida (USA). The stocking of conservation dams with tilapias was a common practice in Central East African countries, and lately Israel has adopted stocking in irrigation reservoirs of 15−24 ha area (Sarig, 1983).

19.3 Spawning and production of seed stock

It is somewhat paradoxical that mass production of seed stock of a group of species that spawns easily, early and frequently should present problems. But in point of fact, one of the major constraints to large-scale commercial tilapia farming is the scarcity of fry. Some of the species attain maturity as early as three months old, and under favourable temperature conditions breed in successive cycles at 4−6 week intervals. This results in continuous production of fry, but the farmer finds it difficult to obtain sufficient fry of the required size at any particular time. The traditional practice of harvesting tilapia ponds at regular intervals during the culture period, to remove larger fish and to allow the fry and young fish (produced by wild spawning) to reach market size before the next harvest, only results in lengthening the culture period and allowing heavy wild spawning in the ponds. This system also has the disadvantage that a good proportion of the unharvested stock would be individuals selected for slow growth. When used for further fry production, there is a likelihood of this characteristic of slow growth being passed on genetically to the offspring.

In sub-tropical or temperate regions, spawning may be limited to the summer. Even here, depending on the preferred spawning temperature range (usually 20−30°C), the fish would spawn several times and overpopulate ponds with fry of different sizes. Thus, the problem of overpopulation and stunting applies also to these regions, though not to the same extent as in the tropics.

19.3.1 Methods of spawning

Attempts have been made to develop a suitable fry production system for tilapias. Campbell (1978) described a relatively simple method of producing large numbers of *T. nilotica* fry using 600 m^2 earth ponds of about 0.4 m depth. Female fish of about 700 g weight and males of 200 g are stocked in one pond at an average density of one per 2 m^2 in the sex ratio of one male to four or five females. They are fed on a high-protein diet for about a month, by which time they will have started spawning. The brood fish are then transferred to a second pond, where they are fed in the same way as in the first pond. Feeding is continued in the first pond as well for another month, by the end of which the fry will have reached a size of about 4 cm and on average about 5000 fry are available for harvest. By this time, spawning will have occurred in the second pond and the

Fig. 19.3 A tilapia cage farm in Costa Rica.

Fig. 19.4 A raceway farm for tilapia near Mombasa, Kenya.

brood fish can be transferred back to the first pond for further spawning. The production per month by this method is reported to be about 4.2 fry per m² or 10.4 fry per female.

In Israel, ponds ranging in size from a few square metres to 5 ha, with gently sloping bottoms, are used for spawning *T. nilotica* and *T. aurea*. The ponds are dried prior to spawning to eradicate weed fish and pests. They are filled to a depth of 50−60 cm, which is the preferred depth for spawning of these species. As the number of eggs per spawning depends on the size of the females, the stocking rate is varied according to their size. While a 100 g *T. nilotica* spawns about 100 eggs, a 600−1000 g fish will spawn about 1000−1500 eggs. A female *T. aurea* of about 1000 g weight may spawn about 2000 eggs each time. The stocking rate for males is generally 100−250 per ha.

In the Philippines, land-based spawning ponds as well as open-water-based cages or hapas are used for spawning and fry rearing. Many farms use hapas made of nylon mosquito netting to breed *T. nilotica* and hybrids of *T. nilotica* × *T. mossambica*. The brood fish are maintained in hapas installed in ponds with about 1 m depth of water. The fish continue to breed throughout the year. A 1:3 male to female sex ratio has been found to be suitable. The fry are collected at intervals of about a month and grown to fingerling stage in nursery ponds or cloth tanks. For cross-breeding, the best sex ratio has been found to be one male to three females. The average monthly production is computed to be about 1466/m³.

Open-water-based cage hatcheries used in the Philippines consist of double-walled net cages very much like the double-walled hapas used for carp hatching in India. The inner coarse-mesh (30 mm) net measures 10 × 2 × 1 m, and the outer fine-mesh net, 12 × 4 × 1.5 m. They are installed in protected calm areas of lakes, such as the ones found in Laguna de Bay. Breeders are stocked at a density of four per m² with the same 1:3 sex ratio as in ponds, and fed with fine rice bran at 3 per cent body weight per day. Spawning occurs at regular intervals and the fry are collected and stocked in rearing hapas (10 × 2 × 1.5 m) at the rate of 1000 per m². Fine rice bran is used for feeding the fry at the rate of about 6−8 per cent of body weight. After two weeks of rearing in hapas, the fingerlings are transferred to larger-meshed (6.5 mm) cages at the rate of 250−500 per m² and fed with fine rice bran at the rate of 4−6 per cent of body weight per day.

Systems that allow a high degree of environmental control make year-round spawning of tilapias possible in temperate climates. Removal of eggs from incubating females and hatching and rearing them separately in special containers helps to increase spawning frequency and thereby overall fry production. Another important advantage of spawning under controlled conditions is that genetic purity of lines can be maintained and this is of special importance in hybrid production. The sex ratio of females to males is generally 3:1 or 4:1. As tilapia spawn at frequent intervals, harvesting has to be carried out every fortnight, when the fry are about 0.5 g in weight.

The aggressive behaviour of the male in an aquarium or tank manifests when mature fish are introduced at a size of about 100 g. Long aquarium tanks (200 × 50 × 40 cm) are stocked with immature, 4−5 months-old fish. One male and seven to ten females form a 'family' in each aquarium. When they become sexually mature, the males of mouth-brooding tilapia species dig nests at the bottom if there is sand or gravel there. Even if the bottom is bare, they exhibit digging movements and the male chooses the ripest female and, after a period of courtship which may last several days, spawning and fertilization take place in the nest or the bottom of the aquarium. Soon after, the female picks up the eggs in her mouth. The male then chooses another ripe female for courtship and spawning. The eggs are removed from the female's mouth after 3−5 days for further incubation. This helps in preventing cannibalism and early preparation of the female for further spawning. Zuge jars or containers placed on a shaking platform (for keeping the eggs separate and in continuous movement) are used in incubation. The eggs hatch out in about 50 hours at temperatures of 25−27°C. The larvae remain in the incubating containers until the yolk sacs are absorbed, which may take about 8−10 days.

Nursing of normal fry or mono-sex hybrids is carried out in nursery or rearing ponds. Stocking densities vary from 50 000 to 100 000

per ha, depending on the size of fingerlings to be raised. When manual sexing of fry is needed, it is necessary to grow them to a size of at least 20–50 g so as to distinguish secondary sexual characteristics with ease. Even a larger size of 100 g is often recommended, but it has to be ensured that the stock are removed before they reach maturity.

19.3.2 Mono-sex seed stock and hybrids

One of the methods in controlling wild spawning of tilapias is mono-sex culture, and since the male tilapia grows faster and attains a larger size, interest has been focused on producing all-male seed stock. Obviously the simplest means is to sort out males from unsorted stocks of fry, commonly known among fish culturists as hand sexing (or manual sexing). The sexes can be distinguished by visual examination of the urino-genital papillae. In the females, the papilla has two orifices whereas the male has only one. Often the female has a smaller genital papilla. No doubt this needs some personal skill and carefulness and can be done reliably only with fingerlings of 20–50 g size. Even at this size, there is likely to be a certain percentage of error in sorting, and even a small number of females in the stock can initiate wild spawning in production ponds. It also involves waste of female fish, although some farmers use the sorted females for preparing feeds for males to be grown to market size. Despite the skilled manual labour involved, sorting of males for commercial scale culture is practised in a number of countries. Growth and production are substantially increased, even though some wild spawning does take place in the ponds. Draining of ponds after harvesting makes it possible to start the next crop with sorted seed stock.

Another approach to the production of mono-sex stock has been by the use of steroid hormones to achieve sex reversal. As mentioned in Chapter 8, it has been possible to reverse the sex of genotypic females by the administration of methyltestosterone or ethynyltestosterone in *T. mossambica*, *T. nilotica* and *T. aurea* (Guerrero, 1982). The degree of success varied between 90 and 100 per cent. The steroids were incorporated in the feed of fry at rates ranging from 10 to 60 mg per kg fish, for durations varying from 18 to 60 days. Feminization of genotypic males was achieved in 90–100 per cent of the males of the above mentioned species, by the administration of oestrogens (ethynyloestradiol, oestrone and diethylstilboestrol). The dose consisted of 50–100 mg/kg of ethynyloestradiol (in experiments with *T. aurea* a 100 mg/kg dose of methallubure was included), 200 mg/kg of oestrone and 100 mg/kg diethylstilboestrol and the duration of treatment varied between 19 and 56 days. From comparative experiments, Hanson *et al*. (1983) found that the sex-reversed male populations have a higher growth rate than hybrids and females.

Rothbard *et al*., (1983) have described the procedure adopted in Israel to produce hormonally sex-inverted all-male tilapia. The fry are placed in outdoor concrete circular tanks of 28 m^3 capacity (diameter 6 m). Commercial high-protein trout starter feed or eel feed is mixed with the androgen 17 a-ethynyltestosterone dissolved in 95 per cent ethanol (technical grade) for feeding the fry. The ethanol is evaporated by drying the mixture in the sun for several hours and the fry are fed at the rate of 12 per cent of their body weight per day. The tanks are protected from sunlight and the water temperature maintained between 21 and 22.5°C. The treatment lasts about 28–29 days. Treatment of hybrids of *T. nilotica* males and *T. aurea* females, F-1 hybrids of *T. nilotica* and *T. aurea*, and the red tilapia yield populations containing 98–100 per cent males. Studies showed the level of testosterone in the plasma of sex-inverted fish to be only 11.1 ± 4.3 ng/ml, compared to sexually active males of *T. nilotica* and *T. hornorum* with 37.8 ± 9.1 ng/ml and 41.7 ± 4.6 ng/ml respectively. From this it is concluded that androgen treatment of fry has no effect on circulating testosterone levels at post-maturation (Rothbard *et al*., 1982).

Hopes of using inter-specific hybrids as a means of controlling wild spawning were aroused by the production of all-male progeny by crossing *T. mossambica* females with *T. hornorum* males by Hickling (1960). Besides producing mono-sex male populations, crossbreeding could help in improving catchability, growth rate, temperature tolerance and body coloration. These could greatly enhance the value of tilapia as candidate species for large-scale fish culture. Since then a number of all-male or predominantly male hybrids have been produced:

T. nilotica × *T. hornorum* (Pruginin and Kanyike, 1965)
T. nilotica × *T. aurea* (Fishelson, 1962)
T. nilotica × *T. variabilis* (Pruginin, 1967)
T. spilurus niger × *T. hornorum* (Pruginin, 1967)
T. vulcani × *T. hornorum* (Pruginin, 1967)
T. vulcani × *T. aurea* (Pruginin, 1967)
T. nilotica × *T. macrochir* (Lessent, 1968)

One hybrid that has received special attention from fish culturists for some time is the so-called red tilapia, the colour of which is a blend of pink, yellow and gold. It is appreciated in the market in preference to the normally silvery grey or black-coloured tilapia. Red tilapia is known to have a faster growth rate and food conversion ratio. It can grow in both fresh- and brackish-water environments. The origin of this hybrid is not yet fully documented. A reddish-orange F-2 progeny with superior qualities was obtained in Taiwan by crossing a mutant reddish-orange female of *T. mossambica* with a normal-coloured grey male *T. nilotica*. In the Philippines, a similar reddish-orange or golden progeny was obtained by cross-breeding a female hybrid of *T. mossambica* × *T. hornorum* with a strain of *T. nilotica*. Galman and Avatlion (1983) found that the red tilapia is intermediate in several characteristics between *T. mossambica*, *T. hornorum*, *T. nilotica* and *T. aurea*, and speculated that all these species are involved in the hybrid.

Lovshin (1982), who has reviewed experience in tilapia hybridization, pointed out that in spite of the knowledge that all-male or predominantly male populations can be produced by hybridization, commercial culture of such hybrids is limited. One reason for this is the difficulty in maintaining pure genetic lines which are necessary to obtain consistent results in hybridization. In commercial production, varying proportions of females occur as a result of contamination of the brood-stock lines. Electrophoretic comparisons of blood proteins and crossing in aquaria of the brood stock until all-male offspring are consistently produced have been suggested as a means of ensuring pure brood stocks. These procedures are feasible in breeding centres, but there are only a few countries where such facilities are available at present for the production and distribution of selected pure lines of aquaculture species. Further investigations by Majumdar and McAndrew (1983) showed that even crosses between pure lines produce varying sex ratios. In 41 trials, only one cross (*T. mossambica* males × *T. macrochir* females) gave 100 per cent male progeny.

Small aquaria, concrete tanks and plastic pools are generally used for hybrid production. As mentioned earlier, male aggression is a problem in spawning operations in small containers. Cannibalism of fingerlings on newly hatched larvae is an additional problem in aquaria and similar containers. According to Lovshin (1982), some hybrid crosses are difficult to carry out in small confined environments. It has been demonstrated that earthen ponds can be successfully used in cross-breeding of tilapias, following the general procedures for pond spawning.

Lovshin (1982) has described a system developed in Brazil for the production of hybrids of *T. hornorum* and *T. nilotica*. Figure 19.5 illustrates the steps recommended. Fingerlings of the two species are sexed when they have reached weights of 20–30 g, and the males and females are stocked separately in segregated brood-stock preparation ponds, at the rate of two or three per m^2. They are fed at the rate of 5 per cent of their body weight daily, and in 2 to 3 months grow to about 60–100 g and become sexually mature.

Mature male *T. hornorum* and female *T. nilotica* which have swollen genital papillae are introduced in the spawning pond at a ratio of 1:1. The stocking rate is one female to every 7 m^2 pond surface. The brood stock are fed with agricultural byproducts at the same rate as for immature fish. After about two and a half months, the spawning ponds are drained and the hybrid fingerlings collected for rearing in nursery ponds. The ponds are dried after draining and poisoned to eliminate any small fry that may have been produced by back-crossing between all-male hybrids and female *T. nilotica*. The process can then be repeated with brood stock that have produced sufficient number of hybrid fingerlings.

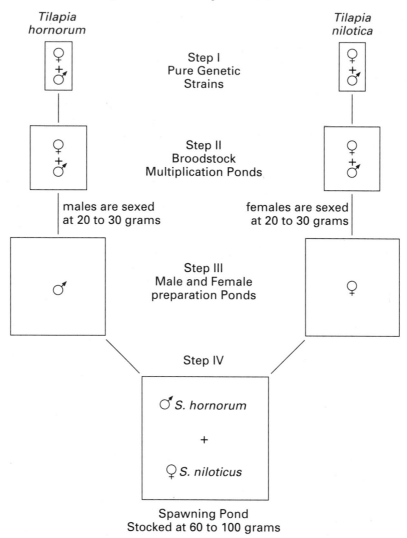

Fig. 19.5 System used for producing all-male tilapia hybrids, *T. nilotica* × *T. hornorum* in Northeast Brazil (after Lovshin, 1980).

According to Hepher and Pruginin (1981), tilapias do not readily hybridize in aquaria and so artificial propagation methods are generally used for hybrid production in Israel. Ripe spawners with swollen papilla and intense pigmentation are selected. The eggs from one of the females of the selected species are stripped into a suitable container and mixed for about 2 minutes with milt from the male of the other species. About 10 ml saline solution is added and the eggs stirred for a further 2 minutes. The eggs can then be rinsed and transferred to an incubator for hatching.

19.4 Grow-out and feeding

19.4.1 Grow-out in ponds

Tilapia culture is generally oriented to producing fish of marketable size of at least 200–300 g. When the grow-out is based on unsorted seed stock, this can be achieved only by low stocking densities and intensive fertilization and feeding, so that the majority of the stock will have reached an acceptable size before they become sexually mature and start breeding. For this purpose, a low stocking density of 3000–5000 per ha is recommended (Hepher and Pruginin, 1981). Two or three crops are raised every year, and after each harvest the ponds are drained completely. The possible use of predators to control fry produced by wild spawning has been referred to earlier, but probably because of insufficient numbers of predator fry, this practice has not yet been widely used.

For successful grow-out it is necessary to stock recently produced fry or fingerlings and not the stunted fingerlings from a previous crop that would breed early and cause overpopulation and stunting of stocks. Commercial culture of tilapia in the Philippines in fresh-water ponds is largely based on *T. nilotica*. Successful farmers stock about 20 000–30 000 fingerlings per ha. The ponds are fertilized with chicken manure and ammonium phosphate. Supplementary feeding is done with fine rice bran, and some farmers mix it with dried broiler manure. Selective harvesting starts after 4 months of culture, and complete harvesting is done after five months. The average production is reported to be about 2 tons/ha per crop.

All-male stocks eliminate many of the problems of tilapia culture, in tropical as well as sub-tropical waters. They can be grown to a larger size of 400–600 g weight to suit consumer acceptance. The longer grow-out period is compensated by the higher prices that large tilapia fetch in the market. Stocking densities of hybrid or all-male populations depend on the level of inputs and management. At even low stocking densities of 3000–5000 per ha, with supplementary feeding of protein-rich feeds, individual weight increases of up to 3–5 g per day have been recorded. All-male *T. nilotica* and all-male hybrids of *T. nilotica* × *T. hornorum* of 60–63 g weight, stocked in densities of 10 000/ha and cultured for a 6-month period, yielded 2.8 tons/ha of *T. nilotica* and 3.2 tons/ha of hybrids (Lovshin, 1977). The average weight of the fish was about 240 g. At stocking levels of 13 000–31 000 per ha of 22–25 g hybrids, annual productions of 5.6–12

tons per ha (of fish weighing up to 495 g) have been recorded (Lovshin, 1982). With stocking densities of 20 000/ha and intensive feeding, production of up to 25 tons/ha per annum has been obtained (Hepher and Pruginin, 1982). In such high-density cultures, it is necessary to aerate the ponds, at least during the night.

In Israel, all-male tilapia (*T. nilotica* and *T. aurea*) or hybrid tilapia are commonly cultured in polyculture systems with common carp, mullet and silver carp. Such a combination is believed to contribute to the maximum utilization of feeds and improvement in the chemical regime (dissolved oxygen, ammonia and nitrite levels) in ponds, due to a proper balance of phytoplankton communities and detritus, and also to the better growth rates of some of the species, especially common carp (Sarig, 1983). The proportion of tilapia varies very much between farms, but it is usually higher than mullet and silver carp, and is second only to common carp. Some farms cull a good percentage of carp towards the end of the rearing period and replace them with tilapia. Tilapia then becomes the predominant species. The annual yield in many farms in polyculture can reach 7.5–10.7 tons per year. Tilapia may account for anything between 9 and 53 per cent of the production, the average being around 25 per cent.

Experiments conducted in Brazil have shown that polyculture of tilapia hybrids (*T. nilotica* × *T. hornorum*) with tambaqui (*Colossoma macropomum*) is feasible. Stocked at the rate of 10 000 tambaqui per ha and 3000–5000 tilapia hybrids per ha, and fed with pelleted chicken diet (17 per cent protein) at 3 per cent of the body weight, an average production of 7.5 tons/ha of tambaqui and 2.9 tons/ha of tilapia have been achieved (Lovshin, 1982). In some areas of China, polyculture of *T. mossambica* and recently also of *T. nilotica* with Chinese carp is practised, but tilapia forms only a subsidiary species. Brood fish and fry are reared indoors in greenhouses at temperatures between 17 and 20°C during winter. In spring they are transferred to well-fertilized outdoor ponds. Yields in 4–5 months of rearing can reach 1100 kg/ha without supplemental feeding, and 1900 kg/ha with feeding.

Although tilapia can be divided into macrophagous and microphagous species, under pond

Fig. 19.6 Catch of tilapia from an intensively cultured pond farm.

culture conditions they have flexible feeding habits. Detritus forms a good percentage of their food. Fertilization of tilapia ponds is aimed at increasing overall food production in ponds. Experience so far seems to indicate greater efficiency of organic manuring in comparison with inorganic fertilizers. Organic manures increase detritus in ponds which stimulates growth of bacteria and zooplankton. In manured ponds, high yields of tilapia have been obtained even with high densities and without supplementary feeding. Ponds in Brazil, stocked with all-male hybrid tilapia at a density of 8000/ha (25 g average weight) and fertilized with 500 kg/ha of chicken manure per week gave an average yield of 1.35 tons/ha after 189-day culture. The average size at harvest was 186 g (Lovshin and Da Silva, 1975). The high performance of tilapia in integrated farming with pig and duck rearing (see Chapter 29) reflects the advantages of organic manuring in tilapia ponds.

19.4.2 Cage and pen culture

Coche (1982) made a very extensive review of cage culture attempts in different countries.

The early interest in cage culture of tilapia was on the assumption that wild spawning would not occur in cages or, if it did, the progeny would not remain in the cages and cause overpopulation, as in ponds. Later the value of cage culture in utilizing open bodies of water, particularly those of eutrophic lakes, coastal areas and running waters was recognized. Though of wide-spread interest, most of the cage culture presently practised is still on an experimental scale, with only few exceptions as in some areas of the Philippines, Ivory Coast, Costa Rica and El Salvador. The commonly used species are *T. mossambica*, *T. nilotica* and *T. aurea*.

Cages are mainly used for grow-out, and the necessary fry or fingerlings are produced in land-based facilities like ponds, cisterns, etc., or in hapas installed in ponds. As described earlier, special double-walled cages can be used for spawning of tilapia in open waters. Fixed and floating cages are used in the open waters of lakes in the Philippines for tilapia grow-out. The fixed cages are used in shallow eutrophic lakes and the floating cages in deep lakes. The stocking density varies with the size of the cage, but in floating cages up to 25 fingerlings/m^2 of 3—4 cm length are common. Artificial feeding is generally not practised, except in waters with low productivity. In 6 months, from February to July, the fingerlings grow to a size of 200—250 g each, and in nine months from August to April, 250—350 g each. The growth rate is largely based on the primary productivity of the lake and the management practices, which include the density of cages in the lake and the distance between cages.

The stocking rate of *T. nilotica* in fixed cages ranges from 15 to 50 fingerlings/m^2, and the duration of culture from 4 to 12 months. The growth rate depends on the productivity of the lake. Without supplemental feeding, 5 cm fingerlings stocked at 15 per m^2 in Laguna de Bay attained 150—180 g in 4 months. The average production of 3.5—7 kg/m^3 of 100—150 g fish has in recent years been greatly reduced as a result of crowding of cages in the lake (Coche, 1982).

Pen culture of tilapia appears to be practised only in the Philippines in the Laguna de Bay. The same materials used for abandoned milkfish pens are used for the construction of tilapia pens, but smaller pens of 0.5—1 ha are preferred. The pens are stocked at the rate of 20—50 fingerlings/m^2 and fed with rice bran or wheat bran at 2—3 per cent of body weight per day. The growth rate varies according to the productivity of the lake, and in productive waters they can grow to a size of 170—250 g in 4—5 months, even without feeding. One of the major problems in pen culture of tilapia is the poor catchability of the species used, namely *T. nilotica*. The harvesting rate with seine and gill nets has been reported to be only about 15—30 per cent.

19.4.3 Tank and raceway culture

The interest in tank and raceway culture of tilapia originated with experiments to determine the suitability of tilapia as bait fish for tuna. In experimental work in Hawaii, it was demonstrated that spawning and fry rearing could be carried out in raceway type 4500 ℓ tanks (6 m × 0.9 m and 0.9 m deep). Later efforts in tank farming were mostly for environmental rehabilitation, as in the case of the Baobab Farm near Mombasa, Kenya, where large limestone quarries were created by the excavation of coral scrublands for cement manufacture. In the Baobab Farm, the fry are stocked in a two-tier raceway system, at the rate of 1000—2000 per m^3. They are regularly graded and the fast-growing ones, comprising 70—90 per cent males, are introduced into the lower tier of the raceways. Fingerlings weighing 50—75 g are transferred to a series of circular production tanks of about 20 m^3 capacity and fed regularly on pelletized feed containing 20—35 per cent protein. A continuous water flow rate of 0.5—1.0 ℓ/min per kg is maintained in the tanks, which is adequate to provide the oxygen requirements and to flush out waste products. The stocking rate is 200—500 per m^3. In about 3 months, around 70 per cent of the stock reach about 250 g and can be marketed. Each tank can produce four crops per year with yields between 100 and 200 kg/m^3 per year. According to Balarin and Haller (1983) the most economic unit under Kenyan conditions is a 75—100 ton facility, with the expected return on capital of nearly 25 per cent of total investment.

19.4.4 Feeds and feeding

The need to grow tilapia to a marketable size in a short time has been pointed out earlier. A variety of feedstuffs have been used in tilapia ponds, including plant leaves, rice bran, oil seeds and oil cakes, copra wastes, manioc and brewery wastes. Culturists have in some cases used chicken diets (often mixed with protein-rich ingredients) or, rarely, the more expensive trout feeds. But in the majority of cases, feeds are prepared on the farm using locally available ingredients. A simple inexpensive formulation used in the Philippines consists of 65 per cent rice bran, 25 per cent fish meal and 10 per cent copra meal. Another formulation tried in the Central African Republic consisted of cotton seed oil cake (82 per cent), wheat flour (8 per cent), cattle blood meal (8 per cent) and bicalcium phosphate (2 per cent). Coche (1982) cited the feed formulations containing 20—22 per cent protein used in the Ivory Coast, consisting of 61—65 per cent rice polishings, 12 per cent wheat middings, 18 per cent peanut oil cake, 4—8 per cent fish meal and 1 per cent oyster shell.

19.5   Diseases and mortality

Comparatively few diseases and mortalities due to infection have been reported in tilapia farms in the tropics. Many of the pathogenic organisms described from the wild stocks only indicate possible infections under culture conditions. Besides a possible natural resistance to disease, the low-density culture practices may have helped to reduce stress and consequent susceptibility to diseases. However, even in low-density culture the high organic loads create conditions suitable for significant bacterial populations to flourish and infect the fish. Several pathogenic protozoans and bacteria have been observed in species of tilapia, but very few of them have been reported to cause major concern. Some of the known diseases seem to occur only in sub-tropical and temperate regions, where over-wintering of fry causes greater stress.

Among mortalities caused by environmental factors, the most important are anoxia following blooms of algae such as *Microcystis, Anabaena, Oscillatoria*, etc. A sudden lowering of the temperature as a result of environmental changes or the entry of very cold water at a temperature below the tolerance level of about 11°C can create problems including mortalities.

19.6   Harvesting and marketing

Harvesting schedules in tilapia culture depend very much on the seed stock used and the climatic conditions in the area. If exclusively mono-sex males are cultured in tropical climates, the duration of rearing can be adjusted according to the preferred size to be marketed. If unsorted stocks are used, or if the hybrids or sorted stock include some females, harvesting is generally carried out before too much wild spawning has occurred.

When ponds and rice fields can be drained, fish harvesting presents few problems. Harvesting from cages is also fairly easy. Partial harvesting in ponds is generally by seines, but significant differences in catchability have been observed between species and hybrids. *T. hornorum* is a species that can be caught easily, whereas *T. nilotica* and *T. aurea* avoid seines by lying on their side on the pond bottom, and repeated seinings are necessary to catch a good proportion of the stock. Catching becomes a major problem in pen culture as indicated in Section 19.4.2, Cage and pen culture. All-male hybrids of *T. nilotica* × *T. hornorum* are reported to be caught much more easily from ponds.

In small-scale rural farms, the marketable surplus catches are generally sold fresh at the farm gate or in the nearby village markets. Larger farms usually transport the catches to urban markets, on ice, and in the case of far away markets sometimes even frozen. In markets where tilapia is not a favoured fish, it has often to be presented in a value-added form under a different name. The demand for red-coloured mutant tilapia was mainly due to the fact that fillets of the fish could be sold under a different name (fresh-water snapper). Cans of processed tilapia have been produced on a limited scale in some countries like Costa Rica. Experience in a number of developing countries, where tilapia have been introduced, seems to show that markets can be developed if fish of at least 200−250 g can be sold at a competitive price. Larger fish of 300−400 g size attract more consumers. The market that has developed in the Philippines for tilapia in the last two decades is illustrative of this.

19.7   Economics

The economics of tilapia farming depend very much on the availability of suitable markets for the product. In extensive and simple small-scale rural farming, where the size of the product is not a major concern and no supplementary feeding is involved, the operation can be profitable when appropriate management measures are employed. In semi-intensive or intensive systems the cost of feeds, labour and, in some cases, water management become quite high and these can be compensated only by appropriate market prices for the product. In most African countries a good percentage of the consumers prefer tilapia, and they can therefore be sold at prices comparable to many other good quality food fish. But in other parts of the world, where tilapias are exotic species, considerable market promotion is required. As mentioned in the previous section, the most important factor in developing a market for tilapia is the size of the fish, and the main thrust of recent improvements in culture technologies has been to obtain marketable-size products in as short a time as possible.

Within the limits of technological constraints, there is considerable variation in the profitability of tilapia farming, between different types and sizes of operations. This is clearly brought out by the costs and returns of sample operations in different provinces of Central Luzon (Philippines) reported by Sevilleja (1985), which are reproduced in Table 19.1. The data relate to land-based fresh-water fish ponds for the calendar year 1982, collected during 1983. The capital investment per hectare varied between the provinces, from 13058 pesos in Bulacan (11.00 pesos for 1 US$ in 1983) to 29661 pesos in Pampanga. The data show the economic viability of both mono- and polyculture of tilapia. The average production obtained by the farms is close to the national average.

Cage farming of tilapia is preferred by producers, mainly because of the lower capital costs involved as well as the lower feeding costs in plankton-rich habitats. However, the cost of seed stock will be higher if it is to be purchased, as normally larger fingerlings are used in cage culture. Aragon *et al.* (1985) studied the economics of tilapia cage culture in the Laguna province of the Philippines. Table 19.2 presents

Table 19.1 Cost and returns (pesos*/ha per year) of tilapia production of sample operators in provinces of the Philippines (from Sevilleja, 1985, reproduced with permission of ICLARM).

Item	Province				Central Luzon Region	
	Bulacan	Nueva Ecija	Pampanga	Tarlac	Amount	%
Monoculture						
Returns						
Cash	23 965	7807	7309	12 633	11 350	89
Non-cash	605	1837	1337	1 100	1 409	11
Total	24 570	9644	8648	13 733	12 759	100
Costs						
Cash	8 184	4709	5347	4 737	5 595	83
Non-cash	2 654	649	840	897	1 130	17
Total	10 838	5358	6187	5 634	6 725	100
Net cash income	15 781	3098	1962	7 896	5 755	95
Net non-cash income	(−2 049)	1188	497	203	279	5
Net earnings	13 732	4286	2459	8 099	6 034	100
Polyculture						
Returns						
Cash	6 222	2651	5181	11 345	8 384	67
Non-cash	4 658	682	1497	5 880	4 045	33
Total	10 880	3333	6678	17 225	12 429	100
Costs						
Cash	6 780	1914	3181	5 778	4 826	84
Non-cash	1 925	93	840	857	924	16
Total	8 705	2007	4021	6 635	5 750	100
Net cash income	558	737	2000	5 567	3 558	53
Net non-cash income	2 733	589	657	5 023	3 121	47
Net earnings	2 175	1326	2657	10 590	6 679	100

* In 1983, 11.00 pesos = 1 US $.

Table 19.2 Costs and returns (in pesos*) per farm per season in tilapia cage culture by farm size and type of operation, for 63 tilapia producers in San Pablo City, Laguna, 1982 (from Aragon *et al.*, 1985, reproduced with permission of ICLARM).

Item	Size of operation					
	Small grow-out operation	Medium grow-out operation	Large grow-out operation[†]	Grow-out operation	Large hatchery[‡]	Total
Costs						
Cash costs						
Fingerlings bought	2 812[a]	6 863[a,b]	20 099[b]	18 816		18 816
Hired labor	732[a]	933[a,b]	3 283[b]	3 235	258	3 493
Interest on capital	798[a]	1 529[a,b]	2 536[b]	2 120	578	2 698
Feed supplies	228[a]	574[a,b]	1 788[b]	1 767	135	1 902
Other costs[§]	434[a]	452[a,b]	469[b]	439		439
Total cash costs	5 004[a]	10 351[a,b]	28 175[b]	26 377	971	27 348
Non-cash costs						
Fingerlings other than bought				15 724		15 724
Unpaid operators' labour	422[a]	444[a,b]	628[b]	600	376	976
Unpaid family labour	395[a]	396[a,b]	453[b]	455	238	693
Brood stock other than bought					30 660	30 660
Depreciation**	2 456[a]	5 052[a,b]	23 841[b]	23 719	1 644	25 363
Total non-cash costs	3 273[a]	5 892[a,b]	24 922[b]	40 498	32 917	73 416
Total costs	8 277[a]	16 243[a,b]	53 097[b]	66 875	33 888	100 763
Returns						
Cash returns						
Fish sold	30 144	66 720	201 179	188 338		188 338
Fingerlings sold					124 704	124 704
Total cash returns	30 144	66 720	201 179	188 338	124 704	313 042
Non-cash returns:						
Fish consumed at home	246	270	1 019	954		954
Fingerlings used by the producers					15 724	15 724
Fish given away	356	472	1 631	1 527		1 527
Total non-cash returns	602[a]	742[a,b]	2 650[b]	2 481	15 724	18 205
Gross returns	30 746[a]	67 462[a,b]	203 829[b]	190 819	140 428	331 247
Net cash farm income[††]	25 140[a]	56 396[a,b]	173 004[b]	161 961	123 733	285 694
Net farm income[‡‡]	22 469[a]	51 219[a,b]	150 732[b]	123 944	106 540	230 484

* In 1982, 8.50 pesos = 1 US $.
[†] Includes farms engaged in grow-out operation only.
[‡] Includes farms engaged in both grow-out and hatchery operations.
[§] Consists of wire, wood, iron, nails and sand.
** Consists of depreciation of bamboo poles, fish net, sinkers, nylon cord, weighing scale and metal containers.
[††] Net cash farm income = total cash returns minus total cash costs.
[‡‡] Net farm income = gross returns minus total costs.
[a,b] Means with the same letter in any given row are not significantly different at the 5% level using the t-test.

cost and returns of different types of cage farm operations in San Pablo City in Laguna. It includes data on farms that undertake only grow-out and those that have their own hatcheries. The total capital investment varies, depending on the number of cages and type of materials used in cage construction. The average capital investment in grow-out operations was 7022, 14 363 and 66 462 pesos for small, medium and large farms respectively. Large farms consisted of cages covering on average 320 m², medium farms on average 314 m², and small ones 280 m². Net farm incomes from all types of cage operations are comparatively high, but the large farms with hatchery operations gave the highest gross and net returns. It should, however, be pointed out that not all cage farms in the country provide similar high incomes: for example, the gross returns from tilapia cage culture in Los Banos amounted to only 3330 pesos per season.

19.8 References

Aragon C.T., de Lim M.M. and Tioseco G.L. (1985) Economics of tilapia culture in Laguna Province, Philippines. In *Philippine Tilapia Economics*, (Ed. by I.R. Smith, E.B. Torres and E.O. Tan). *ICLARM Conf. Proc.*, **12**, 66–82.

Balarin J.D. and Haller R.D. (1983) Commercial tank culture of tilapia. In *Proceedings: International Symposium on Tilapia in Aquaculture*, (Comp. by L. Fishelson and Z. Yaron), pp. 473–83. Tel Aviv University.

Campbell D. (1978) Formulation des aliments destinés à l'élevage de *Tilapia nilotica* (L.) en cages dans le Lac de Kossou. Côte d'Ivoire. *Rapp. Tech.* Authorité Aménagement Vallée du Bandama, Centre Dével. *Pêches Lac Kossou*, **46**.

Chimits P. (1957) The tilapias and their culture. *FAO Fish. Bull.*, **10**(1), 1–24.

Coche A.G. (1982) Cage culture of tilapias. In *The Biology and Culture of Tilapias*, (Ed. by R.S.V. Pullin and R.H. Lowe-McConnell). *ICLARM Conf. Proc.*, **7**, 205–46.

Egusa S. (1976) Some bacterial diseases of freshwater fishes in Japan. *Fish Pathol.*, **10**(2), 103–14.

Fishelson L. (1962) Hybrids of two species of the genus *Tilapia* (Cichlidae, Teleostei). *Fishermen's Bull.*, Haifa, **4**, 14–19. (In Hebrew.)

Fishelson L. and Yaron Z. (Comps) (1983) Classification of tilapias. *Proceedings: International Symposium on Tilapia in Aquaculture*, p. X1. Tel Aviv University.

Galman O.R. and Avatlion R.R. (1983) A preliminary investigation of the characteristics of red tilapias from the Philippines and Taiwan. In *Proceedings: International Symposium on Tilapia in Aquaculture*, (Comp. by L. Fishelson and Z. Yaron), pp. 291–301. Tel Aviv University.

Guerrero R.D. (1975) Use of oral androgens for the production of all-male *Tilapia aurea* (Steindachner). *Trans. Am. Fish. Soc.*, **104**, 342–8.

Guerrero R.D. (1979) Culture of male *Tilapia mossambica* produced through artificial sex reversal. In *Advances in Aquaculture*, (Ed. by T.V.R. Pillay and W.A. Dill), pp. 166–8. Fishing News Books, Oxford.

Guerrero R.D. (1982) Control of tilapia reproduction. In *The Biology and Culture of Tilapias*, (Ed. by R.S.V. Pullin and R.H. Lowe-McConnel). *ICLARM Conf. Proc.*, **7**, 309–16.

Guerrero R.D. (1985) Tilapia farming in the Philippines: practices, problems and prospects. In *Philippine Tilapia Economics*, (Ed. by I.R. Smith, E.B. Torras and E.O. Tan). *ICLARM Conf. Proc.*, **12**, 3–14.

Hanson T.R. *et al.* (1983) Growth comparisons of monosex tilapia produced by separation of sexes, hybridization and sex reversal. In *Proceedings: International Symposium on Tilapia in Aquaculture*, (Comp. by L. Fishelson and Z. Yaron), pp. 570–79. Tel Aviv University.

Henderson-Arzapalo A., Stickney R.R. and Lewis D.H. (1980) Immune hypersensitivity in intensively cultured *Tilapia* species. *Trans. Am. Fish. Soc.*, **109**, 244–7.

Hepher B. and Pruginin Y. (1981) *Commercial Fish Farming*. John Wiley and Sons, New York.

Hepher B. and Pruginin Y. (1982) Tilapia culture in ponds under controlled conditions. In *The Biology and Culture of Tilapias*, (Ed. by R.S.V. Pullin and R.H. Lowe-McConnell). *ICLARM Conf. Proc.*, **7**, 185–203.

Hickling C.F. (1960) The Malacca *Tilapia* hybrids. *J. Genet.*, **57**, 1–10.

Hughes D.G. and Behrends L.L. (1983) Mass production of *Tilapia nilotica* seed in suspended net enclosures. In *Proceedings: International Symposium on Tilapia in Aquaculture*, (Comp. by L. Fishelson and Z. Yaron), pp. 394–401. Tel Aviv University.

Jauncey K. and Ross B. (1982) *A Guide to Tilapia Feeds and Feeding*. University of Stirling.

Jhingran V.G. and Gopalakrishnan V. (1974) Catalogue of cultivated aquatic organisms. *FAO Fish. Technical Paper*, **130**.

King J.E. and Wilson P.T. (1957) Studies on tilapia as skipjack bait. *Sp. Sci. Rep. – Fisheries*, No. 225.

Langford F.H., Ware F.W. and Gasaway R.D. (1978) Status and harvest of introduced *Tilapia aurea* in Florida Lakes. In *Symposium on Culture of Exotic Fishes*. (Ed. by R.O. Smitherman, W.L. Sheldon and J.H. Grover), pp. 102–6. Fish Culture Section, American Fisheries Society, Auburn.

Lessent P. (1968) Essais d'hybridation dans le genre *Tilapia* à la Station de Recherches Piscicoles de Bouaké, Côte d'Ivoire. In *Proceedings of the World Symposium on Warm-water Pond Fish Culture*, Vol. 4, (Ed. by T.V.R. Pillay), pp. 148–59.

Lovshin L.L. (1977) Progress report on fisheries development in north-east Brazil. *Res. Dev. Ser.*

Int. Cent. Aquaculture, Auburn University, **14**.

Lovshin L.L. (1980) Progress report on fisheries development in north-east Brazil. *Res. Dev. Ser. Int. Cent. Aquaculture*, Auburn University, **26**.

Lovshin L.L. (1982) Tilapia hybridization. In *The Biology and Culture of Tilapias*, (Ed. by R.S.V. Pullin and R.H. Lowe-McConnell). *ICLARM Conf. Proc.*, **7**, 279−308.

Lovshin L.L. and da Silva A.B. (1975) Culture of monosex and hybrid tilapia. *FAO/CIFA Technical Paper*, **4** (Suppl. 1), 548−64.

Maar A. (1956) Tilapia culture in farm dams in Southern Rhodesia. *Rhod. Agric. J.*, **53**(5), 667−87.

Majumdar K.C. and McAndrew B.J. (1983) Sex ratios from interspecific crosses within the tilapias. In *Proceedings: International Symposium on Tilapia in Aquaculture*, (Comp. by L. Fishelson and Z. Yaron), pp. 261−9. Tel Aviv University.

NRC (1977) *Nutrient Requirements of Warmwater Fishes*. National Research Council, National Academy of Sciences, Washington.

Paperna I. (1980) Parasites, infections and diseases of fishes in Africa. *CIFA Technical Paper*, **7**.

Popper D. and Lichatovich T. (1975) Preliminary success in predator control of *Tilapia mossambica*. *Aquaculture*, **5**(2), 213−14.

Pruginin Y. (1967) Report to the Government of Uganda on the Experimental Fish Culture Project in Uganda, 1965−66. *FAO UNDP (TA) Reports*, **2446**.

Pruginin Y. and Kanyike E.S. (1965) Mono-sex culture of tilapia through hybridization. Paper presented at Symposium on Fish Farming, Organization of African Unity, Nairobi.

Roberts R.J. and Sommerville C. (1982) Diseases of tilapias. In *The Biology and Culture of Tilapias*, (Ed. by R.S.V. Pullin and R.H. Lowe-McConnell). *ICLARM Conf. Proc.*, **7**, 247−63.

Rothbard S. *et al.* (1983) The technology of mass production of hormonally sex-inversed all-male tilapias. In *Proceedings: International Symposium on Tilapia in Aquaculture*. (Comp. by L. Fishelson and Z. Yaron), pp. 425−34. Tel Aviv University.

Sarig S. (1976) The status of information on fish diseases in Africa and possible means of their control. *CIFA Technical Paper*, **4**, Suppl. 1 to the Report of the Symposium on Aquaculture in Africa 715−21.

Sarig S. (1983) A review on tilapia culture in Israel. *Proceedings: International Symposium on Tilapia in Aquaculture*, (Comp. by L. Fishelson and Z. Yaron), pp. 116−22. Tel Aviv University.

Scott P.W. (1977) *Preliminary studies on diseases in intensively farmed Tilapia in Kenya*. MS thesis, University of Stirling.

Sevilleja R.C. (1985) Tilapia production in freshwater fish ponds of Central Luzon, Philippines. In *Philippine Tilapia Economics*, (Ed. by I.R. Smith, E.B. Torres and E.O. Tan). *ICLARM Conf. Proc.*, **12**, 115−26.

Thys D.F.E. van den Audenaerde (1968) An annotated bibliography of Tilapia (Pisces, Cichlidae). *Mus. R. Afr. Cent. Doc. Zool.*, **14**.

Trewavas E. (1982) Tilapias: taxonomy and speciation. In *The Biology and Culture of Tilapias*, (Ed. by R.S.V. Pullin and R.H. Lowe-McConnell). *ICLARM Conf. Proc.*, **7**, 3−13.

Uchida R.N. and King J.E. (1962) Tank culture of Tilapia. *Fish. Bull.*, Fish and Wildlife Service., 199 (Vol. 62).

Verani J.R. *et al.* (1983) Population control in intensive fish culture associating *Oreochromis* (*Sarotherodon*) *niloticus* with the natural predator *Cichla ocellaris* − quantitative analysis. In *Proceedings: International Symposium on Tilapia in Aquaculture*, (Comp. by L. Fishelson and Z. Yaron), pp. 580−87. Tel Aviv University.

Wohlfarth G.W. and Hulata G.I. (1981) *Applied Genetics of Tilapias*. ICLARM, Manila.

20
Grey Mullets and Milkfish

Grey mullets (family Mugilidae) and the milk-fish (family Chanidae) have been the mainstay of finfish culture in coastal and estuarine impoundments for centuries. In the vallis of the Mediterranean lagoons (especially in Italy), in the so-called coastal 'harbour culture' in Northern China, in the bheris of the Gangetic estuaries in the Indian sub-continent, in the tambaks of Java (Indonesia) and in the coastal fish ponds of Hawaii the grey mullets formed an important group of cultured species. Like many other widely distributed aquaculture species, mullets are relished by consumers in some areas, but considered of poor eating quality in others. This is clearly exemplified in the Mediterranean region, where fresh grey mullet is considered a delicacy in the countries bordering the eastern Mediterranean up to Italy, whereas in countries of the western region there is hardly any market for them. Though considered a high-quality fish in some South American countries like Brazil, they have very little acceptance in North America, except in the southern states and in Hawaii.

In many ways milkfish is similar to the mullet in consumer acceptance. Distributed widely (although not as widely as the mullets), and seldom forming a major capture fishery, milkfish is an important food fish only in Indonesia, the Philippines and Taiwan. However, much of the brackish-water aquaculture experience in Asia has originated from milkfish farming and until recently dominated the scene, until shrimp culture became the major focus of attention in coastal farming.

Although the grey mullets and the milkfish belong to different families, most of them are herbivorous and detritus feeders. In spite of some advances made in induced breeding, culture of these species is still based on wild fry and fingerlings. The main species of aquacultural interest are euryhaline and can be reared both in brackish- and salt-water environments. They can easily be acclimatized for culture in fresh water.

20.1 Grey mullets (Mugilidae)

Though the popular name 'mullets' generally refers to the species of the family Mugilidae, the name grey mullet is used to distinguish them from the red mullets of the family Mullidae. The taxonomic classification of the grey mullets has been rather confusing, and the many revisions of the family have not made it any easier for aquaculturists to identify the various species. Jhingran and Gopalakrishnan (1974) have listed 13 valid species belonging to the genus *Mugil* and one species of the genus *Rhinomugil* which have been used in aquaculture. Though some authors have used the generic name *Liza*, based on the extent of development of the adipose eyelids, others consider this classification invalid as this characteristic is not of diagnostic value (Pillay, 1962). The other genus considered valid, namely *Crenimugil*, is represented by the species *Rhinomugil corsula*, which is of some importance in Indian fish culture. So in the following account of grey mullet culture the candidate species will be treated as belonging to the two genera, *Mugil* and *Crenimugil*.

The most widely distributed and well-known species of grey mullet is *Mugil cephalus*, sometimes referred to as the striped mullet (fig. 20.1). Because of the fast growth rate and the comparatively large size of the adults, this has been the species of choice in all areas. But fry and fingerlings of *M. cephalus* are not as abundantly available as of the other species. So in countries bordering the Mediterranean, *M. capito*, *M. auratus* (fig. 20.2), *M. saliens* (fig. 20.3) and *M. chelo* are also utilized in extensive or intensive farming. Other species used in the Indo-Pacific region are *M. pursia* (= *dussumieri*), *M. tade*, *M. macrolepis*, *M. soiuy* and *R. corsula*. Additional species of importance in South America are *M. curema* and *M. brasiliensis*. Experimental work in West Africa has also included the species *M. falcipinnis* and *M. grandisquamis*.

20.1.1 Culture systems

The traditional extensive culture of mullets together with other euryhaline species in embanked brackish waters still continues to be an important culture system, accounting for a good proportion of present-day production. Culture in more easily manageable ponds is an improvement on the traditional system. The mullets are generally raised together with other species; for example, in Hongkong mullets are cultured in combination with Chinese carps, in Taiwan with Chinese carps and tilapia, in Israel with the common carp and tilapia and in India with *milkfish*, pearl spot (*Etroplus suratensis*) and other estuarine species. The milkfish ponds of the Philippines and tambaks of Indonesia (fig. 20.4) have a certain percentage of grey mullets, although there have been doubts about the suitability of the combination, due to the competing nature of their food habits.

Grey mullets have been transplanted to develop capture fisheries in certain areas. A notable example is the successful transplantation of mullets from the Black Sea into the Caspian Sea. Lake Quaroun and Lake Marut in Egypt and Lake Tiberias (Lake Kinneret) in Israel have been successfully stocked with mullets. *M. capito* has been reported to breed in Lake Quaroun (Wimpenny and Faouzi, 1935).

20.1.2 Fry collection and artificial propagation

As mentioned earlier, most of the fry or finger-

lings used in grey mullet culture the world over are obtained from natural sources. In the extensive type of farming in impoundments like the bheris of India and the Mediterranean vallis, most of the stocking is done by taking advantage of the tidal flow and the habit of mullets to swim against currents. Very small fry may be brought in with the incoming high tide, but larger ones enter at low tides, when there is a slow flow of water from the impoundments. Suitable screens and traps are maintained at the sluice gates to prevent the escape of fish from the impoundments.

Fry collection methods

At present, most farmers supplement the stock obtained through tidal flow with fry caught from the estuaries. The more important species, including *M. cephalus*, breed in the sea and the fry and fingerlings migrate towards the shore and estuaries, where they can be found to congregate in schools. Only fry and small fingerlings below 25 cm in length school in large numbers. In any estuarine area there are certain locations where they congregate in large numbers, but variations in occurrence according to species and location have been observed. As, for example, in Israel *M. saliens* usually concentrate in the lower reaches of rivers, whereas *M. cephalus* and *M. capito* ascend to the higher reaches. On the east coast of India, fry of one or more species can be found in schools in the estuarine waters up to the tidal limits, throughout the year. The most suitable areas for the collection of fry are the marginal areas of rivers, tidal streams, creeks, swamps and inundated fields (Sarojini, 1958). Where a fresh-water stream flows into a river or a brackish-water lagoon, the mullet fry can be observed to congregate and swim against the slow current. Such areas are ideally suited for the collection of large numbers of mullet fry, taking advantage of the tidal flows. According to Hepher and Pruginin (1981), catches of fry are made in Israel after sea storms and rainy periods.

Grey mullets are difficult species to identify even when they reach the adult stage, and this becomes much more difficult in the fry and fingerling stages. Since some of the species are very slow-growing, it is important to sort

Fig. 20.1 Grey mullet, *Mugil cephalus*.

Fig. 20.2 Grey mullet, *Mugil auratus*.

Fig. 20.3 Grey mullet, *Mugil saliens*.

them before stocking in rearing facilities. The seasonality of occurrence of different species in the coastal areas can be of initial help in obtaining the desired species. Perlmutter *et al.*

(1957) observed that the fry of each species of mullet appear regularly at the mouth of rivers on the Israeli coast in particular seasons. Similar seasonality in the occurrence of fry has been

Fig. 20.4 An aerial view of a tambak system in Java, Indonesia (photograph: Michael New).

observed by Sarojini (1958) and Luther (1973) on the east coast of India in West Bengal.

Different types of equipment are used for fry collection, the most common ones being seines and dip nets. Short-bagged drag nets and beach seines are the most common equipment in Taiwan. Fry collectors usually acclimatize the fry through gradual decreases in salinity, when the fry are meant for rearing in fresh water. However, it has been shown experimentally that the fry can be directly transferred to fresh water without any special acclimatization. Fish farmers in Taiwan, Israel and certain parts of India stock rearing ponds directly without any major mortality. It is believed that sudden changes in temperature and low pH affect the survival of fry more than salinity. If the fry are to be transported long distances, it is considered advisable to condition them for a day.

Generally the fry are stocked in production ponds, directly from the collection grounds or after a brief period of conditioning. In Israel the fry are first reared in small ponds for two to three months before stocking in larger production ponds. Mullet fry are stocked at the rate of about 30 000/ha, often with 200–300 young carp per ha to reduce the growth of filamentous algae. The mullets are stocked in larger ponds when they have reached a weight of around 3 g.

Artificial propagation

Considerable effort has been devoted to the artificial propagation of the grey mullets which do not breed in confined waters, especially *M. cephalus*. A number of species have been induced to breed by the administration of pituitary extracts or gonadotropins. Wild *M. cephalus* has been induced to spawn by the administration of mullet pituitary homogenate, often combined with Synahorin (Tang, 1964; Ling 1970). Mature 4- to 6-year-old females are injected intramuscularly with homogenates of two to five pituitaries from the same species, with 10–60 rabbit units of Synahorin (a mixture of chorionic gonadotropin and mammalian hypophyseal extract). Mature males do not require any injections, except towards the end of the breeding season. The best results are obtained by giving two injections at an interval of 24 hours. The treated fish can be stripped easily and a female *M. cephalus* of about 1.5 kg weight is reported to yield 1–1.5 million eggs

(Chen, 1976). The eggs measure 0.9−1 mm in diameter, and are fertilized by the dry or the wet method. At temperatures of 20−24°C the eggs hatch out in 16−30 hours. The larvae are very small, ranging from 2.5−3.5 mm in size, and tend to avoid strong light.

Captive *M. cephalus* have also been spawned by the administration of partially purified salmon gonadotropin, with the potency of 1 mg, equivalent to 2150 IU human chorionic gonadotropin (HCG) (Kuo *et al.*, 1975). The easily available and less expensive HCG can also be used effectively for induced breeding of *M. cephalus*. Female spawners with oocytes of at least 600 μm diameter require a dose of approximately 60 IU HCG per g body weight, administered in two injections. The first injection of about 20 IU per g body weight is followed in 24 hours by an injection of 40 IU per g body weight (Kuo *et al.*, 1973).

Both *M. cephalus* and *M. capito* raised in fresh-water ponds have been induced to breed by injection of carp pituitary homogenates. The brood fish are acclimatized to full sea water for at least two weeks, after which the females are given a series of three injections of homogenized carp pituitaries. The first injection contains one half of a pituitary for every kg of the recipient; the second, given after about 7 hours, contains one pituitary per kg; and the third injection, after 14 hours, contains two pituitaries per kg. Males are also given one half of a pituitary per kg, at the time of the last injection of the females. The females can be stripped 16−24 hours after the third injection and the eggs fertilized with milt stripped from the males. After rinsing in sea water, the fertilized eggs are transferred to incubators. The hatchlings are fed with brine shrimp and zooplankton. The larvae grow to fry stage in about a month and can be acclimatized back to fresh water.

The Chinese mullet *M. so-iuy* is another species which has been successfully bred by the administration of hormones (Zheng, 1987). Three- to four-year-old brood fish are selected for spawning. Mature females with oocytes measuring 600−700 μm are injected with homogenates of carp pituitary, mullet pituitary, HCG or LHRH-A. Generally two injections are given at an interval of 24 hours. When the water temperature is between 15 and 16°C,

spawning takes place in 1−3 days. Incubation can be carried out in sea water or brackish water with salinities above 7 ppt. Hatchlings are reared in indoor concrete tanks, fed with eggs and larvae of oysters and mussels, rotifers and brine shrimp nauplii. Larvae can also be reared in manured nursery ponds with an adequate growth of zooplankton. Soybean milk is given as a supplementary feed initially. Later, soybean or peanutcake is given in the form of a paste. In 30−40 days the fry reach the stocking size.

Mugil macrolepis has been successfully induced to spawn in India by the administration of pituitary of the same species (Sebastian and Nair, 1975). The experiments were conducted with mature wild fish caught in Chinese dip nets. The effective dose was reported to be three to four glands per female of 40−130 g weight, injected intramuscularly at 6-hour intervals. The males did not require any injection. When the injected fish were kept in hapas together with males in salt water of 29−31 ppt salinity, spontaneous spawning occurred, but the rate of fertilization was poor. However, by stripping and artificial fertilization better results were obtained. At temperatures ranging from 26 to 29°C most of the developing eggs hatched in about a day. Three- to four-day-old larvae could be reared in cement tanks in brackish water containing rich growths of chlorella, diatoms, copepods, etc.

Mugil parsia has also been induced to breed with a single low dose of mullet pituitary at the rate of 5 mg per kg body weight of the female (Radhakrishnan *et al.*, 1976). Spontaneous spawning of untreated fish was also observed. Despite experimental success in induced spawning, farmers continue to depend on wild stocks of fry and fingerlings for culture operations.

20.1.3 Grow-out

As indicated earlier, in traditional coastal fish farming mullets are raised along with other species such as milkfish, shrimps, pearlspot, seabass, ten pounders (*Elops*), etc. When the system of tidal stocking is adopted, the stocking rate depends on several extraneous factors and therefore the quantity and composition of the seed stock can seldom be determined. This practice often leads to understocking and the inadvertent presence of slow-growing species. Because of this, fish culturists now often supplement the stocks with fry and fingerlings caught from the wild, as described in the previous section. The proportion of grey mullets in the impoundments is controlled to some extent depending on the market value of the species, but generally in brackish-water areas mullets constitute about 10–40 per cent of the stock. In extensive systems of culture, neither fertilization nor feeding are practised. Regular exchange of water is performed, based on the tidal regime in the area. In estuarine regions, the incoming tidal water often contains large quantities of detritus, besides planktonic organisms. The detritus settles on the bottom of the impoundments and adds to the fertility of the soil. These areas are characterized by rich benthic growths of algal complexes, containing bacterial and microscopic animal populations as well. This is the main food source for the mullets and the milkfish in such waters. In well-managed impoundments, the total production can vary between 150 and 1500 kg/ha (fig. 20.5).

Even in properly designed coastal farms, monoculture of mullets is seldom practised, but mullets can form the main species if selective stocking is carried out. In the coastal fish farms on the west coast of India, mullets, milkfish and the pearlspot are cultured together. In the extensive system adopted in these farms, production seldom exceeds 400 kg/ha. By adopting improved rates of stocking of *M. tade* and *M. parsia* in pilot farms in the Gangetic delta, a production of up to 2200 kg/ha has been obtained (Jhingran, 1982). Production in Hawaiian coastal ponds, before their decline in the last seven to eight decades, used to be around 230 kg/ha, of which about two-thirds were mullets and the rest milkfish.

Mullets form a constituent of stocks in milkfish farms of Southeast Asia, although many farmers consider mullets to be incompatible

Fig. 20.5 A catch of mullets from a Mediterranean valli.

with milkfish in intensive culture. The methods of growing benthic algal pastures in brackish-water ponds, on which both milkfish and mullets feed, is described later in this section. Mullets seldom constitute more than 10 per cent of the stock in such ponds.

The more intensive polyculture of mullets is done in fresh or slightly saline waters in Hong Kong, Taiwan, Israel and on a smaller scale in Egypt and India. In both Hong Kong and Taiwan, they are cultured together with Chinese carp. In Hong Kong ponds, when mullets form the main species, stocking may be done at the rate of 10 000−15 000 mullet fingerlings per ha (about 7.5 cm long) along with 1000−2000 Chinese carp fingerlings per ha in early spring. The stock is thinned to about 3500/ha when the fish have grown to about 12 cm in length. Feeding is generally with rice bran for the first two months and thereafter with a mixture of rice bran and peanut oil cake. Organic manuring of the ponds is also done to increase the production of natural food. Production varies from 2500−3500 kg/ha.

In Taiwan, the stocking rate per ha is reported to be about 3000 mullets, 2000 milkfish, 3250 Chinese carps and 500 common carp. The ponds are fertilized with superphosphates at the rate of 60 kg/ha. The feed comprises rice bran, soybean cake and peanut meal. *Mugil cephalus* attain a size of about 300 g in one year and about 1.2 kg if cultured for the second year. Three-year-old *M. cephalus* reach a weight of 2 kg under pond conditions.

Mullets are reared in polyculture with common carp, silver carp and tilapia. The fry of mullets collected from coastal waters are over-wintered and then grown in nursery ponds to a weight of about 1−2 g for stocking polyculture ponds. Fry of this size are generally stocked at the rate of 5000/ha, when the total stocking density is about 12 300/ha (3000 common carp, 4000 tilapia and 300 silver carp). The mullet reaches around 100 g in weight in about 4 months and about 200 g by the end of the year. As *M. cephalus* fry are not readily available in sufficient numbers, many farmers use *M. capito*. This species grows at a much slower rate and has to be reared for two years to reach a marketable size. Pond fertilization and the type of feeds used in these polyculture ponds are generally the same as in carp ponds.

20.1.4 Diseases

Records of diseases of grey mullets in culture facilities are relatively scarce. Paperna (1975) described a number of parasites and disease conditions occurring in open waters, some of them causing serious fish kills. Fish farms are generally stocked with fry and fingerlings collected from such waters, and so there is every likelihood of these infestations being transmitted to the farms. Heavy infestations of *Ergasilus lizae* in *M. cephalus* have been observed to cause serious losses in brackish-water ponds on the Mediterranean coast of Israel. The crustacean parasite *Pseudocaligus apodus* is believed to infect *M. cephalus*, and *M. capito* is infected by *Caligus pagete* in fish ponds, causing mortality. In fresh-water ponds, *Saprolegnia* infections have been found to cause serious mortalities.

20.2   Milkfish

The milkfish *Chanos chanos* (fig. 20.6), the only species of the family Chanidae, has a wide distribution, though not to the same extent as the grey mullet. It does not form a capture fishery of any significance and its importance is based on the large-scale farming in over 400 000 ha of coastal impoundments in Southeast Asia. Its culture is believed to have originated in Indonesia, during the 15th century, and then spread to the Philippines and Taiwan. According to available information, there are over 183 000 ha of milkfish ponds in Indonesia, about 176 000 ha in the Philippines and about 15 600 ha in Taiwan. The average production per ha in Taiwan is reported to be about 2 tons, in the Philippines 600 kg and in Indonesia about 300 kg. Many individual farms obtain much higher production in all three countries. Some small-scale culture is attempted in peninsular India and Sri Lanka, but the total production is very small.

Though essentially marine fish of the Indian and Pacific oceans, the young ones spend their life in inshore estuarine areas and ascend rivers to the fresh-water zones. They are known to be highly euryhaline and can live in fresh to hyper-saline waters and can tolerate low oxygen levels. Temperature tolerance limits are said to

Fig. 20.6 Milkfish, *Chanos chanos*.

range from 15 to 40°C, but the optimum temperature is between 20 and 33°C. They become sluggish below 20°C and mortality occurs at 12°C.

20.2.1   Culture systems

The most common system for the culture of milkfish is in brackish-water coastal pond farms (fig. 20.7). The farm may include nurseries and rearing ponds, with wintering ponds where the fingerlings have to be over-wintered, as in Taiwan. Some farmers specialize in the production of fry and fingerlings only, in which case rearing ponds may be omitted. Another type of farm has only rearing ponds of different sizes. The farmer buys fry or fingerlings from fry producers and devotes his farm only to raising marketable fish.

Though intended to be monoculture, brackish-water milkfish ponds become polyculture systems as the tidal water brings in early stages of a number of other species, the more important of which are the grey mullets, shrimps and seabass (*Lates calcarifer*).

As in the case of mullets, milkfish are sometimes grown in fresh-water ponds or stocked in lakes and reservoirs. But the more important milkfish farming in fresh waters, is the pen farming that has developed in lakes in the Philippines (Laguna de Bay and Lake Sampaloc).

20.2.2   Fry collection and induced spawning

As in the case of major grey mullet species, the milkfish do not mature and spawn naturally in confined waters. They seem to spawn in the sea near the coast and the small larvae (12–15 mm in length) occur periodically along the sandy coasts and in the estuaries. The collection and rearing of fry from these areas for sale to farmers has become an industry of importance, employing a large number of people in Indonesia, the Philippines and Taiwan.

There seem to be some differences in the periods of availability of larvae and fry along the coasts of these countries. In the Philippines and Taiwan, the season for collection extends from March or April to August, but the peak season in the Philippines is May/June and in Taiwan it is April/May. In Indonesia there are two seasons, one from March to May and the second from September to December. The peak period for collection is October/November.

Fry collection methods

The most common collecting equipment comprises different types of dip nets, such as the triangular scissor net in the Philippines and the scoop net in Taiwan. Dip nets are particularly suited for areas with large concentrations of fry. Seines, drag nets and traps are also used by

Fig. 20.7 A typical milkfish farm in the Philippines. Note the catching ponds on both sides of the central canal.

some fishermen. In traditional methods of fry capture in Indonesia, special fry-congregating devices are used, such as rock walls or lure lines made of fibre ropes strung with plaited strips of coconut and banana leaves (fig. 20.8). In areas where the concentration of fry is low, such lure lines may be deployed in a circle with one end tied to a post. When fry are observed under the lure, the fishermen reduce the circle by pulling the free end of the line and dip out the fry with small dip nets made of coarse cloth. The best collections are made at creek mouths, the leeward side of sandbars in estuarine areas, etc., at high tides during full and new moon periods.

The fry captured are between 10 and 30 mm in length and need to be handled carefully. In Indonesia and the Philippines they are transferred to earthenware jars and acclimatized to lower salinity conditions by gradually diluting the water with fresh water. In Taiwan, the fry are stored temporarily in wooden buckets or cement troughs in sea water. They are packed in plastic bags containing water of lower salinity (10−15 ppt) and filled with oxygen for transport. In the Philippines, fry used to be trans-

ported in earthenware jars of 15−30 ℓ capacity, each containing 1500−3500 fry sorted according to size (fig. 20.9). Now, plastic bags filled with oxygen are commonly used in the Philippines as well. Flat bamboo baskets coated with cement or tar are the traditional containers for fry transport in Indonesia. Each basket carries about 20000−40000 fry in dilute sea water, with daily changes of water. During long-distance transport or storage, the fry are fed on slightly roasted rice flour or wheat flour twice a day, and occasionally on mashed hard-boiled eggs.

Induced spawning

Attempts have been made to develop a hatchery technology for the production of milkfish fry, in order to meet the increasing demands created by intensified farming techniques. The sexes can be distinguished by external characteristics. Females are distinguishable by the presence of three visible pores in the urogenital region, whereas the males show only two pores externally (Chaudhuri *et al.*, 1976). Mature females collected from the sea with ova of

Fig. 20.8 Collecting milkfish fry with a scoop net along a lure line.

Fig. 20.9 The traditional system of transporting acclimatized fry in earthenware jars or pots for shipment.

about 0.7—0.8 mm diameter can be induced to spawn by the administration of carp pituitary homogenate, semi-purified salmon gonado-tropin (SG-G 100) in combination with HCG (Vanstone *et al.*, 1977; Chaudhuri *et al.*, 1978; Kuo *et al.*, 1979; Liao *et al.*, 1979). The average number of eggs spawned annually is estimated to be 2 million/kg body weight. The eggs can be fertilized with milt from untreated males. Where necessary, the free flow of milt is induced by the injection of androgen or salmon pituitary preparations.

Even though the fertilized eggs, which have a diameter of 1.1—1.25 mm, can be hatched in salt water (30—34 ppt) containers in 25—28.5 hours at temperatures of 26.4—29.9°C, it has been difficult to rear the post-larvae to the fry stage in any significant numbers. Survival rates under experimental conditions of 9—47 per cent have been reported during a 21-day rearing period (Liao *et al.*, 1979). Fed initially on fertilized oyster eggs for 14 days and thereafter on a combination of copepods, Artemia nauplii, flour and prepared feed, the fry could be grown to a mean size of 14.5 mm in 20 days.

A system of propagation that appears to hold great possibilities is spawning in open waters. It has been shown that milkfish can attain sexual maturity and spawn in cages installed in protected bays. Floating cages filled with fine-meshed hapa nets to retain spawned milkfish eggs have been successfully used by research institutions for large-scale spawning. The eggs are collected from the cages with a specially designed conical egg-sweeping net, with a rigid frame. This device is reported to have been very successful in the recovery of fertilized eggs and thereby in the production of hatchery produced fry.

Rearing of fry

Considerable attention is paid to fry rearing in the Philippines. Although fry can be introduced into rearing ponds after a brief period of acclimatization in small nursery pools, as done elsewhere, many Philippine farms maintain separate nursery and transition ponds for fry rearing. They normally represent about 3—5 per cent of the farm area. The typically shallow nursery ponds ranging in size from 1000 to 4000 m^2 are located close to transition ponds

which are meant for stunting the fry for later (off-season) stocking. They usually average about 1 ha in area. The nursery ponds are provided with catching ponds (see Section 6.1.2, Dike design and construction), sluice gates and a canal system for easy water distribution and transfer of stock. When the nurseries form a unit of the production farm, their preferred location is in the centre of the farm to facilitate transfer of fingerlings. Milkfish fry are usually grown to fingerling size in ponds with rich growths of the benthic biological complex predominated by blue-green algae (Myxophyceae), generally referred to as lablab. It includes large populations of bacteria, diatoms, green algae (Chlorophyceae) and animal components like protozoans, flat worms, larvae and adults of molluscs, polychaete worms, copepods and larval forms of decapods and insects. In recent years, many farmers have adopted the practice of raising mainly planktonic organisms as food for milkfish fry. In shallow ponds, the distinction between benthic growths and plankton is seldom precise. When plankton is the main source of food, the ponds are generally made deeper for better growth of phyto- and zooplanktonic organisms.

Methods of increasing planktonic growth by fertilization and water management are fairly standard and have been described in Chapter 7. The production and maintenance of the benthic algal complex involves considerable skill and attention. The preparation of the ponds starts about two months before the fry are introduced. The ponds are drained completely during low tides. The bottom is levelled, raked with a wooden rake or ploughed to bring the sub-surface soil nutrients to the surface and to eradicate weeds. The pond bottom is levelled in such a way that it slopes gradually towards the deepest portion of the pond at the sluice gate. Often a shallow diagonal canal is made from the gate to the opposite corner to serve as a refuge for fry and fingerlings during hot days, and to facilitate transfer or harvesting of the stock. The pond is then dried and exposed to the sun for two or three days until the layer of surface soil cracks, after which some water is let in. In order to get rid of any predatory fish or other pests burrowing in the mud, this process of drying and draining may be repeated a few times. Besides eradicating pests and predators, drying also helps in the mineralization of organic matter in the soil. The water gates of the pond are protected with fine-meshed screens to prevent the entry of fish or other organisms from outside.

The ponds are then treated with chicken manure at the rate of 2 tons/ha. Water is let in just to cover the pond bottom and 150 kg/ha of 16−20−0 NPK fertilizer, or half that quantity of 18−46−0 NPK fertilizer per ha, are added after 2 or 3 days. In order to speed up the breakdown of chicken manure, urea may be added at the rate of 25 kg/ha. Within a week lablab growth starts. The water level in the pond is then gradually increased to 25−30 cm in a period of one to one and a half months, increasing the level by 3−5 cm each time. Sudden increases in water level can result in the detachment of lablab from the bottom.

The maintenance of this benthic complex requires proper water management and grazing levels. If overgrowth of the complex occurs, it has to be controlled by additional stocking of fry. Detached lablab is not allowed to accumulate and disintegrate in the ponds, and is removed and dried for later use as feed for milkfish. Organisms that feed on or disturb the growth of lablab are detected and eradicated as far as possible to maintain the algal pasture at an optimum abundance. Further applications of NPK fertilizers are made, if necessary, at intervals of 1 to 2 weeks to maintain the growth of lablab.

Lowering the salinity in the ponds by admixture with fresh water induces growths of filamentous algae which are known in the Philippines as 'lumut'. This is avoided, not only because the fry are not able to feed on them, but also because they become entangled in the filaments.

If the fry are to be reared on plankton, the pond water is maintained at a depth of 75−100 cm, and chemical fertilizers are applied at the same rate as for lablab growth. The fertilizer is placed on a platform from which it can dissipate into the pond. In a few days a plankton bloom develops and the visibility under water is about 15−40 cm. In case of poor growth, a further application of fertilizer is made. If there is an excessive growth of plankton, fertilization is not suspended but a part of the pond water is replaced.

Fry can be stocked directly in the nursery ponds if the salinity of the water in which they are transported is approximately the same as the salinity of the pond water. If there is a difference of over 5 ppt, it is considered preferable to acclimatize them before transfer.

The density of fry in the nurseries is generally 30−50 per m². Nursery management involves the maintenance of suitable conditions for the growth of fry and its natural food (lablab). In order to avoid salinity increases during the summer months, some exchange of water may be needed. It is reported that the growth of milkfish fry is retarded at salinities above 45 ppt. Favourable temperatures for growth are above 23°C. If the growth of natural food is not adequate for the stock, artificial feeds like rice bran or dried lumut are provided. In about one and a half to two months, the fry have grown to a weight of 1−3 g and are then either stocked in the rearing ponds or transferred to transition or stunting ponds.

The transition or stunting pond, as the name implies, is meant to hold the fingerlings in a stunted condition for stocking later, during the off-season for fry. The stocking density in the transition ponds is about 10−15 fingerlings/m², and they subsist on lablab or plankton for one to two months. Fertilization may be carried out to increase natural food production, but this may not be enough to keep the fish in a healthy but stunted condition. Supplementary daily feeding with rice bran at the rate of 5 per cent of the body weight may be required, if the fingerlings have to be held for prolonged periods (up to 6 months or more), or if they become too thin.

In Taiwan, where over-wintering of fry is required, the fry are kept in shallow (20−40 cm deep) ponds with 1.5 m deep wintering ditches protected on the windward side by windbreaks of thatched bamboo frames. For producing stunted fingerlings, fry are stocked at the rate of 300 000−500 000 per ha and fed on benthic algae and rice bran, peanut meal or soybean meal. These wintering ponds may also be used for over-wintering undersized fish from the previous harvest.

20.2.3 Grow-out

Schuster (1952), Djajadiredja and Daulay

(1982), and Bandie *et al.* (1982) have described designs of tambaks that have been used for milkfish farming in different regions of Indonesia. Though the basic principles of pond farm design have not changed, improved designs have been evolved in the Philippines and Taiwan, which are now being introduced by progressive farmers in Indonesia as well (Wardoyo *et al.*, 1982). Basically, each farm has nursery, transition and rearing ponds, which can be independently drained or filled through a canal system. As mentioned earlier, Taiwanese transition ponds are also used as wintering ponds. The rearing ponds generally form 85−90 per cent of a farm, and in modern farms where intensive culture systems are employed individual ponds seldom measure over 4−5 ha. Usually they are rectangular in shape and located on either side of the canal system for water supply and drainage.

Pond preparation

The overall configuration and operation of the rearing ponds are very similar to those described for nursery ponds. The majority of farms depend on the production of benthic organisms for raising milkfish, and so basically the same pond management methods are followed. Even though in pen culture in eutrophic lakes milkfish have been grown to market size on plankton, in actual practice the farmers have not yet been able to obtain consistent results with plankton feeding in pond culture.

In the early days of milkfish farming in the Philippines, lumut or the algal complex dominated by filamentous algae such as *Chaetomorpha, Cladophora* and *Enteromorpha* was considered to be the best natural food to be raised in ponds. It was later observed that filamentous algae like *Chaetomorpha* are too coarse and fibrous to be suitable food for milkfish fingerlings and it is only the decaying algae in the detritus that the fish are able to utilize. Because of this, ponds with lumut growth can be stocked with only 1000−1500 fingerlings per ha and the yield expected would be only 200−300 kg/ha per crop. Experience seems to indicate that lablab is also the best natural food for growing milkfish to market size. Fertilized ponds with good growths of lablab can yield 500−700 kg/ha per crop in a period of two to

three months.

Procedures for preparing rearing ponds are generally the same as for nursery ponds. The ponds are drained and the bottom dried. If required, the soil pH may be adjusted by the application of lime ($CaCO_3$). Initial fertilization is done with a combination of organic and inorganic fertilizers. After the ponds are stocked, fertilization with urea and NPK fertilizer is continued at about half the initial dose, at fortnightly intervals, taking care to exchange the water regularly.

In traditional milkfish ponds, stocking is carried out with milkfish only, but during the course of culture other species, especially grey mullets, shrimps (mainly *Penaeus* and *Metapenaeus* spp.) gain access, converting it into a polyculture system. Stocking density is left to chance; however, in recent times, with the increased demand for and price of shrimps, milkfish farmers are undertaking deliberate stocking of Penaeid shrimps and in some cases even converting milkfish ponds into shrimp ponds. The combination of milkfish with shrimps is not entirely based on compatible feeding habits, as there is obviously some overlap.

In view of the fluctuations in the benthic growth, the success of milkfish production in ponds is largely dependent on the timing and efficiency of stocking. Consequently, a number of systems have been developed for better utilization of the food resources and increased yields in milkfish ponds. The simplest system is to stock rearing ponds at rates that the food resources can sustain, and harvest them when they have reached the marketable size. Since marketable size can be reached in two to four months, three to four crops can be raised every year if fingerlings are available. The usual practice is to stock a single size group of fingerlings (10−15 cm) at the rate of about 2000/ha, and completely harvest when they have grown to marketable size. The main disadvantage of this system is that there is a wastage of food when the fish are small, as they cannot utilize all the food produced; and when they have grown the food produced in the pond may be insufficient because of the increased food requirements of the larger biomass.

In order to avoid shortage of food at critical times in rearing ponds, a procedure known as the 'progression method' is practised by many farmers in the Philippines (fig. 20.10). Rearing is carried out in two stages. The fingerlings grow for a certain period in one pond and are then transferred to another pond where they grow to the market size. The food resources in both ponds are not exhausted and several crops can be raised through proper management. This method has been further improved to a so-called 'modular method', which involves a three-stage rearing. Three contiguous ponds form a series, progressively increasing in size at a ratio of 1:2:4. The first pond is stocked at a density of 15 000/ha. After about 6−7 weeks the stock is transferred to the second pond, and after about 4−5 weeks to the third pond, until they reach market size. As soon as a pond is emptied, it is prepared to receive the next stock.

A more intensive system of stocking developed in Taiwan consists of stocking different size groups and repeated selective harvesting, and this is sometimes referred to as the 'multi-size stocking method'. Initial stocking may be with three size groups; for example, 3000 fingerlings of average length 5 cm, 2000 of 15 cm length, and 2000 of 18 cm length. Subsequent stocking may be with smaller fingerlings (for example, of 5 cm length) at about one to two month intervals, at the rate of 2000−3000 per ha each time. Repeated selective harvesting is performed three to five times to remove the market-sized fish.

Feeding and pond management

As milkfish farming is largely based on the production of natural food, artificial feeding is resorted to only in special circumstances, when the natural food production is not adequate. Locally available feedstuffs like rice bran, peanut meal and soybean meal are used for supplementary feeding. The feeding rate in rearing ponds in Taiwan is generally 30 kg rice bran or 25 kg soybean or peanut meal per ha daily.

An important aspect of pond management consists of reducing or eradicating organisms that disturb or feed on benthic growths in ponds. Chironomid larvae, polychaete worms and snails are the most common pests. Taiwanese farmers use different types of pesti-

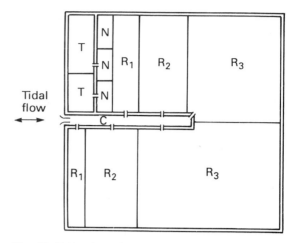

Fig. 20.10 Design of a typical progression or modular pond farm, C=canal, T=transition pond, N=nursery pond, R_1, R_2 and R_3=ponds for the three stages of rearing.

cides to eradicate them. The application of lime and urea for the initial preparation of milkfish ponds usually helps to reduce the growth of these organisms.

There are very few known diseases of milkfish. Schuster (1952) observed a condition described as 'catching cold' when there is a sudden lowering of temperature in shallow ponds. The symptoms are a milky discoloration of the skin and sluggish movements. After two or three days, portions of skin may drop off. No mortality has been observed.

Pen culture

Pen culture of milkfish is practised in the eutrophic lakes of the Philippines, especially in Laguna de Bay. As mentioned in Chapter 6, most of these pens are enclosed by synthetic netting of suitable mesh size, installed on a framework of bamboo poles dug deep into the lake bottom (see fig. 6.34). The size of the pens varies considerably from 1.5 ha to as much as 100 ha. The most common size appears to be between 10 and 20 ha. When fingerlings have to be reared on site, a nursery pen is constructed within the rearing pen. It is made of smaller-meshed netting and usually measures around 20 m × 20 m with a depth of at least 1.5 m.

The stocking rate depends on the density of planktonic blooms in the lake. Generally, it varies between 10 000 and 20 000 fingerlings per ha in the main pen, and 100 fry per m^3 in the nursery pen. Supplementary feeding with rice bran, copra meal, soybean meal, etc., is provided in the nursery, and it may take up to one month for the fry to grow to fingerlings of about 20 g size. Fingerlings stocked in the rearing pen feed on natural food in the lake and no artificial feed is provided. Multiple stocking and harvesting can be practised in pens, as in ponds, if there is a dependable supply of fingerlings. Depending on local conditions, it may take 4 to 5 months for the fingerlings to reach marketable size in pens.

One of the major hazards in such eutrophic lakes is the occurrence of fish kills due to anoxia caused by death and decay of algal blooms.

20.2.4 Harvesting and marketing

Harvesting methods are very much dependent on the rearing techniques and the design of the farm. When thinning of stock during the culture period is involved or when multiple stocking and harvesting is practised, it is necessary to use gill or seine nets. Fish caught by gill netting are likely to be bruised and may lose some scales. The price of such fish in the Philippine markets is low. Even the fish harvested by complete draining of ponds do not command a good price in the Philippines, because the mud adhering to the fish is believed to impart a muddy flavour or taste. This is why milkfish pond farms in the country have special catching ponds. To harvest the fish, the rearing pond is partially drained at low tide and at subsequent high tide water is allowed to flow in through the catching pond. The fish swim against the current and enter the catching pond, from where they are easily gathered with seines or scoop nets. Some farmers use electrical fishing equipment for harvesting.

Taiwanese farmers use large gill nets for harvesting milkfish ponds, along with a scare-line to empty the stomachs of the captured fish. Milkfish with empty stomachs keep better during transport to markets. At the end of the rearing season, the ponds are drained after netting and the remaining fish picked up.

Special care is taken in handling milkfish in the Philippines because of consumer preference for unbruised fish with scales intact (fig. 20.11). The fish are often dipped in iced water before packing to prevent loss of scales during handling. Most of the fish are sold in

Fig. 20.11 Catch of milkfish from a fish pen in Laguna de Bay, the Philippines (courtesy of Robert Gedney).

fresh condition, but there is also an important market for deboned and smoked milkfish.

Available economic data on intensive mono- and polyculture of milkfish in ponds and in pens in the Philippines show high profitability in all these types of systems. Higher returns are obtained in polyculture with quick-growing species of shrimps.

20.3 References

Bandie M.J. *et al.* (1982) Present status of the brackish water fish ponds in East Java, Indonesia with emphasis on engineering related problems. In *Report of Consultation/Seminar on Coastal Fish Pond Engineering*. FAO/UNDP South China Sea Fisheries Development and Coordination Programme. SCS/GEN/82/42, 104–7.

Chaudhuri H. *et al.* (1976) Notes on the external sex characters of *Chanos chanos* (Forskal) spawners. *Fish. Res. J. Philippines*, **1**(2), 76–80.

Chaudhuri H. *et al.* (1978) Observations on artificial fertilization of eggs and the embryonic and larval development of milkfish *Chanos chanos* (Forskal). *Aquaculture*, **13**, 95–113.

Chen T.P., (1976) *Aquaculture practices in Taiwan*. Fishing News Books, Oxford.

Collins R.A. and Delmendo M.N. (1979) Comparative economics of aquaculture in cages, raceways and enclosures. In *Advances in Aquaculture*, (Ed. by T.V.R. Pillay and W.A. Dill) pp. 472–7. Fishing News Books, Oxford.

D'Ancona U. (1954) Fishing and fish culture in brackish-water lagoons. *FAO Fish. Bull.*, **7**(4), 147–72.

Delmendo M.N. and Gedney R.H. (1974) *Fish Farming in Pens — A New Fishery Business in Laguna de Bay*. Laguna Lake Development Authority, *Technical Paper*, **2**.

Djajadiredja R. and Daulay T. (1982) Aspects of design and construction of coastal ponds for milkfish seed production. In *Report of Consultation/ Seminar on Coastal Fish Pond Engineering*. FAO/ UNDP South China Sea Fisheries Development and Coordination Programme, SCS/GEN/82/42, 92–100.

Hepher B. and Pruginin Y. (1981) *Commercial Fish Farming*. John Wiley & Sons, New York.

Hora S.L. and Nair K.K. (1944) Suggestions for the development of salt-water bheris or bhasa-badha fisheries in the Sunderbans. *Fish. Dev. Pamphlet*, Calcutta, **1**.

Jhingran V.G. (1982) *Fish and Fisheries of India*. Hindustan Publishing Corporation (India), Delhi.

Kuo C.M., Shehadeh Z.H. and Nash C.E. (1973) Induced spawning of captive grey mullet (*Mugil*

cephalus L.) females by injection of human chorionic gonadotropin (HCG). *Aquaculture*, **1**, 429–32.

Kuo C.M., Nash C.E. and Shehadeh Z.H. (1974) A procedural guide to induce spawning in grey mullet (*Mugil cephalus* L.). *Aquaculture*, **3**, 1–4.

Kuo, C.M. and Nash, C.E. (1975) Recent progress on the control of ovarian development and induced spawning of the grey mullet (*Mugil cephalus* L.). *Aquaculture*, **5**, 19–29.

Kuo C.M., Nash C.E. and Watanabe W.O. (1979) Induced breeding experiments with milkfish, *Chanos chanos* Forskal, in Hawaii. *Aquaculture*, **18**, 95–105.

Liao I.C. *et al.* (1979) On the induced spawning and larval rearing of milkfish, *Chanos chanos* (Forskal). *Aquaculture*, **18**, pp. 75–93.

Lin S.Y. (1968) *Milkfish Farming in Taiwan*. The Taiwan Fisheries Research Institute. *Fish Culture Report*, **3**.

Ling S.W. (1970) A brief review of the work done on the induced breeding of *Mugil cephalus* in Taiwan. *J. Inland Fish. Soc. India*, **1**, pp. 1–12.

Luther G. (1973) The grey mullet fishery resources of India. In *Proceedings, Symposium on Living Resources of the Seas Around India, CMFRI Special Publication*, pp. 455–60.

Macintosh D.J. (1982) Fisheries and aquaculture significance of mangrove swamps with special reference to the Indo-Pacific region. In *Recent Advances in Aquaculture*, Vol. I (Ed. by J.F. Muir and R.J. Roberts), pp. 4–85. Croom Helm, London.

Madden W.D. and Paulsen C.L. (1977) *The Potential for Mullet and Milkfish Culture in Hawaiian Fish Ponds*. Department of Planning and Economic Development, Hawaii.

Pakrasi B. and Alikunhi K.H. (1952) On the development of the grey mullet, *Mugil corsula* Hamilton. *J. Zool. Soc. India*, **4**(2), 123–40.

Paperna I. (1975) Parasites and diseases of the grey mullet (Mugilidae) with special reference to the seas of the Near East. *Aquaculture*, **5**, 65–80.

Perlmutter A., Bograd L. and Pruginin J. (1957) Use of estuarine and sea fish of the family Mugilidae (grey mullets) for pond culture in Israel. *Proc. Tech. Pap. Gen. Fish. Coun. Mediterr.*, **4** pp. 289–304.

Pillay S.R. (1962) A revision of Indian Mugilidae. Part I. *J. Bombay Nat. Hist. Soc.*, **59**(1), 254–70.

Pruginin Y and Cirlin B. (1975) Techniques used in controlled breeding and production of larvae and fry in Israel. In *Workshop on Controlled Reproduction of Cultivated Fishes. EIFAC Technical Paper*, **25** 90–100.

Rabanal H.R. (1949) *The Culture of Lab-lab, the Natural Food of the Milkfish Fry and Fingerlings under Cultivation*. Department of Agriculture and Natural Resources, Philippines, *Tech. Bull.*, **18**.

Rabanal H.R., Montalban H.R. and Villaluz D.K. (1951) The preparation and management of bangos fishpond nursery in the Philippines. *Philippine J. Fish.*, **1**(1), 3–44.

Rebanal H.R. and Shang Y.C. (1979) The economics of various management techniques for pond culture of fin fish. In *Advances in Aquaculture*, (Ed. by T.V.R. Pillay and W.A. Dill), pp. 224–35. Fishing News Books, Oxford.

Radhakrishnan S. *et al.* (1976) Breeding of mullets by hormone stimulation. *Matsya*, **2**, 28–31.

Sarojini K.K. (1958) On the collection, acclimatisation and transport of mullet seed in West Bengal (India). *J. Bombay Nat. Hist. Soc.*, **55**(1), pp. 42–53.

Schuster W.H. (1952) Fish-culture in brackish-water ponds of Java. *Spec. Publ. Indo-Pac. Fish. Coun.*, **1**.

Sebastian M.J. and Nair V.A. (1975) The induced spawning of the grey mullet, *Mugil macrolepis* (Aguas) Smith and the large-scale rearing of its larvae. *Aquaculture*. **5**, pp. 41–52.

Tang Y.A. (1964) Induced spawning of striped mullet by hormone injection. *Jap. J. Ichthyol.*, **12**, pp. 23–8.

Tang Y.A. (1964) Stock manipulation of coastal fish farms. In *Coastal Aquaculture in the Indo-Pacific Region*, (Ed. by T.V.R. Pillay), pp. 438–53. Fishing News Books, Oxford.

Tang Y.A. (1979) Planning, design and construction of a coastal milkfish farm. In *Advances in Aquaculture*, (Ed. by T.V.R. Pillay and W.A. Dill) pp. 104–17. Fishing News Books, Oxford.

Vanstone W.E. *et al.* (1977) Breeding and larval rearing of the milkfish *Chanos chanos* (Pisces: Chanidae). SEAFDEC Aquaculture Department, *Tech. Rep.*, **3**, 3–17.

Villadolid D.V. and Villaluz D.K. (1950) *A Preliminary Study on Bangos Cultivation and its Relation to Algae Culture in The Philippines*. Department of Agriculture and Natural Resources, Philippines, *Pop. Bull.*, **30**, 16.

Wardoyo S.E., Pirzan M. and Djajadiredja R. (1982) Observations on pond design and engineering of improved privately-owned tambaks in Maros, South Sulawesi Province, Indonesia. In *Report of Consultation/Seminar on Coastal Fish Pond Engineering*. FAO/UNDP South China Sea Fisheries Development and Coordination Programme, SCS/GEN/82/42, 101–3.

Wimpenny R.S. and Faouzi H. (1935) The breeding of a grey mullet *Mugil capito* Cuv., in Lake Quarun, Egypt. *Nature*, **135**, 1041.

Zheng C.W. (1987) Cultivation and propagation of mullet (*Mugil so-iuy*) in China. *Naga*, **10**(3), 18.

21
Yellowtail

The yellowtail, *Seriola quinqueradiata* (family *Carangidae*) is the only carangid that contributes significantly to aquaculture production at present, and its culture is restricted to Japan. Attempts have been made to culture another species of the family, the pompano (*Trachinotus carolinus*), in the southwestern USA, but this has not resulted in any large-scale culture operations. Despite the limited geographical importance, yellowtail culture is of special significance, because it was the first instance of a large-scale culture of a marine fish and it contributes not less than 90–95 per cent of the total finfish mariculture in Japan. Probably this is the only case where a farm-raised fish is unanimously considered superior in quality to fish caught from the sea and fetches a much higher price in the market. Being a high-valued carnivorous species, they are fed in culture facilities with less expensive fish like sardine, mackerel and sand eel. Almost 60 per cent of the commercial landings of these species are used as feed for yellowtail, and it is said that their capture fisheries will collapse in the absence of yellowtail culture, as there is no other demand for them in Japan.

The Japanese name 'hamachi' originally referred to young one-year-old yellowtails; but with the rapid expansion of their culture, all farm-raised yellowtails have come to be known as 'hamachi', irrespective of their age. It is a highly-priced fish in Japanese markets and grows to a length of 80–100 cm and a weight of 5–8 kg in natural waters. The marketable size is at a weight of 1–1.5 kg (fig. 21.1). The capture fishery production had diminished, and this encouraged attempts at farming the species. Though farming started in 1928, significant production was achieved only by about 1960.

21.1 Culture systems

Originally yellowtail was cultured in diked coastal lagoons or lakes and ponds; but later pen culture in coastal areas, fenced off by nets, was started in a number of places. The real boost to yellowtail culture came with the successful development of the cage culture system, starting from about 1954. Most of the present-day production comes from cage culture and some from ponds.

Floating and submersible types of cages are used for yellowtail culture (see Chapter 6). Floating type cages are used in areas protected from high winds and waves (fig. 21.2). The framework of the floating cages is generally made of bamboo or cedarwood (10–15 cm in diameter) or sometimes 3–5 cm steel pipes. Styrofoam buoys encased in polythene (to prevent fouling) are used as floats. In recent years much larger and more durable modern cages have been introduced for operations even in less protected areas.

As described in Chapter 6, the submerged cages have helped in extending the sites for cage culture, and it is now possible to use more exposed open sea areas. The cage can be lowered to the sea bottom by dropping the mooring ropes in rough weather conditions. The weights at the four corners of the cage bottom help to maintain the shape of the cage. The fish can be fed through the opening of the feeding net attached to the top of the cage, which can be drawn to the surface when the cage is lowered.

21.2 Production of seedlings

Although methods of artificial propagation of

Fig. 21.1 One-kilogram yellowtail harvested from a floating cage (from Bardach *et al.* 1972. By permission of John Wiley & Sons).

the Kuroshio current along with floating seaweeds. It is from these coastal areas that they are caught with large encircling nets from fishing boats, from about May to June. The boats carry storage tanks in which the larvae are kept until the boats return to port. The larvae measure 25–40 mm in length.

Artificial propagation of yellowtail has been carried out with mature fish caught in drift nets from the sea. Eggs stripped from mature females have been fertilized with milt from captured males. Experimental work on the use of hormones for maturing brood fish and inducing spawning has also been performed. Maturation can be induced by the injection of Synahorin at the rate of 4 IU/kg of fish. Fertilized eggs hatch out in about 51–68 hours in water temperatures ranging from 18 to 24°C. As starter feed, the hatchlings are given eggs of sea urchins or oysters. Oyster larvae, enriched *Artemia* and *Brachionus* form the main food when they start more active feeding, and later they take to copepods. The fry can be fed on minced juvenile fish and fish meal. They grow around 10 mm in about 20 days, 17 mm in a month and about 63 mm in two months. A month-old fry, which will have developed the characteristic stripes on the sides, becomes cannibalistic.

yellowtail have been developed, the farmers depend largely on seedlings collected from natural sources. Because of the increasing number of seedlings required by the farming industry and the possible adverse effects of the removal of large numbers of larvae, the government enforces restrictions on the quantity that can be caught every season. Yellowtail are migratory fish which move into the offshore waters from March to May, where they spawn. The larvae are brought towards the shore by

Because of the cannibalistic habits of the fry, the seedlings collected from natural sources or propagated artificially are carefully sorted

Fig. 21.2 Floating net cages used for culturing yellowtail in Japan.

out according to size for rearing. The sorted fry can be reared separately in small floating, fine-mesh net-pens. The pens are generally 2−50 m^2 in area and 1−3 m deep. The fry are fed on minced fish (such as sand eel and horse mackerel) and shrimps. A high fat content in the feed and feeding above 80 per cent satiation are avoided. Best results have been obtained by feeding crustaceans or white-meat fish. Shrimp flour made into a paste forms an excellent food for yellowtail fry. Some farmers feed the fry with zooplankton and lights are often hung above the cages to attract the zooplanktonic organisms. The fry rearing usually takes a period of 4−6 weeks, during which time they grow to approximately 8−10 cm in length and 25−50 g in weight.

21.3   Grow-out

As mentioned earlier, most yellowtail culture is now carried out in floating or submersible cages. These are installed in areas where the salinity does not go below 16 ppt. The optimal temperature range is between 24 and 29°C. Temperatures below 9°C and above 31°C are unsuitable for yellowtail culture.

Stocking densities range from 80−200 fingerlings/m^3 in cages, depending on the size of the fingerlings. In net enclosures and ponds, much lower stocking rates of one or two fingerlings/m^3 are adopted. They are generally fed on sliced or whole fish flesh. The same species used for feeding fry are used for feeding fingerlings and adult yellowtails. Even though some farmers use anchovies it is not recommended, because continuous feeding with this fish can cause mortality as a result of the oxidation of the unsaturated fatty acids in anchovy flesh. Thiamine in anchovy flesh has been found to destroy vitamin B$_1$ in yellowtail. Feeding at the rate of 10 per cent body weight or 75−80 per cent satiety gives the best growth rates. The daily ration is given in two feedings for larger fish and three or four for smaller fish. The feed conversion is about 7 for fish that weigh up to 1.5 kg and 10 for fish up to 4 kg, on a wet weight basis. On a dry weight basis, it is 2.1 and 3.0 respectively.

Though artificial diets are commercially available, very few farmers use them because of the high cost and the comparatively poor growth rates obtained. A suitable diet can be made from white fish meal making up about 70 per cent of the feed with 5−10 per cent gluten as a binder and a vitamin and mineral premix. Though the growth rate with this diet is less than with fresh fish, comparatively higher growth rates can be obtained by alternating the two types of feeds. Recent reports indicate that many farmers are now using a moist diet, prepared by mixing fish meat with a formulated powder.

Yellowtail grow rapidly in cages. Fingerlings stocked in May/June grow to 200−700 g by August and 600−1600 g by October. By the end of December they reach a weight of 700−2000 g. In some areas the smaller fish may be kept in the cages for a second year's growth to about 2−3 kg.

21.4   Diseases

With the expansion of cage culture and intensification of culture techniques, an increasing number of diseases and large-scale mortalities are occurring in cage farms. As mentioned earlier, some of the mortalities are caused by the use of poor quality feedstuffs, such as fatty fish. Nutritional diseases are not uncommon. Deterioration of water quality due to accumulation of waste matter from cages also contributes to the occurrence of diseases and mortality. Bacterial and parasitic diseases have become more common and serious.

The most severe disease of yellowtail appears to be due to the infestation of the monogenetic trematodes *Benedenia seriolae* and *Axine heterocerca*. The mature *Benedenia* is about 6 to 7 mm in length and in sea water it reproduces all the year round at temperatures over 12°C. The larva attaches itself to the fish as soon as it comes into contact with it. As many as 50 or more of them may be attached to a fish, and if untreated the fish will lose appetite, stop feeding and eventually die. As the parasite reproduces in polluted waters, the preventive measure is to improve the water quality. The best cure is a bath in fresh water as the parasite has little resistance against low salinity.

Axine may infest the gills of yellowtails throughout the year and cause anaemia resulting in mortality. The best cure is a bath in highly saline water (9−10 per cent) for three minutes, when the parasite will fall off. The fish should not be kept in this water too long as the salinity affects the gills, causing bleeding.

Another important parasite of yellowtail is the copepod *Caligus*, which attaches to the gills and mouth of the fish. Heavy infestation causes weakening of the fish as it affects feeding and eventually causes mortality. The recommended cure is a bath in a 20 ppm solution of Neguvon in fresh water.

Vibriosis and icthyosporidiasis (caused by the phycomycete *Ichthyosporidium hoferi*) are two major diseases believed to be transmitted through the use of infected fish as feed in culture facilities. Both diseases cause significant losses. Nocardial infections of the internal organs such as the spleen and the kidney caused by the bacillus *Nocardia kampachi* are widespread in yellowtail farms and very often occur at the same time as vibriosis, resulting in heavy mortality.

Pseudotuberculosis, caused by *Pasteurella piscida*, is one of the diseases identified in recent years. It is believed to be transmitted through fry and fingerlings introduced from infected areas. Oral administration of sulphur drugs or antibiotics have been used to control the disease. Experimental studies seem to show that prophylactic immunization of juveniles can be achieved by oral spray and immersion methods.

21.5 Harvesting and marketing

The minimum market size of yellowtail is about 300 g and that size can be harvested from August, after a growth of about 4 months. But the preferred size is about 1–1.2 kg and this can be harvested only from October to December. Seines and drag nets have to be used for harvesting from ponds and enclosures, but harvesting from cages needs only dip nets and similar simple devices. Dead fish are packed in ice for transport to markets. As there is a greater demand for live fish and they fetch a higher price, the cage culturists are at a greater advantage as most of the fish can be caught and transported alive. The fish are usually transported to markets by boat or trucks, in canvas tanks.

There are very few detailed records of the economics of yellowtail culture, but like all other types of farming, the cost and earnings are very much dependent on local conditions, the technology employed, and the skill and experience of the farmer.

Yellowtail culture in Japan is usually a family enterprise, and with the income from a production of 10–30 tons per cage farm unit, a family will have an income equivalent to that of a middle-class family in the country. Almost half the production cost consists of the price of feed and the cost of fingerlings. Labour and management costs will be around 10–11 per cent each.

21.6 References

Bardach J.E., Ryther J.H. and McLarney W.D. (1972) *Aquaculture*, pp. 557–65. John Wiley & Sons, New York.

Brown E.E. (1977) Production and culture of yellowtail (*Seriola quinqueradiata*) in Japan. *Proc. World Maricul. Soc.*, **8**, 765–71.

Fujiya M. (1979) Coastal culture of yellowtail (*Seriola quinqueradiata*) and red seabream (*Pagrus major*) in Japan. In *Advances in Aquaculture*, (Ed. by T.V.R. Pillay and W.A. Dill). Fishing News Books, Oxford.

Furukawa A. (1972) Present status of Japanese marine aquaculture. In *Coastal Aquaculture in the Indo-Pacific Region*, (Ed. by T.V.R. Pillay), pp. 29–47. Fishing News Books, Oxford.

Honma A. (1971) *Aquaculture in Japan*, pp. 50–80. Japan FAO Association, Tokyo.

Nose T. (1986) Recent developments in aquaculture in Japan. In *Realism in Aquaculture: Achievements, Constraints, Perspectives*, (Ed. by M. Bilio, H. Rosenthal and C.J. Sindermann), pp. 39–58. European Aquaculture Society, Bredene.

Schmittou H.R. (1973) *Aquaculture Survey in Japan*. Research and Development Series, No. 5. International Center of Aquaculture, Auburn.

22
Sea-basses and sea-breams

Aquaculturally important sea-bass species belong to the families Serranidae and Centropomidae. The European sea-bass common in the Mediterranean and in the eastern Atlantic is *Dicentrarchus labrax*. The Asian sea-bass also known as kakap (cockup) and giant perch, is *Lates calcarifer* (family Centropomidae), distributed in the littoral waters from Iran to Australia. The sea-breams of aquaculture interest are the gilthead sea-bream, *Sparus aurata*, common in the Mediterranean (and occurring also in the Black Sea and the eastern Atlantic), and the red sea-bream, *Pagrus major* of Japan, which occurs also in the East China Sea and Southeast Asian waters.

Common to all sea-basses and sea-breams are their euryhaline and carnivorous habits. They fetch high prices in the markets. Landings from capture fisheries are reported to be declining and there is an increasing and unsatisfied market in major consuming areas. Consequently, aquaculture-produced sea-basses and sea-breams, have the potential for enhancing both domestic and export trade.

Historically, the European and Asian species of sea-bass and the gilthead sea-bream have been constituents of stocks in the Mediterranean vallis (fig. 22.1), Indian bheris and coastal fish farms in Southeast Asian countries. In spite of their predation on other finfish and crustaceans, their market and culinary values made them acceptable species in culture systems. Their stocks in these impoundments were derived from eggs, larvae or fry brought in by the incoming tides. Small-scale efforts in recent years to grow wild and hatchery-reared fry and fingerlings in cages and ponds have shown potential for intensive culture of these species (fig. 22.2). Methods of artificial propagation have been developed and commercial-scale production is becoming established in a number of areas.

22.1 Sea-basses

The main handicap to the introduction of intensive culture of the European sea-bass *Dicentrarchus labrax* and the Asian sea-bass *Lates calcarifer* has been the lack of dependable source of fry and fingerlings. Though wild fry can be collected from natural habitats, the supply is highly inconsistent and inadequate. Because of this, many recent scientific studies on these species have been focused on developing methods of induced spawning and larval rearing.

22.1.1 European sea-bass

Dicentrarchus labrax has been induced to spawn by the administration of different hormone preparations. Natural spawning has also been obtained in captivity without any hormone injections. Girin (1976) reported spontaneous spawning of mature fish in sea-water tanks (salinity 35 ppt), with complete renewal of water every 10 hours. Injections of carp pituitary homogenates and HCG have been successfully used to induce spawning. According to Barnabe (1976), HCG in doses of 800 IU/kg body weight is much more effective than carp pituitary in induced breeding of this species. As in many other species, the best results are obtained by administration of the dose in two injections at intervals of one or two days. Spawning usually takes place about three

Fig. 22.1 Catch from a Mediterranean valli. European sea-bass can be seen below and gilt-head bream in the centre.

Fig. 22.2 Cage culture of European sea-bass in Yugoslavia (courtesy of CENMAR).

days after treatment. The average number of eggs spawned annually is reported to be 300 000/kg body weight and each egg measures 1.2–1.4 mm in diameter. Different types of incubators have been used for hatching the eggs. The salinity of the water supply is maintained around 34–35 ppt and the temperature around 13°C, when the pelagic eggs become demersal. The incubation period naturally depends on temperature and varies from 166 to 47 hours at temperatures ranging from 11 to 19°C, but the optimum temperature is believed to be 13°C. Very high hatching rates, nearing 90 per cent, have been reported (Barnabe, 1976).

The newly hatched larvae measure about 3.5 mm in length and the yolk sac is absorbed by the fourth or fifth day, depending on temperature. If sufficiently large incubators are used, the larvae can be reared for longer periods, with adequate aeration and feeding.

Alternatively, they can be transferred to larger larval rearing tanks, with controlled water quality (fig. 22.3). A temperature of 14–20°C, salinity of 34–37 ppt, pH of 7.9–8.2 and oxygen levels of 6–9 m/ℓ are the recommended environmental requirements. Aeration and regular water renewal together with a temperature slightly higher than ambient help improve larval growth. Various types of larval feeds have been tried in experimental hatcheries, with varying results (fig. 22.4). It appears clear that in the early stages the larvae require live food, such as *Brachionus, Artemia* and copepods. A feeding sequence suggested by Girin (1976) comprises *Brachionus* from the fourth to the fourteenth day, *Artemia* nauplii from the eleventh to the fiftieth day, 1 mm size *Artemia* from the twenty-fifth to the fiftieth day, 2 mm size *Artemia* from the fortieth to the fiftieth day, frozen *Artemia* from the fiftieth to the sixtieth day, thereafter mixed food including

Fig. 22.3 Larval-rearing tanks in the sea-bass hatchery of CENMAR in Yugoslavia (courtesy of CENMAR).

Fig. 22.4 Culture of live food for sea-bass in CENMAR hatchery in Yugoslavia (courtesy of CENMAR).

lyophilized *Artemia* to the seventy-sixth day. A simpler schedule of feeding consisting of live food (*Brachionus* and *Artemia*) up to 35 days, followed by a 52 per cent protein pelleted feed containing about 10 per cent *Artemia*, has been found to be adequate (Barahona and Girin, 1976). Cannibalism among fry can be a major problem, and so the fry have to be frequently graded and separated according to size to minimize losses.

Seventy-five- to eighty-day-old fry, which measure about 40 mm, are transferred to cages for on-growing. The optimum density is reported to be below 20 kg/m^3. The minimum market size is about 250 g, and in extensive systems in temperate areas it may take as long as two years to reach that size. During the third year they can reach 500 g, and 1 kg in the fourth year. Much better growth rates have been obtained in floating cages, when fed on balanced compound feeds. In a pilot farm in Yugoslavia an average weight of 300 g was achieved in 18 months. In Israel, where *D. labrax* has been tried as a predator in fresh-water tilapia ponds, fish of 230–300 g size grew to 650–780 g in about 7 months. In experimental recirculating systems in Denmark, with controlled temperature (between 22 and 25°C), the fish are reported to have grown to 300–500 g within one year.

There are few records of disease causing high mortality in sea-bass culture systems. Outbreaks of vibrio infections are reported to have caused mortality in a pilot farm in Yugoslavia.

22.1.2 Asian sea-bass

The Asian sea-bass *Lates calcarifer*, (fig. 22.5), which occurs in the tropical and sub-tropical areas of Asia, is a highly euryhaline species that lives in brackish-water estuaries and in fresh waters. For spawning they seem to require saline water, but larvae occur in fresh waters, including rice fields. The adult sea-bass is a

Fig. 22.5 The Asian sea-bass *Lates calcarifer* (from *Fisheries Handbook 1*, PPD, Singapore).

voracious carnivore, but juveniles are omnivorous. One of the major problems in culturing them in ponds is their cannibalistic
habits.

Induced spawning and larval rearing

In nature, *L. calcarifer* spawns all the year
round, with the peak season from April to
August. Though it does not spawn normally in
confined areas, methods of induced spawning
have been developed. Brood fish can be obtained from culture ponds or from open waters
and reared in special earthen brood ponds,
cement tanks or floating cages. It is reported
that males predominate among smaller size
groups (1.5–2.5 kg body weight), but 3- to
4-year-old fish from culture ponds show a normal distribution of sexes (Kungvankij *et al.*,
1986). Three-year-old females, weighing 3.5–
5 kg, and two-year-old males, weighing about
2.5–5 kg, are preferred for artificial spawning.
Spawning can be carried out in concrete tanks
(of about 150 ton capacity) with a suitable
supply of saline water (salinity of about 28–32
ppt), with periodic water exchange and aeration.
The brood fish are introduced into the
spawning tanks at the rate of about 10–12
pairs per tank, at least one month before the

spawning. The mature female is recognized by
the red-pink papilla extending out at the urinogenital aperture and the soft belly. The male
is usually more slender, with slightly curved
snout and, when mature, milt oozes out on
slight pressure on the abdomen. Females with
oocytes of about 0.5 mm diameter are suitable
for induced spawning.
The hormones usually used to induce spawning are HCG with pituitary gland of carp and
Puberogen. Puberogen contains 63 per cent
follicle stimulating hormone (FSH) and 34 per
cent leutinizing hormone (LH). The spawners
are usually given two intra-muscular injections at the base of the pectoral fin, the first
one of 50 IU HCG and 0.5–1 pituitary gland,
and the second after 12 hours, of 100–200 IU
HCG and 1.5–2 pituitary glands. Within about
10–12 hours after the second injection, spawning occurs. Repeated spawnings occur in
batches over a period of 3–5 days. The fecundity ranges from 2–17 million eggs, depending on the size of the spawner. Fertilized eggs
float on the surface of the tanks and can be
siphoned out for hatching.
If the brood fish do not spawn in the tank
after the second injection, they are stripped
and the eggs artificially fertilized.
When Puberogen is used to induce spawning,

the dosage is usually 50—200 IU/kg body weight for the female and 20—25 IU/kg fish for the male. If spawning does not occur within 36 hours, a second injection is given and the dosage doubled. This leads to spawning within 12—15 hours.

Induced spawning by environmental manipulation involves changing the salinity and temperature, simulating conditions in the natural spawning areas during the lunar phases. The salinity in the brood fish tank is gradually increased from 20—25 ppt to 30—32 ppt after the spawners are stocked, simulating the increased salinity to which the fish are exposed during migration from coastal areas to the sea. The pre-spawning behaviour of the spawners is carefully monitored. Segregation of the sexes may be done about a week before spawning. Spawning normally takes place during the full moon and new moon periods. At this time, the temperature of the water in the spawning tank containing the females and males is raised to 31—32°C by lowering the water level to about 30 cm and exposing the water to the sun for 2—3 hours. Then the water temperature is suddenly lowered to 27—28°C by the addition of filtered sea water. This induces the fish to spawn during the succeeding night. In case of failure the procedure is repeated. The fish usually spawn intermittently for about 3—7 days.

The fertilized eggs are incubated in 50 ℓ capacity hatching jars or fibreglass tanks. Such containers can hold 50 000—100 000 eggs. A one-minute bath in 5 ppm acriflavine followed by repeated rinsing in salt water is recommended before the eggs are introduced for hatching. The best hatching rates have been observed in salinities between 20 and 30 ppt. With proper aeration and salinity, the eggs hatch out in about 17—18 hours at temperatures of 26—28°C. The hatchlings are about 1.5 mm in length.

The hatchlings can be reared in large nursery tanks, supplied with water of about 20 ppt salinity. The stocking density varies with the age and size of the larvae. The density is gradually decreased from 40 000—50 000 per m^3 during the first week to 2000—5000 per m^3 in the fourth week.

Experience indicates that sea-bass larvae require live food in their early stages. In hatcheries, the first food given to 3-day-old larvae consists mainly of rotifers (*Brachionus plicatilis*), with a small percentage of *Chlorella* sp. and *Tetraselmis* sp., at the rate of 5—10 per ml. This may continue until the fourteenth day, with the addition of *Artemia* from the eighth day to the twentieth day. The suggested density of *Artemia* is 1 or 2 per ml. Usually the larval density in the tanks is reduced to about 20—40 larvae/ℓ, about a week after feeding starts. From the sixteenth day *Daphnia* or *Moina* can be added, at a density of 1 or 2 per ml, several times a day. After about 3 weeks, the fry are fed on minced fish. Generally the fry are graded during rearing to separate out the fast-growing ones from the others. Sorting and separation of fry according to size and thinning of stock help in reducing cannibalism among the fry. After about a month, the fry attain a size of about 12 mm and are then used for grow-out in production ponds or cages.

Grow-out

Lates calcarifer has been cultivated for many years in brackish-water ponds, and in recent years in floating cages, but there is a lack of documented information on grow-out practices.

The main problems of grow-out are feeding and prevention of cannibalism among young fish. In order to reduce losses due to cannibalism, grow-out is performed in two phases. In the first phase the fry are grown to a weight of about 20 g in special nursery-type ponds of up to 2000 m^2. Fry are stocked at the rate of 20—30/m^2. Besides the natural food produced by fertilization, the fry are fed with supplementary feed consisting of adult *Artemia* and ground trash fish twice a day. Exchange of water at the rate of 30 per cent daily is maintained. The rearing period is about 30—45 days. By frequent sorting, fish of similar size are separated and stocked in separate grow-out facilities for growing to market size.

Grow-out to market size lasts for 3—4 months in countries like the Philippines, where 300—400 g fish are acceptable, and 8—12 months in other countries where 700—1200 g fish are preferred. Floating and stationary cages of different sizes (usually 50 m^3) are used. The stocking density in the cages is about 40—50 fish/m^3, but after a growth of about 3 months the stock is thinned out to 10−20 fish/m³. The usual feed is chopped trash fish, fed twice daily at the rate of 10 per cent of body weight initially, which is reduced to only once a day at 50 per cent of the body weight after about two months. When insufficient trash fish is available, rice bran or broken rice is added as a partial substitute.

Both monoculture and polyculture of sea-bass are practised. In intensive polyculture with the sea-bass as the main species, the subsidiary species are forage fish like tilapia. In such polyculture, the ponds are first stocked with the forage fish, which reproduces rapidly. When a sufficient stock of fry and juveniles of the forage has developed in the pond, sea-bass juveniles are stocked at the rate of 3000−5000 per ha. In monoculture systems the stocking rate is usually 10 000−20 000 per ha of uniform-size juveniles which are fed daily with trash fish.

In traditional ponds, the sea-bass attain sizes around 500 g in about 12 months. A gross production of about 2.76 tons/ha in 8 months has been reported (Jhingran, 1977). It has been estimated that in monoculture in ponds with multiple stocking and harvesting, a production of about 3.3 tons/ha can be obtained.

Intensive grow-out of sea-bass has proved to be economical, as can be seen from Table 22.1 which compares the costs and earnings of pond and cage culture in private farms in Thailand.

22.2   Sea-breams

22.2.1   Gilthead sea-bream

The gilthead sea-bream (*S. aurata*) is a highly-priced species in the Mediterranean and neigh-bouring countries, and because of diminishing catches from open waters there is considerable interest in its intensive culture. The traditional polyculture of the species in coastal impound-ments or vallis is based on wild fry, as it does not breed naturally in confined waters. Much of recent research on the species has been directed towards developing a suitable system of artificial propagation to produce fry and fingerlings.

The gilthead sea-bream are characterized by protandric hermaphroditism, which oc-curs from the second to the fourth year of their life. The natural breeding season in the Med-iterranean region is between October and December, when the water temperature varies from 13 to 17°C.

Induced spawning and larval rearing

Maturation of captive fish can be induced by the injection of HCG, the effective dose varying with the state of gonadal maturity. Spawning has been successfully induced with a single injection, but more frequently with a series of two to nine injections of 800−2000 IU/kg body weight. The brood fish are kept in salinities between 35 and 37 ppt at temperatures of 17−21°C. The injection may be given at intervals of 2−3 days, and spawning occurs 4−5 days after the first treatment. If spontaneous spawning does not occur, stripping and artificial fertiliz-ation are carried out. The eggs measure 0.9−1.1 mm in diameter and the estimated annual fecundity is around 500 000/kg. It is reported that mature gilthead sea-bream can also spawn naturally without any hormone treatment in tanks with a constant exchange of sea water of the required temperature (as observed in sea-bass). Incubation of the eggs can be done in any standard incubator or even in 200−600 ℓ tanks with circulating sea water of about 36 ppt salinity and partial aeration. At temperatures around 15−20°C, hatching takes place in about 50 hours. The hatchlings are about 2.5−3 mm in length and start feeding from the third or fourth day, depending on the water temperature.

Different types and sizes of tanks have been used for larval rearing. Indoor tanks of about 200−600 ℓ capacity with circulating sea water of salinity in the range 26−37 ppt, artificial illumination (600−3500 lux) for 12−16 hours a day and aeration have given satisfactory results. The recommended larval density in the tanks is 10 per ℓ. The first food of sea-bream larvae is live organisms (as in the case of sea-bass larvae). Proper larval feeding is important not only in promoting good survival rates, but also in preventing deformities in fry. A suitable sequence can be 20−25 *Brachionus* per ml from the fourth day, 8−10 *Artemia* nauplii per ml from the sixteenth to the fortieth day and 5−8 *Artemia* metanauplii and juveniles

Table 22.1 Comparison of costs and return (in US$) between pond (1 ha) and cage (10 × 50 m^3) culture in Thailand (from Kungvankij *et al.*, 1986).

Item	Pond		Cages
A. Income			
Marketable fish 14 000 kg × 3 US$	42 000	8000 kg × 3 US$	24 000
Sub-total A	42 000		24 000
B. Fixed cost			
Land cost (5000 × 18% interest)	900 (2.7%)	Lease	10
Pond construction (5000 × 20% depreciation)	1 000 (3.0%)	Cage construction	1 667 (8.9%)
Interest (30 000 × 18%)	5 400 (16.0%)	5000 × 33.3 depreciation	1 667 (8.9%)
Property tax (1.5%)	75 (0.2%)	Boat and engine	
Sales tax (1%)	240 (1.3%)	1000 × 20% depreciation	200 (1.1%)
		Interest 2000 × 18%	3 600 (19.2%)
		Sales tax	240 (1.3%)
Sub-total B	7 795		5 717
C. Operating cost			
Seed (1000/15 USD) 60 × 150	9 000 (27.0%)	20 × 150	3 000 (6.0%)
Feed	14 000 (42.1%)		8 000 (42.8%)
Labour 80 × 12	960 (2.9%)		960 (5.1%)
Fuel and lubrication	500 (1.5%)		500 (2.7%)
Maintenance and miscellaneous	1 000 (3.0%)		500 (2.7%)
Sub-total C	25 460		12 960
D. Total cost (B + C)	33 255		18 677
E. Net operating income (A − C)	16 540		11 040
F. Net Income (A−B−C)	8 745		5 323
G. Net income over cost	26.3%		28.5%

per ml from the fortieth day to the fry stage. Rotifers and copepods are also used as early food and chopped mussels or finely minced fish are given from about the fifty-fifth day. About 90−100-day-old fry are used for on-growing. However, the maximum survival rate reported is only about 16 per cent.

Grow-out

Information available on the results of mono-culture of gilthead sea-bream is limited. Experimental work carried out in tanks shows that they can be raised to marketable size of 200−400 g on artificial feeds (trout pellets) in one to two years. Growth rates in floating cages are similar to those in ponds. A pilot farm in Turkey is reported to have grown 5 g juveniles to a weight of over 250 g in 12−18 months in cages, fed on commercial pellets and trash fish. Fingerlings of about 80 g, when stocked in cages at a low density of 36−78 per m^3, are reported to have grown to an average weight of 300 g in 6−7 months, in Israel. At a density of 180/m^3 the fish reached an average weight of 315 g at an age of 15−16 months, when fed on a high-protein pelleted feed. Nutritional studies have shown that feed with a 40 per cent protein level gives optimal utilization. Pellets of 10 mm diameter have been found to be efficient in growing fish of 100 g and above.

Fresh sea-bream fetches the best price and so harvests from ponds or cages are, as far as possible, sold fresh on ice.

22.2.2 Red sea-bream

The red sea-bream *Pagrus major*, also known as the red porgy, is a very valuable species in Japan, because it is considered a symbol of good fortune and eaten on all auspicious occasions. Commercial catches of the species are reported to be declining, and so its propagation is undertaken not only for production of

market-size fish in captivity, but also for stocking open waters for enhancing natural populations. As a result of several years of scientific
effort, it is now possible to induce spawning
and artificially propagate the species.

Induced spawning and larval rearing

Mature brood fish can be obtained from commercial catches or from captive stocks. The
sexes are easily distinguished, especially during
the spawning season. The male has darker
coloration and a more angular head. Maturation can take place in net cages or concrete
tanks. The spawning season extends from April
to June, with a peak in early May. Three- to
four-year-old fish are used for breeding. Females of that age, which weigh about 1 kg,
spawn approximately 300 000 eggs. According
to Kittaka (1977), the average number of eggs
spawned per female has been recorded as
4 800 000 for 6–13-year-old fish and 2 700 000
for 3–4-year-old fish. Fully mature females
and males held in tanks with frequent replacements of sea water spawn naturally in temperatures between 15 and 22°C. Spawning continues
for several days and the eggs are collected with
siphons or through overflow arrangements. An
alternative is to strip the mature brood fish and
fertilize them artificially.

A mature egg measures about 1.2 mm in diameter and has a sticky covering, which will
slowly dissolve in sea water. Fertilized eggs are
highly susceptible to changes in temperature
and high-intensity light. Hatching is performed
in 50–100 ton capacity tanks with circulation
of sea water. The stocking rate of eggs in the
tanks is generally 30 000–40 000 per ton. The
eggs will not hatch at temperatures below 10°C.
The black pigment in the large globule
(0.25 mm in diameter) in the eggs apparently
permits light penetration of roughly 100–3000
lux, and so care has to be taken to ensure that
the illumination in the hatching tanks does not
exceed the maximum of 3000 lux. The eggs are
normally pelagic, but if the specific gravity of
the water is below 1.023 (at 15°C) they may
sink to the bottom, in which case their development will be affected and the hatching rate
poor. So it is necessary to have an adequate
supply of water of the required specific gravity,
corresponding to a salinity of 33.5 ppt.

Under favourable conditions, the eggs hatch
out in about 60 hours at a temperature of 15°C
or in 40 hours at a temperature of 18°C. The
larvae become active on the second day, and
the yolk sacs are absorbed by the third day.
They are then transferred to rearing tanks and
fed on live food. The most common means of
feeding larvae is by the addition of the so-
called green water. Green sea water is produced
in outdoor tanks filled with clean sea water and
fertilized with a chemical fertilizer at the rate
of 500 g/ton water. Phytoplankton is innocu-
lated to stimulate planktonic growth and within
10 days the tanks develop a rich green growth
of phyto- and zooplankton. It is reported that
such green water can sustain a population of
about 40 000 larvae/ton water. After a few
days, the water in the tank is freshened gradually with fresh sea water. At this time the
larvae are fed with live food such as oyster
larvae, rotifers, etc., collected from outside
sources. Ten-day old larvae are quite active
and can be fed on brine shrimp (*Artemia*)
nauplii. Like the eggs, the larvae also require
suitable light, but it appears that the light requirement varies according to the individual
and the stage of growth. Because of this, it has
been recommended that the rearing tank should
have zones with different light intensities (for
example, 0 lux to 2000 lux) so that the larvae
can select the preferred light intensity at any
particular time.

By about the twentieth day, and at a length
of about 10 mm, the fry show signs of benthic
life. At this stage they consume small polychaetes and minced shrimp meat. Fry of 20 mm
size are fed on minced white fish and shrimps.
Grow-out facilities are usually stocked with fry
of this size.

Grow-out

The red sea-bream is generally grown in floating
net cages. The stocking density is about 6–8 kg/
m^3 water and in about 12–18 months they
grow to market size. Most farmers feed them
with frozen fish like anchovies, sand eels, etc.
Pelleted diets are now commercially available,
but many farmers seem to prefer to use mash
diets, or minced fish mixed with a formulated
dry powder diet.

As mentioned earlier, the fry and fingerlings

are also released into the sea for enhancing natural populations. For this purpose, fry are acclimatized and grown in floating cages for a few weeks until they attain a size of about 5–7 cm. Then they are released in suitable areas, where protective devices like concrete blocks or plastic strips are placed to serve as shelters for the fry.

22.3 References

Anon (1979) *Manual for Spawning of Seabass Lates calcarifer in Captivity*. FAO/UNDP/MAL/79/018, Reports and Studies.

Barahona-Fernandes M.H. and Girin M. (1976) Preliminary tests on the optimal pellet-adaptation age for sea bass larvae (Pisces, *Dicentrarchus labrax* L. 1758). *Aquaculture*, **8**, 283–90.

Barnabe G. (1976) Rapport technique sur la ponte induite et l'élevage des larves du loup *Dicentrarchus labrax* (L.) et de la dorade *Sparus aurata* (L.), *Stud. Rev. Gen. Fish. Counc. Med.*, **55**, 63–116.

Brown E.E. (1977) Production and culture of yellowtail (*Seriola quinqueradiata*) in Japan. *Proc. World Maricul. Soc.*, **8**, 765–71.

De Angelis R. (1960) Mediterranean brackish-water lagoons and their exploitation. *Stud. Rev. Gen. Fish. Counc. Med.*, **12**.

Fujita S. (1979) Culture of red sea bream, *Pagrus major*, and its food. In *Cultivation of Fish Fry and its Live Food*, (Ed. by E. Styczynska-Jurewicz et al.). *Europ. Maricul. Soc. Spec. Publ.*, **4**, 183–98.

Fujiya M. (1979) Coastal culture of yellowtail (*Seriola quinqueradiata*) and red seabream (*Pagrus major*) in Japan. In *Advances in Aquaculture*, (Ed. by T.V.R. Pillay and W.A. Dill), pp. 453–8. Fishing News Books, Oxford.

Girin M. (1976) Point des techniques d'elevage larvaire du bar en octobre 1975. *Stud. Rev. Gen. Fish. Counc. Med.*, **55**, 133–41.

Jhingran V.G. (1977) *A note on the progress of work under coordinated project on brackish-water fish farming*. Central Inland Fisheries Research Institute, Barrackpore.

Kariya T. *et al.* (1968) Nocardial infection in cultured yellowtails (*Seriola quinqueradiata* and *S. purpurescens*) 1. Bacteriological study. *Fish Pathol.*, **3**, 16–23.

Kawatsu H., Honma A. and Kawaguchi K. (1979) Epidemic fish diseases and their control in Japan. In *Advances in Aquaculture*, (Ed. by T.V.R. Pillay and W.A. Dill), pp. 197–201. Fishing News Books, Oxford.

Kittaka J. (1977) Red seabream culture in Japan. In *Third Meeting of the Working Group on Mariculture of ICES, Brest. Actes de Colloques CNEXO*, **4**, 111–17.

Kungvankij P. *et al.* (1986) *Biology and Culture of Seabass* (Lates calcarifer). NACA Training Manual 3. Network of Aquaculture Centres in Asia, Bangkok.

Lumare F. (1978) *Present state of knowledge on cultivable species in the Mediterranean*. Presented at the Expert Consultation on Aquaculture Development in the Mediterranean Region, FAO (GCFM)-UNEP.

Person-Le Ruyet J. and Verrilaud P. (1980) Techniques d'élevage intensif de la daurade dorée *Sparus aurata* (L.) de la naissance à l'âge de deux mois. *Aquaculture*, **20**, 351–70.

Pitt R., Tsur O. and Gordin H. (1977) Cage culture of *Sparus aurata*. *Aquaculture*, **11**(4), 285–96.

23
Other Finfishes

In the preceding sections an attempt has been made to summarize the culture practices relating to the major groups of finfish which contribute substantially to aquaculture technologies and production. As mentioned earlier, there are several other species which are cultured on a small scale in limited areas or which have shown potential to become important aquaculture species in the near future. Among the first category are the murrels (or snakeheads) and gouramis of Asia, the groupers and rabbitfish of the Indo-Pacific and Middle East regions and the turbot of Western Europe. This section deals briefly with the culture practices of the above-mentioned groups.

Besides these, the South-American species of the genus *Colossoma* and the Florida pompano (*Trachinotus carolinus*) have received considerable attention from aquaculturists, although commercial-scale production has yet to be developed. The cod (*Gadus morhua*) and the halibut (*Hippoglossus hippoglossus*) are being intensively investigated in North European countries, especially Norway. Though the main focus of attention is to mass-produce young cod for enhancing natural populations, small-scale rearing to market size in captivity has been attempted with encouraging results. Though much more experimental work is required, it has been possible to propagate cod and halibut artificially and rear the larvae to juvenile stages.

Some of the pioneering work on marine fish cultivation was focused on the flat fish plaice (*Pleuronectes platessa*) and sole (*Solea solea*). Techniques of controlled reproduction, larval rearing and grow-out have been tried with considerable success. However, commercial farming of these species is at present considered not economically attractive because of the low market value of plaice and the slow growth rate and problems of feeding sole.

Two groups of species that are presently cultured and of importance in restricted areas are the pufferfish (*Fugu rubripes* and *F. vermicularis*) and the sturgeons. The pufferfish are a speciality seafood only in Japan, Korea and China and are of limited interest elsewhere. Artificial propagation and larval rearing of pufferfish are well-advanced and market size (800 g) fish are produced in ponds and cages, fed on trash fish.

The sturgeons are of special importance as food fish and as a source of the high-valued caviar. Economically important species of sturgeons have been reproduced in hatcheries and young ones are released every year to enhance natural populations that have been greatly reduced by overfishing and environmental changes (see Chapter 30). However, small-scale culture of some species is undertaken in the USSR and other East European countries. The sterlet (*Acipenser ruthenus*) is considered a suitable fish for polyculture in carp ponds with adequate water flow and aeration. The sterlet is benthophagous, feeding on chironomids and oligochaetes, but can be grown on artificial feeds containing about 40 per cent protein. The hybrids of beluga (*Huso huso*) and the sterlet are preferred for monoculture in ponds and pens because of their hardiness and adaptability (fig. 23.1). While pond culture is carried out in the southwestern part of the USSR, pen culture is practised in the bays of the Baltic sea and parts of the Azov sea. Since growing seasons are restricted to

Fig. 23.1 Hybrids of beluga (*Huso huso*) and sterlet (*Acipenser ruthenus*) grown in pens.

5–8 months, the fingerlings usually take two growing seasons to reach the market size of 800 g in the south, and three seasons in the colder northern waters. Stocking rates are usually 7000–8000 per ha in ponds and 20–40 per m^2 in pens, with average yields of 10–25/ kg per m^2 in pens and 980 kg/ha in ponds.

Experimental and pilot scale sturgeon culture have also been reported from France, California (USA) and Italy. Experimental culture of the Siberian sturgeon (*Acipenser baeri*) has been reported from France, where fingerlings imported from the USSR were raised in concrete ponds on artificial diets containing 44 per cent protein at temperatures ranging from 16 to 18°C. By the second year they grew to a weight of 1.8 kg and by the third year to an average weight of 2.65 kg.

Hatchery-produced white sturgeon, *A. transmontanus*, has been experimentally grown in tanks in California in temperatures in the range 12–20°C, on a 40 per cent pelleted trout feed. The market size of approximately 1 kg body weight was reached in 18 months. The white sturgeon imported from California is now established in pilot commercial farms in Northern Italy (fig. 23.2). Larvae are raised in tanks fed with specially formulated aromatic feed. Large concrete tanks are used for growout. The water temperature is regulated from an initial 23–

Fig. 23.2 White sturgeon, *Acipenser transmontanus*, raised in ponds in northern Italy (from *Il Pesce*, **2**, 1987).

24°C down to 18−20°C as they reach adult size. They grow rapidly in the tanks, reaching 200−300 g in 6−7 months, and 1 kg by the end of the first year. By the end of the second year they weigh about 2−3 kg and in three years about 5 kg. In Italy the 5 kg size is considered to be the best for marketing, because of the texture and quality of its flesh. Since the growth rate during the third year is more rapid, a three-year cycle of culture has been recommended. The species is now reproduced locally and eyed ova are offered for sale. Large sturgeon have a ready market. So large-scale commercial farming can be expected to develop in Italy within a reasonable period of time.

23.1 Murrels (snakeheads)

Murrels or snakeheads belonging to the family Channidae (= Ophiocephalidae) are highly regarded food fish in the South and Southeast Asian countries. Their ability to breathe atmospheric oxygen makes it possible to keep them alive for long periods out of water and to sell them alive at high prices in the market. Besides the high-quality flavour and texture of their flesh, murrels are especially regarded as diet for invalids and recuperating patients. Though cultivated in many countries of Asia, murrel culture has not yet developed to major commercial importance.

There are over 30 species of murrels or snakeheads distributed in tropical Asia, including Northern China, and in Africa. Among these, the species of aquacultural importance are *Channa* (= *Ophicephalus*) *striatus* (fig. 23.3). *C. marulius*, *C. punctatus*, *C. maculatus* and *C. micropeltes*. *Channa striatus*, *C. marulius* and *C. micropeltes* grow to sizes of 1−1.2 m, whereas the other two species are smaller in size, reaching 22−30 cm. Though adult murrels of any size have a market in Asian countries, the preferred size is between 600 and 1000 g.

Murrels are very hardy and can tolerate unfavourable conditions. If kept moist, they can live out of water for long periods, and are known to survive droughts by aestivating for months in moist mud. The preferred temperature is in the range of 20−35°C, and the upper and lower lethal limits are reported to be 40° and 15°C respectively. Though sensitive to sudden changes in pH, they can survive in both acidic and alkaline waters. They are essentially fresh-water species, but can withstand low salinity, brackish-water conditions.

The most common system of murrel culture is in earthen ponds ranging in size from 800 to 1600 m^3 and in depth from about 0.5 to 2 m. Often the ponds have fine-meshed wire fencing to prevent escape of the fish. In Kampuchea and Vietnam, murrels are usually grown in cages moored near the shore or trailed behind fishermen's boats. The cages vary in size from 40 to 625 m^3. A traditional system of growing murrels in irrigation wells is practised on a small scale in India and neighbouring countries.

Being highly predaceous and cannibalistic, murrels are generally raised in monoculture using, as far as possible, stock of the same size group. But in countries like Taiwan, murrels are stocked in carp and tilapia ponds to forage on unwanted fish. Recent experimental work in India shows the possibility of culturing *C. marulius* and *C. striatus* in swamp ponds together with several species of local forage fish. Farmers in Thailand have in recent years started integrating murrel culture with pig and poultry production.

Fig. 23.3 The murrel, *Channa striatus*.

23.1.1 Spawning and fry production

Murrels attain maturity at an age between one and two years. Mature *C. striatus* are above 25 cm in length and *C. marulius* above 36 cm. As in the case of other species showing parental care, the fecundity of murrels is comparatively low. Depending on the size of the fish and the species, the fecundity has been observed to vary between 2200 and 34 000 among the cultivated murrels. The peak breeding season for *C. striatus* is during the rainy months, but the species seems to breed throughout the year. *Channa maculatus* appears to breed in Taiwan and Hong Kong from April to September.

All species of murrels exhibit parental care and spawn in nests built in shallow marginal areas with cut pieces of aquatic vegetation or similar material. Spawning lasts for 15–45 minutes and the eggs laid by the female in the nest are fertilized by the sperms shed by the male. The golden-yellow or amber-coloured fertilized eggs float in the centre of the nest in a thin film. The eggs hatch out in 20–57 hours in temperatures ranging from 16 to 33°C, depending on the species. The hatchlings measure about 3.5 mm and within about four days, when they measure about 6.8 mm, they are able to swim about freely. Both males and females take part in caring for the newly hatched young for about 15–20 days, that is until the larvae are about 3.5 mm in length. The newly hatched larvae feed on protozoa and algae. The larval development is completed in about 9 weeks, when they move to the bottom and show adult behaviour. The fry feed on animal foods such as crustaceans, insects, young fish and tadpoles.

Induced spawning of most species of *Channa* has been carried out by hypophysation, although it is not practised to any appreciable extent on production farms, except in Taiwan. Selected brood fish are reared in separate brood ponds, fed on live food such as fish and tadpoles for about two to three months before spawning. Female *C. maculatus* spawners of about 1 kg body weight are injected with one or more common carp pituitary together with 20 rabbit units of Synahorin, in two equal doses at intervals of 12 hours. Male fish do not require injection and spawn naturally. The spawners are then kept in 3–4 m^3 cages made of nylon netting placed in ordinary fish ponds, for spawning. Each cage contains a male and female pair. Sometimes five or six pairs may be released in small shallow ponds (7–10 m^3) without using cages. Spawning generally takes place in about a day. For induced breeding of *C. marulius*, *C. striatus* and *C. punctatus*, hypophysation with carp and catfish pituitaries at doses ranging from 40–80 mg gland/kg female in two injections have been found to be adequate under experimental conditions.

23.1.2 Grow-out

For monoculture as practised in Thailand and Hong Kong, shallow ponds of a surface area of 800 m^2 to 0.5 ha are used. Where intensive farming is undertaken, a continuous flow of water or frequent exchange of water is maintained. The usual pond preparation including draining, liming and drying of the pond bottom is necessary. Thai farmers stock 75–460 fry of *C. striatus* per square metre of pond area. Stocking is carried out in the months of July/

August in order to harvest market-size fish in April/May of the following year. The supply from capture fisheries is low during this period and so the fish can be sold at a high price. Trash fish, rice bran and broken rice are fed thrice daily in the ratio of 8:1:1. The ratio of trash fish may sometimes be raised to 13. The fish grow to market size in 7−8 months.

In cage culture of *C. micropeltes* and *C. striatus* in Kampuchea, a stocking rate of 6000−10000 fry is common in cages measuring about 625 m³. Fed on various types of vegetables (cooked pumpkin, banana, rice and rice bran) and animal products (including live and dead fish), they are reported to reach a weight of 1.5−2.5 kg in 9 months. In Vietnam, where cage culture of the above two species is practised, cages of about 125 m³ are stocked with 4−6 cm long fry collected from the wild, at the rate of about 80 per m³. Harvesting is carried out after about 9 months of culture, as in Kampuchea.

In polyculture with Chinese carp in Taiwan, ponds are stocked with fingerlings of about 10 cm at a rate not exceeding 500/ha. When cultured with tilapia, the recommended stocking rate is about 90 000 10 cm fingerlings/ha. In polyculture with Chinese carps, the murrel (*C. maculatus*) feed on weed-fish, and in culture with tilapia, the fry produced by wild spawning of tilapia forms the main source of food. In order to avoid cannibalism, the stock of murrels is graded two or three times during the culture period, reducing the stock density finally to 15 000−24 000/ha.

Though there are several records of parasite infestation of murrels, serious outbreaks of disease or mortality are not as common. However, in the intensive murrel culture in Thailand, mortalities occur more frequently. The farmers try to control this by providing feed containing antibiotics, even though the actual cause of mortality is not known. Fry of *C. micropeltes* and *C. striatus* have been found to be very susceptible to ectoparasites, such as *Costia* spp., *Chilodonella* spp. and *Trichodina* spp.; *Ichthyophthirius multifiliis* has also been recorded from these species. Gill rot caused by *Branchiomyces sanguinis* has been identified in *C. marulius*, and infection by the fungus *Dictyuchus anomalous* causes mortality of *C. punctatus*.

Harvesting of murrels from ponds is achieved by draining and seining. The yield in Thailand is reported to be up to 25 kg/m². Catches are sold in live condition as the price of dead fish is usually 30−40 per cent less than that of live fish. In some areas in Thailand, murrels are salted and dried, and poor quality fish processed by fermentation. Though feeding is a relatively major cost of production, murrel culture has proved to be very profitable in all Asian countries, because of its high price in the markets. Available data on cost and earnings show 25−67 per cent net income on total operational cost.

23.2   Gouramis

Though the overall aquaculture production of the species generally referred to as gouramis (belonging to the family Anabantidae) is not very high, they have nevertheless been important in traditional farming practices in the Southeast Asian countries and continue to be so. Compared to other aquaculture species, the gouramis have not attracted much attention from scientists so far and the culture practices have not undergone any major changes in the recent past.

The three species known as gouramis are the giant gourami, *Osphronemus goramy* (fig. 23.4), the Siamese gourami or sepat siam, *Trichogaster pectoralis*, and the kissing gourami, *Helostoma temmincki*. Though they were all considered as belonging to the same family Anabantidae, in later classifications the first two species are placed in a new family Osphronemidae and the third in Helostomidae. All are tropical species which can live in swampy conditions and breathe atmospheric oxygen through accessory respiratory organs. They spawn easily under pond conditions and are well relished by consumers. They can be described as omnivorous, with plankton and various types of plant matter dominating the diet. Probably because of the slow growth rate, as in the giant gourami, or the relatively smaller size attained by the others, the earlier widespread interest in culturing gouramis appears to have diminished somewhat and is now restricted to small-scale extensive farming in some of the Southeast Asian countries.

The giant gourami is the most highly-priced

Fig. 23.4 The giant gourami, *Osphronemus goramy*.

species and has been introduced into a number of countries in the Indo-Pacific region. It can be grown in fresh and slightly brackish water, but is sensitive to temperatures below 15°C. The optimum temperature is 24–28°C. It is reported to reach a length of 61 cm and a weight of 9 kg. Young of the species feed on planktonic organisms, but the adults show a preference for aquatic vegetation.

The Siamese gourami attains a maximum length of about 25 cm and is raised in some areas as a rotational crop in rice fields. It is usually marketed as a dried product, which is highly esteemed in countries like Thailand.

The kissing gourami, which derives its name from its characteristic labial contacts (suggestive of kissing) is a fresh-water fish that is predominantly a plankton feeder, but it will feed on most artificial fish feeds. It reaches a maximum length of 30 cm and spawns every three months.

The culture systems adopted for the gouramis are generally extensive, and often in polyculture with other species. The practice of growing them in rice fields is referred to in Chapter 29.

23.2.1 Spawning and fry production

In nature, the giant gourami spawns during the dry season but in ponds it spawns throughout the year, in submerged nests made of plant material. Spawners are generally two to three years old, and both female and male parents guard the progeny. The fecundity of the species is not very high and it is reported that a 5-year-old female produces only 3000–5000 fry per year in two or three spawnings at bi-monthly intervals. The simplest system of breeding adopted is the provision of suitable water plants or other material for nest making and ensuring that the ponds contain enough male and female spawners, usually in the ratio of 1:2 or 2:3. Ponds of various sizes with growths of macro-vegetation along the margins are used for spawning, and in certain areas in Java the species is bred with the cyprinid nilem, *Osteochilus hasselti*, in the same pond. In order to make nest-building material available, farmers in Java place palm fibres on bamboo poles in the pond. Some farmers place conical bamboo containers at depths of about 30 cm, inside which nests are made with water plants (fig. 23.5). The nests are spherical and measure 30–35 cm in diameter with the opening at the bottom. It usually takes about 10 days to make the nests and spawning takes place two or three days after the nests are ready. A characteristic fishy smell and an oily substance that emanates from the nest indicate that spawning has taken place. The eggs hatch out and hatching become clearly visible in two to three days. The hatching rate and larval survival can be increased by removing the fertilized eggs and hatching them indoors. While ants are considered to be the best food for larvae, peanut waste is sometimes used as a substitute. At about 1 cm length, the larvae are transferred to rearing ponds and raised till they reach the size of 2 cm. The fry are usually fed on *Azolla* or minced leaves of other plants. Several variations of these methods followed in Java are described by Hora and Pillay (1962).

Fig. 23.5 Conical bamboo containers placed in ponds for gouramis to build nests in (photograph: Marcel Huet).

The Siamese gourami spawns in floating bubble nests made by the male in 70–100 cm deep spawning ponds. Spawners are usually about 7 months old and about 100 g in weight. The fecundity of the species varies between 20 000 and 40 000 eggs per fish of 90–120 g weight. The nest, which consists of a mass of bubbles, takes about 2 days to make and, when ready, the female lays the eggs under it where they float and are immediately fertilized by the male. Hatching may take 1–3 days and about 4000 fry can be expected from each nest.

Farmers who grow the species in rice fields do not spawn them in separate ponds, but the spawners are first kept in small ponds for some time under crowded conditions without any supplementary feeding. This is done to make them lean in the belief that lean fish are better brood fish. They are then released into the peripheral ditches of the rice field. After a certain period of time, the fields are flooded to submerge the central raised portion of the field covered with weeds. The fish migrate to the weed patches, build bubble nests among them and spawn. The eggs hatch out in about 24 hours. On about the sixth day, when the yolk sacs are fully absorbed and the larvae measure 5–6 mm in length, feeding starts. The main source of larval nutrition is the live food produced in the field as a result of green manuring (with the cut weeds that the farmer piles up in the fields).

Spawning of kissing gourami is undertaken in either separate spawning ponds or harvested rice fields. In Indonesia, the spawning ponds vary in size from 30 to 100 m^2 in area, with a supply of clear water. Rice straw or banana leaves are spread on the pond to provide shade for the eggs and to protect the larvae. Twelve- to eighteen-month-old spawners of about 20 cm length and 150 g weight are held in segregation ponds for a period of about a month, after which they are released into the spawning ponds. Spawning takes place in about 18 hours and the larvae hatch out in less than two days. The ponds are manured with decaying plant material about 7–10 days after larvae have hatched out, to develop enough planktonic food for the larvae. The usual larval rearing period is about 30 days and during this period manuring may be repeated several times.

23.2.2 Grow-out

As mentioned earlier, the giant gourami is a slow-growing fish which appears to take two to three years to reach marketable size. However, the experience so far is not based on any systematic culture practices directed to obtain fast growth and high productivity. The feeds used in grow-out ponds are leaves and tender parts of *Ipomoea*, land grass, yam, tapioca, kitchen waste or a mixture of cooked rice, rice bran and trash fish, with no regard to the nutritional requirements or the stocking densities suitable for optimum growth. The rates of growth are reported as 15 cm in the first year, 25 cm in the second year and 30 cm in the third year, and the production per hectare is about 200 kg.

The Siamese gourami reaches lengths of 16−18 cm and weights of 130−150 g in 12 months, and at the higher stocking rate of 40−220 kg spawners per ha a minimum yield of about 500 kg/ha is obtained in monoculture (fig. 23.6). Culture in rice fields with proper stocking and manuring provides higher yields of up to 2.2 tons/ha.

23.3 Groupers

Groupers belonging to the genus *Epinephelus* (family *Serranidae*) are highly-priced fish in the Indo-Pacific, Middle East and Caribbean regions. Among the several species of the genus, the most important for aquaculture are the estuarine or greasy grouper *E. tauvina* (fig. 23.7). It is a hardy species which can stand rapid changes in salinity between 2.5 and 45.5 ppt. The optimum salinity is reported to be 15−26 ppt. It is carnivorous in feeding habits and feeds on smaller fish and shrimps. Juveniles feed on *Acetes* and mysid shrimps in nature. It spawns throughout the year, probably with a peak in the wet months. The red

Fig. 23.6 Harvest of sepat siam (Siamese gourami), *Trichogaster pectoralis*, from a pond in Thailand.

Fig. 23.7 Greasy grouper, *Epinephelus tauvina* (Courtesy of Chen Fooyan).

grouper, *E. akaar*, which has habits very similar to *E. tauvina*, is the important species for culture in Hong Kong.

One of the major problems in the controlled reproduction of the species is that it is a protogynous hermaphrodite. Like other groupers, it matures as a female but becomes a male with advance in size and age. Fish of 45–50 cm length mature as females, while fish of more than 74 cm weighing more than 11 kg become males and develop ripe testes. An intersex condition can be found in fish of length 66–72 cm, with transitional gonads containing male and female gonadal tissues.

Commercial grouper culture is carried out in floating cages on a small scale in Malaysia, Singapore and Hong Kong. Experimental culture has been attempted in Kuwait.

23.3.1 Fry collection and induced spawning

Although it is reported that mature spawners caught from the wild have been observed to spawn in captivity (Hussain *et al.*, 1975) and accelerated sex-reversal and induced spawning have been achieved experimentally (Chen *et al.*, 1977), wild fry are still used for cage farming. Fry measuring 3.5–5 cm are available during the north-east monsoon and fishermen collect them with small seines and transport them for sale to fish culturists.

Observations seem to indicate that a female *E. tauvina* will take more than 5 years to become a functional male. By oral administration of methyltestosterone incorporated into feeds

three times a week over a period of two months, at a dose of 1 mg per/kg body weight, two- to three-year-old fish yield milt on induced breeding (Chen *et al.*, 1977). Induced spawning of three-year-old females and sex-reversed males has been achieved by injection of HCG and pituitary gland extract from chum salmon or white snapper. Complete ovulation was also obtained by a single injection of 5000 IU HCG. The fertilized eggs can be hatched in tanks within 23–25 hours at a temperature of 27°C. Spawning can be carried out throughout the year, depending on the stage of gonadal development. The larvae metamorphose into juveniles of around 25 mm length in about 33 days. Fifty-day-old juveniles measure up to 70 mm in length.

The red grouper has also been induced to spawn in captivity and the larvae reared under experimental conditions, following more or less the same techniques as employed for the estuarine grouper.

23.3.2 Nursing and grow-out

Wild fry or fingerlings are initially held in hapas for a month or more. Hapas measuring 2 × 2 × 2 m are stocked at the rate of 400–600 advanced fry or fingerlings. When they have reached lengths of 12–15 cm, they are transferred to nursery cages (5 × 5 × 3 m), each holding about 1100 fish, and after about 2–3 months they are transferred to production cages. The cages used are of the floating type, usually made of polyethylene netting supported by a wooden framework. The cages are kept afloat with metal or plastic drums and anchored with concrete blocks. Generally, nursery cages have a mesh size of about 2.5 cm and production cages have a mesh size of 2.5–5.0 cm, for growing fish to a market size of 50–75 cm.

In hapas the fry and fingerlings are fed with mysids and small shrimp after a couple of days of acclimatization. Trash fish form the main feed in nursery and production cages, and they are minced or chopped to suit each size group. Finely minced fish can also be used for feeding fry in hapas. Fish in hapas are fed at 10 per cent of body weight daily and those in nursery cages at about 8 per cent of body weight. Feeding rate in the production cages is 5 per cent of body weight and the conversion ratio is around 1:4.5 (wet weight). The estuarine groupers attain a market size of about 800 g within about 6 months and two crops are raised every year.

As the culture of the red grouper in Hong Kong starts with advanced fingerlings, they are stocked directly in production cages. A $4 \times 4 \times 3$ m cage is generally stocked with 800 fingerlings of a total weight of about 120 kg. They are fed with trash fish at a rate of 3 per cent of body weight in summer and half that in winter. As the fish grow in size the stock is thinned and transferred to additional cages. In about 4 months they reach weights of 300–340 g, and in about 6 months a marketable size of 500 g each. Some fish are left to grow to a size of 1 kg.

While monoculture of groupers is generally practised, in cages the red grouper is cultured together with sea-bream (black sea-bream, *Mylio macrocephalus* and red sea-bream, *Pagrus major*). This is to utilize the feed left uneaten by the grouper and to prevent water pollution due to decaying feed.

The common disease encountered in captive groupers is vibriosis, caused by *Vibrio anguillarum*, commonly referred to as 'red boil disease', characterized by inflammation, haemorrhage and ulceration of skin and musculature. Oral administration of the antibiotic oxytetracycline hydrochloride incorporated in the diet, at the rate of 50–60 mg/kg body weight daily for the first two days, and half that for another five days, has been reported to be effective in controlling the disease. Another

less common disease of groupers is gill infestation by the monogenetic trematode *Diplectanum* sp. A daily bath in 70 ppm formalin for half an hour for a period of 3–4 days is reported to be effective. Infection by the protozoan parasite *Cryptocaryon irritans* causes loss of scales and skin, especially in the head region. The usual treatment is a bath in 200 ppm formalin for half to one hour.

23.4 Rabbit fishes

Rabbit fishes belonging to the family Siganidae (= Teuthidae) include a group of potentially important aquaculture species occurring in the Indo-Pacific, Indian Ocean, Red Sea and Eastern Mediterranean regions. Though culture is presently limited to a few areas, rabbit fishes have attracted the attention of aquaculturists because of their predominantly herbivorous feeding habits, fast growth rates and high prices in the markets. They do not normally grow to lengths greater than 35 cm, but are highly relished in many areas, and for the ethnic Chinese in Southeast Asia they symbolize good fortune and are sought after during auspicious periods such as the Chinese New Year.

In nature they are found in reefs among sea grass, among mangroves and in shallow lagoons. *Siganus canaliculatus*, an important candidate species for culture, has been found to tolerate wide ranges of salinity (17–37 ppt) in its natural habitat. It has been acclimatized to a salinity of 5 ppt and grows well in temperatures between 23 and 36°C. *Siganus rivulatus* and *S. luridus* have been observed to grow well in 20 ppt salinity and tolerate up to 50 ppt. They are sensitive to low oxygen concentrations, below 2 ppm, and pH values above 9. Both juvenile and adult siganids are primarily herbivorous, but under captivity show omnivorous habits and will feed on a variety of foodstuffs of both vegetable and animal origin, as well as on feed pellets. In nature, benthic algae form a major part of their food, but there appear to be differences between species in the algae preferred.

The long-established systems of commercial culture of siganids appear to be as subsidiary species, or in monoculture in brackish-water ponds in the Philippines or in embanked lagoons in Mauritius in combination with oysters. Experimental culture of rabbit fish in floating cages and pens, ponds and raceway systems has been attempted in a number of countries, including Malaysia, Singapore, Guam and Palau Islands, Saudi Arabia, Israel and Tanzania. However, according to reports it is only in Palau that it has resulted in pilot-scale production in cages. The main species used in culture are *S. canaliculatus* (= *oramin*), *S. vermiculatus* (fig. 23.8), *S. rivulatus* and *S. luridus*.

23.4.1 Fry collection and induced spawning

Though some progress has been made in controlled spawning, all culture operations so far are based on fry collected from the wild. Juveniles are available from February to May and again from August to October or November (December in Palau). They are found usually around reef flats and are collected with push nets, scoop or dip nets, seines and cast nets. An improved mobile pyke net was found to be an efficient fry collector in sea-grass beds in the Red Sea in Saudi Arabia.

The spawning season appears to be from January or February to April in several areas. According to Lichatowich *et al.* (1984), it is likely to be April to August in the Red Sea. *Siganus canaliculatus* juveniles grow to adult size in about 9 months, when they mature and spawn. In captivity they may mature even earlier. Spawning is influenced by the lunar cycle and the species has been observed to spawn 4—7 days after the new moon during the spawning season. The number of eggs spawned at a time is estimated to be 300 000—400 000.

Siganids can be induced to spawn spontaneously in tanks and aquaria. The sudden transfer of mature fish from a tank with about 90 cm water to a shallow tank with only 18—23 cm water resulted in immediate spawning of *S. canaliculatus* in Palau. It has been observed that the females release the eggs as soon as the males shed milt and fertilization takes place in the tank. *Siganus rivulatus* has been found to spawn in aquaria in Israel when the water is exchanged with fresh sea water. *Siganus canaliculatus* and *S. argenteus* have been induced to ovulate and spermiate by injection of HCG.

Fig. 23.8 Rabbit fish, *Siganus vermiculata*

Larvae of *S. canaliculatus* have been reared to the post-metamorphosis stage, with successive feeding on mixed phytoplankton, rotifers (*Brachionus*), copepods (*Oithona*) and *Artemia* nauplii. Larvae of *S. rivulatus* and *S. luridus* have also been reared by similar feeding and grown to the juvenile stage.

23.4.2 Grow-out

Not much information is available on polyculture of siganids, nor do accounts of recent experimental or pilot-scale grow-out contain many data on stocking rates and growth performance. In coastal ponds in the Philippines, in monoculture or polyculture with milkfish, siganids, especially *S. vermiculatus*, are reported to attain a marketable size of 150 g within 5–7 months. Similar growth is reported also for *S. canaliculatus* in coastal ponds. In tank culture in Israel, fry fed with algae (*Ulva* sp.) and 25 per cent protein commercial fish feed pellets, showed a two-fold increase in length and a ten-fold increase in weight in over 5 weeks. Similar growth rates are reported to have been achieved with *S. canaliculatus* in the Philippines, where they were fed on algae, and in Palau, with commercial chicken feed pellets.

Siganus rivulatus is reported to grow to a weight of 185 g in about 300 days in floating cages. Lichatowich *et al.* (1984b) reported a growth rate of 105 g in 150 days in floating cages, when this species was fed a mixture of soya meal (53 per cent), fish meal (14 per cent), maize (15 per cent), flour (15 per cent) and a vitamin-mineral mix (3 per cent) (from plastic trays suspended in the cages). The feeding rate was approximately 8 per cent of the biomass. Observations in Palau show that *S. canaliculatus* cultured in floating pens grew faster in areas with good water circulation, and when fed with trout chow instead of with algae only.

Though there is a ready market for siganids in most places, there are occasional reports of ciguatera poisoning as a result of eating them, especially in the South Pacific, and recently in Israel. The source of the ciguatoxin is believed to be marine algae such as *Lyngbya majuscula*, *Plectonema terebrans* and *Schizothrix calciolla*, which are eaten by the fish.

23.5 Turbot

Among the flatfishes the turbot *Scophthalmus maximus* (family Scophthalmidae) (fig. 23.9) has so far proved to be the one with the greatest immediate aquaculture potential. As a result of several years of research, especially in the United Kingdom and France, the main elements of a culture technology have emerged and pilot-scale production in these countries, and in Spain, has been initiated in recent years.

Turbot is a highly-priced marine fish for which there is a good demand, especially in Northern European markets. Natural production is reported to be insufficient to meet the demand. The species is known to be hardy and, though carnivorous in nature, can be fed on various feedstuffs and prepared feeds. Growth rates are fast and at a water temperature of about 18°C, 5 cm juveniles reach a marketable size of 300–400 g in one year, and over 1000 g in about 18 months. The maximum recorded length of the species is 100 cm, but fish of about 50 cm are more common in the markets.

Though early attempts at turbot culture were handicapped by inadequate supplies of seed stock, it is now possible to mass produce fry in hatcheries. As culture has to be based on complete artificial feeding, intensive systems of culture are adopted mainly in tanks and cages. Cooling water from power stations has been used to maintain optimum temperatures and obtain accelerated growth rates.

23.5.1 Controlled spawning and hatchery production of juveniles

Two- to three-year-old female turbot (of 2 kg weight) have been observed to reach sexual maturity and spawn in the wild from May to August. Males reach maturity in the second year (1 kg body weight). The annual fecundity is about a million eggs per kg body weight. The diameter of the eggs varies from 0.9 to 1.2 mm. Mature fish will spawn naturally in tanks. Circular indoor tanks up to 2.7 m^3 in size, supplied with warm water and continuous high-intensity illumination (up to 3000 lux), have been successfully used for spawning and larval rearing. The size and colour of the tanks do not seem to have any significant effect. Eggs and

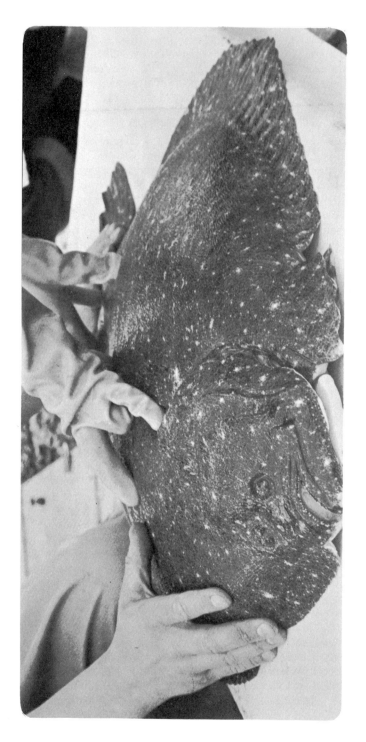

Fig. 23.9 The turbot, *Scophthalmus maximus* (from *Fish Farming International*, **6**(4)).

milt have also been obtained by manual stripping and artificial fertilization has been achieved. Through proper management of brood stock, it is now possible to produce eggs all year round. Temperature is maintained between 10° and 15°C, and the photoperiod is adjusted to obtain spawning at any time of the year. The eggs are incubated at about 12°C in filtered sea water treated with antibiotics. Newly hatched larvae are reared at densities of 30–45 larvae per ℓ, in 60–450 ℓ tanks. Temperatures between 18 and 20°C are maintained with 90 per cent water exchange every day.

The comparatively small size of turbot hatchlings (3.11 mm length and 0.10–0.15 mg weight) makes it necessary to handle them with special care and to feed the right size live foods to obtain reasonable survival rates. The rotifer *Brachionus plicatilis* and nauplii and metanauplii of the brine shrimp *Artemia salina* are the most commonly used larval foods. It has been suggested that the algae used to feed the rotifer affect the growth and survival of the larvae (Howell, 1979). When fed on *Isochrysis galbana* rather than *Dunaliella tertiolecta*, the larvae grew better and mortality was lowered. From the first day of hatching to about the eighth day, *Brachionus* is the preferred food; *Artemia nauplii* are then added and the rotifer reduced, usually terminating by about the eleventh day. Naupliar feeding is continued until the larvae develop into metamorphosed juveniles at a size of 25–30 mm in about 30–40 days. From about the eighteenth day, larger metanauplii of *Artemia* are used.

The juveniles, weighing 55–105 mg, are weaned to artificial diets in less than two weeks. Dry pellet crumbs (400 μm size) give satisfactory results, but greater success has been achieved by the use of moist pellets, when survival rates have been increased to about 50 per cent. Experimental work (Person-Le Ruyet *et al.*, 1983) seems to show that expanded pellets enriched with inosine (a chemical attractant) increase food intake during the beginning of the weaning period, but the economics have yet to be determined. After weaning, dry pellets are normally used for feeding, and in about three months after hatching the fry attain a weight of around 2 g.

In areas where the temperature is much below the optimum range of 18–20°C, the fry are initially reared indoors, in heated water, and then transferred outside. If they are to be transferred during the summer months, they are grown indoors up to a weight of 5 g, but if they are to be transferred during winter, they are grown to at least 20 g size. Survival below 5–6°C is very low. In well-oxygenated water they can tolerate temperatures between 25 and 30°C, and salinities ranging from brackish to 40 ppt. For rearing up to 20 g size, dry pellet feed is commonly used, but for growth beyond that moist pellets (containing 25 per cent trash fish, as well as fish and meat meal, wheat middlings, brewer's yeast, cod liver oil and a vitamin premix) are recommended. Moist pellets made from a mixture of industrial fish and dry compound meal have been used to feed 5 g fish at the rate of 2.6 per cent of body weight at 10°C and at 4.5 per cent at 15°C (Jones, 1981). Juveniles of 50 g weight are fed at 1.1 per cent of body weight at 10°C and at 2 per cent of body weight at 15°C. Dry pellets are not easily accepted by larger fish, but trash fish forms an excellent feed. After a period of one year's growth under favourable conditions, turbot attain weights of 175–350 g in indoor tanks in areas with higher water temperatures, while a weight of about 120 g only may be reached in outdoor tanks.

23.5.2 Grow-out

As commercial grow-out is still in the early stages, there is only limited information available. However, early trials have shown that turbot young can be grown to market size with high levels of survival. Marketable-size fish can be raised in 12–14 months with proper feeding and environmental control, especially of temperature. The optimal temperature appears to be 15° to 19°C. Though heated water may have to be used in growing tanks, the quantity of water required is not large as turbot are reported to require tanks with a surface area only equivalent to their own.

In tanks fed with cooling water from a nuclear power plant in North Wales, UK, a stocking density of 25–55 kg/m^3 of 10–1500 g fish has been maintained, but this required continuous re-oxygenation using special equipment (Jones, 1981). Several foodstuffs and compound diets have been used for on-growing. Chopped or whole trash fish and industrial fish are well accepted by turbot, but it is believed that when such feed is used for a long period the survival rates are adversely affected. So moist compound pellets are recommended. However, fish like mackerel, sprat or mysids stimulate the appetite of turbot and so incorporation of these fish in a ground form to the extent of 20–25 per cent dry weight (as suggested for juvenile growing) has been recommended to improve feeding and growth rates, especially at lower temperatures.

Experimental grow-out of young turbot collected from the wild and raised in hatcheries in floating sea cages of 2.9 m^3 size in Scotland (Hull and Edwards, 1979) has shown that they can be grown successfully in cages, with very high survival rates. To obtain high growth rates, the juveniles had to be reared in land-based nurseries over the first winter period, to a size above 50 g, and then transferred to the cages. The maximum density tried was 41 kg/m^3 or 240 fish/m^3 and this did not depress growth. Moist pellet preparations with a protein content of about 36–39 per cent were used as feed. In the ambient temperature ranging from 6 to 16°C, a market size of 450–500 g was reached in 18 months of on-growing and after another 12 months a weight of 1.2 kg could be attained.

Vibriosis is the most common bacterial disease recorded for young turbot and this can cause high levels of mortality. The antibiotic oxytetracyclin, administered through feed at a concentration of 75 mg/kg body weight per day and also as a bath at 53 mg/ℓ has been found to be effective. In addition, infestation by *Trichodina* spp. has been observed. This is controlled by one-hour baths of formalin at a concentration of 1:6000. In rearing facilities using power-station cooling water, gas bubble disease can occur, but is controlled by changing to ambient sea water.

23.6 References

Abdullah M.A.S. *et al.* (1983) Refinement of spawning and larval rearing techniques in hamoor (*Epinephalus tauvina*). *Ann. Res. Rep. KISR*, 55–7.

Alcántara F., Guerra H. and Wilheim E. (1983) Ensayo preliminar de cultivo de gamitana, *Colossoma macropomum* (Cuvier, 1818) associado a la cria de cerdos. *Revista Latinoamericana de Acuicultura*, Lima, Peru, SELA No. 18, 39–46.

Alikunhi K.H. (1953) Notes on the bionomics, breeding and growth of the murrel. *Ophicephalus striatus* (Bloch). *Proc. Indian Acad. Sci. (B)*. **38**(1), 10–20.

Anon (1981) *Manual on Floating Net-cage Fish Farming in Singapore's Coastal Waters*. Primary Production Department, Singapore. *Fisheries Handbook*, **1**.

Banerji S.R. (1974) Hypophysation and life history of *Channa punctatus* (Bloch). *J. Inland Fish Soc. India*, **6**, 62–73.

Bardach J.E., Ryther J.H. and McLarney W.O. (1972) *Aquaculture*, pp. 218–26. John Wiley & Sons, New York.

Barracund M. *et al.* (1979) Alimentation artificielle de l'esturgeon (*Acipenser baeri*). In *Finfish Nutrition and Fish Feed Technology*, Vol. 1 (Ed. by J.E. Halver and K. Tiews), pp. 411–22. Schriften der Bundesforschungsanstalt fur Fischerei, Berlin.

Ben-Tuvia A., Kissil G.W. and Popper D. (1973) Experiments in rearing rabbitfish (*Siganus rivulatus*) in sea water. *Aquaculture*, **1**, 359–64.

Boonsom J. (1983) *Trichogaster* farming and stomach contents of *Trichogaster* fry. In *Lecture Notes on Aquaculture Practices and Extension in Thailand*, pp. 3–16. Network of Aquaculture Centres in Asia, Bangkok.

Bromley P.J. (1980) Effect of dietary protein lipid and energy content on the growth of turbot (*Scophthalmus maximus* L.). *Aquaculture*, **19**, 359–69.

Bwathondi P.O.J. (1981) *The Culture of Rabbitfish Siganus spp. in Tanzania*. IFS/University of Tanzania.

Caceros-Martinez C., Cadena-Roa M. and Métailler R. (1984) Nutritional requirements of turbot (*Scophthalmus maximus*). I. Preliminary Study of protein and lipid utilization. *J. World Maricul. Soc.*, **15**, 191–202.

Chen F.Y. (1979) Progress and problems of netcage culture of grouper (*Epinephelus tauvina* F.) in Singapore. *Proc. World Maricul. Soc.*, **10**, 260–71.

Chen F.Y. *et al.* (1977) Artificial spawning and larval rearing of the grouper *Epinephelus tauvina* (Forskal) in Singapore. *Singapore J. Pri. Ind.*, **5**(1), 1–21.

Chen T.P. (1976) *Aquaculture Practices in Taiwan*, pp. 63–9. Fishing News Books, Oxford.

Chua T.E. and Keh T.S. (1977) Floating Fish Pens for Rearing Fishes in Coastal Waters, Reservoirs and Mining Pools in Malaysia. *Fisheries Bulletin*, **20**, Ministry of Agriculture, Malaysia.

Darmont M. and Salaya J.J. (1983) Ensayo de cultivo de la cachama, *Colossoma macropomum* (Cuvier, 1818) en jaulas flotantes rigidas. *Simposio de la*

Associacion Latinoamericana de Acuicultura, Valdivia, Chile, **5**.

Fielding J.R. (1968) New systems and new fishes for culture in the United States. In *Proceedings of the World Symposium on Warm-water Pond Fish Culture*, (Ed. by T.V.R. Pillay), *FAO Fish Rep.*, *44*, **5**, 143–61.

Finucane J.H. (1970) Progress in pompano mariculture in the United States. *Proc. World Maricul. Soc.*, **1**, 69–72.

Gomez A. and Scelzo M. (1982) Polyculture experiments of pompano, *Trachinotus carolinus* and spotted red shrimp, *Penaeus brasiliensis* in concrete ponds, Margarita Island, Venezuela, *J. World Maricul. Soc.*, **13**, 146–53.

Herzberg A. (1973) Toxicity of *Siganus luridus* (Ruppell) on the Mediterranean coast of Israel. *Aquaculture*, **2**, 89–91.

Hora S.L. and Pillay T.V.R. (1962) Handbook on fish culture in the Indo-Pacific region. *FAO Fish. Biol. Tech. Paper*, **14**.

Howell B.R. (1979) Experiments on the rearing of larval turbot, *Scophthalmus maximus* L. *Aquaculture*, **18**, 215–25.

Huet M. (1986) *Textbook of Fish Culture*, 2nd edn, pp. 257–8. Fishing News Books, Oxford.

Hull S.T. and Edwards R.D. (1979) Progress in farming turbot *Scophthalmus maximus*, in floating sea cages. In *Advances in Aquaculture*, (Ed. by T.V.R. Pillay and W.A. Dill), pp. 466–72. Fishing News Books, Oxford.

Hussain N., Saif M. and Ukawa M. (1975) *On the culture of* Epinephelus tauvina *(Forskal)*. Kuwait Institute for Scientific Research.

Hussain N.A. and Higuchi H. (1980) Larval rearing and development of brown spotted grouper, *Epinephelus tauvina* (Forskal). *Aquaculture*, **19**(3), 339–50.

Iversen E.S. (1976) *Farming the Edge of the Sea*, pp. 260–63. Fishing News Books, Oxford.

Jhingran V.G. (1982) *Fish and Fisheries of India*, 2nd edn, pp. 454–8. Hindustan Publishing Corporation (India), Delhi.

Jones A. (1974) Sexual maturity, fecundity and growth of the turbot *Scophthalmus maximus* L. *J. Mar. Biol. Assoc. UK*, **54**, 109–25.

Jones A. (1981) Recent developments in techniques for rearing marine flat fish larvae, particularly turbot (*Scophthalmus maximus* L.) on a pilot commercial scale. *Réun Cons. Int. Explor. Mer*, **178**, 522–6.

Jones A. and Houde E.D. (1986) Mass rearing of fish fry for aquaculture. In *Realism in Aquaculture, Achievements, Constraints, Perspectives*, (Ed. by M. Bilio H. Rosenthal and C.J. Sindermann), pp. 351–73. European Aquaculture Society, Bredene.

Jones A., Alderson R. and Howell B.R. (1973) Progress towards the development of a successful rearing technique of larvae of turbot, *Scophthalmus maximus* L. Symposium on the Early Life History of Fish, Oban, *FAO Fish Rep.*, **141**, 38.

Jones A. *et al.* (1980) Progress towards developing methods for the intensive farming of turbot (*Scophthalmus maximus* L.) in cooling water from a nuclear power station. In *Aquaculture in Heated Effluents and Recirculation Systems*, Vol. II (Ed. by K. Tiews), pp. 481–96. Schriften der Bundesforschungsanstalt für Fischerei, Berlin.

Jory D.E., Iversen E.S. and Lewis R.H. (1985) Culture of fishes of the genus *Trachinotus* (Carangidae) in the Western Atlantic: Prospects and problems. *J. World Maricul. Soc.*, **16**, 87–94.

Juario J. *et al.* (1984) *Breeding and Larval Rearing of the Rabbitfish*, Siganus guttatus *(Bloch)*. SEAFDEC Aquaculture Department, Iloilo.

Kingwell S.J., Duggan M.C. and Dye J.E. (1977) Large scale handling of the larvae of the marine flatfish turbot *Scophthalmus maximus* L., and dover sole, *Solea solea* L., with a view to their subsequent fattening under farming conditions. *Actes de Colloques du CNEXO*, **4**, 27–34.

Kuhlmann D., Quantz G. and Witt V. (1981) Rearing of turbot larvae (*Scophthalmus maximus* L.) on cultured food organisms and post metamorphosis growth on natural and artificial food. *Aquaculture*, **23**, 183–96.

Lam T.J. (1944) Siganids: their biology and mariculture potential. *Aquaculture*, **3**, 325–54.

Lichatowich T. *et al.* (1984a) Growth of *Siganus rivulatus* reared in cages in the Red Sea. *Aquaculture*, **40**(3), 273–5.

Lichatowich T. *et al.* (1984b) Spawning cycle, fry appearance and mass collection techniques for fry of *Siganus rivulatus* in the Red Sea. *Aquaculture*, **40**(3), 269–71.

Ling S.W. (1977) *Aquaculture in Southeast Asia*, pp. 21–2. Washington Sea Grant Publication, College of Fisheries, University of Washington, Seattle.

Lovshin L. (1980) Situacion del cultivo de *Colossoma* spp. en Sudamerica. *Revista Latinoamericana de Acuicultura*, Lima, Peru, No. 5.

Martinez M.E. (1984) *El Cultivo de las Especies del Genero* Colossoma *en America Latina*. FAO Regional Office, Santiago, Chile RLAC/84/41-PES-5.

Martyshev F.G. (1983) *Pond Fisheries*, pp. 148–51. A.A. Balkema, Rotterdam.

May R.C., Popper D. and McVey J.P. (1974) Rearing and larval development of *Siganus canaliculatus* (Park) (Pisces: Siganidae). *Micronesica*, **10**(2), 285–98.

Muratori V. (1985) Storioni doc a Calvisano, *Il.*

Pesce, **2**(2), 18–21.

Murugesan V.K. (1978) The growth potential of murrels, *Channa marulius* (Hamilton) and *Channa striatus* (Bloch). *J. Inl. Fish. Soc. India*, **10**, 169–70.

Nunez J.M. and Salaya J.J. (1983) Cultivo de cachama, *Colossoma macropomum* (Cuvier, 1818) en jaulas flotantes no rigidas en la represa de Guanapito, Estado Guarico, Venezuela. *Symposio de la Associacion Latinoamericano de Aciucultura*, Valdivia, Chile, No. 5.

Palma A.L. (1978) Induced spawning and larval rearing of rabbitfish, *Siganus guttatus*. *Philipp. J. Fish.*, **16**(2), 95–104.

Pandian T.J. (1967) Food intake, absorption and conversion in the fish *Ophiocephalus striatus*. *Helgolander Wiss. Meeresunters*, **15**, 637–47.

Pantulu V.R. (1979) Floating cage culture of fish in the lower Mekong basin. In *Advances in Aquaculture*, (Ed. by T.V.R. Pillay and W.A. Dill), pp. 423–7. Fishing News Books, Oxford.

Parameswaran S. and Murugesan V.K. (1976) Observations on the hypophysation of murrels (Ophiocephalidae) *Hydrobiologia*, **50**(1), 81–2.

Person-Le Ruyet J. (1981) Research on rearing turbot (*Scophthalmus maximus*): results and perspectives. *J. World Maricul. Soc.*, **12**(2), 143–52.

Person-Le Ruyet J. and Noel T. (1982) Effects of moist pelleted foods on the growth of hatchery turbot (*Scophthalmus maximus*) juveniles. *J. World Maricul. Soc.*, **13**, 237–45.

Person-Le Ruyet J. *et al.* (1983) Use of expanded pellets supplemented with attractive chemical substances for weaning of turbot (*Scophthalmus maximus* L.) *J. World Maricul. Soc.*, **14**, 676–8.

Pillai T.G. (1962) Fish Farming Methods in the Philippines, Indonesia and Hong Kong. *FAO Fish. Biol. Tech. Paper*, **18**, 51–2.

Popper D., Gordin H. and Kissil G.W. (1973) Fertilization and hatching of rabbitfish *Siganus rivulatus*. *Aquaculture*, **2**, 37–44.

Purdom C.E., Jones A. and Lincoln R.F. (1972) Cultivation trials with turbot (*Scophthalmus maximus*). *Aquaculture*, **1**(2), 213–30.

Qasim S.Z. and Bhatt V.S. (1966) The growth of freshwater murrel, *Ophicephalus punctatus* Bloch. *Hydrobiologia*, **27**, 289–316.

Qayyum A. and Qasim S.Z. (1962) Behaviour of the Indian murrel, *Ophicephalus punctatus*, during brood care. *Copeia*, **2**, 465–7.

Raj B.S. (1946) Notes on the freshwater fish of Madras. *Rec. Ind. Mus.*, **12**, 249–94.

Randall J.E. (1958) A review of ciguatera, tropical fish poisoning, with a tentative explanation of its cause. *Bull. Mar. Sci. Gulf Carib.*, **8**(3), 236–67.

Scott A.P. and Middleton C. (1979) Unicellular algae as food for turbot (*Scophthalmus maximus* L.) larvae – the importance of dietary long-chain polyunsaturated fatty acids. *Aquaculture*, **18**, 227–40.

Smith P.L. (1979) The development of a nursery technique for rearing turbot *Scophthalmus maximus*, from metamorphosis to ongrowing size – progress since 1970 by the British White Fish Authority. In *Advances in Aquaculture*, (Ed. by T.V.R. Pillay and W.A. Dill), pp. 143–9. Fishing News Books, Oxford.

Smith T.I.J. (1973) The commercial feasibility of rearing pompano, *Trachinotus carolinus* (Linnaeus) in cages. *Florida Sea Grant Technical Bulletin*, No. 26.

Soh C.L. and Lam T.J. (1973) Induced breeding and early development of the rabbitfish, *Siganus oramin* (Schneider). *Proc. Symp. Biol. Res. Nat. Dev.*, 49–56.

Tan S.M. and Tan K.S. (1974) Biology of tropical grouper *Epinephelus tauvina* (Forskal) I. A preliminary study on hermaphroditism in *E. tauvina*. *Singapore J. Pri. Ind.*, **2**(2), 123–33.

Tseng W.Y. (1983) Prospect for commercial netcage culture of red grouper (*Epinephelus akara* T. and S.) in Hong Kong. *J. World Maricul. Soc.*, **14**, 650–60.

Tseng W.Y. and Ho S.K. (1979a) Induced breeding of red grouper (*Epinephelus akaara* Temmink and Schlegal) in Hong Kong – embryonic and larval development. *Sci. Fish. Anim. Prod.*, **7**(1), 9–20.

Tseng W.Y. and Ho S.K. (1979b) Cage culture of red grouper (*Epinephelus akaara* T. and S.) in Hong Kong. *China Fish. Month.*, **324**, 9–11.

Von Westernhagen H. (1974) Food preferences in cultured rabbitfishes (Siganidae). *Aquaculture*, **3**, 109–17.

Wee K.L. (1981) *Snakehead* (Channa striatus) *Farming in Thailand*. UNDP/FAO Network of Aquaculture Centres in Asia, Bangkok.

Wee K.L. (1982) Snakeheads – their biology and culture. In *Recent Advances in Aquaculture*, Vol. 1 (Ed. by J.F. Muir and R.J. Roberts), pp. 180–213. Croom Helm, London.

24
Shrimps and Prawns

The popular names shrimps and prawns have been used variously to denote crustaceans of the families Penaeidae and Palaemonidae. Even though there is still some confusion in the use of these names, in most recent aquaculture literature the name prawn appears to be used for fresh-water forms of Palaemonids and shrimp for the others, particularly the marine species.

Shrimps form a group of subsidiary species in most types of fish culture in coastal impoundments and ponds in Asia; and in countries like India, rice fields have been used for a form of extensive culture of shrimps (see Chapter 29) for centuries. However, intensive and semi-intensive culture of these crustaceans are of recent origin. Like the marine finfish referred to in earlier sections, interest in their culture, particularly of shrimps, was triggered by the recent increased market demand and the inadequacy of the capture fishery landings to meet the demand. As the expanding markets were in economically advanced countries like Japan and the USA, the prospects of an export market and opportunities for earning foreign exchange attracted the support of the governments of developing countries and led to investment by private industry. In fact, shrimps and prawns became high-value commodities in many developing countries, mainly because of their export market. There is as much interest in private investment in shrimp farming in tropical countries today as there is for salmon farming in countries in the colder climates, for the very reason of prospects associated with exports.

24.1 Major cultivated species of shrimps and prawns

Attention has so far been directed to the culture of tropical and sub-tropical species of shrimps,

and the so-called giant fresh-water prawn, *Macrobrachium rosenbergii*. Spawning and larval rearing of the kuruma shrimp in captivity in Japan about four decades ago aroused considerable interest in intensive farming of shrimps and this species became the focus of attention for a number of years. It was introduced in many countries in Asia, southern Europe, West Africa, the southern USA and Central and South America. Very soon, attention turned to some of the larger local species of shrimps, which were better adapted to prevailing temperature conditions, and the larvae and juveniles of which were readily available to supplement inadequate production from hatcheries. In Asia, the more important species are the tiger shrimp *P. monodon* and the Indian or white shrimp *P. indicus*. The banana shrimp *P. merguiensis*, the green tiger or bear shrimp *P. semisulcatus* and the oriental shrimp *P. orientalis* (= Chinensis) are also of commercial interest in some countries of the region. The red-tailed shrimp *P. penicillatus* is a species cultured in Taiwan. *Metapenaeus monoceros*, *M. brevicornis* and *M. ensis* form subsidiary species in shrimp farms in several Asian countries.

Besides the imported *P. japonicus*, the main interest in the Mediterranean countries of Europe has been in the local Mediterranean shrimp (triple-grooved shrimp) *P. kerathurus*. Present efforts in establishing shrimp farming in Africa mainly involve the culture of *P. indicus* on the East Coast and *P. notialis* on the West Coast. The most important species in Central and South America are the white-leg shrimp or camaron langostino *P. vannamei* and the blue shrimp *P. stylirostris*. Besides the blue shrimp, there are at least four species that have reached commercial-level culture in the countries bordering the Atlantic Coast of

425

Central and South America, namely the brown shrimp *P. aztecus*, pink shrimp *P. duorarum* and *P. setiferus*, known as the common or white shrimp. As well as these, production on an experimental scale has been undertaken for the southern white shrimp *P. schmitti*. Though all the above species are of potential importance in commercial shrimp farming, and several others are being investigated for their suitability, the bulk of the present production comes from the farming of *P. monodon, P. indicus, P. merguiensis, P. japonicus* and *P. vannamei*.

Much of the research effort on shrimp culture has been concentrated on the development of hatchery techniques for controlled spawning and larval rearing, as in the case of other marine aquaculture species. In the early years of investigations, considerable problems were faced in rearing and feeding the hatchlings through the different stages of development and obtaining reasonable survival rates. The search for species that may be easier to reproduce and have a shorter larval history resulted in investigations on the giant fresh-water prawn, *Macrobrachium rosenbergii*. The success achieved in the mass production of post-larvae of this species in the 1970s led to wide-spread interest in its culture, and it has been imported into an impressive number of countries in the tropical, sub-tropical and even temperate climates, in almost every continent. Some experimental work has been done on the culture of other species of *Macrobrachium*, such as *M. lanchesteri, M. acanthurus, M. carcinus* and *M. malcolmsonii*, but this has not resulted in any commercial production.

Though there are considerable similarities in the culture technologies for the various species, there are also a number of differences brought about by environmental requirements, breeding and feeding behaviour and compatibility with other species. Within the general requirements of water salinity (10−40 ppt), temperature tolerance (18−33°C), the character of soil substrates in the culture facilities, feed quality and response to high-density culture, Penaeid shrimps have specific requirements and limitations. Growth rates and harvest sizes (based on the commercial size of 10−45 g weight) also vary considerably. These factors greatly influence the compatibility of different species of Penaeids in polyculture, but can be advantageously used in rotational production of different species in the same facility, in accordance with seasonal changes of salinity and temperature. In temperate climates, different species can be used for culture in summer and winter. In tropical monsoon areas subject to marked changes of salinity, the production of a species preferring low salinity can be alternated with one that requires high-salinity water. The short duration of culture periods makes such rotation very feasible in many areas.

The fresh-water prawn *M. rosenbergii* is commercially important because of its size, as well as its eating qualities. The males can attain a size of about 25 cm and the females about 15 cm. Though adults are found in fresh- and brackish-water areas, the species requires water of about 12 ppt salinity for larval rearing. This requirement has created problems in siting hatcheries near grow-out facilities, but it has now been shown that this can be overcome by the use of sea water or brine trucked in from the nearest source, as performed by operators of small back-yard hatcheries in Thailand, or by the use of artificial sea water as in a commercial farm in Zimbabwe. The adults are omnivorous and feed on a variety of foods of animal and vegetable origin.

Among the species so far studied, *P. japonicus* (fig. 24.1), *P. orientalis* and *P. setiferus* are considered to be most suited for production in temperate climates. *Penaeus japonicus* is cultured in Japan, Taiwan and in a less intensive way in Brazil, France, Spain and Italy. *Penaeus orientalis* is cultured in Korea and China. *Penaeus setiferus* is the species of interest in the temperate regions of the USA. Much of the available knowledge on modern shrimp culture originated with intensive studies on *P. japonicus*. It is a hardy species, but cannot tolerate low salinity or high temperatures. It requires diets containing about 60 per cent protein for satisfactory growth and grow-out ponds or tanks should have a sandy bottom.

The tiger shrimp *P. monodon* (fig. 24.2), is the fastest growing species used in aquaculture in Asia. The species is euryhaline and can tolerate almost fresh-water conditions, even though 10−25 ppt is considered optimum. It cannot tolerate temperatures below 12°C and

Fig. 24.1 Kuruma shrimp, *Penaeus japonicus* (courtesy of M.N. Mistakidis).

Fig. 24.2 Tiger shrimp, *Penaeus monodon* (photograph: M. Pedini).

the upper limit of tolerance is around 37.5°C.

Penaeus indicus and *P. merguiensis* (figs 24.3 and 24.4) have very similar habits in many respects, but in aquaculture the former species exhibits a preference for sandy substrates and the latter for muddy ones. Both species require high salinities (20−30 ppt) for good growth and cannot tolerate salinities outside the range 5−40 ppt. The lethal temperature is above 34°C. Under the current pond management systems, the duration of culture cannot exceed 3 months, as heavy mortalities occur after that period.

The three Metapenaeid species, *M. monoceros*, *M. brevicornis* and *M. ensis*, are easier to culture, as they mature readily in captivity and their larval culture presents fewer problems. *Metapenaeus monoceros* and *M. brevicornis* are known to breed in ponds. They are tolerant of low salinities and high temperatures and can therefore be cultured in a wider variety of sites. Harvestable size is attained in a shorter time of two to three months, and survival rates are high. But their final size is smaller, generally about 14 cm for *M. monoceros* and *M. ensis* and 7.5−12.5 cm for *M. brevicornis*.

Penaeus semisulcatus (fig. 24.5) grows to a large size and fetches a good price in the markets in Asia and the Middle East, but it requires high salinities and its growth in ponds is slow. Though easy to propagate, survival rates in grow-out facilities are reported to be very low. Among the shrimps cultured in Central and South America, *P. vannamei* (fig. 24.6) is highly euryhaline and can withstand salinities ranging from 0 to 50 ppt and temperatures ranging from 22 to 32°C. Low salinities and warmer temperatures are characteristic of the rainy season (December to April) in countries like Ecuador and higher salinities and cooler temperatures prevail during the remaining months. This partly accounts for the higher survival rate of *P. vannamei* compared to *P. stylirostris*, and its preference in pond farming in these countries.

24.2   Shrimp culture systems

Traditional and modern shrimp culture are carried out mainly in ponds. In traditional systems, where natural stocking was achieved through the intake of tidal water carrying large numbers of shrimp larvae, pond designs were simple and were meant to serve largely as trap ponds; many farmers releasing larvae directly into the rearing or production ponds. It is only in recent years that nursery ponds have been incorporated for growing larvae to an advanced juvenile stage, before transfer to production ponds.

With the adoption of techniques of controlled propagation, many shrimp farms now include hatchery units, together with nursery facilities. There is also greater specialization in the rearing of post-larvae or juveniles for sale to farmers for grow-out. Such nursery farms may maintain brood ponds, hatcheries and nurseries, together with facilities for growing natural food for larvae. When the larvae are collected from the wild, only nursery and live-food growing facilities may be maintained. Even though earth ponds are widely used, many farms adopting semi-intensive systems of culture have nurseries and even rearing ponds with cement concrete dykes. *Macrobrachium* ponds in Taiwan are often made of cement or bricks, or lined with plastic sheets, but sandy loam bottoms containing clay are preferred as they contribute to natural food production. Different kinds of shelters and artificial substrates are provided in ponds, including water plants, hollow bricks, framed nets, plastic pipes, styrofoam sheets, etc.

As mentioned earlier, traditional shrimp culture was necessarily a polyculture system because of the inability to control the composition of the seed stock. In coastal ponds or impoundments, shrimps formed only a small percentage of the harvest. Obviously the species combinations were not always compatible, and the culture procedures were not favourable for high survival rates for shrimps. Recent attempts at polyculture of milkfish and shrimps have shown the conflict of requirements between the species. For example, the shallow depths of milkfish ponds are not favourable for shrimps and do not allow high stocking rates. There is considerable disparity in the time required to grow the species to marketable size. The transfer of stock from pond to pond, practised in milkfish culture, is not very easy for many species of shrimps, and every transfer generally results in injuries or deaths. In view of these practical problems, it is often preferable to

Fig. 24.3 The Indian shrimp or white shrimp, *Penaeus indicus* (Courtesy of PSBR James).

Fig. 24.4 Banana shrimp, *Penaeus merguiensis*.

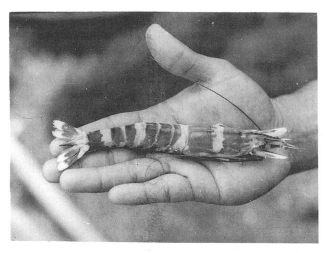

Fig. 24.5 Tiger or bear shrimp, *Penaeus semisulcatus*.

Fig. 24.6 White-leg shrimp, *Penaeus vannamei*.

adopt monoculture methods. Even combinations with other shrimp species do not seem to be very useful in many situations, in view of the differences in environmental requirements between species.

On the other hand polyculture of *Macro-*

brachium with certain species of finfish is believed to be beneficial. Although not so common, *M. rosenbergii* is cultured together with Chinese carps (bighead and grass carp) in Taiwan and Malaysia. There are also reports of successful experimental and commercial culture with grey mullets, tilapia and red swamp crayfish, *Procambarus clarkii*.

Intensive systems of shrimp and prawn culture are generally practised in tank farms. Water in the tanks is frequently exchanged to maintain high oxygen levels and to discharge metabolic products. When the water exchange is low, heavy aeration is adopted to keep organic particles in suspension. These suspended particles serve as biological filters because of the dense colonies of nitrifying bacteria that grow on them. Circular tanks, up to 2000 m^2 surface area, with adequate water circulation and drainage facilities are in use. Raceway systems with a total exchange of water several times a day (sometimes built under greenhouses for better environmental control) have proved particularly efficient for certain species like *P. stylirostris*.

The traditional system of shrimp production in rice fields on the west coast of India is described in Chapter 29. On the east coast of India, in the Gangetic delta, many farmers now raise crops of shrimps in rotation with rice. Similar practices have developed in Bangladesh and in the Mekong delta area in Vietnam. After the shrimps have been harvested, the fields are stocked with carp during the rainy season, when the salinity in the fields is low.

Open-water stocking of shrimps carried out in Japan to enhance natural populations is referred to in Chapter 30. It is reported that at least 300 million post-larvae are released every year, and it is claimed to have formed the basis for new 'sea ranching fisheries' (Uno, 1985). Open-water stocking of the fresh-water prawn *M. rosenbergii* has been carried out in lakes and reservoirs in Thailand, and is reported to have become a source of food and revenue to local fishermen. Small-scale stocking of a river and of dams (reservoirs) has also been attempted in Taiwan (Chao, 1979).

24.2.1 Production of seed stock

Though hatchery techniques have been developed for the main species of shrimps farmed commercially, only a small proportion of the global requirement of seed-stock is presently produced in hatcheries. There are, of course, countries like Japan and Taiwan, where virtually all the required seed are produced in hatcheries. Other major shrimp-culturing countries have depended on the collection of naturally occurring post-larvae and juveniles when available, rather than waiting for hatchery systems of production to be established. However, the collection of wild seed stock is not devoid of problems. There are considerable difficulties in sorting out the required species from the mixed collections, which may contain the larvae of slow-growing, undesirable species of shrimps and also of predatory or weed-fish. Secondly, there may be marked fluctuations in their availability. Thirdly, the shrimp fishermen in the area are more than likely to ascribe poor commercial catches of shrimps to the fishing of the larvae and juveniles for farming, and this can cause social conflicts.

24.2.2 Collection of wild seed stock

The traditional system of stocking ponds with post-larvae and fry brought in by incoming tidal water is still practised in many Asian countries. To eliminate unwanted fish larvae and fry and to estimate the number of shrimp post-larvae and fry stocked, an improved system of stocking has been proposed. This starts by filling nursery ponds with as many larvae and fry as possible by manipulation of the tidal flow. After about a month the nurseries are treated with tea-seed cake (containing 10–15 per cent saponin) at a rate of 10–25 ppm, to kill all the fish without affecting the shrimps. The shrimps can then be transferred to the rearing or production ponds.

Wild shrimp fry can be collected with different types of nets and lure lines. Push nets and scissor nets (fig. 24.7) are probably the most common equipment used. Lure lines, very much like the ones used for milkfish fry collection, are used to gather the fry, which are then removed using scoop nets. Fine-meshed bag nets (similar to the shooting nets used for carp spawn collection in India), with a receptacle at the cod-end, are also placed against tidal currents at high tides in creeks, canals or

Fig. 24.7 Collection of shrimp fry with a scissor net.

in sluice gates for catching incoming fry. The fry collect in the cod-end receptacle and are removed at frequent intervals.

The sorting of shrimp fry according to species requires considerable experience. The nature and location of pigmentation, body shape and mode of locomotion are some of the main identifying characters. ASEAN (1978) gives some distinguishing features for identification of the post-larvae and quotes a provisional key for *P. indicus*, *P. semisulcatus* and *P. monodon*.

24.2.3   Hatchery production of seed stock

Following on the early success of hatchery production of post-larval *P. japonicus*, considerable research effort has been directed towards the controlled maturation, spawning and larval rearing of a number of shrimps and of the fresh-water prawn *M. rosenbergii*. Though initial attempts were directed towards mass production with gravid females caught from the

fishing grounds, success has since been achieved in the maturation and mating of shrimps in captivity.

Berried females of fresh-water prawns can be obtained from natural habitats or from pond farms. Alternatively, breeding stock maintained in tanks and aquaria can be mated after the mature female undergoes prenuptial moulting. Though the basic principles of seed-stock production for shrimps and the fresh-water prawn are similar in many respects, there are some differences in detailed procedures and so they are summarized here separately. Detailed descriptions of seed-stock production can be found in McVey (1983), Huner and Brown (1985) and New and Singholka (1985).

24.3   Reproduction and larval rearing of shrimps

The controlled spawning and larval rearing of shrimps was initiated by Hudinaga (1942) with wild spawners of *P. japonicus* caught from fishing grounds. Since then, as many as 24 *Penaeus* species and seven *Metapenaeus* species are reported to have been fully or partially propagated artificially. Among these the more important species for which methods of commercial-scale propagation are available are *P. aztecus, P. duorarum, P. indicus, P. japonicus, P. kerathurus, P. merguiensis, P. monodon, P. orientalis, P. setiferus, P. stylirostris, P. vannamei* and *M. ensis*.

24.3.1 Brood stock

Spawners of *P. japonicus, P. aztecus, P. duorarum* and *P. setiferus* can be collected in large numbers, whereas spawners of species like *P. monodon* are more difficult to obtain. Therefore, maturation of captive stock of wild-caught or pond-reared adults is necessary for large-scale hatchery production of adequate numbers of larvae of these species.

Some species, like *P. merguiensis* and *P. japonicus*, mature, mate and spawn freely in response to controlled environmental conditions. Unilateral eye-stalk ablation (see Chapter 8) is adopted for species which otherwise do not mature in captivity, like *P. aztecus, P. duorarum, P. monodon* and *P. orientalis*. Even for species that would mature without such treatment, ablation helps speedy maturation and better spawning rates. By eye-stalk ablation it is possible to reduce the interval between spawnings to 3–15 days, from the normal interval of 10–67 days.

The technique of ablation or extirpation involves the removal of either eye and the partial or total removal of the eye stalk by cutting with surgical scissors, cautery (using a soldering iron or clamps or by electrocautery), ligation, squeezing or crushing the eye stalk tissue, or manual pinching. It is important to prevent excessive loss of eye fluids and infection. The interval between ablation and the onset of maturation and subsequent spawning varies from three days to more than two months, depending on a number of factors including the age of the shrimp and the stage of the moulting cycle. It is considered best to undertake ablation during the intermoult, for maturation to follow in less than a week. Ablation during the premoult period can lead

to immediate moulting and a prolonged latency period. Maturation and viability of eggs seem to depend on the water quality (salinity, temperature and pH), light intensity and nutrition. Spawning stock is given high-quality feed, preferably natural foodstuffs like polychaete worms, squids, mussels, clams or cockle meat, at the rate of about 10 per cent of the biomass. A continuous flow of water is maintained in the maturation tanks and a daily exchange of 60–70 per cent of the water is recommended.

24.3.2 Hatchery systems

Spawning and larval rearing are generally carried out in tanks made of cement concrete, ferrocement, fibreglass, plastic, etc. (see Chapter 6) (fig. 24.8). Maturation cages and pens have been used on an experimental basis, but are seldom used on a commercial scale. The hatchery tanks used for spawning and larval rearing in Japan are large, ranging from 100 up to 2000 ton capacity. They are suitable for spawning a large number of spawners at a time and for rearing the resultant hatchlings by what is referred to as the 'community culture method' (see fig. 6.38). Larval foods are raised by fertilizing the tanks directly every day, producing diatoms and zooplankton, which form the food of larval shrimps. Spawning, larval rearing and fry nursing are all done in the same tank.

Many new hatcheries follow a different system, developed in the USA (Galveston, Texas), in which separate smaller tanks made of fibreglass or plastic are used for spawning and larval rearing. Facilities for culture of live food are maintained separately (fig. 24.9). Larval rearing tanks vary in capacity from 1000 to 2000 ℓ and the spawning tanks from 100 to 250 ℓ. As high densities of larvae (200–300 nauplii/ℓ) are reared, they cannot be grown beyond the early post-larval stages (for about 5 days) in the original tanks and further rearing has to be done in nursery tanks or ponds, before grow-out. This type of hatchery system seems to be better suited for species like *P. monodon*, where the availability of spawners is very much limited and so community culture may not prove efficient.

Kungvankij (1982) described a third system, which combines the advantages of the above two systems. It includes spawning tanks with

Fig. 24.8 An indoor hatchery system in Hawaii (courtesy of Amorient Aquafarm and Marine Culture Ent.).

capacities of 1000−2000 ℓ, larval rearing tanks of 1000−3000 ℓ capacity and nursery tanks with a capacity of 30−100 tons for rearing post-larvae to the P_{30} stage (fig. 24.10). This system is reported to maximize tank utilization in spawning and larval rearing of species like *P. monodon*.

Liao (1985) referred to a recently developed 'ladder system' hatchery (fig. 24.11), consisting of four inter-connected tanks built on sloping ground, with the algal culture tank at the top, followed below by the rearing tank for nauplii and zoea larvae, then another rearing tank for mysis to the P_{1-5} stages and finally a larger tank for post-larvae, all built one below the other, with descending water levels.

24.3.3 Spawning and larval rearing

The minimum age of spawning females varies between species and according to the environmental conditions, as can be seen from the following:

P. aztecus	8−9 months
P. indicus	4−8 months (weight 6−8 g)
P. japonicus	7−12 months
P. merguiensis	4−8 months (weight 6−8 g)
P. monodon	9−15 months (weight 32−45 g)
wild spawners	18 months (weight 75 g)
P. orientalis	8−9 months
P. stylirostris	8−9 months (mean length 176 mm, weight 40−50 g)
P. vannamei	8−9 months (mean length 157.9 mm, weight above 30 g)

In general, spawners from captive brood stock are smaller than those from the wild. Farmers usually believe that spawners from

Fig. 24.9 Culture of diatoms and *Artemia* for feeding shrimp larvae.

wild stock are superior to captive ablated spawners, and that the quality and quantity of their eggs are higher.

The maturity of the males can be determined by examination of the petasma on the first pair of pleopods. In mature males these accessory organs are joined together by means of interlocking hooks. Swelling and whitish coloration of the terminal ampoules near the fifth pair of pereiopods indicate gonadal maturity.

The spawning season in nature varies according to species and location. When larvae have to be collected from the wild, or when wild spawners are used for spawning and larval rearing, it is essential to know beforehand the period and locations of their occurrence. On the other hand, captive stocks can be matured and spawned almost throughout the year under controlled conditions. Though tropical species spawn throughout the year, most Penaeids have peak periods of spawning.

In closed thelycum species (i.e. species with lateral plates that lead to a seminal receptacle, where the spermatophores can be inserted), the mating occurs soon after the females have moulted. In species with an open thelycum (with only ridges and protuberances for spermatophore attachment) mating can occur soon after the eggs become mature. In the latter group of species, spermatophores can easily be lost or fail to be affixed before spawning. The spermatophores deposited during a single moulting are generally enough, irrespective of the moult cycle, to fertilize up to three successive spawns.

For controlled spawning, gravid females and males in advanced stages of maturity are stocked in spawning tanks. Spawners obtained from commercial catches during the winter are likely to be infected and are therefore usually treated with 3 ppm $KMnO_4$, 25 ppm formalin or the commercial product Treflan[R] at concentrations of 3–5 ppm. In large tank systems used for community culture, several

Fig. 24.10 A hatchery system with spawning, nursery and rearing tanks.

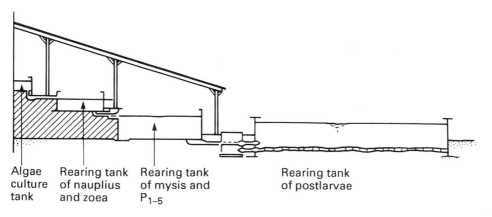

Algae culture tank | Rearing tank of nauplius and zoea | Rearing tank of mysis and P_{1-5} | Rearing tank of postlarvae

Fig. 24.11 A ladder system hatchery in Taiwan (from Liao, 1985).

spawners are introduced into the community tank, whereas in the other systems individual spawners or batches of spawners are placed in separate spawning tanks each time. In large tanks, the density of spawners are generally:

P. japonicus:	1 spawner/2 m^3
P. monodon:	1 spawner/5 m^3
P. indicus:	1 spawner/1 m^3
P. merguiensis:	1 spawner/1 m^3

Generally a 1:1 sex ratio is maintained in spawning tanks, but a ratio of two females to one male has produced higher spawning rates and egg production. There is usually a time lag between mating and spawning, as the eggs may still not be fully mature at the time of mating. *Penaeus japonicus* and *P. indicus* females have been observed to eat their own spawned eggs and so it is advisable to install mesh trays or plates on the bottom of the spawning tanks to protect the eggs. The salinity in the tanks generally ranges from 28 to 35 ppt and the temperature from 23 to 33°C. Spawning usually takes place at night. Fertilization is external and at the above temperature range the embryonic development is rapid.

The nauplius passes through three to six-sub-stages (N_1-N_6) and subsists on its own yolk material. In about 2−3 days it metamorphoses into protozoea with three sub-stages (PZ_1-PZ_3) during which period the larva starts feeding on unicellular algae. This stage, which lasts for 3−6 days is succeeded by the mysis stage with three sub-stages (M_1-M_3). During this stage the larva retains the filtering mechanism for feeding on algal cells. The mysis metamorphoses into post-larva in about 3−5 days. At this stage it ceases to be a filter-feeder and becomes capable of capturing and eating zooplankton. Development from the post-larval stage to the juvenile stage is very gradual and a PL_5 (P_1-P_5 or PL_1-PL_5 denotes the post-larval age in days) may take 15−20 days to reach a size of 20−25 mm, suitable for stocking production ponds.

In larval culture of most Penaeids, the main difficulty is in rearing the protozoeal stage when they start feeding. At this stage the larva is highly light-sensitive, and so the tanks should be properly covered to ensure darkness. The key to the success of the pioneer experiments of Hudinaga in Japan was the method developed for the culture of the diatom *Skeletonema costatum*, which formed a suitable food for the larvae. Since then several other types of live foods and feedstuffs have been tried, but cultured phytoplankton appears to be still the most efficient food for larvae at this stage. Since the size of larvae of different species of Penaeids are not the same, they require phytoplankton of different sizes. They start feeding on zooplankton when they reach the last sub-stage of protozoea. Both mysis and post-larvae up to the fifth day prefer zooplankton, but after that stage they will consume larger food and may feed at the bottom. They can then be fed on polychaetes, chopped mussels, clams, cockles and artificial compound diets.

The more important phytoplankters suitable as food for shrimp larvae are species of *Chaetoceros*, *Skeletonema* and *Tetraselmis*. Algal culture methods have been described in Chapter 7. Among the zooplanktonic organisms, the rotifer *Brachionus plicatilis* is probably the most important as larval food. Many hatcheries depend largely on the brine shrimp, *Artemia salina*, the nauplii of which form excellent food for shrimp larvae. Methods of culturing *Brachionus* and hatching *Artemia* cysts have been described in Chapter 7. In large tank hatcheries practising community culture, the tanks are fertilized soon after hatching, at the rate of 3 ppm KNO_3 and 0.3 ppm Na_2HPO_4 to produce the phytoplankton needed to feed the larvae when they reach the protozoea stage. In some hatcheries pure cultures of diatoms are inoculated before fertilizers are applied. The production of phytoplankton is maintained through additional fertilization if needed. If the density of plankton is inadequate, supplementary feed in the form of soybean cake, soybean curd, egg yolk or fertilized eggs of oysters may be given. When the larvae reach the mysis stage, they are fed on *Brachionus* or brine shrimp nauplii. In most hatcheries, post-larval stages are fed on brine shrimp and after the P_6 stage on minced mussels, clam meat or formulated larval feeds, partly replacing brine shrimp nauplii until they reach the P_7 stage. Beyond this stage, the post-larvae are fed only minced mussel, clam meat or artificial diets, three or four times daily.

As indicated earlier, in hatcheries with separate small hatchery tanks, algae (*Skeletonema costatum* and *Tetraselmis* spp.) are cultured in separate algal tanks or in plastic bags and the required quantities are introduced daily during the protozoea stage. *Artemia* nauplii hatched in special tanks are fed to mysis and early post-larval stages. In hatcheries with intermediate size tanks, fertilization of the tank water and introduction of pure algal cultures are combined, and the post-larvae are reared up to the P_{25} stage.

Hatcheries that produce only P_5 or P_6 stage post-larvae use concrete tanks, earthen ponds or net cages for larval rearing. Small tanks with a filtered sea-water supply and aeration are stocked at a density of up to $150/\ell$. Diatom cultures are introduced to feed the larvae and often a substrate such as polyethylene netting is provided for the larvae to rest on. Early post-larvae are fed with chopped mussel and cockle meat together with young and adult *Artemia*. Daily exchange of water is maintained for the duration of culture (about 30 days).

Earthen nursery ponds range in area from 500 to 2000 m^2 with an average depth of 40–70 cm. The larvae are stocked at the P_9–P_{10} stage at densities of 100–150 per m^2. The ponds are prepared by eradicating predators and ferti-lizing with a combination of organic manures (such as 1000 kg/ha of chicken manure) and inorganic fertilizer (such as 50 kg/ha of ammonium sulphate). Supplementary feeding is done with chopped mussel or cockle meat, at about 10 per cent of the total biomass. The larvae can be reared in such ponds up to the P_{40} or P_{60} stages.

Twenty-one- to twenty-five-day-old post-larvae are suitable for stocking grow-out ponds. In some farms, particularly in Ecuador and Taiwan, post-larvae are grown (sometimes referred to as pre-growing) for 30–60 days at densities of 50–200 per m^2. With daily ex-change of water (10–40 per cent) and sup-plementary feeding with compound feeds, they reach a mean weight of 0.5–2 g, with a survival of 80 per cent depending on species and pond conditions. These fry are then stocked in grow-out facilities.

Nursery cages are used only rarely, as the very small mesh sizes required can become rapidly choked by bio-fouling. The cages, when used, are rectangular in shape (1–2 m × 5 m × 1 m) and are of the floating or stationary type installed in protected bays, lagoons or ponds. Post larvae (P_6–P_7) are stocked at higher den-sities of 1000–2000 per m^3. Feeding is carried out in the same manner as in earthen ponds.

24.4 Grow-out of shrimps

As stated earlier, pond culture is the most common grow-out system, and extensive cul-ture methods are gradually giving way to semi-intensive production, in both Asia and South America. The traditional coastal shrimp ponds of Asia, which were stocked with wild seed stock that gains entrance during tidal water exchange, are now being stocked with sorted fry collected from the wild or bought from fry collectors. The stocking rate is still at a low level of 3000–5000 fry/ha. During the rearing period of about two months, water exchange is maintained using tidal flow. In improved sys-tems, the ponds are carefully prepared before stocking.

The extensive system of rice field culture of shrimps in India has also undergone some changes, such as the introduction of controlled stocking of fry and enhanced production of food organisms through better water manage-ment and manuring by rice stubble. Production may now reach 400 kg/ha per crop under very favourable conditions, but on an average it is around 200 kg/ha per crop.

Modern large-scale shrimp farming is carried out in specially designed pond farms (figs 24.12 and 24.13) following mainly semi-intensive forms of culture, which have generally proved to be more economical. Such farms usually have a hatchery and nursery and rearing ponds. Some pond layouts also include a set of tran-sition ponds, to enable the progression method of culture, involving transfer of stocks from one pond to another as the food resources decrease and the biomass increases. The sluice gates are so located as to create a good circu-lation with the incoming water. The ponds are usually rectangular and about 1–3 ha in area, with a depth of 0.8–1.2 m. Ponds in Central and South American shrimp farms are some-times larger, up to 20 ha in area. Stocking rates vary considerably between 28 000 and 50 000 fry/ha. Natural food is produced by fertilization as mentioned above, and supplementary feed-ing with either fresh feedstuffs or formulated feeds is performed from one to five times daily. Feed rations vary between farms, but generally decrease from 25 per cent in the early juvenile stage to 2–4 per cent before harvest. Some farms in South America do not fertilize the ponds either at the nursery or production stage, and depend entirely on artificial feeding. Japanese shrimp culture has depended very considerably on feeding with the short-necked clam (*Ruditapes (= Venerupis) philippinarum*)

Fig. 24.12 A large-scale commercial shrimp farm in Hawaii (courtesy of Amorient Aquafarm, Kahuku, Hawaii).

Fig. 24.13 A large-scale shrimp farm in Taiwan (photograph: M. New).

and the mussel (*Mytilus edulis*). Formulated moist and dry feeds are also used very widely.

Many farms, particularly those in Central and South America, use diesel pumps to supply water. Farms in Thailand often use 'push pumps'. Though pumping increases the cost of production, regular water exchange and maintenance of good water quality are greatly facilitated. A large water exchange is required in culture ponds from time to time to accelerate and synchronize the moulting cycle of the stock. Yields in this type of culture can be 1.5−2 tons/ha per crop.

Intensive types of culture generally utilize cement tanks, although smaller earthen ponds of 0.5−1 ha size and 60−150 cm depth are also used. Inlets and outlets are arranged in such a way as to effect proper water circulation. Many pond farm designs have a drainage system radiating to a central outlet with a harvest basin. Tanks are generally provided with substrates. Aeration equipment like paddle wheels and air blowers are provided, especially in larger pond systems (fig. 24.14). Stocking densities ranging from 200 to 250 fry/m^2 are common. High-protein formulated diets are fed daily and frequent exchange of water is ensured. The yield from a 1000 ton tank is reported to be about 1.5−3 tons per crop, and in special ponds up to 28 tons/ha per year. However, very intensive culture using high stocking densities and heavy feeding can sometimes result in serious environmental and health problems, leading to large-scale mortalities, as has happened in Taiwan.

Pen culture of shrimps is practised in Japan and has also been carried out on an experimental scale elsewhere, such as in lagoons in southern India. Suitable intertidal areas are enclosed by net fencing. Japanese pens are made of vertical walls of concrete, constructed to a height of about 1 m for holding water during low tide, with a wooden frame with nylon netting set on top of the concrete wall to prevent the escape of shrimps and to facilitate water exchange during high tide (Kungvankij, 1985). In certain respects these pens resemble tidal ponds and the area enclosed may extend to 10 000 m^2 with a depth of 1−1.5 m. The stocking rate is between 20 and 30 per m^2. The average production is reported to be about 3−4 tons/ha per year. In ordinary net enclos-

ures tried elsewhere, the production has seldom reached that level.

In spite of the general belief that polyculture yields higher production, most shrimp culturists seem to prefer monoculture of the fastest growing species available, and they resort to polyculture with other shrimp species mainly because of the shortage of seed stock of the preferred species as for example *P. monodon* in Asia. Even polyculture with milkfish as practised in some Philippine farms is usually due to the scarcity of shrimp fry or because of market demands. It is claimed that polyculture of *P. vannamei* and *P. stylirostris* is beneficial, and if stocked in the ratio of 2:1 respectively, the harvest can be more than doubled. But it is known that *P. stylirostris* will not grow well if stocked at high densities, and the stocking rate should not be more than two per m^2, whereas *P. vannamei* can be stocked at the rate of four to five per m^2. However, behavioural disparities and the differences in salinity and substrate preferences, as well as temperature and feed requirements, can more effectively be made use of by rotating species to ensure continuous use of rearing facilities and an overall increase in yields and income. From an economic point of view, monoculture has been shown to be more profitable, mainly because of the higher market price of shrimps. According to Shang (1983) the average gross revenue per unit area of monoculture farms growing tiger shrimps is about double that of polyculture farms growing the same shrimps together with milkfish and crabs.

24.4.1 Food and feeding

In Chapter 7 the paucity of information on the nutritional requirements of shrimps has been emphasized. Much of the existing information relates to a couple of species and most of it is proprietary and not readily available. It is, however, known that there are considerable differences in dietary requirements between species, particularly with regard to protein levels. Some of the marine shrimps seem to require relatively high protein levels. For example, the protein requirement of *P. japonicus* is between 48 and 60 per cent. The protein requirement of *P. monodon* is about 35−39 per cent, of *P. setiferus* 20−32 per cent,

Fig. 24.14 Paddle-wheels used for aeration in intensive pond culture of shrimps in Taiwan (courtesy of M. New).

of *P. aztecus* 23–40 per cent, of *P. vannamei* 30 per cent, of *P. stylirostris* 35 per cent and of *P. indicus* 43 per cent. Shrimp feeds require sterols and also fatty acids of the linoleic and linolenic series, as *de novo* syntheses of these do not take place in crustaceans (see Chapter 7). Dietary lipids are provided mainly by fish oils with high levels of polyunsaturated fatty acids. Very little is known about the requirements of vitamins and minerals, although standard premixes are added in all diet formulations.

In the present state of knowledge on shrimp nutrition, fresh food continues to be important in larval and fry rearing, as well as adult grow-out. Commercial feeds are becoming available in many areas, but their acceptance in commercial farming is rather slow. When used, many farmers supplement them with natural food and feedstuffs. Water-stable pellets of different shapes and sizes (worm-like or crumbles) are prepared using finely ground ingredients and different kinds of binders, by cooking-extrusion or dry or wet pelletizing.

As is evident from the description of shrimp hatchery operations, the production of adequate quantities of the required type of live food for larval and post-larval stages is a major problem, and because of this several efforts have been made to develop microparticulate or microencapsulated larval diets. However, these have not so far resulted in products which have wide commercial application. Crustacean wet tissue suspension is reported to be used as larval feed successfully in small-scale hatchery operations in India (Hameed Ali *et al.*, 1982). *Mysis* and *Acetes*, blended into a fine particulate suspension and graded by fine-meshed sieves, have been used as the only feed during the entire larval phase and an average larval survival of 44 per cent has been reported. This type of larval feeding resembles the use of fish flesh suspension in the larval rearing of *Macrobrachium* see Section 24.6.2.

24.4.2 Diseases

With the expansion and intensification of shrimp culture, and consequent increased research efforts, a number of diseases that affect shrimps in captivity have been identified. Many of them are associated with sanitary conditions in larval and fry rearing facilities, and some are caused by nutritional deficiences. Among the infectious diseases of shrimps are virus, bacterial, fungal and protozoan diseases; the more important non-infectious diseases are those caused by epibionts.

The virus disease caused by *Baculovirus penaei* has been reported in *P. duorarum*, *P. aztecus*, *P. setiferus*, *P. vannamei* and *P. styliferus*. *Penaeus monodon* has been found to be infected by a baculovirus, referred to as the Monodon baculovirus (MBN) and *P. japonicus* by one that causes mid-gut gland necrosis (baculoviral midgut gland necrosis virus – BMNV). Though the virus infects adult shrimps as well, mortality occurs mainly in the post-larval or early juvenile stages.

In the baculovirus diseases the hepatopancreatic tubule epithelium is affected. In the larval stages the anterior mid-gut epithelium may also be affected. Mortality seems to result from the loss of the infected epithelium. There appear to be no external signs and diagnosis requires histological examination to detect tetrahedral inclusion bodies in the hypertrophied nuclei of affected cells. Diagnosis of BMNV in *P. japonicus* is made by the greatly hypertrophied nuclei within the hepatopancreatic epithelial cells undergoing necrosis. In MBV the polyhedral inclusion bodies tend to be multiple and spherical.

A recently discovered virus disease is the infectious hypodermal and haematopoietic necrosis (IHHN) in *P. stylirostris* (later also found in *P. monodon*), diagnosed by the small particles (16–28 mm) of cubic symmetry in affected tissues. Fry weighing 0.5–2 g are affected most seriously, and it is now known to affect older shrimps as well. In acute cases resulting in death, massive destruction of the cuticular hypodermis and often of the haematopoietic organs, glial cells in the nerve cord and loose connective tissues can be observed. Penaeid shrimps surviving IHHNV infections become carriers of the virus and transmit them to their offspring.

Another new viral infection caused by a hepatopancreatic parvo-like virus (HPV) has been diagnosed in *P. merguiensis*, *P. monodon*, *P. orientalis* and *P. semisulcatus*. The symptoms are poor growth, anorexia, reduced preening capacity, increased surface fouling and occasional opacity of the tail musculature. In all the affected species, necrosis and atrophy of the hepatopancreas can be observed. Heavy mortality occurs during the juvenile stage.

Bacterial diseases in shrimps may occur as 'shell disease', characterized by localized pits in the cuticle, or as localized infections and generalized septicaemias affecting all life stages. In all reported cases, motile, gram-negative, oxidase-positive, fermentative rods, mostly of vibrio species, have been found. Successful therapy includes addition of antibiotics to the tank water in hatcheries and the incorporation of antibiotics in the ration at the grow-out stage. Disinfection of all culture facilities helps to reduce the incidence of the disease.

Systematic non-inflammatory mycoses of larval stages and the generally localized mycoses of the juvenile and adult stages accompanied by inflammation caused by fungus infection are common among most Penaeid species. *Lagenidium* and *Siropidium* are the best known phycomycetes affecting shrimps. Infected individuals become immobile as a result of the profuse growth of the mycelium in the host, replacing most of the muscle and other soft tissues. By using only pretreated or filtered sea water in hatcheries, the entry of zoospores into the water supply system can be prevented. A multiple 6-hour application of Trefla[R] in the parts-per-billion range is also reported to be effective in preventing the disease. Malachite green oxylate at 0.006 ppm concentration is useful in arresting or preventing epizootics, if added prior to their establishment. A single application of 0.01 ppm trifuralin has been reported to be adequate to kill *Lagenedium* and *Siropidium* zoospores.

Another fungal disease of cultured shrimps is caused by *Fusarium solani*. The infection may occur through pond bottom muds and detritus or sea water. Wounds or abrasions on the host can easily be infected. Lesions may occur in the gills or at the bases of appendages or on the cuticle and well-developed lesions are darkly melanized. The black gill disease of *P. japonicus* is caused by this fungus. Mortalities of the whole stock can occur in highly susceptible species, and no effective methods of prevention or cure are known at present.

Among the non-infectious diseases caused by epibionts, Leucothrix disease and ciliate gill diseases are important. Leucothrix disease, caused by the bacterium *Leucothrix mucor*, occurs in juvenile and adult shrimps. The bacteria attach themselves to the body of the host, particularly the gills and accessory gill structures. Larval and post-larval shrimps may become covered by the filaments of the pathogen, affecting respiration, feeding, locomotion and moulting. Severe losses may occur sporadically and, if not controlled, can also cause continuous low-level losses. Treatment with a sea water soluble copper compound (commercial product: Cutrine−Plus[R]) at concentrations of 0.2−0.5 mg Cu/ℓ for 4−6 hours in static conditions has been found to be effective in preventing and curing the disease.

The ciliate gill diseases are caused by protozoans of the genera *Zoothamnium*, *Epistylis* and *Vorticella*. When the surface of the gills are covered with these organisms, hypoxia and death occur in the same manner as in Leucothrix disease. Formalin is reported to be effective in controlling the infection.

The cotton or milk shrimp disease denotes a group of diseases caused by at least four species of microsporidians (Protozoa). Infected shrimps have opaque musculature and ovaries and dark blue or blackish discoloration due to expanded chromatophores in the cuticle. Multiple infections have been reported. There are no proven methods of cure for this disease.

Three types of environmental diseases have been reported from Penaeid shrimps, namely muscle necrosis (spontaneous necrosis), cramped tail and gas-bubble disease. Muscle necrosis is characterized by whitish opaque areas in the striated musculature, especially of the distal segments of the abdomen. It is believed to be caused by severe stress from overcrowding, sudden temperature and salinity changes and low dissolved oxygen or rough handling. If large areas are affected, the disease may prove fatal. The chronic and typically septic form of the disease is known as 'tail rot', when the abdomen or its appendages becomes completely necrotic, red in colour and begins to decompose. In the initial stages the disease can be controlled by reducing stress.

The 'cramped tail' condition generally occurs during summer. It is characterized by a dorsal flexure of the abdomen which is rigid and cannot be straightened. It is believed that this condition is brought about by elevated water and air temperatures and stress due to handling in warm weather.

Gas-bubble disease of shrimps, caused by supersaturation of atmospheric gases and oxygen, is very similar to the one described in finfishes. Early signs of the disease are rapid and erratic swimming, which may soon be followed by a state in which the shrimp floats near the water surface. If supersaturation is due to oxygen, the condition can be controlled by reducing the level, but if caused by nitrogen or other gases it is usually lethal.

Chronic soft-shell syndrome occurs in *P. monodon* in brackish-water ponds with poor soil and water conditions. The affected shrimps show high levels of calcium and phosphorus in the hepatopancreas, and lower levels of phosphorus in the exoskeleton. Soft-shelling could be induced by exposure to pesticides, and reversed by improved diets containing 14% mussel meat.

24.4.3 Harvesting and marketing

Methods of harvesting shrimps have been briefly described in Chapter 11. Efficient harvesting has to be based on the habits of the particular species under culture and the culture procedures. Shrimps graze at night and are generally attracted by light. They also respond actively to water movements, especially when water is let into a pond or is drained. Most Penaeid species are more active during the full moon and new moon, and the maximum activities are shortly after sunset and shortly before sunrise. These are the best times for harvest.

Total harvest is usually carried out at the end of each crop by draining and placing a bag net at the sluice gate to catch the shrimps as they swim from the pond. Partial harvesting can be done by seine nets after partial draining, from peripheral canals or from harvest basins. Some farmers catch the shrimps in large traps built in the outlet canals outside the sluice gates, as the pond is drained. In Taiwan, cast nets are sometimes employed for partial harvesting, using sea snails as bait.

As mentioned in Chapter 11, it is very profitable to sell shrimps alive in countries like Japan, and so the harvesting technique has to be adapted for capturing them without injury. The pound net used in Japan to trap shrimps is well-suited for this purpose. During cool seasons, when the shrimps do not move so actively at the bottom, a pump net or an electric shock is more effective (fig. 24.15). These methods are suitable for *P. japonicus*, which burrows at the bottom. Electric fishing is successfully employed in harvesting tiger shrimps in Taiwan. Harvesting is avoided when most of the shrimps are moulting, as soft-shelled shrimps will not stand handling. Live shrimps are transferred to containers placed in cold water tanks. Methods of packing and transport of live shrimps have been described in Chapter 11.

Since shrimp culture in many developing countries is export-oriented, modern methods of chilling, freezing and packing are practised to meet the requirements and regulations of importing countries. Whole shrimps, shrimp tails or partially or completely shelled products are exported.

24.5 Economics of shrimp farming

Even though reliable economic data on actual commercial production continue to be scarce, as in most other aquaculture systems, greater efforts have been made to assess the economic viability of shrimp farming, because of the investment interests of industries and the somewhat indifferent results of some early ventures. Comparisons and conclusions are made difficult by the wide variations in culture procedures, the climatic conditions of the farming areas and the shrimp species which are used for farming. Hirasawa (1985) analysed in detail the economics of shrimp culture in Asia, based on the type of production system, investment costs and the present and expected future markets. Griffin *et al.* (1985) discussed the investment and production costs of a semi-intensive farm in the USA and compared it with a similar farm in Ecuador. Shang (1983) made a general survey of shrimp farming in Asian countries, the USA and Ecuador. All these studies have confirmed the wide variability of investment and production costs in different areas.

Table 24.1 illustrates the variation in costs and returns in selected countries. Though the average rate of return on operating costs ranges

Fig. 24.15 Electric fishing in shrimp ponds in Taiwan (courtesy of M. New).

Table 24.1 Summary of costs and returns (per ha) of shrimp farming in selected countries in 1982, based on case studies (from Shang, 1983).

	Ecuador	Hawaii	Malaysia	Taiwan	Texas (USA)	Thailand
Size of farm (ha)	203	100	8.4	1.2	162	7
Species cultured	*P. vannamei*	*P. stylirostris* or *P. vannamei*	*P. monodon*	*P.monodon*	*P. vannamei*	*P. monodon*
Crops/year	2	2	3	2.5	1	2
Initial costs (US$)	2838	36 125	28 780	65 000	5370	4416
Yield/year (US$)	1818	2 246	5 000	15 000	1571	2 000
Gross revenue (US$)	9032	25 000	29 545	120 000	9333	14 000
Net revenue (US$) (before taxes)	3734	2 510	9 662	14 566	4174	7 589
Ratio between net revenue and operating cost (%)	70	11	49	14	81	118
Cost of production (US$)/kg	2.91	10.00	3.98	7.0	3.28	3.21

from 11 to 118 per cent, all the farms surveyed were found to be profitable and the lowest returns were from areas where labour costs, capital investments and management costs were high. This is further confirmed by the comparison of estimated costs and returns of similar-sized farms rearing *P. vannamei* in the USA and Ecuador (Griffin *et al.*, 1985) (Table 24.2). The internal rate of return from a 200 ha farm in Ecuador was found to be 2.8 times that of a similar one in the USA.

Investment costs are comparatively high in the USA, where land costs range from $1500–8000/ha and pond construction cost is about 40 per cent of the total investment of about $2 million. The growing season is shorter and only one crop is produced. Feed and labour and harvesting costs account for 36 and 22 per cent respectively, of the variable costs.

Hirasawa's (1985) analysis of shrimp farming shows that the best return on investment is obtained by the use of extensive culture methods in Asia. By a small increase in productivity the relative cost of production can be reduced significantly in extensive systems and this ensures the survival of the farm, even if the market price of shrimp goes down. This may not be possible for intensive systems, as the ratio of variable costs to total costs is rather high and the variable costs do not change as the productivity increases. This is probably the

rationale behind the prevailing interest in semi-intensive systems which also respond to the need for increased production of shrimps to meet the requirements of export markets.

Reproduction and larval rearing of fresh-water prawns

Mature *M. rosenbergii* (Fig. 24.16) easily mates and spawns in captivity throughout the year, even though in nature there are seasonal peaks associated with the onset of the rainy season. Berried females are collected and used for spawning purposes. The general practice is to select gravid females carrying almost ripe (brown-coloured) eggs, from commercial harvests. Brood stock of larger individuals above 45 g in weight are preferred as the quantity of eggs spawned is comparatively higher. Brood prawns are reared in fresh or slightly brackish water (salinity 2–8 ppt) and fed on mussels, cockles, fresh trash fish or compounded feeds. Where brood ponds are maintained, they are stocked at a lower density of about 12 500/ha and the females raised to a weight of 100 g to increase the production of eggs and larvae. The eggs change in colour from orange to greyish-brown as they mature. Mature male prawns are considerably larger than the females, their second walking legs much larger and thicker, the cephalothorax proportionately larger and

Table 24.2 Economic comparison (per ha) of a 200 surface ha shrimp farm using 20-ha ponds by intensive culture in the USA and Ecuador, in 1984 (from Griffin *et al.*, 1985).

Item	USA	Ecuador		
	Semi-intensive	Semi-intensive	Semi-extensive	Extensive
kg/ha/yr (heads off)	1159	1 323	554	232
$/kg	8.47	9.00	10.00	11.00
Value/ha ($)	9798	11 908	5544	2553
Total variable cost ($)				-
Post-larvae	1800	480	180	62
Wages	663	317	190	78
Fuel	225	106	75	40
Feed	2040	1 995	334	0
Fertilizer	54	269	269	0
Repairs	138	311	234	179
Packing	548	448	188	79
Miscellaneous	120	687	339	129
Total	5588	4 613	1809	567
Total fixed cost ($)				
Overhead	404	230	130	100
Depreciation	595	396	268	192
Miscellaneous	175	91	57	50
Total	1174	717	455	342
Total cost ($)	6762	5 330	2264	909
Revenue before taxes ($)	3036	6 578	3280	1644
Taxes ($)	1518	3 289	1640	822
Revenue after taxes ($)	1518	3 289	1640	822
B-E price/kg (heads off) ($)	5.83	4.03	4.09	3.91
IRR (%)	21	59	39	25
Total investment ($\times$ $1000)	1915	1 243	937	715

the abdomen narrower. The male can also be distinguished by the presence of a lump or hard point in the centre on the ventral side of the first segment of the abdomen.

24.6.1 Hatchery systems

There are various hatchery systems in use in experimental centres and commercial farms, but the more important ones can be described as belonging to the green water (see fig. of 3), 'clear water' and recirculation systems. The so-called backyard hatcheries use one of these systems, more commonly the green water system. In many hatcheries, mating, spawning and larval rearing occur in the same tank. Larger tanks are required for water storage,

post-larval rearing and for hatching *Artemia*. The tanks may be in the open in tropical areas, but are usually kept under a roof to prevent the water from getting too warm during summer days and too cold at nights. In temperate climates it is necessary to house the tanks in a greenhouse type construction, to maintain the water temperature at the required level during the cold season.

As mentioned in Chapter 6, hatchery tanks can be made in various shapes and sizes with plastic, fibreglass, reinforced concrete, etc. Circular (with flat or conical bottom), rectangular and square-shaped tanks are in use. Though each has its advantages, rectangular tanks are more convenient and space-saving. A 10 m^3 rectangular tank, with water and air

Fig. 24.16 The giant fresh-water prawn, *Macrobrachium rosenbergii.*

intakes at one end and the drain at the other, has been found to work very efficiently. However, some culturists prefer the circular tanks (fig. 24.17), because they are easy to clean and better water circulation can be maintained in them. Many culturists recommend the tanks to be painted with a dark colour, to enable the larvae to see the feed well. An aeration system at the bottom of the tank helps proper mixing of the water and even distribution of the food particles and the larvae, thereby reducing cannibalism.

Sandifer *et al.* (1983) described the use of artificial 'habitat units' that increase the surface area available to the larvae in rearing systems. In the presence of stacked solid layers, the larvae exhibit what is termed an 'edge effect', which is a pronounced preference for the larger edges. The units consist of a rigid frame to which layers of plastic mesh are attached. The mesh layers increase the area of surface edges significantly, both vertically and horizontally. Strips of screening placed on different layers provide feeding spots, where the food settles. Tanks for holding larvae before distribution and for mixing brackish water are of the same

design as larval tanks, but are larger, about 50 m^3. The water supply, as far as possible from bore wells, provides unpolluted fresh and salt water. A pH in the range of 7.0–8.5, a temperature near to the optimum of 28–31°C and nitrite and nitrate levels not higher than 0.1 ppm (NO$_2$-N) and 20 ppm (NO$_3$-N), have been recommended.

The terms green water, clear water and recirculating systems of larval culture refer largely to the water management procedures. The green water system, as the name implies, involves the production of a mixed phytoplankton culture dominated by *Chlorella* at a cell density of 750 000–1 500 000 cells/ml. A solution of a fertilizer mixture of four parts urea to 1 part NPK (15:15:15) in tap water is added to the tank at least once a week, to maintain the plankton bloom. The production of rotifers in the tank is controlled by the application of CuSO$_4$ at the rate of about 0.6 ppm. The growth of filamentous algae may be controlled by holding *T. mossambica* in the tanks at the rate of 1 fish per 400 ℓ. The salinity of the green water is not more than 12 ppt, and it is used as replacement water during water exchange of

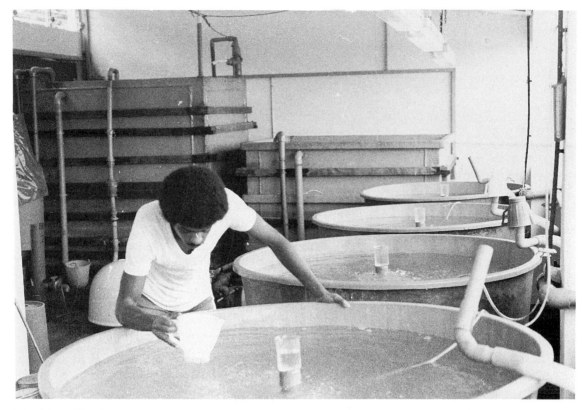

Fig. 24.17 Circular tanks using the clear-water system in a prawn hatchery in Martinique.

the larval tanks. Only green water less than three days old is used. For a continuous production of such water, without causing phytoplankton 'crashes', it is diluted regularly and fertilized as required. The function of the green water is to maintain the water quality in the larval tanks by recycling the waste products of larval metabolism, and to provide food for live food organisms such as *Artemia*. The effect of the green water is very much dependent on meteorological conditions and the quality of fresh and sea water. So, though basically simple, its efficiency depends very much on local conditions. Furthermore, in this system only a lower stocking density of about $30-50$ larvae/ℓ, yielding on average $5-25$ post-larvae/ℓ, is feasible.

The clear-water system can be adopted for different levels of larval production. Circular tanks with conical bottoms or rectangular tanks are used. The system developed by Aquacop (1977a) for high-density larval culture involves the use of conical bottom tanks of 800 ℓ capacity with total water exchange and environmental control, including temperature and light intensity. The water quality is maintained by chlorination, followed by dechlorination with UV light, strong aeration and the use of antibiotics. Higher stocking levels of $100-200$ larvae/ℓ yielding $25-100$ post-larvae/ℓ are feasible, but require very careful management.

Recirculation systems are valuable in water conservation and energy use. The simplest type of this system recirculates the tank water through a graded sand/gravel filter, using mechanical or air lift pumps. More complicated, completely closed systems involving mechanical and biological filtration, chlorination and dechlorination or the use of ozone, have also been developed. Though the value of such a system in controlling the water quality is recognized and is utilized in experimental work, commercial application on any significant scale has yet to be realized. On the other hand, efforts are made in commercial hatcheries to operate both green-water and clear-water systems with only limited water exchange.

24.6.2 Spawning and larval rearing

Mating takes place between hard-shelled males and ripe soft-shelled females which have completed their pre-mating moult. Semen is deposited in a gelatinous mass on the underside of the thoracic region of the female, between the walking legs. Within a few hours, the female extrudes eggs through the gonopore and the eggs are fertilized by the semen attached to the abdomen and transferred to a brood chamber located under the abdomen between the pleopods. It is believed that an enzymatic reaction is involved in the release of sperm cells from the spermatophore. The fertilized eggs become attached to each other and the setae of the first four pleopods by a cementing substance, which hardens into an attachment membrane. The vigorous movements of the pleopods keep the eggs well aerated. The number of eggs laid varies with the size of the prawn, but it is reported to be up to 80 000–100 000 per spawn. However, a one-year old female may spawn only 5000–20 000 eggs during its first spawning. The incubation period ranges from 18 to 23 days at temperatures of about 28°C. Even during the incubation period of fertilized eggs, ovarian eggs start maturing and a second pre-mating moult can occur within a period of about 3 weeks.

Selected healthy berried females from natural sources, rearing ponds, brood tanks or aquaria are introduced into the larval tanks. In community rearing tanks, it is necessary to ensure that all of them have black or grey eggs, so that they will all hatch out within two or three days and the larvae in the tanks will be of the same age. This helps in appropriate feeding and reducing cannibalism among the larvae.

Brackish water with a salinity of about 5 ppt is provided in the tanks, as hatching rates are lower in fresh water. Water of the required salinity is obtained from mixing tanks in the hatchery. It is, however, possible to hatch the eggs in fresh water and raise the salinity after hatching. After the eggs have hatched, the spent females are removed from the tank with a coarse-mesh dip net. Usually larval release is over a four-day period, with a peak between 24 and 72 hours. So the females have to remain in the tanks up to four days. The salinity of the water can then be raised to around 12 ppt. The optimum temperature range is 26–31°C, and sudden changes of salinity or temperature are avoided. Aeration is maintained in the tanks in order to maintain oxygen levels near saturation. In clear-water and recirculating system hatcheries, the tanks are not exposed to direct sunlight, and at least part of the tanks are under cover, providing some suitable source of indirect light.

Macrobrachium undergoes eleven stages during larval development (fig. 24.18) and the time taken for a batch of larvae to metamorphose varies according to temperature and feeding conditions. Under favourable conditions, the post-larval stage is reached in 16–28 days. The post-larvae resemble juvenile prawns and rest or crawl on the tank surfaces. At this stage they are harvested from the tanks for nursing or on-growing.

There is considerable variation in the density of larvae reared in larval tanks. A final density of $40/\ell$ is recommended; but the culture can start with a higher density of $80–120/\ell$, and split the stock on about the twelfth day (stages V–VIII) in order to reach the final desired density. In small rearing ponds, it may be possible to carry out the rearing without any division of stock. In hatcheries following the green-water system, water in the tanks is lowered daily to a 50 per cent level and replaced with green water. In clean-water systems, after removal of about 50 per cent of the water clean water is used for refilling. Some hatcheries in Hawaii practise a combination of these systems. They follow the green-water system during the early part of the culture (3–6 days) and then switch to clean water.

Appropriate feeding is a key factor in successful larval rearing. Different types of feeds are in use in hatcheries, the more important of which are nauplii of *Artemia*, fish flesh, fish roe, egg custard, egg and mussel mixtures and compound feeds. As *Macrobrachium* larvae do not actively search for food, feeding is generally started with swimming nauplii of *Artemia* twice a day and continued till the fifth day after hatching. The actual quantity to be fed is determined by visual examination of the tank,

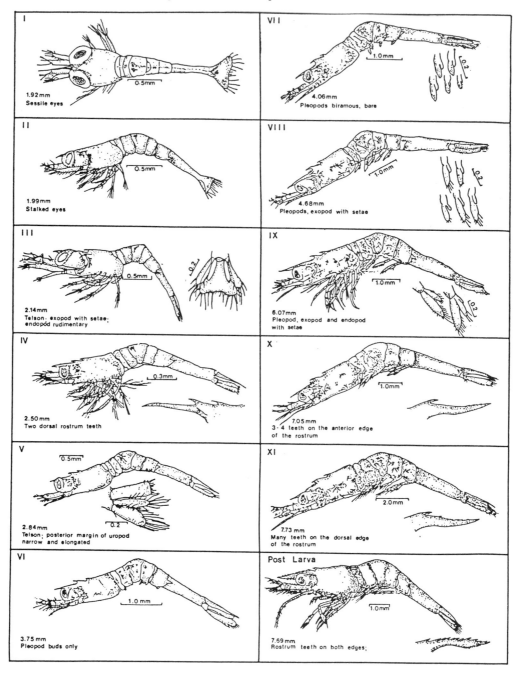

Fig. 24.18 Larval stages of the giant fresh-water prawn (from Malecha, 1983).

the main consideration being to provide an adequate distribution of the nauplii near each larva, to facilitate feeding. New and Singholka (1985) roughly calculated the requirement of nauplii for a 10 m³ larval tank to be about 10−50 million per day. Prepared feeds are generally given only from the fifth day, even though weaning can be started a couple of days earlier. These are given four or five times during the day, and it is recommended that the night feeding should be with *Artemia* nauplii only. The quantity of prepared feed is also determined by the need for a suspension of it in the tank, to make it easy for the larvae to locate it. The quantity needed at each feeding has been calculated to be about 30–60 g per tank initially, gradually increasing up to 200 g. Vigorous aeration helps to keep the feed in suspension near the larvae.

Fish such as skipjack tuna, bonito and pollock, used as feed for the larvae, are filleted, ground and forced through stainless steel sieves with strong jets of water to obtain particles of the required size. Particles of 0.3 mm size are suitable for feeding until the tenth day and from then on 0.3–1 mm size until matamorphosis.

Egg custard for feeding larvae is prepared by cooking scrambled whole chicken eggs and passing it through stainless steel sieves of the required mesh size. A mixture of eggs and mussels has been found to be an efficient feed for larvae. Shelled mussels are minced to fine particles, strained through a coarse sieve and blended with whole beaten egg. The mixture is steamed until it solidifies and then sieved to produce particles of the required size.

As mentioned earlier, the time taken for a larval batch to finish metamorphosis depends on the environmental conditions, but it is usually completed in less than 28 days and it is best to harvest them at this stage. The post-larvae can then be transferred to fresh water, but it is advisable to acclimatize them through a gradual lowering of the salinity by draining part of the brackish water and replacing it with fresh water. The production from a 10 m³ tank is between 100 000 and 200 000 larvae.

The post-larvae are stocked directly in production ponds or grown for a period of one or two months before on-growing to market size. The nursery ponds are stocked at higher densities of about 1000/m². Besides the normal advantages of nursery rearing, this practice helps the farmer to count the number of juveniles and assess their state of health before release into production ponds. Nursery rearing is considered essential in temperate climates, but even in tropical climates this practice can be beneficial. An increasing number of farms seem to be incorporating a nursery stage in the culture operations.

24.7 Grow-out of fresh-water prawns

The preferred size of ponds for grow-out appears to be 0.2–0.6 ha, with an average depth of 0.9 m. The ponds have an adequate supply of fresh or slightly brackish water, not exceeding 4 ppt in salinity. Water quality management is generally through water exchange, though paddle wheels or other aeration devices are maintained for use in emergencies. High growth rates are obtained at temperatures between 26 and 30°C. In semi-tropical areas, the ponds are stocked only when the temperatures are not likely to drop below 20°C. Many farmers do not fertilize their ponds except when they are newly built and lack nutrients in the soil. Prawn ponds in Taiwan with earth or sandy bottoms are dried by exposure to sun and treated with lime and teaseed cake. Filtered water is used to fill the ponds, in order to prevent the entry of eggs of unwanted fish. The water is 'seasoned' with green water and aerated for several days. Fermented chicken manure, pig manure and inorganic fertilizers are applied to stimulate phytoplankton production.

Ponds are normally stocked with 1–4 week-old post-larvae. The stocking rate depends on the preferred market size and the length of the growing season. New and Singholka (1985) recommended a density of five one-month old post-larvae per m² (50 000 per ha) for a growing season of 8 months to harvest prawns of about 70 g (head on). Higher stocking rates are adopted in commercial operations in Hawaii, averaging 16 per m² (160 000 per ha) allowing for a mortality that would bring the stock density to 11 per m² (110 000 per ha). The recommended rate under the temperature conditions in South Carolina (USA) is 4.3–6.5 per m² of nursed juveniles or a mixture of nursed juveniles and post-larvae (43 000–65 000 per ha), for a growing period of 5–6 months, yielding 700–1200 kg per ha.

The most common grow-out practice has been described as the 'continuous culture' or 'continuous stocking and harvesting system', in which the ponds are stocked once or several times a year and are never drained, except for repairs. Harvesting is carried out selectively for marketable prawns about 30 or 45 g size, at regular intervals. A considerable disparity in growth rates between individual prawns (especially between males and females) occurs in ponds. So selective fishing is performed as a means of stock management and to grow under-sized individuals from previous stockings to marketable size. If the whole stock is harvested together, there will be a number of small prawns which may not be acceptable in the market. Prawn ponds following this system in Hawaii are reported to produce about 276 kg/ha per month by selective harvesting, yielding about 3314 kg/ha per year (Malecha, 1983). Some farms in Hawaii have started incorporating a nursery phase or an intermediate grow-out phase, to avoid some of the problems of continuous stocking. This requires specially constructed ponds with harvest sumps.

24.7.1   Food and feeding

The types of feed used in fresh-water prawn farming vary considerably from locally available feedstuffs to compound chicken feeds and specially formulated prawn feeds. Compared to Penaeid shrimps, prawns are reported to have a lower dietary protein requirement of 20−25 per cent. However, some farms use shrimp feeds containing higher protein levels. Also, the results of research seem to indicate that juveniles may need higher protein levels,

and a 40 per cent protein diet has been reported to yield higher weight gains. Asian farmers use rice byproducts, trash fish, molluscs, prawn waste, poultry waste and meal of ipil ipil (*Leucaena* sp.) leaves.

Many farms use commercially available compound chicken feed, sometimes re-extruded after mixing with trash fish or prawn meal. Tables 24.3−24.7 present the ingredient composition of a number of practical compound feeds used in commercial production. At least some of them can be prepared on the farm, if commercial feeds are not available. In most cases, feeding is done manually, but some farmers use automatic blower feeders, as in Hawaii. Many Hawaiian and Thai prawn farms maintain rich blooms of phytoplankton in the grow-out ponds to supplement the compound chicken feed or prawn feed.

The feeding rate is adjusted according to assessments of daily consumption and biomass changes. Recommended rates rise from 2.7−7.4 kg/ha per day in the early rearing period to as much as 37.5 kg/ha per day towards the end of the rearing, in about 6−8 months (New and Singholka, 1985). Feed conversion ratios generally range from 2:1 to 4:1 for commercial diets, giving growth rates averaging around 1−2 cm/month.

Since *M. rosenbergii* is a benthic omnivore, it is considered a suitable candidate for polyculture. Experiments in polyculture with *Tilapia aureus* (Brick and Stickney, 1979) did not reveal

Table 24.3 Ingredient composition of various feeds utilized by Hawaiian prawn farmers (from Corbin *et al.*, 1983).

Ingredient	Waldron's broiler starter (%)	Waldron's game cock pellets (%)	Waldron's prawn No. 1 (%)	Waldron's prawn No. 2 (%)
Alfalfa	—	—	4.00	4.00
Corn	53.25	50.25	56.75	56.75
Cottonseed meal	10.00	15.75	—	—
Soybean meal	24.25	20.50	27.00	25.00
Meat and bone meal	7.00	7.00	11.00	8.00
Tuna meal	—	—	—	5.00
Vitamin mix	1.25	1.25	1.25	1.25
Mineral mix	1.25	1.25	—	—
Molasses	3.00	4.00	—	—

Source: Waldron's Feed Mill, personal communication, 1981.

Table 24.4 Composition of vitamin and mineral mixes used in certain prawn feeds manufactured in Hawaii*

Vitamin mix ingredient	Amount/kg diet[†]	Mineral mix ingredient	Amount/kg diet[‡] (mg)
Vitamin A	5500.0 IU	Zinc oxide	55.1
Vitamin D	1237.0 IU	Ferrous sulphate and carbonate	59.5
Vitamin E	4.1 IU	Manganese oxide	56.0
Vitamin K	0.8 IU	Copper oxide	4.5
Vitamin B_2	3.3 mg	Ethylenediaminedihydroiodide	0.25
Pantothenic acid	4.9 mg	Cobalt sulphate	0.50
Niacin	24.7 mg	Sodium selenite	0.10
Choline chloride	67.1 mg	Sodium chloride	2646.0
Vitamin B_{12}	8.2 mg		
Folic acid	0.3 mg		

* Source: Waldron's Milling Co., Honolulu, Hawaii.
[†] Vitamin mix is added to the prepared diet at 0.025%.
[‡] Mineral mix is added to the prepared diet at 0.29%.

Table 24.5 Ingredient composition of experimental unsupplemented prawn feeds (25% protein) used in Thailand (from New and Singholka, 1985).

Ingredient	Dry pellets	
	1 (%)	2 (%)
Fish oil	3.0	3.0
Shrimp meal	25.0	10.0
Fish meal	10.0	4.0
Peanut meal	5.0	2.0
Soybean meal	5.0	2.0
Broken rice	25.5	39.0
Rice bran	25.5	39.0
Guar gum	1.0	1.0
	100.0	100.0

Table 24.6 Ingredient composition of prawn feed formulated for use in Malaysia (from Chow, 1984).

Ingredient	Semi-moist (%)	Dry (%)
Fish meal (55% CP)	8.00	13.00
Wheat pollards	30.00	47.00
Rice bran	11.75	19.00
Soybean meal	12.00	19.00
Vitamin mix No. 3*	1.25	2.00
Water	37.00	–
	100.00	100.00

* See Table 24.7.

Table 24.7 Composition of vitamin premix No. 3.

Ingredient	Amount/g premix
Vitamin A	500 IU
Vitamin D_3	100 IU
Vitamin B_1	0.1 mg
Vitamin B_2	0.3 mg
Pyridoxine	0.2 mg
Vitamin B_{12}	0.001 mg
Nicotinic acid	2.0 mg
Calcium pantothenate	0.6 mg
Folic acid	0.05 mg
Vitamin K	0.2 mg
Vitamin C	5.0 mg

any significant interaction between the species. Relatively high combined yields were obtained in polyculture trials with tilapia, channel catfish and Chinese carp in Alabama (Behrends *et al.*, 1985). However, the yields and individual harvest weights of the prawns seemed to indicate interspecific competition. In initial experimental polyculture of the prawn with Chinese and common carp in ponds without supplementary feeding, Malecha *et al.* (1981) found that weight gains by the prawns compared favourably with those in monoculture, indicating that they could utilize heterotrophic productivity in manured ponds. Pond fertilization under monoculture is likely to reduce the need for complete compound feeds, making natural food available to the prawns. In Taiwan, where polyculture is practised with grass carp, bighead carp, grey mullet or milkfish, the species combinations are believed to result in more efficient use of pond productivity and the control of the growth of filamentous algae. Polyculture of the prawn with the crayfish *Procambarus clarkii* and the channel catfish appears to lower the survival and growth rates of the prawns (Huner *et al.*, 1983).

Diseases

In fresh-water prawn culture, diseases have a greater occurrence in hatcheries than in grow-out ponds. Several diseases have been identified in larvae, juveniles and adults, but many of them are of undetermined aetiology. Even among those for which the aetiology is known, appropriate prophylactic and curative measures have not yet been developed. But most of the diseases appear to be directly or indirectly due to poor sanitation, inadequate water exchange, poor feeds or low oxygen levels.

Among the diseases identified in larval prawns, those caused by microbial epibionts appear to be more common. The epibionts are mainly filamentous and non-filamentous bacteria, algae or aquatic protozoa. The protozoan agents include the genera *Epistylis*, *Zoothamnium* and *Vorticella*. These organisms attach themselves to the epicuticular surfaces, but do not cause any inflammatory responses. *Zoothamnium* appears to attach itself to the gill lamellae, while other protozoans do not exhibit any site specificity. Bacterial fouling of proximal appendages, gills or the body surface often results in severe mortality. Antibiotic treatment is likely to control the disease. As the disease is triggered by poor biological conditions in hatchery systems, the best control measure is the identification and control of the primary cause.

Brown-spot disease, also known as black-spot or shell disease, occurs commonly in juveniles and adult *Macrobrachium*. It can be recognized by the presence of brown to black, ulcerative to raised lesions on the body surface

or appendages. This disease occurs only in animals which have developed prior cuticular damage due to other causes, including aggression between prawns. Both adults and larvae have been found to develop melanized brown spot lesions. The only control measure suggested is improvement of culture conditions and correction of nutritional deficiencies.

Exuvia entrapment is a disease primarily of late larvae and early post-larvae, with mortality occurring at the time of metamorphosis moult. Affected larvae are not able to free their pereiopods, anterior appendages, eyes or rostrum from the exuvia during ecdysis and consequently die. The aetiology of the disease has not been determined, and the prevention and control measures are limited to the use of algal supplements in larval culture and the maintenance of good quality water conditions in hatchery tanks.

Idiopathic muscle necrosis of prawns, also known as muscle opacity or spontaneous muscle necrosis, is associated with environmental stresses such as salinity, temperature, hypoxia, overcrowding, etc. The aetiology of the disease is not known. It also occurs in Penaeid shrimps and the only preventive measure presently known is reduction of environmental stresses during culture, especially at times of handling and transfer.

24.7.2 Harvesting and marketing

As indicated earlier, harvesting schedules are dictated by culture practices and market requirements. The simplest system, which prawn culturists call 'batch culture', consists of stocking a pond and allowing the stock to grow until they attain marketable size, after which the whole stock is harvested. For harvesting the stock effectively in this system, the ponds have to be drained.

The other techniques involve 'continuous culture' and 'continuous stocking and harvesting'. In continuous culture, the ponds are stocked generally once a year at a comparatively higher rate, and harvesting is done by seines on a continuing basis. After about 5–7 months, market-sized prawns are culled at regular intervals. The ponds are never drained. In the continuous stocking and harvesting system, ponds are restocked up to six times a year after cull-

Fig. 24.19 Harvest of giant fresh-water prawn from a pond farm in Thailand.

ing. Some farms try to combine the main features of the different systems. About 5 months after stocking the post-larvae, regular cull-harvesting is done until about 8 months, when the pond is drained and the whole remaining stock is harvested.

Except in cases where it is possible to drain the whole stock into a harvest sump and remove it by dip nets or mechanical devices like pumps, seining is the most common method of pond harvesting. Seine nets have to be operated with special care, so that the bottom of the seine rides on the sump bottom when in operation, to ensure that prawns do not escape beneath it. In continuous or multiple harvesting, sometimes only one half of the pond is seined at a time, every two or four weeks, in order to avoid disturbing the whole pond each time. Another precautionary measure is to catch only what can easily be removed from the net, to avoid the prawns being crushed to death when hauled out of the water.

Modified seine nets have been designed specifically for cull-harvesting (Hanson and Goodwin, 1977). The head rope is made of light polypropylene which does not sag, and the foot rope is made of soft nylon which rides the contours of the pond. Sufficient floats are used to keep the head rope stretched above water to prevent prawns from crawling over the net. The seine has a bag similar to that of a beach-seine to hold the catch. A mechanical harvesting system described by Williamson and Wang (1982) is a modification of the traditional seine, but uses a tractor or truck to pull the net. The seining time is greatly reduced and harvesting efficiency is reported to be at least as great as in manual seining.

In many areas, live prawns fetch the highest price and so every effort is made to keep the harvested prawns alive. Besides sorting them according to size, soft-shelled (newly moulted) and egg-bearing prawns are separated out. Live prawns can be hauled to markets in trucks in live tanks with proper aeration. Dead prawns are transported on ice, but some producers chill-kill the prawns and blanch them in water at about 65°C for 15–30 seconds, before pack-

Table 24.8 Estimated average costs and returns (in $) of fresh-water prawn production per 0.40 ha pond in Hawaii and in Thailand, in 1980 (from Shang, 1982).

Cost items	Farm size in Hawaii (ha)			Farm size in Thailand (ha)		
	0.4	4	8	0.4	4	8
Initial costs						
Pond construction	6 969.00	6 497.00	6 780.00	1 898.00	1 771.00	1 898.00
Equipment	10 205.00	2 698.00	1 435.00	30.00	253.00	380.00
Total	17 174.00	9 195.00	8 215.00	1 928.00	2 024.00	3 278.00
Annual operating costs						
Labour	1 184.00	3 144.00	2 094.00	356.00	140.00	122.00
Post-larvae	520.00	520.00	520.00	268.00	313.00	455.00
Feed	1 660.00	1 660.00	1 660.00	435.00	283.00	418.00
Electricity	–	207.00	165.00	23.00	43.00	58.00
Land lease	882.00	708.00	708.00	19.00	19.00	19.00
Gasoline and oil	163.00	81.00	81.00	35.00	8.00	21.00
Maintenance	382.00	305.00	286.00	64.00	54.00	43.00
Interest	2 605.00	1 406.00	1 281.00	154.00	118.00	146.00
Depreciation	1 652.00	493.00	389.00	264.00	243.00	273.00
Tax	40.00	40.00	40.00	–	–	–
Insurance	–	334.00	232.00	–	–	–
Miscellaneous	454.00	445.00	373.00	162.00	122.00	156.00
Total	9 542.00	9 343.00	7 829.00	1 780.00	1 343.00	1 711.00
Production and revenue						
Production (kg)	909.00	909.00	909.00	445.00	455.00	380.00
Price ($/kg)	8.80	8.80	8.80	6.50	8.00	7.50
Revenue ($)	7 999.00	7 999.00	7 999.00	2 893.00	3 640.00	2 850.00
Profit ($)	−1 543.00	−1 344.00	170.00	1 113.00	2 297.00	1 139.00
Cost of production/kg	10.50	10.28	8.61	4.00	2.95	4.50
Cost of labour/kg	1.30	3.46	2.30	0.80	0.31	0.32
Cost of feed/kg	1.83	1.83	1.83	0.98	0.62	1.10
Cost of post larvae/kg	0.57	0.57	0.57	0.60	0.69	1.20

ing them on ice. This process seems to help in extending the shelf life of the prawn to 4−6 days.

Experience indicates that the fresh-water prawn can best be marketed in the fresh 'shell-on' form in domestic markets, or in export markets which can be reached by rapid means of transport such as air-freight. The tough exoskeleton and long appendages make peeling difficult and so it is generally sold whole. Further, the frozen prawn is reported to undergo a rapid deterioration in quality. According to Nip and Moy (1979), prawns frozen in still-air and brine solutions, as well as liquid nitrogen, lose elasticity and viable bacterial counts. However, no significant losses of flavour and texture were noticed, and they were of the view that carefully frozen prawns are of good and acceptable quality. Hale and Waters (1981) reported that tails as well as whole prawns can be frozen and stored, up to 10 months in the case of tails and 7 months for whole prawns.

24.8   Economics of fresh-water prawn farming

There are very few reports on cost and earnings of fresh-water prawn culture, and the available estimates are not recent. The profitability of small-scale prawn farming, like that practised by many farmers in Thailand, is never in doubt as this is often the main source of income for the families, and incomes have shown signs of significant increases. The economic performance of large-scale commercial farms appears to vary widely, and there are cases of notable successes and complete failures. Some of the commercial hatcheries and farms have changed hands more than once. In the major *Macrobrachium* farming areas, there does not appear to be any lack of demand for the product, at least for now, since the production is still relatively small (4000 tons in Thailand, 136 tons in Hawaii, 3500 tons in Taiwan). But there is considerable variation between these areas in initial investment and production costs, depending on local conditions, which affects the profitability of the operations.

Shang (1982) compared the average costs and returns of different size prawn farms in Hawaii and Thailand, based on data collected under a UNDP/FAO aid project in Thailand

(see Table 24.8). The data clearly show that the initial costs in starting a farm in Hawaii are 2.5−8.9 times higher than in Thailand. Similarly, the annual operating cost is about 5.6 times higher than in Thailand, even though farmers obtained post-larvae at subsidized prices. The cost of production per kg of prawn is much higher in Hawaii (2.3−2.9 times), even though the yield per ha is more than double. Obviously the cost of inputs in Thailand is lower. As the selling price of prawns is not very different, the profit derived by farmers in Thailand (for all sizes of farms) is much higher.

Shang's data seem to suggest that under conditions in Hawaii, only larger farms can be profitable, whereas large farms (more than 4 ha in area) are less profitable in Thailand. Though it is not possible to make definite conclusions from this, the likely differences in economic performance in relation to farm size and input levels (represented by intensity of farming), should be taken into consideration when large-scale farms are planned in developing countries.

24.9   References

AQUACOP (1977a) *Macrobrachium rosenbergii* (De Man) culture in Polynesia: progress in developing a mass intensive larval rearing technique in clear water. *Proc. World Maricul. Soc.*, **8**, 311−19.

AQUACOP (1977b) Observations on diseases of crustacean cultures in Polynesia. *Proc. World Maricul. Soc.*, **8**, 685−703.

AQUACOP (1985) Overview of Penaeid culture research: impact on commercial culture activity. In *Proceedings First International Conference on the Culture of Penaeid Prawns/Shrimps*, (Ed. by Y. Taki, J.H. Primavera and J.A. Llobrera), pp. 3−10. Aquaculture Department, SEAFDEC, Iloilo City, Philippines.

ASEAN (1978) *Manual on Pond Culture of Penaeid Shrimp*. ASEAN National Coordinating Agency of the Philippines, Manila.

Behrends L.L. *et al.*, (1985) Polyculture of fresh-water prawns, tilapia, channel catfish and Chinese carps. *J. World. Maricul. Soc.*, **16**, 437−50.

Brick. R.W. and Stickney R.R. (1979) Polyculture of *Tilapia aurea* and *Macrobrachium rosenbergii* in Texas. *Proc. World Maricul. Soc.*, **10**, 222−8.

Brock J.A. (1983) Diseases (infectious and non-infectious), metazoan parasites, predators, and public health considerations in *Macrobrachium* culture and fisheries. In *Handbook of Mariculture*,

Vol. I, Crustacean Aquaculture, (Ed. by J.P. McVey), pp. 329—70. CRC Press, Boca Raton.

Caillouet A.C. Jr (1972) Ovarian maturation induced by eyestalk ablation in pink shrimp, *Penaeus duorarum* Burkenroad. *Proc. World Maricul. Soc.*, **3**, 205—25.

Chamberlain G.W., Hutchins D.L. and Lawrence A.L. (1981) Mono- and polyculture of *Penaeus vannamei* and *P. stylirostris* in ponds. *J. World Maricul. Soc.*, **12**(1), 251—70.

Chao N.H. (1979) Freshwater prawn farming in Taiwan — the patterns, problems and prospects. *Proc. World Maricul. Soc.*, **10**, 51—67.

Chiang P. and Liao I.C. (1985) The practice of grass prawn (*Penaeus monodon*) culture in Taiwan from 1968 to 1984. *J. World Maricul. Soc.*, **16**, 297—315.

Chow K.W. (1984) *Artificial Diets for Seabass, Macrobrachium and Tiger Shrimp*. Consultant's report, MAL/79/018, FAO, Rome.

Corbin J.S., Fujimoto M.M. and Iwai T.Y. Jr (1983) Feeding practices and nutritional considerations for *Macrobrachium rosenbergii* culture in Hawaii. In *Handbook of Mariculture, Vol. I, Crustacean Aquaculture*, (Ed. by J.P. McVey), pp. 391—412. CRC Press, Boca Raton.

Griffin W., Lawrence A. and Johns M. (1985) Economics of Penaeid culture in the Americas. In *Proceedings First International Conference on the Culture of Penaeid Prawns/Shrimps*, (Ed. by Y. Taki, J.H. Primavera and J.A. Llobrera), pp. 151—60. SEAFDEC Aquaculture Department, Iloilo City, Philippines.

Hale M.B. and Waters M.E. (1981) Frozen storage stability of whole and headless freshwater prawns, *Macrobrachium rosenbergii*. *Marine Fish. Rev.*, **43**, 18—21.

Hameed Ali K., Dwivedi S.N. and Alikunhi K.H. (1982) A new hatchery system for commercial rearing of penaeid prawn larvae, *Bull. Central Inst. Fish. Education, Bombay*, **2**—3(82).

Hanson J.A. and Goodwin H.L. (Eds) (1977) *Shrimp and Prawn Farming in the Western Hemisphere*. Dowden Hutchinson and Ross, Stroudsburg.

Hirasawa Y. (1985) Economics of shrimp culture in Asia. In *Proceedings First International Conference on the Culture of Penaeid Prawns/Shrimps*, (Ed. by Y. Taki, J.H. Primavera and J.A. Llobrera), pp. 131—50. SAEFDEC Aquaculture Department, Iloilo City, Philippines.

Hudinaga M. (1942) Reproduction, development and rearing of *Penaeus japonicus* Bate. *Japan J. Zool.*, **10**(2), 305—93.

Huner J.V. *et al.* (1983) Interactions of freshwater prawns, channel catfish fingerlings and crayfish in earthen ponds. *Progr. Fish-Culturist.*, **45**(1), 36—40.

Johnson S.K. (1978) *Handbook of Shrimp Diseases*. Texas Agricultural Extension Service, Texas A and M University, College Station.

Kungvankij P. (1985) Overview of Penaeid shrimp culture in Asia. In *Proceedings First International Conference on Culture of Penaeid Prawns/Shrimps*, (Ed. by Y. Taki, J.H. Primavera and J.A. Llobrera), pp. 11—21. SEAFDEC Aquaculture Department, Iloilo City, Philippines.

Liao I.C. (1985) Brief review of the larval rearing techniques of Penaeid prawns. In *Proceedings First International Conference on culture of Penaeid Prawns/Shrimps*, (Ed. by Y. Taki, J.H. Primavera and J.A. Llobrera), Iloilo City, Philippines.

Liao I.C. and Chao N.H. (1982) Progress of *Macrobrachium* farming and its extension in Taiwan. In *Giant Prawn Farming*, (Ed. by M.B. New), pp. 357—9. Elsevier Scientific Publishing, Amsterdam.

Liao I.C., Su H.M. and Lin J.H. (1983) Larval foods for prawns. In *Handbook of Mariculture, Vol. I, Crustacean Aquaculture*. (Ed. by J.P. McVey) pp. 43—69. CRC Press, Boca Raton.

Lightner D.V. (1983) Diseases of cultured Penaeid shrimp. In *Handbook of Mariculture, Vol. I, Crustacean Aquaculture*, (Ed. by J.P. McVey), pp. 289—320. CRC Press, Boca Raton.

Lightner D.V. (1985) A review of the diseases of cultured Penaeid shrimps and prawns with emphasis on recent discoveries and developments. In *Proceedings First International Conference on the Culture of Penaeid Prawns/Shrimps*, (Ed. by Y. Taki, J.H. Primavera and J.A. Llobrera), pp. 79—103. SEAFDEC Aquaculture Department, Iloilo City, Philippines.

Lim L.C., Heng H.H. and Cheong L. (1987) *Manual on Breeding of Banana Prawn*. Fisheries Handbook, No. 3, Primary Production Department, Singapore.

Ling S.W. (1969) The general biology and development of *Macrobrachium rosenbergii*, *FAO Fish. Rep.*, **57**(3), 589—606.

Ling S.W. and Costello T.J. (1979) The culture of freshwater prawns: a review. In *Advances in Aquaculture*, (Ed. by T.V.R. Pillay and W.A. Dill), pp. 299—305. Fishing News Books, Oxford.

Lumare F. (1979) Reproduction of *Penaeus kerathurus* using eye-stalk ablation. *Aquaculture*, **18**, 203—14.

McVey J.P. (Ed.) (1983) *Handbook of Mariculture, Vol. 1, Crustacean Aquaculture*. CRC Press, Boca Raton.

Malecha S.R. (1983a) Commercial seed production of the freshwater prawn, *Macrobrachium rosenbergii*, in Hawaii. In *Handbook of Mariculture, Vol. 1, Crustacean Aquaculture*, (Ed. by J.P.

McVey), pp. 205–30. CRC Press, Boca Raton.

Malecha S.R. (1983b) Commercial pond production of the freshwater prawn, *Macrobrachium rosenbergii*, in Hawaii. In *Handbook of Mariculture, Vol. 1, Crustacean Aquaculture*, (Ed. by J.P. McVey), pp. 231–59. CRC Press, Boca Raton.

Malecha S.R. *et al.* (1981) Polyculture of the freshwater prawn, *Macrobrachium rosenbergii*, Chinese and common carps in ponds enriched with swine manure I. Initial trials. *Aquaculture*, **25**, 101–16.

Menasveta P. and Piyatitivokul S. (1982) Effects of different culture systems on growth, survival and reproduction of the giant freshwater prawn (*Macrobrachium rosenbergii* De Man). In *Giant Prawn Farming*, (Ed. by M.B. New), pp. 175–89. Elsevier Scientific Publishing, Amsterdam.

Menon M.K. (1955) On the paddy-field prawn fishery of Travancore-Cochin and an experiment in prawn culture. *Proc. Indo-Pac. Fish. Council.*, **5**(II and III), 131–5.

New M.B. (1987) *Feed and Feeding of Fish and Shrimp — A Manual on the Preparation and Presentation of Compound Feeds for Shrimp and Fish in Aquaculture.* FAO of the UN, Rome ADCP/REP/87/26.

New M.B. and Singholka S. (1985) Freshwater prawn farming. A manual for the culture of *Macrobrachium rosenbergii. FAO Fish. Tech. Paper*, **225** (Rev. 1).

Nip W.K. and Moy J.H. (1979) Effect of freezing methods on the quality of the prawn, *Macrobrachium rosenbergii. Proc. World Maricul. Soc.*, **10**, 761–8.

Primavera J.H. (1985) A review of maturation and reproduction in closed thelycum Penaeids. In *Proceedings First International Conference on the Culture of Penaeid Prawns/Shrimps*, (Ed. by Y. Taki, J.H. Primavera and J.A. Llobrera), pp. 47–64. SEAFDEC Aquaculture Department, Iloilo City, Philippines.

Primavera J.H., Young T. and de los Reyes C. (1982) Survival, maturation, fecundity and hatching rates of unablated and ablated *Penaeus indicus* H.M. Edwards from brackishwater ponds. *Proc. Symp. Coastal Aquaculture*, **1**, 48–54.

Radhakrishnan E.V. and Vijayakumaran M. (1982) Unprecedented growth induced in spiny lobsters. *Mar. Fish. Infor. Serv. T. E. Ser.*, **43**, 6–8.

Rao P.V. *et al.* (Eds) (1983) *Proceedings of the National Symposium on Shrimp Seed Production and Hatchery Management, Cochin.* The Marine Products Export Development Authority, Cochin.

Roberts K.J. and Bauer L.L. (1978) Costs and returns for *Macrobrachium* grow-out in South Carolina, U.S.A. *Aquaculture*, **15**, 383–90.

Sandifer P.A. *et al.* (1983) Seasonal culture of fresh-water prawns in South Carolina. In *Handbook of Mariculture, Vol. 1, Crustacean Aquaculture*, (Ed. by J.P. McVey), pp. 189–204. CRC Press, Boca Raton.

Shang Y.C. (1972) *Economic Feasibility of Fresh Water Prawn Farming in Hawaii.* University of Hawaii, Honolulu.

Shang Y.C. (1981) *Freshwater Prawn (*Macrobrachium rosenbergii*) Production in Hawaii: Practices and Economics.* Department of Land and Natural Resources, and University of Hawaii Sea Grant College Programme, Honolulu.

Shang Y.C. (1982) Comparison of freshwater prawn farming in Hawaii and in Thailand: culture practices and economics. *J. World Maricul. Soc.*, **13**, 113–19.

Shang Y.C. (1983) *The Economics of Marine Shrimp Farming: A Survey.* Presented at the Brigham Young University Conference.

Shang Y.C. and Mark C.R. (1982) The current state-of-the-art of freshwater prawn farming in Hawaii. In *Giant Prawn Farming*, (Ed. by M.B. New), pp. 351–6. Elsevier Scientific Publishing, Amsterdam.

Shigueno K. (1975) *Shrimp Culture in Japan.* Association for International Technical Promotion, Tokyo.

Silas E.G. *et al.* (1985) Hatchery production of Penaeid prawn seed: *Penaeus indicus. CMFRI Special Publication*, No. 23.

Sindermann C.L. (1974) *Diagnosis and Control of Mariculture Diseases in the United States.* Technical Series Report, No. 2, Middle Atlantic Coast Fisheries Center, Highlands, New Jersey.

Sindermann C.J. (Ed.) (1977) *Disease Diagnosis and Control in North American Marine Aquaculture.* Elsevier, New York.

Smith T.I.J. and Sandifer P.A. (1979) Observations on the behaviour of the Malaysian prawn, *Macrobrachium rosenbergii* (de Man), to artificial habitats. *Mar. Behav. Physiol*, **6**, 131.

Taki Y., Primavera J.H. and Llobrera J.A. (Eds) (1985) *Proceedings of the First International Conference on the Culture of Penaeid Prawns/Shrimps*, (Ed. by Y. Taki, Y.H. Primavera and Y.A. Llobrera). SEAFDEC Aquaculture Department, Iloilo City, Philippines.

Uno Y. (1985) An ecological approach to mariculture of shrimp: shrimp ranching fisheries. In *Proceedings of the First International Conference on the Culture of Penaeid Prawns/Shrimps*, (Ed. by Y. Taki, J.H. Primavera and J.A. Llobrera), pp. 37–45. SEAFDEC Aquaculture Department, Iloilo City, Philippines.

Williamson M.R. and Wang J.K. (1982) An improved harvesting net for freshwater prawns. *Aquacul. Eng.*, **1**, 81–91.

25
Crayfishes and Crabs

High consumer demand and market prices have led aquaculturists to devote considerable attention to the culture of crayfishes (craw fishes), lobsters and crabs. However, it is only the crayfishes that presently accounts for any significant production through culture. Some small-scale production of crabs is also reported from tropical countries. For a number of years, scientists have been involved in investigations on the possibilities of culturing the homarid lobsters, *Homarus americanus* and *H. gammarus*. Preliminary efforts have also been made in rearing the post-larval stages of the spiny lobsters, *Panulirus* spp. and *Jasus* spp. The protracted larval development, the nature of the food required by the different stages of the phyllosoma larvae, the long time the juveniles take to grow to market size and pronounced cannibalism at both larval and adult stages, have made available culture technologies uneconomical for commercial culture. Efforts are presently directed towards large-scale production of juveniles for stocking protected open waters (see Chapter 30).

25.1 Crayfishes

Crayfishes belonging to the families Cambaridae and Astacidae are widely distributed over all the continents, including Africa (where they have been introduced in Uganda and Kenya), and are highly priced in several countries of Europe. Besides being a delicacy, small crayfish are also in demand as bait for anglers in the USA. Among the 300 or so species of crayfish, only four appear to have been used in some form of aquaculture. The most important among these is the red swamp crayfish *Procambarus clarkii* (belonging to the family Cambaridae (fig. 25.1). The other is the signal crayfish, *Pacifastacus leniusculus* (belonging to the family Astacidae) (fig. 25.2). The red swamp crayfish cultured in the southern USA is reported to form at least 85 per cent of all crayfish raised in the country. The other species cultured in the USA are the white river crayfish *P. acutus* and the paper-shell crayfish *Oreonectes immunis*. *Pacifastacus leniusculus* is not yet cultured on a commercial scale under confinement, but has been introduced into Europe from the USA to replace stocks of the European noble crayfish *Astacus astacus*, which were decimated by the crayfish 'plague' caused by the fungus *Aphanomyces astaci*. As *P. leniusculus* is immune to the disease, it has established itself in Scandinavian lakes (which provides favourable environmental conditions). It has also found a ready market, as it was already known in European markets through regular imports from the USA.

Small-scale farms do exist in France for growing the European crayfish, and hatcheries produce juvenile *P. leniusculus* to stock the lakes in Scandinavian countries. In the cold northern waters, the noble crayfish mature when 4−5 years old. Both the signal and noble crayfish mate in the autumn and the eggs hatch out the following spring. By rearing in warm water at 13−16°C, the hatchlings grow to a length of 3−4 cm in less than half a year. They are generally stocked in lakes at this stage, but in lakes with high densities of predatory species stocking with larger juveniles is preferred. Experimental culture of this species to adult size in tanks has also been attempted. The optimal temperature of water in the tanks is 13−16°C,

Fig. 25.1 The red swamp crayfish, *Procambarus clarkii* (right) and the river crayfish, *P. acutus* (left) (from *Fish Farming International*, **2**(3)).

Fig. 25.2 The signal crayfish, *Pacifastacus leniusculus* (from *Fish Farming International*, **4**(2)).

and the food used is a mixture of fresh fish (herring), liver and chalk. The main cause of mortality in culture tanks is cannibalism of the newly-moulted soft animals by the hard-shelled ones.

25.1.1 Culture of *Procambarus clarkii*

As indicated above, it is only *P. clarkii* that are actually farmed on a large scale. In nature they are abundant in open waters during wet periods, but the rest of the time they shelter in burrows in swampy and marshy areas. The crayfish farming practices in the southern USA, particularly in Louisiana, are adapted to the long wet period from autumn to spring when the animal is active. Although crayfish can withstand a wide range of environmental conditions, they thrive best under water temperatures of 20–25°C, 3 ppm dissolved oxygen, salinity of less than 5 ppt, pH of 6.5–8.5 and less than 1 ppm of total ammonia (Avault and Huner, 1985).

Pond culture methods

Procambarus clarkii is farmed in rice fields as a rotation crop or on a continuous basis in ordinary ponds. For growing crayfish, the levees of rice fields are raised to about 50 cm. The recommended size of ponds is about 8 ha for easy management, but some consider a size of 12–16 ha optimal. Crayfish ponds are very similar to those used in finfish culture, except that the levees need not be higher than 50–75 cm. Anti-seap collars for drain pipes are recommended to prevent crayfish and rodents from burrowing and causing leaks. It is useful to have baffle-levees in the pond to facilitate water circulation. Wooded and marshy areas can be embanked with a ring levee for the crayfish to establish themselves naturally, with some supplemental stocking if required.

Depending on the environmental conditions, it takes 3–9 months and a minimum of eleven moults for a crayfish to mature. Brood animals are stocked from April to June at the rate of 20–65 kg per ha, after the pond has been filled to a depth of at least 30 cm. A sex ratio of 1:1 is maintained. After about two weeks the water is slowly drained, which may take another two

weeks to complete. Mating may take place during this period or sometimes even before stocking. Unlike other crayfish which reproduce only once a year, *P. clarkii* reproduces throughout the year. The male transfers sperm to a seminal receptacle between the female's walking legs, where they remain viable for up to 6 months. Eggs are extruded from oviducts and are fertilized by the sperm. A sticky substance that is excreted by glands on the ventral side of the abdominal segments, known as glair, helps to keep the fertilized eggs in place on the paired abdominal appendages (swimmerets). As the water is drained, the crayfish burrows into the levees, and occasionally in exposed portions of the pond bottom. The burrows are sealed with clay plugs to prevent loss of water by evaporation and to protect against predators. Some of the females may lay eggs only after burrowing. The number of eggs varies with the size of the animal, but usually ranges between 100 eggs for a 7.5–8.5 cm crayfish and 600–700 eggs for larger ones of about 12.5 cm length. The eggs develop in the humid burrows and, after hatching, the larvae undergo two moults in the next two weeks. After the second moult they are able to fend for themselves.

Ponds are flooded in about September or October and the rising waters entice the females to release the young and to come out of the burrows. Crayfish are omnivorous, but the bulk of their diet consists of microbially-enriched detritus. Vascular plants and epiphytic growths also form highly relished food items. Animal matter, such as worms, insect larvae, molluscs and zooplankton, are specially important food for juvenile crayfish. Though many farmers depend on natural vegetation, including grasses and sedges, in the pond as food for crayfish, it is beneficial to plant rice or millet in the ponds after they are drained in May or June. The decomposing straw covered with micro-organisms, including fungi and bacteria, forms a greater source of nutrients than the green rice plants on which the crayfish feed. By about March, the straw will be depleted and if no additional source of food is provided growth will be affected and stunting of the population may occur. By using additional substrates, such as hay, it has been possible to prevent this and continue the detrital food chain, obtaining yields of up to 4000 kg/ha in stagnant ponds.

An alternative is to use commercial feeds after the rice-derived detritus is exhausted.

Unlike the European crayfish, the red swamp crayfish appears to be free from any major disease problems. Two bacterial diseases have been encountered (Ambroski *et al.*, 1975a,b) in red swamp crayfish, but they do not seem to cause any harm in pond-raised stocks. Cracked or broken parts of the exoskeleton may be attacked by chitinovorous bacteria, and this may enable other lethal bacteria in the environment to invade the body and uropods. If erosion has not reached the cuticle, moulting will eliminate the affected part. Bacteria producing pathogenic endotoxins occur in gut flora, but do not cause any major problems in low-density cultures. Under crowded conditions in a nutrient-rich environment, blooms of *Flavobacterium* sp. can occur and the accumulated endotoxins can cause mortality. Besides these, protozoan epibionts can interfere with respiration if dense growths occur on the gills. Spent females emerging from the burrows often suffer from 'hollow tail' or 'wasting disease', which is probably because they have used up their body reserves during life in the burrows. According to de La Bretonne (1977) these females moult and develop normally succulent tail meat.

Harvesting is usually done with baited wire-mesh traps or with lift nets from boats. Two types of traps are used: a pillow trap, which has one or more funnel openings and is laid submerged on the pond bottom and the stand-up trap which has two or more funnels at the bottom, with the top open and reaching above the water surface. A metal strip is placed around the inside of the trap to prevent the crayfish from climbing out. Because of the cost and difficulties in obtaining bait, most farmers trap only for about 100 days during the harvest season. Since trapping is labour-intensive and not too efficient, several improved harvesting techniques have been developed, such as a 'crayfish combine' and electro-trawls. Avault and Huner (1985) described the equipment and harvesting methods. The crayfish combine or boat has a motor-powered metal wheel mounted on one end, guided by foot wheels, which digs into the pond bottom to propel the boat. One man can operate the boat and at the same time remove the catch from traps and rebait them. The electric trawl uses an electrical current to guide the crayfish into a surface trawl.

Most of the crayfish harvested are sold alive, but about 30 per cent are processed. They are killed by immersion in boiling water and then the tails are hand peeled for sale in the fresh state. When the product has to be frozen and stored before sale, the hepatopancreas, or what is called the 'fat', is removed.

Double cropping

Following more or less the same techniques as in pond culture, the red swamp crayfish can be produced in rice fields in a double-cropping system in rotation with rice. Crayfish are stocked in the fields in May. When the fields are drained about two weeks before harvest in about August, the crayfish are forced to burrow into the levees and this facilitates the rice harvest without affecting the crayfish. Four to six weeks after the rice harvest, the fields are reflooded. Crayfish then come out of the burrows and release their young. The rice stubbles left in the field give rise to ratoon growths of green leaves and form the substrate for the growth of periphyton. These, as well as detritus in the field, are fed on by the crayfish. Harvesting is carried out from October to May by means of traps. Many farmers practise a triple-crop rotation, including also a crop of soybeans after the crayfish. One of the main problems with this type of farming is the use of pesticides in rice and soybean growing (see Chapter 29).

Techniques of intensive culture of crayfish have been developed, and are very similar to those used for culture of other crustaceans. However, the economics of intensive culture of crayfish are not considered attractive at present under conditions in the USA. There is an established market in the USA for soft-shelled crayfish as bait, and recently a market has developed for large soft-shelled crayfish weighing about 10–30 g. Harvested hard-shelled crayfish can be caught at the pre-moult stage and held in tanks until they moult. Individuals which do not moult readily can be induced to do so by the ablation of both eyestalks at their bases. Another means is to catch the crayfish immediately after they moult, using an active fishing method such as a seine or an electric trawl.

The extensive type of farming as practised in

Louisiana, which involves minimum investment and is generally a subsidiary activity of farmers, has proved to be profitable. Competitive returns have been obtained by skilled farmers, and with the introduction of more cost-effective harvesting methods the returns are expected to increase.

25.2 Crabs

The feasibility of culturing crabs (family Brachyura) has received attention in some countries because of high market demand and decreasing availability. In recognition of the problems of intensive culture of crabs, which exhibit pronounced cannibalism during the larval and adult stages, some of the early attempts at propagating the economically important species were for the purpose of stocking open waters with larvae.

A more profitable operation that has been in existence for over two centuries in the USA is the production of soft-shell crabs popularly known as 'shedding crabs', on a commercial scale. Pre-moult blue crabs (*Callinectes sapidus*) captured from wild stocks are held until they moult and the moulted crabs are sold as soft-shell crabs, which fetch higher prices in the market. Wild-caught peeler crabs (premoult crabs) which can be distinguished by colour changes associated with the formation of the new shell, are held for shedding in either a floating box or on land-based tables with a flow-through or recirculating water supply. The peelers are generally sorted according to colour (which indicates the pre-moult stage) and kept in separate floats or tanks. This process of segregation according to colour stage is continued at intervals of four to six hours. Crabs which have shed the shells are allowed to remain in water until they have expanded to their full size, after which they are removed to stop further hardening of the shell.

25.2.1 Culture of *Scylla serrata*

Opinions may differ on whether shedding crabs can be considered a form of aquaculture, but apart from this the only commercial-scale culture of crabs appears to be that of the serrated crab or mud crab *Scylla serrata* (fig. 25.3), in some of the Asian countries.

In most coastal fish ponds and impoundments in Asia, the serrated crab forms a subsidiary crop. The young enter with the tidal water and grow there for a period of about 6 months, when they reach market size. This species is highly priced in local markets, particularly the females with well-developed gonads. Mating and spawning take place in ponds, but the embryonic development and hatching appear to be retarded by low salinities. Even though the holes that crabs make weaken the earthen dikes of coastal ponds, farmers are compensated by the high price they obtain for the crab harvest.

Scylla serrata is reported to spawn throughout the year, with the peak season from May to September. According to Ong (1966) it attains sexual maturity when about 11 months old (carapace width of about 114 mm) under laboratory conditions, but in nature this probably takes place earlier, as observed by Arriola (1940), about 5 months after hatching. Pair formation can be observed a few days before the pre-copulatory moult, and a few hours after the moulting mating occurs. The spermatophores deposited in the spermatheca of the female remain viable for months. The time of actual spawning of the eggs varies, but three spawns in a period of 5 months have been observed, without further moulting and mating. About 2 million eggs may be found attached to the pleopods of a female, but it is suspected that up to half of them fail to attach to the setae. The larvae hatch as planktonic zoeae and, after passing through a number of zoea stages and a megalopa stage, they metamorphose into juveniles which take on a benthic life. Brine-shrimp nauplius has been used as larval food, but does not seem to be the best, as survival rates are low.

Though the techniques of controlled propagation have been demonstrated in laboratories, crab farmers have depended on wild-caught juveniles for seed stock. In the Philippines, juveniles measuring about 2 cm in carapace length are collected by means of bamboo traps, lift nets or scissor nets. Some farmers buy the seed stock from professional seed collectors. They are stocked in milkfish ponds at a low rate of 500–1000 per ha, and repeated stocking and harvesting are practised. In ponds they feed on natural food such as algae, crustaceans

Fig. 25.3 Mud crab, *Scylla serrata* (photograph: H.R. Rabanal).

and other animal matter. Some farmers feed them with trash fish or other available animal products, at the rate of about 5−7 per cent of body weight. Before moulting, the crab buries itself in the mud or in holes until its shell becomes hard. Juveniles reach a harvesting size of 11−14 cm carapace width in about 6 months. Selective harvesting is carried out with traps and any small crabs caught are released back for further growth.

Taiwanese farmers grow the serrated crab in monoculture or in polyculture with milkfish or the seaweed *Gracilaria*. For monoculture, specially designed tidal ponds with a sandy bottom and a supply of salt water of about 15−30 ppt salinity are used. Ponds with dikes of about 1 m height, made of concrete or bricks with extended crowns to prevent the escape of crabs are preferred. As only a small number of juveniles are stocked at a time, it is more convenient to have a large number of small ponds (about 350 m^2) or larger square ponds divided into four smaller square ponds by means of a central tank (1.5 m^2). This central tank is connected by means of water gates with each of the four sections, and serves as both a water tank and a catching tank, very much like the catching ponds in milkfish farms. If the

ponds have earthern dikes, bamboo screens are placed obliquely towards the inside of the pond to prevent escape of the stock.

Monoculture of crabs in Taiwan is essentially to fatten them to a stage of full gonadal development. The rearing period is therefore short and generally only one to two months. Mated female crabs are stocked between April and September, at the rate of three crabs (about 7−12 cm carapace width) per m^2 water surface. They are fed on soft-shelled snails, trash fish, fish offal and other animal matter. Snails are considered to be the preferred food for maturing crabs. The usual rate of feeding is about 5 per cent of body weight daily, and feeding is usually done after dark.

Harvesting can be done with dip nets baited with trash fish. Plastic or concrete pipes placed on the pond bottom are used as shelter by crabs, and when the pipes are lifted at intervals the crabs can be caught. The most efficient means of harvesting is by making use of their habit of congregating in the central tank when water is let in. It is easy to catch them from the tank with dip nets. Only crabs with fully developed ovaries are removed. For transportation to market, each individual live crab is bound with a wetted heavy straw rope to facilitate handling and also to keep the animal moist.

For polyculture in brackish-water ponds with milkfish, shrimp or seaweeds, larger ponds (0.5–2.0 ha) are used, but oblique fences of bamboo or plastic have to be erected on the dikes to prevent escape of the crabs. Male and female juveniles (1.5–3 cm carapace width) are stocked at rates not exceeding 10 000/ha and are fed the same type of food as in mono-culture. Adequate feeding is reported to reduce cannibalism. The crabs reach market size in about 6 months with a survival rate of 50–70 per cent, depending on the size of juveniles stocked. In certain parts of Taiwan the serrated crab is often reared in rice fields together with shrimps, in rotation with rice after its harvest.

25.2.2 Propagation of crabs in Japan

Seed stock of the Japanese blue crab *Neptunus pelagicus* is regularly produced in hatcheries for stocking open waters. As with other species of crabs, cannibalism has been a major problem in the grow-out of this species to adult size. The provision of various types of shelters on the pond bottom has been tried with some success in improving survival rates, but com-mercial culture has not yet become successful. The larval rearing of this species, as well as of *Portunus trituberculatus*, is carried out in hatcheries. Juveniles have been experimentally raised to adult size in 8–10 months using fresh fish and shellfish as food. But the main interest has been in growing them just to the juvenile stage for stocking the open sea. Berried females collected from the wild are kept in tanks filled with natural sea water of about 33–34 ppt salinity. The hatched early larvae are fed on cultured marine *Chlorella*, but later larval stages are fed on *Artemia* nauplii. In about 25 days after hatching the metamorphosis is com-plete. The juveniles are fed on fresh fish, such as anchovies, and after a further three weeks of rearing are stocked in coastal areas.

The king crab, *Paralithodes camtschatica*, has also been successfully spawned and the larvae reared in the laboratory. As in the case of other crabs, egg-bearing females are used for propagation. The eggs hatch out in about a week at temperatures below 10°C. The zoeae are reared in nylon net cages in running water and fed on *Artemia* nauplii supplemented with shrimp juice, clams and the brown seaweed *Laminaria*. Net cages are comparatively more efficient for rearing post-larvae, but the survival rate has been very low at this stage, probably due to cannibalism and the unsuitability of the food used.

25.3 References

Abrahamsson S. (1972a) Fecundity and growth of some populations of *Astacus astacus* Linné in Sweden, with special regard to introductions in Northern Sweden. *Rep. Inst. Freshwater Res. Drottingholm*, **52**, 23–37.

Abrahamsson S. (1972b) Methods for restoration of crayfish waters in Europe. The development of an industry for production of young of *Pacifastacus leniusculus*. In *Freshwater Crayfish*, (Ed. by S. Abrahamsson). Studentilitteratur Lund, Austria.

Allen P.G. and Johnston W.E. (1976) Research di-rection and economic feasibility. An example of systems analysis for lobster aquaculture. *Aqua-culture*, **9**, 155–80.

Ambroski R.L., Glorioso J.C. and Ambroski G.F. (1975a) Common potential pathogens of crayfish, frogs and fish. *Int. Symp. Freshwater Crayfish*, **2**, 317–26.

Ambroski R.L. *et al.* (1975b). A disease affecting the shell and soft tissues of Louisiana crayfish, *Procambarus clarkii*. *Int. Symp. Freshwater Cray-fish*, **2**, 299–316.

Arrignon J. (1981) *L'ecrevisse et son Elevuge*. Gauthier-villars, Bordas, Paris.

Arriola F.J. (1940) A preliminary study of the life history of *Scylla serrata* Forskal. *Philipp. J. Sci.*, **73**(4), 437–56.

Avault J.W., de la Bretonne L.W. and Huner J.V (1974) Two major problems in culturing crayfish in ponds: oxygen depletion and overcrowding. *Proc. Second International Crayfish Symposium*, 139–144.

Avault J.W. and Huner J.V. (1985) Crawfish culture in the United States. In *Crustacean and Mollusc Aquaculture in the United States*, (Ed. by J.V. Huner and E.E. Brown), pp. 1–61. Avi Publishing Company, Westport.

Cabantous M.A. (1975) Introduction and rearing of *Pacifastacus* at the research center of Les Cloui-zious 18450 Brinon S/Saudre France. *Int. Symp. Freshwater Crayfish*, **2**, 49–56.

Chen T.P. (1976) *Aquaculture Practices in Taiwan*, pp. 123–8. Fishing News Books, Oxford.

Clark D.F., Avault J.W. and Meyers S.P. (1974) Effects of feeding, fertilization and vegetation on production of red swamp crayfish *Procambarus clarkii. Proc. Second International Crayfish Symposium*, 125–38.

Cobb J.S. and Phillips B.J. (Eds) (1980) *The Biology and Management of Lobsters*, Vols I and II. Academic Press. New York.

de la Bretonne L. (1977) A review of crawfish culture in Louisiana. *Proc. World Maricul. Soc.*, **8**, pp. 265–9.

Escritor G.L. (1972) Observations on the culture of the mud crab, *Scylla serrata*. In *Coastal Aquaculture in the Indo-Pacific Region*, (Ed. by T.V.R. Pillay), pp. 355–61, Fishing News Books, Oxford.

Huner J.V. and Avault J.W. (1981) *Producing Crawfish for Fish Bait*. Sea Grant Publ. No. LSU-T1–76001. Center for Wetland Resources, Louisiana State University, Baton Rouge.

Huner J.V. and Lindquist O.V. (1987) Freshwater crayfish culture in Finland. *Aquaculture Magazine*, **13**(1), 22–5.

Lowery R.S. and Mendes A.J. (1977) *Procambarus clarkii* in Lake Naivasha, Kenya, and its effects on established and potential fisheries. *Aquaculture*, **11**, 111–21.

Nakanishi T. (1979) Rearing larvae and post-larvae of the king crab (*Paralithodes camtschatica*). In *Advances in Aquaculture*, (Ed. by T.V.R. Pillay and W.A. Dill), pp. 319–21. Fishing News Books, Oxford.

Ong K.S. (1966a) The early developmental stages of *Scylla serrata* Forskal (Crustacea: Portunidae) reared in the laboratory. *Proc. Indo-Pac. Fish. Coun.*, **11**(II), 135–46.

Ong K.S. (1966b) Observations on the post larval history of *Scylla serrata* Forskal reared in the laboratory. *Malays. agric. J.*, **45**(4), 429–43.

Pagcatipunan R. (1972) Observations on the culture of alimango, *Scylla serrata* at Camarines Norte (Philippines). In *Coastal Aquaculture in the Indo-Pacific Region*, (Ed. by T.V.R. Pillay), pp. 362–5. Fishing News Books, Oxford.

Ting R.Y. (1973) Culture potential of spiny lobster. *Proc. World Maricul. Soc.*, **4**, 165–70.

Varikul V., Phumiphol S. and Hongpromyart M. (1972) Preliminary experiments in pond rearing and some biological studies of *Scylla serrata* (Forskal). In *Coastal Aquaculture in the Indo-Pacific Region*, (Ed. by T.V.R. Pillay), pp. 366–74. Fishing News Books, Oxford.

Westman K. (1973a) Cultivation of the American crayfish *Pacifastacus leniusculus*. In *Freshwater Crayfish*, (Ed. by S. Abrahamsson, pp. 211–20, Studentilitteratur Lund.

Westman K. (1973b) The population of the crayfish *Astacus astacus* in Finland and the introduction of the American crayfish *Pacifastacus leniusculus* Dana. In *Freshwater Crayfish*, (Ed. by S. Abrahamsson), pp. 41–56; Lund.

26
Oysters and Mussels

Aquaculture of molluscs, especially of bivalve molluscs, is unique from several points of view. It is one of the earliest forms of marine aquaculture practised in the western hemisphere, starting from early Roman times. Being sessile and low trophic level filter-feeders for most of their lives, molluscs can be raised at relatively low cost. Culture is carried out mainly in open waters, often in their natural habitats with young ones (spat) collected from the very same area. Historically, culture has contributed substantially in the maintenance of capture fisheries for bivalves like oysters, and it is for the enhancement of oyster populations that fisherman have readily taken to aquatic farming methods. Despite favourable climatic conditions, the availability of inexpensive labour and the need to produce protein food at low cost in tropical countries, most of the bivalve culture has developed in sub-tropical and temperate climates, as in the USA, Japan, Korea, France, Spain, the Netherlands and Italy, where they form high-priced luxury foods.

Molluscs account for over 35 per cent of the total aquaculture production, but in recent years the global rate of increase has declined significantly. This could be due to pollution in growing areas, limitations of domestic markets and stringent import regulations affecting some of the major exporting countries. Since some of the bivalves, like oysters, are eaten raw, and their filter-feeding habit can result in the accumulation of contaminants, they become first suspects in times of epidemics of human intestinal disorders. The occurrence of red tide can severely damage all bivalve production and cause considerable economic loss to the industry. One other unique feature of molluscan aquaculture is that it involves transfer of juveniles from areas of high spat-fall to areas of good growth. Similarly, there have been several introductions and transplantations of exotic species, especially oysters, for enhancing or replacing stocks, and it is believed that these account for the occurrence of a number of exotic diseases among molluscs.

There are several species of bivalves and a smaller number of gastropods which are cultivated. Among these the more important are the oysters (family Ostreidae), mussels (family Mytilidae and Aviculidae), clams (family Mercenaridae), scallops (Pectenidae), the abalone (family Haliotidae) and the cockles (family Arcidae). The group that accounts for the largest production of molluscs through aquaculture is the oysters and several species are cultured in many parts of the world. The culture of mussels and clams is more restricted, but the production of mussels in some countries like Spain has reached high levels, which give rise to marketing problems. Other species are cultured only on a small scale or are in the experimental stages. The farming of pearl oysters (family Pteriidae), though highly developed in Japan and initiated in a number of other countries, does not fall within the scope of this chapter, as it does not contribute to food production. The gastropod topshell *Trochus cornatus* is cultured on a small scale in Japan and Korea, and *T. niolticus* on an experimental scale in the Caroline Islands. The queen conch *Strombus gigas* of the Caribbean has been bred in captivity and the larvae reared through metamorphosis. As the feasibility of commercial culture is uncertain, greater attention is now being devoted to the possibility of enhancing the diminishing natural populations through stocking with hatchery-produced young. Considerable progress has been reported on the controlled breeding and rearing of the giant clam, *Tridacna gigas*, in Australia and other Pacific Islands, indicating the possibility of commercial farming of the species.

A review of mollusc culture methods followed in different countries shows that large-scale commercial farming in most areas adopts extensive systems, depending largely on wild seed stock and natural food production. Natural reproduction is often augmented by concentrating brood stocks and providing substrates for spat settlement. The use of suitable and improved sites for different phases of growth and fattening and the eradication or control of pests and predators are the essential elements of the system. Besides the design of more efficient harvesting and depuration of the harvested molluscs to make them safe for human consumption, the major innovations in production technology relate to the development of off-bottom culture and methods of hatchery spawning and larval rearing. Larvae and adults can be reared on selected algae, but so far no inert feeds have been found to be practical for commercial growing (see Chapter 7), even though experiments have been conducted with fine particles of corn starch and micro-encapsulated feeds (the only exception being the gastropod abalone). In oyster hatcheries the brood stock is sometimes fed on corn starch.

26.1 Oysters

Cultivated oysters belong to two genera: *Crassostrea* (the cupped oysters) and *Ostrea* (flat oysters). Though aquaculture production of cupped oysters is much higher than that of flat oysters, the latter are held in greater esteem to be served on the half-shell, and fetch a much higher price in many countries. The more important species of cultivated oysters are:

Crassostrea gigas	(Pacific oyster)
C. virginica	(American oyster)
C. angulata	(Portuguese oyster)
C. commercialis	(Sydney rock oyster)
C. glomerata	(Auckland rock oyster)
C. plicatula	(Chinese oyster)
C. rivularis	(Chinese oyster)
Ostrea edulis	(Flat oyster; European oyster)
Ostrea chilensis	(Chilean oyster)

Some of these are illustrated in figs 26.1 and 26.2. The mangrove oyster, *C. rhizophorae*, has been cultivated on a relatively smaller scale

in Cuba and Venezuela. A number of other species of mangrove oysters have been experimentally cultivated in tropical areas, such as *C. tulipa* in Sierra Leone, *C. brasiliana* in Brazil and *C. belcherii* in Malaysia, but no commercial production has so far been reported. The slipper oyster *C. eradelie* is cultured in some areas of the Philippines. Most of the cupped oysters grow well at temperatures between 10 and 30°C, though for spawning and larval development temperatures around 20°C are considered optimal. Tropical species like the mangrove oyster, the Chinese oyster (*C. rivularis*) and the slipper oyster tolerate higher temperatures up to 34°C. The flat oyster is less tolerant of high temperatures and grows well only between 10 and 24°C. High mortality occurs above 26°C. Unlike the cupped oysters, the flat oyster cannot withstand continued freezing temperatures. Spawning occurs between 13 and 18°C, which is the optimal temperature range for larval development as well. On the whole, flat oysters are more sensitive to temperature variation.

Most oysters attach themselves to hard substrates and can tolerate wide variations in salinity, often between 5 and 32 ppt. The mangrove and slipper oyster can tolerate up to 40−45 ppt, even though the optimum is below 37 ppt. Lower salinities between 15 and 16 ppt are considered preferable for larval development. The flat oyster is reported to thrive best at salinities above 25 ppt. Being filter-feeders, they draw their food, consisting of phytoplankton and other organic matter, through an inhalant current of water into the mantle. From there the selected food passes into the alimentary tract through the mouth, and the material rejected by the buccal palps is expelled.

Both groups of oysters are protandrous and they generally develop first as males and then may change to females. Change of sex may occur between spawning seasons, as in *Crassostrea*, or even within the same season, as in *Ostrea* species. The spawning behaviour of *Crassostrea* is different from that of *Ostrea*. Fertilization and larval development are external in *Crassostrea*, and males and females release gametes into the water. In *Ostrea*, the female draws the sperm expelled by the male into the pallial cavity near the gills by the inhalant current, and the eggs are fertilized

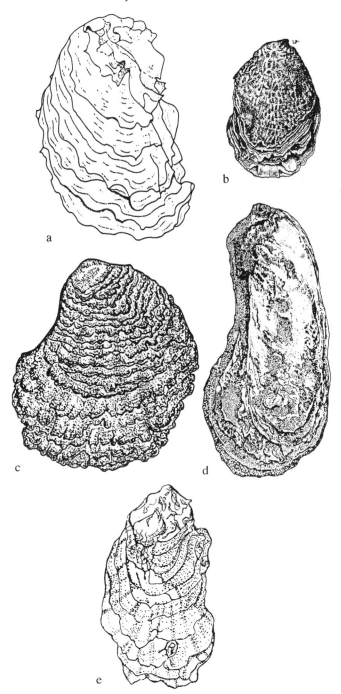

Fig. 26.1 Some of the more important cultivated species of oysters: (a) *Crassostrea gigas* (Pacific oyster); (b) *C. virginica* (American oyster); (c) *Ostrea edulis* (flat oyster); (d) *C. angulata* (Portuguese oyster); (e) *C. rhizophorae* (Mangrove oyster).

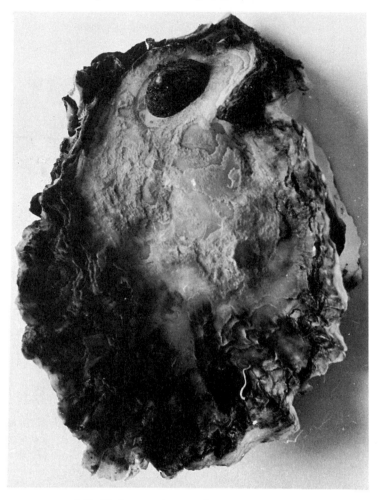

Fig. 26.2 *Crassostrea commercialis* (Sydney rock oyster).

there. The development of eggs and hatching take place in the pallial cavity. When they reach the stage of free-swimming veligers, they are expelled into the surrounding water.

Larval development usually takes up to three weeks, starting with the motile trochophore, followed by the swimming veliger stage. Once the shell hardens and a foot is developed, the larva is able to crawl around and find a substrate for attachment. Though any hard surface may be suitable for attachment, the flat oyster appears to prefer calcareous substrates.

26.1.1 Culture systems

The earliest system of oyster culture, which still prevails in many places, is referred to as bottom culture, where oysters are grown on the sea bottom either in intertidal or sub-tidal areas. This may involve collection of spat from areas of abundant spat-fall and planting them on suitable beds for on-growing. In areas which are suitable for both spat collection and on-growing, the main intervention by the oyster farmer may only be the provision of molluscan shells to stiffen the bottom and to serve as cultch for the attachment of spat, and possible protection from pests and predators. Though this system is the least productive, it continues to be practised because of legal restrictions on the use of off-bottom floating structures in the coastal areas, or because off-bottom culture has not proved economical under local conditions.

Off-bottom culture systems include stake/ stick culture (fig. 26.3), suspended culture and rack culture. Among these, the earliest is probably the stake or stick culture system, where bamboo, wooden or cement stakes or sticks are driven into the bottom or set out horizontally on racks to catch spat. Grow-out may be in the spat-catching area itself or, more often, in separate grow-out areas. The stake or stick system is particularly useful in intertidal areas with soft mud bottoms.

The stone-bridge method of culture employed in China for growing oysters (*C. rivularis*) in muddy areas involves the collection of spat on cement slabs placed in the form of a series of inverted Vs (∧) (fig. 26.4). The seed oysters are generally allowed to grow in the same area where they are caught.

In typical rack culture systems, racks of different designs are used to suspend trays (fig. 26.5) or strings/ropes carrying oyster cultch in or near intertidal zones. They are generally built to about 1−2 m height. Strings of shells or

other types of cultch can be either hung or placed horizontally on the racks for growing.

In suspended culture systems, strings/ropes or trays are suspended from floating rafts or long lines (fig. 26.6). While rafts are used in protected areas, long lines anchored at both ends and supported by floats can be laid out in more exposed areas. Different designs of trays have been developed from the earlier ones made of wood and wire screens. As well as rubber, plastic-coated wire mesh and polypropylene, special metal alloys are also being used to reduce biofouling.

Plastic mesh bags containing shell-cultch are often used in certain areas for spat collection. Elastic mesh bags are also sometimes used to grow adult oysters on intertidal racks (fig. 26.7). Polyethylene mesh bags made in the shape of Chinese lanterns, suspended from long lines, are especially suitable for growing cultchless oysters.

There are a number of other types of containers and substrates which are being tried for off-bottom culture of oysters. Other systems under trial are raceway and pond culture (Lee

Fig. 26.3 The stick culture system of oysters in New Zealand (from Curtin, 1971).

Fig. 26.4 The stone-bridge method of oyster growing in New Zealand (from Curtin, 1971).

et al., 1981) and recirculating systems (Pruder *et al.*, 1977).

26.1.2 Production of seed oysters

As stated earlier, most present-day production of oysters is from collected wild spat. Methods for hatchery production of seed oysters have been developed and are being practised on a commercial scale by a small number of producers. Though the value of and need for hatchery-produced seed oysters are well recognized, it would appear that production costs have stood in the way of wider application of this technique. Oyster larvae normally settle in sites with low current velocities, but these areas may not be rich in plankton and so may not be conducive to good growth rates. As spat-fall often occurs in areas away from environments suitable for oyster growing, the collection, transport and sale of oyster spat has developed into a separate industry. Similarly, hatchery production of seed oysters is a separate specialized activity, and oyster farmers often start their operations with purchased seed. Countries like Japan export large quantities of seed oysters to other oyster-growing countries.

There is considerable variation in the time and abundance of spat-fall in any area, depending on a number of environmental factors including temperature and salinity. For successful spat collection, suitable collecting devices have to be set in the proper place at the proper level and time. Even though spat may settle at a wide range of depths, their survival is greatest in the inter-tidal zone, relatively safe from predators. Early setting of the spat collectors will result in the substrate becoming covered with fouling organisms, whereas late setting may result in poor collection of spat. Regular examination of plankton in the area will help to determine the actual time and place of spat-fall. Generally, spawning occurs at temperatures of 15−20°C in summer and autumn, but tropical species spawn throughout the year at higher temperatures.

As well as the time and place of spat collection, the type of spat collector used is also of

Fig. 26.5 Tray culture of oysters in Brazil.

Fig. 26.6 Ropes suspended from floating rafts used for the culture of oysters.

importance. As mentioned earlier, different types of collectors are in use, and they may be placed on the bottom or on raised structures. Though the main criteria in selecting the type of collector are easy availability, ease of hand-ling and the surface area offered for the attachment of spat, the larvae of some species appear to exhibit preferences for certain types of sub-strates. For example, the European oyster seems to have a preference for materials

Fig. 26.7 Mesh bags used for growing oysters on racks (from *Il Peche*, **1**, 1988).

containing calcium carbonate, whereas the American oyster may settle on any hard surface, including wood, plastic or glass. It has been reported that shells treated with highly chlorinated benzenes, such as polystream, collected about two or three times as many oyster spat as untreated ones, and fouling and drilling of young oysters were significantly reduced (MacKenzie *et al.*, 1961). Though Castagna *et al.* (1969) reported the commercial feasibility of using treated collectors, this practice does not appear to have been widely employed.

Collectors used in France and some other European countries for the European flat oyster and the Portuguese oyster generally consist of semi-cylindrical ceramic roof tiles (10−12 cm in diameter and 30 cm long), stacked in pairs (fig. 26.8). The tiles are coated with lime and this facilitates the subsequent removal of spat which have settled on the tiles. Each stack of three to six pairs is tied together for easy handling and placed on wooden platforms, at least 15−30 cm above the bottom. Collectors for Portuguese oyster spat can be placed nearer to the shore, as they can tolerate higher temperatures and exposure to the sun at low tides. In Italy and Yugoslavia, oyster farmers sometimes use branches of trees or bushes such as juniper as spat collectors and they are suspended from ropes in the littoral areas.

The most common collector used in Japan and many other countries consists of scallops or oyster shells strung on a wire, with cut pieces (1−2 cm long) of tubes or small hollow bamboos as spacers to keep the shells apart and expose more surface for spat attachment (fig. 26.9). Such strings of shells are suspended from rafts, long lines or specially constructed bamboo frames. In some cases, the strings are laid horizontally on the bamboo frames, instead of hanging. When oyster seed is to be used for bottom growing, Japanese spat producers use only oyster shells as collectors, without spacers. Seed oysters attached to the underside of scallop shells, when sown on the sea bottom, are not likely to survive. The Japanese producers subject the collected spat to a hardening process before export. About a month after setting, when the seed oysters measure about 5−10 mm in diameter, the collectors are transferred to hardening racks and laid horizontally

Fig. 26.8 Lime-coated tiles used as spat collectors (from *Fish Farming International*, **7**(4)).

Fig. 26.9 Spat collectors made of scallop shells.

on the platforms. The spat are exposed at each low tide for at least 4 hours. Hardened spat are better able to survive long-distance transport and have less mortality during their growth to seed oysters.

As mentioned earlier, some farmers use chicken wire or expandable plastic mesh bags of various shapes for catching spat and also for on-growing. This type of collector has the advantage that when there is heavy setting it is fairly easy to thin and separate out the spat. This is particularly important in growing oysters of uniform size and shape for consumption on the half shell.

In areas where stake or stick culture is practised, spat are generally collected on cement-dipped wooden, reinforced concrete, or asbestos cement sticks, and their on-growing is also carried out on the same substrate. Plastic or fibreglass sticks can be used (fig. 26.10), if the spat are to be removed from the collectors for growing in trays or other containers.

The spat of mangrove oysters, as the name implies, attach themselves in nature to mangrove roots. Nikolic *et al.* (1976) described the use of terminal branches of the red mangrove suspended from a wooden framework for spat collection (fig. 26.11) as well as on-growing. Other types of collectors, especially strings of oyster or scallop shells and wooden sticks, have also been used successfully for collecting mangrove oyster spat.

Fig. 26.10 Sticks used for spat collection.

Fig. 26.11 Terminal branches of red mangroves used for spat collection.

Hatchery production of seed oysters

In recognition of the need for alternative means of producing seed oysters, other than uncertain collection from the wild, efforts have been made to develop hatchery techniques for the more important species. Declining stocks of oysters in traditional oyster grounds and the consequent reduction in the availability of spat, as also the need to import spat every year, served as further inducement to these efforts. In the absence of controlled breeding, oyster farming had no access to the benefits of genetic selection and hybridization. As a result of many years of research, starting from the early 19th century, controlled reproduction and larval rearing of the more important oysters and several other bivalve species have become possible (Loosanoff and Davis, 1963). Hatchery technologies for commercial application are suitable for the Pacific, Japanese, the American and the European oysters. The basic techniques used in hatcheries for these species, and in fact for all bivalves, are very similar, but variations and modifications are made to suit local conditions and species. According to Chew (1986) 'no two hatcheries are alike, and anyone who starts one can expect several years of experimentation before having any assurance of success'. The economics of hatchery operation also depend very much on local conditions and on whether the farmers have access to wild seed stock. Most hatchery production of seed oysters is performed in the USA, followed by the UK and France on a smaller scale. Small-scale hatcheries are reported to have been established in Chile, Mexico and Australia as well (Chew, 1986).

An adequate supply of unpolluted, clear water of the required salinity, and if possible optimum temperature, is a major criterion for site selection for an oyster hatchery, as well as the normal site requirements for all aquaculture hatcheries (see Chapters 4 and 6). A salinity above 20 ppt and water temperature not exceeding 20°C are recommended for the Pacific oyster, and 11−17 ppt salinity and a 19−25°C temperature range for the American oyster. However, the purpose of an oyster hatchery is to control and manipulate the temperature regime so that mature oysters will be available for spawning at any time, through a process of brood conditioning. For this purpose it is essential to select sufficiently fat spawners in good condition, grown in areas suitable for rapid growth and with similar salinity regimes. At least 30 per cent of the brood stock should be 1.5−2 years old, as a good percentage of them will be males, and the rest can be about 2.5 years old and will have a preponderance of females. It is advisable to have a mixture of spawners grown in different localities, to ensure a varied gene pool.

The brood stock is conditioned for spawning by placing the required number (usually about 50−150 oysters) in flow-through conditioning flumes or trays, supplied with water of the right temperature for maturation. The conditioning may take up to 8 weeks, during which period they are fed on algae. Supplementary feeding with corn starch is recommended to increase the glycogen reserve of the oysters. When fully mature, the European oyster spawns spontaneously at temperatures of 16−26°C. For species of *Crassostrea* it is necessary to manipulate the water temperature to induce spawning. The temperature is first raised to 25°C and then to 30°C over a half-hour period. After this, the temperature is made to fluctuate between 25 and 30°C. These temperature variations normally induce spawning, but if they fail the conditioning trays or flumes are drained and then refilled with fresh sea water of the required temperature. This is likely to stimulate spawning. An additional means of stimulation is a suspension of gametes (eggs or sperm) spawned by other oysters or gametes taken from conditioned oysters. It has been observed that there are distinct differences in the response of spawners from different geographic populations and in their temperature requirements. In such cases a series of temperature shocks is tried, and if even this does not

succeed, eggs and sperm are stripped from mature brood stock and fertilized.

As indicated earlier, there is a major difference in the fertilization and incubation of eggs in *Crassostrea* and *Ostrea*, and so they have to be treated differently in the hatchery. *Crassostrea* which have begun to spawn or show signs of spawning are transferred to separate containers, in order to facilitate separate collection of the spawned eggs and sperm. As soon as spawning ceases, the eggs and sperm are removed and the spawners transferred to cold running sea water. After the eggs are sieved to remove all extraneous matter, they are fertilized with a suspension of sperm. Usually in commercial operations, the eggs of several females (at least two individuals) and the sperm of several males are mixed, to ensure a mixed gene pool. A ratio of 2−4 ml dense sperm suspension for every 4 ℓ egg suspension (containing about a million eggs) has been recommended. Too much sperm can result in abnormal embryonic development due to polyspermy caused by the penetration of the eggs by more than one sperm. Too few sperms mean lower rates of fertilization. The fertilized eggs are diluted with salt water to a concentration of about 200 eggs per ml and allowed to develop for about 24 hours at 25°C into straight-hinged veligers measuring 75−80 μm.

The European oyster, *Ostrea edulis*, is larviparous and retains the eggs and larvae within the mantle cavity for about 7−10 days after spawning. Piles of eggs around the shell margin show that spawning has occurred. The larvae are released in swarms when they measure about 170 μm, much larger than the straight-hinge stage of *Crassostrea*. *Ostrea chilensis* has a prolonged larval incubation and the larvae are released only when ready to settle (Korringa, 1976). This reduces the need for rearing larvae for long periods.

The larvae of *Crassostrea* as well as *Ostrea* can be reared in suitable culture tanks and fed on algae. A number of methods of producing and feeding algae to oyster larvae have been tried, but most hatcheries produce pure cultures of selected algae and feed them singly or in combination. Methods of algal culture and the species preferred by oysters have been discussed in Chapter 7. *Isochrys, Phaeodactylum, Platymonas, Monochrysis, Dunaliella* and

Chlorella are some of the algae which have been found to be efficient larval foods at different stages of growth. The batch culture technique is commonly followed, in which large quantities of green water or algae-rich water are produced in a series of steps, starting with pure cultures inoculated into small quantities of sterilized sea water and nutrient solution which are then used to create progressively larger cultures. In the Wells-Glancey method (Wells, 1927), which is an earlier method of algal production, zooplankton and large algae are first removed from the sea water, using a milk separator, and then incubated for 12−24 hours to produce large numbers of small algae. Probably because of inadequate control of the species of algae grown and the deficiencies of the separating mechanisms, this method has not always given satisfactory results (Loosanoff, 1971).

Various types and sizes of tanks are used for larval rearing. Most of them are large, of at least 500 ℓ capacity. The number of larvae released in the tanks are estimated by counting samples from a suspension. The tanks are stocked at the rate of about 10 per ml. The water supply is generally sand-filtered and UV-sterilized. A temperature of about 25°C, salinity of 25−30 ppt and heavy aeration are maintained. Depending on local conditions, the rearing water is changed once to three times a week. Most hatcheries grade the larvae according to size, and only the best growing ones are allowed to reach settlement. The feeding rate is carefully monitored and the species of algae are varied as the larvae grow to ensure optimum growth at each stage. A starting concentration of 30 000 algal cells/ml water for the first week, followed by 50 000 cells/ml during the second week and 80 000 cells/ml during the third week has been suggested (Breese and Malouf, 1975). The larvae are fed at this concentration once a day during the first week, and twice a day thereafter.

Crassostrea are ready to settle after three weeks and *Ostrea* after two weeks of rearing, at 24−28°C. Hatcheries produce spat settled on cultch or the so-called cultchless seed. The most common cultch used in hatcheries are molluscan shells. Many hatcheries use plastic bags or baskets containing cleaned oyster shells for spat setting in large tanks.

Several methods of producing cultchless spat are in practice. The spat may be allowed to settle on a flexible sheet of smooth plastic, in which case they can be easily removed soon after setting or after a period of growth. An alternative is to use crushed shell chips or calcium carbonate particles as substrates, as each piece will have only one or two spat attached. A third method is the use of vinyl-coated wire trays dipped in concrete and sprinkled with oyster shell chips to collect the spat. Such trays are said to be easier to clean and allow better use of space. Cultchless oysters have to be carefully reared in flumes or trays until they grow to a diameter of at least 2.5 cm.

Spat are sold by hatcheries when they are 4−6 mm in diameter and about 3 months old. Until then, they are reared in water temperatures of 25−30°C. The feeding rate is increased a week after setting to 100 000−150 000 cells/ml per day. Before the spat are removed from the hatchery for sale or planting, the temperature is changed gradually to avoid a sudden temperature shock.

Burrell (1985) described a system developed on the west coast of the USA for transport of eyed larvae of the Pacific oyster. The larvae are concentrated on a screen, placed on a damp cloth and wrapped in wet paper towels for transport in plastic coolers. At temperatures between 1 and 4°C, the larvae can be held for about 7 days without much loss. At the destination they can be set on substrates in the same way as in hatcheries.

26.1.3 Grow-out

In almost all commercial oyster grow-out systems, the main techniques consist of planting spat in protected areas with suitable temperature conditions and high primary production, grading and replanting when needed, providing protection from predators and pests and fattening for the market. The different systems described in Section 26.1.1 are mainly designed to facilitate the use of available sites and to enable better utilization of primary productivity for the nourishment of the oysters.

Flat oysters

Compared to cupped oysters, flat oysters are less hardy and thrive best well below the low-water line. Their growth rates are lower and they require 3−4 years to reach market size. They normally suffer greater mortality during the long culture period and so the overall production of flat oysters is relatively less. Korringa (1976a) has given detailed descriptions of grow-out methods followed by selected enterprises in Europe and the USA and these show significant differences based on local hydrographical and market conditions. Some of the common basic practices are summarized below.

As indicated earlier, the spat-collecting area is not always the best area for on-growing. Besides the water quality and primary production, the nature of the bottom, depth and tidal regimes are of considerable importance, particularly in view of the length of time required for flat-oyster culture and the need for over-wintering. Though in principle off-bottom culture can be practised on muddy beds, a firm bottom is preferred. In oyster parks in France, sand or fine gravel is spread every year after the bottom is levelled to maintain the sub-soil structure. The shallower areas of the park used for rearing spat to young oysters are protected by net fencing.

Spat collected during the summer over-winter in the collecting ground and are brought to the park for rearing in the succeeding spring. The spat are detached from the collectors (usually lime-coated tiles from which they can easily be scraped off) manually or by special mechanical devices. During early spring, the detached spat are planted in the deeper areas of the park (about 80 cm water at low tide), since predation by crabs is not a major problem then. Later in the season the spat are planted in the fenced area and special care taken to eliminate crabs with baited crab traps and other devices. In order to reduce excessive algal fouling, some farmers introduce periwinkles (*Littorina*) which graze on the algae. The young oysters over-winter the second time in the park. After the second winter they are dredged out and replanted in fenced parks in intertidal zones, which are exposed for only short periods during spring and neap tides. A stocking rate of 10 tons/ha of second-year oysters (weighing 6−7 kg per 1000) is believed to be ideal in typical areas in Brittany (France). Some farmers allow the young oysters to grow in the same park, without replanting. Constant care is essential to maintain the parks and protect the stock. After the third over-wintering, the oysters are gathered early in spring and replanted in a clean plot. The third-year oysters can be marketed, but after a further over-wintering the oysters fetch a higher price. The oysters are harvested by dredging followed by hand picking, and are then carefully graded according to size.

Newly detached spat often suffer high mortalities due to predation and therefore many farmers, particularly in the Netherlands, grow them first in trays in areas with a rich supply of food. The trays are placed on racks and filled densely with spat. Under favourable conditions, the growth rate is fast and thinning is necessary to maintain their growth. When they have grown to a weight of 3−4 kg per 1000 by the end of the summer, they are ready to be planted in the growing plots. In some areas, as in Italy and Yugoslavia, the spat may be fastened to ropes for suspended culture. Instead of the old practice of setting spat on pieces of wood or twigs and inserting them in ropes for hanging, some farmers in Yugoslavia cement the spat on wooden boards (7.5 cm × 4 cm) in groups for on-growing. A common method of oyster culture (for both flat and cupped oysters) on the Atlantic coast of France at present is rack culture. The spat are held in synthetic bags (1 m long and 0.5 m wide), which are fastened by rubber bands to wooden or metal racks standing 0.5 m above ground. A density of not more than 6000−7000 bags (each containing 5 kg in the case of 1.5−2-year-old flat oysters and 5−10 kg cupped oysters of the same age) per ha is considered suitable for satisfactory growth.

In Maine, in the USA, the flat-oyster farmers usually purchase hatchery-produced spat and grow them in trays or nets off-bottom from rafts or long lines. The oysters are repeatedly graded during the summer. The trays are lowered to the bottom or suspended low from rafts for over-wintering and then raised in spring to continue growth.

The hanging method of culture has become very popular on the Mediterranean coast of

France in recent years. Ropes laden with oysters are suspended in protected areas from metal or wooden frames, in such a way that the oysters are totally submerged. Seed oysters are stuck on synthetic ropes or specially made wooden poles, using quick-setting cement. On a 2 m long pole or rope, about 75–80 oysters are stuck. Fouling organisms are regularly removed by hand to allow the oysters free access to the water flow. Harvesting is fairly easy, as the ropes or poles can be brought ashore and the oysters detached. Though the growth rate and yield (5 kg per rope or pole) are high, the shell is often fragile and tends to open after harvesting.

The technique of fattening oysters in special oyster ponds called 'claires' in France has been briefly described in Chapter 11. This practice is of special importance in areas where there is a scarcity of fattening grounds but is now used mainly for greening Portuguese oysters.

Predators, pests and diseases

As indicated earlier, once the spat are planted the major effort of the farmer is to protect the stock from predators and pests to the extent that is feasible under open-sea conditions. The vulnerability of spat and young oysters to shore crabs (*Carcinidas* sp.), makes it especially important to provide all possible protection from them. Another major predator is the starfish (*Asterias* sp.), which may settle on the collectors and prey on the spat. Repeated hand picking is the common method of control, although application of quicklime on the oyster bed can also be effective.

Barnacles and other fouling organisms not only compete for space on collectors, trays, nets, etc., but also affect the growth and appearance of the oysters. In addition to manual clearing, which is commonly done, Korringa (1976) described spraying of a low concentration of DDT on spat collectors as a means to prevent fouling.

Large-scale mortalities of the European oyster occurred on the coasts of France from 1968 for over a decade. These epizootics are believed to have been caused by three or four protistan parasites, starting with *Marteilia refringens* and then *Bonamia ostreae*. The only means of control appears to be to avoid planting seed oysters

during the period when infections occur, which is reported to be July and August for *M. refringens*.

Shell disease caused by the fungus *Ostracoblabe implexa* has caused serious losses of young oysters in the Zeeland oyster area of the Netherlands in the years following 1930 (Korringa, 1976). Another reported disease is the 'pit disease', which is described as a congestion caused by the rapid multiplication of the flagellate *Hexamita*. It generally occurs when oysters are kept too long in a storage basin at low temperatures.

Cupped oysters

Crassotrea spp. inhabit a wider range of ecological conditions and can grow well in areas of lower salinities, free from some of the common predators and parasites. Though they require higher temperatures (above 20°C) during the larval development and settling stages, they can withstand very low temperatures better than flat oysters. Because of these advantages, the overall production of cupped oysters worldwide is much higher than that of flat oysters. Farming of *Crassostrea* started in Japan, which is still one of the major producers. Though not considered to be of such high gastronomical qualities as the flat oysters, half-shell, cupped oysters in fattened prime condition are well-relished.

The basic principles of grow-out of the various cupped oysters are similar and so the following account gives only the salient features of the grow-out practices for the more important species, namely the Pacific oyster, *C. gigas*, the American oyster, *C. virginica*, the Portuguese oyster, *C. angulata*, and the Sydney rock oyster, *C. commercialis*.

Both on-bottom and off-bottom grow-out methods are followed, depending on local conditions. While in Japan, in some areas of the USA and on a smaller scale in many other countries, cupped oysters are grown on racks and long lines, bottom culture is the common practice in most parts of the USA because of legal restrictions on the use of floating structures in coastal areas. The availability of hatchery-produced spat has made it possible to adopt off-bottom culture in trays, net bags, etc., of the Pacific oyster, even in places where there are no breeding populations. As spat collection occupies only limited space, suspended collectors can often be legally used in areas where spat-fall occurs.

The traditional system of rearing oysters in parks is widely followed in France, especially in the Arcachon bay, using spat collected on lime-coated tiles. The Sydney rock oyster and the Auckland oyster, *C. glomerata*, are generally farmed on sticks laid on off-bottom racks. The traditional bamboo stick method of culturing the Japanese (Pacific) oyster continues to be a major source of production in China.

The culture method for the American oyster in the USA consists of catching seed oysters in areas where settling occurs and planting them in areas where conditions are suitable for rapid growth. Unpolluted areas with adequate currents and food production are selected for grow-out. Although many growing areas are in the inter-tidal zones, the major production comes from sub-tidal beds, often in estuarine areas with low salinities which have few predators and diseases. Harvesting is usually carried out by hand tongs and box and mechanical dredges.

The main predators of the American oyster are starfish, predatory finfish, oyster drills (*Gastropods*), flat worms and crabs. The starfish are controlled by the application of quicklime on the oyster beds or by capturing them by dragging special mops and killing by hot water dips. Oyster drills (*Urosalpinx*), conches (*Stylochus*) and blue crabs (*Callinectes*) are major predators of young oysters.

The diseases of American oysters have been extensively investigated and Burrell (1985) listed a number of recent reviews, including that of Sindermann (1977). Some of the more serious diseases have been caused by protozoans; for example, *Minchinia nelsonii* causes the salinity-dependent 'Delaware Bay disease' and *M. costalis* causes the 'seaside disease'. Viral infections and mycosis of larvae have been reported to cause major losses in hatcheries.

Although the most common methods of culturing Pacific oysters in Japan today are suspended culture from rafts, long lines or racks, on-bottom sowing is still continued in certain areas (Koganezawa, 1979). The rafts are made mostly of bamboo, or sometimes cedarwood, and measure about 16 m × 8 m. Floats made of styrofoam (covered in polyethylene bags for protection against fouling) are commonly used, although drums or similar material are also suitable. A series of rafts are anchored at distances of 5–10 m and the shell strings or wires bearing the spat are hung from them. For on-growing, the shells are cleaned of all fouling matter and the settled spat thinned where necessary. The spacers between shells are lengthened to provide adequate space for growth and better circulation of water and food organisms. Sometimes the shells are restrung on new wires. The number of floats is adjusted during the rearing period according to the increased weight of the growing oysters.

The long-line system is suited for more exposed areas and can better withstand wind and waves. Essentially, it consists of a series of two parallel wires or fibre ropes, buoyed up with suitable floats (generally wooden or styrofoam). The oyster strings are suspended on ropes between the floats. As in the case of rafts, the floats or the number of strings between floats have to be adjusted as the oysters grow and gain weight. The rack method is adopted in shallow areas and, as described in the earlier section on Production of seed oysters, can be used for hanging culture or stake or stick culture.

The cycle of oyster culture in Japan consists of one or two years. In the one-year cycle, the spat collection lasts about two months, starting in about June, and soon after settling they are transferred for growing on rafts. Harvesting can start from about February to May, giving a grow-out period of 8–9 months. The two-year cycle starts with spat collected later in the season, and the collectors with the spat are transferred to racks in shallow water areas until the following summer, for a period of almost one year, for hardening (see earlier section on Production of seed oysters). During this period there is only limited growth, as the young oysters are usually placed in areas under unsuitable conditions such as low salinity, low tidal flow and fluctuating temperature. For final growing they are moved to rafts during the early summer, and harvesting can start from about the succeeding February. The hardened oysters show higher survival rates, although the size at harvest may be relatively small. In order to satisfy the market for large oysters (10—20 cm shell height and meat weight of 10—30 g), a small percentage of the selected one-year-cycle oysters may be grown individually in net cages with separate cells for each oyster and suspended from rafts for 6—8 months.

Mass mortality in oyster farms in Japan is believed to be related to intensification of farming methods and consequent eutrophication of coastal waters. Starfish and oyster drills (*Thais spp.*, *Tritonalia*, *Ocenebra*, *Rapana*, and *Ceratostoma*) are major enemies of oysters, particularly in rack and bottom culture. According to Fujiya (1970) the parasite *Polydra ciliata* infests 60—70 per cent of the oysters in Japan. On the West Coast of the USA, the parasites of the Pacific oyster include the copepod *Mytilicola* and ciliates such as *Ancistrocama* and *Trichodina*.

The Portuguese oyster (*C. angulata*), which is very similar and considered by some to be the same as or derived from the Japanese or Pacific oyster (*C. gigas*), has become an important species in France, especially after the decline of flat-oyster stocks. The Portuguese oyster grows well in a wider range of salinities and can tolerate salinities as low as 15 ppt, making it easier to grow in estuarine areas. It can thrive in rather turbid water and soft bottoms, and grows well at low temperatures, even though it requires a temperature above 20°C for larval development and settling.

Spat collected on shell strings, slabs of slate, scrap iron or plastic tubes are used for on-growing in specially prepared oyster parks. The park is fenced in with galvanized wire netting and often partitioned into several compartments by plastic-covered wire netting. Shells carrying the spat are planted very closely and sometimes rows of stones or sticks are provided as shelter. The grow-out period is generally one year, when the oysters weigh about 10—30 kg per 1000. After the oysters are separated out from the substrate with a heavy knife, they are sorted according to size and held in trays in a basin for some days before replanting.

Replanting is carried out in parks located in areas that are exposed only at spring tides. The oysters are replanted very densely at the rate of about 10 kg/m². The park is fenced with wire-mesh netting to prevent the oysters from being washed away and to protect them from predatory fish like sting rays (*Trygon* sp.). As an additional protection from predators, masses of willow twigs are also strewn in the park. All necessary maintenance work, including spreading the oysters, placing willow twigs, removing starfish, scraping the silt and sand from the oysters and harrowing the plot, is carried out during spring tides. Oysters planted in spring are generally ready for harvest in the autumn of the same year, at a weight of about 50—60 kg per 1000. The harvested oysters are manually cleaned of all growths on the shells and sorted according to size. The cleaned oysters are fattened in claires. Only a few oysters (three or four per m²) are planted in each claire, to ensure uniformly high quality. They are fattened there for the whole summer season and if a bloom of *Navicula ostrearia* develops, the oysters acquire the desired green hue and excellent flavour. By about September the fattened oysters are harvested.

The Portuguese oysters are also grown on trays placed on supporting frameworks. The trays may be made of wood with wire or plastic netting, or made of only wire netting or plastic, divided into a number of compartments. The oysters are spread in the tray one layer thick. Usually the larger oysters weighing 40 kg per 1000 are grown in these trays and, when harvested, they fetch a relatively high price.

Starfish, rays, skates and oyster-eating birds (*Haematopus* spp.) are some of the predators to be avoided in the oyster parks. Major losses of Portuguese oysters in farms in France were attributed to the so-called gill disease, and another disease of unknown origin. The loss could be made up only by the importation of the Japanese (Pacific) oysters.

Rock oysters

The Sydney rock oyster, *C. commercialis*, and the Auckland rock oyster, *C. glomerata* are cultivated on stakes or sticks in Australia and New Zealand respectively. Like all other oysters, spat of rock oysters are available for collection in areas with low current velocities, which are not the best areas for growing them. Areas with a better tidal flow at levels slightly below extreme low-water neaps are selected for on-growing on racks of the same design as used for spat collection. The sticks containing the spat from the collection grounds are transferred to the growing area when the spat are about 10 months old. The sticks are arranged parallel to the racks in several layers close together, to protect the spat from predators and from the hot sun at low tides.

By the time the oysters are about 18 months old, the sticks are unfastened and rearranged across the rack, leaving more space between them for better circulation of water and food organisms. They are left to grow there until the third year, when they are about 29–30 months old. At that time, if the sticks are fully covered with oysters, it may be necessary to give them more space for growing. A further rearrangement of the sticks is done leaving spaces between of 15–16 cm. Three- to four-year-old oysters are harvested and graded according to size. Large oysters (plate oysters) weighing about 70 kg per 1000 are sold to be served on half-shell (fig. 26.12), and the smaller ones (bottle oysters), weighing about 40 kg per 1000, are shucked and packed in bottles with water, for use in oyster soups, stews, etc.

The Auckland rock oysters, grown on sticks, are ready for harvest when about 3 years old or even less in especially favourable conditions. An alternative practice is to grow the oysters for about 33 months on sticks and then mature or fatten them on trays. Tray farms are built in sheltered areas, where racks are built to support the trays at the optimum growing height. An improvised breakwater is sometimes necessary on one side of the farm for extra protection.

The dreaded oyster predator in Australian rock oyster farms is the porcupine fish (*Dicotylichthys* sp.). Others, like the sting ray, toad fish (*Sphaeroides* sp.), etc., are also major predators. The oyster borer (*Lepsiella*) and the mud worm (*Polydora*) are common enemies. The rack culture system and the practice of arranging the sticks close together when the oysters are young limit exposure to predators to a great extent.

Rock oyster farming in Australia and New Zealand suffers considerably from the so-called 'opening disease' or 'winter mortality' of adult oysters, which actually occurs in early spring. The cause of the disease is not definitely known, but it is suspected that it may be caused by the Haplosporidian *Minchinia*, which also affects the American oysters. Oysters show a characteristic ulceration in the body. Farmers in New Zealand are advised to raise cultivation to mean high-water level or move the oysters well upstream, if possible.

A number of tropical species of oysters associated with mangroves have been cultured on an experimental or small scale. The mangrove roots form excellent substrates for spat setting. In the mangrove environment, the oysters grow best at lower mud levels, with longer periods of immersion. So most attempts to culture these oysters have been based on the principles of collecting spat on stakes or shell-strings and off-bottom growing with the spat suspended from submerged racks or rafts.

The spat of the slipper oyster (*C. eradilei*) are caught on bamboo sticks or on shell-strings suspended from bamboo platforms in intertidal areas. The oysters grow very rapidly and reach a marketable size of about 7.5 cm in 6–9 months. In culture trials of *C. rhizophorae*, mangrove branches supported on a wooden framework were found to be the most cost-effective substrate for spat collection and on-growing. With proper care and seasonal adjustment of the collectors and under favourable conditions, harvesting can be started after 5–6 months, when the oysters weigh 10 kg per 1000 (Nikolic *et al.*, 1976). Seed of *C. tulipa* grown on rafts in Sierra Leone are reported to have grown to lengths of up to 10 cm in 7–8 months (Kamara and McNeil, 1975). *C. brasiliana* grows less rapidly, and in the subtropical climate of Southern Brazil it takes 18 months to grow to 6 cm and 30 months to grow to 8 cm (Wakamatsu, 1973). The mangrove oyster *C. belcherii*, when cultured by a combination technique of seed collection on asbestos strips held in trays and grow-out in trays suspended from rafts in Sabah (Malaysia), were found to grow to a weight of 14–21 kg per 1000 in one year (Chin and Lim, 1975).

26.1.4 Harvesting, handling and marketing

Harvesting and depuration of molluscs have been described briefly in Chapter 11. Handling and marketing of oysters assume special importance when they are meant to be served on half shells. The rearing technique itself is suited for producing oysters in excellent condition with high levels of glycogen and regular-shaped

Fig. 26.12 Large rock oysters suitable to be served on half shells.

shells. Procedures of greening oysters to cater to the gourmet market in France have been described earlier. Besides catering to consumer preferences, it is also of importance to ensure the safety of the products, particularly since they are consumed raw and contamination with pathogenic organisms is more likely in filter-feeding animals such as the oysters. Oysters grown in areas suspected to be contaminated have to be depurated by methods described in Chapter 11, before they are marketed.

When oysters are not eaten raw, the shucked oyster meat is chilled, frozen or dried for marketing. In China and some of the Southeast Asian countries, oyster sauce is a common culinary item and a good part of the production is dried and converted into sauce.

26.2 Mussels

Mussels have been traditionally a well-accepted seafood in many countries of Europe, particularly France, Italy and the Netherlands. But it is only in the last few decades that the potential of their cultivation has attracted any significant attention. The very high levels of production obtained by raft culture in the rias (submerged river valleys or fjords) of Galicia in Spain aroused the expectation that mussel culture may be a quick means of solving the animal protein needs of the populations of the Third World. Not withstanding the fact that the productivity of the Galician Coast cannot be duplicated everywhere, and that many years of work will be needed to promote and popularize mussels in many parts of the world, there

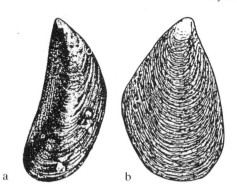

a　　　　b

Fig. 26.13 (a) The blue mussel, *Mytilus edulis*; (b) the Mediterranean mussel, *M. galloprovincialis*.

is undoubtedly very considerable potential for substantial increases in the production of mussels for direct human consumption or as feed for other valued aquaculture species.

There are a number of species of mussels distributed in most parts of the world, but the most important from the point of view of aquaculture production is the blue mussel *Mytilus edulis* (fig. 26.13). It is widely distributed and adapts itself to a variety of ecological conditions. The Mediterranean mussel, *M. galloprovincialis* (fig. 26.13), is considered by some workers to be the same as *M. edulis*, and the morphological differences are well within the geographic variations observed in the species. Irrespective of the taxonomic identity, there is very little difference in the biological characteristics relevant to cultivation. Another group of mussels of importance in aquaculture includes species of *Perna*. The culture of the green mussel *Perna viridis* (= *M. smaragdinus*; *M. viridis*) (fig. 26.14) has been practised in the Philippines for several years. The Chinese mussel culture is largely based on the blue mussel, but the green mussel and the black mussel, *M. crassitesta*, are also of some import-

Fig. 26.14 The green mussel, Perna viridis, grown in the Philippines (from *Fish Farming International*, 3(4)).

ance. Experimental culture of mussels in India includes the brown mussel, *P. indica*. The green mussel cultured in New Zealand belongs to the species *P. canaliculus* (fig. 26.15), and the main species in Venezuela is *P. perna*.

Like other bivalve molluscs, mussels are filter-feeders, feeding on phytoplankton and suspended organic particles. In nature they live in shallow water, attached to hard substrates.

However, they can displace themselves both horizontally and vertically with the aid of the extensible foot, and attach to new substrates by the production of new byssus threads. They often attach themselves to the shells of other mussels, forming large mussel congregations.

The sexes are separate and spawning occurs throughout or for long periods of the year, depending largely on temperature conditions.

Fig. 26.15 The green mussel, *Perna canaliculus*, grown in New Zealand.

Millions of eggs spawned by the female are fertilized by the sperms of males spawning simultaneously. Embryonic development is fairly fast, and in the blue mussel the ciliated trochophore stage is reached within about 24 hours after fertilization. The planktonic veliger stage is reached in 48 hours and at a shell length of 0.25–0.30 mm they can attach themselves with the byssus thread to any filamentous object, including weeds and hydroids. It is reported that in the absence of a suitable substrate, the final metamorphosis can be delayed for up to 40 days in temperatures of about 10°C (Bayne, 1965). The blue mussel reaches the market size of about 5–6 cm shell length in one to three years. Its life span is believed to be not more than 4 years.

26.2.1 Culture systems

The systems adopted for the culture of mussels are very similar to those used in oyster culture. Bottom culture is widely practised in the Netherlands and Germany; 'bouchet' (posts or stockade) or stick culture in France, Italy and other Mediterranean countries and The Philippines; and raft and longline culture in Spain, Sweden and New Zealand.

On very muddy coasts with gentle slopes and large tidal oscillations, bouchet and stick culture give good results. Bottom culture is suitable for coastal areas with stable and hard bottoms which are submerged for long periods. In shallow estuaries and bays with a low tide coefficient, cultivation on hanging ropes is preferred. The raft method is generally practised in protected areas, with steep coastal profiles and considerable tidal oscillations, at depths of at least 3 m at the lowest neap tides.

Bottom culture is entirely based on seed produced in natural beds, and the mussels are more exposed to predators and pests. The mussel parks are located in shallow, enclosed or partially diked areas.

The bouchet system of culture (fig. 26.16) originated in France in the early 13th century and an Irishman, Patrick Walton, who was shipwrecked in the Bay of Aiguillon, appears to have devised the technique based on his observations of the settling behaviour of the blue mussel on the poles of the equipment he used to capture fish. Rows of wooden poles of about 3–6 m length, called bouchets, are driven into the bottom of inter-tidal zones, to form the park for growing mussels on ropes. The rows are normally at right angles to the coastline and spaced about 15–20 m apart, the distance between poles in each row being about 20 cm to 1 m. The lower part of each pole above the sea bottom is covered with smooth plastic sheets to prevent predators like crabs from reaching the mussels. In stick culture, bamboo or wooden sticks replace the bouchets and are used for spat collection as well as for on-growing.

The raft system of culture is generally practised in protected areas like the rias of Spain, in depths of at least 3 m at the lowest low tide (fig. 26.17). Different types and sizes of rafts are in use, ranging from those constructed with the hulls of old fishing vessels to concrete and steel platforms or pontoons with styrofoam and fibreglass floats. There is a wooden lattice framework over the floats, to which ropes containing young mussels are attached. The outside of the wooden platform and floats is protected from wood-boring organisms (mainly *Teredo* and *Limnoria*) by coating with cement, anti-fouling paints, strips of fibreglass or other synthetic material. The size of a raft varies from about 400 m², which can carry some 500 ropes of mussels, to 700 m² or more, carrying 1000 or more ropes. They are built in such a way as to create minimum resistance to surf. The rafts are anchored by long, strong chains (six or seven times longer than the depth of water at the site), facilitating adequate movements and providing strength to withstand bad weather conditions.

The long-line system of culture developed in Sweden (Lutz, 1985) consists of a series of horizontal lines of about 10 m length, buoyed up by a number of suitable floats and anchored down with concrete blocks or other anchoring systems. On these lines a number of vertical 'substrate' lines are hung. The length of these lines can vary, but in the shallow fjords (less than 10 m depth) where they are operated in Sweden, it seldom exceeds 0.5 m. The lengths are standardized, in order to mechanize harvesting. The system is reported to work well under the freezing conditions and low tidal ranges in Swedish fjords.

The long-line culture system has been tried

Fig. 26.16 Bouchet culture of mussels in France, the mussels are ready for harvest (from Bardach *et al.* 1972. By permission of John Wiley & Sons).

on an experimental basis in France to improve the quality of the mussels grown in lagoons and other similar sites. One-year-old mussels are raised on long-lines in the sea for about 8 months, where they achieve faster growth and better taste.

26.2.2   Seed stock

There are several methods of collecting seed stock, but many of them are only local variations. Bottom culture of mussels, as practised in the Netherlands, depends on seed stock which has settled on the inter-tidal and deeper waters, which is collected manually or by dredging from special mussel boats fitted with a number of dredges.

In areas where bouchet or stick culture is practised, the same substrate is also used for collection of seed stock. The poles or sticks are erected in rows near natural mussel beds where spat setting normally occurs regularly, a few months before the spat-fall season (from May to about July in France). During the period before the settling of mussel larvae, barnacles usually settle on the poles and hydroids like *Tubularia* settle in turn on the barnacle shells and these constitute excellent substrates for the

settling of mussel larvae. As the young mussels grow they leave the original filamentous support and attach themselves to the poles. Harvesting of the mussel seed is performed before the poles become overcrowded, by scraping with special tools.

Another means of collecting mussel seed in France is on layers of fibre ropes (mainly coir ropes) of about 1 cm thickness, stretched horizontally on supports of wooden beams built on poles driven in the deeper portions of the inter-tidal zone. The fibre surface is a very suitable substrate for mussel seed, and facilitates early settlement and quick growth. As the mussels grow fast, the ropes have to be transferred for on-growing before they become too heavy with grown mussels. In the Philippines, green mussel seed are collected on bamboo stakes and they grow on the same stakes until they reach market size. Seed are also sometimes collected on bivalve shells or coconut shells and then transferred to the stakes or rafts for on-growing.

For raft culture in Spain, seed mussels are collected on the same rafts used for grow-out. Small amounts are also gathered from sub-littoral areas by raking. Traditionally, ropes made of esparto grass are used for seed collec-

Fig. 26.17 A section of a large raft system used for growing mussels in Spain.

tion. Other fibre ropes of sisal or nylon are also presently used. The ropes vary in length from 2.5 to 6 m, according to the depth of water in the area. The ropes have wooden stays of about 1.5 cm diameter and about 24 cm length, at intervals of about 40 cm, to prevent the mussels sliding down the ropes during bad weather. Though larvae occur in the water throughout the year, the peak settlement is in spring or early summer, with a second peak in autumn. While the larvae of the spring spawn attach to the ropes and growing mussels on the rafts, the autumn larvae seem to attach only to rocks and boulders on the coasts. Seed of autumn spawning show much higher growth rates, reaching a marketable size of 8–9 cm in 15 months after settling, while those of the spring spawning require over one and a half years to reach that size. So there is a greater demand for seed stock from the autumn spawning.

As can be seen from the above, seed for commercial mussel culture is obtained solely from natural reproduction. Methods of induced spawning have been developed, but only for experimental purposes. Sexually mature animals, when transferred to water at a high temperature after mechanical stimulation, will release gametes. Breaking off a chip from the valvular shell at the level of the visceral ganglion and then transferring the animal to a higher temperature of 18–20°C will also induce spawning. Mussels transferred from winter temperatures of 0–1°C to a controlled environment of about 18°C have responded well and released gametes (Hrs-Brenko, 1973).

26.2.3 Grow-out

As mentioned earlier, grow-out is on the sea bed itself, on bouchets or stakes, or suspended on rafts or long lines on ropes or mesh bags. In the majority of cases, the seed have to be transferred from the collecting area to growing areas. In bottom culture, the seed are planted as evenly as possible. During the growing period, the duration of which depends on local conditions, care is taken to avoid overcrowding and if sample dredge hauls show more than about 8 kg mussels per m^2 on the beds, the stock is redistributed. Predators like starfish are regularly caught from the growing area and accumulations of silt are removed from the beds. The mussels may be reared in the same bed until they reach a marketable size of about 5–5.5 cm or in some cases they may be transplanted to deeper plots for fattening. High quality mussels may measure 6–7 cm in shell length and contain about 25–30 per cent of flesh by weight.

In bouchet culture, the seed collected are transferred to bouchets erected for grow-out in the higher regions of the inter-tidal zone. They are placed in cylindrical nylon nets forming bags of about 5 cm length and 10 cm diameter. The net bags are wound round the bouchets. The loosely packed mussel seed rearrange themselves in the bag with their siphons directed outwards. Eventually they crawl through the meshes, but remain attached to the nets or the shells of other mussels. The nets gradually disintegrate and the mussels remain clustered on the bouchets until they reach market size.

In raft culture as practised in Spain, the seed collected on ropes usually appear like strings, formed by the young mussels attaching themselves to each other. They are carefully stripped off and wound and bound to new growing ropes. A very fine synthetic fibre mesh net is used to enclose ropes with the mussel seed (fig. 26.18). Though the net is strong enough to keep the mussels together, the meshes are easily broken by growing mussels. Seed mussels attached to one seed rope are generally enough to fill four or five growing ropes. Some farmers use rolls of meshed tissue to hold young mussels round the ropes. As the mussels are continually submerged in water with an abundant growth of phytoplankton and small particles of detritus, they grow quite rapidly. By the following spring, after a growth of about 6 months, the mussels are again transferred to new growing ropes to provide adequate space for growth to commercial size. Mussels from one rope are

generally enough to cover three new ropes. During such thinning, all fouling organisms are eliminated and then the ropes are suspended on the raft. The mussels are harvested after about 4 months, when they have grown to the market size of 5–8 cm. In semi-tropical climates it seldom takes more than 14 months to attain that size. To produce mussels of uniform size the ropes will have to be turned over during the growing period. In Spain it is usual to have in the same park ropes for collecting seed, ropes covered with seed for growing young mussels and ropes for growing marketable mussels. If the producer operates only one raft, he will have separate ropes with different size groups of mussels on the same raft.

In the off-bottom type of culture practised in Italy, mussel parks consist of networks of poles planted on the shallow muddy bottoms of protected areas, connected with horizontal ropes. The strings of mussels (collected on separate seed-collecting ropes from the park itself, or from nearby areas) are suspended from the horizontal ropes and permanently submerged in sea water. The grow-out method in long-line culture is also very similar.

For culture of the green mussel in Venezuela, wooden rafts very similar to the Spanish mussel rafts are used, in highly productive seas like Golfo de Cariaco (Estado Sucre), near areas of upwelling. In New Zealand, where there are restrictions on the use of coastal areas, submerged platforms and rafts have been used for culturing the green mussel. Plastic receptacles strung on ropes have been used to keep the mussels for on-growing. With the expansion of export markets for mussels, long line culture is now permitted in the country.

The production of marketable mussels varies according to the characteristics of the location, especially the tidal flow, primary production, temperature conditions and stock density. Suspended culture is normally more productive than bottom culture. In the rias of Spain where primary productivity is, on average, 10.5 mg carbon/ℓ per hour, the average annual production per raft of 600 culture ropes is reported to be about 50 tons. In the bouchet type of culture, production per pole is around 10 kg per year and about 4500 kg per ha. The yield from suspended culture in the Mediterranean is about 100 kg/m^2, when only one-quarter of

Fig. 26.18 Enclosing the ropes containing mussel seed with mesh nets in Spain.

the total area of a mussel park is used for on-growing. Observations in Spain show that sunlight has a major influence on the behaviour of the blue mussel. Changes of light intensity affect the mobility of the mussels. Mussels living in darkness are reported to weigh more than 25 per cent of those living in semi-darkness and more than 69 per cent of those exposed to light. The degree of pigmentation is also influenced by light.

Most of the predators and pests of oysters affect mussels as well, especially in bottom culture. Young mussels and weak adults are attacked by shore crabs, starfish, sting rays (*Trygon*), ducks, sea gulls and other birds. Barnacles, shipworm (*Teredo*) and polychaete worms (*Polydora*) are serious pests. The most disastrous losses of mussels have been caused by the intestinal parasite (copepod) *Mytilicola intestinalis*, and the infection is more pronounced in areas with dense populations.

The occurrence of red tide affects growing mussels very seriously. When blooms of dino-flagellates (*Dinophysis, Porocentrum*) occur, the sale of mussels has to be suspended as consumption of mussels from the affected areas may cause gastro-intestinal disorders. Dino-flagellates of the genus *Goniaulax* are more dangerous, as they cause paralytic shellfish poisoning.

26.2.4 Harvesting, depuration and marketing

Harvesting mussels grown on rafts is relatively easy. The ropes are hauled out of the water, often using mechanical winches. The mussels can be shaken off the ropes and sorted according to size. Harvesting from bouchet poles and sticks is normally performed by hand from small boats and the mussels are removed, washed and sorted. The bigger ones (5−7 cm shell length) are sent for depuration or directly to markets. The big mussels are reserved for canning in areas where mussel canning is undertaken. Small mussels are re-attached to ropes and transferred to the park for further

Fig. 26.19 Plastic receptacles used for growing mussels in New Zealand.

growing. In bottom culture, harvesting is carried out using dredges from special boats. Harvested mussels are handled the same way as those from suspended culture, except that the dredged mussels are likely to contain some sand and silt in the shell cavity. The mussels are cleansed in special diked cleansing plots, well protected from winds. They are spread evenly in clean sea water of uniform salinity and high oxygen concentration. Within about 24 hours the mussels are free from sand, but they are kept in the cleansing plots for 3−8 days. In areas where the mussels are consumed raw, it is essential to ensure that they are free from any pathogenic organisms. So harvests from areas exposed to contamination have to be depurated, as in the case of oysters (see

Chapter 11). Besides live mussels, mussel meat is also marketed in the frozen (individually or block frozen), canned, smoked or pickled form.

The presence of pearls in the mantle epithelium is a problem in the utilization of mussels in certain areas, especially on the north-eastern coast of the USA. Although the exact cause and circumstances of pearl formation are still not fully understood, it is known to be associated with infection by a parasitic trematode, probably of the genus *Gymnophallus* (Dubois, 1901; Lutz, 1978). There is also some evidence to show that the incidence of pearls is related to the age of the mussels. Raft-cultured mussels which have been in water for less than 5 years usually provide a high-quality product (Lutz, 1985).

26.3 References

Alfsen C. (1987) *Shellfish Culture in France.* SEAFDEC, Bangkok.

Andreu B. (1968) Pesqueria y cultivo de mejillones y ostras en Espana. *Publ. Tec. Junta Estud. Pesca Madr.*, **7**, 303−20.

Bardach J.E., Ryther J.H. and McLarney W.O. (1972) *Aquaculture*, pp. 674−742, 760−76. John Wiley and Sons, New York.

Bayne B.L. (1965) Growth and delay of metamorphosis of the larvae of *Mytilus edulis* (L.). *Ophelia*, **2**, 1−47.

Burrell V.G. (1985) Oyster culture. In *Crustacean and Mollusc Aquaculture in the United States*, (Ed. by J.V. Huner and E.E. Brown), pp. 235−72. Avi Publishing Company, Westport.

Breese W.P. and Malouf R.E. (1975) *Hatchery Manual for the Pacific Oyster*. Oregon State University Sea Grant College Programme/Agriculture Experiment Station. *Spec. Rep.*, 443.

Castagna M., Haven D.S. and Whitcomb J.B. (1969) Treatment of shell cultch with polystream to increase the yield of seed oysters, *Crassostrea virginica*. *Proc. Nat. Shellfish Assn.*, **59**, 84−90.

Chew K.K. (1986) Review of recent molluscan culture. In *Realism in Aquaculture: Achievement, Constraints, Perspectives*, (Ed. by M. Bilio, H. Rosenthal and C.J. Sindermann), pp. 173−95. European Aquaculture Society, Bredene.

Chin P.K. and Lim L.A. (1975) *Some aspects of oyster culture in Sabah*. Report, Sabah Fisheries Department.

Curtin L. (1971) Oyster Farming in New Zealand. *Fish. Tech. Rep., New Zealand Mar. Dep.*, **72**.

Curtin L. (1972) Development of rock oyster farming in New Zealand. In *Coastal Aquaculture in the*

Indo-Pacific Region, (Ed. by T.V.R. Pillay), pp. 384–93. Fishing News Books, Oxford.

Davy F.B. and Graham M. (Eds) (1982) *Bivalve Culture in Asia and the Pacific*. IDRC, Ottawa.

Dubois R. (1901) Sur la mécanisme de la formation des perles fines dans le *Mytilus edulis*. *C.R. Hebdomad Séances Acad. Sci.*, **133**, 603–5.

Dupuy J.L., Windsor N.T. and Sutton C.E. (1977) *Manual for Design and Operation of an Oyster Seed Hatchery*. VIM/Sea Grant Program. Virginia Institute of Marine Science. *Sp. Rep. Appl. Mar. Sci. Ocean Eng.*, **142**.

Figueras A. (1970) Flat oyster cultivation in Galicia. *Helgoländer Wiss. Meeresunters.*, **20**, 480–85.

Figueras A. (1979) Cultivo del mejillon, *Mytilus edulis*, y posibilidades para su expansion. In *Advances in Aquaculture*, (Ed. by T.V.R. Pillay and W.A. Dill), pp. 361–71. Fishing News Books, Oxford.

Fujiya M. (1970) Oyster farming in Japan. *Helgoländer Wiss. Meersunters.*, **20**, 464–79.

Furfari S.A. (1979) Shellfish purification. A review of current technology. In *Advances in Aquaculture*, (Ed. by T.V.R. Pillay and W.A. Dill), pp. 385–94. Fishing News Books, Oxford.

Hidu H. and Richmond M.S. (1974) Commercial oyster aquaculture in Maine. *Mar. Sea Grant Bull.*, **2**, 1–59.

Hidu H., Chapman S.R. and Dean D. (1981) Oyster mariculture in sub-boreal (Maine, United States of America) waters. Cultchless setting and nursery culture of European and American oysters. *J. Shellfish Res.*, **1**(1), 57–67.

Honma A. (1971) *Aquaculture in Japan*. Japan FΛO Association, Tokyo.

Hrs-Brenko M. (1973) Gonad development, spawning and rearing of *Mytilus* sp. larvae in the laboratory. *Stud. Rev. GFCM*, **52**, 53–65.

Imai T. (1978) Aquaculture in shallow seas. In *Progress in Shallow Sea Culture*, pp. 205–60. A.A. Balkema, Rotterdam.

Juntarashote K., Bahromtanarat S. and Grizel H. (1987) *Shellfish Culture in Southeast Asia*. SEAFDEC, Bangkok.

Kamara A.B. and McNeil K.B. (1975) *Preliminary oyster culture experiments in Sierra Leone*. Ministry of Agriculture and Natural Resources, Freetown, Sierra Leone (unpublished).

Kamara A.B., McNeil K.B. and Quayle D.B. (1979) Tropical mangrove oyster culture: problems and prospects. In *Advances in Aquaculture*, (Ed. by T.V.R. Pillay and W.A. Dill), pp. 344–48. Fishing News Books, Oxford.

Koganezawa A. (1979) The status of Pacific Oyster culture in Japan. In *Advances in Aquaculture*, (Ed. by T.V.R. Pillay and W.A. Dill), pp. 332–7. Fishing News Books, Oxford.

Korringa P. (1976a) *Farming the Flat Oysters of the Genus* Ostrea. Elsevier Scientific Publishing, Amsterdam.

Korringa P. (1976b) *Farming the Cupped Oysters of the Genus* Crassostrea. Elsevier Scientific Publishing, Amsterdam.

Korringa P. (1976b) *Farming Marine Organisms Low in the Food Chain*. Elsevier Scientific Publishing, Amsterdam.

Labrid C. (1968) *Apercu sur l'Ostreiculture*. Arcachonnaise Institut de Biologie Marine, Arcachon.

Lee K., Corbin J. and Brewer W. (1981) *Oyster Culture in Hawaii and Various United States Pacific Island Territories*. North American Oyster Workshop, Seattle, Washington, March 1981.

Lizárraga M. (1977) Técnicas aplicadas an el cultivo de moluscos en América Latina. *FAO Inf. Pesca*, **159**(2), 96–105.

Loosanoff V.L. (1971) Development of shellfish culture techniques. In *Proceedings of the Conference on Artificial Propagation of Commercially Valuable Shellfish*, 9–40. College of Marine Studies, University of Delaware, Newark, October 1969.

Loosanoff V.L. and Davis H.C. (1963) Rearing of bivalve molluscs. In *Advances in Marine Biology*, Academic Press, London, Vol. I, pp. 1–136.

Lutz R.A. (1978) Pearl incidence in *Mytilus edulis* and its commercial raft cultivation implications. *Proc. World Maricul. Soc.*, **9**, 509–22.

Lutz R.A. (1985) Mussel aquaculture in the United States. In *Crustacean and Mollusc Aquaculture in The United States*, (Ed. by J.V. Huner and E.E. Brown), pp. 311–63. Avi Publishing Company, Westport.

McFarlane S. (1971) New Zealand commercial mussel farming experiments in the Hauraki Gulf. In *Report on Mussel Cultivation*, pp. 10–13. Fishing Industry Board, Wellington, New Zealand, 5th Seminar, October 1971.

Mackenzie C.L., Loosanoff V.L. and Gnevuch W.T. (1961) Use of chemically-treated cultch for increased production of seed oysters. *Bur. Commer. Fish. Biol. Lab., Milford Conn. Bull.*, **5**(25), 1–9.

Mandelli E.F. and Acuna A. (1975) The culture of the mussel *Perna perna* and the mangrove oyster, *Crassostrea rhizophorae* in Venezuela. *Mar. Fish. Rev.*, **37**(1), 15–18.

Mason J. (1972) The cultivation of the European mussel, *Mytilus edulis* Linnaeus. *Oceanogr. Mar. Biol.*, **10**, 437–60.

Mathiessen G.C. (1971) *A Review of Oyster Culture and the Oyster Industry in North America*. Woods Hole Oceanographic Institution, Woods Hole.

Meixner R. (1979) Culture of Pacific oysters *Crassostrea gigas* in containers in German coastal waters. In *Advances in Aquaculture*, (Ed. by T.V.R. Pillay and W.A. Dill), pp. 338–9. Fishing News Books, Oxford.

Nikolic M., Bosch A. and Alfonso S. (1976) A system for farming the mangrove oyster (*Crassostrea rhizophorae* Guilding, 1828). *Aquaculture*, **9**, 1–18.

Padilla M. (1973) Observaciones biológicas relacionadas con el cultivo de *Mytilus edulis chilensis* en Aysen. *Publ. Inst. Fom. Pesq. Santiago*, **54**.

Pruder G., Bolton E. and Faunce S. (1977) *System Configuration and Performance in Bivalve Molluscan Aquaculture*. University of Delaware Sea Grant Publication DEL-SG-9–76.

Raimbault R. and Tournier M. (1973) Les Cultures Marines sur le Littoral Francais de la Méditerranée. *Bull. Inf. Doc. Inst. Scien. Tech. Pech. Mar. Nantes*, **223**.

Salaya J.J. (1973) Estudio sobre la biologia pesqueria y cultivo del mejillon, *Perna perna* (L.), en Venezuela. Officina Nacional de Pesca, *Inf. Tec.*, **62**.

Santa Cruz S.G. and Hojas F.C. (1975) Possibilidades de dessarrolo para el cultivo artificial de Mitilidos y Ostras en Chile. *Seminario Agroindustrial Santiago, Chile*, 13–39.

Scalfati G. (1970) La Mitilicoltura, il suo ambiente, l'organizzazione tecnico-economica e la disciplina giuridice. *Mem. Ministr. Mar. Mercant. Ital.*, **29**.

Sebastio C. (1968) *Contributo alla conoscenza della biologia dell'ostrica ed allo sviluppo della ostricoltura razionale in Italia*. Istituto Sperimentale per il Controllo Veterinario dei Prodotti della Pesca, Taranto.

Sindermann C.J. (1977) *Disease Diagnosis and Control in North American Marine Aquaculture*. Elsevier Scientific Publishing, New York.

Sindermann C.J. (1986) The role of pathology in aquaculture. In *Realism in Aquaculture: Achievements, Constraints, Perspectives*, (Ed. by M. Bilio, H. Rosenthal and C.J. Sindermann), pp. 395–419. European Aquaculture Society, Bredene.

Wells W.F. (1927) Report of the Experimental Shellfish Station. *New York State Conservation Department Report*, **16**, 1–22.

Wakamatsu T. (1973) *A ostra de Cananéia e seu cultivo*. SUDELPA-Instituto Oceanographico da Universidade de Sao Paulo.

Walne P.R. (1979) *Culture of Bivalve Molluscs*, 2nd edn. Fishing News Books, Oxford.

Zong-Quing N. (1982) Country reports – China. In *Bivalve Culture in Asia and the Pacific*, (Ed. by F.B. Davy and M. Graham), pp. 21–8. IDRC, Asia Regional Office, Singapore.

27
Clams, Scallops and Abalones

27.1 Clams

Clams, or cockles as they are called in some areas, form a valued item of food in many countries, although they are not so well recognized in others. The natural resources of clams are believed to be over-exploited in countries like Italy, where there is greater demand for it as an ingredient in the normal diet of the population. Simple methods of clam cultivation have been practised for centuries in China and Japan, and in recent years considerable research has been carried out, particularly in the USA and the UK, to develop more sophisticated methods of hatchery production of seed stock and on-growing. For various reasons, these methods have not yet found wide commercial application, and most of the present-day clam production is obtained from the simple systems which have been in existence for a long time, with only marginal improvements.

From among the many species of clams, belonging to some six families, at least eighteen species have been cultured experimentally or commercially. Of these, the more important ones in commercial production are: species of *Venerupis* (= *Tapes*) in Japan, Korea, the Philippines and the USA; *Meretrix* in Japan and Taiwan; *Mercenaria* and *Protothaca* in the USA and *Anadara* in several Southeast Asian countries (fig. 27.1). The species used in commercial culture are *Venerupis japonica* (= *Tapes semidecussata*, *T. japonica*) (Japanese little-neck; Manila clam); *Mercenaria mercenaria* (hard clam, quahog); *Meretrix meretrix* (big clam); *Meretrix lusoria* (clam) and *Anadara granosa* (blood cockle).

Most of these species are marine (found in salinities above 20 ppt) and burrow in sand or mud. Though hermaphroditism and protandry have been observed, the clams are generally dioecious. Fertilization is external and the embryo passes through the free-swimming trochophore and veliger larval stages before metamorphosis. They reach sexual maturity within about a year and readily respond to environmental stimuli, particularly temperature changes, and spawn easily. Larvae of most clams begin to set within 7–14 days. Though the growth rate depends very much on food availability and environmental conditions, most species of clams are reported to grow rapidly for one to four years and thereafter at a slower rate.

27.1.1 Culture systems

Clam culture systems closely follow those described earlier for other bivalves. This also applies to the hatchery methods, which appear to be employed commercially only in the USA. Like other bivalves, clams also seem to spawn in areas which are not very suitable for on-growing. So the simplest system of culture involves the transplantation of seed clams from the spawning areas to growing beds with sandy bottoms in shallow inter-tidal areas, which are not exposed for long periods. This system is widely adopted in Japanese clam culture. Before the seed are planted, the bottom soil is loosened by a harrow and left undisturbed for about a week.

Seedlings are planted by hand, and in temperate climates the best growth is obtained when they are planted in the spring. Seed is also available in the autumn, and if necessary

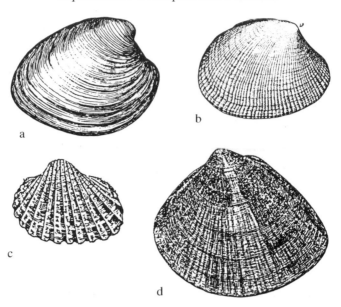

Fig. 27.1 Some of the important cultured clams: (a) *Mercenaria mercenaria*; (b) *Tapes (Ruditapes) philippinarum*; (c) *Anadora granosa*; (d) *Meretrix lusoria*.

planting can be done at this time as well. The stocking density is roughly $1-4\ \ell$ per m^2, but the main consideration is even distribution and avoidance of crowding. Under Japanese conditions, the market size of about 10 g is reached in between one and two years. Among the major predators are ducks, starfish, octopus and crabs. Harvesting is performed throughout the year, removing only the fully grown clams and leaving the smaller ones for further growth. Hand tools as well as dredges operated from boats are used for harvesting.

In the USA, where the hard clam or quahog is commercially cultivated, culture systems vary from intensive recirculating systems to transplantation of seed clams to suitable shellfish beds for on-growing. The hard clam has high temperature tolerance, but grows only in salinities above 20 ppt. Seed clams are produced in several commercial hatcheries in the USA and a clam farmer can obtain his requirements of seed from them, although many hatcheries use a good part of their production for on-growing in their own grow-out facilities. A somewhat unique system of artificial propagation of the small-necked clam (*Tapes philippinarum*) is carried out in China (Zong-Qing, 1982). Shallow ponds in the lower inter-

tidal region are prepared by eradicating pests and predators and fertilizing as in normal fish ponds. Selected phytoplankton, such as *Chaetoceros*, are introduced to produce enough food for the larvae. Sperm and eggs obtained by induced spawning are introduced into the ponds, where fertilization and development of embryos take place. Soyabean milk may be added to feed the larvae for faster development. It is reported that up to 15 million seed clams are thus produced in one hectare of ponds every breeding season.

27.1.2   Hatchery production of seed

The technologies of hatchery production are essentially the same as those employed for oysters. Many commercial hatcheries use salt water from wells to avoid the need for filtration and sterilization. Such hatcheries are entirely dependent on cultured algae for larval rearing while others use cultured algae to rear early larvae and the Wells-Glancy method for growing spat. As mentioned earlier, mature adults spawn in response to stimulation by changes of temperature, but if necessary a further stimulation can be provided by suspensions of sperm or dense phytoplankton infusions. Generally, commercial hatcheries use fibreglass tanks with conical bottoms and circular tops, of about $400-1600\ \ell$ capacity. The density of the larvae in the tanks is about $1-15$ per ml and the concentration of algae between 10000 and 100000 cells/ml per day. The hard clam larvae start metamorphosing in $6-8$ days at temperatures of about 25°C. The set larvae or spat grow to a size of 600 µm in two to three weeks, under favourable conditions. Then they are transferred to nurseries for growing to a size of about $4-6$ mm.

Nurseries and grow-out facilities employ more or less the same methods, namely bottom culture, tray or rack culture or suspended culture. Bottom nurseries in inter-tidal areas are covered by plastic mesh to protect the seed from predators. Baffles or similar devices are provided to prevent siltation of the nurseries. Tray or rack rearing can be carried out in either inter-tidal or sub-tidal areas. Trays are placed directly on the bottom or stacked in tiers. The seed in trays are well protected from predators. For suspended rearing, rafts, suspended plastic trays and lantern nets or pearl nets are commonly used. All of them are intended for high-density rearing in the warmer surface waters, free from benthic predators.

Land-based nurseries often use raceway systems, consisting of shallow trays or tanks made of fibreglass, cement or wood. Besides the shallow raceways, where the seed is spread on the bottom and water flows along the length of the raceway, deeper raceways with tiered racks formed by layers of mesh sheets are also used. Upflow systems, which cause vertical water flow through the seed containers, are becoming common in land-based nurseries. The conditions created by the upflow result in rapid growth.

27.1.3 Grow-out

In the final grow-out to market size, different systems of varying degrees of sophistication are used, ranging from simple sowing on the bottom to recirculation systems. A cost-effective system developed in the USA involves the use of covered sub-tidal areas in which baffles, pens or net tents are used together with crushed stones for protection of the seed and prevention of siltation. A stocking density of up to 4300

seed m^2 is possible in such beds. However, thinning of the stock to about half the original density is necessary as the clams grow in size. The grow-out period is generally two to three years, depending on local conditions, and the market size is 4 to 4.5 cm shell length. Tray culture allows higher densities of stock and many farms in the USA have used them successfully in the grow-out of *M. mercenaria*.

Clams of the genus *Meretrix*, especially *M. meretrix* and *M. lusoria*, are cultured in the Far East. The common species cultured in Taiwan is *M. lusoria*. Its culture is somewhat unique and different from those described above in that it is often grown in association with milkfish in the inlet and outlet channels of pond farms. It is also cultivated on sandy flats and tidal estuaries.

Seed clams are collected from coastal areas with rakes operated from small boats, mainly during summer, although seed can be found throughout the year. The seed clams of about 0.5 mm length are sold to seed clam growers, who rear them in shallow brackish-water ponds. If the ponds are not fertile enough, organic fertilizers are applied to promote growth of phytoplankton. The water is exchanged every three to four days. The stocking rate is about $30-50$ million seed/ha. Where possible, the growing clams are sorted according to size and replanted separately. In about 6 months they reach a size equivalent to $800-1000$ clams per kg and are ready for sale to farmers who grow them to market size.

Grow-out may be in sandy tidal flats, estuarine areas or in fish pond facilities. Beds with a high content of sand (at least 50 per cent) are selected as such bottoms provide for the burrowing of the clam and also seem to promote the attractive pink coloration which is important in marketing. In such areas it is usual to install fences of net mesh to prevent the escape of clams and the entry of predators. The stocking rate generally varies from $2000-5000$ kg seed clams of 600 per kg size in ponds, to as low as 100 kg per ha on sandy flats. They are spread evenly on the bottom. A size of 35 per kg may be reached in about 18 months. Harvesting from ponds is done by hand and from estuaries and tidal flats with rakes fitted with a net bag for holding the collected clams.

The clam that is important in several

countries of Asia is the cockle or the blood cockle *Anadora granosa*. It is cultured on a limited scale in most countries of the region, including China and Taiwan, and is found on muddy estuarine flats and bays with weak tidal currents and waves. The spawning season varies with the locality, e.g. in China from July to September and in Taiwan from January to April. The spat settle on fine, sandy mud flats in the lower inter-tidal areas.

Cockle culture is relatively simple and mainly consists of collecting natural spat and planting them in protected beds for rapid growth. The sites generally selected have soft muddy bottoms with 2−2.5 m water at high tide. In Thailand, shrimp farmers often use elevated parts of their ponds for cockle growing. Exposure of the bed for more than about 6 hours and sudden changes in salinity due to heavy rains can cause serious mortalities. Areas under cockle culture are often fenced in to prevent poaching. In Taiwan, the spat are generally nursed in specially prepared mud flats enclosed by fencing of nylon netting. They are grown to a size of 5000 per kg and then sold to farmers who grow them to market size. Some of the seed producers operate their own grow-out farms. Stocked at the above size, the cockles grow to 500−600 per kg in one year in Taiwan. The minimum market size is 120 per kg and to reach that size it usually takes two to three years or more. During this period the farmer tries to eradicate predators and pests, such as wild ducks, crabs, sea snails and puffer fish. Harvesting is generally done manually.

Relatively fewer clam or cockle diseases have been reported. Many of the known diseases of juveniles and adult clams are caused by the haplosporidian *Perkinsus marinus*, the coccidia *Hyaloklossia* and *Pseudoklossia*, the gregarine *Nematopsis* and ciliates like *Trichodina* and *Ancistrocoma*.

27.2　Scallops

Scallops form another group of bivalve molluscs of considerable importance in the seafood industry. Declining landings from natural beds have aroused interest in increasing production through application of aquaculture techniques. Though significant research on the biology of a number of species of scallops has been done in many countries, commercial-scale culture is limited to only a few, mainly Japan and China. The most important species in Japan is the deep-sea scallop or the giant ezo scallop *Patinopecten yessoensis* and the most important one in China is *Chlamys farreri*. Commercial culture of a larger species, *C. nobilis*, has been started in recent years in some areas of southern China. In the USA, the bay scallop *Argopecten irradians* has been spawned in hatcheries and the seed grown to market size in pens (Castagna, 1975). The giant scallop, *Placopecten magellanicus*, has also been investigated in terms of its potential for farming. Though many species seem to prefer fairly hard substrates with little mud in coastal zones, the European tiger scallop, *Chlamys tigerina*, prefers coarse sandy mud, gravel or stones. The European king scallop, *Pectinopecten maximus*, prefers bottoms of clean firm sand, fine gravel or sandy gravel. The feeding habits of scallops are very similar to those of other bivalves and the main components of food are reported to be diatoms, protozoa and considerable amounts of detritus. The deep-sea scallops show abnormally high levels of fecundity and a fully grown 4-year-old is capable of producing 160 million eggs. The sexes are separate, although hermaphrodites may sometimes be found. In Japan this species spawns between March and July, depending on the location.

The systems of culture adopted for experimental and commercial production of scallops are very similar to those used for other bivalves. Early efforts involved the collection of seed scallops and their planting in either depleted beds or in areas suited for rapid growth. This led to the next stage of rearing the seed in cages or ponds and releasing them in natural beds after they had grown to a shell length of about 3 mm, to improve survival rates. The present trend is to raise the seed to market size using suspended systems. In China, scallops are often cultured in association with seaweed in foreshore areas, in lantern or pearl nets suspended from long lines.

27.2.1　Spat collection

A variety of substrates have been used for collecting seed scallops. The traditional collectors, like cedar leaves and scallop shells, have been replaced in Japan by polyethylene-mesh bags attached to long-line systems. The long lines vary in length from 50 to 100 m and the number of bags attached to each line depends on the water depth and the number of branch lines. The bags are filled with artificial substrates such as mesh monofilament, soft netlon, nylon and plastic meshes and rubberized fibre. Any substrate with a clean surface can be used for collecting seed of the European scallops *C. maximus* and *C. opercularis*. The mesh size of the bag depends on the local conditions including the silt load in the area. The mesh is selected with a view to facilitating free water flow, at the same time preventing the escape of the settled spat and clogging of the meshes by silt. The collectors are laid out at times of spatfall, which are forecast on the basis of close monitoring of water temperature, spawning and larval abundance.

27.2.2 Hatchery production of spat

As in the case of oysters, mature scallops can be induced to spawn by thermal shock and larvae are grown in hatcheries by techniques similar to those for other bivalves. However, this method of seed production has not become very widespread because of the easy availability of natural seed. On the other hand, most of the seed used in culture of *C. farreri* and *C. nobilis* in China are reported to be hatchery-produced. In North America, too, the trend is to use hatchery-produced spat.

Sexually mature adults can be induced to spawn by increases in temperature and by the addition of milt from mature males in spawning tanks. Most species of scallop can be spawned by temperature shock, when properly conditioned brood stock are used. The bay scallop, *A. irradians*, the Chilean scallop, *A. purpuratus*, the deep sea scallop, *P. yessoensis*, and the Chinese scallops mentioned above have all been spawned in hatcheries by this method. Other stimuli which can be used are ultraviolet rays and hydrogen peroxide in very low concentrations.

Scallops show high fecundity, for example a fully mature *P. yessoensis* can produce at least 100 million eggs. In spawning hermaphrodite individuals, care has to be taken to avoid polyspermy by collecting the sperm and eggs in separate containers and then artificially fertilizing the eggs. Fertilization can be achieved by adding 2−6 ml sperm suspension for every litre of egg suspension. Embryonic development is rapid and free-swimming trochophores can be obtained in 8−24 hours, depending on water temperature. The optimum temperature for the development of the bay scallop is reported to be 26−28°C and gentle aeration and frequent water exchange are usually maintained in the tanks. The larvae are fed on phytoplanktonic organisms such as *Monochrysis*, *Chaetoceros*, *Isochrysis*, *Phaeodactylum*, *Dunaliella* and *Tetraselmis*. The larvae normally settle within two weeks.

27.2.3 Grow-out

The simplest system of grow-out is by planting in suitable beds. The spat are released from boats during the summer months. The density of spat depends on local conditions, but five or six per m² is considered suitable for obtaining a commercial size of five or six per kg. The survival rate is only about 25−30 per cent. Even in hanging culture, more widely practised now, the mortality rates are high and so a two-phase system of grow-out has been recommended. In the first phase, the spat are grown in pearl nets or lantern nets on long lines (fig. 27.2) until they develop harder shells.

Long lines used for early grow-out are usually longer than those used for spat collection, and may be as long as 200 m. Pearl nets hung from the long lines are conical or pyramidal in shape (fig. 27.3), made of small-mesh (2−7 mm) net, and provide adequate protection to the spat from predators and silt deposition. Free circulation of water is ensured. The maximum number of spat in each net is about 100 and the nets are hung at 2−12 m depth of water. As the growth rate of the spat is rather rapid, the stock has to be thinned out at regular intervals and transferred to new nets.

The traditional Japanese method of grow-out to market size consists of hanging seed scallops on long lines, sometimes referred to as the 'ear hanging' method. Small holes are drilled on the anterior side of the ear of each seed scallop (c.2 mm size), threaded with nylon

Fig. 27.2 Lantern nets used for growing scallops on long lines in China.

and tied to ropes for hanging on long lines (fig. 27.4).

Most modern hanging culture facilities use long-line systems and rafts. While long lines can be used in shallow inshore areas, as well as offshore deep areas, rafts are only suitable for protected areas. In inshore waters, simple long lines of about 50–60 m length are used, whereas more complicated systems of long lines of 100–480 m length, with several branch lines, are used in deeper areas. Using such large long-line systems, as many as half a million scallops are reported to be cultured in an area of 9 ha in sites exposed to strong wave action and wind during winter months. Spat with hardened shells are placed in lantern nets (known also as Adnon baskets) and hung from the long-line system. The lantern nets are made of monofilament netting (of about 12–25 mm mesh), supported by plastic-coated or gal-

Fig. 27.3 Two-year-old Japanese scallops (*Pectinopecten yessoensis*) grown in nylon-mesh, sandwich-type frames in suspended culture (from Bardach *et al*., 1972. By permission of John Wiley & Sons).

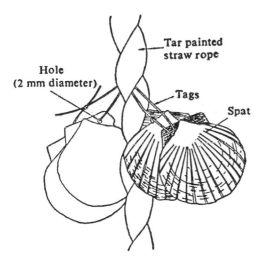

Fig. 27.4 The ear hole hanging method of grow-out of scallops (from Imai, 1970).

vanized wire hoops of about 50 cm diameter which divide the net into several compartments of about 15 cm height (fig. 27.5). Each compartment has a lateral opening through which scallops can be introduced or removed. The stocking density determines the growth rate and a density of 20 scallops per compartment is reported to give good results. These nets are suspended by nylon ropes.

An old system of growing scallops in suspended 'pocket nets' is still practised in some areas of Japan. This type of net is made of polyethylene-mesh attached to a framework (usually 140 cm high and 45 cm wide) made of plastic-coated or galvanized wire. A series of pockets are stitched on the mesh at different levels. Seed scallops are placed in the pockets which are closed by tying with nylon yarn. The net is hung in the water by means of nylon ropes.

It normally takes two to three years for the seed scallops to grow to market size, which for the deep-sea scallop in Japan is 10–11 cm in length.

At temperatures between 12 and 20°C, the Chinese species of scallops reach marketable size in 1.5–2 years. Floating racks and plastic mesh cages are used for grow-out and many growers rear scallops and the seaweed *Laminaria* in the same site, as it increases the

Fig. 27.5 A close view of the scallops growing in lantern nets in *Laminaria* farms in China.

total production and income from the same area. The market size for *C. farreri* is about 6−7 cm and for *C. nobilis* it is 7 cm length.

27.3 Abalones

Abalones belonging to the genus *Haliotis* are the most valuable marine gastropods and probably the most sought-after molluscan seafood in many areas. They are marketed in fresh, frozen, canned and dried forms and are eaten raw or cooked. The valued meat is the very large foot or right shell muscle, one of which can provide several sliced steaks. The shells are also of economic importance as they are used in traditional medicines and also for decorative purposes and jewellery. The major producing countries are Mexico, Japan and Australia.

The USA, New Zealand, South Africa, North and South Korea and Canada also land fair quantities.

Because of the increasing demand and diminishing natural stocks, attention has been directed to enhancing the stocks through transplantation and stocking of open waters with hatchery-reared seed. Significant success in stocking abalone beds with hatchery-produced seed has been achieved in Japan and North Korea and several million seed are annually planted. Recoveries of 15–20 per cent of the stocked abalone have proved the economic viability of this practice under Japanese conditions. Even though the main thrust of abalone culture continues to be natural stock enhancement, concerted efforts are underway in the USA, Japan and China to develop commercially viable methods of growing abalone to market size under controlled conditions. The production so far by aquaculture is rather small, but some of the research findings are of special interest.

There are about 80 species of abalone, distributed in temperate and semi-tropical coasts, which show major differences in thermal requirements and food preferences. These affect not only their growth and survival rates but also their colour and meat quality. The largest abalone world-wide and the most important North American species extending down to Mexico is the red abalone *Haliotes rufescens* (fig. 27.6). The green abalone, *H. fulgens*, and the white abalone, *H. sorenseni*, are known for their high-quality meat, but contribute relatively less to commercial landings. Of about ten species of abalone found in Japan, the most important is *H. discus hannai*. The others of importance are *H. diversicolor*, *H. gigantea* and *H. sieboldi*. *Haliotes discus hannai* is the important species in North and South Korea. The Australian and New Zealand species of abalone are of smaller size and therefore of less commercial value.

Abalones are nocturnal and live in rocky belts. The sexes are separate, but occasionally hermaphroditic individuals have been found. In warm climates they may spawn throughout the year, but in colder areas spawning may be only during the warm months of summer. Spawning of gravid abalones is triggered by sudden changes in water temperature, exposure

Fig. 27.6 The red abalone, *Haliotes rufescens* (from *Fish Farming International*, **7**(1)).

to air or release of gametes by other spawning abalones. A sudden contraction of the foot muscle caused by such factors forces out the eggs and milt. They are highly fecund and a large red abalone may spawn as many as 10 million eggs at a time. Fertilization is external and fertilized eggs sink to the bottom, where embryonic development takes place. In most species, the planktonic trochophore larvae hatch out within about a day under favourable temperature conditions. The veliger stage lasts for about a week, by which time the complete shell and operculum have developed. The larvae then seek out suitable substrates and settle. In the absence of a suitable substrate the larvae can prolong the planktonic stage up to about 3 weeks.

They settle easily on coralline red algae (such as *Lithothamnium* and *Lithophyllum* spp.) and this is reported to be due to a biochemical inducer present on the surface of these algae. Once a suitable substrate is found, the larvae attach themselves by their feet and soon after start feeding on the attached algae, metamorphosing into juvenile abalones. The most interesting aspect of the larval development and metamorphosis of abalone is that the larvae are completely dependent on the egg yolk for their nutrition until they reach the veliger stage. So the difficult problems of feeding early larvae do not occur.

As the juveniles grow in size they feed on epiphytic diatoms and other microscopic algae. Abalones are known for their slow growth rates, but the growth can be enhanced to a certain extent by increasing the water temperature and abundance of the preferred algal species. Each species of abalone has its own preferences for algae. For example, the preferred algae for the North American species include brown algae: giant kelp (*Macrocystis*), bull kelp (*Nereocystis*), feather boa kelp (*Egregia*); red algae: *Gigartina*, *Gelidium* and *Plocamium* and the green alga; sea lettuce (*Ulva*). The young (4−5 mm) of the Japanese species feed on *Undaria*, *Eiosenia*, *Codium*, *Aalymenia* and *Ulva*. The adults show greater selectivity and choose, in order of preference, brown algae, followed by green and then red. The colour and pattern of the abalones' shells are very much dependent on the algae they feed on.

Adult abalones lead a sedentary life in crevices or rock ledges, but the juveniles are more mobile. They forage mainly at night on drifting macroalgae. The typical growth rate is 20−30 mm per year. As light has a depressing effect on feeding and growth, enhanced growth can be obtained by reducing the lighting. Doubling of the growth rate has been achieved on an experimental scale.

27.3.1 Controlled reproduction and larval rearing

Wild or hatchery-reared brood stock can be used for controlled production of seed. If wild brood stock are used, it is considered essential to condition them in holding tanks for two or three weeks. Investigations on the Japanese abalone, *H. discus hannai*, show the importance of conditioning at an optimum temperature and feeding on the preferred food of fresh seaweed (e.g. *Undaria* and *Laminaria*) for successful maturation. According to Kan-No (1975), if the rearing is carried out at 20°C, the species will attain maturity in about 80 days, even in the winter season. Maturity can be maintained for at least three months.

Since the stimuli of elevated water temperature and/or air drying do not provide a reliable means of inducing spawning, the use of ultraviolet irradiated sea water is sometimes adopted to ensure consistent spawning. The gravid animals are exposed to flowing heated sea water, irradiated with uv light. An irradiation of 800 milliwatt hours per litre is reported to be adequate, and spawning occurs in about three hours.

Another method of spawning abalones is the one developed in the USA of exposing brood stock to hydrogen peroxide. This is based on the finding that hormone-like prostaglandins regulate spawning in abalones: hydrogen peroxide activates the natural enzymic synthesis of prostaglandins in gravid animals and so spawning can be induced. Gravid stock are kept in suitable containers of sea water (temperature 12−18°C) made alkaline to a pH of about 9.1 by the addition of sodium hydroxide. A 6 per cent solution (freshly diluted from a 30 per cent stock solution) of hydrogen peroxide is introduced into the container at the rate of 50 ml for each 12 ℓ of water. After an exposure of about 2.5 hours, the water is drained and immediately replaced with isothermal fresh sea water. Spawning can be expected to occur within 2.5−3.5 hours.

If the spawning tank contains both females and males, fertilization of the eggs takes place in the tank itself. A sex ratio of one male to four females is maintained in such tanks. If, however, the males and females are spawned in separate tanks, the gametes can be collected and fertilized separately. The fertilized eggs are washed free from excess sperm and incubated in clean sea water at a temperature of about 14−16°C. The larval trochophores hatch out 18−24 hours after fertilization. In about 7 days they reach the veliger stage and if care is taken to maintain water quality and prevent microbial growth in the culture, very high larval survival rates can be expected. Species like *H. diversicolor* develop faster and may assume a benthic life within 43−46 hours after hatching.

For larval settlement, special tanks made of fibreglass provided with filtered running sea water are commonly used. The tanks are 'seasoned' with a growth of benthic diatoms, bacteria and microalgae, in particular red algae. Fluorescent lighting is provided to promote the growth of diatoms in the tanks. The stocking density of larvae appears to vary very considerably, depending on the water quality and methods of feeding.

Investigations have shown that the abalone requires a specific biochemical inducer for normal settlement, metamorphosis and rapid subsequent development of juveniles, which has been identified as the amino acid gamma-aminobutyric acid (GABA), contained in the red algae. The addition of a low concentration of this inducer causes rapid synchronous and completely normal settlement, metamorphosis and juvenile growth. Crustose coralline red algae, or specific proteins derived from these algae, cause a similar induction of settlement and metamorphosis, but in culture systems predation by microscopic faunae associated with the algae may cause mortalities. The use of GABA is therefore considered more convenient and inexpensive. Another means of inducing larvae to settle is by using the mucus of juveniles and adult animals together with diatoms (Chew, 1986).

27.3.2 Grow-out

Juvenile abalones of 5 mm size are generally transferred to grow-out tanks, and are gradually introduced to macroalgae. For the juveniles of the Japanese species of abalone, particularly *H. discus hannai*, the most suitable food for growth are *Undaria* and *Eisenia*, followed by *Codium*, *Ulva*, *Grateloupia* and *Rhodymenia*. Benthic diatoms are very suitable food until the juveniles reach a shell length of about 20 mm. *Egregia* and *Macrocystis* are commonly fed to abalones in the USA. When they reach a size of 15−20 mm, a diet of macroalga is required. The stocking density in grow-out tanks with gravel substrates in Japan is about 2000−2500 per m^2. The mortality of young abalones of 5 mm size in commercial culture is very high (as high as 99 per cent), but it is much less (about 10 per cent) from the 5 to 30 mm stage.

On the Californian coast in the USA, abalone growers concentrate on growing to the speciality market size of 5−10 cm, rather than the normal commercial size of 18 cm harvested from natural stocks. It takes 2 to 5 years to grow them to the gourmet size. Considerable

Fig. 27.7 *Haliotes discus hannai* cultured in plastic cylinders, in association with *Laminaria*, suspended from long lines in China (photograph: P. Buino).

research is presently underway to enhance growth rates by artificial feeds and by using intensive culture in tanks, raceways or ponds or containment systems in the open ocean or protected bays. In raceways supplied with heated water from power plants, abalones are reported to have grown four or five times faster than in the natural environment. Artificial diets for young abalones, containing sodium alginate extracted from the giant kelp *Macrocystis*, which also serves as a feeding stimulant and binder, have been in use for some time now. A crude protein content above 20 per cent has been found to be adequate for normal growth.

In China, hatchery-raised seed abalones are grown in onshore areas, inside plastic cylinders covered with close-meshed sieves (fig. 27.7). They are suspended from long lines in a polyculture system with scallops and seaweeds (see Chapter 28). The sea water is fertilized regularly by the spraying of inorganic fertilizers and the algal production induced by this adds to the availability of natural food for the abalone. They are reported to grow to a marketable size of about 6 cm in one to two years.

As mentioned earlier, the most successful system so far is the planting of hatchery-reared seed in protected areas of the sea. Several ways of increasing survival of the planted seed have been tried. One is the provision of portable habitats made of concrete blocks or shelves, to serve as substrates and refuges on the sea bed. Another is the use of transplantation cages to acclimatize the seed and to reduce initial mortality. Planting larger juveniles (about 22 mm size) has been found to give an average recovery of 23–31 per cent.

Diseases and parasitic infections have not yet become a problem in abalone culture. Post-larval mortality is generally caused by predation by small worms and crustaceans hiding on algal substrates. Several species of fish, crustaceans, starfish and other molluscs prey on juvenile abalones. Octopus, starfish, rock fish, rays and the sea otter, *Enhydra lutris*, have been observed to be particularly important predators of adult abalones.

27.4 References

Anderson G.J., Miller M.B. and Chew K.K. (1982) *A Guide to Manila Clam Aquaculture in Puget Sound*. Sea Grant Report, University of Washington, Seattle.

Bardach J.E., Ryther J.H. and McLarney W.O. (1972) *Aquaculture*. John Wiley & Sons, New York.

Brand A.R., Paul J.D. and Hoogesteger J.N. (1980) Spat settlement of the scallops *Chlamys opercularis* (L.) and *Pecten maximus* (L.) on artificial collectors. *J. Mar. Biol. Assoc.*, **60**, 379–90.

Castagna M. (1975) Culture of the bay scallop *Argopecten irradians* in Virginia. *Mar. Fish. Rev.*, **37**, 19–24.

Castagna M. and Duggan W. (1971) Rearing of the bay scallop, *Aequipecten irradians*. *Proc. Natl. Shellfish Assoc.*, **61**, 80–85.

Castagna M. and Kraeuter J.N. (1981) *Manual for Growing the Hard Clam* Mercenaria. Virginia Institute of Marine Science, Virginia, *Special Report*, No. 249.

Chen T.P. (1976) *Aquaculture Practices in Taiwan*, pp. 95–102. Fishing News Books, Oxford.

Chew K.K. (1986) Review of recent molluscan culture. In *Realism in Aquaculture: Achievements, Constraints, Perspectives*, (Ed. by M. Bilio, H. Rosenthal and C.J. Sindermann), pp. 173–95. European Aquaculture Society, Bredene.

Comely C.A. (1972) Larval culture of the scallop *Pecten maximus*. *J. Cons. Int. Expl. Mer.*, **34**, 365–78.

Culliney J.L. (1974) Larval development of the giant scallop, *Placopecten magellanicus* (Gmelin). *Biol. Bull. Mar. Biol. Lab. Woods Hole*, **147**, 321–32.

Daggan W.P. (1973) Growth and survival of the bay scallop, *Argopecten irradians*, at various locations in the water column at various densities. *Proc. Natl Shellfish Assoc.*, **63**, 68–71.

Hadley N.H. and Manzi J.J. (1984) Growth of seed clams (*Mercenaria mercenaria*) at various densities in a commercial scale nursery system. *Aquaculture*, **36**, 369–78.

Hooker N. and Morse D.E. (1985) Abalone: The emerging development of commercial cultivation in the United States. In *Crustacean and Mollusc Aquaculture in the United States*, (Ed. by J.V. Huner and E.E. Brown), pp. 365–413. Avi Publishing Company, Westport.

Illanes-Bucher J.E. (1987) Cultivation of the northern scallop of Chile [*Chlamys (Argopecten) purpurata*] in controlled and natural environments. In *6th International Pectinid Workshop, Menai Bridge, Wales*, (Ed. by A.R. Beaumont and J. Mason). ICES.

Imai T. (Ed.) (1978) *Aquaculture in the Shallow Seas*, pp. 263–364. A.A. Balkema, Rotterdam.

Ino T. (1980) Biological studies on the propagation of the Japanese Abalone (Genus *Haliotis*). *Bull.*

Tokai Reg. Fish Res. Lab., **5**, 1−102.

Inoue M. (1976) Mass production and transplantation of abalone. *Bull. Kanagawa Exp. Fish. Stn*, **131**, 295−307.

Ito S., Kanno H. and Takashashi K. (1975) Some problems on the culture of the scallop in Mutsu Bay. *Bull. Mar. Biol. Stn. Asamushi, Tohoku University*, **15**, 89−100.

Iverson E.S. (1976) *Farming the Edge of the Sea*, pp. 174−7. Fishing News Books, Oxford.

Kan-No H. (1975) Recent advances in abalone culture in Japan. *Proc. 1st Int. Conf. Aquaculture Nutr.*, pp. 195−211. University of Delaware, Newark.

Kikuchi S. and Uki N. (1974) Technical study on artificial spawning of abalone, genus *Haliotis. II.* Effects of seawater irradiated with ultraviolet rays in induction of spawning. *Bull. Tohoku Reg. Fish. Res. Lab.*, **33**, 79−86.

Kikuchi S. *et al.* (1967) Food values of certain marine algae for the growth of the young abalone, *Haliotis discus hannai. Bull. Tohoku Reg. Fish. Res. Lab.*, **27**, 93−100.

Leighton D.L. (1971) Observations of the effect of diet on shell coloration in the red abalone *Haliotis rufescens* Swainson. *Veliger*, **4**, 104−13.

Leighton D.L. *et al.* (1981) Acceleration of development and growth in green abalone, *Haliotis fulgens*, using warmed effluent seawater. *Proc. World Maricul. Soc.*, **12**, 170−80.

Lovatelli A. (1987) *Status of Scallop Farming: A Review of Techniques.* Network of Aquaculture Centres in Asia, NACA-SF/WP/87/1.

MacKenzie C.L. (1977) Predation on hard clam (*Mercenaria mercenaria*) populations. *Trans. Am. Fish. Soc.*, **106**(6), 530−37.

Manzi J.J. (1985) Clam aquaculture. In *Crustacean and Mollusc Aquaculture in the United States*, (Ed. by J.V. Huner and E.E. Brown), pp. 275−310. Avi Publishing Company, Westport.

Miyamoto T. *et al.* (1982) Experimental studies on the release of the cultured seeds of abalone *Haliotis discus hannai* Ino in Oshoro Bay, Hokkaido. *Sci. Rep. Hokkaido Fish. Exp. Stn*, **24**, 59−89.

Morse D.E. (1980) Recent advances in biochemical control of reproduction, settling, metamorphosis and development of abalones and other molluscs: applicability for more efficient cultivation and reseeding. *Proc. Natl. Shellfish. Assoc.*, **70**, 132−3.

Morse A.N.C. and Morse D.E. (1984) Recruitment and metamorphosis of *Haliotis* larvae are induced by molecules uniquely available at the surfaces of crustose red algae. *J. Exp. Mar. Biol. Ecol.*, **75**, 191−215.

Morse D.E., Hooker N. and Morse A. (1978) Chemical control of reproduction in bivalve and gastropod molluscs. III: An inexpensive technique for mariculture of many species. *Proc. World Maricul. Soc.*, **9**, 543−7.

Morse D.E. *et al.* (1977) Hydrogen peroxide induces spawning in molluscs with activation of prostaglandin endoperoxide synthetase. *Science*, **196**, 298−300.

Mottett M. (1979) A review of the fishery biology and culture of scallops. *Tech. Rep. Wash. Dept. Fish.*, **39**, 1−100.

Naidu K.S. and Scaplen R. (1979) Settlement and survival of the giant scallop *Placopecten magellanicus* larvae on enclosed polyethylene film collectors. In *Advances in Aquaculture*, (Ed. by T.V.R. Pillay and W.A. Dill), pp. 379−81. Fishing News Books, Oxford.

Paul J.D., Brand A.R. and Hoogesteger J.N. (1981) Experimental cultivation of the scallop *Chlamys opercularis* (L.) and *Pecten maximus* (L.) using naturally produced spat. *Aquaculture*, **24**, 31−44.

Sastry A.N (1963) Reproduction of the bay scallop *Aequipecten irradians* Lamarck: influence of temperature on maturation and spawning. *Biol. Bull.*, **125**(1), 146−53.

Sastry A.N. (1965) The development and external morphology of pelagic larval and post-larval stages of the bay scallop *A. irradians concentricus* reared in the laboratory. *Bull. Mar. Sci.*, **15**(2), 417−35.

Shaw W.N. (1972) Aquaculture of sea scallops and abalone in Japan. *Proc. World Maricul. Soc.*, **3**, 303−8.

Tanaka K. (1978) Development of culture-based abalone fishery. *Fish. J.*, **4**(6), 9−97.

Ventilla R.F. (1982) The scallop industry in Japan. *Adv. Mar. Biol.*, **20**, 310−82.

Wallace J.C. and Reinsnes T.G. (1985) The significance of various environmental parameters for the growth of the Iceland scallop *Chlamys islandica* (Pectinidae) in hanging culture. *Aquaculture*, **44**, 229−42.

Yamamoto G. (1975) Recent advances in the ecological studies on the Japanese scallops. *Bull. Biol. Stn Asamushi, Tohoko University*, **15**, 53−8.

Zhong-Qing N. (1982) Country Report − China. In *Bivalve culture in Asia and the Pacific*, (Ed. by F.B. Davy and M. Graham). pp. 21−28. IDRC Asia Regional Office, Singapore.

28
Seaweeds

Though aquaculture as practised today is largely based on vertebrate and invertebrate animal species, plants contribute a substantial proportion of world production through aquatic farming. It was estimated that the total seaweed aquaculture production in 1983 was about 2.4 million tons (wet weight), which was about 22.8 per cent of the overall aquaculture production that year. According to Tseng (1984) China is now able to produce more than 1 million tons of fresh seaweeds by phycoculture. So from the point of view of output, seaweed farming forms a significant part of aquaculture, although it is largely confined to about five countries of Asia (Japan, China, Korea, the Philippines and Taiwan).

In the major producing countries, namely Japan, China and Korea, seaweeds are grown mainly for human consumption. Seaweeds are also used as fodder and in the manufacture of agar, carrageenan, alginates, mannitol and iodine. China now produces on a commercial scale analogue foods like 'shredded jellyfish' from seaweeds. Many other countries, including the USA, Canada and some Carribbean Islands, are now undertaking experimental and pilot-scale culture to produce raw material for industrial uses. Small-scale farming of freshwater aquatic plants like the water chestnut (*Trapa* spp.), water cress (*Nasturtium* spp.) and the water spinach (*Ipomoea* spp.) has been undertaken by Asian farmers for many centuries, but large-scale farming of aquatic plants has only been done in the marine environment. It originated in Japan about three centuries ago with the culture of 'nori' or the laver (*Porphyra* spp.), which continues to be the most important species cultivated for human consumption. Sea-weed culture, particularly of *Laminaria*, has advanced rapidly in China.

Doubts have been expressed about the value of seaweeds in human nutrition. Their use as a condiment or vegetable is limited to oriental countries and among certain ethnic groups elsewhere. Though the total per capita consumption is not very high, its protein content is not low (35.6 per cent in dried nori, Bardach *et al.*, 1972). The amino acid composition is reported to be $10-30$ per cent of the dry weight, and the contents of vitamins A, B_1, B_2, B_6, B_{12}, C and niacin are very high. In addition, these edible seaweeds have higher contents of the important minerals calcium and iron than vegetables and fruits (Fujiwara−Arasaki *et al.*, 1984). Irrespective of the direct nutritional value, it has an important role in the overall food consumption of the people and therefore obviously in their nutrition. An unsatisfied demand for good quality seaweed products for food, additives for food products and other industrial uses provided the rationale for the increased interest in introducing or expanding seaweed culture in several countries.

The main groups of seaweeds cultivated for human food are the following:

Red algae (Rhodophyceae)
    Porphyra spp.
Brown algae (Phaeophyceae)
    Undaria pinnatifida
    Laminaria spp.
Green algae (Chlorophyceae)
    Enteromorpha compressa
    Monostroma spp.

All are typically marine species, but there

are differences between species in their salinity and temperature tolerance. Many of them cannot withstand exposure to wide variations in salinity. Many of the edible seaweeds require lower temperatures, between 10 and 20°C, for rapid growth. They are largely inter-tidal and sub-tidal species, and the lower limits of vertical distribution are governed by the levels of light intensity. Reproduction can be both sexual and asexual. Some species of red algae exhibit a biphasic (gametophyte, carposporophyte) type of alternation of generation, while others are triphasic (gametophyte, carposporophyte, tetrasporophyte). The discovery of the microscopic conchocelis phase in *Porphyra* spp. has been a landmark in the understanding of the summer phase of the reproductive cycle of these seaweeds. Asexual reproduction by means of asexual (neutral) spores occurs in the rather young stage of the leafy plants and this often accounts for the heavy settlement on collectors in culture operations. Vegetative propagation is also common among most of the cultivated species.

28.1 Culture systems

The nori culture system used to consist of bundles of bamboo or twigs, or rocks or concrete blocks, placed on the seabed for the monospores to settle on. The materials with their attached monospores were then transferred to suitable sites such as inshore areas near estuaries, for the development of thalli to the desired size. But the most common method now is the use of nets with a large mesh, of 15 cm x 15 cm, and 'blinds' made of split bamboo, strung with ropes at intervals of about 10–15 cm generally known as 'hibi'. The spores are caught on these and transferred to suitable places for grow-out.

Blasting rocky reefs or rock surfaces to alter depth and to expose additional surface for propagation is a common practice, particularly in *Laminaria* and *Undaria* culture. The scattering of broken pieces of rock also helps to increase substrate area and eliminate unwanted weeds. Artificial seeding was started when more was known about the life history of the cultivated seaweeds, and various types of nets and frameworks of braided strings are now used for this purpose. In commercial culture of *Echeuma* in China, cuttings of the plant are inserted in sublittoral reefs by divers. A new method adopted in recent years consists of fastening cuttings to coral branches with rubber rings and dropping them on the reefs. Divers rearrange them if required.

Raft and rack culture are practised in Japan and China. Brown algae, particularly *Laminaria*, are often cultured on long lines of ropes. An ingenious way of culturing *Laminaria* in China is the use of basket rafts, consisting of a series of cylindrical bamboo baskets tied together, each containing an amount of fertilizer which can be replaced when used up. Sporophytes are attached to the ropes tied along the sides of the basket. Instead of baskets, single rows of bamboo poles or synthetic tubing are also used to suspend young sporophytes, with earthenware jars containing fertilizer hung at intervals of about 6 m.

A system of polyculture of seaweeds (mainly *Laminaria*) has been developed in recent years in China on the Qingdao coast. Scallops are grown in lantern or pearl nets and abalones in plastic drums suspended from long lines between the seaweeds (fig. 28.1). Inorganic fertilizer is sprayed over the sea to improve the growth of seaweeds. The nutrients not used by the seaweeds serve to increase the production of algae which forms the food of scallops and abalones.

In some of the Asian countries, like the Philippines and Taiwan, *Gracilaria* and *Caulerpa* are grown in ponds, following many of the procedures common in fish culture such as pond fertilization, water management and disease and pest control. The culture of these species are largely for industrial use. Large-scale seaweed culture for waste recycling and industrial uses in North America has concentrated on growing unattached masses in raceways and greenhouse tanks flushed with sea water. Raceways shaped in the form of a 'V' in cross-section or the usual ones with a single sloping bottom are used, with compressed air circulation. Experimental culture of the carrageenan-producing red alga *Hypnea* in artificial upwelling systems has given encouraging results. Nutrient-rich water pumped from depths of 870 m into ponds was first used for filter feeding shellfish, and seaweed formed a secondary crop.

Fig. 28.1 Long-line method of culturing *Laminaria* in China.

28.2 Culture practices

28.2.1 Porphyra culture

In terms of magnitude of production and the history of technological developments, the red alga, nori or amanore (*Porphyra*) is the most important among edible seaweeds in Japan. There are over twenty species of the genus, but only four species are actually cultivated to any appreciable extent. *Porphyra tenera* is the predominant species, the others being *P. angusta*, *P. kuniedai* and *P. yezoensis*. *Porphyra tenera* is androdioecious, *P. kuniedai* and *P. yezoensis* are monoecious and *P. angusta* is dioecious. The growth periods of *P. tenera* and *P. angusta* are relatively short and that of *P. yezoensis* is somewhat long. The slowest growth is found in *P. kuniedai*. The species of *Porphyra* cultivated in China are mainly *P. haitanensis* in the southern part of the country and *P. yezoensis* in the north.

Culture using natural seedlings

The traditional system of catching monospores of *Porphyra* on bamboo or twigs of other trees is seldom practised now. Nets made of palm fibre or synthetic twine are laid flat at a suitable level below the sea surface, supported by a series of bamboo poles or wooden stakes driven into the seabed along the length of the net at intervals of 2.5–3 m. Instead of nets, hibi (blinds) made of split bamboo tied together by ropes, are also used in the same manner as nets. The size of these hibi in Japan varies with locality, but the common size is 18, 36 or 45 m long and 1.2, 1.8 or 2.4 m wide. To utilize deeper offshore waters, floating systems consisting of synthetic ropes supported by buoys and held by anchors are used.

The nets or hibi are spread to catch seed at different times and localities (usually in September and October), when the water temperature is about 22–23°C. The best catches are reported to be made on the second to the fourth day after the 1st or the 15th of the lunar month, or after a storm. Four or five hibi may be placed one over the other for spore collection. As in the case of molluscs, the areas suited for seed collection are not always the best sites for on-growing of nori. The spores are developed in areas of higher salinity near the open sea, but areas of lower salinity near river mouths are much more suitable for subsequent growth.

About a month after the hibi are spread, small buds can be observed on them and at this time they can be transplanted to the growing area. During the growing period, it may be necessary to adjust the level of the net or hibi according to the temperature and tide conditions and also considering possible exposure to diseases such as the fungus disease caused by *Pythicum* spp. At lower levels the nori grow well, but are more exposed to disease. By exposing them to air, fungal growth can be partially prevented, but overexposure may affect growth and toughen the thallus, reducing the value of the product. In about 50−60 days, the thalli grow and multiply in temperatures between 5 and 10°C. Lengths of about 15−20 cm are considered a suitable size for harvest. Smaller ones are left for further growth. Harvesting can be done from the same substrate three or four times, and the final harvest contains plants of different sizes.

Controlled production of seedlings

Controlled production of seedlings is now widely practised by the use of conchocelis. After a period of active growth during winter months, the thalli of *Porphyra* become progressively reduced in size, and some of them develop carpogonia and others spermatia. Sexual fusion between the contents of carpogonia and spermatia gives rise to carpospores. In summer the thalli disintegrate and the liberated carpospores sink to the seabed. They settle on mollusc shells and start germinating, giving rise to the microspic filamentous plant known as the conchocelis stage or the summer phase of *Porphyra* (fig. 28.2). The filaments burrow and grow beneath the surface of the shell substrate, eventually forming a darkly-stained area consisting of a plant mass on the shell surface. In autumn, when the water temperature drops and the photoperiod shortens, the spore-bearing branches or sporangia of the conchocelis mature and release non-motile monospores (conchospores). They are carried around by tidal currents until they attach themselves to suitable substrates like rocks or, in culture situations, to nettings or hibi spread out to catch them. The cultivated species of *Porphyra* reproduce asexually as well, and the young plants release monospores.

Taking advantage of the above described patterns of reproduction, it is now possible to produce seed indoors in open containers. At the end of the growing season (February to early April) ripe *Porphyra* leaves with well-developed sporangia along the margins are collected and placed in containers of about 20 ℓ capacity, filled with sea water. To stimulate the release of spores, the leaves are lightly squeezed and within a few minutes the carpospores can be observed on the water surface. Other methods of obtaining carpospores are also employed. Parent thalli, dried overnight and immersed in sea water for 4−5 hours, will induce spore formation. An alternative is to use pulverized thalli in sea water, filtered to produce a carpospore suspension. To set the spores, clean oyster shells are placed in shallow concrete tanks with the inner side up. Water containing carpospores is then evenly poured over the shells. About 20 000 shells are needed to set the spores from about 1 kg mature leaves. At temperatures of about 10−15°C, the carpospores germinate and burrow into the shells. According to Saito (1979), transparent vinyl films covered with calcite granules are now used as a substitute for shells.

For proper growth of conchocelis the shells are strung on nylon cords after seeding and suspended in large sea-water tanks (2 × 3 × 0.7 m) placed indoors with a suitable means to control sunlight. The strings of shells are suspended from bamboo sticks in such a way that each shell is freely bathed in sea water. The temperature is maintained around 23−25°C and the light intensity at 500 lux or less to prevent premature liberation of spores. By manipulation of the temperature and light intensity, the time of maturation of the spores can be varied to meet the requirements of farming. The nutrients in the water and the substrate are

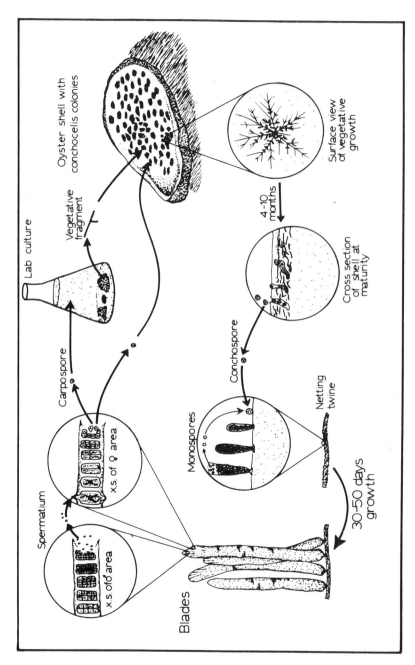

Fig. 28.2 Porphyra/conchocelis life cycle (from Melvin *et al.*, 1986).

usually enough for the growth of the algae, but the addition of small doses of phosphate and nitrogen fertilizers improves the rate of development. The water in the tanks is renewed only if its quality deteriorates. The shells remain in the tanks for about 5 months and, with adequate growth, the algae become visible to the naked eye as greyish-purple spots. If the shells are overgrown with diatoms or other organisms they can be brushed clean, as the conchocelis is safe inside the shell material. By about September or October the conchocelis have their sporangia well developed. When the temperature falls below 22°C and the days become shorter, maturation becomes rapid and the release of spores occurs.

A system for the production of free-living conchocelis, without any substrate, has been developed. Mature thalli are kept in containers with enriched sea water for the release of carpospores. The carpospores grow into globular colonies on the bottom of the containers in about two months. These colonies are divided into smaller portions and transferred to other containers and cultured in aerated sea water. As the colonies grow in size, they are again divided and transferred to new containers and the process repeated several times. Thus a large number of free-living conchocelis are produced and the culture is maintained for a long time with temperatures ranging from 24 to 28°C and 750 lux illumination for 8 hours a day. The conchospores are released by cutting the conchocelis into small pieces to enable seeding of the shell substrates.

To seed the hibi (nets, blinds or ropes), several methods are employed, but the basic requirement is to ensure maximum exposure to the hibi substrate and to keep the spores in motion to facilitate contact with it. This can be done on shore or in the sea. By immersing the hibi in conchocelis culture tanks for a few minutes, enough spores can be collected. Spore collection on nets is made easier by rolling them on rotating drums fitted above the tanks (fig. 28.3). A number of nets are rolled around a drum which is slowly rotated to immerse the nets in the tank, so the spores can settle on them. Single or double drums can be used. The settling of spores can also be facilitated by creating a curtain of air bubbles from pipes placed at the bottom of the tank. The nets are stretched horizontally below the surface, and the air bubbles drive the spores towards the net.

For seeding at sea, nets or bamboo blinds are stretched between poles in areas which are not exposed at low tide. Vinyl bags containing the conchocelis-bearing shells are hung under the nets or blinds. The spores, when released, settle on the substrates. This type of seeding allows proper regulation of spore density on the substrate and protects from adverse weather conditions.

Another method of seeding hibi in the sea is by spreading conchocelis-bearing shells in the inter-tidal seabed and placing five or six layers of hibi on top of them. In a couple of days the spores will have settled on the hibi, which can be transferred to racks for rearing.

Although the seeded hibi are generally used immediately for on-growing, they can be kept if necessary for several weeks in a cold store, covered with polyethylene sheets to keep them moist. *Porphyra* buds can also be stored for over 6 months by first drying them to bring the moisture content down to 20−30 per cent and then packing them in polyethylene bags for storage at −20°C. This helps to make seeded hibi available for replacing the old ones which have ceased to be productive.

Grow-out of conchocelis seedlings

The grow-out procedures for conchocelis-derived seedlings are essentially the same as for natural seedlings. In about 50 days after germination the plants grow to lengths of 15−20 cm and harvesting is then started (otherwise the plants are torn off by waves). Subsequent harvests are carried out at 15−20 day intervals. After repeated harvesting, when the size and quantity of thalli on the hibi decrease, they can be replaced with new ones stored under refrigeration. In areas where the nutrients are scarce or depleted, the cultivating grounds are fertilized to improve harvest. Fertilizer pellets containing about 90 per cent nitrogen and 10 per cent potassium are suspended from porous containers to last for about a fortnight at a time. Fertilization is also adopted in some areas just before harvesting to improve the colour of the product. Diseases associated with poor weather conditions, poor water quality

Fig. 28.3 Artificial seeding of *Porphyra* in Japan (from *Fish Farming International*, **3**(3)).

and crowding can sometimes occur. These can generally be controlled by reducing the number of hibi. Harvesting machines suited for different culture systems are available now and can be used to save labour at sea.

Methods of processing *Porphyra* have been described in Chapter 11. The drying of the thalli has to be carried out within two to three hours at a temperature below 50°C, to enhance the gloss, taste and flavour of the product.

28.2.2 Culture of *Undaria*

Undaria, known popularly as 'wakame', is an important group of cultivated edible seaweeds. Its culture is important in Japan, Korea and China.

The main species of *Undaria* cultivated is *U. pinnatifida*, which is in great demand in Japan for the traditional as well as the salted products made from it. Production by culture has outstripped harvests from natural sources. Two other species, *U. undarioides* and *U. peterseniana*, are also presently cultivated on a smaller scale in the southern parts of Japan. These are cold-water, open-sea species and can be grown only in areas where winter temperatures remain below 22°C. Optimal temperatures are different for different stages of the life cycle. Salinities between 30 and 33 ppt are reported to be optimum. The life history of *Undaria* includes an alternation of sexual and asexual forms. The macroscopic plant is the asexual form or sporophyte, which grows during the winter months at temperatures between 10 and 15°C. Asexual zoospores are produced by the sporophyte during this season and are released when the water temperature rises above 14°C. The planktonic zoospores settle within a short time on solid substrates like rocks, shells, etc., and germinate at temperatures between 15 and 20°C, to produce microscopic gametophytes or the sexual plants. During summer the gametophytes develop and by the end of the season, around September, the sperms are released and fertilize the ova within the oogonuim to form the zygote, which develops into the sporophyte.

The systems of culture are essentially the same as for *Porphyra*. The spores are collected and reared on so-called 'seeding twines', made of synthetic yarns of 2–3 mm diameter. These twines, which are usually about 100 m in length, are wound round a square (50 cm × 50 cm) plastic frame, leaving intervals of about 1 cm. Seeding is performed in large concrete or plastic tanks kept in cool shaded positions. Mature sporophylls (sporophyte plants) are partially dried and then placed in the tanks filled with fresh sea water. The twine frames are arranged in layers inside the tank to collect the zoospores released from the sporophylls. After about two hours they are transferred to large culture tanks with a depth of about 1 m, in which the frames are hung vertically. The gametophytes and the young sporophytes develop on the strings throughout summer. The light intensity in the tanks is regulated to facilitate rapid growth and high survival rates. If the water quality deteriorates it is exchanged; fertilizers may be added if the growth is poor. In areas with short periods of seed rearing and a lower likelihood of fouling, tank culture of gametophytes can be avoided, but most seaweed growers find it beneficial to perform nursery rearing in indoor tanks.

Grow-out of thalli starts in autumn, between September and November, when the water temperature falls below 20°C and there is less likelihood of fouling by epiphytic organisms. Twines containing the seed are set out on cultivation ropes in water depths up to 5 m, depending on local conditions. The cultivation ropes are made of synthetic fibre of 10–20 mm diameter and the seed twines are attached to them at intervals of about 15 cm. Sometimes the twines are cut into 5–6 cm pieces and inserted into the braided strands of the cultivation rope. In exposed areas the cultivation rope is set out with heavy anchors and a sufficient number of floats. This system of floating ropes has made it possible to culture *Undaria* in deeper waters (up to 50 m depth) exposed to heavy seas. In protected areas, like bays, the ropes can be stretched out on rafts.

Undaria grows rapidly and can be ready for harvest in about three months. Though heavy seas may damage the floating rope system, they aid rapid growth of the plants. Water temperature is a major factor affecting growth. The optimum is between 15 and 17°C, and below 5°C growth will be greatly retarded. The harvesting length of the thalli is greater than 50 cm.

Harvesting is done according to the growth pattern of the thalli which is itself very much dependent on the length of the growing period. When there is dense growth and the growing period is long, the well-developed thalli are harvested first. This helps the remaining ones to grow faster and enables repeated harvesting. In areas with poor growth, the thalli are harvested by cutting the upper portion first and allowing the remaining portion to grow. In areas with short growing periods, all the thalli may be harvested at a time, when they have grown to the maximum size.

Undaria is marketed in the dry state. Drying may be done in the sun or in a dryer. Salted *Undaria* is preferred by many consumers because it is convenient to use. Some are sold fresh in certain areas and, if required, can be stored under refrigeration.

28.2.3 Culture of *Laminaria*

Laminaria japonica, known as 'Kombu', is an important edible seaweed and a source of alginic acid in Japan. China was a major importer of Japanese and Korean *Laminaria* until about six decades ago, when a local source of the alga was discovered, and in recent years its cultivation has expanded very rapidly.

The reproductive cycle of the species involves a long sporophyte generation and a very short gametophyte generation of about two weeks. The mature sporophytes form sori from which motile zoospores are released which very soon attach themselves and germinate into male and female gametophytes. The zygotes formed by the fusion of the eggs and sperm released by the gametophytes develop into the sporophytes.

The earlier system of sowing spore-bearing rocks into the sea has been partially replaced by the more intensive techniques of spore collection and on-growing. In late autumn, special collectors made of split bamboo in the form of short ladders are suspended from floating rafts. The zygotes which attach themselves to the collectors develop into young sporophytes by about January. Temperature and light intensity affect the formation and release of spores in nature and, by controlling these factors, it is possible to propagate the species indoors with consistent results (fig. 28.4). Temperatures between 10 and 15°C have been found to be favourable for the maturation of sporophytes and formation of zoospores. At temperatures above 20°C, the production of new sori and release of zoospores cease. Temperatures of 10–20°C have been found to be suitable for the germination of spores and the formation of gametophytes. Periodic darkness is necessary for release of eggs and sperm and successful fertilization, but illumination is required for the development of the gametophytes and the growth of the sporophytes. So periods of illumination of 500–4000 m candles, alternated with periods of darkness, have been recommended in indoor propagation facilities.

Major increases in *Laminaria* production in China are attributable to the use of the raft systems. Spores attached to ropes are suspended from rafts either vertically or in a cradle-like device made of bamboo, hung between two parallel rows of the bamboo rafts. The main advantage of this arrangement is that the plants grow upwards and therefore have better exposure to sunlight. Immersion of the sporophytes in a fertilizer solution, such as ammonium nitrate, before they are attached to the rafts promotes an early start to rapid growth. The use of spores produced in summer under controlled conditions and cultured in mid-autumn when temperatures fall below 20°C has greatly improved growth rates and reduced fouling. The yield of a basket raft is reported to be only about 1 kg per kg fertilizer used, whereas the bamboo rafts yield over 3 kg per kg fertilizer. However, the quality of the basket raft plants is said to be higher.

In more exposed areas of the sea, long lines are now commonly used for the cultivation of *Laminaria* in China. As mentioned earlier, polyculture with scallops and abalones is presently practised. Application of fertilizer is by spraying and this helps in the faster growth of *Laminaria* as well as scallops and abalones.

Predation and fouling as well as rot diseases have been oberved, but no practical methods of prevention are known at present. Bottom culture of *Laminaria* on stones and ropes is still practised in China, but the zoospores are first attached to the substrates before being deposited on the sea bed.

Laminaria japonica grows under favourable conditions to lengths of over 3 m in 4–5

Fig. 28.4 An indoor facility for propagating *Laminaria* in China.

months (fig. 28.5). This size is suitable for harvesting, although the plants can grow up to 6 m on rafts. While harvesting of raft-grown *Laminaria* is performed from boats, efficient harvesting in bottom cultures is possible only by diving. Periodic harvesting of the distal one-third of the sporophytes has been found to facilitate better growth and higher yields. The farmer is especially interested in the dry weight rather than the wet weight of the harvested plants, and the dry weight is reported to increase at water temperatures of 20−21.5°C, even when the wet weight shows a decline (fig. 28.6). The reported yield of *Laminaria* in bottom culture is about 2.4 tons/ha. A single bamboo raft can culture about 72 000−134 000 plants per ha.

28.2.4 Culture of other edible seaweeds

The green algae *Monostroma* and *Entero-* *morpha* are cultivated in Japan. *Monostroma* is a highly-priced species and may be cultured together with *Porphyra* or separately. The methods of culture are the same as for *Porphyra*, involving the collection of spores on hibi and growing them in inter-tidal areas near river mouths. *Enteromorpha* appears to be grown only along with *Porphyra*.

Another species cultured in the Philippines is the green alga *Caulerpa*, which is consumed locally as a fresh vegetable and also exported to Japan in fresh or dried form. Its culture is generally carried out in ponds with clay-loam bottoms, and free from pollution as the plant is eaten fresh. The depth of water varies from 60 to 100 cm, depending on the clarity of the water, and the optimal salinity is 30−32 ppt. The water temperature is maintained between 27 and 30°C for good growth and, since the alga is rather sensitive to intense sunlight, the water depth is adjusted so that the plants are only just visible from the surface. The water

Fig. 28.5 *Laminaria* grown on long lines ready for harvest.

management required to maintain the necessary temperature and light intensity makes it necessary to site *Caulerpa* ponds where exchange of water is possible on most days of the tidal cycle. A water pH above the optimum of 7–8 can result in stunted growth and tough thalli, and so is avoided.

Cuttings or fragments of the plant are generally spread on the pond surface, but many farmers embed one end of the cuttings in balls of clay so that they sink to the bottom rather than float around in the pond for long periods. About 1.5 tons of cuttings are planted for every ha of pond surface. Frequent fertilization with small quantities of inorganic fertilizers helps to improve growth. The plants can be harvested in about two to three months and if enough stock is left in the ponds, further planting will not be required.

28.2.5 Culture of seaweeds for industrial use

Descriptions of seaweed culture practices in this chapter are mainly focused on edible seaweeds. However, a considerable amount of research and development is underway on establishing viable practices of intensive culture of seaweeds as sources of agar, alginates and carrageenans. Mathieson (1986) has reviewed the work done on a number of species and in this section experience in commercial farming of some of the more important species is summarized.

Methods of enhancing the production of *Gelidium amansii* and related species (used in agar production) have been practised in Japan for many years. Cuttings of the plants are sown in protected bays, where they generate new fronds. Rope cultivation techniques, very

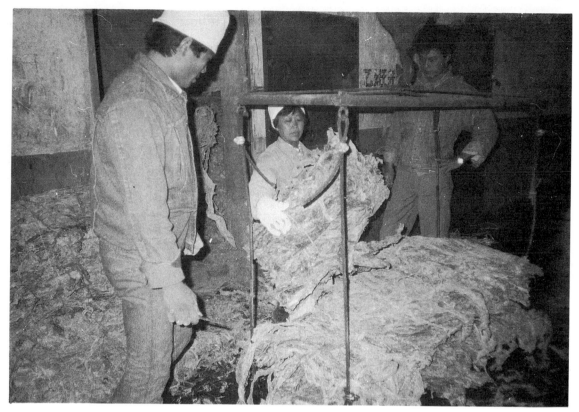

Fig. 28.6 Dried *Laminaria* ready for processing to produce agar in a factory in China.

similar to those used for *Undaria*, are presently employed and the growth is enhanced with fertilizer pellets.

Species of *Gracilaria*, especially *G. confervoides* and *G. gigas*, are cultured in Taiwan in old milkfish ponds. The species can adapt to wide variations in environmental conditions; for example, it can withstand salinities ranging from 8 to 25 ppt and temperatures up to 25°C. Cuttings are planted uniformly on the pond bottom at the rate of about 3000−5000 cuttings per ha of pond, supported by bamboo sticks dug into the bottom. The depth of water in the pond is increased to about 60−80 cm, with the rise in temperature in June. Exchange of tidal water is maintained to provide additional nutrients and to adjust salinity levels. Inorganic or organic fertilizers are applied at the rate of about 3 kg/ha of urea every week, or 120−180 kg/ha of fermented pig manure every two or three days at times of water exchange. If the water temperature falls below 8°C, either the depth of the pond water is increased or the plants are transferred to protected wintering ponds. Major pests and competitors in the ponds are overgrowths of algae, including *Enteromorpha* and *Chaetomorpha*. The common control measures are to lower the water level and reduce the water exchange; plant additional large *Gracilaria* to utilize nutrients in the pond water; and stock adult milkfish or tilapia (about 150 g weight) at the rate of 500−1000 per ha to feed on the green algae.

Harvesting of *Gracilaria* from the ponds is done by hand or with scoop nets, from June to November. The annual yield is around 10 tons/ha. The main costs of production are labour (about 53 per cent) and seed (18 per cent) and the rate of return on initial investment is 50 per cent according to Shang (1976). When poly-culture with crabs is practised, the production of *Gracilaria* is reduced to about 9 tons/ha and the operating costs increased more than four times, but an additional production of 6.3 tons/ha of crabs and shrimps is obtained. The harvested plants are sun-dried on bamboo screens or plastic sheets for export.

The culture of the red algae (*Eucheuma* spp.) has become a commercial operation in the Philippines for production of kappa carra-geenan. The most successful species so far is *E. cottonii*, but the related species *E. muricatum* (= *spinosum*) has also been tried. For the culture of *E. cottonii*, protected coral reef areas, with a good water flow and temperatures between 26 and 32°C are selected. Nets and long lines are used for suspended culture. A major factor in the success of commercial cul-ture of *E. cottonii* in the Philippines is the use of a clone of the species known as 'tambalang', produced by natural selection. It grows and multiplies rapidly and can survive in a wide variety of environmental conditions. Vegeta-tive parts of the plant are attached to polye-thylene nets and strung parallel to the bottom between poles. The seed is attached with thin plastic strips which allow movement and do not cut the fronds. When the fronds reach a weight of 1200–1500 g, harvesting is done by pruning about one-third. Further harvesting is done when the fronds grow to the desired size, and this cycle is repeated. Long lines are preferred to nets for grow-out as they are easy to main-tain, although the productivity per unit area may be lower.

Experimental culture of several other species of seaweeds is being carried out by many institutions and private enterprises. *Chondrus crispus* (Irish moss) has been culti-vated in greenhouse tanks flushed with sea water. *Eucheuma isoforme* has been cultured in tanks with slanted bottoms and circulating water to keep the plants in suspension. Studies on the growth of *Hypnea musciformis* in an artificial upwelling system in the US Virgin Islands have been referred to earlier. Mass culture techniques using sporophytes as seed have been developed for enhancing natural stocks of the kelp *Macrocystis* (North, 1972).

28.3   References

Bardach J.E., Ryther J.H. and McLarney W.O. (1972) *Aquaculture*, pp. 790–814. John Wiley & Sons, New York.

Chen T.P. (1976) *Aquaculture Practices in Taiwan*, pp. 145–9. Fishing News Books, Oxford.

Cheng T.H. (1969) Production of kelp – a major aspect of China's exploitation of the sea. *Econ. Bot.*, **23**, 215–36.

Deveau L.E. and Castle J.R. (1979) The industrial development of farmed marine algae: the case history of *Eucheuma* in the Philippines and USA. In *Advances in Aquaculture*, pp. 410–15. Fishing News Books, Oxford.

Fujiwara-Arasaki T., Mino N. and Kuroda M. (1984) The protein value in human nutrition of edible marine algae in Japan. *Hydrobiologia*, **116/117**, 513–16.

Hasegawa Y. (1976) Progress of *Laminaria* culti-vation in Japan, *J. Fish. Res. Bd Can.*, **33**, 1002–6.

Huguenin J.E. (1976) An examination of problems and potentials for future large-scale intensive sea-weed culture systems. *Aquaculture*, **9**, 313–42.

Korringa P. (1976) 'Nori' farming in Japan. In *Farming Marine Organisms Low in the Food Chain*, pp. 17–48. Elsevier Scientific Publishing, Amsterdam.

Kurogi M. (1963) Recent laver cultivation in Japan. *Fish. News Int.*, **2**, 269–74.

Mathieson A.C. (1986) A comparison of seaweed mariculture programs – activities. In *Realism in Aquaculture: Achievements, Constraints, Perspec-tives*, (Ed. by M. Bilio, H. Rosenthal and C.J. Sindermann), pp. 107–40. European Aquaculture Society, Bredene.

Melvin D.G. *et al.* (1986) Conchocelis culture – *Equipment and Techniques for Nori Farming in Washington State*, Vol. I.

Miura A. (1975) Porphyra cultivation in Japan. In *Advance of Phycology in Japan*, (Ed. by J. Tokida and H. Hirose), pp. 304–20. Fischer, Jena.

Neish I.C. (1979) Developments in the culture of algae and seaweeds and the future of the industry. In *Advances in Aquaculture*, (Ed. by T.V.R. Pillay and W.A. Dill), pp. 395–402. Fishing News Books, Oxford.

North W.J. (1972) Mass-cultured *Macrocystis* as a means of increasing kelp stands in nature. In *Proceedings of the 7th International Seaweed Symposium*, (Ed. by K. Nisizawa), pp. 394–9. University of Tokyo Press.

Rao G.N.S. (Com.) (1965) Use of seaweeds directly as human food. *IPFC Regional Studies*, **2**.

Ren G., Wang J. and Chen M. (1984) Cultivation of

Gracilaria by means of low rafts. *Hydrobiologia*, **116/117**, 72−6.

Saito Y. (1979) Seaweed aquaculture in the Northwest Pacific. In *Advances in Aquaculture*, (Ed. by. T.V.R. Pillay and W.A. Dill), pp. 402−10. Fishing News Books, Oxford.

Shang Y.C. (1976) The economic aspects of *Gracilaria* culture in Taiwan. *Aquaculture*, **8**, 1−7.

Sijian L. and Ping Z. (1984) The commercial cultivation of *Eucheuma* in China. *Hydrobiologia*, **116/117**, 243−5.

Tseng C.K. (1981) Marine phycoculture in China. In *Proceedings of the 10th International Seaweed Symposium, Gooteborg*, (Ed. by T. Levring), pp. 123−52. De Gruyter, Berlin.

Tseng C.K. (1984) Phycological research in the development of the Chinese seaweed industry. *Hydrobiologia*, **116/117**, 7−18.

29
Integration of Aquaculture with Crop and Livestock Farming

29.1 Rationale of integrated farming

The peasant economies, under which farming of fish originated, probably accounted for the practice of integrating it with crop and animal production. It is an ancient practice in China and the immigrant Chinese have introduced it into several Southeast Asian countries. Historically, fish farming has been a part-time activity of peasant farmers, who developed it as an efficient means of utilizing farm resources to the maximum extent. Farm ponds and reservoirs had to be constructed and maintained as sources of water supply for farm animals and plants, and it was only logical that non-consumptive uses for these water bodies would be developed in the course of time.

Fish culture can be carried out in these waters without a great deal of additional expense and with minimum adverse effects on crop and animal farming. It is a means of diversifying farm outputs and producing food for the peasant families and neighbourhood populations. The labour required can easily be shared between family members or even hired labour, with much of the hard labour being carried out when there is less demand for other farm activities. Farm wastes can be used for fertilizing and feeding the fish and accumulations of silt in the ponds can be used for fertilizing agricultural crops, vegetables and fruit trees grown around the pond farms. The embankments of the farm can be used very conveniently for the cultivation of different cash crops. The ponds, together with their embankments, also provide suitable areas for raising ducks.

The ponds are generally built on low-lying areas not suited for agriculture and therefore they do not in any way affect crop production. In fact, it is a means of land reclamation in certain areas and the relatively wide embankments built in such pond farms serve to increase the total land area available for crop and animal farming. Integrated farming of fish and ducks has been developed as a means of reclaiming sodic soils for agriculture in countries such as Hungary.

Because of the role that such integrated farming can play in increasing the employment opportunities, nutrition and income of rural populations, it has received considerable attention in recent years. Besides many developing countries of Asia, some in Africa (Madagascar, Central African Republic, Zambia) and South America (Panama, Brazil) have introduced this system on a pilot or larger scale. Some of the East European countries (Hungary, Czechoslavakia, Poland) have expanded and improved in recent years the practice of integrating animal production with fish culture. The rationale of rice-field aquaculture is also very similar. Flooded fields which lie fallow after harvest were used to raise crops of fish. The benefits derived from this practice led to the cultivation of fish along with rice. Such farming flourished under circumstances where rice farming and fish culture were truly complementary and there were no conflicts in the farming practices.

The basic principles involved in integrated farming are the utilization of the synergetic effects of inter-related farm activities, and the conservation, including the full utilization, of farm wastes. It is based on the concept that 'there is no waste', and 'waste is only a misplaced resource which can become a valuable material for another product' (FAO, 1977). This would seem to imply as well a certain amount of self-reliance and the minimum use of inputs from outside the farm. In an integrated farm, the importance of different crops may not be the same, from the point of view of production capacity, inputs involved and benefits gained. In China there are farms where fish are the main crop and livestock and agriculture of secondary importance. In others, livestock or agricultural crops form the mainstay. The allocation of land and water for fish, crops and livestock also varies. For example, in a state farm in China, about 60 per cent of the land was devoted to fish culture, 14 per cent to pigs and cattle, 14 per cent to the cultivation of fodder and 10 per cent to growing rice and wheat (UNDP/FAO, 1979). The use of animal manure in fish ponds for promoting live food production is an ancient practice in Asia (see Chapter 7) and there is as yet no evidence that this causes the transmission of pathogenic organisms to humans through fish.

29.2 Integrated farming of fish and livestock

Although fish farming is integrated with the husbandry of most domesticated animals, pig and duck raising appear to have been most successful in this respect. In the agricultural economy of China, pigs have a special significance. They are considered as 'costless fertilizer factories, moving on hooves', and pig manure forms the main source of home-made fertilizer. So the farmer has to raise them to meet the needs of crop farming, as well as fish farming. The pig sties can be built on fish farm embankments and aquatic plants from the ponds and associated water bodies, along with crop wastes, can be used for feeding the pigs. Part of the pig manure, raw or fermented, can be used very conveniently for fertilizing the ponds and the rest used for fertilizing agricultural fields. From the point of view of utilization of manure, the production of cattle and chicken are also important. As mentioned earlier, duck raising in association with fish farming has proved to be mutually beneficial. The ponds and their embankments provide the space for duck rearing and the ducks fertilize the ponds with their excreta, as well as feeding on unwanted organisms growing in ponds, such as snails, which may be carriers of water-borne diseases.

29.2.1 Pig and fish farming

In almost all the well-established integrated systems referred to above, herbivorous or omnivorous species of fish are used. The most common species are the common and Chinese carps. The catfish *Pangasius*, Indian carps and species of tilapia have also been used on a limited scale. The major benefit to fish farming of integration is easy access to the manure which fertilizes the pond and produces plankton and other micro-organisms to feed the fish. The feeding habits of the fish, particularly carps, make such integrated farming possible and beneficial.

As mentioned earlier, it is a common practice to have wider embankments in fish ponds to facilitate the building of pig sties and also planting of fruit trees, vegetables or other crops. In China, embankments in integrated farms may be over 10 m wide and planted with groundnuts, vegetables, colza, corn, sugar cane, mulberry, bananas, castor, etc. (fig. 29.1). The slopes are planted with grass, which can be used as fodder for grass carp and also for other farm animals like cattle. Feeder channels and irrigation ditches associated with the pond farms are used for growing aquatic plants, such as azolla, duck weed, water hyacinth (*Eichhornia crassipes*), and *Pistia stratiotis*, which are used for feeding the pigs. The control of macrovegetation in tropical fish ponds is a major problem, but in the integrated farming as practised in China, such vegetation is considered a valuable food resource.

Fairly high rates of stocking are practised, as the productivity is generally high. Because of the variety of food materials which become available in the ponds, polyculture is more commonly practised. A total of about 60 000 early fingerlings of different species weighing 20–30 g are stocked per ha of ponds.

Pig sties are built on the pond embankments or on neighbouring land. The number of pigs reared in relation to the pond area differs (figs 29.1 and 29.2). In China about 45–75 pigs per ha are generally raised, but some farms have up to 90 pigs per ha. The average production of manure (faeces and urine) per pig is around 7.8–8 tons per annum. This amounts to 351–600 tons of manure/ha per year and a very high loading of organic matter in the

Fig. 29.1 An integrated fish farm in China, with a large pig sty on the embankment. Note the fodder grass planted on the slopes.

Fig. 29.2 Small-scale integrated farming of fish and pigs in the Central African Republic. Only 20 to 30 pigs are reared per ha of pond surface and the washings go directly into the ponds.

ponds. According to Chen and Li (1980), even higher rates of 150–300 pigs per ha of pond area are maintained in Taiwan. Considerable care and water management skills are required to prevent pollution of the water and mortality of the stock. It has to be remembered that the number of pigs raised per ha and the manuring rates are worked out empirically, based on years of experience. Experimental work seems to indicate that satisfactory fish production can be obtained with much lower manuring (Buck *et al.*, 1979). Under Hungarian conditions, the maximum loading possible is reported to be 600 kg/ha per day, when manure is placed in localized heaps in the ponds (Woynarovich, 1980). The output of manure depends on the size and age of the pig. While a piglet gives about 3.4 kg manure a day, a one-year-old pig will produce about 12.5 kg a day. It is now a common practice to avoid direct washing of the wastes into the ponds. The wastes are conveyed to a specially built tank, where sedimentation and fermentation of the manure take place. At regular intervals, the supernatant liquid from the tank is allowed to flow into the ponds. The sludge that remains is removed for fertilizing agricultural crops. Thus the loading of decomposable organic matter in the ponds is reduced. The chemical composition of pig wastes is presented in Chapter 7 (see Table 7.18). The waste output and chemical composition depend on the quality and quantity of food and water consumed by the pigs.

Traditionally, the pigs in integrated farms depend on feedstuffs produced on the farm. As stated earlier, several aquatic plants such as the water hyacinth, *Ipomoea*, *Pistia*, *Wolffia*, *Lemna* and *Azolla* are grown in the canals and associated water bodies near the farm. These as well as the foliage of several terrestrial plants, such as vegetables, corn, rice and leguminous plants, are utilized as feedstuffs for pigs. Over 10 tons of aquatic plants can be produced in one ha of water area and this is considered enough to feed ten pigs. These plant materials are generally mixed with rice bran, bananas, coconut meal, soyabean wastes, fish meal, etc., for feeding to the pigs.

The duration of the culture of fish and pigs varies, but generally it is about one year. The yield of fish is reported to vary from 2 to 18 tons/ha per annum. The pigs are generally sold when about 90–100 kg in weight. The overall economics of combined fish and pig raising depend on the local conditions. It has however, been clearly demonstrated in many areas that the adoption of such integrated farming increases the productivity per area and input, and it also increases the farmer's income by a factor of two or more. Table 29.1 presents the annual operating costs and returns of a pig/fish farm in Malaysia, which illustrates the economic benefits derived.

29.2.2 Duck and fish farming

Ducks have been raised on fish ponds in Eastern Europe and parts of China for several centuries. Though the compatibility of ducks and fish have long been recognized, the interaction and benefits of the association have been understood only in recent years. Considerable progress has been made in developing suitable methods of raising ducks on fish ponds in East

Table 29.1 The annual operating costs and income for a pig/fish farm on disused mining land at Taiping, Perak, Malaysia, from 1977–8 (after Tan and Khoo, 1980; reproduced with permission of ICLARM).

Item	Value (M$*)	
Operating costs		
Pig food	61 000	
Labour	3 600	
Maintenance and operation of equipment		1 800
Subtotal for pigs (A)		66 400
Purchase of fish fingerlings (B)		3 650
Total		70 050
Gross income		
Sale of 480 pigs at M$170/pig (C)	81 600	
Sale of bighead carp at M$1360/ton (D)		15 000
Total		96 600
Net income		
Pig production (C−A)	15 200	
Fish production (D−B)		11 350
Total		26 550

* US$1.00 = M$2.06.

European countries, such as Hungary, Czechoslovakia, Poland and the German Democratic Republic. Such integrated farming is practised widely in Taiwan and on a limited scale in a number of other tropical countries.

As presently practised, the combination of duck and fish farming is seen as a means of reducing the cost of feed for ducks and a convenient and inexpensive way of fertilizing ponds for the production of fish food. Obviously, more animal protein can be produced per unit area by such a combination. The ducks search for and feed on a variety of organisms, including tadpoles, frogs, insects, insect larvae, snails and water weeds, which need to be eradicated from ponds. The protein content of supplementary feeds which are necessary to achieve high production rates for ducks can be reduced to 10 or 15 per cent, when the ducks are raised on ponds. In addition, the pond provides a clean and healthy environment for the ducks. Special strains of ducks suited for pond raising have been developed. If a suitable strain is used, approximately 50−60 per cent of their droppings will fall into the pond and act as fertilizer to the fish food organisms.

The barrage type of ponds, made by damming shallow valleys, are believed to be the most suitable ones for duck/fish farming, as the ducks can lie on the natural slopes (Woynarovich, 1980). There is a likelihood of ducks damaging earthen dikes while foraging for food, but with proper reinforcement and maintenance, this problem can be solved.

Timely and reliable supplies of good quality ducklings of the required strain are of critical importance in successful farming. While small-scale producers may have to depend on outside sources, larger farms will find it more convenient and profitable to have their own breeding centres. The breeders are selected after the first egg laying, which commences when they are about 6−7 months old. A protein-rich feed, suitable drinking water and appropriate temperature are some of the essential requirements for breeding stocks. They are fed at the rate of 9−10 per cent of the body weight, i.e. about 240−300 g per day. About 120−140 eggs are produced by a female every year. The incubation time for the eggs is about 28 days, and a survival rate of 75 per cent of one-day-old ducklings can be expected.

These one-day-old ducklings require special care and have to be reared in a controlled environment to the age of 10−14 days. A temperature of 30−32°C is maintained in the rearing room, which can hold 50−55 ducklings per m². A screen floor is provided to allow manure to fall through. Ducklings are fed on pelleted starter feed and tepid water is provided for drinking from special troughs. After the third or fourth day, the ducklings are released to shallow splashing pools to become accustomed to water. During the second week they are allowed to swim in small indoor conditioning ponds and are fed on prepared feeds. The ducklings are ready for release into fish ponds in 14−20 days, depending on weather conditions.

In East European countries, about 300−500 ducks are raised per ha of ponds during the summer season. The rapid growing strains of ducks presently used reach a marketable size when about 42−58 days old. During the period of about 5 weeks when the ducks are on the ponds, they contribute approximately 2.1−3.5 tons of droppings to fertilize them. The composition of duck droppings is given in Chapter 7, Section 7.4.2, Organic fertilizers.

There are two basic ways of keeping ducks on fish ponds. One is to allow them free access to the whole pond area and the other is to confine them to enclosures (figs 29.3 and 29.4). When the ducks are allowed to swim around freely on the pond surface, a good proportion of their droppings fall directly into the pond and are distributed more or less uniformly. The ducks are able to forage around the whole pond for food organisms. Small duck houses are built on or near the ponds with facilities for providing the ducks with prepared feeds. In such a system, considerable energy is used up by the ducks in swimming around and this is believed to affect the growth rate and feed conversion. The second option of confining them in enclosures is preferred by many farmers, who use selected strains of ducks for maximum growth. Wire fences are built, enclosing part of the pond area and adjacent banks, and suitable feeding and resting areas are provided. About one-quarter of the enclosure will be on land and the rest in water. Some of the droppings fall directly into the pond, and the rest have to be washed into it.

Fig. 29.3 A fish pond in Hungary, where ducks have access to the whole pond.

Fig. 29.4 Ducks confined to enclosures on the banks of a fish pond.

Wave action and water circulation ensure distribution of the manure and manured water in the whole pond. It is reported that fish production under this system is nearly equivalent to the free-range system.

As in pig/fish culture, the most common species of fish used in duck/fish culture are herbivores and omnivores. In East European countries the common carp was traditionally the main species, but now the Chinese carps are also included to make full use of the food resources in integrated polyculture systems. As supplementary feeding is generally carried out, fairly high stocking rates can be adopted. A fish yield per ha of 500−600 kg silver carp, 150−200 kg bighead carp and 1000−1200 kg common carp of marketable size is common in Hungary. The same production can be achieved without supplementary feeding, if the proportion of Chinese carps is increased. However, since the consumers prefer the common carp, it continues to be the main species. Duck production is about 1000−1200 kg (each weighing 2−2.4 kg).

In Taiwan, ducks are raised on fish ponds in the southern part of the island for the production of eggs as well. A hybrid of the native mallard and the drake of the muscovy is raised for meat, and the native mallard is raised for eggs. About 2000−4000 ducks are raised per ha of ponds, depending on the depth and abundance of the water supply. Four crops of an average of about 3200 ducks can be obtained. The egg-laying ducks start producing eggs in 4−5 months when they reach a weight of 1.2−1.5 kg and they continue to lay for two years, after which they are sold. Each duck lays about 250 eggs per year; 1500 egg-laying ducks are raised per ha of ponds and so the annual output of ducks' eggs per ha is about 375 000. Both the hybrids and the native strain are fed pelleted supplementary diets.

As well as the three species of Chinese carps (grass carp, silver carp and bighead) and common carp, other important species used in Taiwan are the hybrid tilapia (mainly *Tilapia nilotica* male × *T. mossambica* female) and the grey mullet. Small numbers of eel, Asian catfish and sea perch are also added in polyculture, making roughly 11 000 advanced fingerlings and young fish (Chen and Li, 1980). Tilapia and the mullet together make up over

70 per cent of the fish stock. Some farmers stock only male tilapia.

29.2.3 Farming of other animals in association with fish

As well as pigs and ducks, a number of other animals are farmed in association with fish, although the association is not as close as with ducks and pigs. Cattle, chicken, silk worms, etc., when grown in the same farm in close proximity, make it easy and inexpensive to utilize their waste materials for fertilizing the ponds. For example, in some of the farms in China the cattle sheds are situated very near fish ponds and wastes and washings from the sheds are conveyed through pipes directly into the ponds. Part of the fodder required for the cattle can be grown on the pond banks. Greater efforts have been made to combine chicken farming with fish culture. Chicken manure is a very efficient fertilizer for fish ponds and thus it is greatly advantageous to have easy access to it. Chickens can be housed over the ponds and the droppings can fall directly into them (figs 29.5 and 29.6). However, in present-day poultry farming there are only limited benefits for the chickens in being close to the fish ponds. The farmer, of course, is likely to gain by increased income and diversification of activities. Integrated farming of fish and chicken appears to be more prevalent in Indonesia and has been adopted by some farmers in Thailand.

In many farms in China, mulberry plants are grown on fish farm dikes and in neighbouring fields for silkworm production. The mulberry wastes and silkworm pupae (after removal of the silk) are used to feed the fish directly, and also serve partly as fertilizers for the ponds.

On a very limited scale, the culture of geese is conducted along similar lines to ducks in combination with fish culture in East European countries and in Hong Kong. The low egg production (30−60 eggs per year) and the high juvenile mortality have hindered large-scale farming, even though the growth rate and feed conversion ratio are better than for both ducks and chickens. Fast-growing European geese grown for meat eat green fodder and attain marketable size (4−4.5 kg) in about 50 days. Acclimatized and water-habituated young geese can be stocked in a pond when 20 days

Fig. 29.5 Combined farming of fish and chicken in Indonesia. Note the chicken house close to the ponds.

Fig. 29.6 Chicken and vegetables grown on an experimental fish farm in the Philippines.

old and weighing around 1500 g. It is reported that the beneficial effects of geese rearing on a pond are far lower than those of duck rearing (Woynarovich, 1980).

29.3　Rice-field aquaculture

Inundated rice fields always have a small population of fish which gain access with the water, and this probably gave rise to the practice of deliberate stocking and harvesting. The trapping of shrimp larvae in fallow rice fields and growing to market size is an age-old practice and still exists in parts of India. It is believed that rice-field fish culture was introduced into Southeast Asia from India about 1500 years ago (Tamura, 1961). It seems to have started in the 19th century in Indonesia, where it became an important peasant activity. Scarcity of food during the Second World War impelled farmers to devote greater attention to this type of integrated farming in countries like Japan. As will be discussed later in this section, improvements in rice farming techniques led to its general decline, but efforts are now being made to revive the practice.

29.3.1　Objectives and types of rice-field aquaculture

Rice-field aquaculture has been practised mainly to improve the income of the farmers and to make available an essential item in the diet of rural people in areas where 'rice and fish' form the staple food. In its traditional form, it required only very little extra input and provided off-season employment to the farmers and farm labour. Even though complete evaluations have not always been made, observations do indicate that the combination of rice and fish farming is mutually beneficial. The fish feed on organisms which grow in the fields and on many of the noxious insects and their larval stages, thus promoting better rice production. When a strongly herbivorous fish is cultured, the weed growths can be controlled to a considerable extent, as the fish will feed on them. When there is proper water management it is possible to control the growth of molluscs and the breeding of mosquitoes, thus reducing public health hazards. The movements of fish in inundated rice fields

cause increased tillering, which can result in higher rice production. The greater depth of water maintained in the fields is reported to prevent pests like rats digging holes in the bunds, and will also flood any holes that exist. Despite these benefits, the combination of these cultures entails additional costs for the farmer, particularly in management and labour, which will have to be offset by income from fish production.

Three major types of rice-field aquaculture are practised. The first and probably the simplest form consists of using flooded rice fields after harvest, to raise one or more crops of fish or shrimps. The second is growing fish along with the rice and harvesting the rice and fish at the end of the rice-growing season. The third and more complicated system, which ensures a prolonged period of fish culture, involves transferring the stock to specially prepared ditches, channels or pools at the time of the rice harvest, and restocking them in the field for a further growing period. By this system, the fish are grown to a larger size than is possible in the short duration of one rice crop.

In rice/fish farming the main crop is rice and therefore fish farming techniques have to be modified to make them compatible with rice farming. It may become necessary in certain cases to reinforce and increase the height of bunds to prevent escape of fish, but this will not affect rice farming. The construction of ditches and canals will reduce the area available for rice planting, as they may occupy 5–10 per cent of the land. Higher levels of water have to be maintained (10–25 cm) for growing fish together with rice. In areas where the water supply is limited, this may prove to be a major handicap. Also the short-stemmed, high-yielding varieties presently used by farmers may tolerate only moderate water depths, even when the water supply is not a constraint. The duration of cultivation of such varieties is shorter (105–125 days) and may not be long enough to grow fish to a marketable size. Deep water (floating) rice will be more suited for combined farming with fish. Fields with a high soil percolation may be unsuitable for rice/fish culture. The fertility of the soil is equally important to rice farming and to fish culture. The small additional fertilization that may be necessary to stimulate adequate growth of fish food may not affect production costs very much. The water quality in the fields has to be maintained at a level which is suitable for the fish and its food organisms. The very serious problem which has affected the combined culture of rice and fish, and contributed to its decline in many countries, is the intensive use of pesticides which create lethal conditions for fish life. There is also the risk of accumulation of pesticides in the fish and their effect on consumers.

The recommended dose of insecticides used in rice fields depends on the stage of rice growth, severity of attack, method of application and the pest species. Foliar sprays and the broadcast of granules are the most common methods of application. Because of the practical problems in penetrating the rice canopy and the short duration of the effectiveness of these methods capsule application in the root zone has been developed. This has provided higher efficiency and residual action. Carbofuran is a typical systemic insecticide which can be used in gelatine capsules for root-zone application, but capsule production is costly and its application is too laborious for general acceptance. Because of this, a method of application of carbofuran using a liquid band injector has been devised. If this method is used by farmers or if pesticides are incorporated otherwise into the soil, one of the major constraints to rice/fish culture can be overcome. Experimental work has shown that if fish are stocked in fields treated with carbofuran (by broad-casting, root-zone application or spraying of 18–15 kg Furadan, 3G with a basal fertilizer) after 7 days, no mortality of fish will occur. The use of pest-resistant rice varieties will also reduce the need for insecticide application. Seiber and Argente (1976a,b) reported that carbofuran is not accumulated in the fatty tissues of tilapia and so tilapia grown in fields treated by this pesticide is safe for human consumption. Carbofuran appears to be completely converted to water-soluble metabolites.

29.3.2 Rice–fish rotation

The constraints and conflicts mentioned above do not apply when fish culture is practised in rotation with rice cultivation. Even then the possible risk of pesticide residue in fish has to be considered. However, the interval between rice farming and fish stocking is long enough to allow degradation of pesticides. Infestation by insect pests is also reduced, as their life cycles are disrupted by the alteration of crops and associated practices. If chemicals like carbofuran are used, the risk is further reduced. In countries like the Philippines where the yield of the wet-season rice crop is rather low, it is logical to rear fish as an alternating crop.

When fish are to be raised in rotation, the fields are prepared after the rice harvest. The bunds surrounding the field have to be raised and reinforced where necessary, to maintain the required depth of water. The water level depends on the habits of the species and their size. As the effect of water level on the rice is not a constraint in this type of farming, an adequate water level can be maintained if there is a suitable supply.

It is advantageous to flood the fields soon after harvesting the rice, without removing the stubble. The submerged stubble provides the substrate for the development of fish food organisms. When decomposed, they fertilize the water and stimulate higher productivity. After the fish harvest, the residues remaining in the soil serve as fertilizers for the rice crop.

The selection of species for culture depends to a large extent on the likely duration of culture and the quality of the water. The practice of shrimp production in rice fields on the west coast of India is carried out in areas where generally only one crop (July to September) of a salt-resistant variety of rice is grown. After the rice is harvested, the bunds are strengthened and suitable sluices installed to control the water supply. The brackish-water lagoons nearby, from where water is obtained to irrigate the fields, have large numbers of shrimp larvae at this time. The fields are filled with tidal water at high tides and the larvae gain access to the fields with the water, where they find shelter and food. Lamps may be hung above the inlets to attract the larval shrimps. The natural process of stocking continues with every high tide for two or three months. The sluice gates are provided with conical bag nets to prevent escape of larvae and juvenile shrimps at low tides. Harvesting of shrimps starts in December, by when the early stock will have reached marketable size. Regular harvesting helps to thin the stock, leading to a better growth rate and a higher percentage of larger shrimps. Several species of shrimps are grown in the fields: *Penaeus indicus*, *Macrobrachium rude* and *Palaemon styliferus*. Incidental species are *Caridina gracilirostris*, *Acetes* sp. and the finfishes, grey mullets and pearlspot (*Etroplus* sp.). The total yield per ha is reported to be around 780–2100 kg. With increasing interest in shrimp farming and the high price of shrimps, farmers now devote greater attention to water management and stimulation of primary production in the fields. Where possible, controlled stocking with sorted larvae is undertaken. In the deltaic areas of eastern India flooded rice fields are stocked with larvae of quick-growing shrimps, particularly *Penaeus monodon* and *P. semisulcatus*.

In rice fields irrigated with fresh water, either mono- or polyculture of finfish is practised. The most common species are probably common carp, tilapia and *Trichogaster*. The snakehead (murrel) and the catfish, *Clarias*, are also used. The use of goldfish (*Carassius auratus*) and tench in Italy appears to have been discontinued. Similarly, the production of buffalo fish (*Ictiobus cyprinellus*) and channel catfish (*Ictalurus punctatus*) as a rotational crop in rice field reservoirs in Arkansas appears to be only on a very limited scale. Since the fields, when flooded after rice harvest, serve as shallow ponds, some of the pond culture practices such as fertilization and supplementary feeding can be adopted. Through proper water management, a suitable water temperature and oxygen content have to be maintained. Depending on the period available for fish farming, the stocking rate and size can be determined. The duration of culture is generally three to four months. Some farmers use the rice fields to grow fry to late fingerling stage, or from late fingerling stage to marketable size. When tilapia are cultured, special efforts are made to grow them to market size during this period. Naturally the fish yield varies very considerably with species, culture practices, etc. In well managed fields a yield of up to 700 kg/ha can be expected. De la Cruz (1980) gave the data (shown in Table 29.2) on costs and returns for a rotational crop of tilapia and common carp in the Philippines, for a culture period of about 116 days and a stocking rate of 10 000 tilapia (*Tilapia nilotica*) and common carp. Although the economics of the practice vary from place to place, these data give some indication of the income that can be expected when fields left fallow are used for fish culture. Available records show that the income from fish farming is approximately the same as it would be if the fields were used for rice production during the period it is left fallow. But under the circumstances it adds to the income of the farmer and is a more efficient use of land and farm resources.

In the traditional system of shrimp production, the production cost is minimal when operated by the owner of the field. However, in a number of cases the fields are leased from owners for shrimp growing and the cost of the lease is relatively high. This influences the net return from shrimps. In improved systems where sorted larvae and juveniles are stocked and fertilization or feeding are adopted, the operational expenses are higher but the net income is compensatingly high.

In rice fields in Louisiana (USA) the crayfish *Procambarus clarkii* is raised as a rotational crop (See Chapter 25). The culture techniques

Table 29.2 Costs and returns of polyculture of *Tilapia nilotica* and common carp, *Cyprinus carpio* with supplemental feeding in a 1 ha rice field* (from De la Cruz, 1980, reproduced with permission of ICLARM).

Item	Value (P)
Returns	
470 kg marketable tilapia at P 7.50/kg	3525.00
222 kg common carp at P 6.50/kg	1443.00
Total (A)	4968.00
Costs	
Fingerlings	
10 000 *S. niloticus* fingerlings at P 0.08 each	800.00
5 000 common carp at P 0.05 each	250.00
Feed: 1270 kg fine rice bran at P 0.75/kg	952.92
Fertilizer: 386.5 kg 16:20:0 (N:P:K) at P 1.71/kg	660.92
Labour: 18.7 man-days at P 11.00/man-day	205.70
Total (B)	2869.12
Net returns (A−B)	2098.88

* (US$1.00 = Philippine P 7.33)

have been intensified in recent years. Adults are stocked at the rate of 6−12 kg per ha in fields flooded after the rice is harvested, when the rice stubbles start to sprout. A depth of about 15−45 cm is maintained in the field and the crayfish feed on the rice stubble and various aquatic plants found in the field. After about 6 months they reach the marketable size of 10−15 g. Larger ones weighing 40−45 g fetch better prices and it takes about 8−14 months to reach that size. The usual yield is about 400−700 kg per ha.

29.3.3 Combined culture of rice and fish

When combined culture of rice and fish is planned, some additional constructions will be needed. As in the case of rotational culture, the bunds around the field have to be strengthened and the height increased. Straw may be embedded along the inside walls of the bunds to make them water-tight. A height of 25−60

cm is required, depending on the water level required and the species to be cultured. Usually the inlets and outlets are provided with pipes and screens to prevent escape of fish and ingress of undesirable animals. It is a common practice to dig a series of trenches to serve as fish refuges when the water in the field gets too cold or too warm, or when the water level in the rest of the field has to be reduced (fig. 29.7). They may be built along the peripheries or across the field. A width of 50 cm and a depth of 30 cm would normally be enough, but for extreme temperature conditions deeper trenches (up to 90 cm) are recommended. Depending on the type of soil, a side slope of 30−45 degrees may be necessary. It will also be useful to have a sump connected to the trenches near the inlet to facilitate harvesting of the fish and provide additional shelter during the cultivation period.

Although there are several species cropped from rice fields, the main species presently

Fig. 29.7 A trench along the peripheral bund of the rice field, which serves as a fish refuge.

used in combined rice/fish culture in fresh waters are the common carp, tilapia, nilem carp (*Osteochilus hasseltii*), kissing gourami (*Helostoma temmincki*) and sepat siam (*Trichogaster pectoralis*) and less frequently the Java carp (*Puntius gonionotus*), snakehead (*Channa*) and catfish, (*Clarias*). Limited experimental work in India has shown the suitability of Indian carps for such integrated farming. As mentioned earlier, the use of the goldfish *Carassius* and the tench *Tinca* in Italy appears to have died out now, largely because of changes in rice farming methods. In the extensive system of polyculture of shrimps and brackish-water species, practised in the estuarine regions of the river Ganges in eastern India, all the locally important shrimps and prawns (*Penaeus semisulcatus*, *Metapenaeus monoceros*, *M. brevicornis*, *Palaemon carcinus* and *P. rudis*) and brackish-water fish (*Mugil parsia*, *M. tade*, *Rhinomugil corsula*, *Lates calcarifer* and *Mystus gulio*) are used.

The most successful species in rice fields are those which can thrive in shallow waters and tolerate fairly high turbidity and high temperatures. Since the duration of culture is rather limited, they should have high growth rates and reach marketable size in a few months.

The literature on rice-field aquaculture shows that a variety of techniques and stocking rates have been employed in different countries. As many of them are presently not employed, because of declining interest or conflict in production practices, it may not be of much use to summarize them here. As, for example, the very intensive system of common carp farming in rice fields developed in Japan is not practised now, if at all, on any significant scale. One of the few countries where it still exists on a wide scale is Indonesia, where through government legislation it has been possible to restrict the types of pesticides that can be used in rice fields. Another factor that has promoted the continued practice of rice-field fish culture is the acceptance of small fish by Indonesian consumers.

Basically, the methods of culture adopted in Indonesia are for production of either fingerlings or fish for consumption. Monoculture of common carp is more popular than polyculture. Tilapia, kissing gourami, nilem carp and Java

carp form only minor components in species combinations. Many farmers continue to use local varieties of rice, which take up to 6 months to be harvested. This makes it possible to have a longer period for fish culture and to grow more than one crop per year. Fertilizers are used in the preparation of the rice fields. Although organic manures are generally preferred in this type of integrated farming, most farmers at present also use chemical fertilizers. A higher dose of fertilizer, sometimes twice as much as normal, is used when fish culture is combined with rice farming. Normally an increase in fertilizer of about 50% is recommended. As excessive use of fertilizers may reduce rice yields due to lodging and more severe attacks by pests, it is recommended that nitrogen fertilizers should be applied in split doses and incorporated into the soil to reduce nitrogen losses.

In general, fish are stocked no earlier than 5 days after the transplantation of rice seedlings, to give enough time for the seedlings to root properly. It is recommended that stocking should be done only after 10 days if fry are used, and in the case of fingerlings about 3 weeks after transplantation of rice (fig. 29.8). The rate of stocking varies considerably, depending on species and the size or age of the fry or fingerlings used. According to recent records, when small fry are stocked for rearing to fingerling size, 2.2–7.6 kg per ha of fry are stocked. If the duration of culture is one month, 6.5–15 kg per ha of fingerlings can be harvested. When larger fry or fingerlings are stocked and reared to consumption size a yield of about 100 kg/ha, after two or three months, has been reported. This is a comparatively low production, but is probably due to lack of any supplementary feeding and less intensive management. The yield per ha is generally higher when fish are grown alone in the rice fields as a rotational crop. In Thailand, a yield of 210–250 kg per ha is obtained when common carp are reared for about 6 months with supplementary feeding. When advanced fingerlings are held over in ponds or reservoirs after the rice harvest and reintroduced in the field for a second period of rearing with the rice crop, much higher yields of up to 1800 kg/ha per year have been reported from Japan. In Taiwan, with monoculture of tilapia, and in

Fig. 29.8 Combined farming of rice and fish in Indonesia. The farmer is releasing fry in the prepared field (photograph: M. Huet).

Fig. 29.9 Combined farming of shrimps with rice in deltaic areas of eastern India. Note the canals around the field, where larvae and juveniles remain during dry months.

Madagascar, with monoculture of carp, yields of 200–250 kg/ha per year are reported.

In the culture of brackish-water shrimps and fish in rice fields in deltaic areas of India, the stocking rate is not regulated, but the larvae and juveniles which gain access at the end of winter remain in the canal system (0.6–1.5 m deep) which runs all round the field or part of it. The water level in the canals is maintained at about 30 cm below the level of the field (fig. 29.9). The fish grow in the canals up to about June, when the flow of tidal water into the canal is stopped. The fields are fertilized with organic manure and planted with a salt-resistant variety of rice seedlings in about July. During the rainy season, rainwater floods the fields and the canals and then the shrimps and fish migrate to the rice-growing area and forage among the rice. By about October or November, when the water level in the field goes down, the fish and shrimps move back into the canal. Partial harvesting may start at this time, but the final harvest coincides with the rice harvest. The estimated production of shrimps and fish is about 100−200 kg/ha per year.

Reliable economic evaluations of the effect of such combined farming on rice yields are not readily available. There is some reduction in the area available for rice cultivation because of the canals, trenches and sumps which are constructed as fish refuges, but it is believed that this is compensated for by the beneficial effects of the fish on rice production and the overall income derived by the farmer. Rice seedlings are planted with a distance of about 20 cm between rows and 15 or 20 cm between plants, in Asian countries. This allows enough space for the movement of fish and so there is no need to reduce the number of seedlings planted, to make room for the fish. It has been

Table 29.3 Annual inputs and returns from 7 rice/fish farms in Indonesia excluding any depreciation costs, in Indonesian Rupiahs*) (after Djajadiredja *et al.* 1980; reproduced with permission of ICLARM).

	Paddy areas (m²)						
	1400	1890	2800	2800	4200	4200	6300
Fish							
Inputs							
Labour	12 000	12 400	4 800	8 400	21 800	22 400	23 200
Fingerlings/fry	38 000	20 800	8 850	11 200	52 400	86 000	41 000
Fertilizers	—	—	—	—	—	—	—
Feed	20 000	10 000	—	—	—	—	15 000
Total (A)	70 000	43 200	13 650	19 600	74 200	108 400	79 200
Output (B)	170 00	84 000	25 200	44 000	168 800	296 000	115 000
Net return (B−A)	100 000	40 800	11 550	24 400	94 600	187 600	35 800
Rate of return as a % of A	143	94	85	124	127	173	45
Rice							
Inputs							
Field construction	40 333	34 100	40 000	50 500	75 768	150 000	113 666
Buildings and equipment	2 200	2 200	2 200	2 200	2 200	2 200	2 200
Labour	16 000	34 500	39 450	52 600	68 990	107 161	75 200
Seedlings	800	2 500	1 050	3 600	4 500	2 000	7 500
Fertilizers	4 320	7 000	5 250	12 600	14 700	17 500	24 360
Pesticides	500	—	—	—	—	1 400	—
Taxes	1 400	1 900	2 800	2 800	4 200	4 200	6 300
Total (C)	65 553	82 200	90 750	124 300	170 358	284 461	229 226
Output (D)	110 000	143 922	150 000	164 920	322 400	396 610	448 000
Net return (D−C)	44 447	61 722	59 250	40 620	151 042	112 149	218 774
Rate of return as a % of C	68	75	65	33	89	39	95
Fish + Rice							
Total costs (A+C)	135 553	125 400	104 400	143 900	244 558	392 861	308 426
Total returns (B+D)	280 000	227 922	175 200	208 920	491 200	692 610	563 000
Total net return	144 447	102 522	70 800	65 020	246 642	299 749	254 574
Total rate of return as a % of total costs	107	82	68	45	101	76	83

* (US$ 1.00 = Rp 627)

suggested that even more space can be made available for fish movement by increasing the distance between rows to 25−30 cm and reducing the distance between plants, without affecting the number of seedlings planted and the yield of rice (Singh *et al.*, 1980).

Table 29.3 presents cost and return data on seven rice/fish farms in Indonesia, to illustrate the relationship between inputs and outputs in this type of farming. As is obvious the economics are site-specific and highly variable.

When fish have to be grown to a larger size than is possible in the limited duration of a rice crop, the farm should have additional holding or rearing facilities for the period between rice harvest and the planting of new seedlings. The use of the canal system for the culture of brackish-water shrimps and fish in India, described earlier, is one type of facility that would enable prolonged culture. The procedure of culture of one- and two-year-old carp in rice fields described by Kuronuma (1954) involves rearing harvested fingerlings from rice fields in separate ponds and later in wintering ponds, before stocking them in the rice fields for a further period of growth along with the rice. With adequate feeding the carp can be grown to the required size. Obviously, this involves higher inputs, including labour.

29.4 References

Agricultural Experiment Station, University of Arkansas, (1959) Use of reservoirs for production of fish in the rice areas of Arkansas. Agricultural Experiment Station, University of Arkansas. *Special Report*, 9.

Ardiwinata R.O. (1957) Fish culture on paddy fields in Indonesia. *Proc. Indo-Pacific Fish. Council*, **7**(II−III), 119−54.

Bardach J.E., Ryther J.H. and McLarney W.O. (1972) *Aquaculture*. John Wiley & Sons, New York.

Buck D.H., Baur R.J. and Rose C.R. (1979) Experiments in recycling swine manure in fish ponds. In *Advances in Aquaculture*, (Ed. by T.V.R. Pillay and W.A. Dill), pp. 489−92. Fishing News Books, Oxford.

Chen T.P. (1954) The culture of tilapia in rice paddies in Taiwan. *Jt Comm. Rural Reconstr. China Fish. Ser.*, **2**.

Chen T.P. and Li Y. (1980) Integrated agriculture-aquaculture studies in Taiwan. In *Integrated Agriculture-Aquaculture Farming Systems*, (Ed. by R.S.V. Pullin and Z.H. Shehadeh), *ICLARM Conf. Proc.*, **4**, 239−41.

Coche A.G. (1967) Fish culture in rice fields − a world-wide synthesis. *Hydrobiologia*, **30**, 1−44.

De la Cruz, C.R. (1980) Integrated agriculture-aquaculture farming systems in the Philippines, with two case studies on simultaneous and rotational rice-fish culture. In *Integrated Agriculture-Aquaculture Farming Systems*, (Ed. by R.S.V. Pullin and Z.H. Shehadeh). *ICLARM Conf. Proc.*, **4**, 209−23.

Delmendo M.N. (1980) A review of integrated live-stock-fowl-fish farming systems. In *Integrated Agriculture-Aquaculture Farming Systems*, (Ed. by R.S.V. Pullin and Z.H. Shehadeh). *ICLARM Conf. Proc.*, **4**, 59−71.

Djajadiredja R and Jangkaru Z. (1978) *Small-scale fish/crop/livestock/home industry integration: a preliminary study in West Java, Indonesia*. Report No. 7, Inland Fisheries Research Institute, Bogor.

Djajadiredja R., Jangkaru Z. and Junus M. (1980) Freshwater aquaculture in Indonesia with special reference to small-scale agriculture-aquaculture integrated farming systems in West Java. In *Integrated Agriculture-Aquaculture Farming Systems*, (Ed. by R.S.V. Pullin and Z.H. Shehadeh). *ICLARM Conf. Proc.*, **4**, 143−65.

Estores R.A., Laigo F.M. and Adordionisio C.I. (1980) Carbofuran in rice-fish culture. In *Integrated Agriculture-Aquaculture Farming Systems*, (Ed. by R.S.V. Pullin and Z.H. Shehadeh). *ICLARM Conf. Proc.*, **4**, 53−7.

FAO (1977) China: recycling of organic wastes in agriculture. *FAO Soils Bull.*, **40**.

FAO (1980) Freshwater Aquaculture Development in China. *FAO Fish. Tech. Paper*, **215**.

Hora S.L. and Pillay T.V.R. (1962) Handbook on fish culture in the Indo-Pacific region. *FAO Fish. Biol. Tech. Paper*, **14**.

Kuronuma K. (1954) *Carp Culture in Rice Fields as a Side Work of Japanese Farmers*. Ministry of Agriculture and Forestry, Japanese Government.

Menon M.K. (1955) On the paddy-field prawn fishery of Travancore-Cochin and an experiment in prawn culture. *Proc. Indo-Pac. Fish. Counc.*, **5**(II and III), 131−5.

Pillay T.V.R. and Bose B. (1958) Observations on the culture of brackishwater fishes in paddy fields in West Bengal (India). *Proc. Indo-Pac. Fish. Counc.*, **7**(II and III), 187−92.

Pillay T.V.R. (1958b) Land reclamation and fish culture in deltaic areas of West Bengal, India. *Prog. Fish Cult.*, **20**, 99−103.

Seiber J.N. and Argente A. (1976a) *Carbofuran residues in paddy-reared* Tilapia mossambica *from CLSU. Trial 1*. IRRI Research Report, International Rice Research Institute, Los Baños, Laguna, Philippines.

Seiber J.N. and Argente A. (1976b) *Carbofuran residues in paddy-reared* Tilapia mossambica *from CLSU. Trial 2.* IRRI Research Report, International Rice Research Institute, Los Baños, Laguna, Philippines.

Sin A.W. and Cheng K.W.J. (1977) Management systems of inland fish culture in Hong Kong. *Proc. Indo-Pac. Fish Counc.,* Sect. III, 390—98.

Singh V.P., Early A.C. and Wickham T.H. (1980) Rice agronomy in relation to rice-fish culture. In *Integrated Agriculture-Aquaculture Farming Systems*, (Ed. by R.S.V. Pullin and Z.H. Shehadeh). *ICLARM Conf. Proc.*, **4**, 15—34.

Tamura T. (1961) Carp cultivation in Japan. In *Fish as Food* (Ed. by G. Borgstrom). Academic Press, New York.

Tan E.S.P. and Khoo K.H. (1980) The integration of fish farming with agriculture in Malaysia. In *Integrated Agriculture-Aquaculture Farming Systems*, (Ed. by R.S.V. Pullin and Z.H. Shehadeh). *ICLARM Conf. Proc.*, **4**, 175—87.

Tapiador D.D. *et al.* (1977) Freshwater Fisheries and Aquaculture in China. *FAO Fish Tech. Paper*, **168**.

UNDP/FAO (1979) *Aquaculture Development in China.* ADCP/REP/79/10, FAO of the UN, Rome.

Woynarovich E. (1979) The feasibility of combining animal husbandry with fish farming, with special reference to duck and pig production. In *Advances in Aquaculture*, (Ed. by T.V.R. Pillay and W.A. Dill), pp. 203—8. Fishing News Books, Oxford.

Woynarovich E. (1980) Utilization of piggery wastes in fish ponds. In *Integrated Agriculture-Aquaculture Farming Systems*, (Ed. by R.S.V. Pullin and Z.H. Shehadeh). *ICLARM Conf. Proc.*, **4**, 125—8.

30
Stocking of Open Waters and Ranching

The technologies of aquaculture described in Chapters 15–27 relate to growing aquatic animals in confinement. Some of those techniques can be adopted for building up populations of selected species or for enhancing existing populations in open waters such as streams, lakes, reservoirs, lagoons and sea areas. If a suitable anadromous species is selected, the homing behaviour can be utilized to ensure that the surviving stock from releases return to the home waters for spawning, after ranching in the open seas. The main advantage in stocking and ranching is the elimination of the controlled grow-out phase and consequent savings on artificial feeding and stock maintenance, besides the capital costs of grow-out facilities. In both open-water stocking and ranching, the released animals feed on natural food in the environment and are exposed to predators and other causes of mortality which cannot be controlled. The term ranching is used here only when a species is allowed to forage around freely in extensive water areas like the sea, and is able to return to its home waters at a certain stage of its life.

The need for and value of human intervention in enhancing fishery resources have been demonstrated by the state of natural stocks of a number of aquatic species. Populations of salmon and trout in many rivers which were decimated by excessive fishing or environmental degradation could be rehabilitated only through continued stocking of hatchery-raised young. Many multi-purpose reservoirs formed by damming rivers, where populations of economically important fish species had declined or disappeared due to environmental changes, have developed into major resources of important species by stocking. Even though it is not always easy to measure very precisely the economic return in every case, there is evidence from experience in several instances that adequately planned release of spawners of hatchery-raised young in sufficient numbers for the required periods of time has resulted in remarkable increases in commercial catches. However, it involves considerable expenditure of both money and organized effort for a number of years to yield noticeable results. It is also necessary to grow the animal before release to a size when it can fend for itself, in order to reduce mortality due to predation. The lack of success of some of the earlier efforts of population enhancement could have been because these requirements were not fulfilled.

30.1 Open-water stocking

30.1.1 Reservoirs

Some of the very successful stocking operations in open waters are those carried out to build up fish populations in reservoirs formed by the construction of dams across rivers. Large land areas are inundated by the construction of dams and as a result very spectacular changes take place in the fauna above the dams. Due to increased water fertility caused by decaying vegetation and the flooded soils, explosive increases in fish fauna occur, but generally they are of the uneconomic species, considered as weed or trash fish. This is usually followed in a few years by a trophic depression, which results in the reduction of the fish populations. In the succeeding phase the productivity stabilizes, depending on the rate of growth of the biota and the amount of organic substances accumulated in the bottom soil. The fish populations in these reservoirs can be manipulated to provide a lucrative fishery by judicious stocking.

If a species that can breed in the reservoir is selected, initial stocking of an adequate number of spawners or adults may prove to be of considerable value. If the conditions in the reservoir are favourable, a breeding population of the species can be expected to develop in a reasonable period of time, depending on the magnitude of the initial stock, spawning success and environmental conditions, including production of food organisms and protection from predators.

Where there is a lack of a suitable spawning habitat, or when the species selected will not breed in the lentic environment of reservoirs, it will be necessary to establish hatcheries and nursery farms to produce fingerlings or yearlings for stocking purposes. When non-indigenous species are transplanted to fill ecological niches or to build up a dominant fish population, there is usually the need for a steady source of fingerlings and yearlings for at least a number of years. This necessitates access to hatchery and nursery facilities.

A considerable amount of experience has accumulated in different parts of the world in the establishment and management of fish populations in reservoirs. It is now widely accepted that pre-impoundment studies for reservoir construction should include detailed investigations of the fish fauna, the possible effects of the dam and reservoir on the fishery resources and the possibilities of preserving them as well as developing new resources. Some of the largest reservoirs utilized for fishery resource development are in the USSR and very impressive efforts have been made there to stock and manage several species for commercial fishing. Reservoirs on the Volga and other rivers in the European part of the USSR are stocked with bream (*Abramis brama*), common carp, white fish (*Coregonus lavaretus* and *C. albula*) and pike perch (*Stizostedion lucioperca*). Southern reservoirs are also stocked with silver carp for effective utilization of phytoplankton. Reservoirs created by hydro-electric dams on the Black, Caspian and Aral Sea basins have been stocked with young of migratory cyprinid species like roach (*Rutilus rutilus*), bream, vimba (*Vimba vimba*) and shemoia (*Chalcalburnus chalcoides*). Propagation and continued stocking in the reservoirs of the Volga are reported to have been largely responsible for the maintenance of the fisheries of the Caspian sturgeons, namely beluga (*Huso huso*), spiny sturgeon (*Acipenser nudiventris*), Russian sturgeon (*A. guldenstadti*) and sevryuga (*A. stellatus*).

Enhancement of fish resources forms an integral part of the hydro-electric reservoir management in the USSR. It starts as soon as filling of the reservoir commences, and spawners are transplanted to build up a spawning stock. Hatchery and nursery facilities are established near the reservoir to propagate valuable species for stocking, to add to those which are produced by natural breeding. Fishing is prohibited in the reservoir until a satisfactory population level is achieved. Very often weed fish are captured and destroyed, in order to reduce competition for food and space. Food organisms of the stocked species are transplanted and provided with favourable conditions for growth during the period of trophic depression. Special care is taken to maintain the optimum water levels required for breeding and survival of the stocked species. Yields of fish from the reservoirs vary depending to a large extent on the climatic conditions in the area. Yields in the order of 25–45 kg per ha have been reported from the southern part of the country, but the production is much lower in the north.

Hydro-electric and irrigation reservoirs in a number of other countries also, especially in India and China, have been stocked regularly to develop commercial fishing. Most of the Indian reservoirs are regularly stocked with fry and fingerlings of Indian carps grown in nearby nursery farms. Although there is evidence that these carps can breed in the reservoirs or in streams draining into the reservoirs, the production of fry and fingerlings appears to be low and so stocking is required to maintain the stock size. Some of the reservoirs in the north are stocked with varieties of common carp and some in southern India with tilapia, along with other local species. Breeding populations of tilapia have established in some of the lakes. Experience has shown the need for using large fingerlings or yearlings for stocking to obtain satisfactory survival, and presently sizes above 15 cm are used for this purpose (Jhingran, 1982). The rate of stocking varies very considerably, but where regular stocking is practised it ranges between about 3700 and 5000 fingerlings per ha. The yield per ha of reservoir also varies considerably and is reported to be between 6.2 and 39 kg.

In China, reservoirs and lakes are managed very much on the lines of large fish ponds, to derive maximum production. Those under about 100 ha in area are managed more intensively with supplementary feeding and heavy stocking. Management of larger ones of over 10 000 ha in area involves stocking, protection of natural spawning sites and creation of additional spawning grounds, as well as regulation of fishing through restrictions on fishing equipment and fishing seasons. The more common species stocked are the Chinese carps (bighead, silver carp and black carp). Wuchang fish (*Megalobrama amblycephala*), common carp and the crucian carp (*Carassius auratus*) are also used. Fingerlings about five months old, of 15–20 cm length, are preferred for stocking and the stocking rate may be as much as 2250 per ha. The average production in large shallow lakes is reported to be about 60 kg/ha, whereas a production of up to 750 kg/ha has been reported from smaller reservoirs.

Many of the South American countries, particularly Mexico, Brazil and Argentina have been stocking their inland waters, including reservoirs, with indigenous and introduced species of fish. The large-mouth bass (*Micropterus salmoides*), trout (*Salmo gairdnerii, Salvelinus fontinalis*), tilapia (*Tilapia* spp.) and the common carp are the more important non-indigenous species used. Hatchery production of the commercially important local species and their regular stocking in reservoirs have been an accepted practice in hydro-electric projects in Brazil. Reservoirs in north-eastern Brazil have now established populations of species such as *Prochilodus cearensis, P. argenteus, Pimelodus clarias, Salminus maxillosus, S. brevidens, Cichla ocellaris* and *C. temensis*. The pejerry (*Bacilichthys bonariensis*) is one of the common indigenous species used for stocking reservoirs in Argentina, and this species has been introduced successfully into reservoirs in Brazil and Chile. In North America, stocking

of reservoirs has been largely confined to selected species of sport fish. Besides salmon and trout, the fry of white fish (*Coregonus clupeaformis*) have been regularly stocked for over six decades in the Canadian waters of Lake Ontario, but there does not appear to be any conclusive evidence of its contribution to the fishing of the lake. The striped bass (*Morone saxatilis*) has been stocked in reservoirs, lakes and rivers all over the USA to enhance sport fisheries.

Accidental and intentional stocking of lakes and reservoirs have occurred in the African continent. A recent organized stocking took place in the Kariba reservoir, where *Tilapia macrochir* has been stocked regularly for a period of about four years.

One of the major problems in the exploitation of fish stocks from reservoirs is the difficulty in operating nets due to the uneven nature of reservoir bottoms. Large rocks and tree stumps covered with water during the formation of the reservoir, hinder the operation of seines and gill nets. Specially designed gill and entangling nets are only partially effective when there are so many obstructions. So, it has been an accepted policy in the USSR to level the reservoir bottom for commercial fishing by trawling and clear hauling areas for seine nets. Reservoir bottoms are also graded in China to enable the operation of pair trawls, seines, gill nets and encircling nets. Enforcement of strict regulations relating to the size and periods of fishing has also contributed substantially to the success of stocking programmes.

Through a coordinated programme of reservoir management by engineers and fishery biologists in both China and the USSR, it has been possible to maintain the required water levels for fish production and spawning. The relatively shallow lakes and reservoirs in China are often partitioned by means of dykes and artificial islands, to enable more intensive management, often combining fry and fingerling rearing in sections of the reservoir itself.

30.1.2 Lakes and streams

The main characteristic of lakes and streams, in so far as stocking is concerned, is that they have significant autochthonous populations of fish, and stocking is generally intended to enhance the economically important ones or to occupy ecological niches in the fauna. Unlike newly-formed reservoirs, these natural water bodies have more or less stabilized ecosystems and therefore call for appropriate studies to determine the need and desirability for stocking. Although there are several instances where natural lakes have been stocked and managed along the same lines as reservoirs, as described above, it would appear that the majority of enhancement programmes in lakes and streams have been for establishing or augmenting sport fisheries.

The introduction of the rainbow and brown trout in streams in several countries for sport fisheries has already been referred to in Chapter 16. In order to maintain an adequate population of the species in the trout streams, hatcheries and nursery farms are established in suitable sites with a plentiful supply of clean water of the right temperature. The improvement of streams to ensure good quality water and the production of the natural food of trouts and implementation of the regulations regarding the size and number of fish that an angler can catch are some of the management measures taken to maintain the trout stocks. Though there are not many cases of established self-perpetuating populations of trout outside their natural range of distribution, it has been possible to maintain small populations through continued stocking of hatchery-produced young in many streams. Exceptions are some of the rivers in Canada and the Falkland Islands, where breeding populations have established.

Large-mouth bass, *Micropterus salmoides*, is an important sport fish in the USA and among the management measures adopted are regular stocking and balancing of its populations with those of forage species. When forage species such as the black crappie (*Pomoxis nigromaculatus*) blue gill (*Lepomis macrochirus*) and golden shiners (*Notemigonus crysoleucas*) overcrowd a lake, corrective restocking has to be undertaken. This involves the stocking of fingerling large-mouth bass, following a marginal or sectional treatment with a fish toxicant like rotenone or Fintrol-S. A stocking rate of 60–120 fingerlings per ha is generally adopted. If overcrowding of black crappie or golden shiners (which spawn before large-mouth bass) occur, 'early spawned' large-mouth bass may

be stocked at the rate of 125–250 per ha. An alternative management measure, when the lakes show a trend towards a crowded forage condition or have been heavily fished, is to stock large-mouth bass of 20–25 cm length. The lakes are closed to bass fishing for at least one month after stocking, as the hatchery-reared bass are extremely susceptible to hook and line.

Generally an annual harvest quota is enforced on the basis of catch records and population estimates. The introduction of additional forage species like the threadfin shad, *Dorosoma petense*, has very often resulted in increased bass fishing in lakes deficient in forage fish. The channel catfish (*Ictalurus punctatus*) and sometimes the white catfish (*I. catus*) are stocked in the lakes as additional sport fish to take some of the fishing pressure off large-mouth bass. They are usually stocked at the rate of 250 per ha.

The pike *Esox lucius* is another important game fish, especially in Europe (fig. 30.1), and its artificial propagation is practised in a number of countries for stocking lakes and other water bodies. Besides being a sport fish, the pike is considered to be a valuable fish to reduce coarse fish populations in lakes, as it is a voracious carnivore. Pikes spawn from February to the end of May, when the water temperature is about 8–10°C. In nature they spawn in shallow grassy waters. Brood fish can be caught from the neighbourhood of such areas, held in special brood ponds and fed on coarse fish. In small sheltered ponds, supplied with spring water of constant temperature, male and female pikes readily mature and spawn. Normally a female spawns about 20 000–25 000 eggs per kg. Incubation and hatching of eggs are generally carried out in large Zoug jars. Each jar (60–70 cm in height and 15–20 cm diameter) will carry between 1 and 5 ℓ eggs. Other types of hatchery jars can also be used. The best temperature for hatching pike eggs is between 8 and 15°C. Artificial fertilization of pike eggs can be carried out in more or less the same way as for salmonids. Pike eggs are smaller in size (2.5–3 mm in diameter) and more delicate and should be handled with special care. The incubation period is about 120 degree-days. Just before hatching, the embryos are removed to hatching trays similar to those used for

Fig. 30.1 Pike, *Esox lucius* (from Huet, 1986).

salmonids, but with holes less than 2 mm in diameter. The absorption of the yolk sac takes about 160–180 degree-days. When the yolk sac is fully absorbed, the larvae are released into the lake.

It is reported that if post-larvae are stocked, a survival of ten per cent can be expected. In order to minimize mortality after stocking, it is preferable to stock fingerlings, as with all other species used for repopulation. But the main problem is that pikes require live food from the post-larval stage onwards and from the time they reach 6–10 cm they are highly carnivorous. Because of this, the fry are reared in troughs for 3 to 4 weeks, fed on zooplankton collected from natural waters in sufficient quantities to avoid cannibalistic tendencies. Troughs made of concrete or other durable material can be stocked with up to 2500 larvae per m³ and will give about 60 per cent survival. The fry are then transferred to rearing ponds and are harvested in 6–8 weeks, by the end of April. The rearing ponds are similar to carp ponds. Sometimes carp ponds themselves are used for pike rearing in rotation with carp. Under normal conditions, 10 000–25 000 fry are stocked per ha and the average survival is about 20 per cent.

Pike perch is another important carnivorous fish of equal importance as a sport fish and also as a food fish in Europe (fig. 30.2). They are well-suited for stocking lakes and grow rapidly, reaching a weight of up to 1 kg in about three years. They spawn in the spring, when the temperature is 12–16°C. The male digs shallow nests on sandy or gravel beds in shallow areas. Eggs are laid on roots of aquatic plants placed in the nest. A female weighing about 1 kg can

lay about 200 000 eggs, which stick to the substratum. The brood fish watch over the fertilized eggs, which hatch after 8–10 days at temperatures of 11–14°C.

Under culture conditions, selected brood fish are liberated in spawning ponds (very often spawning ponds of carp are used) where artificial nests are placed. These nests are made of wire netting on a wooden frame and attached to it are small roots of willow, alder or water plants on which the fish can deposit the eggs. Several designs of nests, in particular one with a canopy of roots, have been used. Soon after the eggs are laid, the nests are transferred to hatching ponds. Eyed eggs can be transported long distances. Fertilized eggs can also be transferred to open waters in perforated hatching boxes, from which the hatched larvae can escape directly into the lake. However, it is necessary to grow them for a few weeks at least to obtain a higher survival rate. The larvae should be reared in nursery ponds with a dense growth of zooplankton, such as rotifers and cyclops. Fry of 15 mm will eat insect larvae, plankton and benthic fauna. In the absence of forage fish, they become cannibalistic at a size of about 4 cm. Since feeding them after this stage is difficult, they are usually stocked in open waters at this point.

Besides pike and pike perch, species of coregonids, especially *Coregonus albula* and *Coregonus lavaretus*, are propagated on a limited scale for stocking alpine oligotrophic lakes in Europe. The methods of propagation are very similar to those of pike. The brood fish are collected from the natural spawning grounds in the lake. *Coregonus albula* breeders may weigh between 30 and 200 g, while

Fig. 30.2 Pike perch, *Lucioperca lucioperca* (from Huet, 1986).

C. lavaretus can vary from 1 to 3 kg. A female of *C. albula* gives from 2000 to 10 000 eggs and *C. lavaratus* 10 000 to 19 000 eggs per kg weight. Artificial fertilization is carried out in exactly the same way as for salmonids and the incubation is performed in Zoug or MacDonald jars. The eggs can be left to hatch in the jars and take between 300 and 360 degree-days, at a low hatching temperature of 5°C. The hatchlings measure only about 10 mm in length and 1 mm in diameter. Coregonid fry have very small yolk sacs, which are rapidly absorbed in about three to five days after hatching. They are usually stocked soon after the yolk sacs are absorbed, and so the survival rate is low, between 1 and 10 per cent. It is not too clear whether such stocking operations are economically justifiable.

Though an anadromous species, the transplantation of ayu (*Plecoglossus altivelis*) as practised in Japan is essentially a stocking operation, rather than ranching. No reproduction is expected from the stocked fish as most surviving fish are caught as adults by sport or commercial fishermen during the year of release. This is largely because its entire life span, including the migration to the sea and return, is completed within a year and the rate of recapture is very high.

The construction of a series of dams across rivers has prevented the upward migration of ayu and greatly reduced its living areas. This, along with pollution, has resulted in a considerable decline in ayu populations. Being a very popular object of river fishing and a highly-priced food fish, continued efforts have been made to resuscitate the stocks and maintain the catch at an increasing level. This has been accomplished by regular stocking of fry. According to one estimate, not less than 266 tons of fry are stocked in the rivers. In the past, Lake Biwas was the main source of fry, but now fry from the foreshore areas near the spawning grounds of the fish, and also fry from the river when they start ascending, are caught for stocking. When stocked, the lake form of fry gives the highest survival rate, the next best being the river-run fry (those collected from the river). However, the collection of fry from all these sources are still not able to fulfil the increasing demand. Techniques for artificial propagation and rearing of ayu fry have therefore been developed and this is expected to help in meeting the increased demand, including the requirements of pond culture of the species.

The sea-run fry have to be conditioned for several days to acclimatize them for transfer to fresh water. This is done by holding them initially in ponds filled with 1 part sea water to 4 parts fresh water, and after a day or two transferring them to fresh water. They are held in fresh water for 5−7 days before shipment for release. The river and lake-run forms are also held in fresh water for a time to empty their guts for transportation. Special transport containers are used to transport the fry in water saturated with oxygen. It is believed that a medicated bath during transport will result in higher survival rates.

The rate of recapture after stocking varies widely (depending on the nature of the river), ranging from 10 to 80 per cent, but usually between 40 and 60 per cent. In most rivers the fish reach lengths of 20−22 cm about 5 months after stocking.

30.1.3 Coastal and inland seas

The stocking or repopulation described earlier in this chapter was restricted largely to freshwater environments. The limitations of human control on aquatic populations in the sea areas and the immensity of efforts required to overcome them have discouraged some of the early attempts to enhance marine stocks. Probably the first large-scale attempt to stock the sea was the one started in 1950 with the cod (*Gadus morhua*) larvae in the Oslofjord in Norway, which was eventually abandoned in 1971 as the benefits of stocking could not be demonstrated. It is only in recent years that more organized intensive research started to examine more closely the viability of stocking this species.

The most organized and ambitious programme of sea stocking today takes place in Japan, particularly in the Seto Inland Sea area. It is an effort to establish what is referred to as culture-based fisheries, and forms part of a policy to increase the marine resources of the country for future harvests. It is carried out by fishery cooperatives and is promoted by both the national and prefectural governments. Besides stocking young ones, the programme includes environmental improvements, including improvements to the sea bottom, to facilitate the interchange of sea and fresh water, the creation of special nursery and growing areas, provision of shelters, measures to reduce wave velocity, etc. The term 'fish farming' is used in Japan to denote this type of programme, which is considered to be a stage intermediate between ordinary fishing from natural resources and extensive fish culture. Although the breeding techniques are intensive, the grow-out takes place in the open sea, with feeding on natural food.

Probably the most important species presently stocked is the kuruma shrimp, *Penaeus japonicus*. Several millions of fry of about 1 cm length are released every year. After about four or five months, the shrimps have grown to the commercial size of 11−12 cm. The survival rate of the released fry depends very much on the techniques of release, environmental conditions during growth and fishing conditions. Survival rates of 0.1−10 per cent have been reported. As a result of considerable research, several measures are now being adopted to increase the survival rate of released shrimps. Being a typical burrowing species, the released fry burrow into the substrate and become more or less sedentary within a well-defined area, until they gradually move to offshore areas. The hatchery-reared juveniles acquire this burrowing habit only when about 10 mm in size, as against 7−9 mm in the natural stocks. This has led to the practice of releasing them only after they reach 10 mm in size. Selection of the stocking site is an equally important factor. The most suitable sites are usually inter-tidal sand flats between the mean low-water neap tide and mean sea level, where there are shallow pools which will only contain a few metres of water at high tides. Hatchery-reared shrimp fry are planted at low tide into such pools, which may sometimes be fenced in for an initial period.

It has been shown experimentally that initial severe mortality occurs within 24 hours after release of the fry. Even though a combination of factors such as unfavourable temperature, salinity, oxygen content or turbulence can kill the fry, predation by invertebrates such as the hermit crab (*Pagurus dubius*) and the gastropod *Niotha livescens* and by inter-tidal species of fish such as the goby (*Gobius gymnauchen*) is the main cause of mortality during this period. While the invertebrates attack the shrimp fry only when their activity is impaired by adverse environmental conditions, such as low salinity or high temperatures, the predation by gobies is severe, even when the fry are very active. Almost 60 per cent of the initial mortality is ascribed to predation by fish. To reduce this initial mortality the fry are released first into a fenced enclosure or an artificial lagoon in the inter-tidal area. Since such fenced enclosures can be subject to frequent damage, an artificial tide-land is devised to serve as stocking site. The main objective is to control the en-

vironmental conditions in order to prevent foraging by predatory fish and yet to permit a sedentary life for the hatchery-reared shrimp fry.

The artificial tide-land (fig. 30.3) consists of (i) a stocking zone at the mean high-water neap tide level, inclined seaward at a slope of 1/200 and divided into rectangular blocks by concrete septa buried in the substrate, leaving the upper 5 cm above sand, and (ii) a transfer zone at mean sea level, linking the stocking zone to the natural sand flats fronting the tide-land. The two zones are considered necessary as it has been observed that the initial loss is much less at higher elevations of about 40 cm above

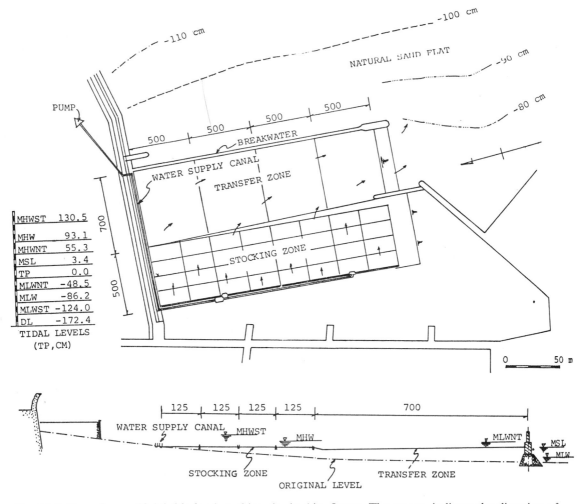

MHWST	130.5
MHW	93.1
MHWNT	55.3
MSL	3.4
TP	0.0
MLWNT	-48.5
MLW	-86.2
MLWST	-124.0
DL	-172.4

TIDAL LEVELS
(TP,CM)

Fig. 30.3 Plan of an artificial tide-land stocking site in Aio, Japan. The arrows indicate the direction of water flow. A cross-section of the tide-land is shown below (from Kurata and Shigueno, 1979).

mean sea level, if the level of water in the pool is maintained by pumping. However, for free dispersion of fry as they grow in size they would have to be transferred to a transfer zone at a lower level. If the area is exposed at low tide, a pump is used to inundate it to a depth of less than 5 cm. The fry are released into the stocking zone at the rate of about 100 per m². As they grow, the fry move from the stocking zone to the transfer zone and then to the natural sand flats, to grow into adult shrimps. By the time they attain an average body length of about 25–30 mm, in about two to three weeks after release, the majority of the survivors will have left the stocking zone for deeper waters. The stocking zone can then be used again for further release.

It is obviously very difficult to determine conclusively the survival rate and cost benefit ratio of kuruma shrimp release, when the commercial fisheries are also dependent on natural recruitment in the area. If increased catches are observed they can also be ascribed to natural fluctuations of the fisheries. Despite all this, there appears to be a general agreement among certain sectors of the public, as well as among shrimp fishermen in Japan, that the stocking of kuruma is not only capable of substantially increasing the local shrimp production, but can also support a considerable fishery, if adequate numbers of fry are stocked according to an organized plan.

Although attempts have been made to stock the seas with shrimp larvae in certain other areas, as on the Kuwaiti coast, the only really successful case of developing a fishery by stocking hatchery-reared juveniles of shrimps appears to be in Italy in the lagoons of Lessina and Venice. Introduced *Penaeus japonicus* has been propagated in hatcheries and, as a result of stocking, small populations of the species have developed and are commercially exploited. The rearing and release of the lobster *Homarus gammarus* has a long history in Norway, starting from around 1928. It is now possible to produce seed stock in hatcheries, using low intensity lighting, adequate feeding, and continuous sorting according to size. Larvae can be grown individually in specially designed partitioned troughs up to the juvenile stage, to avoid cannibalism. Preliminary experiments in ranching the American lobster

(*H. americanus*) have been conducted in Massachusetts, USA, using concrete-filled tyres as shelters (Hruby, 1979), but because lobsters are continually moving, stable populations could not be established. It is reported that half a million young lobsters are released on the Massachusetts coast every year. Experimental stocking is being done also in the state of Maine (USA), Norway and France.

Many of the on-bottom mollusc culture methods described in Chapters 26 and 27 are essentially stocking operations to increase populations in foreshore areas and to develop new resources. In northern Japan fisheries of the common scallop (*Pectinopecten*) have been enhanced by the collection and stocking of natural seed and by the eradication of predators such as starfish and sea urchins. Attempts have also been made to improve the production of abalones, by planting hatchery-reared seed and developing beds of seaweeds to increase their food resources. Preliminary experiments in the repopulation of the queen conch (*Strombus gigas*) have been made in some of the islands of the Caribbean area. It is possible to grow young conch from egg cases hatched under hatchery conditions. However, without any control of fishing for the released juveniles, their survival rates cannot be assessed.

As mentioned earlier, several efforts have been made to stock the seas with hatchery-reared fish species. Other than anadromous species, which are discussed in Section 30.2, none of the others appear to have proved successful in enhancing populations or developing new ones. Progress is being made in the release of the red seabream (*Pagrus major*) in Japan, and the possibility of repopulating cod is being re-examined in Norway.

30.2 Ranching of anadromous species

The strong homing behaviour during spawning migration of anadromous species makes them the most suitable aquatic animals for ranching. Successful ranching of anadromous species on a sufficiently large scale, however, relates only to two groups of temperate-zone fishes: the salmon and sturgeons. The maintenance or revival of salmon fisheries in countries of the northern hemisphere has, to a very large extent, been due to the successful hatchery rearing of smolts, regular release of smolts into the open seas and their subsequent return to the home streams for spawning. Attempts have been made to introduce certain species of salmon in some countries of the southern hemisphere. In recognition of the merits of ranching, as opposed to intensive rearing of species in confinement, there is interest in identifying a suitable species for ranching in the tropics. Though the shad *Hilsa ilisha* has often been considered a possible candidate, its short-range migrations and ability to propagate and form riverine populations above dams and other obstructions in rivers, as well as the lack of a well-established technology for the hatchery rearing of young *Hilsa*, have stood in the way of even experimental work.

30.2.1　Salmons

Among the anadromous species, it is the Pacific salmons which have received the maximum attention so far for transplantation to new areas. The chinook salmon (*Oncorhynchus tshawytcha*) (fig. 30.4), coho salmon (*O. kisutch*) (fig. 30.5), sockeye salmon (*O. nerka*) (fig. 30.6) and the pink salmon (*O. gorbuscha*) (fig. 30.7) have all been transplanted to different locations in North America and other parts of the world, starting as early as the 1870s. Beginning with the establishment of an egg collecting station on the McCloud river in California, in the USA in 1872, efforts were made to establish runs of *Oncorhynchus* in eighteen states of the USA and fifteen other countries until 1930.

Fig. 30.4 Chinook salmon, *Oncorhynchus tshawytcha*; (a) adult female; (b) adult male (from US Bureau of Fisheries Bulletin, XXVI, 1906).

Fig. 30.5 Coho salmon, *Oncorhynchus kisutch*; (a) adult male; (b) breeding female (from US Bureau of Fisheries Bulletin, XXVI, 1906).

McNeil (1979) briefly reviewed the history of transplantation and pointed out that successful transplantations have usually been the exception in the earlier years. With increased knowledge of the ecological requirements of the transplanted species and improvements in the techniques of artificial propagation and rearing, more organized and persistent efforts have been made to establish breeding populations of different species of Pacific salmon and, to a lesser extent, the Atlantic salmon (*Salmo salar*).

Ocean ranching of the Pacific salmon has developed into a major activity in the northwest USA, with Alaska allowing private parties to produce juvenile salmon and undertake ranching operations. The chinook salmon is reported to account for about 70 per cent of the juveniles produced, with coho forming most of the remainder. The most consistent success in ranching appears to have occurred when stocks native to the hatchery stream, and with short fresh-water rearing period, were used. Transplanted stocks may need continuous artificial recruitment to maintain satisfactory returning runs. Despite intermingling of stocks in marine waters, it is believed that in fresh waters each stock is adapted to a unique combination of environmental factors. The homing behaviour and genetic control over the time of maturation result in segregation into discrete stocks in space and/or time for reproduction. Though the final recognition of home waters is in response to environmental cues to which the fish become conditioned early in life, there is evidence to show that homing in salmonids is

Fig. 30.6 Sockeye salmon, *Oncorhynchus nerka*; (a) adult female; (b) adult male (from US Bureau of Fisheries Bulletin, XXVI, 1906).

partly inherited (Raleigh, 1971; Brannon, 1972).

Salmon enhancement programmes in the northwest USA have been based on hatchery production of juveniles, as well as the provision of spawning channels (fig. 30.8). The artificial spawning channels were intended to supplement the natural spawning of the salmon in the rivers, where conditions were not optimum for successful spawning of larger populations. They are relatively less expensive to construct and maintain. Efforts are made to simulate conditions in the natural spawning areas of the species and to provide the prerequisites for successful spawning, in as far as they are known. A typical channel may be around 7000 m in length, 4–11 m in width and 10–75 cm in depth. Stability of the spawning beds is ensured by the provision of suitable substrate material of medium or fine gravel and the provisions made for diversion of flood water. The water quality, particularly oxygen content, is maintained by the selection of the correct gradient for the channel and the construction of baffles or other such devices. In order to avoid drying up of the spawning beds and dehydration of laid eggs during periods of drought, provision is made for pumping water into the channel or for diverting water from the main stream. The adverse effects of freezing conditions are minimized by maintaining good circulation and constant water depth and flow. Predation and fluctuation of temperature are also controlled to a very great extent.

Fig. 30.7 Pink salmon, *Oncorhynchus gorbuscha*, (a) adult female; (b) adult male (from US Bulletin of Fisheries Bulletin, XXVI, 1906).

Hatchery production of juveniles has a longer history, and although initial attempts of stocking gave inconclusive results, satisfactory returns were obtained later in a number of areas, as a result of improved knowledge of the biology of the salmon and better technologies of propagation. Hatchery propagation and ranching is now accepted as an effective fishery management technique for most of the Pacific salmons. The hatchery techniques followed are described in Chapter 16. For stocking to be effective, it is considered necessary to grow the alevins to fry or fingerling stage before release.

Several successful transplantations have been made in North America and elsewhere, and evaluations have shown that many of them have been highly successful. A strikingly suc-

cessful example is the stocking of coho salmon in Lakes Michigan and Superior, starting in 1966. As a result of infestation by the parasitic sea lamprey (*Petromyzon marinus*), many of the economically important species of fish from the lakes had diminished. Eradication of the lamprey gave rise to the abundance of another unwanted fish, alewife (*Alosa pseudoharengus*). The introduction of coho salmon was to control the alewife and to develop a new sport fishery. Chinook salmon were also subsequently introduced into Lake Michigan.

Several economic evaluations have been attempted to determine the benefits of stocking hatchery-reared young. The benefit values do vary, but almost all well organized stocking programmes have given positive results, with

Fig. 30.8 Spawning channels for salmon in British Columbia, Canada (courtesy of W.R. Hoursten).

cost benefit ratios ranging from 1:2.3 to 1:5.5.

Ranching of the chum salmon (*O. keta*) (fig. 30.9) and the pink salmon in northern Japan is a spectacular example of the success of continued and patient efforts. Since the latter half of the 19th century, the scale of releases of hatchery juveniles has steadily increased. Hatcheries of pink and chum salmon established on Sakhalin Island by the USSR have been releasing an equally impressive number of juveniles of the two species. At least 1.5 billion chum and 0.5 billion pink salmon are released annually from the hatcheries of northern Japan and the Sakhalin Islands. The USSR alone is planning to increase the release of Pacific salmon juveniles to 5 billion by the year 2000. These hatcheries rely on substrate incubators, where alevins repose on gravel in shallow channels with a horizontal flow of water. Stocking densities generally range from 15 000 to 30 000 alevins per m^2 gravel surface. The raceways are darkened to simulate conditions in the natural spawning beds.

The marine survival of hatchery-raised chum salmon from Hokkaido (northern Japan) ranges between 2 and 2.5 per cent. The rate of exploitation of hatchery fish returning to coastal waters ranges from 80 to 90 per cent and ave-rages about 87 per cent. Adult chum salmon returning to the Sakhalin hatcheries average about 2.42 per cent of the number released. Available statistics show that an average of 81.7 per cent of the total returning run is caught in Soviet coastal fisheries. The estimated 2.3 per cent return of hatchery chum salmon to Sakhalin is very similar to that of Hokkaido Island. The average rate of return of pink salmon to hatcheries in Sakhalin Island is about three times higher than of chum salmon. The rate of exploitation of pink salmon in USSR waters is held between 50 and 70 per cent in order to conserve the stocks. Marine survival of hatchery pink salmon is estimated to be in the order of 2−5 per cent. Overall economics of the ranching operation indicate that every dollar invested in hatchery production yields up to $50 in fresh salmon. The total increase of Pacific salmon landings from released juveniles is estimated to be between 50 000 and 70 000 tons.

The establishment of chum salmon in the Caspian Sea and of chinook and Atlantic salmon in New Zealand waters are other notable examples of successful transplantations. Hatcheries located in the Baltic, White and Barents Sea basins regularly release millions of

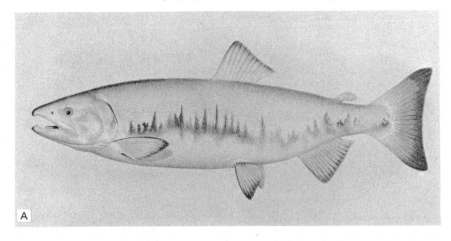

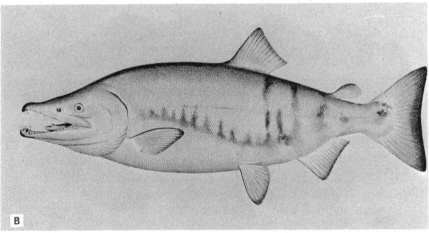

Fig. 30.9 Chum salmon, *Oncorhynchus keta*; (a) breeding female (b) breeding male (from US Bureau of Fisheries Bulletin, XXVI, 1906).

young ones to enhance the salmon fisheries of the area. Chum and pink salmon transplanted from Sakhalin Island to the Barent and White Seas have become fully established and pink salmon are reported in catches also from neighbouring north European countries. Several efforts are now under way to introduce Atlantic salmon in temperate countries, including Australia, but the main objective is to develop intensive cage culture of the species based on experience in Norway and Scotland. Even in north European countries, the emphasis is turning towards intensive culture rather than the release of smolts for open-water ranching. However, countries like Sweden and Iceland have succeeded in establishing homing populations of the Atlantic salmon in their rivers. Farwell and Porter (1979) described successful efforts to establish spawning populations of Atlantic salmon within inaccessible watersheds of Newfoundland in Canada, by systematic stocking with fry produced in an artificial spawning channel and upwelling incubation boxes.

A recent effort in the transplantation of salmon that has received considerable international attention is of the Pacific salmon in Chile. Encouraged by the successful introduction of chinook salmon in New Zealand in the southern hemisphere, the Government of Chile, with the cooperation of institutions and private agencies in the USA and the Government of Japan, have been trying to transplant Pacific salmon in the country. The initial attempts to transplant chinook and coho salmon and the cherry salmon (*Oncorhynchus masou*) were not successful. The most encouraging results so far obtained are by a private sector company which has released several hatchery-raised coho and chinook salmon yearlings. There were good returns for chinook and only marginal returns for the coho. It is believed that the difference in return is due to the migratory patterns of the species. Chinooks are generally a shelf-hugging species, whereas coho has a more wide-ranging pattern of migration. Ocean current patterns in the area may also be affecting the survival and return of the released fish. Recently Atlantic salmon have been transplanted to Chile from Norway through private efforts, but it is mainly intended to develop intensive cage farming rather than ranching.

30.2.2 Sturgeons

Sturgeons are of considerable economic importance in the USSR and Iran, particularly as the source of true caviar. From among the thirteen or so species occurring in the area, five are of major importance. These are the stellate sturgeon (*Acipenser stellatus*), Russian sturgeon (*A. güldenstädti*), thorn sturgeon (*A. nudiventris*), sterlet (*A. ruthenus*) and beluga (*Huso huso*). Acipenseridae are slow-growing, slow-maturing and long-living fish. The beluga lives up to 100 or more years, the Russian sturgeon 50−60 years and the stellate sturgeon up to 30 years. But the older fish are usually rare in the rivers. Among the sexually mature fish which ascend River Volga in the USSR, beluga are usually between 12 and 27 years old, Russian sturgeon 10−20 years old and stellate sturgeon 8−17 years old. The maximum weight of a beluga may be as much as one ton, but the average is 80−100 kg. The average weight of Russian sturgeon is about 15 kg and of the stellate sturgeon 8−9 kg. These species do not spawn every year, but after certain time intervals, ranging from three to six years for females, and somewhat less for males. Sexually mature fish which ascend the river for spawning form a relatively small percentage of the stock belonging to about a dozen generations.

The sturgeon populations of the Caspian, Black and Azor Seas have been severely affected by the construction of dams across the rivers for power generation and by excessive fishing. In order to rehabilitate the stocks and maintain them at an optimum level, the state has established a programme of hatchery production and release of young sturgeons, together with measures for the regulation of fishing and provision of facilities for upstream migration of the required number of brood fish. In order to enhance the natural populations of the species, large-scale transplantations of the spawners caught in the tail waters of Volga were carried out in live fish containers. During a period of six years, 223 000 spawners were released into the Volgograd reservoir from the dam of the hydro-electric station, either by transporting them in containers or with the help of fish elevators. Though these transplantations were very effective in improving natural propagation, the construction of additional dams made hatchery production the only major means of maintaining the resource.

Several hatcheries have been established, especially in the Astrakhan region in the Volga delta, for artificial propagation of sturgeons. Although spawning channels have also been tried, most of the propagation is now done by hypophysation. Brood fish are collected from the rivers during the breeding season, generally between April and August, and held in long rectangular maturing ponds that have a strong flow of clean fresh water. The upstream portions of the ponds have bottoms paved with pebbles and the downstream portions are of earth. In order to hasten maturity and spawning, sturgeon pituitary gland extract is injected at the rate of 2−2.5 mg dry weight per kg body weight in females and half of that in males. Carp pituitary injection at the rate of 4−5 mg per kg of body weight is also reported to be effective. Injected fish are held in tanks until they are ripe. As the fish are too large and difficult to handle, and the structure of the oviduct allows only partial stripping, they have to be killed for egg collection. A ripe female is stunned and suspended from a hook. The abdomen is slit open to remove the eggs, taking care to prevent loss of loose eggs through the genital opening when pressure is exerted. After washing, the eggs are fertilized with milt. Once

Fig. 30.10 An accumulation of sturgeon in a fish lock on the Volga hydroelectric station in the USSR.

the shells of the eggs have hardened, the excess sperm and the sticky coating on the eggs are washed off with running water containing 10 per cent chalk or powdered clay for 20–30 minutes. Salmon incubators can be used for hatching the fertilized eggs. Simple incubators which are just boxes with perforations or incubators with one or more hatching trays in a hatching trough are used. Incubation can be carried out in the river itself, by floating the trays in a series across the river. The hatching time varies depending on the temperature and the species, but is seldom more than 6 days at 15°C. The yolk sac is absorbed in 5–10 days.

In order to increase survival after release, the larvae are reared to fry stage for 2 to 3 weeks or 4 to 6 weeks. For growing to early fry in 2 to 3 weeks, shallow concrete tanks are used. The stocking rate is high and the larvae are fed on daphnia, chironomids and oligochaetes grown in special ponds. They are protected from light as far as possible, as light is reported to affect the development of all sturgeon species, except the stellate sturgeon. To ensure availability of live food for early growth and to enhance survival rate after release, it is considered preferable to raise them in confinement for 4 to 6 weeks. For this, hatchlings or early fry are reared in larger ponds (0.5–2 ha and 0.5–2 m deep), stocked at the rate of four to six per m^2. The ponds are fertilized with mineral or organic fertilizers to increase the production of food organisms, especially cladocerans, copepods and benthic fauna, including chironomids. In about 4 to 6 weeks they reach a weight of 1.5–3.5 g and are ready for release. Live boats are used to carry the fingerlings to brackish-water areas for release.

According to McNeil (1979) the release of juvenile sturgeons was approaching 100 million fish annually in the USSR and 5 million in Iran. The artificial recruitment of sturgeon juveniles into the Caspian Sea alone was reported to have reached 70 million. Bardach *et al.* (1972) estimated the survival of released fish to be about 3 per cent on the basis of the available information.

30.3 Ownership and regulation of fishing

Open-water stocking and ranching have come a long way from the days when they were considered a mere waste of effort and money. Greater knowledge of the behaviour and environmental requirements of the species has contributed considerably to the development of suitable methods. Of special importance have been improvements in techniques of artificial propagation and rearing of fry and fingerlings, accompanied by control of predators and environmental improvements where possible. In many cases, economic evaluations have been made and these have shown the favourable cost benefit ratios of the operations. It is, however, obvious that indirect methods have to be adopted to estimate survivals of released stocks, when the fisheries are based on combined stocks of released and resident individuals. Data obtained by experimental marking may not always be accepted as representative of what happens in a commercial fishery. Because of all this, some scepticism is still expressed about the value of certain types of stocking, as, for example, the sea stocking of shrimps in Japan or the continued stocking of reservoirs in some of the South American countries.

Although the subject may be controversial, one of the more important constraints to the expansion of stocking and ranching is the problem of regulating the fishing of released stocks. The majority of successful stockings described in this chapter have been undertaken by, or under the auspices of, the State. Exceptions are some of the smaller operations undertaken by sport fishery associations or agencies for the benefit of anglers. Under administrative systems where the State can undertake such work on a continuing basis for the benefit of fisheries, there may not be much difficulty in justifying such programmes. The system may also permit strict adherence to regulations of fishing periods and fishing quota. But in a large majority of cases it is extremely difficult and costly to implement such regulations for cultured stocks in common property waters. It may be possible to allocate ownership rights to the releasing agency for returning spawners to a home stream in the case of anadromous species. But if fishing in the seas cannot be regulated and harmonized with the release operations, the profitability and success of the programme can be adversely affected. It is, therefore, necessary to consider in advance the economic and organizational management of the resulting fishery, before undertaking large-scale stocking or ranching.

The emergence of cage and pen culture has introduced another option for the use of open waters, at least in protected areas. It may well be possible to produce in such areas at least as many fish by such intensive culture methods as could be expected by open-water stocking. No comparative cost/benefit ratios have been worked out, but it is not unlikely that intensive culture would prove to be economically more attractive. Not all species presently used for stocking and ranching may be suitable for such intensive culture, as for example the large sturgeons or species meant for sport fishing, but at least for the species which can be cultured in cages or pens for human consumption, that option has to be considered against release into open waters.

30.4 References

Atkinson C.E. (1976) Salmon aquaculture in Japan, the Koreas, and the U.S.S.R. In *Proceedings of the Conference on Salmon Aquaculture and the Alaskan Fishing Community*, (Ed. by D.H. Rosenberg. *University of Alaska Sea Grant Report*, **76**(2), 79−154.

Bardach J.E., Ryther J.H. and McLarney W.O. (1972) *Aquaculture*. John Wiley & Sons, New York.

Braaten B. (1985) Status and prospects on aquaculture world-wide − the situation in Norway. In *Status and Prospects on Aquaculture Worldwide*, 51−9. Aquanor '85, Trondheim.

Brannon E.L. (1972) Mechanisms controlling migration of sockeye salmon fry. *Bul. IPSFC*, **21**.

Brown E.E. (1969) *The Fresh Water Cultured Fish Industry of Japan*. Research Report, University of Georgia, No. 41.

Cleaver F. (1969) Recent advances in artificial culture of salmon and steelhead trout of the Columbia River. *US Fish. Wildl. Serv. Fish. Leafl.*, **623**.

Doroshov S.I. (1982) Biology and culture of sturgeon. In *Recent Advances in Aquaculture* Vol. 2 (Ed. by J.F. Muir and R.J. Roberts pp. 251–74. Croom Helm, London.

FAO (1977) La Acuicultura en America Latina Vol. 3 – Informes Nacionales. *FAO Inf. Pesca*, **159**.

Farwell M.K. and Porter T.R. (1979) Atlantic salmon enhancement techniques in Newfoundland. In *Advances in Aquaculture*, (Ed. by T.V.R. Pillay and W.A. Dill), pp. 560–63. Fishing News Books, Oxford.

Fraser J.M. (1972) Recovery of planted brook trout, splake and rainbow trout from selected Ontario Lakes. *J. Fish. Res. Board. Can.*, **29**, 129–42.

Frey D.G. (1967) Reservoir research objectives and practices with an example from the Soviet Union. In *Reservoir Fishery Resources Symposium* pp. 26–36. Presented by the Reservoir Committee of the Southern Division, American Fisheries Society at the University of Georgia.

Hanamura N. (1979) Advances and problems in culture-based fisheries in Japan. In *Advances in Aquaculture*, (Ed. by T.V.R. Pillay and W.A. Dill), pp. 541–7. Fishing News Books, Oxford.

Harding E. (1966) Lake Kariba: the hydrology and development of fisheries. In *Man-made Lakes*, (Ed. by R.H. Lowe-McConnell), pp. 7–20. Academic Press, London.

Hickling C.F. (1961) *Tropical Inland Fisheries*. Longmans, London.

Honma A. (1971) *Aquaculture in Japan*. Japan FAO Association, Tokyo.

Hora S.L. and Pillay T.V.R. (1962) Handbook on fish culture in the Indo-Pacific region. *FAO Fish. Biol. Tech. Paper*, **14**.

Hruby T. (1979) Experimental lobster ranching in Massachusetts. *Proc. World Maricul. Soc.*, **10**, 194–202.

Huet M. (1986) *Textbook on Fish Culture*, 2nd edn. Fishing News Books, Oxford.

Ishida R. (1979) Stocking of ayu *Plecoglossus altivelis*, in the rivers of Japan. In *Advances in Aquaculture* (Ed. by T.V.R. Pillay and W.A. Dill), pp. 563–7. Fishing News Books, Oxford.

Japan Fisheries Association (1975) *Fish Farming in Japan*. Fisheries Association, Tokyo.

Japan Fisheries Resources Conservation Association (1966) *Propagation of the Chum Salmon in Japan*. Japan Fisheries Resource Conservation Association, Tokyo.

Jhingran V.G. (1982) *Fish and Fisheries of India*, 2nd edn. Hindustan Publishing Corporation (India,) Delhi.

Jhingran V.G. and Natarajan A.V. (1979) Improvement of fishery resources in inland waters through stocking. In *Advances in Aquaculture*, (Ed. by T.V.R. Pillay and W.A. Dill), pp. 532–41. Fishing News Books, Oxford.

Joyner T. and Mahnken C. (1975) Toward a planetary aquaculture – the seas as a range and cropland. *Mar. Fish. Rev.*, **37**(4), 5–10.

Kanid'yev A.N., Kostyunin G.M. and Salmin S.A. (1970) Hatchery propagation of the pink and chum salmons as a means of increasing the salmon stocks of Sakhalin. *J. Ichthyol.*, **10**, 249–259.

Karpova E.I. and Kutyanina L.G. (1969) Prospects for the acclimatization of fish and other aquatic organisms. *Rybn. Khoz.*, **1**, 24–6.

Kurata H. and Shigueno K. (1979) Recent progress in the farming of kuruma shrimp (*Penaeus japonicus*). In *Advances in Aquaculture*, (Ed. by T.V.R. Pillay and W.A. Dill), pp. 258–68. Fishing News Books, Oxford.

Lapitzky I.I. (1968) Development of fisheries and the ways of raising fish productivity of big reservoirs. In *First and Second Group Fellowship Study Tours on Inland Fisheries Research Management and Fish Culture in the Union of Soviet Socialist Republics – Lectures Rep. FAO/UNDP(TA)*, **2547**, 124–34.

Legeza M.I. (1971) Features of the biology of Caspian sturgeons (Fam. *Acipenseridae*) and their use in the propagation of stocks. *J. Ichthyol.*, **11**(3), 354–62.

Lindbergh J.M., Noble R.E. and Blackburn K.M. (1981) Salmon ranching in Chile. *ICLARM Newsletter*, **4**(4), 9–10.

Lumare F. (1985) L'allevamento della mazzancolla (*Penaeus japonicus*) – Una realtà emergente dell' acquacoltura nazionale. (In Italian.) *Il Pesce*, **2**(4), 19–29.

MacCrimmon H.R. (1971) World distribution of brown trout, *Salmo trutta*, *J. Fish Res. Board Can.*, **15**(12), 2527–48.

MacCrimmon H.R. (1968) World distribution of rainbow trout (*Salmo gairdnerii*). *J. Fish. Res. Board Can.*, **28**(5), 663–704.

McNeil W.J. (1979) Review of transplantation and recruitment of anadromous species. In *Advances in Aquaculture*, (Ed. by T.V.R. Pillay and W.A. Dill), pp. 547–54. Fishing News Books, Oxford.

Magomedov G.M. (1968) Some data on the chum salmon (*Oncorhynchus keta* (Walb.) acclimatized in the Caspian Sea. *J. Ichthyol.*, **10**(4), 552–5.

Nagasawa A. (1981) Salmon ranching in Chile. *ICLARM Newsletter*, **4**(4), 6–7.

Nakamura M. and Uekita M. (1975) Structure and function of the artificial tideland – a measure for controlling environment for the kuruma shrimp fry stocking. *Ann. Rep. Shallow Sea Farming Proj. Bingo Nada 1974(5)*, 9–17 (in Japanese).

Nikol'skii G.V. (1961) *Special Ichthyology*, 3rd edn, translated from Russian, Israel Program of Scientific Translation, Jerusalem.

Pillay T.V.R. (1973) The role of aquaculture in fishery development and management. *J. Fish. Res. Board Can.*, **30**, 2202–17.

Powell D.H. (1975) Management of largemouth bass in Alabama's State-owned public fishing lakes. In *Black Bass Biology and Management*, pp. 386–90. Sport Fishing Institute, Washington.

Raleigh R.F. (1971) Innate control of migrations of salmon and trout fry from natal gravels to rearing areas. *Ecology*, **52**(2), 291–7.

Shaverdrov R.S. *et al.* (1962) Steps toward acclimatization of pink salmon (*Oncorhynchus gorbuscha*) in the Black Sea Basin. *Tr. Nauchno−Issled. Rybokhoz. Stn. Gruzii*, **7**, 61–7.

Szalay M. (1975) Controlled reproduction and rearing of *Lucioperca lucioperca*. *EIFAC Tech. Paper*, **25**, 174–80.

Tapiador D.D. *et al.* (1977 Freshwater Fisheries and Aquaculture in China. *FAO Fish. Tech. Paper*, **168**.

Thorpe J.E. (Ed.) (1980) *Salmon Ranching*. Academic Press, London.

UNDP/FAO (1979) *Aquaculture Development in China.* FAO ADCP/REP/79/10. FAO of the UN, Rome.

UNDP/FAO (1979) *Aquaculture Development in Mexico.* ADCP/MR/79/4. FAO of the UN, Rome.

Vovk F.I. (1968) The biological basis of reproduction of migratory fish when the river is controlled. In *Lectures of the First and Second Group Fellowship Study Tours on Inland Fisheries Research Management and Fish Culture in the Union of Soviet Socialist Republics − Lectures. Rep. FAO/UNDP (TA)*, **2547**, 152–60.

Index

Books published by
Fishing News Books

Free catalogue available on request from Fishing News Books, Blackwell Scientific Publications Ltd, Osney Mead, Oxford OX2 OEL, England

Advances in fish science and technology
Aquaculture in Taiwan
Aquaculture: principles and practice
Aquaculture training manual
Aquatic weed control
Atlantic salmon: its future
Better angling with simple science
British freshwater fishes
Business management in fisheries and aquaculture
Cage aquaculture
Calculations for fishing gear designs
Carp farming
Commercial fishing methods
Control of fish quality
Crab and lobster fishing
The crayfish
Culture of bivalve molluscs
Design of small fishing vessels
Developments in electric fishing
Developments in fisheries research in Scotland
Echo sounding and sonar for fishing
The economics of salmon aquaculture
The edible crab and its fishery in British waters
Eel culture
Engineering, economics and fisheries management
European inland water fish: a multilingual catalogue
FAO catalogue of fishing gear designs
FAO catalogue of small scale fishing gear
Fibre ropes for fishing gear
Fish and shellfish farming in coastal waters
Fish catching methods of the world
Fisheries oceanography and ecology
Fisheries of Australia
Fisheries sonar
Fisherman's workbook
Fishermen's handbook
Fishery development experiences
Fishing and stock fluctuations
Fishing boats and their equipment
Fishing boats of the world 1
Fishing boats of the world 2
Fishing boats of the world 3
The fishing cadet's handbook
Fishing ports and markets
Fishing with electricity
Fishing with light

Freezing and irradiation of fish
Freshwater fisheries management
Glossary of UK fishing gear terms
Handbook of trout and salmon diseases
A history of marine fish culture in Europe and North America
How to make and set nets
Inland aquaculture development handbook
Intensive fish farming
Introduction to fishery by-products
The law of aquaculture: the law relating to the farming of fish and shellfish in Great Britain
The lemon sole
A living from lobsters
The mackerel
Making and managing a trout lake
Managerial effectiveness in fisheries and aquaculture
Marine fisheries ecosystem
Marine pollution and sea life
Marketing in fisheries and aquaculture
Mending of fishing nets
Modern deep sea trawling gear
More Scottish fishing craft
Multilingual dictionary of fish and fish products
Navigation primer for fishermen
Net work exercises
Netting materials for fishing gear
Ocean forum
Pair trawling and pair seining
Pelagic and semi-pelagic trawling gear
Penaeid shrimps — their biology and management
Planning of aquaculture development
Refrigeration of fishing vessels
Salmon and trout farming in Norway
Salmon farming handbook
Scallop and queen fi   ries in the British Isles
Seine fishing
Squid jigging from small boats
Stability and trim of fishing vessels and other small ships
Study of the sea
Textbook of fish culture
Training fishermen at sea
Trends in fish utilization
Trout farming handbook
Trout farming manual
Tuna fishing with pole and line